Vladimir G. Ivancevic and Tijana T. Ivancevic

Computational Mind: A Complex Dynamics Perspective

Studies in Computational Intelligence, Volume 60

Editor-in-chief
Prof. Janusz Kacprzyk
Systems Research Institute
Polish Academy of Sciences
ul. Newelska 6
01-447 Warsaw
Poland
E-mail: kacprzyk@ibspan.waw.pl

Further volumes of this series
can be found on our homepage:
springer.com

Vol. 37. Jie Lu, Da Ruan, Guangquan Zhang (Eds.)
E-Service Intelligence, 2007
ISBN 978-3-540-37015-4

Vol. 38. Art Lew, Holger Mauch
Dynamic Programming, 2007
ISBN 978-3-540-37013-0

Vol. 39. Gregory Levitin (Ed.)
Computational Intelligence in Reliability Engineering, 2007
ISBN 978-3-540-37367-4

Vol. 40. Gregory Levitin (Ed.)
Computational Intelligence in Reliability Engineering, 2007
ISBN 978-3-540-37371-1

Vol. 41. Mukesh Khare, S.M. Shiva Nagendra (Eds.)
Artificial Neural Networks in Vehicular Pollution Modelling, 2007
ISBN 978-3-540-37417-6

Vol. 42. Bernd J. Krämer, Wolfgang A. Halang (Eds.)
Contributions to Ubiquitous Computing, 2007
ISBN 978-3-540-44909-6

Vol. 43. Fabrice Guillet, Howard J. Hamilton (Eds.)
Quality Measures in Data Mining, 2007
ISBN 978-3-540-44911-9

Vol. 44. Nadia Nedjah, Luiza de Macedo Mourelle, Mario Neto Borges, Nival Nunes de Almeida (Eds.)
Intelligent Educational Machines, 2007
ISBN 978-3-540-44920-1

Vol. 45. Vladimir G. Ivancevic, Tijana T. Ivancevic
Neuro-Fuzzy Associative Machinery for Comprehensive Brain and Cognition Modeling, 2007
ISBN 978-3-540-47463-0

Vol. 46. Valentina Zharkova, Lakhmi C. Jain
Artificial Intelligence in Recognition and Classification of Astrophysical and Medical Images, 2007
ISBN 978-3-540-47511-8

Vol. 47. S. Sumathi, S. Esakkirajan
Fundamentals of Relational Database Management Systems, 2007
ISBN 978-3-540-48397-7

Vol. 48. H. Yoshida (Ed.)
Advanced Computational Intelligence Paradigms in Healthcare, 2007
ISBN 978-3-540-47523-1

Vol. 49. Keshav P. Dahal, Kay Chen Tan, Peter I. Cowling (Eds.)
Evolutionary Scheduling, 2007
ISBN 978-3-540-48582-7

Vol. 50. Nadia Nedjah, Leandro dos Santos Coelho, Luiza de Macedo Mourelle (Eds.)
Mobile Robots: The Evolutionary Approach, 2007
ISBN 978-3-540-49719-6

Vol. 51. Shengxiang Yang, Yew Soon Ong, Yaochu Jin Honda (Eds.)
Evolutionary Computation in Dynamic and Uncertain Environment, 2007
ISBN 978-3-540-49772-1

Vol. 52. Abraham Kandel, Horst Bunke, Mark Last (Eds.)
Applied Graph Theory in Computer Vision and Pattern Recognition, 2007
ISBN 978-3-540-68019-2

Vol. 53. Huajin Tang, Kay Chen Tan, Zhang Yi
Neural Networks: Computational Models and Applications, 2007
ISBN 978-3-540-69225-6

Vol. 54. Fernando G. Lobo, Cláudio F. Lima and Zbigniew Michalewicz (Eds.)
Parameter Setting in Evolutionary Algorithms, 2007
ISBN 978-3-540-69431-1

Vol. 55. Xianyi Zeng, Yi Li, Da Ruan and Ludovic Koehl (Eds.)
Computational Textile, 2007
ISBN 978-3-540-70656-4

Vol. 56. Akira Namatame, Satoshi Kurihara and Hideyuki Nakashima (Eds.)
Emergent Intelligence of Networked Agents, 2007
ISBN 978-3-540-71073-8

Vol. 57. Nadia Nedjah, Ajith Abraham and Luiza de Macedo Mourella (Eds.)
Computational Intelligence in Information Assurance and Security, 2007
ISBN 978-3-540-71077-6

Vol. 58. Jeng-Shyang Pan, Hsiang-Cheh Huang, Lakhmi C. Jain and Wai-Chi Fang (Eds.)
Intelligent Multimedia Data Hiding, 2007
ISBN 978-3-540-71168-1

Vol. 59. Andrzej P. Wierzbicki and Yoshiteru Nakamori (Eds.)
Creative Environments, 2007
ISBN 978-3-540-71466-8

Vol. 60. Vladimir G. Ivancevic and Tijana T. Ivancevic
Computational Mind: A Complex Dynamics Perspective, 2007
ISBN 978-3-540-71465-1

Vladimir G. Ivancevic
Tijana T. Ivancevic

Computational Mind: A Complex Dynamics Perspective

With 108 Figures and 4 Tables

Springer

Vladimir G. Ivancevic
Human Systems Integration
Land Operations Division
Defence Science & Technology Organisation
PO Box 1500
75 Labs
Edinburgh SA 5111
Australia
E-mail: vladimir.ivancevic@dsto.defence.gov.ac

Tijana T. Ivancevic
The University of Adelaide
Department of Applied Mathematics
School of Mathematical Sciences
SA 5005
Australia
E-mail: tijana.ivancevic@adelaide.edu.au

Library of Congress Control Number: 2007925682

ISSN print edition: 1860-949X
ISSN electronic edition: 1860-9503
ISBN 978-3-540-71465-1 Springer Berlin Heidelberg New York

Springer is a part of Springer Science+Business Media
springer.com

Cover design: deblik, Berlin
Typesetting by the editors using a Springer LaTeX macro package
Printed on acid-free paper SPIN: 11977551 89/SPi 5 4 3 2 1 0

Dedicated to Nitya, Atma and Kali

Preface

Computational Mind: A Complex Dynamics Perspective is a graduate–level monographic textbook in the field of Computational Intelligence. It presents a modern dynamical theory of the computational mind, combining cognitive psychology, artificial and computational intelligence, and chaos theory with quantum consciousness and computation. The book has three Chapters. The first Chapter gives an introduction to human and computational mind, comparing and contrasting main themes of cognitive psychology, artificial and computational intelligence. The second Chapter presents brain/mind dynamics from the chaos theory perspective, including sections on chaos in human EEG, basics of nonlinear dynamics and chaos, techniques of chaos control, synchronization in chaotic systems and complexity in humanoid robots. The last Chapter presents modern theory of quantum computational mind, including sections on Dirac–Feynman quantum dynamics, quantum consciousness, and quantum computation using Josephson junctions. The book is designed as a one–semester course for computer scientists, engineers, physicists and applied mathematicians, both in industry and academia. It includes a strong bibliography on the subject and detailed index.

Adelaide,
Now 2006

V. Ivancevic, Defence Science & Technology Organisation, Australia, e-mail: Vladimir.Ivancevic@dsto.defence.gov.au

T. Ivancevic, School of Mathematics, The University of Adelaide, e-mail: Tijana.Ivancevic@adelaide.edu.au

Contents

Acknowledgments

The authors wish to thank Land Operations Division, Defence Science & Technology Organisation, Australia, for the support in developing the *Human Biodynamics Engine* (HBE) and all the HBE–related text in this monograph.

We also express our gratitude to *Springer* book series *Studies in Computational Intelligence* and especially to the Editor, Professor Janusz Kacprzyk.

1

Introduction: Human and Computational Mind

In this Chapter we compare and contrast human and computational mind, from psychological, AI and CI perspectives.

1.1 Natural Intelligence and Human Mind

Recall that the word *intelligence* (plural *intelligences*) comes from Latin *intellegentia*.[1] It is a property of *human mind* that encompasses many related *mental abilities*, such as the capacities to *reason*, *plan*, solve problems, think abstractly, comprehend ideas and *language*, and learn. Although many regard the concept of intelligence as having a much broader scope, for example in *cognitive science* and *computer science*, in some schools of *psychology*,[2] the

[1] *Intellegentia* is a combination of Latin *inter* = *between* and *legere* = *choose, pick out, read.* Inter–lege–nt–ia, literally means 'choosing between.'

Also, note that there is a is a scientific journal titled 'Intelligence', dealing with intelligence and psychometrics. It was founded in 1977 by Douglas K. Detterman of Case Western Reserve University. It is currently published by Elsevier and is the official journal of the International Society for Intelligence Research.

[2] Recall that *psychology* is an academic and applied field involving the study of the human mind, brain, and behavior. Psychology also refers to the application of such knowledge to various spheres of human activity, including problems of individuals' daily lives and the treatment of mental illness.

Psychology differs from anthropology, economics, political science, and sociology in seeking to explain the mental processes and behavior of individuals. Psychology differs from biology and neuroscience in that it is primarily concerned with the interaction of mental processes and behavior, and of the overall processes of a system, and not simply the biological or neural processes themselves, though the subfield of neuropsychology combines the study of the actual neural processes with the study of the mental effects they have subjectively produced.

The word psychology comes from the ancient Greek 'psyche', which means 'soul' or 'mind' and 'ology', which means 'study'.

V.G. Ivancevic and T.T. Ivancevic: *Introduction: Human and Computational Mind*, Studies in Computational Intelligence (SCI) **60**, 1–269 (2007)
www.springerlink.com

study of intelligence generally regards this trait as distinct from *creativity*, *personality*, *character*, or *wisdom*.

Briefly, the word *intelligence* has five common meanings:

1. Capacity of human mind, especially to understand principles, truths, concepts, facts or meanings, acquire knowledge, and apply it to practise; the ability to learn and comprehend.
2. A form of life that has such capacities.
3. Information, usually secret, about the enemy or about hostile activities.
4. A political or military department, agency or unit designed to gather such information.
5. Biological intelligent behavior represents animal's ability to make productive decisions for a specific task, given a root objective; this decision is based on learning which requires the ability to hold onto results from previous tasks, as well as being able to analyze the situation; the root objective for living organisms is simply survival; the 'specific task' could be a choice of food, i.e., one that provides long steady supply of energy as it could be a long while before the next mealtime; this is in perfect harmony with the root biological objective – survival.

According to Encyclopedia Britannica, *intelligence* is the *ability to adapt effectively to the environment, either by making a change in oneself or by changing the environment or finding a new one.* Different investigators have emphasized different aspects of intelligence in their definitions. For example, in a 1921 symposium on the definition of intelligence, the American psychologist Lewis Terman emphasized the *ability to think abstractly*, while another American psychologist, Edward Thorndike, emphasized *learning* and the ability to give good responses to questions. In a similar 1986 symposium, however, psychologists generally agreed on the importance of adaptation to the environment as the key to understanding both what intelligence is and what it does. Such adaptation may occur in a variety of environmental situations. For example, a student in school learns the material that is required to pass or do well in a course; a physician treating a patient with an unfamiliar disease adapts by learning about the disease; an artist reworks a painting in order to make it convey a more harmonious impression. For the most part, adapting involves making a change in oneself in order to cope more effectively, but sometimes effective adaptation involves either changing the environment or finding a new environment altogether. Effective adaptation draws upon a number of cognitive processes, such as perception, learning, memory, reasoning, and problem solving. The main trend in defining intelligence, then, is that it is not itself a cognitive or mental process, but rather a selective combination of these processes purposively directed toward effective adaptation to the environment. For example, the physician noted above learning about a new disease adapts by perceiving material on the disease in medical literature, learning what the material contains, remembering crucial aspects of it that are needed to treat the patient, and then reasoning to solve the problem of how to apply the information to the needs of the patient. Intelligence, in sum, has come to be regarded as not a single ability but an effective drawing together of many abilities. This has not always been obvious to investigators of the subject, however, and, indeed, much of the history of the field revolves around arguments regarding the nature and abilities that constitute intelligence.

Now, let us quickly reflect on the above general *intelligence–related keywords.*

Reason

Recall that in the philosophy of arguments, *reason* is the ability of the human mind to form and operate on concepts in abstraction, in varied accordance with rationality and logic —terms with which reason shares heritage. Reason is thus a very important word in Western intellectual history, to describe a type or aspect of mental thought which has traditionally been claimed as distinctly human, and not to be found elsewhere in the animal world. Discussion and debate about the nature, limits and causes of reason could almost be said to define the main lines of historical philosophical discussion and debate. Discussion about reason especially concerns:

(a) its relationship to several other related concepts: language, logic, consciousness etc,
(b) its ability to help people decide what is true, and
(c) its origin.

The concept of reason is connected to the concept of language, as reflected in the meanings of the Greek word 'logos', later to be translated by Latin 'ratio' and then French 'raison', from which the English word derived. As reason, rationality, and logic are all associated with the ability of the human mind to predict effects as based upon presumed causes, the word 'reason' also denotes a ground or basis for a particular argument, and hence is used synonymously with the word 'cause'.

It is sometimes said that the contrast between reason and logic extends back to the time of Plato[3] and Aristotle[4]. Indeed, although they had no

[3] Plato (c. 427 — c. 347 BC) was an immensely influential ancient Greek philosopher, a student of Socrates, writer of philosophical dialogues, and founder of the Academy in Athens where Aristotle studied. Plato lectured extensively at the Academy, and wrote on many philosophical issues, dealing especially in politics, ethics, metaphysics, and epistemology. The most important writings of Plato are his dialogues, although some letters have come down to us under his name. It is believed that all of Plato's authentic dialogues survive. However, some dialogues ascribed to Plato by the Greeks are now considered by the consensus of scholars to be either suspect (e.g., First Alcibiades, Clitophon) or probably spurious (such as Demodocus, or the Second Alcibiades). The letters are all considered to probably be spurious, with the possible exception of the Seventh Letter. Socrates is often a character in Plato's dialogues. How much of the content and argument of any given dialogue is Socrates' point of view, and how much of it is Plato's, is

separate Greek word for logic as opposed to language and reason, Aristotle's *syllogism* (Greek 'syllogismos') identified logic clearly for the first time as a distinct field of study: the most peculiarly reasonable ('logikê') part of reasoning, so to speak.

---

heavily disputed, since Socrates himself did not write anything; this is often referred to as the 'Socratic problem'. However, Plato was doubtless strongly influenced by Socrates' teachings.

Platonism has traditionally been interpreted as a form of metaphysical dualism, sometimes referred to as Platonic realism, and is regarded as one of the earlier representatives of metaphysical objective idealism. According to this reading, Plato's metaphysics divides the world into two distinct aspects: the *intelligible world* of 'forms', and the *perceptual world* we see around us. The perceptual world consists of imperfect copies of the intelligible forms or ideas. These forms are unchangeable and perfect, and are only comprehensible by the use of the intellect or understanding, that is, a capacity of the mind that does not include sense-perception or imagination. This division can also be found in Zoroastrian philosophy, in which the dichotomy is referenced as the *Minu* (intelligence) and *Giti* (perceptual) worlds. Currently, in the domain of mathematical physics, this view has been adopted by Sir Roger Penrose [Pen89].

[4] Aristotle (384 BC — March 7, 322 BC) was an ancient Greek philosopher, a student of Plato and teacher of Alexander the Great. He wrote books on divers subjects, including physics, poetry, zoology, logic, rhetoric, government, and biology, none of which survive in their entirety. Aristotle, along with Plato and Socrates, is generally considered one of the most influential of ancient Greek philosophers. They transformed Presocratic Greek philosophy into the foundations of Western philosophy as we know it. The writings of Plato and Aristotle founded two of the most important schools of Ancient philosophy.

Aristotle valued knowledge gained from the senses and in modern terms would be classed among the modern empiricists. He also achieved a 'grounding' of dialectic in the Topics by allowing interlocutors to begin from commonly held beliefs (Endoxa), with his frequent aim being to progress from 'what is known to us' towards 'what is known in itself' (Physics). He set the stage for what would eventually develop into the empirical scientific method some two millennia later. Although he wrote dialogues early in his career, no more than fragments of these have survived. The works of Aristotle that still exist today are in treatise form and were, for the most part, unpublished texts. These were probably lecture notes or texts used by his students, and were almost certainly revised repeatedly over the course of years. As a result, these works tend to be eclectic, dense and difficult to read. Among the most important ones are Physics, Metaphysics (or Ontology), Nicomachean Ethics, Politics, De Anima (On the Soul) and Poetics. These works, although connected in many fundamental ways, are very different in both style and substance.

Aristotle is known for being one of the few figures in history who studied almost every subject possible at the time, probably being one of the first polymaths. In science, Aristotle studied anatomy, astronomy, economics, embryology, geography, geology, meteorology, physics, and zoology. In philosophy, Aristotle

No philosopher of any note has ever argued that logic is the same as reason. They are generally thought to be distinct, although logic is one important aspect of reason. But the tendency to the preference for 'hard logic', or 'solid logic', in modern times has incorrectly led to the two terms occasionally being

---

wrote on aesthetics, ethics, government, metaphysics, politics, psychology, rhetoric and theology. He also dealt with education, foreign customs, literature and poetry. His combined works practically constitute an encyclopedia of Greek knowledge. According to Aristotle, everything is made out of the five basic elements:

1. Earth, which is cold and dry;
2. Water, which is cold and wet;
3. Fire, which is hot and dry;
4. Air, which is hot and wet; and
5. Aether, which is the divine substance that makes up the heavenly spheres and heavenly bodies (stars and planets).

Aristotle defines his philosophy in terms of essence, saying that philosophy is 'the science of the universal essence of that which is actual'. Plato had defined it as the 'science of the idea', meaning by idea what we should call the unconditional basis of phenomena. Both pupil and master regard philosophy as concerned with the universal; Aristotle, however, finds the universal in particular things, and called it the essence of things, while Plato finds that the universal exists apart from particular things, and is related to them as their prototype or exemplar. For Aristotle, therefore, philosophic method implies the ascent from the study of particular phenomena to the knowledge of essences, while for Plato philosophic method means the descent from a knowledge of universal ideas to a contemplation of particular imitations of those ideas. In a certain sense, Aristotle's method is both inductive and deductive, while Plato's is essentially deductive from a priori principles.

In the larger sense of the word, Aristotle makes philosophy coextensive with reasoning, which he also called 'science'. Note, however, that his use of the term science carries a different meaning than that which is covered by the scientific method. "All science (dianoia) is either practical, poetical or theoretical." By practical science he understands ethics and politics; by poetical, he means the study of poetry and the other fine arts; while by theoretical philosophy he means physics, mathematics, and metaphysics.

Aristotle's conception of logic was the dominant form of logic up until the advances in mathematical logic in the 19th century. Kant himself thought that Aristotle had done everything possible in terms of logic. The Organon is the name given by Aristotle's followers, the Peripatetics, for the standard collection of six of his works on logic. The system of logic described in two of these works, namely On Interpretation and the Prior Analytics, is often called Aristotelian logic.

Aristotle was the creator of syllogisms with modalities (modal logic). The word modal refers to the word 'modes', explaining the fact that modal logic deals with the modes of truth. Aristotle introduced the qualification of 'necessary' and 'possible' premises. He constructed a logic which helped in the evaluation of truth but which was difficult to interpret.

seen as essentially synonymous or perhaps more often logic is seen as the defining and pure form of reason.

However machines and animals can unconsciously perform logical operations, and many animals (including humans) can unconsciously, associate different perceptions as causes and effects and then make decisions or even plans. Therefore, to have any distinct meaning at all, 'reason' must be the type of thinking which links language, consciousness and logic, and at this time, only humans are known to combine these things.

However, note that reasoning is defined very differently depending on the context of the understanding of reason as a form of knowledge. The logical definition is the act of using reason to derive a conclusion from certain premises using a given methodology, and the two most commonly used explicit methods to reach a conclusion are deductive reasoning and inductive reasoning. However, within idealist philosophical contexts, reasoning is the mental process which informs our imagination, perceptions, thoughts, and feelings with whatever intelligibility these appear to contain; and thus links our experience with universal meaning. The specifics of the methods of reasoning are of interest to such disciplines as philosophy, logic, psychology, and artificial intelligence.

In deductive reasoning, given true premises, the conclusion must follow and it cannot be false. In this type of reasoning, the conclusion is inherent in the premises. Deductive reasoning therefore does not increase one's knowledge base and is said to be non–ampliative. Classic examples of deductive reasoning are found in such syllogisms as the following:

1. One must exist/live to perform the act of thinking.
2. I think.
3. Therefore, I am.

In inductive reasoning, on the other hand, when the premises are true, then the conclusion follows with some degree of *probability*.[5] This method of

[5] Recall that the word *probability* derives from the Latin 'probare' (to prove, or to test). Informally, probable is one of several words applied to uncertain events or knowledge, being closely related in meaning to likely, risky, hazardous, and doubtful. Chance, odds, and bet are other words expressing similar notions. Just as the theory of mechanics assigns precise definitions to such everyday terms as work and force, the theory of probability attempts to quantify the notion of probable.

The scientific study of probability is a modern development. Gambling shows that there has been an interest in quantifying the ideas of probability for millennia, but exact mathematical descriptions of use in those problems only arose much later. The doctrine of probabilities dates to the correspondence of Pierre de Fermat and Blaise Pascal (1654). Christiaan Huygens (1657) gave the earliest known scientific treatment of the subject. Jakob Bernoulli's 'Ars Conjectandi' (posthumous, 1713) and Abraham de Moivre's 'Doctrine of Chances' (1718) treated the subject as a branch of mathematics.

reasoning is ampliative, as it gives more information than what was contained in the premises themselves. A classical example comes from David Hume:[6]

1. The sun rose in the east every morning up until now.
2. Therefore the sun will also rise in the east tomorrow.

A third method of reasoning is called abductive reasoning, or inference to the best explanation. This method is more complex in its structure and can involve both inductive and deductive arguments. The main characteristic of abduction is that it is an attempt to favor one conclusion above others by either attempting to falsify alternative explanations, or showing the likelihood of the favored conclusion given a set of more or less disputable assumptions.

A fourth method of reasoning is analogy. Reasoning by analogy goes from a particular to another particular. The conclusion of an analogy is only plausible. Analogical reasoning is very frequent in common sense, science, philosophy and the humanities, but sometimes it is accepted only as an auxiliary method. A refined approach is *case–based reasoning*.

Pierre–Simon Laplace (1774) made the first attempt to deduce a rule for the combination of observations from the principles of the theory of probabilities. He represented the law of probability of errors by a curve $y = \varphi(x)$, x being any error and y its probability, and laid down three properties of this curve: (i) it is symmetric as to the $y-$axis; (ii) the $x-$axis is an asymptote, the probability of the error being 0; (iii) the area enclosed is 1, it being certain that an error exists. He deduced a formula for the *mean* of three observations. He also gave (1781) a formula for the law of facility of error (a term due to Lagrange, 1774), but one which led to unmanageable equations. Daniel Bernoulli (1778) introduced the principle of the maximum product of the probabilities of a system of concurrent errors.

The *method of least squares* is due to Adrien–Marie Legendre (1805), who introduced it in his 'Nouvelles méthodes pour la détermination des orbites des comètes' (New Methods for Determining the Orbits of Comets). In ignorance of Legendre's contribution, an Irish–American writer, Robert Adrain, editor of 'The Analyst' (1808), first deduced the law of facility of error,

$$\phi(x) = ce^{-h^2x^2}$$

where c and h are constants depending on precision of observation. He gave two proofs, the second being essentially the same as John Herschel's (1850). Carl Friedrich Gauss gave the first proof which seems to have been known in Europe (the third after Adrain's) in 1809. Further proofs were given by Laplace (1810, 1812), Gauss (1823), James Ivory (1825, 1826), Hagen (1837), Friedrich Bessel (1838), W. F. Donkin (1844, 1856), and Morgan Crofton (1870).

[6] David Hume (April 26, 1711 – August 25, 1776)[1] was a Scottish philosopher, economist, and historian, as well as an important figure of Western philosophy and of the Scottish Enlightenment.

Plan

Recall that a *plan* represents a proposed or intended method of getting from one set of circumstances to another. They are often used to move from the present situation, towards the achievement of one or more objectives or goals.

Informal or ad–hoc plans are created by individual humans in all of their pursuits. Structured and formal plans, used by multiple people, are more likely to occur in projects, diplomacy, careers, economic development, military campaigns, combat, or in the conduct of other business.

It is common for less formal plans to be created as abstract ideas, and remain in that form as they are maintained and put to use. More formal plans as used for business and military purposes, while initially created with and as an abstract thought, are likely to be written down, drawn up or otherwise stored in a form that is accessible to multiple people across time and space. This allows more reliable collaboration in the execution of the plan.

The term planning implies the working out of sub–components in some degree of detail. Broader–brush enunciations of objectives may qualify as metaphorical road–maps.

Planning literally just means the creation of a plan; it can be as simple as making a list. It has acquired a technical meaning, however, to cover the area of government legislation and regulations related to the use of resources.

Planning can refer to the planned use of any and all resources, as for example, in the succession of Five–Year Plans through which the government of the Soviet Union sought to develop the country. However, the term is most frequently used in relation to planning for the use of land and related resources, for example in urban planning, transportation planning, and so forth.

Problem Solving

The *problem solving* forms part of thinking. Considered the most complex of all intellectual functions, problem solving has been defined as higher–order cognitive process that requires the modulation and control of more routine or fundamental skills. It occurs if an organism or an artificial intelligence system does not know how to proceed from a given state to a desired goal state. It is part of the larger problem process that includes problem finding and problem shaping.

The nature of human problem solving has been studied by psychologists over the past hundred years. There are several methods of studying problem solving, including: *introspection*,[7] *behaviorism*,[8] computer simulation and experimental methods.

[7] Introspection is contemplation on one's self, as opposed to extrospection, the observation of things external to one's self. Introspection may be used synonymously with self–reflection and used in a similar way. Cognitive psychology accepts the use of the scientific method, but rejects introspection as a valid method

Beginning with the early experimental work of the Gestaltists in Germany (e.g., [Dun35], and continuing through the 1960s and early 1970s, research on problem solving typically conducted relatively simple, laboratory tasks that appeared novel to participants (see, e.g. [May92]). Various reasons account for the choice of simple novel tasks: they had clearly defined optimal solutions, they were solvable within a relatively short time frame, researchers could trace participants' problem–solving steps, and so on. The researchers made the underlying assumption, of course, that simple tasks such as the Tower of Hanoi captured the main properties of 'real world' problems, and that the cognitive processes underlying participants' attempts to solve simple problems were representative of the processes engaged in when solving 'real world' problems. Thus researchers used simple problems for reasons of convenience, and thought generalizations to more complex problems would become possible. Perhaps the best–known and most impressive example of this line of research remains the work by Newell and Simon [NS72].

See more on problem solving below.

Learning

Recall that learning is the process of acquiring knowledge, skills, attitudes, or values, through study, experience, or teaching, that causes a change of behavior that is persistent, measurable, and specified or allows an individual to formulate a new mental construct or revise a prior mental construct (conceptual knowledge such as attitudes or values). It is a process that depends on

---

of investigation. It should be noted that Herbert Simon and Allen Newell identified the 'thinking–aloud' protocol, in which investigators view a subject engaged in introspection, and who speaks his thoughts aloud, thus allowing study of his introspection.
Introspection was once an acceptable means of gaining insight into psychological phenomena. Introspection was used by German physiologist Wilhelm Wundt in the experimental psychology laboratory he had founded in Leipzig in 1879. Wundt believed that by using introspection in his experiments he would gather information into how the subject's minds were working, thus he wanted to examine the mind into its basic elements. Wundt did not invent this way of looking into an individual's mind through their experiences; rather, it can be dated back to Socrates. Wundt's distinctive contribution was to take this method into the experimental arena and thus into the newly formed field of psychology.

[8] Behaviorism is an approach to psychology based on the proposition that behavior can be studied and explained scientifically without recourse to internal mental states. A similar approach to political science may be found in Behavioralism. The behaviorist school of thought ran concurrent with the psychoanalysis movement in psychology in the 20th century. Its main influences were Ivan Pavlov, who investigated classical conditioning, John B. Watson who rejected introspective methods and sought to restrict psychology to experimental methods, and B.F. Skinner who conducted research on operant conditioning.

experience and leads to long–term changes in behavior potential. Behavior potential describes the possible behavior of an individual (not actual behavior) in a given situation in order to achieve a goal. But potential is not enough; if individual learning is not periodically reinforced, it becomes shallower and shallower, and eventually will be lost in that individual.

Short term changes in behavior potential, such as fatigue, do not constitute learning. Some long–term changes in behavior potential result from aging and development, rather than learning.

Education is the conscious attempt to promote learning in others. The primary function of 'teaching' is to create a safe, viable, productive learning environment. Management of the total learning environment to promote, enhance and motivate learning is a *paradigm shift*[9] from a focus on teaching to a focus on learning.

---

[9] Recall that an *epistemological paradigm shift* was called a *scientific revolution* by epistemologist and historian of science Thomas Kuhn in his 1962 book 'The Structure of Scientific Revolutions', to describe a change in basic assumptions within the ruling theory of science. It has since become widely applied to many other realms of human experience as well.

A scientific revolution occurs, according to Kuhn, when scientists encounter anomalies which cannot be explained by the universally accepted paradigm within which scientific progress has thereto been made. The paradigm, in Kuhn's view, is not simply the current theory, but the entire worldview in which it exists, and all of the implications which come with it. There are anomalies for all paradigms, Kuhn maintained, that are brushed away as acceptable levels of error, or simply ignored and not dealt with (a principal argument Kuhn uses to reject Karl Popper's model of falsifiability as the key force involved in scientific change). Rather, according to Kuhn, anomalies have various levels of significance to the practitioners of science at the time. To put it in the context of early 20th century physics, some scientists found the problems with calculating Mercury's perihelion more troubling than the Michelson–Morley experiment results, and some the other way around. Kuhn's model of scientific change differs here, and in many places, from that of the logical positivists in that it puts an enhanced emphasis on the individual humans involved as scientists, rather than abstracting science into a purely logical or philosophical venture. When enough significant anomalies have accrued against a current paradigm, the scientific discipline is thrown into a state of crisis, according to Kuhn. During this crisis, new ideas, perhaps ones previously discarded, are tried. Eventually a new paradigm is formed, which gains its own new followers, and an intellectual 'battle' takes place between the followers of the new paradigm and the hold–outs of the old paradigm. Again, for early 20th century physics, the transition between the Maxwellian electromagnetic worldview and the Einsteinian Relativistic worldview was not instantaneous nor calm, and instead involved a protracted set of 'attacks', both with empirical data as well as rhetorical or philosophical arguments, by both sides, with the Einsteinian theory winning out in the long–run. Again, the weighing of evidence and importance of new data was fit through the human sieve: some scientists found the simplicity

The stronger the stimulation for the brain, the deeper the impression that is left in the neuronal network. Therefore a repeated, very intensive experience perceived through all of the senses (audition, sight, smell) of an individual will remain longer and prevail over other experiences. The complex interactions of neurons that have formed a network in the brain determine the direction of flow of the micro–voltage electricity that flows through the brain when a person thinks. The characteristics of the neuronal network shaped by previous impressions is what we call the person's 'character'.

The most basic learning process is *imitation*, one's personal repetition of an observed process, such as a smile. Thus an imitation will take one's time (attention to the details), space (a location for learning), skills (or practice), and other resources (for example, a protected area). Through copying, most infants learn how to hunt (i.e., direct one's attention), feed and perform most basic tasks necessary for survival.

The so–called *Bloom's Taxonomy*[10] divides the learning process into a six–level hierarchy, where knowledge is the lowest order of cognition and evaluation the highest [Blo80]:

---

of Einstein's equations to be most compelling, while some found them more complicated than the notion of Maxwell's aether which they banished. Some found Eddington's photographs of light bending around the sun to be compelling, some questioned their accuracy and meaning. Sometimes the convincing force is just time itself and the human toll it takes, Kuhn pointed out, using a quote from Max Planck: "A new scientific truth does not triumph by convincing its opponents and making them see the light, but rather because its opponents eventually die, and a new generation grows up that is familiar with it." After a given discipline has changed from one paradigm to another, this is called, in Kuhn's terminology, a scientific revolution or a paradigm shift. It is often this final conclusion, the result of the long process, that is meant when the term paradigm shift is used colloquially: simply the (often radical) change of worldview, without reference to the specificities of Kuhn's historical argument.

[10] Benjamin Bloom (21 February 1913 – September 13, 1999) was an American educational psychologist who made significant contributions to the classification of educational objectives and the theory of mastery learning.
Bloom's classification of educational objectives, known as Bloom's Taxonomy, incorporates cognitive, psychomotor, and affective domains of knowledge. While working at the University of Chicago in the 1950s and '60s, he wrote two important books, Stability and Change in Human Characteristics and Taxonomy of Educational Objectives (1956). Bloom's taxonomy provides structure in which to categorize test questions. This taxonomy helps teachers pose questions in such a way to determine the level of understanding that a student possesses. For example, based upon the type of question asked, a teacher can determine that a student is competent in content knowledge, comprehension, application, analysis, synthesis and/or evaluation. This taxonomy is organized in a hierarchal way to organize information from basic factual recall to higher order thinking. This data table below is from the article written by W. Huitt titled, "Bloom *et al.*'s Taxonomy of

1. Knowledge is the memory of previously–learnt materials such as facts, terms, basic concepts and answers.
2. Comprehension is the understanding of facts and ideas by organization, comparison, translation, interpretation, and description.
3. Application is the use of new knowledge to solve problems.
4. Analysis is the examination and division of information into parts by identifying motives or causes. A person can analyze by making inferences and finding evidence to support generalizations.
5. Synthesis is the compilation of information in a new way by combining elements into patterns or proposing alternative solutions.
6. Evaluation is the presentation and defense of opinions by making judgments about information, validity of ideas or quality of work based on the following set of criteria:

- *Attention* – the cognitive process of selectively concentrating on one thing while ignoring other things. Examples include listening carefully to what someone is saying while ignoring other conversations in the room (e.g. the cocktail party problem, Cherry, 1953). Attention can also be split, as when a person drives a car and talks on a cell phone at the same time. Sometimes our attention shifts to matters unrelated to the external environment, this is referred to as mind-wandering or 'spontaneous thought'. Attention is one of the most intensely studied topics within psychology and cognitive neuroscience. Of the many cognitive processes associated with the human mind (decision–making, memory, emotion, etc), attention is considered the most concrete because it is tied so closely to perception. As such, it is a gateway to the rest of cognition. The most famous definition of attention was provided by one of the first major psychologists, William James[11] in

the Cognitive Domain". The table below describes the levels of Bloom's Taxonomy, beginning with the lowest level of basic factual recall. Each level in the table is defined, gives descriptive verbs that would foster each level of learning, and describes sample behaviors of that level. Bloom's taxonomy helps teachers better prepare questions that would foster basic knowledge recall all the way to questioning styles that foster synthesis and evaluation. By structuring the questioning format, teachers will be able to better understand what a child's weaknesses and strengths are and determine ways to help students think at a higher–level.

[11] William James (January 11, 1842 — August 26, 1910) was a pioneering American psychologist and philosopher. He wrote influential books on the young science of psychology, educational psychology, psychology of religious experience and mysticism, and the philosophy of pragmatism. He gained widespread recognition with his monumental Principles of 'Psychology' (1890), fourteen hundred pages in two volumes which took ten years to complete. 'Psychology: The Briefer Course', was an 1892 abridgement designed as a less rigorous introduction to the field. These works criticized both the English associationist school and the Hegelianism of

his 1890 book 'Principles of Psychology': "Everyone knows what attention is. It is the taking possession by the mind in clear and vivid form, of one out of what seem several simultaneously possible objects or trains of thought ... It implies withdrawal from some things in order to deal effectively with others." Most experiments show that one neural correlate of attention is enhanced firing. Say a neuron has a certain response to a stimulus when the animal is not attending to that stimulus. When the animal attends to the stimulus, even if the physical characteristic of the stimulus remains the same the neurons response is enhanced. A strict criterion, in this paradigm of testing attention, is that the physical stimulus available to the subject must be the same, and only the mental state is allowed to change. In this manner, any differences in neuronal firing may be attributed to a mental state (attention) rather than differences in the stimulus itself.

- *Habituation* – an example of non–associative learning in which there is a progressive diminution of behavioral response probability with repetition of a stimulus. It is another form of integration. An animal first responds to a sensory stimulus, but if it is neither rewarding nor harmful the animal learns to suppress its response through repeated encounters. One example of this can be seen in small song birds – if a stuffed owl (or similar predator) is introduced into the cage, the birds react to it as though it were a real predator, but soon realise that it is not and so become habituated to it. If another stuffed owl is introduced (or the same one removed and re–introduced), the birds react to it as though it were a predator, showing that it is only a very specific stimulus that is being ignored (namely, one particular unmoving owl in one place). This learned suppression of response is habituation. Habituation is stimulus specific. It does not cause a general decline in responsiveness. It functions like an average weighted history wavelet interference filter reducing the responsiveness of the organism to a particular stimulus. Frequently one can see opponent processes after the

---

his day as competing dogmatisms of little explanatory value, and sought to re–conceive of the human mind as inherently purposive and selective.

James defined *true beliefs* as those that prove useful to the believer. Truth, he said, is that which works in the way of belief. "True ideas lead us into useful verbal and conceptual quarters as well as directly up to useful sensible termini. They lead to consistency, stability and flowing human intercourse" but "all true processes must lead to the face of directly verifying sensible experiences somewhere," he wrote.

Pragmatism as a view of the meaning of truth is considered obsolete by many in contemporary philosophy, because the predominant trend of thinking in the years since James' death in 1910 has been toward non–epistemic definitions of truth, i.e., definitions that don't make truth dependent upon the warrant of a belief. A contemporary philosopher or logician will often be found explaining that the statement 'the book is on the table' is true iff the book is on the table.

stimulus is removed. Habituation is connected to associational reciprocal inhibition phenomenon, opponent process, motion after effect, color constancy, size constancy, and negative image after effect. Habituation is frequently used in testing psychological phenomena. Both infants and adults look less and less as a result of consistent exposure to a particular stimulus. The amount of time spent looking to a presented alternate stimulus (after habituation to the initial stimulus) is indicative of the strength of the remembered percept of the previous stimulus. It is also used to discover the resolution of perceptual systems, for example, by habituating a subject to one stimulus, and then observing responses to similar ones, one can detect the smallest degree of difference that is detectable by the subject.

Closely related to habituation is *neural adaptation* or *sensory adaptation* – a change over time in the responsiveness of the sensory system to a constant stimulus. It is usually experienced as a change in the stimulus. For example, if one rests one's hand on a table, one immediately feels the table's surface on one's skin. Within a few seconds, however, one ceases to feel the table's surface. The sensory neurons stimulated by the table's surface respond immediately, but then respond less and less until they may not respond at all; this is neural adaptation. More generally, neural adaptation refers to a temporary change of the neural response to a stimulus as the result of preceding stimulation. It is usually distinguished from memory, which is thought to involve a more permanent change in neural responsiveness. Some people use adaptation as an umbrella term that encompasses the neural correlates of priming and habituation. In most cases, adaptation results in a response decrease, but response facilitation does also occur. Some adaptation may result from simple fatigue, but some may result from an active re–calibration of the responses of neurons to ensure optimal sensitivity. Adaptation is considered to be the cause of perceptual phenomena like afterimages and the motion aftereffect. In the absence of fixational eye movements, visual perception may fade out or disappear due to neural adaptation.

- *Sensitization* – an example of non–associative learning in which the progressive amplification of a response follows repeated administrations of a stimulus [BHB95]. For example, electrical or chemical stimulation of the rat hippocampus causes strengthening of synaptic signals, a process known as long–term potentiation (LTP). LTP is thought to underlie memory and learning in the human brain. A different type of sensitization is that of kindling, where repeated stimulation of hippocampal or amygdaloid neurons eventually leads to seizures. Thus, kindling has been suggested as a model for temporal lobe epilepsy. A third type is central sensitization, where nociceptive neurons in the dorsal horns of the spinal cord become sensitized by peripheral tissue damage or inflammation. These various types indicate that sensitization may underlie both pathological and adaptive functions in the organism, but whether they also share the same physiological and molecular properties is not yet established.

- *Classical Pavlovian conditioning* – a type of associative learning. Ivan Pavlov described the learning of conditioned behavior as being formed by pairing two stimuli to condition an animal into giving a certain response. The simplest form of classical conditioning is reminiscent of what Aristotle would have called the law of contiguity, which states that: 'When two things commonly occur together, the appearance of one will bring the other to mind.' Classical conditioning focuses on reflexive behavior or involuntary behavior. Any reflex can be conditioned to respond to a formerly neutral stimulus. The typical paradigm for classical conditioning involves repeatedly pairing a neutral stimulus with an unconditioned stimulus. An unconditioned reflex is formed by an unconditioned stimulus, a stimulus that elicits a response–known as an unconditioned response–that is automatic and requires no learning and are usually apparent in all species. The relationship between the unconditioned stimulus and unconditioned response is known as the unconditioned reflex. The conditioned stimulus, is an initially neutral stimulus that elicits a response–known as a conditioned response–that is acquired through learning and can vary greatly amongst individuals. Conditioned stimuli are associated psychologically with conditions such as anticipation, satisfaction (both immediate and prolonged), and fear. The relationship between the conditioned stimulus and conditioned response is known as the conditioned (or conditional) reflex. In classical conditioning, when the unconditioned stimulus is repeatedly or strongly paired with a neutral stimulus the neutral stimulus becomes a conditioned stimulus and elicits a conditioned response.
- *Operant conditioning* – the use of consequences to modify the occurrence and form of behavior. Operant conditioning is distinguished from Pavlovian conditioning in that operant conditioning deals with the modification of voluntary behavior through the use of consequences, while Pavlovian conditioning deals with the conditioning of involuntary reflexive behavior so that it occurs under new antecedent conditions. Unlike reflexes, which are biologically fixed in form, the form of an operant response is modifiable by its consequences. Operant conditioning, sometimes called instrumental conditioning or instrumental learning, was first extensively studied by Edward Thorndike,[12] who observed the

[12] Edward Lee Thorndike (August 31, 1874 - August 9, 1949) was an American psychologist who spent nearly his entire career at Teachers College, Columbia University. His work on animal behavior and the learning process led to the theory of connectionism.

Among Thorndike's most famous contributions were his research on how cats learned to escape from puzzle boxes, and his related formulation of the law of effect. The law of effect states that responses which are closely followed by satisfy-

behavior of cats trying to escape from home-made puzzle boxes. When first constrained in the boxes, the cats took a long time to escape. With experience, ineffective responses occurred less frequently and successful responses occurred more frequently, enabling the cats to escape in less time over successive trials. In his Law of Effect, Thorndike theorized that successful responses, those producing satisfying consequences, were 'stamped in' by the experience and thus occurred more frequently. Unsuccessful responses, those producing annoying consequences, were stamped out and subsequently occurred less frequently. In short, some consequences strengthened behavior and some consequences weakened behavior. Burrhus Skinner[13] built upon Thorndike's ideas to construct a more detailed

ing consequences are associated with the situation, and are more likely to reoccur when the situation is subsequently encountered. Conversely, if the responses are followed by aversive consequences, associations to the situation become weaker. The puzzle box experiments were motivated in part by Thorndike's dislike for statements that animals made use of extraordinary faculties such as insight in their problem solving: "In the first place, most of the books do not give us a psychology, but rather a eulogy of animals. They have all been about animal intelligence, never about animal stupidity." (Animal Intelligence, 1911).

Thorndike meant to distinguish clearly whether or not cats escaping from puzzle boxes were using insight. Thorndike's instruments in answering this question were 'learning curves' revealed by plotting the time it took for an animal to escape the box each time it was in the box. He reasoned that if the animals were showing 'insight,' then their time to escape would suddenly drop to a negligible period, which would also be shown in the learning curve as an abrupt drop; while animals using a more ordinary method of trial and error would show gradual curves. His finding was that cats consistently showed gradual learning.

Thorndike interpreted the findings in terms of associations. He asserted that the connection between the box and the motions the cat used to escape was 'strengthened' by each escape. A similar, though radically reworked idea was taken up by B.F. Skinner in his formulation of Operant Conditioning, and the associative analysis went on to figure largely in behavioral work through mid-century, now evident in some modern work in behavior as well as modern *connectionism*.

[13] Burrhus Frederic Skinner (March 20, 1904 – August 18, 1990) was an American psychologist and author. He conducted pioneering work on experimental psychology and advocated behaviorism, which seeks to understand behavior as a function of environmental histories of experiencing consequences. He also wrote a number of controversial works in which he proposed the widespread use of psychological behavior modification techniques, primarily operant conditioning, in order to improve society and increase human happiness; and as a form of social engineering.

Skinner was born in rural Susquehanna, Pennsylvania. He attended Hamilton College in New York with the intention of becoming a writer and received a B.A. in English literature in 1926. After graduation, he spent a year in Greenwich Village attempting to become a writer of fiction, but he soon became disillusioned with his literary skills and concluded that he had little world experience, and no

theory of operant conditioning based on: (a) *reinforcement* (a consequence that causes a behavior to occur with greater frequency), (b) *punishment* (a consequence that causes a behavior to occur with less frequency), and (c) *extinction* (the lack of any consequence following a response). There are four contexts of operant conditioning:
(i) *Positive reinforcement* occurs when a behavior (response) is followed by a favorable stimulus (commonly seen as pleasant) that increases the frequency of that behavior. In the Skinner box experiment, a stimulus such as food or sugar solution can be delivered when the rat engages in a target behavior, such as pressing a lever.
(ii) *Negative reinforcement* occurs when a behavior (response) is followed by the removal of an aversive stimulus (commonly seen as unpleasant) thereby increasing that behavior's frequency. In the Skinner box experiment, negative reinforcement can be a loud noise continuously sounding inside the rat's cage until it engages in the target behavior, such as pressing a lever, upon which the loud noise is removed.
(iii) *Positive punishment* (also called 'Punishment by contingent stimulation') occurs when a behavior (response) is followed by an aversive stimulus, such as introducing a shock or loud noise, resulting in a decrease in that behavior.
(iv) *Negative punishment* (also called 'Punishment by contingent withdrawal') occurs when a behavior (response) is followed by the removal of a favorable stimulus, such as taking away a child's toy following an undesired behavior, resulting in a decrease in that behavior.

- *Observational (or social) learning* – learning that occurs as a function of observing, retaining and replicating behavior observed in others. It is most associated with the work of psychologist Albert Bandura,[14] who implemented some of the seminal studies in the area and initiated social learning theory. Although observational learning can take place at any stage in life, it is thought to be particularly important during childhood, particularly as authority becomes important. Because of this, social learning theory has influenced debates on the effect of television violence and parental role models. Bandura's Bobo doll experiment is widely cited in

strong personal perspective from which to write. During this time, which Skinner later called 'the dark year,' he chanced upon a copy of Bertrand Russell's book 'An Outline of Philosophy', in which Russell discusses the behaviorist philosophy of psychologist John B. Watson. At the time, Skinner had begun to take more interest in the actions and behaviors of those around him, and some of his short stories had taken a 'psychological' slant. He decided to abandon literature and seek admission as a graduate student in psychology at Harvard University (which at the time was not regarded as a leading institution in that field).

[14] Albert Bandura (born December 4, 1925 in Mundare, Alberta) is a Canadian psychologist most famous for his work on social learning theory (or Social Cognitivism) and self efficacy. He is particularly noted for the Bobo doll experiment.

psychology as a demonstration of observational learning and demonstrated that children are more likely to engage in violent play with a life size rebounding doll after watching an adult do the same. Observational learning allows for learning without any change in behavior and has therefore been used as an argument against strict behaviorism which argued that behavior change must occur for new behaviors to be acquired. Bandura called the process of social learning modelling and gave four conditions required for a person to successfully model the behavior of someone else: (i) attention to the model (a person must first pay attention to a person engaging in a certain behavior – the model); (ii) retention of details (once attending to the observed behavior, the observer must be able to effectively remember what the model has done); (iii) motor reproduction (the observer must be able to replicate the behavior being observed; e.g., juggling cannot be effectively learned by observing a model juggler if the observer does not already have the ability to perform the component actions, i.e., throwing and catching a ball); (iv) motivation and opportunity (the observer must be motivated to carry out the action they have observed and remembered, and must have the opportunity to do so; e.g., a suitably skilled person must want to replicate the behavior of a model juggler, and needs to have an appropriate number of items to juggle to hand). Social learning may affect behavior in the following ways: (i) teaches new behaviors; (ii) increases or decreases the frequency with which previously learned behaviors are carried out; (iii) can encourage previously forbidden behaviors; (iv) can increase or decrease similar behaviors (e.g., observing a model excelling in piano playing may encourage an observer to excel in playing the saxophone).

- *Communication* – the process of symbolic activity, sometimes via a language. Specialized fields focus on various aspects of communication, and include: (i) *mass communication* (academic study of various means by which individuals and entities relay information to large segments of the population all at once through mass media); (ii) *communication studies* (academic discipline that studies communication; subdisciplines include argumentation, speech communication, rhetoric, communication theory, performance studies, group communication, information theory, intercultural communication, interpersonal communication, intrapersonal communication, marketing, organizational communication, persuasion, propaganda, public affairs, public relations and telecommunication); (iii) *organizational communication* (the study of how people communicate within an organizational context, or the influence of, or interaction with organizational structures in communicating/organizing), (iv) *conversation analysis* (commonly abbreviated as CA, is the study of talk in interaction; CA generally attempts to describe the orderliness, structure and sequential patterns of interaction, whether this is institutional, in the school, doctor's surgery, courts or elsewhere, or casual conversation); (v) *linguistics* (scientific study of human language and speech; usually is conducted along two major axes: theoretical vs. applied, and autonomous vs. contextual); (vi) *cognitive linguistics* (commonly abbreviated as CL, refers to the school of linguistics that views the important essence of language as innately based in evolutionary–developed and speciated faculties, and seeks explanations that advance or fit well into the current understandings of the human mind); (vii) *sociolinguistics* (the study of the effect of any and all aspects of society, including cultural norms, expectations, and context, on the way language is used); (viii) *pragmatics* (concerned with bridging the explanatory gap between sentence meaning and speaker's meaning – how context influences the interpretation is crucial); (ix) *semiotics* (the study of signs, both individually and grouped in sign systems; it includes the study of how meaning is made and understood); and (x) *discourse analysis* (a general term for a number of approaches to analyzing written, spoken or signed language use; includes: discourse grammar, rhetoric and stylistics). Communication as a named and unified discipline has a history of contestation that goes back to the Socratic dialogues, in many ways making it the first and most contestatory of all early sciences and philosophies. Seeking to define 'communication' as a static word or unified discipline may not be as important as understanding communication as a family of resemblances with a plurality of definitions as Ludwig Wittgenstein[15] had put forth. Some definitions are broad, recognizing that animals can communicate, and some are more narrow, only including human beings within the parameters of human symbolic interaction. Nonetheless, communication is usually described along three major dimensions: content, form, and destination. In the advent of 'noise' (internal psychological noise and/or physical realities) these three components of communication often become skewed and inaccurate. (between parties, communication content include acts that declare knowledge and experiences, give advice and commands, and ask questions. These acts may take many forms, including gestures (nonverbal communication, sign language and body language), writing, or verbal speaking. The form depends on the symbol systems used. Together, communication content and form make messages that are sent towards a destination. The target can be oneself, another person (in interpersonal communication), or another entity (such as a corporation or group). There

[15] Ludwig Josef Johann Wittgenstein (April 26, 1889 – April 29, 1951) was an Austrian philosopher who contributed several ground-breaking works to contemporary philosophy, primarily on the foundations of logic, the philosophy of mathematics, the philosophy of language, and the philosophy of mind. He is widely regarded as one of the most influential philosophers of the 20th century.

are many theories of communication, and a commonly held assumption is that communication must be directed towards another person or entity. This essentially ignores intrapersonal communication (note intra–, not inter–) via diaries or self–talk. Interpersonal conversation can occur in dyads and groups of various sizes, and the size of the group impacts the nature of the talk. Small–group communication takes place in settings of between three and 12 individuals, and differs from large group interaction in companies or communities. This form of communication formed by a dyad and larger is sometimes referred to as the psychological model of communication where in a message is sent by a sender through channel to a receiver. At the largest level, mass communication describes messages sent to huge numbers of individuals through mass media, although there is debate if this is an interpersonal conversation.

Language

Recall that a language is a system of signals, such as voice sounds, gestures or written symbols that encode or decode information.

Human spoken and written languages can be described as a system of symbols (sometimes known as lexemes) and the grammars (rules) by which the symbols are manipulated. The word 'language' is also used to refer to common properties of languages.

Language learning is normal in human childhood. Most human languages use patterns of sound or gesture for symbols which enable communication with others around them. There are thousands of human languages, and these seem to share certain properties, even though many shared properties have exceptions.

Languages are not just sets of symbols. They also often conform to a rough grammar, or system of rules, used to manipulate the symbols. While a set of symbols may be used for expression or communication, it is primitive and relatively unexpressive, because there are no clear or regular relationships between the symbols.

Human languages are usually referred to as natural languages, and the science of studying them is *linguistics*, with Ferdinand de Saussure[16] and Noam Chomsky[17] as the most influential figures.

[16] Ferdinand de Saussure (November 26, 1857 – February 22, 1913) was a Geneva–born Swiss linguist whose ideas laid the foundation for many of the significant developments in linguistics in the 20th century. He is widely considered the 'father' of 20th–century linguistics.

Saussure's most influential work, 'Course in General Linguistics', was published posthumously in 1916 by former students Charles Bally and Albert Sechehaye on the basis of notes taken from Saussure's lectures at the University of Geneva. The Course became one of the seminal linguistics works of the 20th century, not primarily for the content (many of the ideas had been anticipated in the works

Humans and computer programs have also constructed other languages, including constructed languages such as Esperanto, Ido, Interlingua, Klingon, programming languages, and various mathematical formalisms. These

---

of other 19th century linguists), but rather for the innovative approach that Saussure applied in discussing linguistic phenomena. Its central notion is that language may be analyzed as a formal system of differential elements, apart from the messy dialectics of realtime production and comprehension.

Saussure's famous quotes are:

"A sign is the basic unit of language (a given language at a given time). Every language is a complete system of signs. Parole (the speech of an individual) is an external manifestation of language."

"A linguistic system is a series of differences of sound combined with a series of differences of ideas."

[17] Noam Avram Chomsky (born December 7, 1928) is the Institute Professor Emeritus of linguistics at the MIT. Chomsky is credited with the creation of the theory of generative grammar, considered to be one of the most significant contributions to the field of theoretical linguistics made in the 20th century. He also helped spark the cognitive revolution in psychology through his review of B.F. Skinner's 'Verbal Behavior', in which he challenged the behaviorist approach to the study of mind and language dominant in the 1950s. His naturalistic approach to the study of language has also affected the philosophy of language and mind. He is also credited with the establishment of the so–called *Chomsky hierarchy*, a classification of formal languages in terms of their generative power.

'Syntactic Structures' was a distillation of Chomsky's book 'Logical Structure of Linguistic Theory' (1955) in which he introduces transformational grammars. The theory takes utterances (sequences of words) to have a syntax which can be (largely) characterised by a formal grammar; in particular, a *context–free grammar* extended with transformational rules. Children are hypothesised to have an innate knowledge of the basic grammatical structure common to all human languages (i.e. they assume that any language which they encounter is of a certain restricted kind). This innate knowledge is often referred to as universal grammar. It is argued that modelling knowledge of language using a formal grammar accounts for the 'productivity' of language: with a limited set of grammar rules and a finite set of terms, humans are able to produce an infinite number of sentences, including sentences no one has previously said.

Chomsky's ideas have had a strong influence on researchers investigating the acquisition of language in children, though some researchers who work in this area today do not support Chomsky's theories, often advocating emergentist or connectionist theories reducing language to an instance of general processing mechanisms in the brain.

Chomsky's work in linguistics has had major implications for modern psychology. For Chomsky linguistics is a branch of cognitive psychology; genuine insights in linguistics imply concomitant understandings of aspects of mental processing and human nature. His theory of a universal grammar was seen by many as a direct challenge to the established behaviorist theories of the time and had major consequences for understanding how language is learned by children and what,

languages are not necessarily restricted to the properties shared by human languages.

Some of the areas of the human brain involved in language processing are: Broca's area, Wernicke's area, Supramarginal gyrus, Angular gyrus, Primary Auditory Cortex.

Mathematics and computer science use artificial entities called *formal languages* (including programming languages and markup languages, but also some that are far more theoretical in nature). These often take the form of character strings, produced by some combination of formal grammar and semantics of arbitrary complexity.

The classification of natural languages can be performed on the basis of different underlying principles (different closeness notions, respecting different properties and relations between languages); important directions of present classifications are:

1. Paying attention to the historical evolution of languages results in a genetic classification of languages—which is based on genetic relatedness of languages;
2. Paying attention to the internal structure of languages (grammar) results in a typological classification of languages—which is based on similarity of one or more components of the language's grammar across languages; and
3. Respecting geographical closeness and contacts between language-speaking communities results in areal groupings of languages.
4. The different classifications do not match each other and are not expected to, but the correlation between them is an important point for many linguistic research works. (Note that there is a parallel to the classification of species in biological phylogenetics here: consider monophyletic vs. polyphyletic groups of species.)

The task of genetic classification belongs to the field of historical–comparative linguistics, of typological—to linguistic typology. The world's languages have been grouped into families of languages that are believed to have common ancestors. Some of the major families are the Indo–European languages, the Afro–Asiatic languages, the Austronesian languages, and the Sino–Tibetan languages. The shared features of languages from one family can be due to shared ancestry.

An example of a typological classification is the classification of languages on the basis of the basic order of the verb, the subject and the object in a sentence into several types: SVO, SOV, VSO, and so on, languages. (English, for instance, belongs to the SVO language type.)

---

exactly, is the ability to use language. Many of the more basic principles of this theory (though not necessarily the stronger claims made by the principles and parameters approach described above) are now generally accepted in some circles.

The shared features of languages of one type (= from one typological class) may have arisen completely independently. (Compare with analogy in biology.) Their cooccurence might be due to the universal laws governing the structure of natural languages—language universals.

The following language groupings can serve as some linguistically significant examples of areal linguistic units, or sprachbunds: Balkan linguistic union, or the bigger group of European languages; Caucasian languages. Although the members of each group are not closely genetically related, there is a reason for them to share similar features, namely: their speakers have been in contact for a long time within a common community and the languages converged in the course of the history. These are called 'areal features'.

Mathematics and computer science use artificial entities called formal languages (including programming languages and markup languages, but also some that are far more theoretical in nature). These often take the form of character strings, produced by some combination of formal grammar and semantics of arbitrary complexity.

Abstraction

Recall that *abstraction* is the process of reducing the information content of a concept, typically in order to retain only information which is relevant for a particular purpose. For example, abstracting a leather soccer ball to a ball retains only the information on general ball attributes and behavior. Similarly, abstracting an emotional state to happiness reduces the amount of information conveyed about the emotional state.

Abstraction typically results in complexity reduction leading to a simpler conceptualization of a domain in order to facilitate processing or understanding of many specific scenarios in a generic way.

In philosophical terminology, abstraction is the thought process wherein ideas are distanced from objects.

Abstraction uses a strategy of simplification, wherein formerly concrete details are left ambiguous, vague, or undefined; thus effective communication about things in the abstract requires an intuitive or common experience between the communicator and the communication recipient.

Abstractions sometimes have ambiguous referents; for example, 'happiness' (when used as an abstraction) can refer to as many things as there are people and events or states of being which make them happy. Likewise, 'architecture' refers not only to the design of safe, functional buildings, but also to elements of creation and innovation which aim at elegant solutions to construction problems, to the use of space, and at its best, to the attempt to evoke an emotional response in the builders, owners, viewers and users of the building.

Abstraction in philosophy is the process (or, to some, the alleged process) in concept–formation of recognizing some set of common features in individuals, and on that basis forming a concept of that feature. The notion of abstraction is important to understanding some philosophical controversies surrounding empiricism and the problem of universals. It has also recently become popular in formal logic under predicate abstraction.

Some research into the human brain suggests that the left and right hemispheres differ in their handling of abstraction. One side handles collections of examples (e.g., examples of a tree) whereas the other handles the concept itself.

Abstraction in mathematics is the process of extracting the underlying essence of a mathematical concept, removing any dependence on real world objects with which it might originally have been connected, and generalizing it so that it has wider applications.

Many areas of mathematics began with the study of real world problems, before the underlying rules and concepts were identified and defined as abstract structures. For example, geometry has its origins in the calculation of distances and areas in the real world; statistics has its origins in the calculation of probabilities in gambling; and algebra started with methods of solving problems in arithmetic.

Abstraction is an ongoing process in mathematics and the historical development of many mathematical topics exhibits a progression from the concrete to the abstract. Take the historical development of geometry as an example; the first steps in the abstraction of geometry were made by the ancient Greeks, with Euclid being the first person (as far as we know) to document the axioms of plane geometry. In the 17th century Descartes introduced Cartesian coordinates which allowed the development of analytic geometry. Further steps in abstraction were taken by Lobachevsky, Bolyai and Gauss[18]

[18] *Gauss–Bolyai–Lobachevsky space* is a non–Euclidean space with a negative Gaussian curvature, that is, a *hyperbolic geometry*. The main topic of conversation involving Gauss–Bolyai–Lobachevsky space involves the impossible process (at least in Euclidean geometry) of squaring the circle. The space is named after Carl Gauss, János Bolyai, and Nikolai Lobachevsky.

Carl Friedrich Gauss (30 April 1777 – 23 February 1855) was a German mathematician and scientist of profound genius who contributed significantly to many fields, including number theory, analysis, differential geometry, geodesy, magnetism, astronomy and optics. Sometimes known as 'the prince of mathematicians' and 'greatest mathematician since antiquity', Gauss had a remarkable influence in many fields of mathematics and science and is ranked among one of history's most influential mathematicians. Gauss was a child prodigy, of whom there are many anecdotes pertaining to his astounding precocity while a mere toddler, and made his first ground–breaking mathematical discoveries while still a teenager. He completed Disquisitiones Arithmeticae, his magnum opus, at the age of twenty–one (1798), though it would not be published until 1801. This work was fundamental in consolidating number theory as a discipline and has shaped the field to the present day. One of his most important results is his 'Theorema Egregrium', establishing an important property of the notion of curvature as a foundation of differential geometry.

János Bolyai (December 15, 1802–January 27, 1860) was a Hungarian mathematician. Between 1820 and 1823 he prepared a treatise on a complete system

who generalized the concepts of geometry to develop non–Euclidean geometries. Later in the 19th century mathematicians generalized geometry even further, developing such areas as geometry in n dimensions, projective geometry, affine geometry, finite geometry and differential geometry. Finally Felix Klein's 'Erlangen program'[19] identified the underlying theme of all of these geometries, defining each of them as the study of properties invariant under a given group of symmetries. This level of abstraction revealed deep connections between geometry and abstract algebra.

The advantages of abstraction are:

(i) It reveals deep connections between different areas of mathematics;

(ii) Known results in one area can suggest conjectures in a related area; and

(iii) Techniques and methods from one area can be applied to prove results in a related area.

An abstract structure is a formal object that is defined by a set of laws, properties, and relationships in a way that is logically if not always historically independent of the structure of contingent experiences, for example, those involving physical objects. Abstract structures are studied not only in logic and mathematics but in the fields that apply them, as computer science, and in the studies that reflect on them, as philosophy and especially the philosophy of mathematics. Indeed, modern mathematics has been defined in a very general sense as the study of abstract structures by the *Bourbaki* group.[20]

---

of non–Euclidean geometry. Bolyai's work was published in 1832 as an appendix to a mathematics textbook by his father. Gauss, on reading the Appendix, wrote to a friend saying "I regard this young geometer Bolyai as a genius of the first order." In 1848 Bolyai discovered not only that Lobachevsky had published a similar piece of work in 1829, but also a generalisation of this theory.

Nikolai Ivanovich Lobachevsky (December 1, 1792–February 24, 1856 (N.S.) was a Russian mathematician. Lobachevsky's main achievement is the development (independently from János Bolyai) of non–Euclidean geometry. Before him, mathematicians were trying to deduce Euclid's fifth postulate from other axioms. Lobachevsky would instead develop a geometry in which the fifth postulate was not true.

[19] Felix Christian Klein (April 25, 1849, Düsseldorf, Germany – June 22, 1925, Göttingen) was a German mathematician, known for his work in group theory, function theory, non-Euclidean geometry, and on the connections between geometry and group theory. His 1872 Erlangen Program, classifying geometries by their underlying symmetry groups, was a hugely influential synthesis of much of the mathematics of the day.

[20] Nicolas Bourbaki is the collective allonym under which a group of (mainly French) 20th-century mathematicians wrote a series of books presenting an exposition of modern advanced mathematics, beginning in 1935. With the goal of founding all of mathematics on set theory, the group strove for utmost rigour and generality, creating some new terminology and concepts along the way.

While Nicolas Bourbaki is an invented personage, the Bourbaki group is officially known as the Association des collaborateurs de Nicolas Bourbaki

The main disadvantage of abstraction is that highly abstract concepts are more difficult to learn, and require a degree of mathematical maturity and experience before they can be assimilated.

In computer science, abstraction is a mechanism and practice to reduce and factor out details so that one can focus on a few concepts at a time.

The concept is by analogy with abstraction in mathematics. The mathematical technique of abstraction begins with mathematical definitions; this has the fortunate effect of finessing some of the vexing philosophical issues of abstraction. For example, in both computing and in mathematics, numbers are concepts in the programming languages, as founded in mathematics. Implementation details depend on the hardware and software, but this is not a restriction because the computing concept of number is still based on the mathematical concept.

Roughly speaking, abstraction can be either that of control or data. Control abstraction is the abstraction of actions while data abstraction is that of data structures. For example, control abstraction in structured programming is the use of subprograms and formatted control flows. Data abstraction is to allow for handling data bits in meaningful manners. For example, it is the basic motivation behind data–type. Object–oriented programming can be seen as an attempt to abstract both data and code.

Creativity

Now, recall that *creativity* is a mental process involving the generation of new ideas or concepts, or new associations between existing ideas or concepts. From a scientific point of view, the products of creative thought (sometimes referred to as divergent thought) are usually considered to have both originality and

('association of collaborators of Nicolas Bourbaki'), which has an office at the École Normale Supérieure in Paris.

The emphasis on rigour may be seen as a reaction to the work of Jules–Henri Poincaré, who stressed the importance of free–flowing mathematical intuition, at a cost in completeness (i.e., proof) in presentation. The impact of Bourbaki's work initially was great on many active research mathematicians world–wide.

Notations introduced by Bourbaki include: the symbol $\varnothing$ for the *empty set*, and the terms *injective*, *surjective*, and *bijective*.

Aiming at a completely self-contained treatment of most of modern mathematics based on set theory, the group produced the following volumes (with the original French titles in parentheses):

I Set theory (Théorie des ensembles);

II Algebra (Algèbre);

III General Topology (Topologie générale);

IV Functions of one real variable (Fonctions d'une variable réelle);

V Topological vector spaces (Espaces vectoriels topologiques);

VI Integration (Intégration);

VII Commutative algebra (Algèbre commutative); and

VIII Lie groups and algebras (Groupes et algèbres de Lie).

appropriateness. An alternative, more everyday conception of creativity is that it is simply the act of making something new. Although intuitively a simple phenomenon, it is in fact quite complex. It has been studied from the perspectives of behavioral psychology, social psychology, psychometrics, cognitive science, artificial intelligence, philosophy, history, economics, design research, business, and management, among others. The studies have covered everyday creativity, exceptional creativity and even artificial creativity. Unlike many phenomena in science, there is no single, authoritative perspective or definition of creativity. Unlike many phenomena in psychology, there is no standardized measurement technique.

Creativity has been attributed variously to divine intervention, cognitive processes, the social environment, personality traits, and chance ('accident', 'serendipity'). It has been associated with genius, mental illness and humor. Some say it is a trait we are born with; others say it can be taught with the application of simple techniques. Although popularly associated with art and literature, it is also an essential part of innovation and invention and is important in professions such as business, economics, architecture, industrial design, science and engineering.

Despite, or perhaps because of, the ambiguity and multi–dimensional nature of creativity, entire industries have been spawned from the pursuit of creative ideas and the development of creativity techniques. This mysterious phenomenon, though undeniably important and constantly visible, seems to lie tantalizingly beyond the grasp of scientific investigation.

More than 60 different definitions of creativity can be found in the psychological literature (see [Tay88]). The etymological root of the word in English and most other European languages comes from the Latin 'creatus', which literally means 'to have grown'. Perhaps the most widespread conception of creativity in the scholarly literature is that creativity is manifested in the production of a creative work (for example, a new work of art or a scientific hypothesis) that is both novel and useful. Colloquial definitions of creativity are typically descriptive of activity that results in producing or bringing about something partly or wholly new; in investing an existing object with new properties or characteristics; in imagining new possibilities that were not conceived of before; and in seeing or performing something in a manner different from what was thought possible or normal previously.

A useful distinction has been made by [Rho61], between the creative person, the creative product, the creative process, and the creative 'press' or environment. Each of these factors are usually present in creative activity. This has been elaborated by [Joh72], who suggested that creative activity may exhibit several dimensions including sensitivity to problems on the part of the creative agent, originality, ingenuity, unusualness, usefulness, and appropriateness in relation to the creative product, and intellectual leadership on the part of the *creative agent*.

Boden [Bod04] noted that it is important to distinguish between ideas which are psychologically creative (which are novel to the individual mind which had the idea), and those which are historically creative (which are novel with respect to the whole of human history). Drawing on ideas from artificial intelligence, she defines psychologically creative ideas as those which cannot be produced by the same set of generative rules as other, familiar ideas.

Often implied in the notion of creativity is a concomitant presence of inspiration, cognitive leaps, or intuitive insight as a part of creative thought and action [Koe64]. Popular psychology sometimes associates creativity with right or forehead brain activity or even specifically with lateral thinking. Some students of creativity have emphasized an element of chance in the creative process. Linus Pauling,[21] asked at a public lecture how one creates scientific theories, replied that one must endeavor to come up with many ideas — then discard the useless ones.

The formal starting point of the scientific study of creativity is sometimes considered to be J. Joy Guilford's[22] address to the American Psychological Association in 1950, which helped to popularize the topic (see [SL99]). Since then, researchers from a variety of fields have studied the nature of creativity

---

[21] Linus Carl Pauling (February 28, 1901 – August 19, 1994) was an American quantum chemist and biochemist, widely regarded as the premier chemist of the twentieth century. Pauling was a pioneer in the application of quantum mechanics to chemistry (quantum mechanics can, in principle, describe all of chemistry and molecular biology), and in 1954 was awarded the Nobel Prize in chemistry for his work describing the nature of chemical bonds. He also made important contributions to crystal and protein structure determination, and was one of the founders of molecular biology. Pauling is noted as a versatile scholar for his expertise in inorganic chemistry, organic chemistry, metallurgy, immunology, anesthesiology, psychology, debate, radioactive decay, and the aftermath of nuclear weapons, in addition to quantum mechanics and molecular biology.

Pauling received the Nobel Peace Prize in 1962 for his campaign against above-ground nuclear testing, becoming the only person in history to individually receive two Nobel Prizes (Marie Curie won Nobel Prizes in physics and chemistry, but shared the former and won the latter individually; John Bardeen won two Nobel Prizes in the field of physics, but both were shared; Frederick Sanger won two Nobel Prizes in chemistry, but one was shared).

Later in life, he became an advocate for regular consumption of massive doses of vitamin C, which is still regarded as unorthodox by conventional medicine.

[22] Joy Paul Guilford (1897–1988) was a US psychologist, best remembered for his psychometric study of human intelligence.

He graduated from the University of Nebraska before studying under Edward Titchener at Cornell. He then held a number of posts at Nebraska and briefly at the University of Southern California before becoming Director of Psychological Research at Santa Ana Army Air Base in 1941. There he worked on the selection and ranking of air–crew trainees.

Developing the views of L. L. Thurstone, Guilford rejected Charles Spearman's view that intelligence could be characterized in a single numerical parameter and proposed that three dimensions were necessary for accurate description: (i) content, (ii) operations, and (iii) productions. He made the important distinction between convergent and divergent production.

from a scientific point of view. Others have taken a more pragmatic approach, teaching practical creativity techniques. Three of the best–known are Alex Osborn's[23] *brainstorming* techniques, Genrikh Altshuller's[24] 'Theory of Inventive Problem Solving' (TIPS), and Edward de Bono's[25] *lateral thinking* (1960s to present).

The *neurology of creativity* has been discussed by F. Balzac in [Bal06]. The study found that creative innovation requires *coactivation and communication between regions of the brain that ordinarily are not strongly connected.* Highly creative people who excel at creative innovation tend to differ from others in three ways: they have a high level of specialized knowledge, they are capable of divergent thinking mediated by the frontal lobe, and they are able to modulate neurotransmitters such as norepinephrine in their frontal lobe. Thus, the frontal lobe appears to be the part of the cortex that is most important for creativity. The study also explored the links between creativity and sleep, mood and addiction disorders, and depression.

J. Guilford's group developed the so–called 'Torrance Tests of Creative Thinking'. They involved simple tests of divergent thinking and other problem–solving skills, which were scored on [Gui67]:

1. Fluency: the total number of interpretable, meaningful, and relevant ideas generated in response to the stimulus;
2. Flexibility: the number of different categories of relevant responses;

---

[23] Alex Faickney Osborn (May 24, 1888 – May 4, 1966) was an advertising manager and the author of the creativity technique named *brainstorming*.

[24] Genrikh Saulovich Altshuller (October 15, 1926 - September 24, 1998), created the Theory of Inventive Problem Solving (TIPS). Working as a clerk in a patent office, Altshuller embarked on finding some generic rules that would explain creation of new, inventive, patentable ideas.

[25] Edward de Bono (born May 19, 1933) is a psychologist and physician. De Bono writes prolifically on subjects of lateral thinking, a concept he is believed to have pioneered and now holds training seminars in. Dr. de Bono is also a world-famous consultant who has worked with companies like Coca-cola and Ericsson. In 1979 he co–founded the School of Thinking with Dr Michael Hewitt–Gleeson.

De Bono has detailed a range of 'deliberate thinking methods' – applications emphasizing thinking as a deliberate act rather than a reactive one. His writing style is simple and clear, though often criticized for being dry and repetitive. Avoiding academic terminology, he has advanced applied psychology by making theories about creativity and perception into usable tools. A distinctive feature of De Bono's books is that he never acknowledges or credits the ideas of other authors or researchers in the field of creativity.

De Bono's work has become particularly popular in the sphere of business – perhaps because of the perceived need to restructure corporations, to allow more flexible working practices and to innovate in products and services. The methods have migrated into corporate training courses designed to help employees and executives 'think out of the box' / 'think outside the box'.

3. Originality: the statistical rarity of the responses among the test subjects; and
4. Elaboration: the amount of detail in the responses.

Personality

On the other hand, *personality* is a *collection of emotional, thought and behavioral patterns unique to a person* that is consistent over time. Personality psychology is a branch of psychology which studies personality and individual different processes – that which makes us into a person. One emphasis is on trying to create a coherent picture of a person and all his or her major psychological processes. Another emphasis views it as the study of individual differences. These two views work together in practice. Personality psychologists are interested in broad view of the individual. This often leads to an interest in the most salient individual differences among people.

The word *personality* originates from the Latin *persona*, which means 'mask'.[26] In the History of theater of the ancient Latin world, the mask was not used as a plot device to disguise the identity of a character, but rather was a convention employed to represent, or typify that character.

There are several theoretical perspectives on personality in psychology, which involve different ideas about the relationship between personality and other psychological constructs, as well as different theories about the way personality develops. Most theories can be grouped into one of the following classes.

Generally the opponents to personality theories claim that personality is 'plastic' in time, places, moods and situations. Changing personality may in fact resulting from diet (or lack of), medical effects, historical or subsequent events, or learning. Stage managers (of many types) are especially skilled in changing a person's resulting 'personality'. Most personality theories will not cover such flexible nor unusual people situations. Therefore, although personality theories do not define personality as 'plastic' over time like their opponents, they do imply a drastic change in personality is highly unusual.

According to the Diagnostic and Statistical Manual of Mental Disorders of the American Psychiatric Association, personality traits are 'prominent aspects of personality that are exhibited in a wide range of important social and personal contexts.' In other words, persons have certain characteristics which partly determine their behavior. According to the theory, a friendly

---

[26] A *persona*, in the word's everyday usage, is a social role, or a character played by an actor. The word derives from the Latin for 'mask' or 'character', derived from the Etruscan word 'phersu', with the same meaning.

For instance, in Dostoevsky's novel, Notes from Underground (generally considered to be the first existentialist novel), the narrator ought not to be conflated with Dostoevsky himself, despite the fact that Dostoevsky and his narrator may or may not have shared much in common. In this sense, the persona is basically a mouthpiece for a particular world–view.

person is likely to act friendly in any situation because of the traits in his personality. One criticism of trait models of personality as a whole is that they lead professionals in clinical psychology and lay–people alike to accept classifications, or worse offer advice, based on superficial analysis of one's profile.

The most common models of traits incorporate four or five broad dimensions or factors. The least controversial dimension, observed as far back as the ancient Greeks, is simply extraversion vs. introversion (outgoing and physical–stimulation–oriented vs. quiet and physical–stimulation–averse).

Gordon Allport[27] delineated different kinds of traits, which he also called dispositions. Central traits are basic to an individual's personality, while secondary traits are more peripheral. Common traits are those recognized within a culture and thus may vary from culture to culture. Cardinal traits are those by which an individual may be strongly recognized.

Raymond Cattell's[28] research propagated a two–tiered personality structure with sixteen 'primary factors' (16 Personality Factors) and five 'secondary factors' (see Table 1.1). Cattell referred to these 16 factors as *primary factors*, as opposed to the so–called 'Big Five' factors which he considered *global factors*. All of the primary factors correlate with global factors and could therefore be considered subfactors within them.

---

[27] Gordon Willard Allport (November 11, 1897 - October 9, 1967) was an American psychologist. He was born in Montezuma, Indiana, the youngest of four brothers. One of his older brothers, Floyd Henry Allport, was an important and influential psychologist as well. Gordon W. Allport was a long time and influential member of the faculty at Harvard University from 1930-1967. His works include Becoming, Pattern and Growth in Personality, The Individual and his Religion, and perhaps his most influential book The Nature of Prejudice.

Allport was one of the first psychologists to focus on the study of the personality, and is often referred to as one of the fathers of personality psychology. Characteristically for this ecletic and pluralistic thinker, he was also an important contributor to social psychology as well. He rejected both a psychoanalytic approach to personality, which he thought often went too deep, and a behavioral approach, which he thought often did not go deep enough. He emphasized the uniqueness of each individual, and the importance of the present context, as opposed to past history, for understanding the personality.

[28] Raymond Bernard Cattell (20 March 1905 – 2 February 1998) was a British and American psychologist who theorized the existence of fluid and crystallized intelligences to explain human cognitive ability. He was famously productive throughout his 92 years, and ultimately was able to claim a combined authorship and co–authorship of 55 books and some 500 journal articles in addition to at least 30 standardized tests. His legacy includes not just that intellectual production, but also a spirit of scientific rigor brought to an otherwise soft science and kept burning by his students and co–researchers whom he was survived by.

In keeping with his devotion to rigorous scientific method, Cattell was an early proponent of the application in psychology of factor analytical methods, in place of what he called mere 'verbal theorizing.' One of the most important results of Cattell's application of factor analysis was the derivation of 16 factors underlying

Table 1.1. Cattell's 16 Personality Factors

Descriptors of Low Range	Primary Factor	Descriptors of High Range
Impersonal, distant, cool, reserved, detached, formal, aloof (Sizothymia)	Warmth	Warm, outgoing, attentive to others, kindly, easy going, participating, likes people (Affectothymia)
Concrete thinking, lower general mental capacity, less intelligent, unable to handle abstract problems (Lower Scholastic Mental Capacity)	Reasoning	Abstract–thinking, more intelligent, bright, higher general mental capacity, fast learner (Higher Scholastic Mental Capacity)
Reactive emotionally, changeable, affected by feelings, emotionally less stable, easily upset (Lower Ego Strength)	Emotional Stability	Emotionally stable, adaptive, mature, faces reality calm (Higher Ego Strength)
Deferential, cooperative, avoids conflict, submissive, humble, obedient, easily led, docile, accommodating (Submissiveness)	Dominance	Dominant, forceful, assertive, aggressive, competitive, stubborn, bossy (Dominance)
Serious, restrained, prudent, taciturn, introspective, silent (Desurgency)	Liveliness	Lively, animated, spontaneous, enthusiastic, happy go lucky, cheerful, expressive, impulsive (Surgency)
Expedient, nonconforming, disregards rules, self indulgent (Low Super Ego Strength)	Rule–Consciousness	Rule–conscious, dutiful, conscientious, conforming, moralistic, staid, rule bound (High Super Ego Strength)
Shy, threat–sensitive, timid, hesitant, intimidated (Threctia)	Social Boldness	Socially bold, venturesome, thick skinned, uninhibited (Parmia)
Utilitarian, objective, unsentimental, tough minded, self–reliant, no–nonsense, rough (Harria)	Sensitivity	Sensitive, aesthetic, sentimental, tender minded, intuitive, refined (Premsia)
Trusting, unsuspecting, accepting, unconditional, easy (Alaxia)	Vigilance	Vigilant, suspicious, skeptical, distrustful, oppositional (Protension)
Grounded, practical, prosaic, solution oriented, steady, conventional (Praxernia)	Abstractedness	Abstract, imaginative, absent minded, impractical, absorbed in ideas (Autia)
Forthright, genuine, artless, open, guileless, naive, unpretentious, involved (Artlessness)	Privateness	Private, discreet, nondisclosing, shrewd, polished, worldly, astute, diplomatic (Shrewdness)

Self–Assured, unworried, complacent, secure, free of guilt, confident, self satisfied (Untroubled)	Apprehension	Apprehensive, self doubting, worried, guilt prone, insecure, worrying, self blaming (Guilt Proneness)
Traditional, attached to familiar, conservative, respecting traditional ideas (Conservatism)	Openness to Change	Open to change, experimental, liberal, analytical, critical, free thinking, flexibility (Radicalism)
Group–oriented, affiliative, a joiner and follower dependent (Group Adherence)	Self–Reliance	Self–reliant, solitary, resourceful, individualistic, self sufficient (Self-Sufficiency)
Tolerated disorder, unexacting, flexible, undisciplined, lax, self-conflict, impulsive, careless of social rules, uncontrolled (Low Integration)	Perfectionism	Perfectionistic, organized, compulsive, self-disciplined, socially precise, exacting will power, control, self-sentimental (High Self–Concept Control)
Relaxed, placid, tranquil, torpid, patient, composed low drive (Low Ergic Tension)	Tension	Tense, high energy, impatient, driven, frustrated, over wrought, time driven. (High Ergic Tension)

A different model was proposed by Hans Eysenck,[29] who believed that just three traits: *extroversion*, *neuroticism* and *psychoticism* – were sufficient to describe human personality. Eysenck was one of the first psychologists to study personality with the method of *factor analysis*, a statistical technique introduced by Charles Spearman[30] and expanded by Raymond Cattell. Eysenck's

human personality. He called these 16 factors source traits because he believed that they provide the underlying source for the surface behaviors that we think of as personality. ('Psychology and Life, 7 ed.' by Richard Gerrig and Philip Zimbardo.) This theory of 16 personality factors and the instruments used to measure them are known collectively as the 16 Personality Factors.

29 Hans Jürgen Eysenck (March 4, 1916 – September 4, 1997) was an eminent psychologist, most remembered for his work on intelligence and personality, though he worked in a wide range of areas. At the time of his death, Eysenck was the living psychologist most frequently cited in science journals.

Hans Eysenck was born in Germany, but moved to England as a young man in the 1930s because of his opposition to the Nazi party. Eysenck was the founding editor of the journal Personality and Individual Differences, and authored over 50 books and over 900 academic articles. He aroused intense debate with his controversial dealing with variation in IQ among racial groups.

30 Charles Edward Spearman (September 10, 1863 - September 7, 1945) was an English psychologist known for work in statistics, as a pioneer of factor analysis, and for Spearman's rank correlation coefficient. He also did seminal work on models for human intelligence, including his theory that disparate cognitive test scores reflect a single general factor and coining the term g factor. Spearman had an unusual background for a psychologist. After 15 years as an officer in the British Army he resigned to study for a PhD in experimental psychology. In Britain

results suggested two main personality factors [Eys92a, Eys92b]. The first factor was the tendency to experience negative emotions, and Eysenck referred to it as 'neuroticism'. The second factor was the tendency enjoy positive events, especially social events, and Eysenck named it 'extraversion'. The two personality dimensions were described in his 1947 book 'Dimensions of Personality'. It is common practice in personality psychology to refer to the dimensions by the first letters, E and N. E and N provided a 2–dimensional space to describe individual differences in behavior. An analogy can be made to how latitude and longitude describe a point on the face of the earth. Also, Eysenck noted how these two dimensions were similar to the four personality types first proposed by the ancient Greek physician Galen[31]:

---

psychology was generally seen as a branch of philosophy and Spearman chose to study in Leipzig under Wilhelm Wundt. Besides Spearman had no conventional qualifications and Leipzig had liberal entrance requirements. He started in 1897 and after some interruption (he was recalled to the army during the South African War) he obtained his degree in 1906. He had already published his seminal paper on the factor analysis of intelligence (1904). Spearman met and impressed the psychologist William McDougall who arranged for Spearman to replace him when he left his position at University College London. Spearman stayed at University College until he retired in 1931. Initially he was Reader and head of the small psychological laboratory. In 1911 he was promoted to the Grote professorship of the Philosophy of Mind and Logic. His title changed to Professor of Psychology in 1928 when a separate Department of Psychology was created. When Spearman was elected to the Royal Society in 1924 the citation read "Dr. Spearman has made many researches in experimental psychology. His many published papers cover a wide field, but he is especially distinguished by his pioneer work in the application of mathematical methods to the analysis of the human mind, and his original studies of correlation in this sphere. He has inspired and directed research work by many pupils."

Spearman was strongly influenced by the work of Francis Galton. Galton did pioneering work in psychology and developed correlation, the main statistical tool used by Spearman. Spearman developed rank correlation (1904) and the widely used correction for attenuation (1907). His statistical work was not appreciated by his University College colleague Karl Pearson and there was long feud between them. Although Spearman achieved most recognition for his statistical work, he regarded this work as subordinate to his quest for the fundamental laws of psychology (see [WZZ03] for details).

31 Galen, (Latin: Claudius Galenus of Pergamum) was an ancient Greek physician. The forename 'Claudius' is absent in Greek texts; it was first documented in texts from the Renaissance. Galen's views dominated European medicine for over a thousand years.

Galen transmitted Hippocratic medicine all the way to the Renaissance. His On the Elements According to Hippocrates describes the philosopher's system of four bodily humours, blood, yellow bile, black bile and phlegm, which were identified with the four classical elements, and in turn with the seasons. He created his own theories from those principles, and much of Galen's work can be seen as building on the Hippocratic theories of the body, rather than being purely innovative. In

1. High N and High E = Choleric type;
2. High N and Low E = Melancholic type;
3. Low N and High E = Sanguine type; and
4. Low N and Low E = Phlegmatic type.

The third dimension, 'psychoticism', was added to the model in the late 1970s, based upon collaborations between Eysenck and his wife, Sybil B.G. Eysenck, the current editor of Personality and Individual Differences (see [Eys69, Eys76]).

The major strength of Eysenck's model was to provide detailed theory of the causes of personality (see his 1985 book 'Decline and Fall of the Freudian Empire'). For example, Eysenck proposed that extraversion was caused by variability in cortical arousal; 'introverts are characterized by higher levels of activity than extraverts and so are chronically more cortically aroused than extraverts'. While it seems counterintuitive to suppose that introverts are more aroused than extraverts, the putative effect this has on behavior is such that the introvert seeks lower levels of stimulation. Conversely, the extravert seeks to heighten their arousal to a more optimal level (as predicted by the *Yerkes–Dodson Law*) by increased activity, social engagement and other stimulation–seeking behaviors.

Differences between Cattell and Eysenck emerged due to preferences for different forms of factor analysis, with Cattell using oblique, Eysenck orthogonal, rotation to analyze the factors that emerged when personality questionnaires were subject to statistical analysis. Today, the Big Five factors have the weight of a considerable amount of empirical research behind them. Building on the work of Cattell and others, Lewis Goldberg[32] proposed a five–dimensional personality model, nicknamed the 'Big Five' personality traits:

Extroversion (i.e., 'extroversion vs. introversion' above; outgoing and physical–stimulation–oriented vs. quiet and physical–stimulation–averse);

---

turn, he mainly ignored Latin writings of Celsus, but accepted that the ancient works of Asclepiades had sound theory.

Galen's own theories, in accord with Plato's, emphasized purposeful creation by a single Creator ('Nature' – Greek 'phusis') – a major reason why later Christian and Muslim scholars could accept his views. His fundamental principle of life was pneuma (air, breath) that later writers connected with the soul. These writings on philosophy were a product of Galen's well rounded education, and throughout his life Galen was keen to emphasize the philosophical element to medicine. Pneuma physicon (animal spirit) in the brain took care of movement, perception, and senses. Pneuma zoticon (vital spirit) in the heart controlled blood and body temperature. 'Natural spirit' in the liver handled nutrition and metabolism. However, he did not agree with the Pneumatist theory that air passed through the veins rather than blood.

[32] Lewis R. Goldberg is an American personality psychologist and a professor emeritus at the University of Oregon. Among his other accomplishments, Goldberg is closely associated with the Big Five taxonomy of personality. He has published well over 100 research articles and has been active on editorial boards.

1. Neuroticism (i.e., emotional stability; calm, unperturbable, optimistic vs. emotionally reactive, prone to negative emotions);
2. Agreeableness (i.e., affable, friendly, conciliatory vs. aggression aggressive, dominant, disagreeable);
3. Conscientiousness (i.e., dutiful, planful, and orderly vs. spontaneous, flexible, and unreliable); and
4. Openness to experience (i.e., open to new ideas and change vs. traditional and staid).

Character

A *character structure* is a system of relatively permanent motivational and other traits that are manifested in the characteristic ways that an individual relates to others and reacts to various kinds of challenges. The word 'structure' indicates that these several characteristics and/or learned patterns of behavior are linked in such a way as to produce a state that can be highly resistant to change. The idea has its roots in the work of Sigmund Freud[33] and several of his followers, the most important of whom (in this respect) is Erich Fromm.[34] Among other important participants in the establishment of this concept must surely be counted Erik Erikson.[35]

Among the earliest factors that determine an individual's eventual character structure are his or her genetic characteristics and early childhood nurture and education. A child who is well nurtured and taught in a relatively benign and consistent environment by loving adults who intend that the child should

[33] Sigmund Freud (May 6, 1856–September 23, 1939) was an Austrian neurologist and the founder of the psychoanalytic school of psychology. Freud is best known for his studies of sexual desire, repression, and the unconscious mind. He is commonly referred to as 'the father of psychoanalysis' and his work has been tremendously influential in the popular imagination–popularizing such notions as the unconscious, defence mechanisms, Freudian slips and dream symbolism – while also making a long-lasting impact on fields as diverse as literature, film, marxist and feminist theories, literary criticism, philosophy, and of course, psychology.

[34] Erich Pinchas Fromm (March 23, 1900 – March 18, 1980) was an internationally renowned German-American psychologist and humanistic philosopher. He is associated with what became known as the Frankfurt School of critical thinkers.

Central to Fromm's world view was his interpretation of the Talmud, which he began studying as a young man under Rabbi J. Horowitz and later studied under Rabbi Salman Baruch Rabinkow while working towards his doctorate in sociology at the University of Heidelberg and under Nehemia Nobel and Ludwig Krause while studying in Frankfurt. Fromm's grandfather and two great grandfathers on his father's side were rabbis, and a great uncle on his mother's side was a noted Talmudic scholar. However, Fromm turned away from orthodox Judaism in 1926 and turned towards secular interpretations of scriptural ideals.

[35] Erik Homburger Erikson (June 15, 1902 – May 12, 1994) was a developmental psychologist and psychoanalyst known for his theory on social development of human beings, and for coining the phrase identity crisis.

learn how to make objective appraisals regarding the environment will be likely to form a normal or productive character structure. On the other hand, a child whose nurture and/or education are not ideal, living in a treacherous environment and interacting with adults who do not take the long–term interests of the child to heart will be more likely to form a pattern of behavior that suits the child to avoid the challenges put forth by a malign social environment. The means that the child invents to make the best of a hostile environment. Although this may serve the child well while in that bad environment, it may also cause the child to react in inappropriate ways, ways damaging to his or her own interests, when interacting with people in a more ideal social context. Major trauma that occurs later in life, even in adulthood, can sometimes have a profound effect. However, character may also develop in a positive way according to how the individual meets the psychosocial challenges of the life cycle (Erikson).

Freud's first paper on character described the anal character consisting of stubbornness, stinginess and extreme neatness. He saw this as a reaction formation to the child's having to give up pleasure in anal eroticism.The positive version of this character is the conscientious, inner directed obsessive. Freud also described the erotic character as both loving and dependent. And the narcissistic character as the natural leader, aggressive and independent because of not internalizing a strong super ego.

For Erich Fromm, character develops as the way in which an individual structures modes of assimilation and relatedness. The character types are almost identical to Freud's but Fromm gives them different names, receptive, hoarding, exploitative. Fromm adds the marketing type as the person who continually adapts the self to succeed in the new service economy. For Fromm, character types can be productive or unproductive. Fromm notes that character structures develop in each individual to enable him or her to interact successfully within a given society, to adapt to its mode of production and social norms may be very counter–productive when used in a different society.

Wisdom

On the other hand, *wisdom* is the ability, developed through experience, insight and reflection, to discern truth and exercise good judgment. It is sometimes conceptualized as an especially well developed form of common sense. Most psychologists regard wisdom as distinct from the cognitive abilities measured by standardized intelligence tests. Wisdom is often considered to be a trait that can be developed by experience, but not taught. When applied to practical matters, the term wisdom is synonymous with prudence. Some see wisdom as a quality that even a child, otherwise immature, may possess independent of experience or complete knowledge. The status of wisdom or prudence as a virtue is recognized in cultural, philosophical and religious sources. Some define wisdom in a utilitarian sense, as foreseeing consequences and acting to maximize the long–term common good.

A standard philosophical definition says that wisdom consists of making the best use of available knowledge. As with all decisions, a wise decision may be made with incomplete information. The technical philosophical term for the opposite of wisdom is folly. For example, in his *Metaphysics*, Aristotle defines wisdom as knowledge of causes: why things exist in a particular fashion.

Beyond the simple expedient of experience (which may be considered the most difficult way to gain wisdom as through the 'school of hard knocks'), there are a variety of other avenues to gaining wisdom which vary according to different philosophies. For example, the so–called *freethinkers*[36]believe that wisdom may come from pure reason and perhaps experience. Recall that *freethought* is a philosophical doctrine that holds that beliefs should be formed on the basis of science and logical principles and not be comprised by authority, tradition or any other dogmatic or otherwise fallacious belief system that restricts logical reasoning. The cognitive application of freethought is known as *freethinking*, and practitioners of freethought are known as freethinkers. Freethought holds that individuals should neither accept nor reject ideas proposed as truth without recourse to knowledge and reason. Thus, freethinkers strive to build their beliefs on the basis of facts, scientific inquiry, and logical principles, independent of the factual/logical fallacies and intellectually-limiting effects of authority, cognitive bias, conventional wisdom, popular culture, prejudice, sectarianism, tradition, urban legend and all other dogmatic or otherwise fallacious principles. When applied to religion, the philosophy of freethought holds that, given presently–known facts, established scientific theories, and logical principles, there is insufficient evidence to support the existence of supernatural phenomena. A line from 'Clifford's Credo' by the 19th Century British mathematician and philosopher William Clifford[37] perhaps best describes the premise of freethought: "It is wrong always,

---

[36] Freethought is a philosophical doctrine that holds that beliefs should be formed on the basis of science and logical principles and not be comprised by authority, tradition or any other dogmatic or otherwise fallacious belief system that restricts logical reasoning. The cognitive application of freethought is known as freethinking, and practitioners of freethought are known as freethinkers.

[37] William Kingdon Clifford, FRS (May 4, 1845 – March 3, 1879) was an English mathematician who also wrote a fair bit on philosophy. Along with Hermann Grassmann, he invented what is now termed *geometric algebra*, a special case being the *Clifford algebras* named in his honour, which play a role in contemporary mathematical physics. He was the first to suggest that gravitation might be a manifestation of an underlying geometry. His philosophical writings coined the phrase 'mind–stuff'.

Influenced by Riemann and Lobachevsky, Clifford studied non–Euclidean geometry. In 1870, he wrote On the space theory of matter, arguing that energy and matter are simply different types of curvature of space. These ideas later played a fundamental role in Albert Einstein's general theory of relativity. Yet Clifford is now best remembered for his eponymous Clifford algebras, a type of associative algebra that generalizes the complex numbers and William Rowan Hamilton's *quaternions*. The latter resulted in the octonions (biquaternions),

everywhere, and for anyone, to believe anything upon insufficient evidence." Since many popular beliefs are based on dogmas, freethinkers' opinions are often at odds with commonly–established views.

On the other hand, there is also a common belief that wisdom comes from *intuition* or, 'superlogic', as it is called by Tony Buzan,[38] inventor of *mind maps*. For example, *holists* believe that wise people sense, work with and align themselves and others to life. In this view, wise people help others appreciate the fundamental interconnectedness of life. Also, some religions hold that wisdom may be given as a gift from God. For example, *Buddha* taught that a wise person is endowed with good bodily conduct, good verbal conduct and good mental conduct and a wise person does actions that are unpleasant to do but give good results and doesn't do actions that are pleasant to do but give bad results; this is called *karma*. According to *Hindu scriptures*, spiritual wisdom – *jnana* alone can lead to liberation. *Confucius* stated that wisdom can be learned by three methods: (i) *reflection* (the noblest), (ii) *imitation* (the easiest) and (iii) *experience* (the bitterest).

1.1.1 Human Intelligence

At least two major 'consensus' definitions of intelligence have been proposed. First, from 'Intelligence: Knowns and Unknowns', a report of a task force convened by the *American Psychological Association*[39] in 1995 (see [APS98]):

---

which he employed to study motion in non–Euclidean spaces and on certain surfaces, now known as Klein–Clifford spaces. He showed that spaces of constant curvature could differ in topological structure. He also proved that a Riemann surface is topologically equivalent to a box with holes in it. On Clifford algebras, quaternions, and their role in contemporary mathematical physics.

[38] Tony Buzan (1942–) is the originator of mind mapping and coined the term mental literacy. He was born in London and received double Honours in psychology, English, mathematics and the General Sciences from the University of British Columbia in 1964. He is probably best known for his book, Use Your Head, his promotion of mnemonic systems and his mind–mapping techniques. Following his 1970s series for the BBC, many of his ideas have been set into his series of five books: Use Your Memory, Master Your Memory, Use Your Head, The Speed Reading Book and The Mind Map Book.

In essence, Buzan teaches "Learn how your brain learns rapidly and naturally." His work is partly based on the explosion of brain research that has taken place since the late 1950s, and the work on the left and right brain by psychologist Robert Ornstein and Nobel Laureate Roger Wolcott Sperry.

[39] The American Psychological Association (APA) is a professional organization representing psychology in the US. It has around 150,000 members and an annual budget of around $70m. The APA mission statement is to "advance psychology as a science and profession and as a means of promoting health, education, and human welfare." The APA was founded in July 1892 at Clark University by a group of 26 men. Its first president was G. Stanley Hall. There are currently 54

Individuals differ from one another in their ability to understand complex ideas, to adapt effectively to the environment, to learn from experience, to engage in various forms of reasoning, to overcome obstacles by taking thought. Although these individual differences can be substantial, they are never entirely consistent: a given person's intellectual performance will vary on different occasions, in different domains, as judged by different criteria. Concepts of 'intelligence' are attempts to clarify and organize this complex set of phenomena.

A second definition of intelligence comes from the 'Mainstream Science on Intelligence', which was signed by 52 intelligence researchers in 1994 (also see [APS98]): Intelligence is a very general mental capability that, among other things, involves the ability to reason, plan, solve problems, think abstractly, comprehend complex ideas, learn quickly and learn from experience. It is not merely book learning, a narrow academic skill, or test–taking smarts. Rather, it reflects a broader and deeper capability for comprehending our surroundings, i.e., 'catching on', 'making sense' of things, or 'figuring out' what to do.

Individual intelligence experts have offered a number of similar definitions:

(i) David Wechsler:[40] " ... the aggregate or global capacity of the individual to act purposefully, to think rationally, and to deal effectively with his environment."
(ii) Cyril Burt:[41] " ... innate general cognitive ability."
(iii) Howard Gardner:[42] " To my mind, a human intellectual competence must entail a set of skills of problem solving, enabling the individual to resolve genuine problems or difficulties that he or she encounters and, when appropriate, to create an effective product, and must also entail the potential for finding or creating problems, and thereby laying the groundwork for the acquisition of new knowledge."

---

professional divisions in the APA. It is affiliated with 58 state and territorial and Canadian provincial associations.

[40] David Wechsler (January 12, 1896, Lespedi, Romania – May 2, 1981, New York, New York) was a leading Romanian-American psychologist. He developed well–known intelligence scales, such as the Wechsler Adult Intelligence Scale (WAIS) and the Wechsler Intelligence Scale for Children (WISC).

[41] Sir Cyril Lodowic Burt (March 3, 1883 — October 10, 1971) was a prominent British educational psychologist. He was a member of the London School of Differential Psychology. Some of his work was controversial for its conclusions that genetics substantially influence mental and behavioral traits. After his death, he was famously accused of scientific fraud.

[42] Howard Gardner (born in Scranton, Pennsylvania, USA in 1943) is a psychologist based at Harvard University best known for his theory of multiple intelligences. In 1981 he was awarded a MacArthur Prize Fellowship.

(iv) Richard Herrnstein[43] and Charles Murray: " ... cognitive ability."
(v) Robert Sternberg:[44] "... goal–directed adaptive behavior."

Psychometric Definition of Intelligence and Its Criticisms

Despite the variety of concepts of intelligence, the most influential approach to understanding intelligence (i.e., with the most supporters and the most published research over the longest period of time) is based on *psychometric testing*,[45] which regards intelligence as *cognitive ability.*

[43] Richard J. Herrnstein (May 20, 1930 – September 13, 1994) was a prominent researcher in comparative psychology who did pioneering work on pigeon intelligence employing the Experimental Analysis of Behavior and formulated the 'Matching Law' in the 1960s, a breakthrough in understanding how reinforcement and behavior are linked. He was the Edgar Pierce Professor of psychology at Harvard University and worked with B. F. Skinner in the Harvard pigeon lab, where he did research on choice and other topics in behavioral psychology. Herrnstein became more broadly known for his work on the correlation between race and intelligence, first in the 1970s, then with Charles Murray, discussed in their controversial best–selling 1994 book, The Bell Curve. Herrnstein described the behavior of hyperbolic discounting, in which people will choose smaller payoffs sooner instead of larger payoffs later. He developed a type of non–parametric statistics that he dubbed ρ.

[44] Robert J. Sternberg (borne 8 December 1949) is a psychologist and psychometrician and the Dean of Arts and Sciences at Tufts University. He was formerly IBM Professor of Psychology and Education at Yale University and the President of the American Psychological Association. Sternberg currently sits on the editorial board of Intelligence. Sternberg has proposed the so–called *Triarchic theory of intelligence* and a triangular theory of love. He is the creator (with Todd Lubart) of the investment theory of creativity, which states that creative people buy low and sell high in the world of ideas, and a propulsion theory of creative contributions, which states that creativity is a form of leadership.

[45] Psychometrics is the field of study concerned with the theory and technique of psychological measurement, which includes the measurement of knowledge, abilities, attitudes, and personality traits. The field is primarily concerned with the study of differences between individuals. It involves two major research tasks, namely: (i) the construction of instruments and procedures for measurement; and (ii) the development and refinement of theoretical approaches to measurement. Much of the early theoretical and applied work in psychometrics was undertaken in an attempt to measure intelligence. The origin of psychometrics has connections to the related field of psychophysics. Charles Spearman, a pioneer in psychometrics who developed approaches to the measurement of intelligence, studied under Wilhelm Wundt and was trained in psychophysics. The psychometrician L.L. Thurstone later developed and applied a theoretical approach to the measurement referred to as the law of comparative judgment, an approach which has close connections to the psychophysical theory developed by Ernst Heinrich Weber and Gustav Fechner. In addition, Spearman and Thurstone both made important contributions to the theory and application of factor analysis, a statistical

Recall that *psychometrics* is the field of study concerned with the theory and technique of psychological measurement, which includes the measurement of knowledge, abilities, attitudes, and personality traits. The field is primarily concerned with the study of differences between individuals. It involves two major research tasks, namely:

(i) the construction of instruments and procedures for measurement; and

(ii) the development and refinement of theoretical approaches to measurement. Much of the early theoretical and applied work in psychometrics was undertaken in an attempt to measure intelligence.

The origin of psychometrics has connections to the related field of psychophysics. Charles Spearman, a pioneer in psychometrics who developed approaches to the measurement of intelligence, studied under Wilhelm Wundt[46] and was trained in psychophysics. The psychometrician Louis

---

method that has been used extensively in psychometrics. More recently, psychometric theory has been applied in the measurement of personality, attitudes and beliefs, academic achievement, and in health-related fields. Measurement of these unobservable phenomena is difficult, and much of the research and accumulated art in this discipline has been developed in an attempt to properly define and quantify such phenomena. Critics, including practitioners in the physical sciences and social activists, have argued that such definition and quantification is impossibly difficult, and that such measurements are often misused. Proponents of psychometric techniques can reply, though, that their critics often misuse data by not applying psychometric criteria, and also that various quantitative phenomena in the physical sciences, such as heat and forces, cannot be observed directly but must be inferred from their manifestations. Figures who made significant contributions to psychometrics include Karl Pearson, L. L. Thurstone, Georg Rasch and Arthur Jensen.

[46] Wilhelm Maximilian Wundt (August 16, 1832–August 31, 1920) was a German physiologist and psychologist. He is generally acknowledged as a founder of experimental psychology and cognitive psychology. He is less commonly recognised as a founding figure in social psychology, however, the later years of Wundt's life were spent working on Völkerpsychologie which he understood as a study into the social basis of higher mental functioning.

Wundt combined philosophical introspection with techniques and laboratory apparatuses brought over from his physiological studies with Helmholtz, as well as many of his own design. This experimental introspection was in contrast to what had been called psychology until then, a branch of philosophy where people introspected themselves. Wundt argued in his 1904 book 'Principles of Physiological Psychology' that "we learn little about our minds from casual, haphazard self–observation... It is essential that observations be made by trained observers under carefully specified conditions for the purpose of answering a well–defined question."

The methods Wundt used are still used in modern psychophysical work, where reactions to systematic presentations of well–defined external stimuli are measured in some way–reaction time, reactions, comparison with graded colors or sounds, and so forth. His chief method of investigation was called *introspection* in the terminology of the time, though *observation* may be a better translation.

Thurstone[47] later developed and applied a theoretical approach to the measurement referred to as the law of comparative judgment, an approach which has close connections to the psychophysical theory developed by Ernst Weber and Gustav Fechner (see below). In addition, Spearman and Thurstone both made important contributions to the theory and application of factor analysis, a statistical method that has been used extensively in psychometrics. More recently, psychometric theory has been applied in the measurement of personality, attitudes and beliefs, academic achievement, and in health–related fields. Measurement of these unobservable phenomena is difficult, and much of the research and accumulated art in this discipline has been developed in an attempt to properly define and quantify such phenomena. Critics, including practitioners in the physical sciences and social activists, have argued that such definition and quantification is impossibly difficult, and that such measurements are often misused. Proponents of psychometric techniques can reply, though, that their critics often misuse data by not applying psychometric criteria, and also that various quantitative phenomena in the physical sciences, such as heat and forces, cannot be observed directly but must be inferred from their manifestations. Figures who made significant contributions to psychometrics include Karl Pearson, Louis Thurstone, Georg Rasch and Arthur Jensen.

Wundt subscribed to a 'psychophysical parallelism' (which entirely excludes the possibility of a mind–body/cause–effect relationship), which was supposed to stand above both materialism and idealism. His epistemology was an eclectic mixture of the ideas of Spinoza, Leibniz, Kant, and Hegel.

[47] Louis Leon Thurstone (29 May 1887–29 September 1955) was a U.S. pioneer in the fields of psychometrics and psychophysics. He conceived the approach to measurement known as the law of comparative judgment, and is well known for his contributions to *factor analysis*. He is responsible for the standardized mean and standard deviation of IQ scores used today, as opposed to the Intelligence Test system originally used by Alfred Binet. He is also known for the development of the Thurstone scale.

Thurstone's work in factor analysis led him to formulate a model of intelligence center around 'Primary Mental Abilities' (PMAs), which were independent group factors of intelligence that different individuals possessed in varying degrees. He opposed the notion of a singular general intelligence that factored into the scores of all psychometric tests and was expressed as a mental age. This idea was unpopular at the time due to its obvious conflicts with Spearman's 'mental energy' model, and is today still largely discredited. Nonetheless, Thurstone's contributions to methods of factor analysis have proved invaluable in establishing and verifying later psychometric factor structures, and has influenced the hierarchical models of intelligence in use in intelligence tests such as WAIS and the modern Stanford–Binet IQ test.

The *seven primary mental abilities* in Thurstone's model were *verbal comprehension, word fluency, number facility, spatial visualization, associative memory, perceptual speed and reasoning.*

Intelligence, narrowly defined by psychometrics, can be measured by intelligence tests, also called *intelligence quotient* (IQ)[48] tests. Such intelligence tests take many forms, but the common tests (*Stanford–Binet*,[49] *Raven's Progressive Matrices*,[50] Wechsler Adult Intelligence

[48] An intelligence quotient or IQ is a score derived from a set of standardized tests of intelligence. Intelligence tests come in many forms, and some tests use a single type of item or question. Most tests yield both an overall score and individual sub–tests scores. Regardless of design, all IQ tests measure the same general intelligence. Component tests are generally designed and chosen because they are found to be predictable of later intellectual development, such as educational achievement. IQ also correlates with job performance, socioeconomic advancement, and 'social pathologies'. Recent work has demonstrated links between IQ and health, longevity, and functional literacy. However, IQ tests do not measure all meanings of 'intelligence', such as creativity. IQ scores are relative (like placement in a race), not absolute (like the measurement of a ruler). The average IQ scores for many populations were rising during the 20th century: a phenomenon called the *Flynn effect*. It is not known whether these changes in scores reflect real changes in intellectual abilities. On average, IQ scores are stable over a person's lifetime, but some individuals undergo large changes. For example, scores can be affected by the presence of learning disabilities.

[49] The modern field of intelligence testing began with the Stanford-Binet IQ test. The Stanford-Binet itself started with the French psychologist Alfred Binet who was charged by the French government with developing a method of identifying intellectually deficient children for placement in special education programs. As Binet indicated, case studies may be more detailed and at times more helpful, but the time required to test large numbers of people would be huge. Unfortunately, the tests he and his assistant Victor Henri developed in 1896 were largely disappointing [Fan85].

[50] Raven's Progressive Matrices are widely used non–verbal intelligence tests. In each test item, one is asked to find the missing part required to complete a pattern. Each Set of items gets progressively harder, requiring greater cognitive capacity to encode and analyze. The test is considered by many intelligence experts to be one of the most **g**–loaded in existence. The matrices are offered in three different forms for different ability levels, and for age ranges from five through adult: (i) Colored Progressive Matrices (younger children and special groups); (ii) Standard Progressive Matrices (average 6 to 80 year olds); and (iii) Advanced Progressive Matrices (above average adolescents and adults). According to their author, Raven's Progressive Matrices and Vocabulary tests measure the two main components of general intelligence (originally identified by Spearman): the ability to think clearly and make sense of complexity, which is known as eductive ability (from the Latin root 'educere', meaning 'to draw out'; and the ability to store and reproduce information, known as reproductive ability. Adequate standardization, ease of use (without written or complex instructions), and minimal cost per person tested are the main reasons for its widespread international use in most countries of the world. It appears to measure a type of *reasoning ability* which is fundamental to making sense out of the 'booming buzzing confusion' in

Scale,[51] *Wechsler–Bellevue I*,[52] and others) all measure the same dominant form of intelligence, **g** or 'general intelligence factor'. The abstraction of **g** stems from the observation that scores on all forms of cognitive tests *positively correlate* with one another. **g** can be derived as the *principal intelligence factor* from *cognitive test scores* using the *multivariate correlation statistical method* of *factor analysis* (FA).

---

all walks of life. Thus, it has among the highest predictive validities of any test in most occupational groups and, even more importantly, in predicting social mobility ... the level of job a person will attain and retain. Although it is sometimes criticized for being costly, this is based on a failure to calculate cost per person tested with re–usable test booklets that can be used up to 50 times each. The authors of the Manual recommend that, when used in selection, RPM scores are set in the context of information relating to Raven's framework for the assessment of Competence. Some of the most fundamental research in cognitive psychology has been carried out with the RPM. The tests have been shown to work–scale–measure the same thing – in a vast variety of cultural groups. There is no truth in the assertion that the low mean scores obtained in some groups arise from a general lack of familiarity with the way of thought measured by the test. Two remarkable, and relatively recent, findings are that, on the one hand, the actual scores obtained by people living in most countries with a tradition of literacy – from China, Russia, and India through Europe to Kuwait – are very similar at any point in time. On the other hand, in all countries, the scores have increased dramatically over time ... such that 50% of our grandparents would be assigned to special education classes if they were judged against today's norms. Yet none of the common explanations (e.g., access to television, changes in education, changes in family size etc.) hold up. The explanation seems to have more in common with those put forward to explain the parallel increase in life expectancy ... which has doubled over the same period of time.

51 Wechsler Adult Intelligence Scale or WAIS is a general IQ test, published in February 1955 as a revision of the Wechsler–Bellevue test (1939), standardized for use with adults over the age of 16. In this test intelligence is quantified as the global capacity of the individual to act purposefully, to think rationally, and to deal effectively with the environment.

52 David Wechsler (January 12, 1896, Lespedi, Romania – May 2, 1981, New York, New York) was a leading Romanian–American psychologist. He developed well–known intelligence scales, such as the Wechsler Adult Intelligence Scale (WAIS) and the Wechsler Intelligence Scale for Children (WISC). The Wechsler Adult Intelligence Scale (WAIS) was developed first in 1939 and then called the Wechsler–Bellevue Intelligence Test. From these he derived the Wechsler Intelligence Scale for Children (WISC) in 1949 and the Wechsler Preschool and Primary Scale of Intelligence (WPPSI) in 1967. Wechsler originally created these tests to find out more about his patients at the Bellevue clinic and he found the then–current *Binet IQ test* unsatisfactory. The tests are still based on his philosophy that intelligence is "the global capacity to act purposefully, to think rationally, and to deal effectively with (one's) environment."

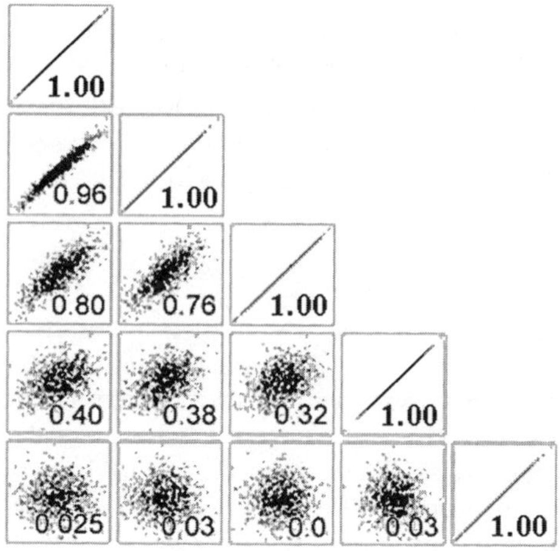

Fig. 1.1. Example of positive linear correlations between 1000 pairs of numbers. Note that each set of points correlates maximally with itself, as shown on the diagonal. Also, note that we have not plot the upper part of the correlation matrix as it is symmetrical.

Correlation and Factor Analysis

Recall that *correlation*, also called *correlation coefficient*, indicates the strength and direction of a linear relationship between two random variables (see Figure 1.1). In other words, correlation is a measure of the relation between two or more statistical variables. In general statistical usage, correlation (or, co–rrelation) refers to the departure of two variables from independence, although correlation does not imply their *functional causal relation*. In this broad sense there are several coefficients, measuring the degree of correlation, adapted to the nature of data. A number of different coefficients are used for different situations. Correlation coefficients can range from -1.00 to $+1.00$. The value of -1.00 represents a perfect negative correlation while a value of $+1.00$ represents a perfect positive correlation. The perfect correlation indicates an existence of functional relation between two statistical variables. A value of 0.00 represents a lack of correlation.Geometrically, the correlation coefficient can also be viewed as the cosine of the angle between the two vectors of samples drawn from the two random variables.

The most widely–used type of correlation simple linear coefficient is Pearson r, also called *linear* or *product–moment* correlation, which assumes that the two variables are measured on at least interval scales, and it determines the extent to which values of the two variables are 'proportional' to each other. The value of correlation coefficient does not depend on the specific measurement units used. Proportional means linearly related using regression line or least squares line. If the correlation coefficient is squared, then the resulting value (r^2, the *coefficient of determination*) will represent the proportion of common variation in the two variables (i.e., the 'strength' or 'magnitude' of the relationship). In order to evaluate the correlation between variables, it is important to know this 'magnitude' or 'strength' as well as the significance of the correlation.

The significance level calculated for each correlation is a primary source of information about the reliability of the correlation. The significance of correlation coefficient of particular magnitude will change depending on the size of the sample from which it was computed. The test of significance is based on the assumption that each of the two variables is normally distributed and that their bivariate ('combined') distribution is normal (which can be tested by examining the 3D bivariate distribution histogram). However, *Monte–Carlo* studies suggest that meeting those assumptions (especially the second one) is not absolutely crucial if our sample size is not very small and when the departure from normality is not very large. It is impossible to formulate precise recommendations based on those Monte–Carlo results, but many researchers follow a rule of thumb that if our sample size is 50 or more then serious biases are unlikely, and if our sample size is over 100 then you should not be concerned at all with the normality assumptions.

Recall that the *normal distribution*, also called *Gaussian distribution*, is an extremely important probability distribution in many fields. It is a family of distributions of the same general form, differing in their location and scale parameters: the *mean* ('average') μ and *standard deviation* ('variability') σ, respectively. The *standard normal distribution* is the normal distribution with a mean of zero and a standard deviation of one. It is often called the *bell curve* because the graph of its *probability density function pdf*, given by the *Gaussian function*

$$pdf = \frac{1}{\sigma\sqrt{2\pi}} \exp\left(-\frac{(x-\mu)^2}{2\sigma^2}\right),$$

resembles a bell shape (here, $\frac{1}{\sqrt{2\pi}}\mathrm{e}^{-x^2/2}$ is the *pdf* for the standard normal distribution). The corresponding *cumulative distribution function cdf* is defined as the probability that a variable X has a value less than or equal to x, and it is expressed in terms of the *pdf* as

$$cdf = \frac{1}{\sigma\sqrt{2\pi}} \int_{-\infty}^{x} \exp\left(-\frac{(u-\mu)^2}{2\sigma^2}\right) du.$$

Now, the correlation $r_{X,Y}$ between two *normally distributed random variables* X and Y with expected values μX and μY and standard deviations σX and σY is defined as:

$$r_{XY} = \frac{\mathrm{cov}(X,Y)}{\sigma_X \sigma_Y} = \frac{E((X-\mu_X)(Y-\mu_Y))}{\sigma_X \sigma_Y},$$

where E denotes the expected value of the variable and cov means covariance. Since $\mu X = E(X)$, $\sigma_X^2 = E(X^2) - E^2(X)$ and similarly for Y, we can write (see, e.g., [CCW03])

$$r_{XY} = \frac{E(XY) - E(X)E(Y)}{\sqrt{E(X^2) - E^2(X)}\ \sqrt{E(Y^2) - E^2(Y)}}.$$

Assume that we have a data matrix $\mathbf{X} = \{x_{i\alpha}\}$ formed out of the *sample* $\{\mathbf{x_i}\}$ of n normally distributed simulator tests called observable–vectors or *manifest variables*, defined on the sample $\{\alpha = 1, \ldots, N\}$ of pilot (for the statistical significance the practical user's criterion is $N \geq 5n$). The *maximum likelihood estimator* of the Pearson correlation coefficient r_{ik} between any two manifest variables $\mathbf{x_i}$ and $\mathbf{x_k}$ is defined as[53]

$$r_{ik} = \frac{\sum_{\alpha=1}^{N}(x_{i\alpha} - \boldsymbol{\mu}_i)(x_{k\alpha} - \boldsymbol{\mu}_k)}{\sqrt{\sum_{\alpha=1}^{N}(x_{i\alpha} - \boldsymbol{\mu}_i)^2}\sqrt{\sum_{\alpha=1}^{N}(x_{k\alpha} - \boldsymbol{\mu}_k)^2}},$$

where

$$\mu_i = \frac{1}{N}\sum_{\alpha=1}^{N} x_{i\alpha}$$

is the arithmetic mean of the variable $\mathbf{x_i}$.[54] Correlation matrix $\mathbf{R}$ is the matrix $\mathbf{R} \equiv \mathbf{R}_{ik} = \{r_{ik}\}$ including $n \times n$ Pearson correlation coefficients r_{ik} calculated between n manifest variables $\{\mathbf{x_i}\}$. Therefore, $\mathbf{R}$ is symmetrical matrix

---

53 A time–dependent generalization $C_{\alpha\beta} = C_{\alpha\beta}(t)$ of the correlation coefficient r_{XY} is the *correlation function*, defined as follows. For the two time–series, $x_\alpha(t_i)$ and $x_\beta(t_i)$ of the same length $(i = 1, \ldots, T)$, one defines the correlation function by

$$C_{\alpha\beta} = \frac{\sum_i (x_\alpha(t_i) - \bar{x}_\alpha)(x_\beta(t_i) - \bar{x}_\beta)}{\sqrt{\sum_i (x_\alpha(t_i) - \bar{x}_\alpha)^2 \sum_j (x_\beta(t_j) - \bar{x}_\beta)^2}},$$

where $\bar{x}$ denotes a time average over the period studied. For two sets of N time–series $x_\alpha(t_i)$ each $(\alpha, \beta = 1, \ldots, N)$ all combinations of the elements $C_{\alpha\beta}$ can be used as entries of the $N \times N$ correlation matrix $\mathbf{C}$. By diagonalizing $\mathbf{C}$, i.e., solving the eigenvalue problem:

$$\mathbf{C}\mathbf{v}^k = \lambda_k \mathbf{v}^k,$$

one gets the eigenvalues λ_k $(k = 1, \ldots, N)$ and the corresponding eigenvectors $\mathbf{v}^k = \{v_\alpha^k\}$.

54 The following algorithm (in pseudocode) estimates bivariate correlation coefficient with good numerical stability:

```
Begin
   sum_sq_x = 0;
   sum_sq_y = 0;
   sum_coproduct = 0;
   mean_x = x[1];
   mean_y = y[1];
```

with ones on the main diagonal. The correlation matrix **R** represents the total variability of all included manifest variables. In other words it stores all information about all simulator tests and all pilot. Now, if the number of included simulator tests is small, this information is meaningful for the human mind. But if we perform one hundred tests (on five hundred pilot), then the correlation matrix contains ten thousand Pearson correlation coefficients. This is the reason for seeking the 'latent' factor structure, underlying the whole co–variability contained in the correlation matrix.

Therefore, the correlation is defined only if both of the standard deviations are finite and both of them are nonzero. It is a corollary of the *Cauchy–Schwarz inequality*[55] that the correlation cannot exceed 1 in absolute value.

```
  for i in 2 to N:
    sweep = (i - 1.0) / i;
    delta_x = x[i] - mean_x;
    delta_y = y[i] - mean_y;
    sum_sq_x += delta_x * delta_x * sweep;
    sum_sq_y += delta_y * delta_y * sweep;
    sum_coproduct += delta_x * delta_y * sweep;
    mean_x += delta_x / i;
    mean_y += delta_y / i ;
  end_for;
  pop_sd_x = sqrt( sum_sq_x / N );
  pop_sd_y = sqrt( sum_sq_y / N );
  cov_x_y = sum_coproduct / N;
  correlation = cov_x_y / (pop_sd_x * pop_sd_y);
End.
```

[55] The Cauchy–Schwarz inequality, named after Augustin Louis Cauchy (the father of complex analysis) and Hermann Amandus Schwarz, is a useful inequality encountered in many different settings, such as linear algebra applied to vectors, in analysis applied to infinite series and integration of products, and in probability theory, applied to variances and covariances. The Cauchy–Schwarz inequality states that if x and y are elements of real or complex inner product spaces then

$$|\langle x,y\rangle|^2 \leq \langle x,x\rangle \cdot \langle y,y\rangle.$$

The two sides are equal iff x and y are linearly dependent (or in geometrical sense they are parallel). This contrasts with a property that the inner product of two vectors is zero if they are orthogonal (or perpendicular) to each other. The inequality hence confers the notion of *the angle between the two vectors* to an inner product, where concepts of *Euclidean geometry* may not have meaningful sense, and justifies that the notion that inner product spaces are generalizations of *Euclidean space*.

An important consequence of the Cauchy–Schwarz inequality is that the inner product is a continuous function.

Another form of the Cauchy–Schwarz inequality is given using the notation of norm, as explained under norms on inner product spaces, as

The correlation is 1 in the case of an increasing linear relationship, -1 in the case of a decreasing linear relationship, and some value in between in all other cases, indicating the degree of linear dependence between the variables. The closer the coefficient is to either -1 or 1, the stronger the correlation between the variables (see Figure 1.1). If the variables are independent then the correlation is 0, but the converse is not true because the correlation coefficient detects only linear dependencies between two variables. For example, suppose the random variable X is uniformly distributed on the interval from -1 to 1, and $Y = X^2$. Then Y is completely determined by X, so that X and Y are dependent, but their correlation is zero; this means that they are uncorrelated. The correlation matrix of n random variables $X_1, \ldots, X_n$ is the $n \times n$ matrix whose ij entry is $r_{X_i X_j}$. If the measures of correlation used are product–moment coefficients, the correlation matrix is the same as the covariance matrix of the standardized random variables X_i/σ_{X_i} (for $i = 1, \ldots, n$). Consequently it is necessarily a non–negative definite matrix. The correlation matrix is symmetrical (the correlation between X_i and X_j is the same as the correlation between X_j and X_i).

As a higher derivation of the correlation matrix analysis and its eigenvectors, the so–called principal components, the *factor analysis* (FA) is a multivariate statistical technique used to explain variability among a large set of observed random variables in terms of fewer unobserved random 'latent' variables, called *factors*. The observed, or 'manifested' variables are modelled as linear combinations of the factors, plus 'error terms'. According to FA, classical bivariate correlation analysis is an artificial extraction from a rial multivariate world, especially in human sciences. FA originated in psychometrics, and is used in social sciences, marketing, product management, operations research, and other applied sciences that deal with large multivariate quantities of data.

For example,[56] suppose a psychologist proposes a theory that there are two kinds of intelligence, 'verbal intelligence' and 'mathematical intelligence'. Note that these are inherently unobservable. Evidence for the theory is sought in the examination scores of 1000 students in each of 10 different academic fields. If a student is chosen randomly from a large population, then the student's 10 scores are random variables. The psychologist's theory may say that the average score in each of the 10 subjects for students with a particular level of verbal intelligence and a particular level of mathematical intelligence is a certain number times the level of verbal intelligence plus a certain number times the level of mathematical intelligence, i.e., it is a linear combination of those two 'factors'. The numbers by which the two 'intelligences' are multiplied

---

$$|\langle x, y \rangle| \leq \|x\| \cdot \|y\|.$$

[56] This oversimplified example should not be taken to be realistic. Usually we are dealing with many factors.

are posited by the theory to be the same for all students, and are called 'factor loadings'. For example, the theory may hold that the average student's aptitude in the field of amphibology is

{ 10 × the student's verbal intelligence } + { 6 × the student's mathematical intelligence }.

The numbers 10 and 6 are the factor loadings associated with amphibology. Other academic subjects may have different factor loadings. Two students having identical degrees of verbal intelligence and identical degrees of mathematical intelligence may have different aptitudes in amphibology because individual aptitudes differ from average aptitudes. That difference is called the 'error' — an unfortunate misnomer in statistics that means the amount by which an individual differs from what is average. The observable data that go into factor analysis would be 10 scores of each of the 1000 students, a total of 10,000 numbers. The factor loadings and levels of the two kinds of intelligence of each student must be inferred from the data. Even the number of factors (two, in this example) must be inferred from the data.

In the example above, for $i = 1, \ldots, 1,000$ the ith student's scores are

$$\begin{array}{ccccccccc} x_{1,i} & = & \mu_1 & + & \ell_{1,1}v_i & + & \ell_{1,2}m_i & + & \varepsilon_{1,i} \\ \vdots & & \vdots & & \vdots & & \vdots & & \vdots \\ x_{10,i} & = & \mu_{10} & + & \ell_{10,1}v_i & + & \ell_{10,2}m_i & + & \varepsilon_{10,i} \end{array}$$

where $x_{k,i}$ is the ith student's score for the kth subject, μ_k is the mean of the students' scores for the kth subject, ν_i is the ith student's 'verbal intelligence', m_i is the ith student's 'mathematical intelligence', $\ell_{k,j}$ are the factor loadings for the kth subject, for $j = 1, 2$; $\epsilon_{k,i}$ is the difference between the ith student's score in the kth subject and the average score in the kth subject of all students whose levels of verbal and mathematical intelligence are the same as those of the ith student. In matrix notation, we have

$$X = \mu + LF + \epsilon,$$

where X is a $10 \times 1,000$ matrix of observable random variables, μ is a 10×1 column vector of unobservable constants (in this case constants are quantities not differing from one individual student to the next; and random variables are those assigned to individual students; the randomness arises from the random way in which the students are chosen), L is a 10×2 matrix of factor loadings (unobservable constants), F is a $2 \times 1,000$ matrix of unobservable random variables, ϵ is a $10 \times \; 1,000$ matrix of unobservable random variables.

Observe that by doubling the scale on which 'verbal intelligence', the first component in each column of F, is measured, and simultaneously halving the factor loadings for verbal intelligence makes no difference to the model. Thus, no generality is lost by assuming that the standard deviation of verbal intelligence is 1. Likewise for 'mathematical intelligence'. Moreover, for similar reasons, no generality is lost by assuming the two factors are uncorrelated with each other. The 'errors' ϵ are taken to be independent of each other.

The variances of the 'errors' associated with the 10 different subjects are not assumed to be equal.

Mathematical basis of FA is *principal components analysis* (PCA), which is a technique for simplifying a dataset, by reducing multidimensional datasets to lower dimensions for analysis. Technically speaking, PCA is a *linear transformation*[57] that transforms the data to a new coordinate system such that the greatest variance by any projection of the data comes to lie on the first coordinate (called the first principal component), the second greatest variance on the second coordinate, and so on. PCA can be used for *dimensionality reduction*[58] in a dataset while retaining those characteristics of the dataset that contribute most to its variance, by keeping lower–order principal components and ignoring higher–order ones. Such low–order components often contain the 'most important' aspects of the data. PCA is also called the (discrete) *Karhunen–Loève transform* (or KLT, named after Kari Karhunen and Michel Loève) or the *Hotelling transform* (in honor of Harold Hotelling[59]). PCA has

---

[57] Recall that a *linear transformation* (also called *linear map* or *linear operator*) is a function between two vector spaces that preserves the operations of vector addition and scalar multiplication. In the language of abstract algebra, a linear transformation is a *homomorphism of vector spaces*, or a *morphism* in the category of vector spaces over a given field.

Let V and W be vector spaces over the same field K. A function (operator) $f : V \to W$ is said to be a *linear transformation* if for any two vectors $x, y \in V$ and any scalar $a \in K$, the following two conditions are satisfied:

$$\text{additivity} : f(x+y) = f(x) + f(y), \qquad \text{and}$$
$$\text{homogeneity} : f(ax) = af(x).$$

This is equivalent to requiring that for any vectors $x_1, \ldots, x_m$ and scalars $a_1, \ldots, a_m$, the following equality holds:

$$f(a_1 x_1 + \cdots + a_m x_m) = a_1 f(x_1) + \cdots + a_m f(x_m).$$

[58] Dimensionality reduction in statistics can be divided into two categories: *feature selection* and *feature extraction*.

Feature selection approaches try to find a subset of the original features. Two strategies are *filter* (e.g., information gain) and *wrapper* (e.g., genetic algorithm) approaches. It is sometimes the case that data analysis such as regression or classification can be carried out in the reduced space more accurately than in the original space. On the other hand, feature extraction is applying a mapping of the multidimensional space into a space of fewer dimensions. This means that the original feature space is transformed by applying e.g., a linear transformation via a *principal components analysis*.

Dimensionality reduction is also a phenomenon discussed widely in physics, whereby a physical system exists in three dimensions, but its properties behave like those of a lower–dimensional system.

[59] Harold Hotelling (Fulda, Minnesota, September 29, 1895 - December 26, 1973) was a mathematical statistician. His name is known to all statisticians because of

the distinction of being the optimal linear transformation for keeping the subspace that has largest variance. This advantage, however, comes at the price of greater computational requirement if compared, for example, to the discrete cosine transform. Unlike other linear transforms, the PCA does not have a fixed set of basis vectors. Its basis vectors depend on the data set.

Assuming zero *empirical mean* (the empirical mean of the distribution has been subtracted from the data set), the principal component $\mathbf{w}_1$ of a dataset $\mathbf{x}$ can be defined as

$$\mathbf{w}_1 = \arg\max_{\|\mathbf{w}\|=1} E\left\{\left(\mathbf{w}^T\mathbf{x}\right)^2\right\}.$$

With the first $k-1$ components, the kth component can be found by subtracting the first $k-1$ principal components from $\mathbf{x}$,

$$\hat{\mathbf{x}}_{k-1} = \mathbf{x} - \sum_{i=1}^{k-1} \mathbf{w}_i \mathbf{w}_i^T \mathbf{x},$$

and by substituting this as the new dataset to find a principal component in

$$\mathbf{w}_k = \arg\max_{\|\mathbf{w}\|=1} E\left\{\left(\mathbf{w}^T\hat{\mathbf{x}}_{k-1}\right)^2\right\}.$$

Therefore, the Karhunen–Loève transform is equivalent to finding the *singular value decomposition*[60] of the data matrix $\mathbf{X}$,

---

Hotelling's T–square distribution and its use in statistical hypothesis testing and confidence regions. He also introduced canonical correlation analysis, and is the eponym of *Hotelling's law*, *Hotelling's lemma*, and *Hotelling's rule* in economics.

[60] Recall that in linear algebra, the *singular value decomposition* (SVD) is an important factorization of a rectangular real or complex matrix, with several applications in signal processing and statistics. The SVD can be seen as a generalization of the *spectral theorem*, which says that normal matrices can be unitarily diagonalized using a basis of eigenvectors, to arbitrary, not necessarily square, matrices.

Suppose M is an $m \times n$ matrix whose entries come from the field K, which is either the field of real numbers, or the field of complex numbers. Then there exists a factorization of the form:

$$M = U\Sigma V^*,$$

where U is an $m \times m$ unitary matrix over K, the matrix Σ is $m \times n$ with nonnegative numbers on the diagonal and zeros off the diagonal, and V^* denotes the conjugate transpose of V, an $n \times n$ unitary matrix over K. Such a factorization is called a singular–value decomposition of M.

The matrix V thus contains a set of orthonormal 'input' or 'analyzing' basis vector directions for M. The matrix U contains a set of orthonormal 'output' basis vector directions for M. The matrix Σ contains the singular values, which can be thought of as scalar 'gain controls' by which each corresponding input is multiplied to give a corresponding output. A common convention is to order the values Σ_{ii} in non–increasing fashion. In this case, the diagonal matrix Σ is uniquely determined by M (although the matrices U and V are not).

$$\mathbf{X} = \mathbf{W}\mathbf{\Sigma}\mathbf{V}^T,$$

and then obtaining the reduced–space data matrix $\mathbf{Y}$ by projecting $\mathbf{X}$ down into the reduced space defined by only the first L singular vectors $\mathbf{W_L}$,

$$\mathbf{Y} = \mathbf{W_L}^T\mathbf{X} = \mathbf{\Sigma_L}\mathbf{V_L}^T.$$

The matrix $\mathbf{W}$ of singular vectors of $\mathbf{X}$ is equivalently the matrix $\mathbf{W}$ of eigenvectors of the matrix of observed covariances $\mathbf{C} = \mathbf{X}\mathbf{X}^T$,

$$\mathbf{X}\mathbf{X}^T = \mathbf{W}\mathbf{\Sigma}^2\mathbf{W}^T.$$

The eigenvectors with the largest eigenvalues correspond to the dimensions that have the strongest correlation in the dataset.

Now, FA is performed as PCA[61] with subsequent orthogonal (non–correlated) or oblique (correlated) *factor rotation* for the simplest possible interpretation (see, e.g., [KM78a]).

---

[61] The alternative FA approach is the so–called *principal factor analysis* (PFA, also called *principal axis factoring*, PAF, and *common factor analysis*, PFA). PFA is a form of factor analysis which seeks the least number of factors which can account for the common variance (correlation) of a set of variables, whereas the more common principal components analysis (PCA) in its full form seeks the set of factors which can account for all the common and unique (specific plus error) variance in a set of variables. PFA uses a PCA strategy but applies it to a correlation matrix in which the diagonal elements are not 1's, as in PCA, but iteratively–derived estimates of the *communalities*.

In addition to PCA and PFA, there are other less–used extraction methods:

1. Image factoring: based on the correlation matrix of predicted variables rather than actual variables, where each variable is predicted from the others using multiple regression.
2. Maximum likelihood factoring: based on a linear combination of variables to form factors, where the parameter estimates are those most likely to have resulted in the observed correlation matrix, using MLE methods and assuming multivariate normality. Correlations are weighted by each variable's uniqueness. (As discussed below, uniqueness is the variability of a variable minus its communality.) MLF generates a *chi–square goodness–of–fit test.* The researcher can increase the number of factors one at a time until a satisfactory goodness of fit is obtained. Warning: for large samples, even very small improvements in explaining variance can be significant by the goodness-of-fit test and thus lead the researcher to select too many factors.
3. Alpha factoring: based on maximizing the reliability of factors, assuming variables are randomly sampled from a universe of variables. All other methods assume cases to be sampled and variables fixed.
4. Unweighted least squares (ULS) factoring: based on minimizing the sum of squared differences between observed and estimated correlation matrices, not counting the diagonal.
5. Generalized least squares (GLS) factoring: based on adjusting ULS by weighting the correlations inversely according to their uniqueness (more unique variables are weighted less). Like MLF, GLS also generates a chi–square goodness–of–fit

FA is used to uncover the latent structure (dimensions) of a set of variables. It reduces attribute space from a larger number of variables to a smaller number of factors and as such is a 'non–dependent' procedure (that is, it does not assume a dependent variable is specified). Factor analysis could be used for any of the following purposes:

1. To reduce a large number of variables to a smaller number of factors for modelling purposes, where the large number of variables precludes modelling all the measures individually. As such, factor analysis is integrated in *structural equation modelling* (SEM),[62] helping create the latent variables modeled by SEM. However, factor analysis can be and is often used on a standalone basis for similar purposes.

---

test. The researcher can increase the number of factors one at a time until a satisfactory goodness of fit is obtained.

[62] Structural equation modelling (SEM) grows out of and serves purposes similar to multiple regression, but in a more powerful way which takes into account the modelling of interactions, nonlinearities, correlated independents, measurement error, correlated error terms, multiple latent independents each measured by multiple indicators, and one or more latent dependents also each with multiple indicators. SEM may be used as a more powerful alternative to multiple regression, path analysis, factor analysis, time series analysis, and analysis of covariance. That is, these procedures may be seen as special cases of SEM, or, to put it another way, SEM is an extension of the general linear model (GLM) of which multiple regression is a part.

SEM is usually viewed as a confirmatory rather than exploratory procedure, using one of three approaches:

a) Strictly confirmatory approach: A model is tested using SEM goodness–of–fit tests to determine if the pattern of variances and covariances in the data is consistent with a structural (path) model specified by the researcher. However as other unexamined models may fit the data as well or better, an accepted model is only a not–disconfirmed model.
b) Alternative models approach: One may test two or more causal models to determine which has the best fit. There are many goodness–of–fit measures, reflecting different considerations, and usually three or four are reported by the researcher. Although desirable in principle, this AM approach runs into the real-world problem that in most specific research topic areas, the researcher does not find in the literature two well-developed alternative models to test.
c) Model development approach: In practice, much SEM research combines confirmatory and exploratory purposes: a model is tested using SEM procedures, found to be deficient, and an alternative model is then tested based on changes suggested by SEM modification indexes. This is the most common approach found in the literature. The problem with the model development approach is that models confirmed in this manner are post–hoc ones which may not be stable (may not fit new data, having been created based on the uniqueness of an initial dataset). Researchers may attempt to overcome this problem by using a cross–validation strategy under which the model is developed using a calibration data sample and then confirmed using an independent validation sample.

2. To select a subset of variables from a larger set, based on which original variables have the highest correlations with the principal component factors.
3. To create a set of factors to be treated as uncorrelated variables as one approach to handling multi–collinearity in such procedures as multiple regression
4. To validate a scale or index by demonstrating that its constituent items load on the same factor, and to drop proposed scale items which cross–load on more than one factor.
5. To establish that multiple tests measure the same factor, thereby giving justification for administering fewer tests.
6. To identify clusters of cases and/or outliers.
7. To determine network groups by determining which sets of people cluster together.

The so–called *exploratory factor analysis* (EFA) seeks to uncover the underlying structure of a relatively large set of variables. The researcher's à priori assumption is that any indicator may be associated with any factor. This is

---

Regardless of approach, SEM cannot itself draw causal arrows in models or resolve causal ambiguities. Theoretical insight and judgment by the researcher is still of utmost importance.

The SEM process centers around two steps: validating the measurement model and fitting the structural model. The former is accomplished primarily through confirmatory factor analysis, while the latter is accomplished primarily through path analysis with latent variables. One starts by specifying a model on the basis of theory. Each variable in the model is conceptualized as a latent one, measured by multiple indicators. Several indicators are developed for each model, with a view to winding up with at least three per latent variable after confirmatory factor analysis. Based on a large ($n > 100$) representative sample, factor analysis (common factor analysis or principal axis factoring, not principle components analysis) is used to establish that indicators seem to measure the corresponding latent variables, represented by the factors. The researcher proceeds only when the measurement model has been validated. Two or more alternative models (one of which may be the null model) are then compared in terms of *model fit*, which measures the extent to which the covariances predicted by the model correspond to the observed covariances in the data. The so–called modification indices and other coefficients may be used by the researcher to alter one or more models to improve fit.

Advantages of SEM compared to multiple regression include more flexible assumptions (particularly allowing interpretation even in the face of multicollinearity), use of confirmatory factor analysis to reduce measurement error by having multiple indicators per latent variable, the attraction of SEM's graphical modelling interface, the desirability of testing models overall rather than coefficients individually, the ability to test models with multiple dependents, the ability to model mediating variables, the ability to model error terms, the ability to test coefficients across multiple between–subjects groups, and ability to handle difficult data (time series with autocorrelated error, non–normal data, incomplete data).

the most common form of factor analysis. There is no prior theory and one uses factor loadings to intuit the factor structure of the data.

On the other hand, the so–called *confirmatory factor analysis* (CFA) seeks to determine if the number of factors and the loadings of measured (indicator) variables on them conform to what is expected on the basis of pre–established theory. Indicator variables are selected on the basis of prior theory and factor analysis is used to see if they load as predicted on the expected number of factors. The researcher's à priori assumption is that each factor (the number and labels of which may be specified à priori) is associated with a specified subset of indicator variables. A minimum requirement of confirmatory factor analysis is that one hypothesize beforehand the number of factors in the model, but usually also the researcher will posit expectations about which variables will load on which factors (see, e.g., [KM78b]). The researcher seeks to determine, for instance, if measures created to represent a latent variable really belong together.

The *factor loadings*, also called component loadings in PCA, are the correlation coefficients between the variables (rows) and factors (columns) in the *factor matrix*. Analogous to Pearson's r, the squared factor loading is the percent of variance in that variable explained by the factor. To get the percent of variance in all the variables accounted for by each factor, add the sum of the squared factor loadings for that factor (column) and divide by the number of variables (note that the number of variables equals the sum of their variances as the variance of a standardized variable is 1). This is the same as dividing the factor's eigenvalue by the number of variables.

The *factor scores*, also called component scores in PCA, factor scores are the scores of each case (row) on each factor (column). To compute the factor score for a given case for a given factor, one takes the case's standardized score on each variable, multiplies by the corresponding factor loading of the variable for the given factor, and sums these products. Computing factor scores allows one to look for factor outliers. Also, factor scores may be used as variables in subsequent modelling.

Rotation serves to make the output more understandable and is usually necessary to facilitate the interpretation of factors. The sum of eigenvalues is not affected by rotation, but rotation will alter the eigenvalues (and percent of variance explained) of particular factors and will change the factor loadings. Since alternative rotations may explain the same variance (have the same total eigenvalue) but have different factor loadings, and since factor loadings are used to intuit the meaning of factors, this means that different meanings may be ascribed to the factors depending on the rotation – a problem some cite as a drawback to factor analysis. If factor analysis is used, the researcher may wish to experiment with alternative rotation methods to see which leads to the most interpretable factor structure.

Varimax rotation is an orthogonal rotation of the factor axes to maximize the variance of the squared loadings of a factor (column) on all the variables (rows) in a factor matrix, which has the effect of differentiating the original variables by extracted factor. Each factor will tend to have either large or small loadings of any particular variable. A varimax solution yields results which make it as easy as possible to identify each variable with a single factor. This is the most common rotation option.

The oblique rotations allow the factors to be correlated, and so a factor correlation matrix is generated when oblique is requested. Two most common oblique rotation methods are:

Direct oblimin rotation – the standard method when one wishes a non–orthogonal solution , that is, one in which the factors are allowed to be correlated; this will result in higher eigenvalues but diminished interpretability of the factors; and

Promax rotation – an alternative non–orthogonal rotation method which is computationally faster than the direct oblimin method and therefore is sometimes used for very large datasets.

FA advantages are:

1. Offers a much more objective method of testing intelligence in humans;
2. Allows for a satisfactory comparison between the results of intelligence tests; and
3. Provides support for theories that would be difficult to prove otherwise.

Charles Spearman pioneered the use of factor analysis in the field of psychology and is sometimes credited with the invention of factor analysis. He discovered that schoolchildren's scores on a wide variety of seemingly unrelated subjects were positively correlated, which led him to postulate that a general mental ability, or g, underlies and shapes human cognitive performance. His postulate now enjoys broad support in the field of intelligence research, where it is known as the g theory.

Raymond Cattell expanded on Spearman's idea of a two–factor theory of intelligence after performing his own tests and factor analysis. He used a multi–factor theory to explain intelligence. Cattell's theory addressed alternate factors in intellectual development, including motivation and psychology. Cattell also developed several mathematical methods for adjusting psychometric graphs, such as his 'scree' test and similarity coefficients. His research lead to the development of his theory of fluid and crystallized intelligence. Cattell was a strong advocate of factor analysis and psychometrics. He believed that all theory should be derived from research, which supports the continued use of empirical observation and objective testing to study human intelligence.

Factor Structure and Rotation

Starting with the correlation matrix **R** including the number of significant correlations, the goal of exploratory factor analysis (FA) is to detect latent underlying dimensions (i.e., the factor structure) among the set of all manifest variables. Instead of the correlation matrix, the factor analysis can start from the covariance matrix (see Figure 4), which is the symmetrical matrix with variances of all manifest variables on the main diagonal and their covariances in other matrix cells. For the purpose of the present project the correlation matrix is far more meaningful starting point. Three main applications of factor analytic techniques are (see [CL71, And84, Har75]):

1. to *reduce* the number of manifest variables,
2. to *classify* manifest variables, and
3. to *score* each individual soldier on the latent factor structure.

Factor analysis model expands each of the manifest variables $\mathbf{x}_i$ with the means $\boldsymbol{\mu}_i$ from the data matrix $\mathbf{X} = \{x_{i\alpha}\}$ as a linear vector–function

$$\mathbf{x}_i = \boldsymbol{\mu}_i + \mathbf{L}_{ij}\,\mathbf{f}_j + \mathbf{e}_i, \quad (i = 1, \dots, n;\ j = 1, \dots, m) \tag{1.1}$$

where n and m denote the numbers of manifest and latent variables, respectively, $\mathbf{f}_j$ denotes the jth common–factor vector (with zero mean and unity–matrix covariance), $\mathbf{L} = \mathbf{L}_{ij}$ is the matrix of factor loadings l_{ij}, and $\mathbf{e}_i$ corresponds to the ith specific–factor vector (specific variance not explained by the common factors, with zero mean and diagonal–matrix covariance).

That portion of the variance of the ith manifest variable $\mathbf{x}_i$ contributed by the m common factors $\mathbf{f}_j$, the sum of squares of the loadings l_{ij}, is called the ith communality.

Now, in the correlation matrix $\mathbf{R}$ the variances of all variables are equal to 1.0. Therefore, the total variance in that matrix is equal to the number of variables. Extraction of factors is based on the solution of eigenvalue problem, i.e., characteristic equation for the correlation matrix $\mathbf{R}$,

$$\mathbf{R}\mathbf{x}_i = \lambda_i \mathbf{x}_i,$$

where λ_i are eigenvalues of $\mathbf{R}$, representing the variances extracted by the factors, and $\mathbf{x}_i$ now represent the corresponding eigenvectors, representing principal components or factors. The question then is, how many factors do we want to extract? Note that as we extract consecutive factors, they account for less and less variability. The decision of when to stop extracting factors basically depends on when there is only very little 'random' variability left. According to the widely used Kaiser criterion we can retain only factors with eigenvalues greater than 1. In essence this is like saying that, unless a factor extracts at least as much as the equivalent of one original variable, we drop it. The proportion of variance of a particular item that is due to common factors (shared with other items) is called communality. Therefore, an additional task facing us when applying this model is to estimate the communalities for each variable, that is, the proportion of variance that each item has in common with other items. The proportion of variance that is unique to each item is then the respective item's total variance minus the communality. A common starting point is to use the squared multiple correlation of an item with all other items as an estimate of the communality. The correlations between the manifest variables and the principal components are called factor loadings.

The first factor is generally more highly correlated with the variables than the second, third and other factors, as these factors are extracted successively and will account for less and less variance overall.

Therefore, the principal component factor analysis of the sample correlation matrix $\mathbf{R}$ is specified in terms of its $m < n$ eigenvalue–eigenvector pairs $(\lambda_j, \mathbf{x}_j)$ where $\lambda_j \geq \lambda_{j+1}$. The matrix of estimated factor loadings l_{ij} is given by

$$\mathbf{L} = \left[\sqrt{\lambda_1}\mathbf{x}_1 | \sqrt{\lambda_2}\mathbf{x}_2 | \dots | \sqrt{\lambda_m}\mathbf{x}_m\right].$$

Factor extraction can be performed also by other methods, collectively called *principal factors*, including: (i) Maximum likelihood factors, (ii) Principal axis method, (iii) Centroid method, (iv) Multiple R^2-communalities, and (v) Iterated Minres communalities. However, we shall stick on the principal components because of their obvious eigen–structure.

In any case, matrix of factor loadings $\mathbf{L}$ is determined only up to an orthogonal matrix $\mathbf{O}$. The communalities, given by the diagonal elements of $\mathbf{L}\mathbf{L}^T$ are also unaffected by the choice of $\mathbf{O}$. This ambiguity provides the rationale for 'factor rotation', since orthogonal matrices correspond to 'coordinate' rotations.

We could plot, theoretically, the factor loadings in a $m-$dimensional scatter–plot. In that plot, each variable is represented as a point. In this plot we could rotate the axes in any direction without changing the relative locations of the points to each other; however, the actual coordinates of the points, that is, the factor loadings would of course change. There are various rotational strategies that have been proposed. The goal of all of these strategies is to get a clear pattern of loadings, that is, factors that are somehow clearly marked by high loadings for some variables and low loadings for others. This general pattern is also sometimes referred to as simple structure (a more formalized definition can be found in most standard textbooks). Typical rotational strategies are Varimax, Quartimax, and Equimax (see Anderson, 1984). Basically, the extraction of principal components amounts to a variance maximizing Varimax–rotation of the original space of manifest–variables. We want to get a pattern of loadings on each factor that is as diverse as possible, lending itself to easier interpretation. After we have found the line on which the variance is maximal, there remains some variability around this line. In principal components analysis, after the first factor has been extracted, that is, after the first line has been drawn through the data, we continue and define another line that maximizes the remaining variability, and so on. In this manner, consecutive factors are extracted. Because each consecutive factor is defined to maximize the variability that is not captured by the preceding factor, consecutive factors are independent of each other. Put another way, consecutive factors are uncorrelated or orthogonal to each other.

Basically, the rotation of the matrix of the factor loadings $\mathbf{L}$ represents its post–multiplication, i.e. $\mathbf{L}^* = \mathbf{LO}$ by the rotation matrix $\mathbf{O}$, which itself resembles one of the matrices included in the classical rotational Lie groups

$SO(m)$ (containing the specific $m-$fold combination of sinuses and cosinuses. The linear factor equation (1.1) represents the orthogonal factor model, provided that vectors $\mathbf{f}_j$ and $\mathbf{e}_i$ are independent (orthogonal to each other, i.e., having zero covariance).

The most frequently used Kaiser's Normal Varimax rotation procedure selects the orthogonal transformation $\mathbf{T}$ that 'spreads out' the squares of the loadings on each factor as much as possible, i.e., maximizes the total 'squared' variance

$$V = \frac{1}{n}\sum_{j=1}^{m}\left[\sum_{i=1}^{n}(l_{ij}^{*})^{4} - \frac{1}{n}\left(\sum_{i=1}^{n}(l_{ij}^{*})^{2}\right)^{2}\right],$$

where l_{ij}^{*} denote the rotated factor loadings from the rotated factor matrix $\mathbf{L}^{*}$.

Besides orthogonal rotation, there is another concept of oblique (non-orthogonal, or correlated) factors, which could help to achieve more interpretable simple structure. Specifically, computational strategies have been developed to rotate factors so as to best represent clusters of manifest variables, without the constraint of orthogonality of factors. Oblique rotation produces the factor structure made from the smaller set of mutually correlated factors. An oblique rotation to the simple structure corresponds to *nonrigid* rotation of the factor-axes (i.e., principal components) in the factor space such that the rotated axes $\mathbf{l}_{\mathbf{j}}^{*} = \mathbf{L}_{\mathbf{obl}}^{*}$ (no longer perpendicular) pass (nearly) through the clusters of manifest variables. Although the purest mathematical background does not exist for the non–orthogonal factor rotation, the *parsimony principle*: "explain the maximum of the common variability of the data matrix $\mathbf{X} = \{x_{i\alpha}\}$ with the minimum number of factors", is fully developed only in this form of factor analysis, and the factor–correlation matrix $\mathbf{L}_{\mathbf{obl}}^{*}$ resembles the correlation matrix between manifest variables in the latent, factor space with double–reduced number of observables.

The linear factor equation (1.1) becomes now the *oblique factor model*

$$\mathbf{x}_i = \boldsymbol{\mu}_i + \mathbf{L}_{\mathbf{obl}}^{*}\,\mathbf{f}_j + \mathbf{e}_i, \quad (i = 1, \dots, n;\ j = 1, \dots, m),$$

where the vectors $\mathbf{f}_j$ and $\mathbf{e}_i$ are interdependent (correlated to each other). With oblique rotation, using common procedures, like Kaiser–Harris Orthoblique, Oblimin, Oblimax, Quartimin, Promax (see [And84]), we could

1. perform a hierarchical (iterated) factor analysis, obtaining second–order factors, third–order factors, etc., finishing with a single general factor (for example using principal component analysis of the factor–correlation matrix $\mathbf{L}_{\mathbf{obl}}^{*}$); and
2. develop the so–called 'cybernetic models': when two factors in the factor–correlation matrix $\mathbf{L}_{\mathbf{obl}}^{*}$ are highly correlated we can assume a linear functional link between them; connecting all correlated factors on the certain hierarchical level, we can make a block–diagram out of them depicting a linear system; this is the real point of the *exploratory* factor analysis.

The factor scores $S_{j\alpha}$ (where j labels factors and α labels individual pilot) are incidental parameters that characterize general performance of the individuals (see [CL71, And84, Har75]). Factor scores with zero mean and unity-matrix covariance are usually automatically evaluated in principal–component, orthogonal and oblique factor analysis, according to the formula:

$$S_{j\alpha} = (\mathbf{L}^T\mathbf{L})^{-1}\mathbf{L}^T(x_{j\alpha} - \bar{x}_{j\alpha}),$$

and replacing $\mathbf{L}$ by $\mathbf{L}^*$, and by $\mathbf{L}^*_{\mathbf{obl}}$, respectively. They represent an objective measure of the general performance of pilot on the battery of psycho–tests.[63]

---

[63] Here is the Mathematica algorithm for calculating the basic factor structure:
$Mean[x\_] := \frac{Plus@@x}{Length[x]}$;
$Variance[x\_] := \frac{Plus@@(mean[x]-x)^2}{Length[x]}$;
$StDev[x\_] := \sqrt{Variance[x]}$;
$Covar[x1\_, x2\_] := \frac{Plus@@((mean[x1]-x1)((mean[x2]-x2))}{Length[x1]}$;
$Corr[x1\_, x2\_] := \frac{Covar[x1,x2]}{StDev[x1]\,StDev[x2]}$;
$CorrMat[X\_] := Table[Corr[X[[1,j]], X[[1,i]]]//N, \{i,m\}, \{j,m\}]$;
Generate random data–matrix (m variables $\times$ n cases):
$NoVars = 10;\ NoCases = 50;\ m = NoVars;\ n = NoCases;$
$data = Array[x, \{NoCases, NoVars\}]//MatrixForm;$
$Table[x[i,j] = Random[Integer, \{1,5\}], \{i, NoCases\}, \{j, NoVars\}];$
Print[“data = ”,data//MatrixForm];
Calculate correlation matrix:
$R = CorrMat[data];\ Print[\text{“R=”}, R//MatrixForm]$
Calculate eigenvalues of the correlation matrix:
$\lambda = Eigenvalues[R]//MatrixForm$
Corresponding eigenvectors:
$vec = Eigenvectors[R];\ Print[vec//Transpose//MatrixForm]$
Determine significant principal components
according to the criterion $\lambda \geq 2$:
$Print[$“PRINCIPAL COMPONENTS”
$\to \{vec[[1]], vec[[2]]\}//Transpose//MatrixForm]$
Define operator matrix:
$NoFact = 2;\ P = Array[p, NoVars, NoFact];$
$Table[p[i,j] = 1, \{i, NoVars\}, \{j, NoFact\}];$
$Table[p[i,j] = 0, \{i, 2, NoVars, 2\}, \{j, 2, NoFact, 2\}];$
$Table[p[i,j] = 0, \{i, 1, NoVars, 2\}, \{j, 1, NoFact, 2\}];$
$Print[\text{“}P = \text{”}, P//MatrixForm];$
Perform oblique rotation:
$Q = Transpose[P];\ S = R.P;\ G = Q.S;$
$Do[k = \frac{1}{\sqrt{G\|i,i\|}}, \{i, NoFact\}];$
$F = Sk;\ Z = kG;\ C = Zk;$
$L = Inverse[C];\ \Phi = F.L;$
Factor structure matrix:
$Print[\text{“}F = \text{”}, F//MatrixForm]$

The factor scores can be used further for multivariate regression in the latent space (instead in the original manifest space) for reducing the number of predictors in the general regression analysis (see [CL71]).

Quantum–Like Correlation and Factor Dynamics

To develop correlation and factor dynamics model, we are using geometrical analogy with *nonrelativistic quantum mechanics* (see [Dir49]). A time dependent state of a quantum system is determined by a normalized (complex), time–dependent, wave psi–function $\psi = \psi(t)$, i.e. a unit Dirac's 'ket' vector $|\psi(t)\rangle$, an element of the Hilbert space $L^2(\psi)$ with a coordinate basis (q^i), under the action of the Hermitian operators, obtained by the procedure of quantization of classical mechanical quantities, for which real eigenvalues are measured . The state–vector $|\psi(t)\rangle$, describing the motion of de Broglie's waves, has a statistical interpretation as the probability amplitude of the quantum system, for the square of its magnitude determines the density of the probability of the system detected at various points of space. The summation over the entire space must yield unity and this is the normalization condition for the psi–function, determining the unit length of the state vector $|\psi(t)\rangle$.

In the coordinate q–representation and the Schrödinger S–picture we consider an action of an evolution operator (in normal units Planck constant $\hbar = 1$)

$$\hat{S} \equiv \hat{S}(t, t_0) = \exp[-i\hat{H}(t - t_0)],$$

i.e., a one–parameter Lie–group of unitary transformations evolving a quantum system. The action represents an exponential map of the system's total energy operator – Hamiltonian $\hat{H} = \hat{H}(t)$. It moves the quantum system from one instant of time, t_0, to some future time t, on the state–vector $|\psi(t)\rangle$, rotating it: $|\psi(t)\rangle = \hat{S}(t, t_0)|\psi(t_0)\rangle$. In this case the Hilbert coordinate basis (q^i) is fixed, so the system operators do not evolve in time, and the system evolution is determined exclusively by the time–dependent Schrödinger equation

$$i\partial_t|\psi(t)\rangle = \hat{H}(t)|\psi(t)\rangle, \qquad (\partial_t = \partial/\partial t), \tag{1.2}$$

with initial condition given at one instant of time t_0 as $|\psi(t_0)\rangle = |\psi\rangle$.

Inter-factor correlation matrix:
$Print[\text{"}C = \text{"}, C//MatrixForm]$
Factor projection matrix:
$Print[\text{"}\Phi = \text{"}, \Phi//MatrixForm]$
Calculate factor scores for individual pilot:
$var[x\_] := x - mean[x];$
$Table[v[i] = var[X[[i]]//N], \{i, n\}];$
$T_F = Transpose[F];\ F_F = Inverse[T_F.F].T_F;$
$Table[F_F.v[i], \{i, n\}]//MatrixForm.$

If the Hamiltonian $\hat{H} = \hat{H}(t)$ does not explicitly depend on time (which is the case with the absence of variables of macroscopic fields), the state vector reduces to the exponential of the system energy:

$$|\psi(t)\rangle = \exp(-iE(t - t_0)|\psi\rangle,$$

satisfying the time–independent (i.e., stationary) Schrödinger equation

$$\hat{H}|\psi\rangle = E|\psi\rangle, \tag{1.3}$$

which represents the characteristic equation for the Hamiltonian operator $\hat{H}$ and gives its real eigenvalues (stationary energy states) E_n and corresponding orthonormal eigenfunctions (i.e., probability amplitudes) $|\psi_n\rangle$.

To model the correlation and factor dynamics we start with the characteristic equation for the correlation matrix

$$\mathbf{R}\mathbf{x} = \lambda\mathbf{x},$$

making heuristic analogy with the stationary Schrödinger equation (1.3). This analogy allows a 'physical' interpretation of the correlation matrix $\mathbf{R}$ as an operator of the 'total correlation or covariation energy' of the statistical system (the simulator–test data matrix $\mathbf{X} = \{x_{i\alpha}\}$), eigenvalues λ_n corresponding to the 'stationary energy states', and eigenvectors $\mathbf{x}_n$ resembling 'probability amplitudes' of the system.

So far we have considered one instant of time t_0. Including the time–flow into the stationary Schrödinger equation (1.3) we get the time–dependent Schrödinger equation (1.2) and returning back with our heuristic analogy, we get the basic equation of the n–dimensional correlation dynamics

$$\partial_t \mathbf{x}(t) = \mathbf{R}(t)\, \mathbf{x}_k(t), \tag{1.4}$$

with initial condition at time t_0 given as a stationary manifest–vectors $\mathbf{x}_k(t_0) = \mathbf{x}_k$ $(k = 1, \dots, n)$.

In more realistic case of 'many' observables (i.e., very big n), instead of the correlation dynamics (1.4), we can use the reduced–dimension factor dynamics, represented by analogous equation in the factor space spanned by the extracted (oblique) factors $\mathbf{F} = \mathbf{f}_i$, with inter–factor–correlation matrix $\mathbf{C} = c_{ij}$ $(i, j = 1, \dots, \text{no. of factors})$

$$\partial_t \mathbf{f}_i(t) = \mathbf{C}(t)\, \mathbf{f}_i(t), \tag{1.5}$$

subject to initial condition at time t_0 given as stationary vectors $\mathbf{f}_i(t_0) = \mathbf{f}_i$.

Now, according to the fundamental existence and uniqueness theorem for linear autonomous ordinary differential equations, if $A = A(t)$ is an $n \times n$ real matrix, then the initial value problem

$$\partial_t \mathbf{x}(t) = A\mathbf{x}(t), \qquad \mathbf{x}(0) = \mathbf{x}_0 \in \mathbb{R}^n,$$

has the unique solution

$$\mathbf{x}(t) = \mathbf{x}_0 \mathrm{e}^{tA}, \qquad \text{for all } t \in \mathbb{R}.$$

Therefore, analytical solutions of our correlation and factor–correlation dynamics equations (1.4) and (1.5) are given respectively by exponential maps

$$\mathbf{x}_k(t) = \mathbf{x}_k \exp[t\,\mathbf{R}],$$
$$\mathbf{f}_i(t) = \mathbf{f}_i \exp[t\,\mathbf{C}].$$

Thus, for each $t \in \mathbb{R}$, the matrix $\mathbf{x}\exp[t\,\mathbf{R}]$, respectively the matrix $\mathbf{f}\exp[t\,\mathbf{C}]$, maps

$$\mathbf{x}_k \mapsto \mathbf{x}_k \exp[t\,\mathbf{R}], \qquad \text{respectively} \qquad \mathbf{f}_i \mapsto \mathbf{f}_i \exp[t\,\mathbf{C}].$$

The sets $g^t_{corr} = \{\exp[t\,\mathbf{R}]\}_{t\in\mathbb{R}}$ and $g^t_{fact} = \{\exp[t\,\mathbf{C}]\}_{t\in\mathbb{R}}$ are 1–parameter families (groups) of linear maps of $\mathbb{R}^n$ into $\mathbb{R}^n$, representing the *correlation flow*, respectively the *factor–correlation flow* of simulator–tests. The linear flows g^t (representing both g^t_{corr} and g^t_{fact}) have two essential properties:

1. identity map: $g^0 = I$, and
2. composition: $g^{t_1+t_2} = g^{t_1} \circ g^{t_2}$.

They partition the state space $\mathbb{R}^n$ into subsets that we call 'correlation orbits', respectively 'factor–correlation orbits', through the initial states $\mathbf{x}_k$, and $\mathbf{f}_i$, of simulator tests, defined respectively by

$$\gamma(\mathbf{x}_k) = \{\mathbf{x}_k g^t | t \in \mathbb{R}\} \qquad \text{and} \qquad \gamma(\mathbf{f}_i) = \{\mathbf{f}_i g^t | t \in \mathbb{R}\}.$$

The correlation orbits can be classified as:

1. If $g^t\mathbf{x}_k = \mathbf{x}_k$ for all $t \in \mathbb{R}$, then $\gamma(\mathbf{x}_k) = \{\mathbf{x}_k\}$ and it is called a *point orbit*. Point orbits correspond to equilibrium points in the manifest and the factor space, respectively.
2. If there exists a $T > 0$ such that $g^T\mathbf{x}_k = \mathbf{x}_k$, then $\gamma(\mathbf{x}_k)$ is called a *periodic orbit*. Periodic orbits describe a system that evolves periodically in time in the manifest and the factor space, respectively.
3. If $g^t\mathbf{x}_k \neq \mathbf{x}_k$ for all $t \neq 0$, then $\gamma(\mathbf{x}_k)$ is called a *non–periodic orbit*.

Analogously, the factor–correlation orbits can be classified as:

1. If $g^t\mathbf{f}_i = \mathbf{f}_i$ for all $t \in \mathbb{R}$, then $\gamma(\mathbf{f}_i) = \{\mathbf{f}_i\}$ and it is called a point orbit. Point orbits correspond to equilibrium points in the manifest and the factor space, respectively.
2. If there exists a $T > 0$ such that $g^T\mathbf{f}_i = \mathbf{f}_i$, then $\gamma(\mathbf{f}_i)$ is called a periodic orbit. Periodic orbits describe a system that evolves periodically in time in the manifest and the factor space, respectively.
3. If $g^t\mathbf{f}_i \neq \mathbf{f}_i$ for all $t \neq 0$, then $\gamma(\mathbf{f}_i)$ is called a non–periodic orbit.

Now, to interpret properly the meaning of (really discrete) time in the correlation matrix $\mathbf{R} = \mathbf{R}(t)$ and factor–correlation matrix $\mathbf{C} = \mathbf{C}(t)$, we can perform a successive time–series $\{t, t + \Delta t, t + 2\Delta t, t + k\Delta t, \cdots\}$ of simulator tests (and subsequent correlation and factor analysis), and discretize our correlation (respectively, factor–correlation) dynamics, to get

$$\mathbf{x}_k(t + \Delta t) = \mathbf{x}_k(0) + \mathbf{R}(t)\,\mathbf{x}_k(t)\Delta t, \qquad \text{and}$$
$$\mathbf{f}_i(t + \Delta t) = \mathbf{f}_i(0) + \mathbf{C}(t)\,\mathbf{f}_i(t)\Delta t,$$

respectively. Finally we can represent the discrete correlation and factor–correlation dynamics in the form of the (computationally applicable) *three–point iterative dynamics equation*, respectively in the manifest space

$$\mathbf{x}_k^{s+1} = \mathbf{x}_k^{s-1} + \mathbf{R}_k^s\,\mathbf{x}_k^s,$$

and in the factor space

$$\mathbf{f}_i^{s+1} = \mathbf{f}_i^{s-1} + \mathbf{C}_i^s\,\mathbf{f}_i^s,$$

in which the time–iteration variable s labels the time occurrence of the simulator tests (and subsequent correlation and factor analysis), starting with the initial state, labelled $s = 0$.

FA–Based Intelligence

In the psychometric view, the concept of intelligence is most closely identified with Spearman's **g**, or Gf ('fluid **g**'). However, psychometricians can measure a wide range of abilities, which are distinct yet correlated. One common view is that these abilities are hierarchically arranged with **g** at the vertex (or top, overlaying all other cognitive abilities).[64]

On the other hand, critics of the psychometric approach, such as Robert Sternberg from Yale, point out that people in the general population have a somewhat different conception of intelligence than most experts. In turn, they argue that the psychometric approach measures only a part of what

[64] Intelligence, IQ, and **g** are distinct terms. As already said above, intelligence is the term used in ordinary discourse to refer to cognitive ability. However, it is generally regarded as too imprecise to be useful for a scientific treatment of the subject. The intelligence quotient (IQ) is an index calculated from the scores on test items judged by experts to encompass the abilities covered by the term intelligence. IQ measures a multidimensional quantity: it is an amalgam of different kinds of abilities, the proportions of which may differ between IQ tests. The dimensionality of IQ scores can be studied by factor analysis, which reveals a single dominant factor underlying the scores on all IQ tests. This factor, which is a hypothetical construct, is called **g**. Variation in **g** corresponds closely to the intuitive notion of intelligence, and thus **g** is sometimes called *general cognitive ability* or *general intelligence*.

is commonly understood as intelligence. Other critics, such as Arthur Eddington,[65] argue that the equipment used in an experiment often determines the results and that proving that e.g., intelligence exists does not prove that current equipment measure it correctly. Sceptics often argue that so much scientific knowledge about the brain is still to be discovered that claiming the conventional IQ test methodology to be infallible is just a small step forward from claiming that *craniometry*[66] was the infallible method for measuring intelligence (which had scientific merits based on knowledge available in the nineteenth century).

A more fundamental criticism is that both the psychometric model used in these studies and the conceptualization of cognitive ability itself are fundamentally off beam. These views were expressed by none other than Charles Spearman, the 'discoverer' of **g** – himself. Thus he wrote: "Every normal man, woman, and child is a genius at something. It remains to discover at what. This must be a most difficult matter, owing to the very fact that it occurs in only a minute proportion of all possible abilities. It certainly cannot be detected by any of the testing procedures at present in current usage. But these procedures are capable, I believe, of vast improvement." In this context he noted that it is more important to ask 'What does this person think about?' than 'How well can he or she think?' Spearman went on to observe that the tests from which his **g** had emerged had no place in schools since they did not reflect the diverse talents of the children and thus deflected teachers from their fundamental educational role, which is to nurture and recognize these diverse talents.

He also noted, as paraphrased here, that the so–called 'cognitive ability' is not primarily cognitive but affective and conative. In constructing meaning out of confusion (Spearman's eductive ability) one first follows feelings that beckon or attract. One then has to engage in 'experimental interactions with the environment' to check out those, largely non–verbal, 'hunches'. This requires determination and persistence — *conation*. Now, all of these are difficult and demanding activities which will only be undertaken whilst one is undertaking activities one cares about. So the first question is: 'What kinds of activity is this person strongly motivated to undertake' (and the kinds of activity which people may be strongly motivated to undertake are legion and mostly unrelated to those assessed in conventional 'intelligence' tests). And the second question is: 'How many of the cumulative and substitutable

---

[65] Sir Arthur Stanley Eddington, OM (December 28, 1882 — November 22, 1944) was an astrophysicist of the early 20th century. The Eddington limit, the natural limit to the luminosity that can be radiated by accretion onto a compact object, is named in his honor. He is famous for his work regarding the Theory of Relativity. Eddington wrote an article in 1919, Report on the relativity theory of gravitation, which announced Einstein's theory of general relativity to the English–speaking world.

[66] Craniometry is the technique of measuring the bones of the skull. Craniometry was once intensively practiced in anthropology/ethnology.

components of competence required to carry out these activities effectively does this person display whilst carrying out that activity?' So one cannot, in reality, assess a person's intelligence, or even their eductive ability, except in relation to activities they care about. What one sees in e.g., the Raven Progressive Matrices is the cumulative effect of how well they do all these things in relation to a certain sort of task. The problem is that this is not — and cannot be — 'cognitive ability' in any general sense of the word but only in relation to this kind of task. As Roger Sperry[67] has observed, what is neurologically localized is not 'cognitive ability' in any general sense but the emotional predisposition to 'think' about a particular kind of thing (for more details, see e.g., papers of John Raven[68] [Rav02]).

Most experts accept the concept of a single dominant factor of intelligence, general mental ability or **g**, while others argue that intelligence consists of a set of relatively independent abilities [APS98]. The evidence for **g** comes from factor analysis of tests of cognitive abilities. The methods of factor analysis do not guarantee a single dominant factor will be discovered. Other *psychological tests*, which do not measure cognitive ability, such as *personality tests*, generate multiple factors.

Proponents of *multiple–intelligence theories* often claim that **g** is, at best, a measure of academic ability. Other types of intelligence, they claim, might be just as important outside of a school setting. Robert Sternberg has proposed a 'Triarchic Theory of Intelligence'. Howard Gardner's theory of multiple intelligences breaks intelligence down into at least eight different components: logical, linguistic, spatial, musical, kinesthetic, naturalist, intra–personal and

---

[67] Roger Wolcott Sperry (August 20, 1913 – April 17, 1994) was a neuropsychologist who, together with David Hunter Hubel and Torsten Nils Wiesel, won the 1981 Nobel Prize in Medicine for his work with *split–brain* research. Before Sperry's experiments, some research evidence seemed to indicate that areas of the brain were largely undifferentiated and interchangeable. In his early experiments Sperry challenged this view by showing that after early development circuits of the brain are largely hardwired. In his Nobel–winning work, Sperry separated the *corpus callosum*, the area of the brain used to transfer signals between the right and left hemispheres, to treat epileptics. Sperry and his colleagues then tested these patients with tasks that were known to be dependent on specific hemispheres of the brain and demonstrated that the two halves of the brain may each contain consciousness. In his words, each hemisphere is "... indeed a conscious system in its own right, perceiving, thinking, remembering, reasoning, willing, and emoting, all at a characteristically human level, and . . . both the left and the right hemisphere may be conscious simultaneously in different, even in mutually conflicting, mental experiences that run along in parallel." This research contributed greatly to understanding the lateralization of brain functions. In 1989, Sperry also received National Medal of Science.

[68] John Carlyle Raven first published his Progressive Matrices in the United Kingdom in 1938. His three sons established Scotland–based test publisher JC Raven Ltd. in 1972. In 2004, Harcourt Assessment, Inc. a division of Harcourt Education acquired JC Raven Ltd.

inter–personal intelligences. Daniel Goleman and several other researchers have developed the concept of *emotional intelligence* and claim it is at least as important as more traditional sorts of intelligence. These theories grew from observations of human development and of brain injury victims who demonstrate an acute loss of a particular cognitive function (e.g., the ability to think numerically, or the ability to understand written language), without showing any loss in other cognitive areas.

In response, **g** theorists have pointed out that **g**'s *predictive validity*[69] has been repeatedly demonstrated, for example in predicting important non–academic outcomes such as job performance, while no multiple–intelligences theory has shown comparable validity. Meanwhile, they argue, the relevance, and even the existence, of multiple intelligences have not been borne out when actually tested [Hun01]. Furthermore, **g** theorists contend that proponents of multiple–intelligences (see, e.g., [Ste95]) have not disproved the existence of a general factor of intelligence [Kli00]. The fundamental argument for a general factor is that test scores on a wide range of seemingly unrelated cognitive ability tests (such as sentence completion, arithmetic, and memorization) are positively correlated: people who score highly on one test tend to score highly on all of them, and g thus emerges in a factor analysis. This suggests that the tests are not unrelated, but that they all tap a common factor.

Cognitive vs. Not–Cognitive Intelligence

Clearly, biologically realized 'cognitive intelligence' is the most complex property of human mind and can be perceived only by itself. Our problem is what we call or may call cognitive intelligence. From the formal, computational perspective, cognitive intelligence is one of ill defined concepts. Its definitions are immersed in numerous scientific contexts and mirrors their historical evolutions, as well as, different 'interests' of researchers. Its weakness is usually based on its abstract multifaces image and, on the other hand, a universal utility character.

---

[69] In psychometrics, *predictive validity* is the extent to which a scale predicts scores on some criterion measure. For example, the validity of a cognitive test for job performance is the correlation between test scores and, say, supervisor performance ratings. Such a cognitive test would have predictive validity if the observed correlation were statistically significant. Predictive validity shares similarities with concurrent validity in that both are generally measured as correlations between a test and some criterion measure. In a study of concurrent validity the test is administered at the same time as the criterion is collected. This is a common method of developing validity evidence for employment tests: A test is administered to incumbent employees, then a rating of those employees' job performance is obtained (often, as noted above, in the form of a supervisor rating). Note the possibility for restriction of range both in test scores and performance scores: The incumbent employees are likely to be a more homogeneous and higher performing group than the applicant pool at large.

The classical behavioral/biologists definition of intelligence reads: "Intelligence is the ability to adapt to new conditions and to successfully cope with life situations." This definition seems to be the best, but 'intelligence' here depends on available physical tools and specific life experience (individual hidden knowledge, preferences and access to information), therefore it is not enough selective to be measured, compared or designed. In general, cognitive intelligence is a human–like intelligence. Unfortunately there are many opinions what human–like intelligence means. For example, (i) cognitive intelligence uses a human mental introspective experience for the modelling of intelligent system thinking; and (ii) cognitive intelligence may use brain models to extract brain's intelligence property.

Therefore, cognitive intelligence can be seen as a product of human self–conscious recognition of efficient mental processes, defined a'priori as intelligent. In order to get a consensus on the notion of cognitive intelligence is useful to have an agreement on which intelligence is not cognitive. A not–cognitive intelligence could be considered as an intelligence being developed using not human analogies; e.g., it is possible to construct very different models of flying objects starting from the observation of storks, balloons, beetles or clouds – maybe this observation can be useful.

The difference between human and artificial intelligence theories is similar to the difference between a birds theory of fly and the airplanes fly theory, the both can lead to a more general theory of fly but this last needs a goal–oriented and a higher abstraction level of the conceptualization/ontology.

According to the *TOGA meta–theory paradigms*,[70] for scientific and practical modelling purposes, it is reasonable to separate conceptually the following five concepts: *information, knowledge, preferences, intelligence and emotions.* If properly defined, all of them can be independently identified and designed.

Such conceptual modularity should enable to construct: *emotional intelligence, social intelligence, skill intelligence, organizational intelligence,* and many other X–intelligences, where X denotes a type of knowledge, preferences or a carrier system involved.

[70] According to the *top–down object–based goal–oriented approach* (TOGA) standard, the Information–Preferences–Knowledge *cognitive architecture* consists of:

Data: everything what is/can be processed/transformed in computational and mental processes. Concept data is included in the ontology of 'elaborators', such as developers of methods, programmers and other computation service people. In this sense, data is a relative term and exists only in the couple (data, processing).

Information: data which represent a specific property of the domain of human or artificial agent's activity (such as: addresses, tel. numbers, encyclopedic data, various lists of names and results of measurements). Every information has always a source domain. It is a relative concept. Information is a concept from the ontology of modeler/problem–solver/decision–maker.

Knowledge: every abstract property of human/artificial agent which has ability to process/transform a quantitative/qualitative information into other information, or into another knowledge. It includes: instructions, emergency procedures,

For example, business intelligence and emotional intelligence, rather are applications of intelligence either for business activities or for the second, under emotional/(not conscious) constrains and 'biological requests'.

In the above context, an *abstract intelligent agent* can be considered as a functional kernel of any natural or artificial intelligent system.

Intelligence and Cognitive Development

Although there is no general *theory of cognitive development*, the most historically influential theory was developed by Jean Piaget.[71] *Piaget theory*

---

exploitation/user manuals, scientific materials, models and theories. Every knowledge has its reference domain where it is applicable. It has to include the source domain of the processed information. It is a relative concept.

Preference: an ordered relation among two properties of the domain of activity of a *cognitive agent*, it indicates a property with higher utility. Preference relations serve to establish an intervention goal of an agent. Cognitive preferences are relative. A preference agent which manages preferences of an intelligent agent can be external or its internal part.

Goal: a hypothetical state of the domain of activity which has maximal utility in a current situation. Goal serves to the choice and activate proper knowledge which process new information.

Document: a passive carrier of knowledge, information and/or preferences (with different structures), comprehensive for humans, and it has to be recognized as valid and useful by one or more human organizations, it can be physical or electronic.

Computer Program: (i) from the modelers and decision-makers perspective: an active carrier of different structures of knowledge expressed in computer languages and usually focused on the realization of predefined objectives (a design-goal). It may include build-in preferences and information and/or request specific IPK as data. (ii) from the software engineers perspective: a data-processing tool (more precise technical def. you may find on the Web).

[71] Jean Piaget (August 9, 1896 – September 16, 1980) was a Swiss natural scientist and developmental psychologist, well known for his work studying children and his theory of cognitive development. Piaget served as professor of psychology at the University of Geneva from 1929 to 1975 and is best known for reorganizing cognitive development theory into a series of stages, expanding on earlier work from James Baldwin: four levels of development corresponding roughly to (1) infancy, (2) pre–school, (3) childhood, and (4) adolescence. Each stage is characterized by a general cognitive structure that affects all of the child's thinking (a structuralist view influenced by philosopher Immanuel Kant). Each stage represents the child's understanding of reality during that period, and each but the last is an inadequate approximation of reality. Development from one stage to the next is thus caused by the accumulation of errors in the child's understanding of the environment; this accumulation eventually causes such a degree of cognitive disequilibrium that thought structures require reorganising. For his development of the theory, Piaget was awarded the Erasmus Prize.

provided many central concepts in the field of developmental psychology. His theory concerned the growth of intelligence, which for Piaget meant the ability to more accurately represent the world, and perform logical operations on representations of concepts grounded in the world. His theory concerns the emergence and acquisition of schemata, schemes of how one perceives the world, in 'developmental stages', times when children are acquiring new ways of mentally representing information. Piaget theory is considered 'constructivist, meaning that, unlike nativist theories (which describe cognitive development as the unfolding of innate knowledge and abilities) or empiricist theories (which describe cognitive development as the gradual acquisition of knowledge through experience), asserts that we construct our cognitive abilities through self–motivated action in the world.

The four development stages are described in Piaget's theory as:

1. Sensorimotor stage: from birth to age 2 years (children experience the world through movement and senses)
2. Preoperational stage: from ages 2 to 7(acquisition of motor skills)
3. Concrete operational stage: from ages 7 to 11 (children begin to think logically about concrete events)
4. Formal Operational stage: after age 11 (development of abstract reasoning).

These chronological periods are approximate, and in light of the fact that studies have demonstrated great variation between children, cannot be seen as rigid norms. Furthermore, these stages occur at different ages, depending upon the domain of knowledge under consideration. The ages normally given for the stages, then, reflect when each stage tends to predominate, even though one might elicit examples of two, three, or even all four stages of thinking at the same time from one individual, depending upon the domain of knowledge and the means used to elicit it. Despite this, though, the principle holds that within a domain of knowledge, the stages usually occur in the same chronological order. Thus, there is a somewhat subtler reality behind the normal characterization of the stages as described above. The reason for the invariability of sequence derives from the idea that knowledge is not simply acquired from outside the individual, but it is constructed from within. This idea has been extremely influential in pedagogy, and is usually termed constructivism. Once knowledge is constructed internally, it is then tested against reality the same way a scientist tests the validity of hypotheses. Like a scientist, the individual learner may discard, modify, or reconstruct knowledge based on its utility in the real world. Much of this construction (and later reconstruction) is in fact done subconsciously. Therefore, Piaget's four stages actually reflect four types of thought structures. The chronological sequence is inevitable, then, because one structure may be necessary in order to construct the next level, which is simpler, more generalizable, and more powerful. It's a little like saying that you need to form metal into parts in order to build machines, and then coordinate machines in order to build a factory.

Piaget divided schemes that children use to understand the world through four main stages, roughly correlated with and becoming increasingly sophisticated with age:

1. Sensorimotor stage (years 0–2),
2. Preoperational stage (years 2–7),
3. Concrete operational stage (years 7–11), and
4. Formal operational stage (years 11–adulthood).

Sensorimotor Stage

Infants are born with a set of congenital reflexes, according to Piaget, as well as a drive to explore their world. Their initial schemas are formed through differentiation of the congenital reflexes (see assimilation and accommodation, below).

The sensorimotor stage is the first of the four stages. According to Piaget, this stage marks the development of essential spatial abilities and understanding of the world in six sub–stages:

1. The first sub–stage occurs from birth to six weeks and is associated primarily with the development of reflexes. Three primary reflexes are described by Piaget: sucking of objects in the mouth, following moving or interesting objects with the eyes, and closing of the hand when an object makes contact with the palm (palmar grasp). Over these first six weeks of life, these reflexes begin to become voluntary actions; for example, the palmar reflex becomes intentional grasping.
2. The second sub–stage occurs from six weeks to four months and is associated primarily with the development of habits. Primary circular reactions or repeating of an action involving only ones own body begin. An example of this type of reaction would involve something like an infant repeating the motion of passing their hand before their face. Also at this phase, passive reactions, caused by classical or operant conditioning, can begin.
3. The third sub–stage occurs from four to nine months and is associated primarily with the development of coordination between vision and prehension. Three new abilities occur at this stage: intentional grasping for a desired object, secondary circular reactions, and differentiations between ends and means. At this stage, infants will intentionally grasp the air in the direction of a desired object, often to the amusement of friends and family. Secondary circular reactions, or the repetition of an action involving an external object begin; for example, moving a switch to turn on a light repeatedly. The differentiation between means also occurs. This is perhaps one of the most important stages of a child's growth as it signifies the dawn of logic. Towards the late part of this sub–stage infants begin to have a sense of object permanence, passing the A–not–B error test.
4. The fourth sub-stage occurs from nine to twelve months and is associated primarily with the development of logic and the coordination between

means and ends. This is an extremely important stage of development, holding what Piaget calls the 'first proper intelligence'. Also, this stage marks the beginning of goal orientation, the deliberate planning of steps to meet an objective.

5. The fifth sub–stage occurs from twelve to eighteen months and is associated primarily with the discovery of new means to meet goals. Piaget describes the child at this juncture as the 'young scientist', conducting pseudo–experiments to discover new methods of meeting challenges.
6. The sixth sub–stage is associated primarily with the beginnings of insight, or true creativity. This marks the passage into the preoperational stage.

Preoperational Stage

The Preoperational stage is the second of four stages of cognitive development. By observing sequences of play, Piaget was able to demonstrate that towards the end of the second year a qualitatively quite new kind of psychological functioning occurs. Operation in Piagetian theory is any procedure for mentally acting on objects. The hallmark of the preoperational stage is sparse and logically inadequate mental operations.

According to Piaget, the Sensorimotor stage of development is followed by this stage (2–7 years), which includes the following five processes:

1. Symbolic functioning, which is characterised by the use of mental symbols words or pictures which the child uses to represent something which is not physically present.
2. Centration, which is characterized by a child focusing or attending to only one aspect of a stimulus or situation. For example, in pouring a quantity of liquid from a narrow beaker into a shallow dish, a preschool child might judge the quantity of liquid to have decreased, because it is 'lower', that is, the child attends to the height of the water, but not to the compensating increase in the diameter of the container.
3. Intuitive thought, which occurs when the child is able to believe in something without knowing why she or he believes it.
4. Egocentrism, which is a version of centration, this denotes a tendency of child to only think from own point of view.
5. Inability to Conserve; Through Piaget's conservation experiments (conservation of mass, volume and number) Piaget concluded that children in the preoperational stage lack perception of conservation of mass, volume, and number after the original form has changed. For example, a child in this phase will believe that a string of beads set up in a 'O–O–O–O–O' pattern will have the same number of beads as a string which has a 'O–O–O–O–O' pattern, because they are the same length, or that a tall, thin 8-ounce cup has more liquid in it than a wide, fat 8–ounce cup.

Concrete Operational Stage

The concrete operational stage is the third of four stages of cognitive development in Piaget's theory. This stage, which follows the Preoperational stage and occurs from the ages of 7 to 11, is characterized by the appropriate use of logic. The six important processes during this stage are:

1. Decentering, where the child takes into account multiple aspects of a problem to solve it. For example, the child will no longer perceive an exceptionally wide but short cup to contain less than a normally-wide, taller cup.
2. Reversibility, where the child understands that numbers or objects can be changed, then returned to their original state. For this reason, a child will be able to rapidly determine that $4 + 4$ which they can answer to be 8, minus 4 will equal four, the original quantity.
3. Conservation: understanding that quantity, length or number of items is unrelated to the arrangement or appearance of the object or items. For instance, when a child is presented with two equally–sized, full cups they will be able to discern that if water is transferred to a pitcher it will conserve the quantity and be equal to the other filled cup.
4. Serialisation: the ability to arrange objects in an order according to size, shape, or any other characteristic. For example, if given different–shaded objects they may make a color gradient.
5. Classification: the ability to name and identify sets of objects according to appearance, size or other characteristic, including the idea that one set of objects can include another. A child is no longer subject to the illogical limitations of animism (the belief that all objects are animals and therefore have feelings).
6. Elimination of Egocentrism: the ability to view things from another's perspective (even if they think incorrectly). For instance, show a child a comic in which Jane puts a doll under a box, leaves the room, and then Jill moves the doll to a drawer, and Jane comes back; a child in this stage will not say that Jane will think the doll is in the drawer.

Formal Operational Stage

The formal operational stage is the fourth and final of the stages of cognitive development of Piaget's theory. This stage, which follows the Concrete Operational stage, commences at around 11 years of age (puberty) and continues into adulthood. It is characterized by acquisition of the ability to think abstractly and draw conclusions from the information available. During this stage the young adult functions in a cognitively normal manner and therefore is able to understand such things as love, 'shades of gray', and values. Lucidly, biological factors may be traced to this stage as it occurs during puberty and marks the entering into adulthood in physiologically, cognitive, moral (Kohlberg), psychosexual (Freud), and social development (Erikson).

Many people do not successfully complete this stage, but mostly remain in concrete operations.

Psychophysics

Recall that *psychophysics* is a subdiscipline of psychology, founded in 1860 by Gustav Fechner[72] with the publication of 'Elemente der Psychophysik', dealing with the relationship between physical stimuli and their subjective correlates, or percepts. Fechner described research relating physical stimuli with how they are perceived and set out the philosophical foundations of the field. Fechner wanted to develop a theory that could relate matter to the mind, by describing the relationship between the world and the way it is perceived (Snodgrass, 1975). Fechner's work formed the basis of psychology as a science. Wilhelm Wundt, the founder of the first laboratory for psychological research, built upon Fechner's work.

The *Weber–Fechner law* attempts to describe the relationship between the physical magnitudes of stimuli and the perceived intensity of the stimuli.

---

[72] Gustav Theodor Fechner (April 19, 1801 – November 28, 1887), was a German experimentle psychologist. A pioneer in experimental psychology.

Fechner's epoch–faking work was his Elemente der Psychophysik (1860). He starts from the Spinozistic thought that bodily facts and conscious facts, though not reducible one to the other, are different sides of one reality. His originality lies in trying to discover an exact mathematical relation between them. The most famous outcome of his inquiries is the law known as *Weber–Fechner law* which may be expressed as follows: "In order that the intensity of a sensation may increase in arithmetical progression, the stimulus must increase in geometrical progression." Though holding good within certain limits only, the law has been found immensely useful. Unfortunately, from the tenable theory that the intensity of a sensation increases by definite additions of stimulus, Fechner was led on to postulate a unit of sensation, so that any sensations might be regarded as composed of n units. Sensations, he argued, thus being representable by numbers, psychology may become an 'exact' science, susceptible of mathematical treatment.

His general formula for getting at the number of units in any sensation is $S = c \log R$, where S stands for the sensation, R for the stimulus numerically estimated, and c for a constant that must be separately determined by experiment in each particular order of sensibility. This reasoning of Fechner's has given rise to a great mass of controversy, but the fundamental mistake in it is simple. Though stimuli are composite, sensations are not. "Every sensation," says William James, "presents itself as an indivisible unit; and it is quite impossible to read any clear meaning into the notion that they are masses of units combined." Still, the idea of the exact measurement of sensation has been a fruitful one, and mainly through his influence on Wilhelm Wundt, Fechner was the father of that 'new' psychology of laboratories which investigates human faculties with the aid of exact scientific apparatus.

Ernst Weber[73] was one of the first people to approach the study of the human response to a physical stimulus in a quantitative fashion. Gustav Fechner later offered an elaborate theoretical interpretation of Weber's findings, which he called simply Weber's law, though his admirers made the law's name a hyphenate. Fechner believed that Weber had discovered the fundamental principle of mind/body interaction, a mathematical analog of the function Rene Descartes once assigned to the pineal gland.

In one of his classic experiments, Weber gradually increased the weight that a blindfolded man was holding and asked him to respond when he first felt the increase. Weber found that the response was proportional to a relative increase in the weight. That is to say, if the weight is 1 kg, an increase of a few grams will not be noticed. Rather, when the mass is increased by a certain factor, an increase in weight is perceived. If the mass is doubled, the threshold is also doubled. This kind of relationship can be described by a linear ordinary differential equation as,

$$dp = k\frac{dS}{S},$$

where dp is the differential change in perception, dS is the differential increase in the stimulus and S is the stimulus at the instant. A constant factor k is to be determined experimentally. Integrating the above equation gives: $p = k\ln S + C$, where C is the constant of integration. To determine C, we can put $p = 0$, which means no perception; then we get, $C = -k\ln S_0$,where S_0 is that threshold of stimulus below which it is not perceived at all. In this way, we get the solution

$$p = k\ln\frac{S}{S_0}.$$

Therefore, the relationship between stimulus and perception is logarithmic. This logarithmic relationship means that if a stimulus varies as a geometric progression (i.e. multiplied by a fixed factor), the corresponding perception is altered in an arithmetic progression (i.e. in additive constant amounts). For example, if a stimulus is tripled in strength (i.e, 3×1), the corresponding perception may be two times as strong as its original value (i.e., $1 + 1$). If the stimulus is again tripled in strength (i.e., $3 \times 3 \times 1$), the corresponding perception will be three times as strong as its original value (i.e., $1 + 1 + 1$). Hence, for multiplications in stimulus strength, the strength of perception

[73] Ernst Heinrich Weber (Wittenberg, June 24, 1795 – January 26, 1878) was a German physician who is considered a founder of experimental psychology. Weber studied medicine at Wittenberg University. In 1818 he was appointed Associate Professor of comparative anatomy at Leipzig University, where he was made a Fellow Professor of anatomy and physiology in 1821.

Around 1860 Weber worked with Gustav Fechner on psychophysics, during which time he formulated Weber's Law. In 1866 Weber retired as professor of physiology and also as professor of anatomy in 1871. Around this time he and his brother, Eduard Weber, discovered the inhibitory power of the vagus nerve.

only adds. This logarithmic relationship is valid, not just for the sensation of weight, but for other stimuli and our sensory perceptions as well.

In case of vision, we have that the eye senses brightness logarithmically. Hence stellar magnitude is measured on a logarithmic scale. This magnitude scale was invented by the ancient Greek astronomer Hipparchus in about 150 B.C. He ranked the stars he could see in terms of their brightness, with 1 representing the brightest down to 6 representing the faintest, though now the scale has been extended beyond these limits. An increase in 5 magnitudes corresponds to a decrease in brightness by a factor 100.

In case of sound, we have still another logarithmic scale is the decibel scale of sound intensity. And yet another is pitch, which, however, differs from the other cases in that the physical quantity involved is not a 'strength'. In the case of perception of pitch, humans hear pitch in a logarithmic or geometric ratio–based fashion: For notes spaced equally apart to the human ear, the frequencies are related by a multiplicative factor. For instance, the frequency of corresponding notes of adjacent octaves differ by a factor of 2. Similarly, the perceived difference in pitch between 100 Hz and 150 Hz is the same as between 1000 Hz and 1500 Hz. Musical scales are always based on geometric relationships for this reason. Notation and theory about music often refers to pitch intervals in an additive way, which makes sense if one considers the logarithms of the frequencies, as $\log(a \times b) = \log a + \log b$.

Psychophysicists usually employ experimental stimuli that can be objectively measured, such as pure tones varying in intensity, or lights varying in luminance. All the senses have been studied:vision, hearing, touch (including skin and enteric perception), taste, smell, and the sense of time. Regardless of the sensory domain, there are three main topics in the psychophysical classification scheme: absolute thresholds, discrimination thresholds, and scaling.

The most common use of psychophysics is in producing scales of human experience of various aspects of physical stimuli. Take for an example the physical stimulus of frequency of sound. Frequency of a sound is measured in Hertz (Hz), cycles per second. But human experience of the frequencies of sound is not the same as the frequencies. For one thing, there is a frequency below which no sounds can be heard, no matter how intense they are (around 20 Hz depending on the individual) and there is a frequency above which no sounds can be heard, no matter how intense they are (around 20,000 Hz, again depending on the individual). For another, doubling the frequency of a sound (e.g., from 100 Hz to 200 Hz) does not lead to a doubling of experience. The perceptual experience of the frequency of sound is called pitch, and it is measured by psychophysicists in mels.

More analytical approaches allow the use of psychophysical methods to study neurophysiological properties and sensory processing mechanisms. This is of particular importance in human research, where other (more invasive) methods are not used due to ethical reasons. Areas of investigation include sensory thresholds, methods of measurement of sensitivity, and signal detection theory.

Perception is the process of acquiring, interpreting, selecting, and organizing sensory information. Methods of studying perception range from essentially biological or physiological approaches, through psychological approaches to the often abstract 'thought–experiments' of mental philosophy.

Experiments in psychophysics seek to determine whether the subject can detect a stimulus, identify it, differentiate between it and another stimulus, and describe the magnitude or nature of this difference [Sno75]. Often, the classic methods of experimentation are argued to be inefficient. This is because a lot of sampling and data has to be collected at points of the psychometric function that is known (the tails). Staircase procedures can be used to quickly estimate threshold. However, the cost of this efficiency, is that we do not get the same amount of information regarding the *psychometric function* as we can through classical methods; e.g., we cannot extract an estimate of the slope (derivative) of the function.

A psychometric function describes the relationship between a parameter of a physical stimulus and the responses of a person who has to decide about a certain aspect of that stimulus. The psychometric function usually resembles a sigmoid function with the percentage of correct responses (or a similar value) displayed on the ordinate and the physical parameter on the abscissa. If the stimulus parameter is very far towards one end of its possible range, the person will always be able to respond correctly. Towards the other end of the range, the person never perceives the stimulus properly and therefore the probability of correct responses is at chance level. In between, there is a transition range where the subject has an above–chance rate of correct responses, but does not always respond correctly. The inflection point of the sigmoid function or the point at which the function reaches the middle between the chance level and 100% is usually taken as sensory threshold. A common example is visual acuity testing with an eye chart. The person sees symbols of different sizes (the size is the relevant physical stimulus parameter) and has to decide which symbol it is. Usually, there is one line on the chart where a subject can identify some, but not all, symbols. This is equal to the transition range of the psychometric function and the sensory threshold corresponds to visual acuity.

On the other hand, a *sensory threshold* is a theoretical concept which states: "A stimulus that is less intense than the sensory threshold will not elicit any sensation." Whilst the concept can be applied to all senses, it is most commonly applied to the detection and perception of flavours and aromas. Several different sensory thresholds have been defined:

1. Absolute threshold: the lowest level at which a stimulus can be detected.
2. Recognition threshold: the level at which a stimulus can not only be detected but also recognised.
3. Differential threshold: the level at which an increase in a detected stimulus can be perceived.
4. Terminal threshold: the level beyond which a stimulus is no longer detected.

In other words, a threshold is the point of intensity at which the participant can just detect the presence of, or difference in, a stimulus. Stimuli with intensities below the threshold are considered not detectable, however stimuli at values close to threshold will often be detectable some proportion of the time. Due to this, a threshold is considered to be the point at which a stimulus, or change in a stimulus, is detected some proportion p of the time. An absolute threshold is the level of intensity of a stimulus at which the subject is able to detect the presence of the stimulus some proportion of the time (a p level of 50% is often used). An example of an absolute threshold is the number of hairs on the back of one's hand that must be touched before it can be felt, a participant may be unable to feel a single hair being touched, but may be able to feel two or three as this exceeds the threshold. A difference threshold is the magnitude of the difference between two stimuli of differing intensities that the participant is able to detect some proportion of the time (again, 50% is often used). To test this threshold, several difference methods are used. The subject may be asked to adjust one stimulus until it is perceived as the same as the other, may be asked to describe the magnitude of the difference between two stimuli, or may be asked to detect a stimulus against a background. Absolute and difference thresholds are sometimes considered similar because there is always background noise interfering with our ability to detect stimuli, however study of difference thresholds still occurs, for example in pitch discrimination tasks (see [Sno75]).

The *sensory analysis* applies principles of experimental design and statistical analysis to the use of human senses (sight, smell, taste, touch and hearing) for the purposes of evaluating consumer products. The discipline requires panels of human assessors, on whom the products are tested, and recording the responses made by them. By applying statistical techniques to the results it is possible to make inferences and insights about the products under test. Most large consumer goods companies have departments dedicated to sensory analysis. Sensory Analysis can generally be broken down into three sub–sections:

1. Effective Testing (dealing with objective facts about products);
2. Affective Testing (dealing with subjective facts such as preferences); and
3. Perception (the biochemical and psychological aspects of sensation).

The *signal detection theory* (SDT) is a means to quantify the ability to discern between signal and noise. It has applications in many fields such as quality control, telecommunications, and psychology (see [Abd06]). The concept is similar to the signal to noise ratio used in the sciences, and it is also usable in alarm management, where it is important to separate important events from background noise. According to the theory, there are a number of psychological determiners of how we will detect a signal, and where our threshold levels will be. Experience, expectations, physiological state (e.g, fatigue) and other factors affect thresholds. For instance, a sentry in wartime will likely detect fainter stimuli than the same sentry in peacetime. SDT is used when psychologists want to measure the way we make decisions under conditions of uncertainty, such as how we would perceive distances in foggy conditions. SDT assumes that 'the decision maker is not a passive receiver of information, but an active decision–maker who makes difficult perceptual judgements under conditions of uncertainty'. In foggy circumstances, we are forced to decide how far an object is away from us based solely upon visual stimulus which is impaired by the fog. Since the brightness of the object, such as a traffic light, is used by the brain to discriminate the distance of an object, and the fog reduces the brightness of objects, we perceive the object to be much further away than it actually is. To apply signal detection theory to a data set where stimuli were either present or absent, and the observer categorized each trial as having the stimulus present or absent, the trials are sorted into one of four categories, depending upon the stimulus and response:

	Respond 'Absent'	Respond 'Present'
Stimulus Present	Miss	Hit
Stimulus Absent	Correct Rejection	False Alarm

1.1.2 Human Problem Solving

Beginning in the 1970s, researchers became increasingly convinced that empirical findings and theoretical concepts derived from simple laboratory tasks did not necessarily generalize to more complex, real–life problems. Even worse, it appeared that the processes underlying creative problem solving in different domains differed from each other [Ste95]. These realizations have led to rather different responses in North America and in Europe.

In North America, initiated by the work of Herbert Simon on learning by doing in semantically rich domains (see, e.g., [AS79, BS77]), researchers began to investigate problem solving separately in different natural knowledge domains – such as physics, writing, or chess playing – thus relinquishing their attempts to extract a global theory of problem solving (see, e.g., [SF91]). Instead, these researchers have frequently focused on the development of problem solving within a certain domain, that is on the development of expertise (see, e.g., [ABR85], [CS73]; [CFG81]).

Areas that have attracted rather intensive attention in North America include such diverse fields as: reading [SC91], writing [BBS91], calculation [SM91], political decision making [VWL91], managerial problem solving [Wag91], lawyers' reasoning [ALL91], personal problem solving [HK87], mathematical problem solving [Pol45, Sch85], mechanical problem solving [Heg91], problem solving in electronics [LL91], computer skills [Kay91], game playing [FS91], and social problem solving [D'Zur86].

In particular, George Pólya's 1945 book 'How to Solve It' [Pol45], is a small volume describing methods of problem–solving. It suggests the following steps when solving a mathematical problem:

1. First, you have to understand the problem.
2. After understanding, then make a plan.

Heuristic	Informal Description	Formal analogue
Analogy	can you find a problem analogous to your problem and solve that?	Map
Generalization	can you find a problem more general than your problem ...?	Generalization
Induction	can you solve your problem by deriving a generalization from some examples?	Induction
Variation of the Problem	can you vary or change your problem to create a new problem (or set of problems) whose solution(s) will help you solve your original problem?	Search
Auxiliary Problem	can you find a subproblem or side problem whose solution will help you solve your problem?	Subgoal
Here is a problem related to yours and solved before	can you find a problem related to yours that has already been solved and use that to solve your problem?	Pattern recognition Pattern matching
Specialization	can you find a problem more specialized?	Specialization
Decomposing and Recombining	can you decompose the problem and "recombine its elements in some new manner"?	Divide and conquer
Working backward	can you start with the goal and work backwards to something you already know?	Backward chaining
Draw a Figure	can you draw a picture of the problem?	Diagrammatic Reasoning
Auxiliary Elements	can you add some new element to your problem to get closer to a solution?	Extension

3. Carry out the plan.
4. Look back on your work. How could it be better?

If this technique fails, Polya advises: "If you cannot solve a problem, then there is an easier problem you can solve: find it." Or, "If you cannot solve the proposed problem try to solve first some related problem. Could you imagine a more accessible related problem?"

His small book contains a dictionary–style set of heuristics, many of which have to do with generating a more accessible problem, like the ones given in the table below:

The technique 'have I used everything' is perhaps most applicable to formal educational examinations (e.g., n men digging m ditches, see footnote below) problems. The book has achieved 'classic' status because of its considerable influence. Marvin Minsky[74] said in his influential paper 'Steps Toward Artificial Intelligence': "And everyone should know the work of George Polya

[74] Marvin Lee Minsky (born August 9, 1927), sometimes affectionately known as 'Old Man Minsky', is an American cognitive scientist in the field of artificial

on how to solve problems." Polya's book has had a large influence on mathematics textbooks. Most formulations of a problem solving framework in U.S. textbooks attribute some relationship to Polya's problem solving stages. Other books on problem solving are often related to less concrete and more creative techniques, like e.g., lateral thinking, mind mapping and brainstorming (see below).

On the other hand, in Europe, two main approaches have surfaced, one initiated by Donald Broadbent in the UK [Bro77, BB95] and the other one by Dietrich Dörner in Germany [Dor75, DV85, DW95]. The two approaches have in common an emphasis on relatively complex, semantically rich, computerized laboratory tasks, constructed to resemble 'real–life' problems. The approaches differ somewhat in their theoretical goals and methodology, however. The tradition initiated by Broadbent emphasizes the distinction between cognitive problem–solving processes that operate under awareness versus outside of awareness, and typically employs mathematically well–defined computerized systems. The tradition initiated by Dörner, on the other hand, has an interest in the interplay of the cognitive, motivational, and social components of problem solving, and utilizes very complex computerized scenarios that contain up to 2,000 highly interconnected variables. Buchner [Buc95] describes the two traditions in detail.

To sum up, researchers' realization that problem–solving processes differ across knowledge domains and across levels of expertise (see, e.g. [Ste95]) and that, consequently, findings obtained in the laboratory cannot necessarily generalize to problem–solving situations outside the laboratory, has during the past two decades led to an emphasis on real–world problem solving. This emphasis has been expressed quite differently in North America and Europe, however. Whereas North American research has typically concentrated on studying problem solving in separate, natural knowledge domains, much of the European research has focused on novel, complex problems, and has been performed with computerized scenarios (see [Fun95], for an overview).

Characteristics of Difficult Problems

As elucidated by Dietrich Dorner and later expanded upon by Joachim Funke, difficult problems have some typical characteristics. Recategorized and somewhat reformulated from these original works, these characteristics can be summarized as follows:

Intransparency (lack of clarity of the situation), including commencement opacity and continuation opacity;

Polytely (multiple goals), including inexpressiveness, opposition and transience;

---

intelligence (AI), co–founder of MIT's AI laboratory, and author of several texts on AI and philosophy.

Complexity (large numbers of items, interrelations, and decisions), including enumerability, connectivity (hierarchy relation, communication relation, allocation relation), and heterogeneity;

Dynamism (time considerations), including temporal constraints, temporal sensitivity, phase effects, and dynamic unpredictability.

The resolution of difficult problems requires a direct attack on each of these characteristics that are encountered.

Some *standard problem–solving techniques*, also known as creativity techniques, include:

1. Trial–and–error;[75]

---

[75] Trial and error (also known in computer science literature as generate and test and as 'guess and check' when solving equations in elementary algebra) is a method of problem solving for obtaining knowledge, both propositional knowledge and know-how.

This approach can be seen as one of the two basic approaches to problem solving and is contrasted with an approach using insight and theory.

In trial and error, one selects (or, generates) a possible answer, applies it to the problem and, if it is not successful, selects (or generates) another possibility that is subsequently tried. The process ends when a possibility yields a solution.

In some versions of trial and error, the option that is a priori viewed as the most likely one should be tried first, followed by the next most likely, and so on until a solution is found, or all the options are exhausted. In other versions, options are simply tried at random.

This approach is most successful with simple problems and in games, and is often resorted to when no apparent rule applies. This does not mean that the approach need be careless, for an individual can be methodical in manipulating the variables in an effort to sort through possibilities that may result in success. Nevertheless, this method is often used by people who have little knowledge in the problem area.

Trial and error has a number of features:

solution-oriented: trial and error makes no attempt to discover why a solution works, merely that it is a solution.

problem-specific: trial and error makes no attempt to generalize a solution to other problems.

non-optimal: trial and error is an attempt to find a solution, not all solutions, and not the best solution.

needs little knowledge: trial and error can proceed where there is little or no knowledge of the subject.

For example, trial and error has traditionally been the main method of finding new drugs, such as antibiotics. Chemists simply try chemicals at random until they find one with the desired effect.

The *scientific method* can be regarded as containing an element of trial and error in its formulation and testing of hypotheses. Also compare *genetic algorithms*, *simulated annealing* and *reinforcement learning* – all varieties of search which apply the basic idea of trial and error.

2. Brainstorming;[76]
3. Morphological box;[77]

Biological Evolution is also a form of trial and error. Random mutations and sexual genetic variations can be viewed as trials and poor reproductive fitness as the error. Thus after a long time 'knowledge' of well–adapted genomes accumulates simply by virtue of them being able to reproduce.

Bogosort can be viewed as a trial and error approach to sorting a list.

In mathematics the method of trial and error can be used to solve formulae – it is a slower, less precise method than algebra, but is easier to understand.

[76] Brainstorming is a creativity technique of generating ideas to solve a problem. The main result of a brainstorm session may be a complete solution to the problem, a list of ideas for an approach to a subsequent solution, or a list of ideas resulting in a plan to find a solution. Brainstorming was originated in 1953 in the book 'Applied Imagination' by Alex Osborn, an advertising executive. Other methods of generating ideas are individual ideation and the morphological analysis approach.

Brainstorming has many applications but it is most often used in:

New product development – obtaining ideas for new products and improving existing products

Advertising – developing ideas for advertising campaigns

Problem solving – issues, root causes, alternative solutions, impact analysis, evaluation

Process management – finding ways of improving business and production processes

Project Management – identifying client objectives, risks, deliverables, work packages, resources, roles and responsibilities, tasks, issues

Team building – generates sharing and discussion of ideas while stimulating participants to think

Business planning – develop and improve the product idea.

Trial preparation by attorneys.

Brainstorming can be done either individually or in a group. In group brainstorming, the participants are encouraged, and often expected, to share their ideas with one another as soon as they are generated. Complex problems or brainstorm sessions with a diversity of people may be prepared by a chairman. The chairman is the leader and facilitator of the brainstorm session.

The key to brainstorming is to not interrupt the thought process. As ideas come to mind, they are captured and stimulate the development of better ideas. Thus a group brainstorm session is best conducted in a moderate–sized room, and participants sit so that they can all look at each–other. A flip chart, blackboard, or overhead projector is placed in a prominent location. The room is free of telephones, clocks, or any other distractions.

[77] Morphological analysis was designed for multi-dimensional, non-quantifiable problems where causal modelling and simulation do not function well or at all. Fritz Zwicky developed this approach to seemingly non-reducible complexity [Zwi69]. Using the technique of cross consistency assessment (CCA) [Rit02], the system however does allow for reduction, not by reducing the number of variables involved, but by reducing the number of possible solutions through the elimination of the illogical solution combinations in a grid box.

4. Method of focal objects;[78]
5. Lateral thinking;[79]

---

[78] The technique of *focal objects* for problem solving involves synthesizing the seemingly non–matching characteristics of different objects into something new.

For example, to generate new solutions to gardening take some ideas at random, such swimming and a couch, and invent ways for them to merge. Swimming might be used with the idea of gardening to create a plant oxygen tank for underwater divers. A couch might be used with the idea of gardening to invent new genes that would grow plants into the shape of a couch. The larger the number of diverse objects included, the greater the opportunity for inventive solutions.

Another way to think of focal objects is as a memory cue: if you're trying to find all the different ways to use a brick, give yourself some random 'objects' (situations, concepts, etc.) and see if you can find a use. Given 'blender', for example, I would try to think of all the ways a brick could be used with a blender (as a lid?). Another concept for the brick game: find patterns in your solutions, and then break those patterns. If you keep finding ways to build things with bricks, think of ways to use bricks that don't involve construction. Pattern–breaking, combined with focal object cues, can lead to very divergent solutions.

[79] Lateral thinking is a term coined by Edward de Bono [Bon73], a Maltese psychologist, physician, and writer, although it may have been an idea whose time was ready; the notion of lateral truth is discussed by Robert M. Pirsig in Zen and the Art of Motorcycle Maintenance. de Bono defines Lateral Thinking as methods of thinking concerned with changing concepts and perception. For example:

It took two hours for two men to dig a hole five feet deep. How deep would it have been if ten men had dug the hole for two hours?

The answer appears to be 25 feet deep. This answer assumes that the thinker has followed a simple mathematical relationship suggested by the description given, but we can generate some lateral thinking ideas about what affects the size of the hole which may lead to different answers:

A hole may need to be of a certain size or shape so digging might stop early at a required depth.

The deeper a hole is, the more effort is required to dig it, since waste soil needs to be lifted higher to the ground level. There is a limit to how deep a hole can be dug by manpower without use of ladders or hoists for soil removal, and 25 feet is beyond this limit.

Deeper soil layers may be harder to dig out, or we may hit bedrock or the water table.

Each man digging needs space to use a shovel.

It is possible that with more people working on a project, each person may become less efficient due to increased opportunity for distraction, the assumption he can slack off, more people to talk to, etc.

More men could work in shifts to dig faster for longer.

There are more men but are there more shovels?

The two hours dug by ten men may be under different weather conditions than the two hours dug by two men.

Rain could flood the hole to prevent digging.

Temperature conditions may freeze the men before they finish.

Would we rather have 5 holes each 5 feet deep?

6. Mind mapping;[80]

The two men may be an engineering crew with digging machinery.

What if one man in each group is a manager who will not actually dig?

The extra eight men might not be strong enough to dig, or much stronger than the first two.

The most useful ideas listed above are outside the simple mathematics implied by the question. Lateral thinking is about reasoning that is not immediately obvious and about ideas that may not be obtainable by using only traditional step–by–step logic.

Techniques that apply lateral thinking to problems are characterized by the shifting of thinking patterns away from entrenched or predictable thinking to new or unexpected ideas. A new idea that is the result of lateral thinking is not always a helpful one, but when a good idea is discovered in this way it is usually obvious in hindsight, which is a feature lateral thinking shares with a joke.

Lateral thinking can be contrasted with critical thinking, which is primarily concerned with judging the truth value of statements and seeking error. Lateral Thinking is more concerned with the movement value of statements and ideas, how to move from them to other statements and ideas.

For example the statement 'cars should have square wheels' when considered with critical thinking would be evaluated as a poor suggestion, as there are many engineering problems with square wheels. The Lateral Thinking treatment of the same statement would be to see where it leads. Square wheels would produce predictable bumps. If bumps can be predicted then suspension can be designed to compensate. Another way to predict bumps would be a laser or sonar on the front of the car examining the road surface ahead. This leads to the idea of active suspension with a sensor on the car that has normal wheels. The initial statement has been left behind.

[80] Recall that a *mind map* is a diagram used to represent words, ideas, tasks or other items linked to and arranged radially around a central key word or idea. It is used to generate, visualize, structure and classify ideas, and as an aid in study, organization, problem solving, and decision making.

It is an image–centered diagram that represents semantic or other connections between portions of information. By presenting these connections in a radial, nonlinear graphical manner, it encourages a brainstorming approach to any given organizational task, eliminating the hurdle of initially establishing an intrinsically appropriate or relevant conceptual framework to work within.

A mind map is similar to a semantic network or cognitive map but there are no formal restrictions on the kinds of links used.

Most often the map involves images, words, and lines. The elements are arranged intuitively according to the importance of the concepts and they are organized into groupings, branches, or areas. The uniform graphic formulation of the semantic structure of information on the method of gathering knowledge, may aid recall of existing memories.

People have been using image centered radial graphic organization techniques referred to variably as mental or generic mind maps for centuries in areas such as engineering, psychology, and education, although the claim to the origin of the mind map has been made by a British popular psychology author, Tony Buzan.

7. Analogy with similar problems;[81] and

The mind map continues to be used in various forms, and for various applications including learning and education (where it is often taught as 'Webs' or 'Webbing'), planning and in engineering diagramming.

When compared with the earlier original concept map (which was developed by learning experts in the 1960s) the structure of a mind map is a similar, but simplified, radial by having one central key word.

Mind maps have many applications in personal, family, educational, and business situations, including note-taking, brainstorming (wherein ideas are inserted into the map radially around the center node, without the implicit prioritization that comes from hierarchy or sequential arrangements, and wherein grouping and organizing is reserved for later stages), summarizing, revising and general clarifying of thoughts. For example, one could listen to a lecture and take down notes using mind maps for the most important points or keywords. One can also use mind maps as a mnemonic technique or to sort out a complicated idea. Mind maps are also promoted as a way to collaborate in color pen creativity sessions.

[81] Recall that analogy is either the cognitive process of transferring information from a particular subject (the analogue or source) to another particular subject (the target), or a linguistic expression corresponding to such a process. In a narrower sense, analogy is an inference or an argument from a particular to another particular, as opposed to deduction, induction, and abduction, where at least one of the premises or the conclusion is general. The word analogy can also refer to the relation between the source and the target themselves, which is often, though not necessarily, a similarity, as in the biological notion of analogy.

Niels Bohr's model of the atom made an analogy between the atom and the solar system.Analogy plays a significant role in problem solving, decision making, perception, memory, creativity, emotion, explanation and communication. It lies behind basic tasks such as the identification of places, objects and people, for example, in face perception and facial recognition systems. It has been argued that analogy is 'the core of cognition'. Specifically analogical language comprises exemplification, comparisons, metaphors, similes, allegories, and parables, but not metonymy. Phrases like and so on, and the like, as if, and the very word like also rely on an analogical understanding by the receiver of a message including them. Analogy is important not only in ordinary language and common sense, where proverbs and idioms give many examples of its application, but also in science, philosophy and the humanities. The concepts of association, comparison, correspondence, homomorphism, iconicity, isomorphism, mathematical homology, metaphor, morphological homology, resemblance, and similarity are closely related to analogy. In cognitive linguistics, the notion of conceptual metaphor may be equivalent to that of analogy.

Analogy has been studied and discussed since classical antiquity by philosophers, scientists and lawyers. The last few decades have shown a renewed interest in analogy, most notable in cognitive science.

With respect to the terms source and target, there are two distinct traditions of usage:

The logical and mathematical tradition speaks of an arrow, homomorphism, mapping, or morphism from what is typically the more complex domain or source

8. Research;[82]

1.1.3 Human Mind

Recall that the word *mind* commonly refers to the collective aspects of *intellect* and *consciousness* which are manifest in some combination of *thought*, *perception*, *emotion*, *will*, *memory*, and *imagination*.

There are many theories of what the mind is and how it works, dating back to Plato, Aristotle and other Ancient Greek philosophers. Modern theories, based on a scientific understanding of the brain, see the mind as a phenomenon of psychology, and the term is often used more or less synonymously with *consciousness*.

The question of which human attributes make up the mind is also much debated. Some argue that only the 'higher' intellectual functions constitute

---

to what is typically the less complex codomain or target, using all of these words in the sense of mathematical category theory.

The tradition that appears to be more common in cognitive psychology, literary theory, and specializations within philosophy outside of logic, speaks of a mapping from what is typically the more familiar area of experience, the source, to what is typically the more problematic area of experience, the target.

[82] Research is often described as an active, diligent, and systematic process of inquiry aimed at discovering, interpreting, and revising facts. This intellectual investigation produces a greater understanding of events, behaviors, or theories, and makes practical applications through laws and theories. The term research is also used to describe a collection of information about a particular subject, and is usually associated with science and the scientific method.

The word research derives from Middle French; its literal meaning is 'to investigate thoroughly'.

Thomas Kuhn, in his book 'The Structure of Scientific Revolutions', traces an interesting history and analysis of the enterprize of research.

Basic research (also called fundamental or pure research) has as its primary objective the advancement of knowledge and the theoretical understanding of the relations among variables. It is exploratory and often driven by the researcher's curiosity, interest, or hunch. It is conducted without any practical end in mind, although it may have unexpected results pointing to practical applications. The terms "basic" or "fundamental" indicate that, through theory generation, basic research provides the foundation for further, sometimes applied research. As there is no guarantee of short-term practical gain, researchers often find it difficult to get funding for basic research. Research is a subset of invention.

Applied research is done to solve specific, practical questions; its primary aim is not to gain knowledge for its own sake. It can be exploratory, but is usually descriptive. It is almost always done on the basis of basic research. Applied research can be carried out by academic or industrial institutions. Often, an academic institution such as a university will have a specific applied research program funded by an industrial partner interested in that program. Common areas of applied research include electronics, informatics, computer science, material science, process engineering, drug design ...

mind: particularly reason and memory. In this view the emotions – love, hate, fear, joy – are more 'primitive' or subjective in nature and should be seen as different in nature or origin to the mind. Others argue that the rational and the emotional sides of the human person cannot be separated, that they are of the same nature and origin, and that they should all be considered as part of the individual mind.

In popular usage *mind* is frequently synonymous with *thought*: It is that private conversation with ourselves that we carry on 'inside our heads' during every waking moment of our lives. Thus we 'make up our minds,' or 'change our minds' or are 'of two minds' about something. One of the key attributes of the mind in this sense is that it is a private sphere. No–one else can 'know our mind.' They can only know what we communicate.

Both philosophers and psychologists remain divided about the nature of the mind. Some take what is known as the substantial view, and argue that the mind is a single entity, perhaps having its base in the brain but distinct from it and having an autonomous existence. This view ultimately derives from Plato, and was absorbed from him into Christian thought. In its most extreme form, the substantial view merges with the theological view that the mind is an entity wholly separate from the body, in fact a manifestation of the soul, which will survive the body's death and return to God, its creator.

Others take what is known as the functional view, ultimately derived from Aristotle, which holds that the mind is a term of convenience for a variety of mental functions which have little in common except that humans are conscious of their existence. Functionalists tend to argue that the attributes which we collectively call the mind are closely related to the functions of the brain and can have no autonomous existence beyond the brain, nor can they survive its death. In this view mind is a subjective manifestation of consciousness: the human brain's ability to be aware of its own existence. The concept of the mind is therefore a means by which the conscious brain understands its own operations.

A leading exponent of the *substantial view* at the mind was George Berkeley, an 18th century Anglican bishop and philosopher. Berkeley argued that there is no such thing as matter and what humans see as the material world is nothing but an idea in God's mind, and that therefore the human mind is purely a manifestation of the soul or spirit. This type of belief is also common in certain types of spiritual non–dualistic belief, but outside this field few philosophers take an extreme view today. However, the view that the human mind is of a nature or essence somehow different from, and higher than, the mere operations of the brain, continues to be widely held.

Berkeley's views were attacked, and in the eyes of many philosophers demolished, by T.H. Huxley,[83] a 19th century biologist and disciple of Charles

[83] Thomas Henry Huxley, FRS (4 May 1825 – 29 June 1895) was an English biologist, known as 'Darwin's Bulldog' for his defence of Charles Darwin's theory of evolution. His scientific debates against Richard Owen demonstrated that there were

Darwin,[84] who agreed that the phenomena of the mind were of a unique order, but argued that they can only be explained in reference to events in the brain. Huxley drew on a tradition of materialist thought in British philosophy dating to Thomas Hobbes,[85] who argued in the 17th century that mental events were ultimately physical in nature, although with the biological knowledge of his day he could not say what their physical basis was. Huxley blended Hobbes with Darwin to produce the modern *functional view*. Huxley's view was reinforced by the steady expansion of knowledge about the functions of the human brain. In the 19th century it was not possible to say with certainty how the brain carried out such functions as memory, emotion, perception and reason. This left the field open for substantialists to argue for an autonomous mind, or for a metaphysical theory of the mind. But each advance in the study of the brain during the 20th century made this harder, since it became more and more apparent that all the components of the mind have their origins in

---

close similarities between the cerebral anatomy of humans and gorillas. Huxley did not accept many of Darwin's ideas, such as gradualism and was more interested in advocating a materialist professional science than in defending natural selection.

A talented populariser of science, he coined the term 'agnosticism' to describe his stance on religious belief. He is credited with inventing the concept of 'biogenesis', a theory stating that all cells arise from other cells and also 'abiogenesis', describing the generation of life from non–living matter.

84 Charles Robert Darwin (12 February 1809 – 19 April 1882) was an English naturalist who achieved lasting fame by producing considerable evidence that species originated through evolutionary change, at the same time proposing the scientific theory that natural selection is the mechanism by which such change occurs. This theory is now considered a cornerstone of biology.

Darwin developed an interest in natural history while studying first medicine, then theology, at university. Darwin's observations on his five-year voyage on the Beagle brought him eminence as a geologist and fame as a popular author. His biological finds led him to study the transmutation of species and in 1838 he conceived his theory of natural selection. Fully aware that others had been severely punished for such 'heretical' ideas, he confided only in his closest friends and continued his research to meet anticipated objections. However, in 1858 the information that Alfred Wallace had developed a similar theory forced an early joint publication of the theory.

His 1859 book 'On the Origin of Species by Means of Natural Selection' established evolution by common descent as the dominant scientific explanation of diversification in nature.

85 Thomas Hobbes (April 5, 1588–December 4, 1679) was an English philosopher, whose famous 1651 book Leviathan set the agenda for nearly all subsequent Western political philosophy. Although Hobbes is today best remembered for his work on *political philosophy*, he contributed to a diverse array of fields, including history, geometry, ethics, general philosophy and what would now be called political science. Additionally, Hobbes's account of human nature as self–interested cooperation has proved to be an enduring theory in the field of philosophical anthropology.

the functioning of the brain. Huxley's rationalism, was disturbed in the early 20th century by Freudian a theory of the unconscious mind, and argued that those mental processes of which humans are subjectively aware are only a small part of their total mental activity.

More recently, Douglas Hofstadter's[86] 1979 Pulitzer Prize–winning book 'Gödel, Escher, Bach – an eternal Golden Braid', is a *tour de force* on the subject of mind, and how it might arise from the neurology of the brain. Amongst other biological and cybernetic phenomena, Hofstadter places tangled loops and recursion at the center of self, self–awareness, and perception of oneself, and thus at the heart of mind and thinking. Likewise philosopher Ken Wilber posits that Mind is the interior dimension of the brain holon, i.e., mind is what a brain looks like internally, when it looks at itself.

Quantum physicist David Bohm[87] had a theory of mind that is most comparable to Neo–Platonic theories. "Thought runs you. Thought, however, gives false info that you are running it, that you are the one who controls thought. Whereas actually thought is the one which controls each one of us ..." [Boh92].

The debate about the nature of the mind is relevant to the development of artificial intelligence (see next section). If the mind is indeed a thing separate from or higher than the functioning of the brain, then presumably it will not be possible for any machine, no matter how sophisticated, to duplicate it. If on the other hand the mind is no more than the aggregated functions of the

[86] Douglas Richard Hofstadter (born February 15, 1945 in New York, New York) is an American academic, the son of Nobel Prize–winning physicist Robert Hofstadter. He is probably best known for his book Gödel, Escher, Bach: an Eternal Golden Braid (abbreviated as GEB) which was published in 1979, and won the 1980 Pulitzer Prize for general non-fiction. This book is commonly considered to have inspired many students to begin careers in computing and artificial intelligence, and attracted substantial notice outside its central artificial intelligence readership owing to its drawing on themes from such diverse disciplines as high-energy physics, music, the visual arts, molecular biology, and literature.

[87] David Joseph Bohm (born December 20, 1917 in Wilkes–Barre, Pennsylvania, died October 27, 1992 in London) was an American–born quantum physicist, who made significant contributions in the fields of theoretical physics, philosophy and neuropsychology, and to the Manhattan Project.

Bohm made a number of significant contributions to physics, particularly in the area of quantum mechanics and relativity theory. While still a post-graduate at Berkeley, he developed a theory of plasmas, discovering the electron phenomenon now known as Bohm–diffusion. His first book, Quantum Theory published in 1951, was well–received by Einstein, among others. However, Bohm became dissatisfied with the orthodox approach to quantum theory, which he had written about in that book, and began to develop his own approach (Bohm interpretation), a non–local hidden variable deterministic theory whose predictions agree perfectly with the nondeterministic quantum theory. His work and the EPR argument became the major factor motivating John Bell's inequality, whose consequences are still being investigated.

brain, then it will be possible, at least in theory, to create a machine with a mind.

Currently, the Mind/Brain/Behavior Interfaculty Initiative (MBB) at Harvard University aims to elucidate the structure, function, evolution, development, and pathology of the nervous system in relation to human behavior and mental life. It draws on the departments of psychology, neurobiology, neurology, molecular and cellular biology, radiology, psychiatry, organismic and evolutionary biology, history of science, and linguistics.

Bohm also made significant theoretical contributions to neuropsychology and the development of the so–called *holonomic brain model*. In collaboration with Stanford neuroscientist Karl Pribram, Bohm helped establish the foundation for Pribram's theory that the brain operates in a manner similar to a hologram, in accordance with quantum mathematical principles and the characteristics of wave patterns. These wave forms may compose hologram–like organizations, Bohm suggested, basing this concept on his application of *Fourier analysis*, a mathematical method for decomposing complex waves into component sine waves. The holonomic brain model developed by Pribram and Bohm posits a lens defined world view, much like the textured prismatic effect of sunlight refracted by the churning mists of a rainbow, a view which is quite different from the more conventional 'objective' approach. Pribram believes that if psychology means to understand the conditions that produce the world of appearances, it must look to the thinking of physicists like Bohm.

Bohm proposes thus in his book 'Thought as a System' a pervasive, systematic nature of thought: "What I mean by 'thought' is the whole thing – thought, 'felt', the body, the whole society sharing thoughts – it's all one process. It is essential for me not to break that up, because it's all one process; somebody else's thoughts becomes my thoughts, and vice versa. Therefore it would be wrong and misleading to break it up into my thoughts, your thoughts, my feelings, these feelings, those feelings ... I would say that thought makes what is often called in modern language a system. A system means a set of connected things or parts. But the way people commonly use the word nowadays it means something all of whose parts are mutually interdependent – not only for their mutual action, but for their meaning and for their existence. A corporation is organized as a system – it has this department, that department, that department. They do not have any meaning separately; they only can function together. And also the body is a system. Society is a system in some sense. And so on. Similarly, thought is a system. That system not only includes thoughts and feelings, but it includes the state of the body; it includes the whole of society – as thought is passing back and forth between people in a process by which thought evolved from ancient times. A system is constantly engaged in a process of development, change, evolution and structure changes ... although there are certain features of the system which become relatively fixed. We call this the structure ... Thought has been constantly evolving and we can't say when that structure began. But with the growth of civilization it has developed a great deal. It was probably very simple thought before civilization, and now it has become very complex and ramified and has much more incoherence than before ...

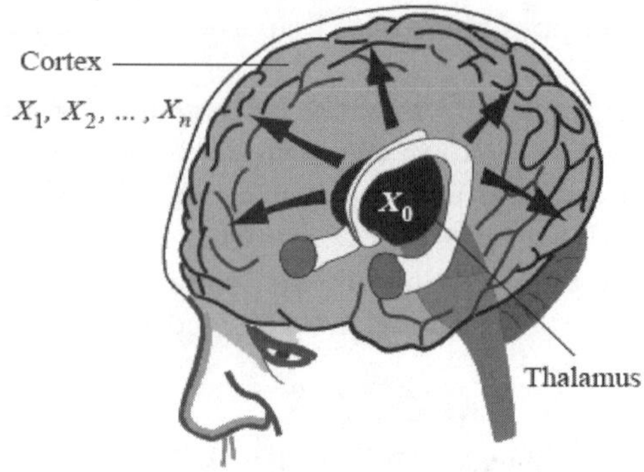

Fig. 1.2. A *possibly chaotic* 1–to–many relation: *Thalamus* $\Rightarrow$ *Cortex* in the human brain (with permission from E. Izhikevich).

On the other hand, human brain has been considered (by E.M. Izhikevich, Editor of the new Encyclopedia of Computational Neuroscience) as a *weakly–connected neural network*, with possibly *chaotic behavior* [Izh99b], consisting of n quasi–periodic cortical oscillators $X_1, \ldots, X_n$ forced by the thalamic input X_0 (see Figure 1.2)

The Mind–Body Problem

The *mind–body problem* is essentially the problem of explaining the relationship between minds, or mental processes, and bodily states or processes (see, e.g., [Kim95a]). Our perceptual experiences depend on stimuli which arrive at our various sensory organs from the external world and that these stimuli cause changes in the states of our brain, ultimately causing us to feel a sensation which may be pleasant or unpleasant. Someone's desire for a slice of pizza will tend to cause that person to move their body in a certain manner in a certain direction in an effort to get what they want. But how is it possible that conscious experiences can arise out of an inert lump of gray matter endowed with electrochemical properties? [Kim95b]. How does someone's desire cause that individual's neurons to fire and his muscles to contract in exactly the right manner? These are some of the essential puzzles that have confronted philosophers of mind at least from the time of René Descartes.[88]

[88] René Descartes (March 31, 1596 – February 11, 1650), also known as Cartesius, was a noted French philosopher, mathematician, and scientist. Dubbed the 'Founder of Modern Philosophy' and the 'Father of Modern Mathematics', he ranks as one of the most important and influential thinkers of modern times. Much of subsequent western philosophy is a reaction to his writings, which have been closely studied from his time down to the present day. Descartes was one of the key thinkers of the Scientific Revolution in the Western World. He is also

Dualism

Recall that *dualism* is a set of views about the relationship between mind and matter, which begins with the claim that mental phenomena are, in some respects, non–physical [Har96]. One of the earliest known formulations of mind–body dualism existed in the eastern *Sankhya school* of Hindu philosophy (c. 650 BCE) which divided the world into *Purusha* (mind/spirit) and *Prakrti* (material substance). In the Western philosophical tradition, we first encounter similar ideas with the writings of Plato and Aristotle, who maintained, for different reasons, that man's *intelligence* could not be identified with, or explained in terms of, his physical body (see, e.g., [RPW97]). However, the best–known version of dualism is due to René Descartes (1641), and holds that the mind is a non–physical substance [Des91]. Descartes was the first to clearly identify the mind with consciousness and self–awareness and to distinguish this from the brain, which was the seat of intelligence. Hence, he was the first to formulate the mind–body problem in the form in which it still exists today.

The main argument in favour of dualism is simply that it appeals to the common–sense intuition of the vast majority of non–philosophically–trained people. If asked what the mind is, the average person will usually respond by identifying it with their self, their personality, their soul, or some other such entity, and they will almost certainly deny that the mind simply is the brain or vice–versa, finding the idea that there is just one ontological entity at play to be too mechanistic or simply unintelligible [Har96]. The majority of modern philosophers of mind reject dualism, suggesting that these intuitions, like many others, are probably misleading. We should use our critical faculties, as well as empirical evidence from the sciences, to examine these assumptions and determine if there is any real basis to them [Har96] Another very important, more modern, argument in favor of dualism consists in the idea that the mental and the physical seem to have quite different and perhaps irreconcilable properties [Jac82]. Mental events have a certain subjective quality to them, whereas physical events obviously do not. For example, what does a burned finger feel like? What does blue sky look like? What does nice

---

honoured by having the *Cartesian coordinate system* used in plane geometry and algebra named after him.

Descartes was a major figure in 17th century continental rationalism, later advocated by Baruch Spinoza and Gottfried Leibniz, and opposed by the empiricist school of thought, consisting of Hobbes, Locke, Berkeley, and Hume. Leibniz, Spinoza and Descartes were all versed in mathematics as well as philosophy, and Descartes and Leibniz contributed greatly to science as well. As the inventor of the Cartesian coordinate system, Descartes founded analytic geometry, that bridge between algebra and geometry crucial to the invention of the calculus and analysis. Descartes' reflections on mind and mechanism began the strain of western thought that much later, impelled by the invention of the electronic computer and by the possibility of machine intelligence, blossomed into, e.g., the Turing test. His most famous statement is "Cogito ergo sum" (I think, therefore I am).

music sound like? Philosophers of mind call the subjective aspects of mental events qualia (or raw feels) [Jac82]. There is something that it is like to feel pain, to see a familiar shade of blue, and so on; there are qualia involved in these mental events. And the claim is that qualia seem particularly difficult to reduce to anything physical [Nag74].

Interactionist dualism, or simply *interactionism*, is the particular form of dualism first espoused by Descartes in the 'Meditations' [Des91]. In the 20th century, its major defenders have been Karl Popper[89] and John Eccles[90]

---

[89] Sir Karl Raimund Popper (July 28, 1902 – September 17, 1994), was an Austrian and British philosopher and a professor at the London School of Economics. He is counted among the most influential philosophers of science of the 20th century, and also wrote extensively on social and political philosophy. Popper is perhaps best known for repudiating the classical observationalist–inductivist account of scientific method by advancing empirical falsifiability as the criterion for distinguishing scientific theory from non–science; and for his vigorous defense of liberal democracy and the principles of social criticism which he took to make the flourishing of the 'open society' possible. In 1934 he published his first book, 'The Logic of Scientific Discovery', in which he criticized psychologism, naturalism, inductionism, and logical positivism, and put forth his theory of potential falsifiability being the criterion for what should be considered science.

Popper coined the term *critical rationalism* to describe his philosophy. This designation is significant, and indicates his rejection of classical empiricism, and of the observationalist-inductivist account of science that had grown out of it. Popper argued strongly against the latter, holding that scientific theories are universal in nature, and can be tested only indirectly, by reference to their implications. He also held that scientific theory, and human knowledge generally, is irreducibly conjectural or hypothetical, and is generated by the creative imagination in order to solve problems that have arisen in specific historico–cultural settings. Logically, no number of positive outcomes at the level of experimental testing can confirm a scientific theory, but a single genuine counterexample is logically decisive: it shows the theory, from which the implication is derived, to be false. Popper's account of the logical asymmetry between verification and falsification lies at the heart of his philosophy of science. It also inspired him to take falsifiability as his criterion of demarcation between what is and is not genuinely scientific: a theory should be considered scientific if and only if it is falsifiable. This led him to attack the claims of both psychoanalysis and contemporary Marxism to scientific status, on the basis that the theories enshrined by them are not falsifiable. His scientific work was influenced by his study of quantum mechanics (he has written extensively against the famous Copenhagen interpretation) and by Albert Einstein's approach to scientific theories.

In his book 'All Life is Problem Solving' (1999), Popper sought to explain the apparent progress of scientific knowledge, how it is that our understanding of the universe seems to improve over time. This problem arises from his position that the truth content of our theories, even the best of them, cannot be verified by scientific testing, but can only be falsified. If so, then how is it that the growth

(see [PE02]). It is the view that mental states, such as beliefs and desires, causally interact with physical states [Har96]. Descartes' famous argument for this position can be summarized as follows: Fred has a clear and distinct idea of his mind as a thinking thing which has no spatial extension (i.e., it cannot be measured in terms of length, weight, height, and so on) and he also has a clear and distinct idea of his body as something that is spatially

---

of science appears to result in a growth in knowledge? In Popper's view, the advance of scientific knowledge is an evolutionary process characterised by his formula:

$$PS_1 \to TT_1 \to EE_1 \to PS_2 \,.$$

In response to a given problem situation, $PS1$, a number of competing conjectures, or tentative theories, TT, are systematically subjected to the most rigorous attempts at falsification possible. This process, error elimination, EE, performs a similar function for science that natural selection performs for biological evolution. Theories that better survive the process of refutation are not more true, but rather, more 'fit', in other words, more applicable to the problem situation at hand, $PS1$. Consequently, just as a species' 'biological fit' does not predict continued survival, neither does rigorous testing protect a scientific theory from refutation in the future. Yet, as it appears that the engine of biological evolution has produced, over time, adaptive traits equipped to deal with more and more complex problems of survival, likewise, the evolution of theories through the scientific method may, in Popper's view, reflect a certain type of progress: toward more and more interesting problems, $PS2$. For Popper, it is in the interplay between the tentative theories (conjectures) and error elimination (refutation) that scientific knowledge advances toward greater and greater problems; in a process very much akin to the interplay between genetic variation and natural selection.

As early as 1934 Popper wrote of the *search for truth* as one of the "strongest motives for scientific discovery." Still, he describes in 'Objective Knowledge' (1972) early concerns about the much–criticised notion of *truth as correspondence*. Then came the *semantic theory of truth* formulated by the logician Alfred Tarski. Popper writes of learning in 1935 of the consequences of Tarski's theory, to his intense joy. The theory met critical objections to truth as correspondence and thereby rehabilitated it. The theory also seemed to Popper to support metaphysical realism and the regulative idea of a search for truth.

Among his contributions to philosophy is his answer to David Hume's 'Problem of Induction'. Hume stated that just because the sun has risen every day for as long as anyone can remember, doesn't mean that there is any rational reason to believe it will come up tomorrow. There is no rational way to prove that a pattern will continue on just because it has before. Popper's reply is characteristic, and ties in with his *criterion of falsifiability*. He states that while there is no way to prove that the sun will come up, we can theorize that it will. If it does not come up, then it will be disproven, but since right now it seems to be consistent with our theory, the theory is not disproven. Thus, Popper's demarcation between science and non–science serves as an answer to an old logical problem as well. This approach was criticised by Peter Singer for masking the role induction plays in empirical discovery.

extended, subject to quantification and not able to think. It follows that mind and body are not identical because they have radically different properties, according to Descartes [Des91]. At the same time, however, it is clear that Fred's mental states (desires, beliefs, etc.) have causal effects on his body and vice-versa: a child touches a hot stove (physical event) which causes pain (mental event) and makes him yell (physical event) which provokes a sense of fear and protectiveness in the mother (mental event) and so on. Descartes' argument obviously depends on the crucial premise that what Fred believes to be 'clear and distinct' ideas in his mind are necessarily true. Most modern philosophers doubt the validity of such an assumption, since it has been shown

[90] Sir John Carew Eccles (January 27, 1903 – May 2, 1997) was an Australian neurophysiologist who won the 1963 Nobel Prize in Physiology or Medicine for his work on the synapse. He shared the prize together with Andrew Fielding Huxley and Alan Lloyd Hodgkin.

In the early 1950s, Eccles and his colleagues performed the key experiments that would win Eccles the Nobel Prize. To study synapses in the peripheral nervous system, Eccles and colleagues used the stretch reflex as a model. This reflex is easily studied because it consists of only two neurons: a sensory neuron (the muscle spindle fiber) and the motor neuron. The sensory neuron synapses onto the motor neuron in the spinal cord. When Eccles passed a current into the sensory neuron in the quadriceps, the motor neuron innervating the quadriceps produced a small excitatory postsynaptic potential (EPSP). When he passed the same current through the hamstring, the opposing muscle to the quadriceps, he saw an inhibitory postsynaptic potential (IPSP) in the quadriceps motor neuron. Although a single EPSP was not enough to fire an action potential in the motor neuron, the sum of several EPSPs from multiple sensory neurons synapsing onto the motor neuron could cause the motor neuron to fire, thus contracting the quadriceps. On the other hand, IPSPs could subtract from this sum of EPSPs, preventing the motor neuron from firing.

Apart from these seminal experiments, Eccles was key to a number of important developments in neuroscience. Until around 1949, Eccles believed that synaptic transmission was primarily electrical rather than chemical. Although he was wrong in this hypothesis, his arguments led himself and others to perform some of the experiments which proved chemical synaptic transmission. Bernard Katz and Eccles worked together on some of the experiments which elucidated the role of acetylcholine as a neurotransmitter.

[91] Pierre Maurice Marie Duhem (10 June 1861 – 14 September 1916) French physicist and philosopher of science. Duhem's sophisticated views on the philosophy of science are explicated in 'The aim and structure of physical theory' (foreword by Prince Louis de Broglie). In this work he refuted the inductivist untruth that Newton's laws can de deduced from Kepler, *et al.* (a selection was published as Medieval cosmology: theories of infinity, place, time, void, and the plurality of worlds. He gave his name to the Quine-Duhem thesis, which holds that for any given set of observations there are an innumerably large number of explanations. Thus empirical evidence cannot force the revision of a theory.

in modern times by Freud (a third–person psychologically–trained observer can understand a person's unconscious motivations better than she does), by Pierre Duhem[91]
(a third–person philosopher of science can know a person's methods of discovery better than she does), by Bronisław Malinowski[92] (an anthropologist can know a person's customs and habits better than he does), and by theorists of perception (experiments can make one see things that are not there and scientists can describe a person's perceptions better than he can), that such an idea of privileged and perfect access to one's own ideas is dubious at best.

Other important forms of dualism which arose as reactions to, or attempts to salvage, the Cartesian version are:

(i) Psycho–physical parallelism, or simply parallelism, is the view that mind and body, while having distinct ontological statuses, do not causally influence one another, but run along parallel paths (mind events causally interact with mind events and brain events causally interact with brain events) and only seem to influence each other [RPW97]. This view was most prominently defended by Gottfried Leibniz.[93] Although Leibniz was actually an ontological monist who believed that only one fundamental substance, monads, exists in the universe and everything else is reducible to it, he nonetheless maintained that there was an important distinction between 'the mental' and 'the physical' in terms of causation. He held that God had arranged things in advance so that minds and bodies would be in harmony with each other. This is known as the doctrine of pre–established harmony [Lei714].

---

[92] Bronisław Kasper Malinowski (April 7, 1884 – May 16, 1942) was a Polish anthropologist widely considered to be one of the most important anthropologists of the twentieth century because of his pioneering work on ethnographic fieldwork, the study of reciprocity, and his detailed contribution to the study of Melanesia.

[93] Gottfried Wilhelm Leibniz (July 1 (June 21 Old Style) 1646 – November 14, 1716) was a German polymath. Educated in law and philosophy, Leibniz played a major role in the European politics and diplomacy of his day. He occupies an equally large place in both the history of philosophy and the history of mathematics. He invented *calculus* independently of Newton, and his notation is the one in general use since. He also invented the *binary system*, foundation of virtually all modern computer architectures. In philosophy, he is most remembered for *optimism*, i.e., his conclusion that our universe is, in a restricted sense, the best possible one God could have made. He was, along with René Descartes and Baruch Spinoza, one of the three great 17th century rationalists, but his philosophy also both looks back to the *Scholastic tradition* and anticipates logic and analysis. Leibniz also made major contributions to physics and technology, and anticipated notions that surfaced much later in biology, medicine, geology, probability theory, psychology, knowledge engineering, and information science. He also wrote on politics, law, ethics, theology, history, and philology, even occasional verse. His contributions to this vast array of subjects are scattered in journals and in tens of thousands of letters and unpublished manuscripts. To date, there is no complete edition of Leibniz's writings, and a complete account of his accomplishments is not yet possible.

(ii) Occasionalism is the view espoused by Nicholas Malebranche which asserts that all supposedly causal relations between physical events or between physical and mental events are not really causal at all. While body and mind are still different substances on this view, causes (whether mental or physical) are related to their effects by an act of God's intervention on each specific occasion [Sch02].

(iii) Epiphenomenalism is a doctrine first formulated by Thomas Huxley [Hux898]. Fundamentally, it consists in the view that mental phenomena are causally inefficacious. Physical events can cause other physical events and physical events can cause mental events, but mental events cannot cause anything, since they are just causally inert by-products (i.e. epiphenomena) of the physical world [RPW97] The view has been defended most strongly in recent times by Frank Jackson [Jac82].

(iv) Property dualism asserts that when matter is organized in the appropriate way (i.e., in the way that living human bodies are organized), mental properties emerge. Hence, it is a sub–branch of emergent materialism [Har96]. These emergent properties have an independent ontological status and cannot be reduced to, or explained in terms of, the physical substrate from which they emerge. This position is espoused by David Chalmers and has undergone something of a renaissance in recent years [Cha97].

Monism

In contrast to dualism, *monism* states that there is only one fundamental substance. Monism, first proposed in the West by Parmenides[94] and in modern times by Baruch Spinoza,[95] maintains that there is only one substance; in the East, rough parallels might be the Hindu concept of *Brahman* or the *Tao* of Lao Tzu [Spi670]. Today the most common forms of monism in Western philosophy are physicalistic [Kim95b]. Physicalistic monism asserts that the only existing substance is physical, in some sense of that term to be clarified

---

[94] Parmenides of Elea (early 5th century BC) was an ancient Greek philosopher born in Elea, a Hellenic city on the southern coast of Italy. Parmenides was a student of Ameinias and the founder of the School of Elea, which also included Zeno of Elea and Melissus of Samos.

[95] Benedictus de Spinoza (November 24, 1632 – February 21, 1677), named Baruch Spinoza by his synagogue elders, was a Jewish–Dutch philosopher. He is considered one of the great rationalists of 17th–century philosophy and, by virtue of his magnum opus the 'Ethics', one of the definitive ethicists. His writings, like those of his fellow rationalists, reveal considerable mathematical training and facility. Spinoza was a lens crafter by trade, an exciting engineering field at the time because of great discoveries being made by telescopes. The full impact of his work only took effect some time after his death and after the publication of his 'Opera Posthuma'. He is now seen as having prepared the way for the 18th century Enlightenment, and as a founder of modern biblical criticism. 20th century philosopher, Gilles Deleuze (1990), referred to Spinoza as "The absolute philosopher, whose Ethics is the foremost book on concepts."

by our best science [Sto05]. Another form of monism is that which states that the only existing substance is mental. Such idealistic monism is currently somewhat uncommon in the West [Kim95b].

Phenomenalism, the theory that all that exists are the representations (or sense data) of external objects in our minds and not the objects themselves, was adopted by Bertrand Russell[96] and many of the logical positivists during

[96] Bertrand Arthur William Russell, (3rd Earl Russell, 18 May 1872 – 2 February 1970), was a British philosopher, logician, and mathematician, working mostly in the 20th century. A prolific writer, Bertrand Russell was also a populariser of philosophy and a commentator on a large variety of topics, ranging from very serious issues to the mundane. Continuing a family tradition in political affairs, he was a prominent liberal as well as a socialist and anti–war activist for most of his long life. Millions looked up to Russell as a prophet of the creative and rational life; at the same time, his stances on many topics were extremely controversial.

Russell was born at the height of Britain's economic and political ascendancy. He died of influenza nearly a century later, at a time when the British Empire had all but vanished, its power dissipated by two debilitating world wars. As one of the world's best–known intellectuals, Russell's voice carried great moral authority, even into his early 90s. Among his political activities, Russell was a vigorous proponent of nuclear disarmament and an outspoken critic of the American war in Vietnam.

In 1950, Russell was made a Nobel Laureate in Literature, "in recognition of his varied and significant writings in which he champions humanitarian ideals and freedom of thought."

Russell is generally recognized as one of the founders of *analytical philosophy*, even of its several branches. At the beginning of the 20th century, alongside G.E. Moore, Russell was largely responsible for the British 'revolt against Idealism', a philosophy greatly influenced by Georg Hegel. This revolt was echoed 30 years later in Vienna by the logical positivists' 'revolt against metaphysics'. Russell was particularly appalled by the idealist doctrine of internal relations, which held that in order to know any particular thing, we must know all of its relations. Russell showed that this would make space, time, science and the concept of number unintelligible. Russell's logical work with Alfred Whitehead continued this project.

Russell had great influence on modern mathematical logic. His first mathematical book, An Essay on the Foundations of Geometry, was published in 1897. This work was heavily influenced by Immanuel Kant. Russell soon realised that the conception it laid out would have made Albert Einstein's schema of space-time impossible, which he understood to be superior to his own system. Thenceforth, he rejected the entire Kantian program as it related to mathematics and geometry, and he maintained that his own earliest work on the subject was nearly without value. Russell discovered that Gottlob Frege had independently arrived at equivalent definitions for 0, successor, and number, and the definition of number is now usually referred to as the *Frege–Russell definition*. It was largely Russell who brought Frege to the attention of the English-speaking world. He did this in 1903, when he published 'The Principles of Mathematics', in which the concept of class is inextricably tied to the definition of number. The appendix to this work detailed a paradox arising in Frege's application of second– and higher–order

the early 20th century [Rus18]. It lasted for only a very brief period of time. A third possibility is to accept the existence of a basic substance which is neither physical nor mental. The mental and physical would both be properties of this neutral substance. Such a position was adopted by Baruch Spinoza [Spi670] and popularized by Ernst Mach[97] [Mac59] in the 19th century. This neutral monism, as it is called, resembles property dualism.

Behaviorism

Behaviorism dominated philosophy of mind for much of the 20th century, especially the first half [Kim95b]. In psychology, *behaviorism* developed as a reaction to the inadequacies of introspectionism. Introspective reports on one's own interior mental life are not subject to careful examination for accuracy and are not generalizable. Without generalizability and the possibility of third-person examination, the behaviorists argued, science is simply not possible [Sto05]. The way out for psychology was to eliminate the idea of an interior mental life (and hence an ontologically independent mind) altogether and focus instead on the description of observable behavior [Ski72].

functions which took first-order functions as their arguments, and he offered his first effort to resolve what would henceforth come to be known as the *Russell Paradox*, which he later developed into a complete theory, the Theory of types. Aside from exposing a major inconsistency in naive set theory, Russell's work led directly to the creation of modern axiomatic set theory. It also crippled Frege's project of reducing arithmetic to logic. The Theory of Types and much of Russell's subsequent work have also found practical applications with computer science and information technology.

Russell continued to defend *logicism*, the view that mathematics is in some important sense reducible to logic, and along with his former teacher, Alfred Whitehead, wrote the monumental 'Principia Mathematica', an *axiomatic system* on which all of mathematics can be built. The first volume of the Principia was published in 1910, and is largely ascribed to Russell. More than any other single work, it established the specialty of mathematical or symbolic logic. Two more volumes were published, but their original plan to incorporate geometry in a fourth volume was never realised, and Russell never felt up to improving the original works, though he referenced new developments and problems in his preface to the second edition. Upon completing the Principia, three volumes of extraordinarily abstract and complex reasoning, Russell was exhausted, and he never felt his intellectual faculties fully recovered from the effort. Although the Principia did not fall prey to the paradoxes in Frege's approach, it was later proven by Kurt Gödel that neither Principia Mathematica, nor any other consistent system of primitive recursive arithmetic, could, within that system, determine that every proposition that could be formulated within that system was decidable, i.e., could decide whether that proposition or its negation was provable within the system (*Gödel's incompleteness theorem*).

[97] Ernst Mach (February 18, 1838 – February 19, 1916) was an Austrian–Czech physicist and philosopher and is the namesake for the 'Mach number" (aka Mach speed) and the optical illusion known as Mach bands.

Parallel to these developments in psychology, a philosophical behaviorism (sometimes called logical behaviorism) was developed [Sto05]. This is characterized by a strong verificationism, which generally considers unverifiable statements about interior mental life senseless. But what are mental states if they are not interior states on which one can make introspective reports? The answer of the behaviorist is that mental states do not exist but are actually just descriptions of behavior and/or dispositions to behave made by external third parties in order to explain and predict others' behavior [Ryl49]. Philosophical behaviorism is considered by most modern philosophers of mind to be outdated [Kim95a]. Apart from other problems, behaviorism implausibly maintains, for example, that someone is talking about behavior if she reports that she has a wracking headache.

Continental Philosophy of Mind

In contrast to Anglo–American *analytic philosophy*[98] there are other schools of thought which are sometimes subsumed under the broad label of *continental philosophy*. These schools tend to differ from the analytic school in

[98] Analytic philosophy is the dominant academic philosophical movement in English–speaking countries and in the Nordic countries. It is distinguished from Continental Philosophy which pertains to most non–English speaking countries. Its main founders were the Cambridge philosophers G.E. Moore and Bertrand Russell. However, both were heavily influenced by the German philosopher and mathematician Gottlob Frege and many of analytic philosophy's leading proponents, such as Ludwig Wittgenstein, Rudolf Carnap, Kurt Gödel, Karl Popper, Hans Reichenbach, Herbert Feigl, Otto Neurath, and Carl Hempel have come from Germany and Austria. In Britain, Russell and Moore were succeeded by C. D. Broad, L. Stebbing, Gilbert Ryle, A. J. Ayer, R. B. Braithwaite, Paul Grice, John Wisdom, R. M. Hare, J. L. Austin, P. F. Strawson, William Kneale, G. E. M. Anscombe, and Peter Geach. In America, the movement was led by many of the above-named European emigres as well as Max Black, Ernest Nagel, C. L. Stevenson, Norman Malcolm, W. V. Quine, Wilfrid Sellars, and Nelson Goodman, while A. N. Prior, John Passmore, and J. J. C. Smart were prominent in Australasia.

Logic and philosophy of language were central strands of analytic philosophy from the beginning, although this dominance has diminished greatly. Several lines of thought originate from the early, language-and-logic part of this analytic philosophy tradition. These include: logical positivism, logical empiricism, logical atomism, logicism and ordinary language philosophy. Subsequent analytic philosophy includes extensive work in ethics (such as Philippa Foot, R. M. Hare, and J. L. Mackie), political philosophy (John Rawls, Robert Nozick), aesthetics (Monroe Beardsley, Richard Wollheim, Arthur Danto), philosophy of religion (Alvin Plantinga, Richard Swinburne), philosophy of language (David Kaplan, Saul Kripke, Richard Montague, Hilary Putnam, W.V.O. Quine, Nathan Salmon, John Searle), and philosophy of mind (Daniel Dennett, David Chalmers, Putnam). Analytic metaphysics has also recently come into its own (Kripke, David Lewis, Salmon, Peter van Inwagen, P.F. Strawson).

that they focus less on language and logical analysis and more on directly understanding human existence and experience. With reference specifically to the discussion of the mind, this tends to translate into attempts to grasp the concepts of thought and perceptual experience in some direct sense that does not involve the analysis of linguistic forms [Dum01]. In particular, in his 'Phenomenology of Mind', G.W. F. Hegel[99] discusses three distinct types of mind: the subjective mind, the mind of an individual; the objective mind, the mind of society and of the State; and the Absolute mind, a unity of all concepts. In modern times, the two main schools that have developed in response or opposition to this Hegelian tradition are *phenomenology* and *existentialism*. Phenomenology, founded by Edmund Husserl,[100] focuses on the contents of

[99] Georg Wilhelm Friedrich Hegel (August 27, 1770 – November 14, 1831) was a German philosopher born in Stuttgart, Württemberg, in present-day southwest Germany. His influence has been widespread on writers of widely varying positions, including both his admirers (F.H. Bradley, J.P. Sartre, Hans Küng, Bruno Bauer), and his detractors (Kierkegaard, Schopenhauer, Heidegger, Schelling). His great achievement was to introduce for the first time in philosophy the idea that History and the concrete are important in getting out of the circle of philosophia perennis, i.e., the perennial problems of philosophy. Also, for the first time in the history of philosophy he realised the importance of the Other in the coming to be of self–consciousness, see slave–master dialectic.

Some of Hegel's writing was intended for those with advanced knowledge of philosophy, although his 'Encyclopedia' was intended as a textbook in a university course. Nevertheless, like many philosophers, Hegel assumed that his readers would be well–versed in Western philosophy, up to and including Descartes, Spinoza, Hume, Kant, Fichte, and Schelling. For those wishing to read his work without this background, introductions to Hegel and commentaries about Hegel may suffice. However, even this is hotly debated since the reader must choose from multiple interpretations of Hegel's writings from incompatible schools of philosophy. Presumably, reading Hegel directly would be the best method of understanding him, but this task has historically proved to be beyond the average reader of philosophy.[citation needed] This difficulty may be the most urgent problem with respect to the legacy of Hegel.

One especially difficult aspect of Hegel's work is his innovation in logic. In response to Immanuel Kant's challenge to the limits of Pure Reason, Hegel developed a radically new form of logic, which he called speculation, and which is today popularly called *dialectics*. The difficulty in reading Hegel was perceived in Hegel's own day, and persists into the 21st century. To understand Hegel fully requires paying attention to his critique of standard logic, such as the *law of contradiction* and the *law of the excluded middle*, and, whether one accepts or rejects it, at least taking it seriously. Many philosophers who came after Hegel and were influenced by him, whether adopting or rejecting his ideas, did so without fully absorbing his new speculative or dialectical logic.

[100] Edmund Gustav Albrecht Husserl (April 8, 1859, Prostějov – April 26, 1938, Freiburg) was a German philosopher, known as the father of phenomenology. Husserl was born into a Jewish family in Prostějov (Prossnitz), Moravia, Czech Republic (then part of the Austrian Empire). A pupil of Franz Brentano and

the human mind and how phenomenological processes shape our experiences. Existentialism, a school of thought led by Jean–Paul Sartre,[101] focuses on the content of experiences and how the mind deals with such experiences [Fly04].

Neurobiology

On the other hand, within the tangible field of *neurobiology*, there are many subdisciplines which are concerned with the relations between mental and physical states and processes [Bea95]:

1. Sensory neurophysiology investigates the relation between the processes of perception and stimulation [Pine97].

Carl Stumpf, Husserl came to influence, among others, Edith Stein (St. Teresa Benedicta of the Cross), Eugen Fink, Martin Heidegger, Jean–Paul Sartre, and Maurice Merleau–Ponty; in addition, Hermann Weyl's interest in *intuitionistic logic* and impredicativity appear to have resulted from contacts with Husserl. Rudolf Carnap was also influenced by Husserl, not only concerning Husserl's notion of essential insight that Carnap used in his Der Raum, but also his notion of *formation rules* and *transformation rules* is founded on Husserl's philosophy of logic. In 1887 Husserl converted to Christianity and joined the Lutheran Church. He taught philosophy at Halle as a tutor (Privatdozent) from 1887, then at Göttingen as professor from 1901, and at Freiburg im Breisgau from 1916 until he retired in 1928. After this, he continued his research and writing by using the library at Freiburg, until barred therefrom because of his Jewish heritage under the rectorship of his former pupil and intended protege, Martin Heidegger.

Husserl held the belief that *truth–in–itself* has as ontological correlate *being–in–itself*, just as meaning categories have formal–ontological categories as correlates. The discipline of logic is a formal theory of judgment, that studies the formal a priori relations among judgments using meaning categories. Mathematics, on the other hand, is formal ontology, it studies all the possible forms of being (of objects). So, in both of these disciplines, formal categories, in their different forms, are their object of study, not the sensible objects themselves. The problem with the psychological approach to mathematics and logic is that it fails to account for the fact that it is about formal categories, not abstractions from sensibility alone. The reason why we do not deal with sensible objects in mathematics is because of another faculty of understanding called *categorial abstraction*. Through this faculty we are able to get rid of sensible components of judgments, and just focus on formal categories themselves. Thanks to 'eidetic (or essential) intuition', we are able to grasp the possibility, impossibility, necessity and contingency among concepts or among formal categories. Categorial intuition, along with categorial abstraction and eidetic intuition, are the basis for logical and mathematical knowledge.

101 Jean–Paul Charles Aymard Sartre (June 21, 1905 – April 15, 1980), was a French existentialist philosopher, dramatist and screenwriter, novelist and critic.

The basis of Sartre's existentialism is found in his 'The Transcendence of the Ego'. To begin with, the thing–in–itself is infinite and overflowing. Any direct consciousness of the thing–in–itself, Sartre refers to as a 'pre–reflective consciousness'. Any attempt to describe, understand, historicize etc. the thing–in–itself,

2. Cognitive neuroscience studies the correlations between mental processes and neural processes [Pine97].
3. Neuropsychology describes the dependence of mental faculties on specific anatomical regions of the brain [Pine97].
4. Lastly, evolutionary biology studies the origins and development of the human nervous system and, in as much as this is the basis of the mind, also describes the ontogenetic and phylogenetic development of mental phenomena beginning from their most primitive stages [Pink97].

Since the 1980's, sophisticated neuroimaging procedures, such as fMRI, have furnished increasing knowledge about the workings of the human brain, shedding light on ancient philosophical problems.The methodological breakthroughs of the neurosciences, in particular the introduction of high–tech neuroimaging procedures, has propelled scientists toward the elaboration of increasingly ambitious research programs: one of the main goals is to describe and comprehend the neural processes which correspond to mental functions [Bea95]. A very small number of neurobiologists, such as Emil Reymond[102] be and John Eccles have denied the possibility of a 'reduction' of mental phenomena to cerebral processes (see [PE02]). However, the contemporary neurobiologist and philosopher Gerhard Roth continues to defend a form of 'non–reductive materialism' [Rot01].

Analytical Psychology

Recall that *analytical psychology* (AP) is part of the *Jungian psychology movement* started by Carl G. Jung[103] and his followers. Although considered to

---

Sartre calls 'reflective consciousness'. There is no way for the reflective consciousness to subsume the pre–reflective, and so reflection is fated to a form of anxiety, i.e., the human condition. The reflective consciousness in all its forms, (scientific, artistic or otherwise) can only limit the thing–in–itself by virtue of its attempt to understand or describe it. It follows therefore that any attempt at self–knowledge (self–consciousness) is a construct that fails no matter how often it is attempted. (self-consciousness is a reflective consciousness of an overflowing infinite) In Sartre's words "Conciousness is consciousness of itself insofar as it is consciousness of a transcendent object." The same holds true about knowledge of the 'Other' (being), which is a construct of reflective consciousness. One must be careful to understand this more as a form of warning than as an ontological statement. However, there is an implication of Solipsism here that Sartre considers fundamental to any coherent description of the human condition.

[102] Emil du Bois–Reymond (November 7, 1818, Berlin, Germany – November 26, 1896), was a German physician and physiologist, discoverer of the nerve action potential and the father of experimental electrophysiology.

[103] Carl Gustav Jung (July 26, 1875 – June 6, 1961) was a Swiss psychiatrist and founder of *analytical psychology*.

Jung's unique and broadly influential approach to psychology emphasized understanding the *psyche* through exploring the worlds of dreams, art, mythology, world religion and philosophy. Though not the first to analyze dreams, he has become perhaps the best–known pioneer in the field of *dream analysis*. Although he was a theoretical psychologist and practicing clinician for most of his life, much of his life's work was spent exploring other realms: Eastern vs. Western philosophy, alchemy, astrology, sociology, as well as literature and the arts. Jung also emphasized the importance of balance. He cautioned that modern humans rely too heavily on science and logic and would benefit from integrating spirituality and appreciation of the unconscious realm. Interestingly, Jungian ideas are not typically included in curriculum of most major universities' psychology departments, but are occasionally explored in humanities departments. Many pioneering psychological concepts were originally proposed by Jung. Some of these are: (i) *archetype*, (ii) *collective unconscious*, (iii) *unconscious complex*, and (iv) *synchronicity*. In addition, the popular career test currently offered by high school and college career centers, the *Myers–Briggs Type Indicator*, is strongly influenced by Jung's theories.

The overarching goal of Jung's work was the reconciliation of the life of the individual with the world of the *supra–personal archetypes*. He came to see the individual's encounter with the unconscious as central to this process. The human experiences the unconscious through symbols encountered in all aspects of life: in dreams, art, religion, and the symbolic dramas we enact in our relationships and life pursuits. Essential to the encounter with the unconscious, and the reconciliation of the individual's consciousness with this broader world, is learning this symbolic language. Only through attention and openness to this world (which is quite foreign to the modern Western mind) are individuals able to harmonize their lives with these supra–personal archetypal forces. In order to undergo the individuation process, the individual must be open to the parts of oneself beyond one's own ego. In order to do this, the modern individual must pay attention to dreams, explore the world of religion and spirituality, and question the assumptions of the operant societal world–view (rather than just blindly living life in accordance with dominant norms and assumptions).

The collective unconscious could be thought of as the DNA of the human psyche. Just as all humans share a common physical heritage and predisposition towards specific physical forms (like having two legs, a heart, etc.) so do all humans have a common psychological predisposition. However, unlike the quantifiable information that composes DNA (in the form of coded sequences of nucleotides), the collective unconscious is composed of archetypes. In contrast to the objective material world, the subjective realm of archetypes can not be fully plumbed through quantitative modes of research. Instead it can be revealed more fully through an examination of the symbolic communications of the human psyche — in art, dreams, religion, myth, and the themes of human relational/behavioral patterns. Devoting his life to the task of exploring and understanding the collective unconscious, Jung theorized that certain symbolic themes exist across all cultures, all epochs, and in every individual.

The *shadow* is an *unconscious complex* that is defined as the diametrical opposite of the conscious self, the ego. The shadow represents unknown attributes and qualities of the ego. There are constructive and destructive types of shadow. On the destructive side, it often represents everything that the conscious person does not wish to acknowledge within themselves. For instance, someone who identifies as being kind has a shadow that is harsh or unkind. Conversely, an individual who is brutal has a kind shadow. The shadow of persons who are convinced that they are ugly appears to be beautiful. On the constructive side, the shadow may represent hidden positive influences. Jung points to the story of Moses and Al–Khidr in the 18th Book of the Koran as an example. Jung emphasized the importance of being aware of shadow material and incorporating it into conscious awareness, lest one project these attributes on others. The shadow in dreams is often represented by dark figures of the same gender as the dreamer. According to Jung the human being deals with the reality of the shadow in four ways: denial, projection, integration and/or transmutation.

Jung identified the *anima* as being the unconscious feminine component of men and the *animus* as the unconscious masculine component in women. However, this is rarely taken as a literal definition: many modern–day Jungian practitioners believe that every person has both an anima and an animus. Jung stated that the anima and animus act as guides to the unconscious unified *Self*, and that forming an awareness and a connection with the anima or animus is one of the most difficult and rewarding steps in psychological growth. Jung reported that he identified his anima as she spoke to him, as an inner voice, unexpectedly one day. Oftentimes, when people ignore the anima or animus complexes, the anima or animus vies for attention by projecting itself on others. This explains, according to Jung, why we are sometimes immediately attracted to certain strangers: we see our anima or animus in them. Love at first sight is an example of anima and animus projection. Moreover, people who strongly identify with their gender role (e.g., a man who acts aggressively and never cries) have not actively recognized or engaged their anima or animus. Jung attributes human rational thought to be the male nature, while the irrational aspect is considered to be natural female. Consequently, irrationality is the male anima shadow and rationality is the female animus shadow.

There are four primary modes of experiencing the world in Jung's *extrovert/introvert model*: two rational functions: *thinking* and *feeling*, and two perceptive functions: *sensation* and *intuition*. Sensation is the perception of facts. Intuition is the perception of the unseen. Thinking is analytical, deductive cognition. Feeling is synthetic, all–inclusive cognition. In any person, the degree of introversion/extroversion of one function can be quite different to that of another function. Broadly speaking, we tend to work from our most developed function, while we need to widen our personality by developing the others. Related to this, Jung noted that the unconscious often tends to reveal itself most easily through a person's least developed function. The encounter with the unconscious and development of the underdeveloped function(s) thus tend to progress together.

Jung had a profesional relationship with the Nobel lauret physicist Wolfgang Pauli. Their work has been published in the books [PJ55, PJ01] as well as in Jung's famous [Jun80].

be a part of *psychoanalysis*, it is distinct from *Freudian psychoanalysis*.[104] While Freudian psychoanalysis assumes that the repressed material hidden in the unconscious is given by repressed sexual instincts, analytical psychology has a more general approach. There is no preconceived assumption about the unconscious material. The unconscious, for Jungian analysts, may contain repressed sexual drives, but also aspirations, fears, etc.

The aim of AP is the personal experience of the deep forces and motivations underlying human behavior. It is related to the so–called *depth psychology* and *archetypal psychology*. Its basic assumption is that the personal unconscious is a potent part, probably the more active part, of the normal human psyche. Reliable communication between the conscious and unconscious parts of the psyche is necessary for wholeness. Also crucial is the belief that *dreams* show ideas, beliefs, and feelings of which individuals may not be readily aware, but need to be, and that such material is expressed in a personalized vocabulary of visual metaphors. Things 'known but unknown' are contained in the unconscious, and dreams are one of the main vehicles for the unconscious to express them.

AP distinguishes between a *personal and* a *collective unconscious*. The collective unconscious contains *archetypes* common to all human beings. That is, individuation may bring to surface symbols that do not relate to the life experiences of a single person. This content is more easily viewed as answers to the more fundamental questions of humanity: life, death, meaning, happiness, fear. Among these more spiritual concepts may arise and be integrated into the personality.

AP distinguishes two main psychological types or temperaments: (i) *extrovert*, and (ii) *introvert*.[105]The attitude type could be thought of as the *energy*

---

[104] For a period of some 6 years, Carl Jung was a close friend and collaborator of Sigmund Freud. However after Jung published his 'Wandlungen und Symbole der Libido' (The Psychology of the Unconscious) in 1913, their theoretical ideas had diverged sharply.

[105] In the context of *personality psychology*, *extroverts and introverts* differ in how they get or lose energy as a function of their immediate social context. In particular, extroverts feel an increase of perceived energy when interacting with large group of people, but a decrease of energy when left alone. Conversely, introverts feel an increase of energy when alone, but a decrease of energy when surrounded by large group of people.

Extroverts tend to be energetic when surrounded by people and depressive when not. To induce human interactions, extroverts tend to be enthusiastic, talkative, and assertive. Extroverts enjoy doing activities that involve other people, such as taking part in community activities and involving in business, religious, political, and scientific affairs; their affinity to large groups allow them to enjoy large social gatherings including parties and marches. As such, an extroverted person is likely to enjoy time spent with people and find less reward in time spent alone.

On the other hand, introverts are 'geared to inspect' rather than to act in social settings. In a large social setting, introverts tend to be quiet, low–key,

flow of *libido*, or *psychic energy* (*ch'i* in Roman–Chinese and 'ki' in Roman–Japanese).[106] The introvert's energy flow is inward to the subject and away

deliberate, and engaged in non–social activities. Conversely, introverts gain energy when alone performing solitary activities. Thus they tend to enjoy reading, writing, watching movies at home, inventing, and designing - and doing these activities in quiet, minimally socially interactive environment such as home, library, labs, and quiet coffee shops. While introverts avoid social situations with large numbers of people, they tend to enjoy intense, one–to–one or one–to–few social interactions. They tend to have small circle of very close friends, compared to the extroverts' typically larger circle of less–close friends.

While most people view being either introverted or extroverted as a question with only two answers, levels of extraversion in fact fall in a normally distributed bell curve, with most people falling in between. The term *ambivert* was coined to denote people who fall more or less directly in the middle and exhibit tendencies of both groups. An ambivert is normally comfortable with groups and enjoys social interaction, but also relishes time alone and away from the crowd.

[106] Freud introduced the term *libido* as the instinctual energy or force that can come into conflict with the conventions of civilized behavior. It is the need to conform to society and control the libido, contained in what Freud defined as the Id, that leads to tension and disturbance in both society and the individual. This disturbance Freud labelled neurosis. Thus, libido has to be transformed into socially useful energy, according to Freud, through the process of 'sublimation'.

Ch'i (or *qi*, or *ki*) is a fundamental concept of traditional Chinese culture. Ch'i is believed to be part of everything that exists, as in 'life force' or 'life energy', something like the 'force' in Lucas' Star Wars. It is most often translated as 'energy flow,' or literally as 'air' or 'breath'.

The nature of ch'i is a matter of controversy among those who accept it as a valid concept, while those who dismiss its very existence ignore it, except for purposes of discussion with its adherents. Disputing the nature of qi is an old controversy in Chinese philosophy. Among some traditional Chinese medicine practitioners, qi is sometimes thought of as a metaphor for biological processes similar to the Western concept of energy flow for *homeostatic balance* in biological regulations. Others argue that qi involves some new physics or biology. Attempts to directly connect qi with some scientific phenonomena have been attempted since the mid–nineteenth century. *Ch'i* is a central concept in many martial arts; e.g., in the Japanese arts, Ki is developed in Aikido and given special emphasis in Ki–Aikido (a classic combat story concerns two opponents who held each others hands before a fight, while doing so each felt the others *ch'i* and the one with the weaker *ch'i* resigned without a blow being struck).

The concept of *quantum tunneling* in modern physics where physical matters can 'tunnel' through energy barriers using quantum mechanics captured some of the similar concepts of *ch'i* (which allows one to transcend normal physical forces in nature). The seemingly impossibility of tunneling through energy barriers (walls) is only limited by the conceptual framework of classical mechanics, but can easily be resolved by the *wave–particle duality* in modern physics. By the same token, this duality is similar to the metaphorical duality of *yin* and *yang*, which is governed by the flow of energy *ch'i*. Examples of quantum tunneling can be found as a mechanism in biology used by enzymes to speed up reactions

from the object, i.e., external relations. The extrovert's energy flow is outward toward the object, ie. towards external relations and away from the inner, subjective world. Extroverts desire breadth, while introverts seek depth. The introversion/extroversion attitude type may also influence mental breakdown. Introverts may be more inclined to catatonic type schizophrenia and extroverts towards manic depression.

Samuels [Sam95] has distinguished three schools of 'post–Jungian' psycho–therapy: the classical, the developmental and the archetypal. The classical school is that which tries to remain faithful to what Jung himself proposed and taught in person and in his 20–plus volumes of work. The developmental school, associated with M. Fordham, B. Feldman etc., can be considered a bridge between Jungian psychoanalysis and M. Klein's *object relations theory*. The archetypal school (sometimes called 'the imaginal school'), with different views associated with the *mythopoeticists*, such as J. Hillman in his intellectual theoretical view of archetypal psychology, C.P. Estés, in her view that ethnic and Aboriginal people are the originators of archetypal psychology and have long carried the maps to the journey of the soul in their songs, tales, dream–telling, art and rituals; M. Woodman who proposes a feminist viewpoint regarding archetypal psychology, and other Jungians like T. Moore and R. Moore, as well. Most mythopoeticists/archetypal psychology innovators either imagine the *Self* not to be the main archetype of the collective unconscious as Jung thought, but rather assign each archetype equal value... Others, who are modern progenitors of archetypal psychology (such as Estés), think of the *Self* as that which contains and yet is suffused by all the other archetypes, each giving life to the other.

1.2 Artificial and Computational Intelligence

1.2.1 Artificial Intelligence

Recall that *artificial intelligence* (AI) is a branch of computer science that deals with *intelligent behavior*, *learning* and *adaptation* in machines. Research in AI is concerned with producing machines to automate tasks requiring intelligent behavior. Examples include control, planning and scheduling, the ability to answer diagnostic and consumer questions, handwriting, speech, and facial recognition. As such, it has become an engineering discipline, focused on providing solutions to real life problems. AI systems are now in routine use in economics, medicine, engineering and the military, as well as being built

---

in lifeforms to millions of times their normal speed [MRJ06]. Other examples of quantum tunneling are found in semiconductor and superconductors, such as field emission used in flash memory and major source of current leakage in *very–large–scale integration* (VLSI) electronics draining power in mobile phones and computers.

into many common home computer software applications, traditional strategy games like computer chess and other video games.

In the philosophy of artificial intelligence, the so–called *strong AI* is the supposition that some forms of artificial intelligence can truly reason and solve problems; strong AI supposes that it is possible for machines to become sapient, or self–aware, but may or may not exhibit human–like thought processes. The term strong AI was originally coined by John Searle [Sea80]: "According to strong AI, the computer is not merely a tool in the study of the mind; rather, the appropriately programmed computer really is a mind." The term 'artificial intelligence' would equate to the same concept as what we call 'strong AI' based on the literal meanings of 'artificial' and 'intelligence'. However, initial research into artificial intelligence was focused on narrow fields such as pattern recognition and automated scheduling, in hopes that they would eventually allow for an understanding of true intelligence. The term 'artificial intelligence' thus came to encompass these narrower fields, the so–called *weak AI* as well as the idea of strong AI.

In contrast to strong AI, weak AI refers to the use of software to study or accomplish specific problem solving or reasoning tasks that do not encompass (or in some cases, are completely outside of) the full range of human cognitive abilities. An example of weak AI software would be a chess program such as *Deep Blue*. Unlike strong AI, a weak AI does not achieve self–awareness or demonstrate a wide range of human–level cognitive abilities, and at its finest is merely an intelligent, more specific problem–solver. Some argue that weak AI programs cannot be called 'intelligent' because they cannot really think.

AI divides roughly into two schools of thought: *Conventional AI* and *Computational Intelligence* (CI). Conventional AI mostly involves methods now classified as machine learning, characterized by formalism and statistical analysis. This is also known as symbolic AI, logical AI, neat AI and good old–fashioned AI (which mainly deals with symbolic problems). Basic Ai methods include:

1. Expert systems: apply reasoning capabilities to reach a conclusion. An expert system can process large amounts of known information and provide conclusions based on them.
2. Case based reasoning
3. Bayesian networks
4. Behavior based AI: a modular method of building AI systems by hand.

On the other hand, CI involves iterative development or learning (e.g., parameter tuning in connectionist systems). Learning is based on empirical data and is associated with non–symbolic AI and soft computing. Methods mainly include:

1. Neural networks: systems with very strong pattern recognition capabilities;

2. Fuzzy systems: techniques for reasoning under uncertainty, have been widely used in modern industrial and consumer product control systems; and
3. Evolutionary computation: applies biologically inspired concepts such as populations, mutation and survival of the fittest to generate increasingly better solutions to the problem. These methods most notably divide into evolutionary algorithms (e.g. genetic algorithms) and swarm intelligence (e.g. ant algorithms).

With hybrid intelligent systems attempts are made to combine these two groups. Expert inference rules can be generated through neural network or production rules from statistical learning such as in ACT–R.

A promising new approach called intelligence amplification tries to achieve artificial intelligence in an evolutionary development process as a side–effect of amplifying human intelligence through technology.

Brief AI History

Early in the 18th century, René Descartes envisioned the bodies of animals as complex but reducible machines, thus formulating the mechanistic theory, also known as the 'clockwork paradigm'. Wilhelm Schickard created the first mechanical digital calculating machine in 1623, followed by machines of Blaise Pascal[107] (1643) and Gottfried Wilhelm von Leibniz (1671), who also invented the binary system. In the 19th century, Charles Babbage and Ada Lovelace worked on programmable mechanical calculating machines.

Bertrand Russell and Alfred Whitehead published their 'Principia Mathematica' in 1910–1913, which revolutionized *formal logic*. In 1931 Kurt Gödel showed that sufficiently powerful consistent formal systems contain true theorems unprovable by any theorem–proving AI that is systematically deriving all possible theorems from the axioms. In 1941 Konrad Zuse built the first working program–controlled computers. Warren McCulloch and Walter Pitts published A Logical Calculus of the Ideas Immanent in Nervous Activity

[107] Blaise Pascal (June 19, 1623 – August 19, 1662) was a French mathematician, physicist, and religious philosopher. Pascal was a child prodigy, who was educated by his father. Pascal's earliest work was in the natural and applied sciences, where he made important contributions to the construction of mechanical calculators and the study of fluids, and clarified the concepts of pressure and vacuum by expanding the work of Evangelista Torricelli. Pascal also wrote powerfully in defense of the scientific method.

He was a mathematician of the first order. Pascal helped create two major new areas of research. He wrote a significant treatise on the subject of projective geometry at the age of sixteen and corresponded with Pierre de Fermat from 1654 on probability theory, strongly influencing the development of modern economics and social science.

(1943), laying the foundations for neural networks. Norbert Wiener's 'Cybernetics or Control and Communication in the Animal and the Machine' (MIT Press, 1948) popularizes the term 'cybernetics'.

The 1950s were a period of active efforts in AI. In 1950, Alan Turing introduced the 'Turing test' as a way of operationalizing a test of intelligent behavior. The first working AI programs were written in 1951 to run on the Ferranti Mark I machine of the University of Manchester: a draughts–playing program written by Christopher Strachey and a chess–playing program written by Dietrich Prinz. John McCarthy coined the term 'artificial intelligence' at the first conference devoted to the subject, in 1956. He also invented the Lisp programming language. Joseph Weizenbaum built ELIZA, a chatterbot implementing Rogerian psychotherapy. At the same time, John von Neumann,[108] who had been hired by the RAND Corporation, developed the *game theory*, which would prove invaluable in the progress of AI research.

During the 1960s and 1970s, Joel Moses demonstrated the power of symbolic reasoning for integration problems in the Macsyma program, the first successful knowledge–based program in mathematics. Leonard Uhr and Charles Vossler published 'A Pattern Recognition Program That Generates, Evaluates, and Adjusts Its Own Operators' in 1963, which described one of the first machine learning programs that could adaptively acquire and modify features and thereby overcome the limitations of simple perceptrons of Frank Rosenblatt.[109] Marvin Minsky and Seymour Papert published their

[108] John von Neumann (Neumann János) (December 28, 1903 – February 8, 1957) was an Austro–Hungarian mathematician and polymath who made contributions to quantum physics, functional analysis, set theory, game theory, economics, computer science, topology, numerical analysis, hydrodynamics (of explosions), statistics and many other mathematical fields as one of world history's outstanding mathematicians. His PhD supervisor was David Hilbert. Most notably, von Neumann was a pioneer of the modern digital computer and the application of operator theory to quantum mechanics, a member of the Manhattan Project and the first faculty of the Institute for Advanced Study at Princeton (along with Albert Einstein and Kurt Gödel), and creator of *game theory* and the concept of *cellular automata*. Along with Edward Teller and Stanislaw Ulam, von Neumann worked out key steps in the nuclear physics involved in thermonuclear reactions and the hydrogen bomb.

[109] Frank Rosenblatt (1928–1969) was a New York City born computer scientist who completed the Perceptron (the simplest kind of feedforward neural network: a linear classifier) on MARK 1, computer at Cornell University in 1960. This was the first computer that could learn new skills by trial and error, using a type of neural network that simulates human thought processes.

Rosenblatt's perceptrons were initially simulated on an IBM 704 computer at Cornell Aeronautical Laboratory in 1957. By the study of neural networks such as the Perceptron, Rosenblatt hoped that "the fundamental laws of organization which are common to all information handling systems, machines and men included, may eventually be understood."

book 'Perceptrons', which demonstrated the limits of simple neural nets. Alain Colmerauer developed the Prolog computer language. Ted Shortliffe demonstrated the power of rule–based systems for knowledge representation and inference in medical diagnosis and therapy in what is sometimes called the first expert system. Hans Moravec developed the first computer–controlled vehicle to autonomously negotiate cluttered obstacle courses.

In the 1980s, neural networks became widely used due to the *backpropagation algorithm*, first described by Paul Werbos in 1974. The team of Ernst Dickmanns built the first robot cars, driving up to 55 mph on empty streets. The 1990s marked major achievements in many areas of AI and demonstrations of various applications. In 1995, one of Dickmanns' robot cars drove more than 1000 miles in traffic at up to 110 mph. Deep Blue, a chess–playing computer, beat Garry Kasparov in a famous six–game match in 1997. DARPA stated that the costs saved by implementing AI methods for scheduling units in the first Persian Gulf War have repaid the US government's entire investment in AI research since the 1950s. Honda built the first prototypes of humanoid robots.

During the 1990s and 2000s AI has become very influenced by probability theory and statistics. Bayesian networks are the focus of this movement, providing links to more rigorous topics in statistics and engineering such as Markov models and Kalman filters, and bridging the old divide between 'neat' and 'scruffy' approaches. The last few years have also seen a big interest in game theory applied to AI decision making. This new school of AI is sometimes called 'machine learning'. After the September 11, 2001 attacks there has been much renewed interest and funding for threat–detection AI systems, including machine vision research and data–mining. The DARPA Grand Challenge is a race for a \$2 million prize where cars drive themselves across several hundred miles of challenging desert terrain without any communication with humans, using GPS, computers and a sophisticated array of sensors. In 2005 the winning vehicles completed all 132 miles of the course.

Cybernetics, General Systems Theory and Bionics

Closely related to AI is *cybernetics*, which is the study of communication and control, typically involving regulatory feedback, in living organisms, in machines and organisations and their combinations, for example, in sociotechnical systems, computer controlled machines such as automata and robots. The term *cybernetics* stems from the Greek 'kybernetes', which means steersman, governor, pilot, or rudder, which has the same root as government. It is an earlier but still–used generic term for many of the subject matters that are increasingly subject to specialization under the headings of adaptive systems, artificial intelligence, complex systems, complexity theory, control systems, decision support systems, dynamical systems, information theory, learning organizations, mathematical systems theory, operations research, simulation, and systems engineering.

Contemporary cybernetics began as an interdisciplinary study connecting the fields of control systems, electrical network theory, logic modeling, and neuroscience in the 1940s. The name cybernetics was coined by Norbert Wiener[110] to denote the study of 'teleological mechanisms' and was popularized through his book 'Cybernetics, or Control and Communication in the Animal and Machine' (MIT, 1948).

The study of *teleological mechanisms* in machines with *corrective feedback* dates from as far back as the late 1700s when James Watt's steam engine was equipped with a governor, a centrifugal feedback valve for controlling the speed of the engine. In 1868 James Clerk Maxwell[111] published a theoretical article on governors. In 1935 Russian physiologist P.K. Anokhin published a book 'Physiology of Functional Systems' on in which the concept of feedback ('back afferentation') was studied. In the 1940s the study and mathematical modelling of regulatory processes became a continuing research effort and two key articles were published in 1943. These papers were 'Behavior, Purpose and Teleology' by Rosenblueth, Wiener and Bigelow; and the paper 'A Logical Calculus of the Ideas Immanent in Nervous Activity' by McCulloch and Pitts.

[110] Norbert Wiener (November 26, 1894 – March 18, 1964) was an American mathematician and applied mathematician, especially in the field of electronics engineering. He was a pioneer in the study of *stochastic processes* (random processes) and noise processes, especially in the field of *electronic communication systems* and *control systems*. He is known as the founder of cybernetics. He coined the term 'cybernetics' in his book 'Cybernetics or Control and Communication in the Animal and the Machine' (MIT Press, 1948), widely recognized as one of the most important books of contemporary scientific thinking. He is also considered by some to be the first American–born–and–trained mathematician on an intellectual par with the traditional bastions of mathematical learning in Europe. He thus represents a watershed period in American mathematics. Wiener did much valuable work in defense systems for the United States, particularly during World War II and the Cold War.

[111] James Clerk Maxwell (13 June 1831 – 5 November 1879) was a Scottish mathematical physicist, born in Edinburgh. Maxwell formulated a set of equations expressing the basic laws of *electricity and magnetism* and developed the Maxwell distribution in the *kinetic theory of gases*. He is also credited with developing the first permanent colour photograph in 1861.

Maxwell had one of the finest mathematical minds of any theoretical physicist of his time. Maxwell is widely regarded as the nineteenth century scientist who had the greatest influence on twentieth century physics, making contributions to the fundamental models of nature. In 1931, on the centennial anniversary of Maxwell's birthday, Einstein described Maxwell's work as the "most profound and the most fruitful that physics has experienced since the time of Newton."

Algebraic mathematics with elements of geometry are a feature of much of Maxwell's work. Maxwell demonstrated that electric and magnetic forces are two complementary aspects of electromagnetism. He showed that electric and magnetic fields travel through space, in the form of waves, at a constant velocity of 3.0×108 m/s. He also proposed that light was a form of electromagnetic radiation.

Wiener himself popularized the social implications of cybernetics, drawing analogies between automatic systems such as a regulated steam engine and human institutions in his best–selling 'The Human Use of Human Beings: Cybernetics and Society' (Houghton–Mifflin, 1950).

In scholarly terms, cybernetics is the study of systems and control in an abstracted sense, that is, it is not grounded in any one empirical field. The emphasis is on the functional relations that hold between the different parts of a system, rather than the parts themselves. These relations include the transfer of *information*, and circular relations (*feedbacks*) that result in emergent phenomena such as *self–organization*. The main innovation of cybernetics was the creation of a scientific discipline focused on goals: an understanding of goal–directedness or purpose, resulting from a *negative feedback loop* which minimizes the deviation between the perceived situation and the desired situation (goal). As mechanistic as that sounds, cybernetics has the scope and rigor to encompass the human social interactions of agreement and collaboration that, after all, require goals and feedback to attain (see, e.g., [Ash56]). Related to cybernetics are: engineering cybernetics, quantum cybernetics, biological cybernetics, medical cybernetics, psychocybernetics, sociocybernetics and organizational cybernetics.

On the other hand, *general systems theory* is an interdisciplinary field that studies the properties of systems as a whole. It was founded by Ludwig von Bertalanffy, Ross W. Ashby, Margaret Mead, Gregory Bateson and others in the 1950s. Also, John von Neumann discovered cellular automata and self–reproducing systems without computers, with only pencil and paper. Aleksandr Lyapunov and Jules Henri Poincaré worked on the foundations of chaos theory without any computer at all. Ilya Prigogine, Prigogine has studied 'far from equilbrium systems' for emergent properties, suggesting that they offer analogues for living systems.

Systems theory brought together theoretical concepts and principles from ontology, philosophy of science, physics, biology and engineering and later found applications in numerous fields including geography, sociology, political science, organizational theory, management, psychotherapy (within family systems therapy) and economics among others. Cybernetics is a closely related field. In recent times systems science, systemics and complex systems have been used as synonyms.

Cybernetics, catastrophe theory and chaos theory have the common goal to explain complex systems that consist of a large number of mutually interacting and interrelated parts in terms of those interactions. Cellular automata (CA), neural networks (NN), artificial intelligence (AI), and artificial life (ALife) are related fields, but they do not try to describe general(universal) complex (singular) systems. The best context to compare the different "C"-Theories about complex systems is historical, which emphasizes different tools and methodologies, from pure mathematics in the beginning to pure computer science now. Since the beginning of chaos theory when Edward Lorenz accidentally discovered a *strange attractor*[112] with his computer, computers have become an indispensable source of information. One could not imagine the study of complex systems without computers today.

In recent years, the field of systems thinking has been developed to provide techniques for studying systems in holistic ways to supplement more traditional reductionistic methods. In this more recent tradition, systems theory is considered by some as a humanistic extension of the natural sciences.

Finally, bionics is the application of methods and systems found in nature to the study and design of engineering systems and modern technology. Also a short form of biomechanics, the word 'bionic' is actually a portmanteau formed from biology and electronic.

The transfer of technology between lifeforms and synthetic constructs is desirable because evolutionary pressure typically forces natural systems to become highly optimized and efficient. A classical example is the development of dirt– and water–repellent paint (coating) from the observation that the surface of the lotus flower plant is practically unsticky for anything (the lotus effect). Examples of bionics in engineering include the hulls of boats imitating the thick skin of dolphins, sonar, radar, and medical ultrasound imaging imitating the echolocation of bats.

In the field of computer science, the study of bionics has produced cybernetics, artificial neurons, artificial neural networks, and swarm intelligence. Evolutionary computation was also motivated by bionics ideas but it took the idea further by simulating evolution in silico and producing well-optimized solutions that had never appeared in nature.

Often, the study of bionics emphasizes imitation of a biological structure rather than just an implementation of its function. The conscious copying of examples and mechanisms from natural organisms and ecologies is a form of applied case–based reasoning, treating nature itself as a database of solutions that already work. Proponents argue that the selective pressure placed on all natural life forms minimizes and removes failures.

Roughly, we can distinguish three biological levels in biology after which technology can be modelled:

1. mimicking natural methods of manufacture of chemical compounds to create new ones;
2. imitating mechanisms found in nature; and
3. studying organizational principles from social behaviour of organisms, such as the flocking behaviour of birds or the emergent behaviour of bees and ants.

[112] *Strange attractor* is an attracting set that has zero measure in the embedding phase–space and has fractal dimension. Trajectories within a strange attractor appear to skip around randomly (see Chapter 2 for details).

Turing Test and General AI

Recall that the *Turing test* is a proposal for a test of a machine's capability to perform human–like conversation. Described by Alan Turing[113] in his 1950 paper 'Computing machinery and intelligence,'[114] it proceeds as follows: a human judge engages in a natural language conversation with two other parties, one a human and the other a machine; if the judge cannot reliably tell which is which, then the machine is said to pass the test. It is assumed that both the human and the machine try to appear human. In order to keep the test setting simple and universal (to explicitly test the linguistic capability of the machine instead of its ability to render words into audio), the conversation is usually limited to a text–only channel such as a teletype machine as Turing suggested or, more recently IRC or instant messaging.

General artificial intelligence research aims to create AI that can *replicate human intelligence completely*, often called an Artificial General Intelligence (AGI) to distinguish from less ambitious AI projects. As yet, researchers have devoted little attention to AGI, many claiming intelligence is too complex to be completely replicated. Some small groups of computer scientists are doing some AGI research, however. By most measures, demonstrated progress towards strong AI has been limited, as no system can pass a full Turing test for unlimited amounts of time, although some AI systems can at least fool some people initially now (see the Loebner prize winners). Few active AI researchers are prepared to publicly predict whether, or when, such systems will be developed, perhaps due to the failure of bold, unfulfilled predictions for AI research progress in past years. There is also the problem of the AI

---

[113] Alan Mathison Turing, OBE (June 23, 1912 – June 7, 1954) was an English mathematician, logician, and cryptographer. Turing is often considered to be the father of modern computer science.

With the Turing test, Turing made a significant and characteristically provocative contribution to the debate regarding artificial intelligence: whether it will ever be possible to say that a machine is conscious and can think. He provided an influential formalisation of the concept of algorithm and computation with the Turing machine, formulating the now widely accepted "Turing" version of the Church–Turing thesis, namely that any practical computing model has either the equivalent or a subset of the capabilities of a Turing machine. During World War II, Turing worked at Bletchley Park, Britain's codebreaking centre and was for a time head of Hut 8, the section responsible for German Naval cryptanalysis. He devised a number of techniques for breaking German ciphers, including the method of the bombe, an electromechanical machine which could find settings for the Enigma machine.

[114] In Turing's paper, the term 'Imitation Game' is used for his proposed test as well as the party game for men and women. The name 'Turing test' may have been invented, and was certainly publicized, by Arthur C. Clarke in the science–fiction novel 2001: A Space Odyssey (1968), where it is applied to the computer HAL 9000.

effect, where any achievement by a machine tends to be deprecated as a sign of true intelligence.

Computer Simulation of Human Brain

This is seen by many as the quickest means of achieving Strong AI, as it doesn't require complete understanding. It would require three things:

1. Hardware: an extremely powerful computer would be required for such a model. Futurist Ray Kurzweil estimates 1 million MIPS. If *Moore's law* continues, this will be available for £1000 by 2020.
2. Software: this is usually considered the hard part. It assumes that the human mind is the central nervous system and is governed by physical laws.
3. Understanding: finally, it requires sufficient understanding thereof to be able to model it mathematically. This could be done either by understanding the central nervous system, or by mapping and copying it. Neuro–imaging technologies are improving rapidly, and Kurzweil predicts that a map of sufficient quality will become available on a similar timescale to the required computing power.

Once such a model is built, it will be easily altered and thus open to trial and error experimentation. This is likely to lead to huge advances in understanding, allowing the model's intelligence to be improved/motivations altered. Current research in the area is using one of the fastest supercomputer architectures in the world, namely the Blue Gene platform created by IBM to simulate a single Neocortical Column consisting of approximately 60,000 neurons and 5km of interconnecting synapses. The eventual goal of the project is to use supercomputers to simulate an entire brain.

In opposition to human–brain simulation, the direct approach attempts to achieve AI directly without imitating nature. By comparison, early attempts to construct flying machines modelled them after birds, but modern aircraft do not look like birds. The main question in the direct approach is: 'What is AI?'. The most famous definition of AI was the operational one proposed by Alan Turing in his 'Turing test' proposal (see footnote above). There have been very few attempts to create such definition since (some of them are in the AI Project). John McCarthy[115] stated in his work 'What is AI?' that we

---

[115] John McCarthy (born September 4, 1927, in Boston, Massachusetts, sometimes known affectionately as Uncle John McCarthy), is a prominent computer scientist who received the Turing Award in 1971 for his major contributions to the field of Artificial Intelligence. In fact, he was responsible for the coining of the term 'Artificial Intelligence' in his 1955 proposal for 1956 Dartmouth Conference.

McCarthy championed expressing knowledge declaratively in mathematical logic for Artificial Intelligence. An alternative school of thought emerged at MIT and elsewhere proposing the 'procedural embedding of knowledge' using high

still do not have a solid definition of intelligence (compare with the previous section).

Machine Learning

As a broad AI–subfield, *machine learning* (ML) is concerned with the development of algorithms and techniques that allow computers to 'learn'. At a general level, there are two types of learning: inductive, and deductive. Inductive machine learning methods create computer programs by extracting rules and patterns out of massive data sets. It should be noted that although pattern identification is important to ML, without rule extraction a process falls more accurately in the field of *data mining*.

Machine learning overlaps heavily with statistics. In fact, many machine learning algorithms have been found to have direct counterparts with statistics. For example, boosting is now widely thought to be a form of stagewise regression using a specific type of loss function.

Machine learning has a wide spectrum of applications including search engines, medical diagnosis, bioinformatics and cheminformatics, detecting credit card fraud, stock market analysis, classifying DNA sequences, speech and handwriting recognition, object recognition in computer vision, game playing and robot locomotion.

Some machine learning systems attempt to eliminate the need for human intuition in the analysis of the data, while others adopt a collaborative approach between human and machine. Human intuition cannot be entirely eliminated since the designer of the system must specify how the data are to be represented and what mechanisms will be used to search for a characterization of the data. Machine learning can be viewed as an attempt to automate parts of the scientific method. Some machine learning researchers create methods within the framework of *Bayesian statistics*.

---

level plans, assertions, and goals first in Planner and later in the Scientific Community Metaphor. The resulting controversy is still ongoing and the subject matter of research.

McCarthy invented the Lisp programming language and published its design in Communications of the ACM in 1960. He helped to motivate the creation of Project MAC at MIT, but left MIT for Stanford University in 1962, where he helped set up the Stanford AI Laboratory, for many years a friendly rival to Project MAC.

In 1961, he was the first to publicly suggest (in a speech given to celebrate MIT's centennial) that computer time–sharing technology might lead to a future in which computing power and even specific applications could be sold through the utility business model (like water or electricity). This idea of a computer or information utility was very popular in the late 1960s, but faded by the mid–1970s as it became clear that the hardware, software and telecommunications technologies of the time were simply not ready. However, since 2000, the idea has resurfaced in new forms.

Machine learning algorithms are organized into a taxonomy, based on the desired outcome of the algorithm. Common algorithm types include:

1. *supervised learning*, where the algorithm generates a function that maps inputs to desired outputs. One standard formulation of the supervised learning task is the classification problem: the learner is required to learn (to approximate the behavior of) a function which maps a vector into one of several classes by looking at several input–output examples of the function.
2. *unsupervised/self–organized learning*, which models a set of inputs: labeled examples are not available.
3. *semi–supervised learning*, which combines both labeled and unlabeled examples to generate an appropriate function or classifier.
4. *reinforcement learning*, where the algorithm learns a policy of how to act given an observation of the world. Every action has some impact in the environment, and the environment provides feedback that guides the learning algorithm.
5. *transduction*, similar to supervised learning, but does not explicitly construct a function: instead, tries to predict new outputs based on training inputs, training outputs, and new inputs.
6. *learning to learn*, where the algorithm learns its own inductive bias based on previous experience.

Symbol–Based Learning

The *symbol–based learning* relies on learning algorithms that can be characterized into the following five dimensions [Lug02]:

- *data and goals*: here the learning problem is described according to the goals of the learner and the data it is initially given;
- *knowledge representation*: using representation languages with programs to store the knowledge learned by the system in a logical way;
- *learning operations*: an agent is given a set of training instances and it is tasked to construct a generalization, heuristic rule or a plan that satisfies its goals;
- *concept space*: the representation language along with the learning operations define a space of possible concept definitions, the learner needs to search this space to find the desired concept. The complexity of this concept space is used to measure how difficult the problem is; and
- *heuristic search*: heuristics are used to commit to a particular direction when searching the concept space.

Connectionist Learning

The *connectionist learning* is performed using artificial neural networks (see subsection below), which are systems comprised of a large number of interconnected artificial neurons. They have been widely used for (see, e.g., [Hay94, Kos92, Lug02]):

- *classification*: deciding the category or grouping where an input value belongs;
- *pattern recognition*: identifying a structure in sometimes noisy data;
- *memory recall*: addressing the content in memory;
- *prediction/forecasting*: identifying an effect from different causes;
- *optimization*: finding the best organization within different constraints; and
- *noise filtering*: separating a signal from the background noise or removing irrelevant components to a signal.

The knowledge of the network is encapsulated within the organization and interaction of the neurons. Specifically, the global properties of neurons are characterized as:

- *network topology*: the topology of the network is the pattern of connections between neurons;
- *learning algorithm*: the algorithm used to change the weight between different connections; and
- *encoding scheme*: the interpretation of input data presented to the network and output data obtained from the network.

Learning is achieved by modifying the structure of the neural network, via adjusting weights, in order to map input combinations to required outputs. There are two general classes of learning algorithms for training neural networks, they are supervised and unsupervised learning. Supervised learning requires the neural network to have a set of training data, consisting of the set of data to be learned as well as the corresponding answer. The data set is repeatedly presented to the neural network, in turn, the network adapts by changing the weights of connections between the neurons until the network output corresponds closely to the required answers. The goal of supervised learning is to find a model or mapping that will correctly associate its inputs with its targets. Supervised learning is suited to applications when the outputs expected from the network are well known. This allows the designer (or another fully trained network) to provide feedback.

In the case of unsupervised learning the target value is not provided and the information in the training data set is continuously presented until some convergence criteria is satisfied. This involves monitoring the output of the network and stopping its training when some desired output is observed. The main difference to supervised learning is that the desired output is not known when the training starts. During training, the network has to continuously adapt and change its output until it demonstrates a useful output behavior at which time it receives a single feedback to stop. The input data provided to the network will need to include sufficient information so that the problem is unambiguous. Unsupervised learning is suitable in situations where there is no clear–cut answer to a given problem.

The biggest problem of using neural networks with agents with that the concepts cannot intuitively fit within the agent oriented paradigm. However, neural networks have been used to implement part of a system such as pattern recognition and classification. It is also believed that neural learning concepts and techniques will play an important role in future research [Lug02].

Computational Learning Theory

The performance and computational analysis of machine learning algorithms is a branch of statistics known as *computational learning theory*. Machine learning algorithms take a training set, form hypotheses or models, and make predictions about the future. Because the training set is finite and the future is uncertain, learning theory usually does not yield absolute guarantees of performance of the algorithms. Instead, probabilistic bounds on the performance of machine learning algorithms are quite common. In addition to performance bounds, computational learning theorists study the time complexity and feasibility of learning. In computational learning theory, a computation is considered feasible if it can be done in polynomial time. There are two kinds of time complexity results (see, e.g., [Ang92]):

1. positive results, showing that a certain class of functions is learnable in polynomial time.
2. negative results, showing that certain classes cannot be learned in polynomial time.

Negative results are proven only by assumption. The assumptions that are common in negative results are:
(i) *computational complexity*: $P \neq NP$,[116] and

[116] The relationship between the complexity classes P and NP is an unsolved question in theoretical computer science. It is generally agreed to be the most important such unsolved problem, and one of the most important unsolved problems in all of mathematics. The Clay Mathematics Institute has offered a US $1,000,000 prize for a correct solution.

In essence, the $P = NP$ question asks: if positive solutions to a YES/NO problem can be verified quickly, can the answers also be computed quickly? Consider, for instance, the subset–sum problem, an example of a problem which is easy to verify, but is believed (but not proved) to be difficult to compute the answer. Given a set of integers, does any subset of them sum to 0? For instance, does a subset of the set $\{-2, -3, 15, 14, 7, -10\}$ add up to 0? The answer is YES, though it may take a little while to find a subset that does – and if the set was larger, it might take a very long time to find a subset that does. On the other hand, if someone claims that the answer is "YES, because $\{-2, -3, -10, 15\}$ add up to zero," then we can quickly check that with a few additions. Verifying that the subset adds up to zero is much faster than finding the subset in the first place. The information needed to verify a positive answer is also called a certificate. So we conclude that given the right certificates, positive answers to our problem can be verified quickly (i.e. in polynomial time) and that's why this problem is in NP.

(ii) *cryptography:*[117] *one–way functions* exist.

Recall that a one–way function is a function that is easy to calculate but hard to invert, i.e., it is difficult to calculate the input to the function given its output. The precise meanings of 'easy' and 'hard' can be specified mathematically. With rare exceptions, almost the entire field of public key cryptography rests on the existence of one–way functions. Formally, two variants of one–way functions are defined: strong and weak one–way functions:

An answer to the $P = NP$ question would determine whether problems like SUBSET–SUM are really harder to compute than to verify (this would be the case if P does not equal NP), or that they are as easy to compute as to verify (this would be the case if $P = NP$). The answer would apply to all such problems, not just the specific example of SUBSET–SUM.

The restriction to YES/NO problems doesn't really make a difference; even if we allow more complicated answers, the resulting problem (whether $FP = FNP$) is equivalent.

[117] Recall that cryptography (or cryptology; derived from Greek 'kryptós–hidden' and 'gráfein–to write') is a mathematical discipline concerned with information security and related issues, particularly encryption, authentication, and access control. Its purpose is to hide the meaning of a message rather than its existence. In modern times, it has also branched out into computer science. Cryptography is central to the techniques used in computer and network security for such things as *access control* and *information confidentiality*. Cryptography is used in many applications that touch everyday life; the security of ATM cards, computer passwords, and electronic commerce all depend on cryptography.

The so–called *symmetric–key cryptography* refers to encryption methods in which both the sender and receiver share the same key (or, less commonly, in which their keys are different, but related in an easily computable way). This was the only kind of encryption publicly known until 1976.

The modern study of symmetric–key ciphers relates mainly to the study of block ciphers and stream ciphers and to their applications (see, e.g., [Gol01]). A block cipher is the modern embodiment of Alberti's polyalphabetic cipher: block ciphers take as input a block of plaintext and a key, and output a block of ciphertext of the same size. Block ciphers are used in a mode of operation to implement a cryptosystem. DES and AES are block ciphers which have been designated cryptography standards by the US government (though DES's designation was eventually withdrawn after the AES was adopted)[8]. Despite its delisting as an official standard, DES (especially its still-approved and much more secure triple–DES variant) remains quite popular; it is used across a wide range of applications, from ATM encryption to e–mail privacy and secure remote access. Many other block ciphers have been designed and released, with considerable variation in quality. Stream ciphers, in contrast to the 'block' type, create an arbitrarily long stream of key material, which is combined with the plaintext bit by bit or character by character, somewhat like the one–time pad. In a stream cipher, the output stream is created based on an internal state which changes as the cipher operates. That state's change is controlled by the key, and, in some stream ciphers, by the plaintext stream as well.

1. Strong one–way functions. A function

$$f : \{0,1\}^* \rightarrow \{0,1\}^*$$

is called (strongly) one–way if the following two conditions hold: (i) easy to compute: there exists a (deterministic) polynomial–time algorithm A, such that for input x algorithm A outputs $f(x)$ (i.e., $A(x) = f(x)$); and (ii) hard to invert: for any probabilistic polynomial–time algorithm A', and any polynomial $p(\cdot)$, and for sufficiently large n,

$$P(A'(f(U_n), 1^n) \in f^{-1}|f(U_n)) < \frac{1}{p(n)},$$

where U_n denotes a random variable uniformly distributed over $\{0,1\}^n$. Hence, the probability in the second condition is taken over all the possible values assigned to U_n and all possible internal coin tosses of A' with uniform probability distribution. In addition to an input in the range of f the inverting algorithm is also given the desired length of the output in unary notation. The main reason for this convention is to rule out the possibility that a function is considered one–way merely because the inverting algorithm does not have enough time to print the output. The left hand part of the comparison is quite easy to understand: it is the probability, that A' finds any value U, with property $f(U) = f(Un)$. So, basically, the hard–to–invert condition requires this probability to be negligibly small.
2. Weak one–way functions only require that all efficient inverting algorithms fail with some non–negligible probability. A function

$$f : \{0,1\}^* \rightarrow \{0,1\}^*$$

is called weakly one–way if the following two conditions hold: (i) easy to compute: as in the definition of strong one–way functionl and (ii) slightly–hard to invert: There exists a polynomial such that for every probabilistic polynomial–time algorithm, A', and all sufficiently large n's,

$$P(A'(f(U_n), 1^n) \notin f^{-1}|f(U_n)) > \frac{1}{p(n)}$$

It is not known whether one–way functions exist. In fact, their existence would imply $P \neq NP$, resolving the foremost unsolved question of computer science. However, it is not clear if $P \neq NP$ implies the existence of one–way functions. It can be proved that weak one–way functions exist if and only if strong one-way functions do. Thus, as far as the mere existence of one–way function goes, the notions of weak and strong one–way functions are equivalent. It is known that the existence of one–way functions implies the existence of many other useful cryptographic primitives, including:

1. Pseudorandom bit generators;
2. Pseudorandom function families;
3. Digital signature schemes (secure against adaptive chosen–message attack).

In particular, a *trapdoor one–way function* (or, trapdoor permutation) is a special kind of one–way function. Such a function is hard to invert unless some secret information, called the trapdoor, is known. RSA is a well known example of a function believed to belong to this class.

Now, there are several different approaches to computational learning theory, which are often mathematically incompatible. This incompatibility arises from using different inference principles: principles which tell us how to generalize from limited data. The incompatibility also arises from differing definitions of probability (see frequency probability, Bayesian probability). The different approaches include:

1. *probably approximately correct learning* (PAC learning),[118] proposed by Leslie Valiant;
2. *statistical learning theory* (or VC theory),[119] proposed by Vladimir Vapnik;

---

[118] Probably approximately correct learning (PAC learning) is a framework of learning that was proposed by Leslie Valiant in his paper 'A theory of the learnable'. In this framework the learner gets samples that are classified according to a function from a certain class. The aim of the learner is to find a bounded approximation (approximately) of the function with high probability (probably). We demand the learner to be able to learn the concept given any arbitrary approximation ratio, probability of success or distribution of the samples. The model was further extended to treat noise (misclassified samples). The PAC framework allowed accurate mathematical analysis of learning. Also critical are definitions of efficiency. In particular, we are interested in finding efficient classifiers (time and space requirements bounded to a polynomial of the example size) with efficient learning procedures (requiring an example count bounded to a polynomial of the concept size, modified by the approximation and likelihood bounds).

[119] Vapnik–Chervonenkis theory (also known as VC theory, or statistical learning theory) was developed during 1960–1990 by Vladimir Vapnik and Alexey Chervonenkis. The theory is a form of computational learning theory, which attempts to explain the learning process from a statistical point of view. VC theory covers four parts:

a) Theory of consistency of learning processes – what are (necessary and sufficient) conditions for consistency of a learning process based on the empirical risk minimization principle?
b) Nonasymptotic theory of the rate of convergence of learning processes – how fast is the rate of convergence of the learning process?
c) Theory of controlling the generalization ability of learning processes – how can one control the rate of convergence (the generalization ability) of the learning process?
d) Theory of constructing learning machines – how can one construct algorithms that can control the generalization ability?

3. *Bayesian inference* (see below), arising from work first done by Thomas Bayes;[120] and
4. *algorithmic learning theory*,[121] from the work of Mark Gold.

The last part of VC theory introduced a well–known learning algorithm: the *support vector machine*. VC theory contains important concepts such as the VC dimension and structural risk minimization.

[120] Thomas Bayes (c. 1702 – April 17, 1761) was a British mathematician and Presbyterian minister, known for having formulated a special case of Bayes' theorem, which was published posthumously. Bayes' solution to a problem of 'inverse probability' was presented in the Essay Towards Solving a Problem in the Doctrine of Chances (1764), published posthumously by his friend Richard Price in the Philosophical Transactions of the Royal Society of London. This essay contains a statement of a special case of Bayes' theorem.

Bayesian probability is the name given to several related interpretations of probability, which have in common the application of probability to any kind of statement, not just those involving random variables. 'Bayesian' has been used in this sense since about 1950.

It is not at all clear that Bayes himself would have embraced the very broad interpretation now called Bayesian. It is difficult to assess Bayes' philosophical views on probability, as the only direct evidence is his essay, which does not go into questions of interpretation. In the essay, Bayes defines probability as follows:

"The probability of any event is the ratio between the value at which an expectation depending on the happening of the event ought to be computed, and the chance of the thing expected upon it's happening."

In modern *utility theory* we would say that expected utility is the probability of an event times the payoff received in case of that event. Rearranging that to solve for the probability, we get Bayes' definition. As Stigler points out, this is a subjective definition, and does not require repeated events; however, it does require that the event in question be observable, for otherwise it could never be said to have 'happened' (some would argue, however, that things can happen without being observable).

The search engine Google, and the information retrieval company Autonomy Systems, employ Bayesian principles to provide probable results to searches. Microsoft is reported as using Bayesian probability in its future Notification Platform to filter unwanted messages.

In statistics, empirical Bayes methods involve:

a) An 'underlying' probability distribution of some unobservable quantity assigned to each member of a statistical population. This quantity is a random variable if a member of the population is chosen at random. The probability distribution of this random variable is not known, and is thought of as a property of the population.
b) An observable quantity assigned to each member of the population. When a random sample is taken from the population, it is desired first to estimate the "underlying" probability distribution, and then to estimate the value of the unobservable quantity assigned to each member of the sample.

[121] *Algorithmic learning theory* (or *inductive inference*) is a framework for machine learning, introduced in E.M. Gold's seminal paper 'Language identification in the

Computational learning theory has led to practical algorithms. For example, PAC theory inspired boosting, VC theory led to *support vector machines*, and Bayesian inference led to Bayesian belief networks (see below).

---

limit' [Gol67]. The objective of language identification is for a machine running one program to be capable of developing another program by which any given sentence can be tested to determine whether it is 'grammatical' or 'ungrammatical'. The language being learned need not be English or any other natural language – in fact the definition of 'grammatical' can be absolutely anything known to the tester.

In the framework of algorithmic learning theory, the tester gives the learner an example sentence at each step, and the learner responds with a hypothesis, which is a suggested program to determine grammatical correctness. It is required of the tester that every possible sentence (grammatical or not) appears in the list eventually, but no particular order is required. It is required of the learner that at each step the hypothesis must be correct for all the sentences so far. A particular learner is said to be able to 'learn a language in the limit' if there is a certain number of steps beyond which its hypothesis no longer changes. At this point it has indeed learned the language, because every possible sentence appears somewhere in the sequence of inputs (past or future), and the hypothesis is correct for all inputs (past or future), so the hypothesis is correct for every sentence. The learner is not required to be able to tell when it has reached a correct hypothesis, all that is required is that it be true.

Gold showed that any language which is defined by a Turing machine program can be learned in the limit by another Turing–complete machine using enumeration. This is done by the learner testing all possible Turing machine programs in turn until one is found which is correct so far; this forms the hypothesis for the current step. Eventually, the correct program will be reached, after which the hypothesis will never change again (but note that the learner does not know that it won't need to change).

Gold also showed that if the learner is given only positive examples (that is, only grammatical sentences appear in the input, not ungrammatical sentences), then the language can only be guaranteed to be learned in the limit if there are only a finite number of possible sentences in the language (this is possible if, for example, sentences are known to be of limited length).

Language identification in the limit is a very theoretical model. It does not allow for limits of runtime or computer memory which can occur in practice, and the enumeration method may fail if there are errors in the input. However the framework is very powerful, because if these strict conditions are maintained, it allows the learning of any program known to be computable. This is because a Turing machine program can be written to mimic any program in any conventional programming language. Other frameworks of learning consider a much more restricted class of function than Turing machines, but complete the learning more quickly (in polynomial time). An example of such a framework is *probably approximately correct learning*.

Social and Emergent Learning

the *social and emergent learning* focuses on learning algorithms using the underlying concept of evolution, in other words, shaping a population $P(t)$ of candidate solutions x_i^t through the survival of the fittest members at time t. $P(t)$ is defined as:

$$P(t) = \{x_1^t, x_2^t, \ldots, x_n^t\}.$$

The attributes of a solution are represented with a particular pattern that is initialized by a *genetic algorithm.* As time passes, solution candidates are evaluated according to a specific fitness function that returns a measure of the candidate's fitness at that time. After evaluating all candidates the algorithm selects pairs for recombination. Genetic operators from each individual are used to produce new solutions that combine components of their parents. The fitness of a candidate determines the extent to which it reproduces. The general form of the genetic algorithm reads [Lug02]:

1. $t \leftarrow 0$;
2. Initialize population $P(t)$;
3. **while** termination condition not met **do**;
4.   **for** each member x_i^t within $P(t)$ **do**;
5. $fitness(member) \leftarrow FitnessFunction(member)$;
6.   **end for**;
7.   select members from $P(t)$ based on $fitness(member)$;
8.   produce offspring of selected members using generic operators;
9.   replace members of $P(t)$ with offspring based on fitness;
10. $t \leftarrow t + 1$;
11. **end while**.

Reinforcement Learning

Recall that *reinforcement learning* (RL) is designed to allow computers to *learn by trial and error*. It is an approach to machine intelligence that combines two disciplines to solve a problem that each discipline cannot solve on its own. The first discipline, *dynamic programming* is a field in mathematics used to solve problems of optimization and control. The second discipline, supervised learning is discussed in section on neural networks below. In most real–life problems the correct answers required with supervised learning are not available, using RL the agent is simply provided with a *reward–signal* that implicitly trains the agent as required, Figure 1.3 illustrates the agent–environment interaction used with RL. The agent and the environment interact in a discrete sequence of time steps $t = 0, 1, 2, 3, ...$, for each time step the agent is presented with the current instance of the state $s_t \in S$ where S is the set of all possible states. The agent then uses the state to select and execute an action $a_t \in A(st)$ where $A(st)$ is the set of all possible actions available in state st. In the next time step the agent receives a reward $r_{t+1} \in R$, and

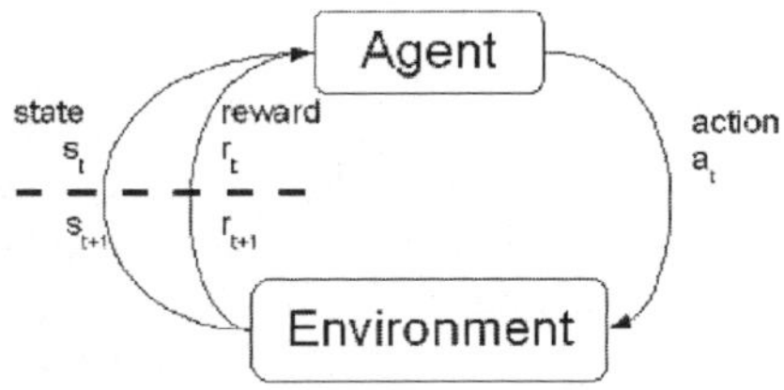

Fig. 1.3. The agent–environment interface in reinforcement learning (adapted from [SB98]).

is presented with a new state s_{t+1}. The system learns by mapping an action to each state for a particular environment. A specific mapping of actions and states is known as a *policy* π where $\pi_t(s, a)$ is the probability that $a_t = a$ if $s_t = s$. Actions available to agents can be separated into three different categories [SB98]:

- Low–level actions (e.g., supplying voltage to a motor);
- High–level actions (e.g., making a decision);
- Mental actions (e.g., shifting attention focus);

An important point to note is that according to Figure 1.3, the reward is calculated by the environment which is external to the agent. This is a confusing concept because at first it seems that the designer of an RL system is required to somehow implement something in the environment in order to provide an agent with appropriate rewards. The RL literature overcome this problem by explaining that the boundary between the agent and the environment need not be distinctively physical. The boundary of the agent is shortened to include only the reasoning process, everything outside the reasoning process which includes all other components of the agent, are treated as part of the environment. In the context of human reasoning, this is analogous to treating the human brain as the agent and the entire human body as part of the environment [Sio05].

Markov property of RL is concerned with the way that the state signal received from the environment is represented. This is an important issue when developing an RL system because all actions are directly dependent on the state of the environment. In a causal system the response of the environment for an action taken at time t will depend on all actions previously taken, formally written as

$$PR\{s_{t+1} = s', \quad r_{t+1} = r | s_t, a_t, r_t, s_{t-1}, a_{t-1}, r_{t-1}, ..., s_0, a_0\}.$$

However, the state signal should not be expected to represent everything about the environment because certain information might be inaccessible or intentionally made unavailable.

When the response of the environment depends only on the state and action representations at time t, is it said to have the Markov property and can be defined as

$$PR\{s_{t+1} = s', \quad r_{t+1} = r | s_t, a_t\}.$$

This means that the state signal is able to summarize all past sensations compactly such that all relevant information is retained for making decisions.

When a reinforcement learning problem satisfies the Markov property it is called a *Markov decision process* (MDP), additionally if the states and actions sets are finite then it is called a finite MDP. In some cases even when a particular problem is non–Markov it may be possible to consider it as an approximation of an MDP for the basis for learning, in such cases the learning performance will depend on how good the approximation is.

Reward function $R^a_{ss'}$ provides rewards depending on the actions of the agent. The sequence of rewards received after time step t is $r_{t+1}, r_{t+2}, r_{t+3}, \ldots$, the agent learns by trying to maximize the sum of rewards received when starting from an initial state and proceeding to a terminal state. An additional concept is the one when an agent tries to maximize the expected discounted return as

$$R_t = r_{t+1} + \gamma r_{t+2} + \gamma^2 r_{t+3} + \ldots = \sum_{k=0}^{\infty} \gamma^k r_{t+k+1},$$

where $0 \le \gamma \le 1$. This involves the agent discounting future rewards by a factor of γ.

There are two important classes of reward functions [HH97]. In the *pure delayed reward functions*, rewards are all zero except at a terminal state where the sign of the reward indicates whether it is a goal or penalty state. A classic example of pure delayed rewards is the cart–pole problem, where the cart is supporting a hinged inverted pendulum and the goal of the RL agent is to learn to balance the pendulum in an upright position. The agent has two actions in every state, move left and move right. The reinforcement function is zero everywhere except when the pole falls or the cart hits the end of the track, when the agent receives a -1 reward. Through such a set–up an agent will eventually learn to balance the pole and avoid the negative reinforcement. On the other hand, using the *minimum–time reward functions* it becomes possible to find the shortest path to a goal state. The reward function returns a reward of -1 for all actions except for the one leading to a terminal state for which the value is again dependent on whether it is a goal or penalty state. Due to the fact that the agent wants to maximize its rewards, it tries to achieve its goal at the minimum number of actions and therefore learns the optimal policy. An example used to illustrate this problem is driving a car up the hill problem, which is caused by the car not having enough thrust to drive up the hill on its own and therefore the RL agent needs to learn to use the momentum of the car climb the hill.

Value function. The issue of how an agent knows what is a good action is tackled using the *value function* $V^\pi(s)$ which provides a value of 'goodness' to states with respect to a specific policy. For MDPs, the information in a value function can be formally defined by

$$V^\pi(s) = E_\pi\{R_t|s_t = s\} = E_\pi\left\{\sum_{k=0}^{\infty}\gamma^k r_{t+k+1}|s_t = s\right\}$$

where $E_\pi\{\}$ denotes the expected value if the agent follows policy π, this is called the *state value function*. Similarly, the *action value function* starting from s, taking action a, and thereafter following policy π is defined by

$$Q^\pi(s,a) = E_\pi\{R_t|s_t = s,\, a_t = a\} = E_\pi\left\{\sum_{k=0}^{\infty}\gamma^k r_{t+k+1}|s_t = s,\, a_t = a\right\}.$$

A value function that returns the highest value for the best action in each state is known as the *optimal value function*. $V^*(s)$ and $Q^*(s,a)$ denote the optimal state and action value functions and are given respectively by

$$V^*(s) = \max_a \sum_{s'} P^a_{ss'}[R^a_{ss'} + \gamma V^*(s')],$$

$$Q^*(s,a) = \sum_{s'} P^a_{ss'}[R^a_{ss'} + \gamma \max_{a'} Q^*(s',a')].$$

Learning algorithms are concerned with how and when to update the value function using provided rewards. The differences in algorithms range depending on the required data that they need to operate, how they perform calculations and finally when this update takes place. Learning algorithms can be divided into three major classes: *dynamic programming*, *Monte–Carlo method* and *time–difference method*.

Dynamic programming (DP) works by assigning blame to the many decisions a system has to do while operating, this is done using two simple principles. Firstly, if an action causes something bad to happen immediately, then it learns not to do that action from that state again. Secondly, if all actions from a certain state lead to a bad result then that state should also be avoided. DP requires a *perfect environment model* in order to find a solution. Therefore the environment must have finite sets of states S and actions $A(s)$, and also finite sets of transition probabilities $P^a_{ss'} = \Pr\{s_{t+1} = s'|s_t = s,\, a_t = a\}$ and immediate rewards $R^a_{ss'} = E\{r_{t+1}|s_{t+1} = s',\, s_t = s,\, a_t = a\}$ for all $s \in S$, $a \in A(s)$. The value function in DP is updated using the equation

$$V^\pi(s) = \sum_a \pi(s,a) \sum_{s'} P^a_{ss'}[R^a_{ss'} + \gamma V^*(s')].$$

Starting from the far right in this equation it can be seen that the reward received for taking an action is added to the discounted value of the resulting state of that action. However, a single action may have multiple effects in a complex environment leading to multiple resulting states. The value of each possible resulting state is multiplied by the corresponding transition probability and all results are added to get the actual value of a single action. In order to calculate the value of the state itself, the value of each action is calculated and added to produce the full value of the state.

The two biggest problems encountered when developing applications using DP are [Sio05]: (i) the requirement of previously knowing all effects of actions taken in the environment, and (ii) the exponential increase in computation required to calculate the value of a state for only a small increase in possible actions and/or effects.

Monte Carlo (MC) methods however do not assume complete knowledge of the environment and require only experience through sampling sequences of states, actions and rewards from direct interaction with an environment. They are able to learn by segmenting sequences of actions into episodes and averaging rewards received as shown by the following algorithm [SB98]:

1: $\pi \longleftarrow$ policy to be evaluated;
2: $V \longleftarrow$ an arbitrary state–value function;
3: $Returns(s) \longleftarrow$ an empty list, for all $s \in S$;
4: **while** true **do;**
5:   Generate an episode using;
6:   **for** each state s appearing in the episode **do**;
7:     $R \longleftarrow$ return following the first occurrence of s;
8:     Append R to $Returns(s)$;
9:     $V(s) \longleftarrow average(Returns(s))$;
10:   **end for**;
11: **end while**;

Note that the algorithm requires the generation of an entire episode (line 5) before performing any updates to the value function.

MC is also able to estimate action values rather than state values, in this case policy evaluation is performed by estimating $Q^{\pi}(s, a)$, which is the expected return when starting in state s, taking action a, and thereafter following policy π. The relevant algorithm has the same structure as above. When MC is used for approximating optimal policies, the *generalized policy iteration* (GPI) is used. GPI maintains an approximate policy and an approximate value function, it then performs policy evaluation[122] and policy improvement[123] repeatedly. This means that the value function is updated to reflect the current policy while the policy is then improved with respect to the value function. Using these two processes GPI is able to maximize its rewards.

Temporal–Difference (TD) learning combines ideas from both MC and DP methods. Similarly to MC, TD methods are able to learn from experiences and do not need a model of the environment's dynamics. Like DP, TD methods update the value function based in part on estimates of future states

[122] Policy evaluation calculates the value function of a given policy.

[123] Policy improvement changes the policy such that it takes the best actions as dictated by the value function.

(this feature is called *bootstraping*) and hence do not require waiting for the episode to finish. An example of TD learning is the *Sarsa* algorithm [SB98]:

1: Initialize $Q(s,a)$ arbitrarily;
2: **for** each episode **do**;
3:   Initialize s;
4:   Choose a from s using policy *derived* from Q;
5:   **for** each state s in episode **do**;
6:     Take action a, observe r, s';
7:     Choose a' from s' using policy *derived* from Q;
8:     $Q(s,a) \longleftarrow Q(s,a) + \alpha[r + \gamma Q(s',a') - Q(s,a)]$;
9:     $s \longleftarrow s'$; $a \longleftarrow a'$;
10:   **end for**;
11: **end for**;

The most important part of the algorithm is line 8 where the action value function is updated according to the rule:

$$Q(s,a) \longleftarrow Q(s,a) + \alpha[r + \gamma Q(s',a') - Q(s,a)],$$

where α is called the step–size parameter and it controls how much the value function is changed with each update. Sarsa is an on–policy TD–algorithm and it requires the agent to select the following action before updating $Q(s,a)$. This is because $Q(s,a)$ is calculated by subtracting $Q(s,a)$ from the discounted value of $Q(s',a')$, which can only be known by selecting a'. Note that actions are selected using a policy that is based on the value function and in turn the value function is updated from the reward received.

Off–policy TD is able to approximate the optimal value function independently of the policy being followed. An example is the *Qlearning* algorithm [SB98]:

Initialize $Q(s,a)$ arbitrarily;
for each episode **do**;
  Initialize s;
  for Each state s in episode **do**;
    Choose a' from s' using policy *derived* from Q;
    Take action a, observe r, s';
    $Q(s,a) \longleftarrow Q(s,a) + \alpha[r + \gamma \max_{a'} Q(s',a') - Q(s,a)]$;
    $s \longleftarrow s'$;
  end for;
end for;

The main difference between Sarsa and Qlearning lies in the calculation that updates the value function, the Qlearning update function is given by

$$Q(s,a) \longleftarrow Q(s,a) + \alpha[r + \gamma \max_{a'} Q(s',a') - Q(s,a)].$$

With Sarsa the value function is updated based on the *next chosen action*, while with Qlearning it is updated based on the *best known future action* even if that action is actually *not selected* in the next iteration of the algorithm.

Exploration versus exploitation. One of the more well known problems within the RL literature is the *exploration/exploitation problem*. During its operation the agent forms the *action estimates* $Q^{\pi}(a) = Q^{*}(a)$. The best known action at time t would therefore be

$$a_t^* = \arg\max_a Q_t(a).$$

An agent is said to be *exploring* when it tries an new action for a particular situation $a \neq a_t^*$. The reward obtained from the execution of that action is used to update the value function accordingly. An agent is said to be *exploiting* its learning knowledge when it chooses the *greedy action* (i.e., best action) indicated by its value function in a particular state $a = a_t^*$. In this case, the agent also updates the value function according to the reward received. This may have two effects, firstly, the reward may be similar to the one expected by the value function, which means that the value function is stabilizing on the problem trying to be solved. Secondly, it may be totally different to the value expected, therefore changing the value function and possibly the ordering of the actions with respect the their values. Hence, another action may subsequently become the 'best' action for that state.

An action selection policy controls the exploitation/exploration that is performed by the agent while learning. There are two types of policies commonly considered. Firstly, the *EGreedy policy* explores by selecting actions randomly but only for a defined percentage of all actions chosen as

$$a_t = \begin{cases} a_t^* & \text{if} \quad PR = (1-\epsilon), \\ \text{random} & \text{if} \quad PR = \epsilon. \end{cases}$$

For example, if $\epsilon = 0.1$ then the agent will explore only 10% of the time, the rest of the time it chooses the greedy action.

Secondly, the *SoftMax action selection* is more complex. It makes its choice based on the relation

$$a_t = \frac{\mathrm{e}^{Q_t(a)/\tau}}{\sum_{b=1}^{n} \mathrm{e}^{Q_t(a)/\tau}},$$

where τ is called the *temperature value*. A high temperature selects all actions randomly, while a low temperature selects actions in a greedy fashion. An intermediate temperature value causes SoftMax to select actions with a probability that is based on their value. This way actions with a high value have a greater chance of being selected while actions with a lower value have less chance of being selected. The advantage of SoftMax is that it tends to select the best action most of the time followed by the second–best, the third–best and so on, an action with a very low value is seldom executed. This is useful when a particular action is known to cause extremely bad rewards. Using SoftMax, that action will always get a very small probability of execution, with EGreedy however, it has the same probability as any other action when exploring.

AI Programming Languages

Lisp

Recall that *Lisp* resps a family of computer programming languages with a long history and a distinctive fully–parenthesized syntax. Originally specified in 1958, Lisp is the second–oldest high–level programming language[124] in widespread use today; only Fortran is older. Like Fortran, Lisp has changed a great deal since its early days, and a number of dialects have existed over its history. Today, the most widely–known general–purpose Lisp dialects are Common Lisp[125] and Scheme.[126]

---

[124] Recall that a high–level programming language is a programming language that, in comparison to low–level programming languages, may be more abstract, easier to use, or more portable across platforms. Such languages often abstract away CPU operations such as memory access models and management of *scope*.

[125] Common Lisp, commonly abbreviated CL, is a dialect of the Lisp programming language, standardised by ANSI X3.226-1994. Developed to standardize the divergent variants of Lisp which predated it, it is not an implementation but rather a language specification. Several implementations of the Common Lisp standard are available, including commercial products and open source software.

Common Lisp is a general–purpose programming language, in contrast to Lisp variants such as Emacs Lisp and AutoLISP which are embedded extension languages in particular products. Unlike many earlier Lisps, Common Lisp (like Scheme) uses lexical variable scope.

Common Lisp is a multi–paradigm programming language that:

(i) Supports programming techniques such as imperative, functional and object-oriented programming.

(ii) Is dynamically typed, but with optional type declarations that can improve efficiency.

(iii) Is extensible through standard features such as Lisp macros (compile–time code rearrangement accomplished by the program itself) and reader macros (extension of syntax to give special meaning to characters reserved for users for this purpose).

[126] Scheme is a multi–paradigm programming language and a dialect of Lisp which supports functional and procedural programming. It was developed by Guy L. Steele and Gerald Jay Sussman in the 1970s. Scheme was introduced to the academic world via a series of papers now referred to as Sussman and Steele's Lambda Papers. There are two standards that define the Scheme language: the official IEEE standard, and a de facto standard called the Revisedn Report on the Algorithmic Language Scheme, nearly always abbreviated RnRS, where n is the number of the revision.

Scheme's philosophy is minimalist. Scheme provides as few primitive notions as possible, and, where practical, lets everything else be provided by

Lisp was originally created as a practical mathematical notation for computer programs, based on Church's[127] *lambda calculus* (which provides a theoretical framework for describing functions and their evaluation; though it is a mathematical abstraction rather than a programming language, lambda calculus forms the basis of almost all *functional programming languages*[128] today).

---

programming libraries. Scheme, like all Lisp dialects, has very little syntax compared to many other programming languages. There are no operator precedence rules because fully nested and parenthesized notation is used for all function calls, and so there are no ambiguities as are found in infix notation, which mimics conventional algebraic notation.

Scheme uses lists as the primary data structure, but also has support for vectors. Scheme was the first dialect of Lisp to choose static (a.k.a. lexical) over dynamic variable scope. It was also one of the first programming languages to support first–class continuations.

[127] Alonzo Church (June 14, 1903 — August 11, 1995) was an American mathematician and logician who was responsible for some of the foundations of theoretical computer science. Born in Washington, DC, he received a bachelor's degree from Princeton University in 1924, completing his Ph.D. there in 1927, under Oswald Veblen. After a postdoc at Göttingen, he taught at Princeton, 1929—1967, and at the University of California, Los Angeles, 1967–1990.

Church is best known for the following accomplishments:

(i) His proof that Peano arithmetic and first–order logic are undecidable. The latter result is known as *Church's theorem.*

(ii) His articulation of what has come to be known as *Church's thesis.*

(iii) He was the founding editor of the Journal of Symbolic Logic, editing its reviews section until 1979.

(iv) His creation of the *lambda calculus.*

The lambda calculus emerged in his famous 1936 paper showing the existence of an 'undecidable problem'. This result preempted Alan Turing's famous work on the halting problem which also demonstrated the existence of a problem unsolvable by mechanical means. He and Turing then showed that the lambda calculus and the Turing machine used in Turing's halting problem were equivalent in capabilities, and subsequently demonstrated a variety of alternative 'mechanical processes for computation'. This resulted in the *Church—Turing thesis.*

The lambda calculus influenced the design of the LISP programming language and functional programming languages in general. The Church encoding is named in his honor.

[128] Recall that *functional programming* is a programming paradigm that conceives computation as the evaluation of mathematical functions and avoids state and mutable data. Functional programming emphasizes the application of functions, in contrast with imperative programming, which emphasizes changes in state and the execution of sequential commands. A broader conception of functional programming simply defines a set of common concerns and themes rather than a list of distinctions from other paradigms. Often considered important are higher–order and first-class functions, closures, and recursion. Other common features of functional programming languages are continuations, *Hindley–Milner type inference systems*, non–strict evaluation, and monads.

Lisp quickly became the favored programming language for artificial intelligence research. As one of the earliest programming languages, Lisp pioneered many ideas in computer science, including tree data structures, automatic storage management, dynamic typing, object–oriented programming, and the self–hosting compiler.

The name Lisp derives from 'List Processing'. Linked lists are one of Lisp languages' major data structures, and Lisp source code is itself made up of lists. As a result, Lisp programs can manipulate source code as a data structure, giving rise to the macro systems that allow programmers to create new syntax or even new 'little languages' embedded in Lisp.

The interchangeability of code and data also give Lisp its instantly recognizable syntax. All program code is written as s–expressions, or parenthesized lists. A function call or syntactic form is written as a list with the function or operator's name first, and the arguments following: $(f\ x\ y\ z)$.

Lisp was invented by John McCarthy in 1958 while he was at MIT. McCarthy published its design in a paper in Communications of the ACM in 1960, entitled 'Recursive Functions of Symbolic Expressions and Their Computation by Machine'.[129] He showed that with a few simple operators and a notation for functions, one can build a Turing–complete language for algorithms. Lisp was first implemented by Steve Russell on an IBM 704 computer. Russell had read McCarthy's paper, and realized (to McCarthy's surprise) that the eval function could be implemented as a Lisp interpreter. The first complete Lisp compiler, written in Lisp, was implemented in 1962 by Tim Hart and Mike Levin at MIT. (AI Memo 39, 767 kB PDF.) This compiler introduced the Lisp model of incremental compilation, in which compiled and interpreted functions can intermix freely. The language used in Hart and Levin's memo is much closer to modern Lisp style than McCarthy's earlier code.

Largely because of its resource requirements with respect to early computing hardware (including early microprocessors), Lisp did not become as popular outside of the AI community as Fortran and the ALGOL–descended C language. Newer languages such as Java have incorporated some limited versions of some of the features of Lisp, but are necessarily unable to bring the coherence and synergy of the full concepts found in Lisp. Because of its suitability to ill–defined, complex, and dynamic applications, Lisp is presently enjoying some resurgence of popular interest.

Functional programming languages, especially 'purely functional' ones, have largely been emphasized in academia rather than in commercial software development. However, notable functional programming languages used in industry and commercial applications include Erlang (concurrent applications), R (statistics), Mathematica (symbolic math), J and K (financial analysis), and domain–specific programming languages like XSLT. Important influences on functional programming have been the *lambda calculus*, APL, Lisp and Haskell.

[129] McCarthy's original notation used bracketed 'M–expressions' that would be translated into S–expressions.

Prolog

Prolog is a *logic programming* language. The name Prolog is taken from 'programmation en logique' (which is French for 'programming in logic'). It was created by Alain Colmerauer and Robert Kowalski[130] around 1972 as an alternative to the American–dominated Lisp programming languages. It has been an attempt to make a programming language that enables the expression of logic instead of carefully specified instructions on the computer. In some ways Prolog is a subset of Planner, e.g., see Kowalski's early history of logic programming. The ideas in Planner were later further developed in the *Scientific Community Metaphor*.[131]

---

[130] Alain Colmerauer (born January 24, 1941) is a French computer scientist. He is the creator of the logic programming language Prolog and Q–Systems, one of the earliest linguistic formalisms used in the development of the TAUM–METEO machine translation prototype. He is a professor at the University of Aix–Marseilles, specialising in the field of constraint programming.

Robert Anthony Kowalski (born May 15, 1941 in Bridgeport, Connecticut, USA) is an American logician who has spent much of his career in the UK. He has been important in the development of logic programming, especially the programming language Prolog. He is also interested in legal reasoning.

[131] The Scientific Community Metaphor is one way of understanding scientific communities. In this approach, a high level programming language called Ether was developed that made use of pattern–directed invocation to invoke high–level procedural plans on the basis of messages (e.g. assertions and goals). The Scientific Community Metaphor builds on the philosophy, history and sociology of science with its analysis that scientific research depends critically on monotonicity, concurrency, commutativity, and pluralism to propose, modify, support, and oppose scientific methods, practices, and theories.

The first publications on the Scientific Community Metaphor (Kornfeld & Hewitt 1981, Kornfeld 1981, Kornfeld 1982) involved the development of a programming language named 'Ether' that invoked procedural plans to process goals and assertions concurrently by dynamically creating new rules during program execution. Ether also addressed issues of conflict and contradiction with multiple sources of knowledge and multiple viewpoints.

According to Carl Hewitt [Hew69], Scientific Community Metaphor systems have characteristics of:

(i) monotonicity (once something is published it cannot be withdrawn),

(ii) concurrency (scientists can work concurrently, overlapping in time and interacting with each other),

(iii) commutativity (publications can be read regardless of whether they initiate new research or become relevant to ongoing research),

(iv) pluralism (publications include heterogeneous, overlapping and possibly conflicting information),

(v) skepticism (great effort is expended to test and validate current information and replace it with better information), and

(vi) provenance (the provenance of information is carefully tracked and recorded).

Prolog is used in many AI programs and in *computational linguistics* (especially natural language processing, which it was originally designed for; the original goal was to provide a tool for computer–illiterate linguists) A lot of the research leading up to modern implementations of Prolog came from spin–off effects caused by the *fifth generation computer systems* project (FGCS) which chose to use a variant of Prolog named *Kernel Language* for their operating system (however, this area of research is now actually almost defunct).

Prolog is based on *first–order predicate calculus*;[132] however it is restricted to allow only *Horn clauses.*[133] Execution of a Prolog program is effectively an application of theorem proving by *first–order resolution.*

'Planner' is a programming language designed by Carl Hewitt at MIT, and first published in 1969. First subsets such as Micro–Planner and Pico–Planner were implemented and then essentially the whole language was implemented in Popler and derivations such as QA–4, Conniver, QLISP and Ether.

132 Recall that *predicate calculus* consists of

1. *formation rules* (i.e. recursive definitions for forming well–formed formulas),
2. *transformation rules* (i.e. inference rules for deriving theorems), and
3. *axioms* or *axiom schemata* (possibly a countably infinite number).

When the set of axioms is infinite, it is required that there be an algorithm which can decide for a given well–formed formula, whether it is an axiom or not. There should also be an algorithm which can decide whether a given application of an inference rule is correct or not.

133 A Horn clause is a clause (a disjunction of literals) with at most one positive literal. A Horn clause with exactly one positive literal is a definite clause; a Horn clause with no positive literals is sometimes called a goal clause, especially in logic programming. A Horn formula is a conjunctive normal form formula whose clauses are all Horn; in other words, it is a conjunction of Horn clauses. A dual–Horn clause is a clause with at most one negative literal. Horn clauses play a basic role in logic programming and are important for constructive logic. For example,

$$\neg p \vee \neg q \vee \cdots \vee \neg t \vee u$$

is a definite Horn clause. Such a formula can be rewritten in the following form, which is more common in logic programming,

$$(p \wedge q \wedge \cdots \wedge t) \to u.$$

The relevance of Horn clauses to theorem proving by *first–order resolution* is that the resolution of two Horn clauses is a Horn clause. Moreover, the resolution of a goal clause and a definite clause is again a goal clause. In automated theorem proving, this can lead to greater efficiencies in proving a theorem (represented as a goal clause). Prolog is a programming language based on Horn clauses. Horn clauses are also of interest in computational complexity, where the problem of finding a set of variable assignments to make a conjunction of Horn clauses true is a *P–complete problem.*

Recall that a *resolution rule* in *propositional logic* is a single valid inference rule producing, from two clauses, a new clause implied by them. The resolution rule takes two clauses – a clause is a disjunction of literals – containing complementary literals, and produces a new clause with all the literals of both except for the complementary ones. The clause produced by the resolution rule is called the resolvent of the two input clauses. When the two clauses contain more than one pair of complementary literals, the resolution rule can be applied (independently) for each such pair. However, only the pair of literals that are resolved upon can be removed: all other pair of literals remain in the resolvent clause.

When coupled with a complete *search algorithm*, the resolution rule yields a sound and complete algorithm for deciding the *satisfiability* of a propositional formula, and, by extension, the validity of a sentence under a set of axioms. This resolution technique uses *proof by contradiction* and is based on the fact that any sentence in propositional logic can be transformed into an equivalent sentence in *conjunctive normal form*. Its steps are:

1. All sentences in the knowledge base and the negation of the sentence to be proved (the conjecture) are conjunctively connected.
2. The resulting sentence is transformed into a conjunctive normal form (treated as a set of clauses, S).
3. The resolution rule is applied to all possible pairs of clauses that contains complementary literals. After each application of the resolution rule, the resulting sentence is simplified by removing repeated literals. If the sentence contains complementary literals, it is discarded (as a *tautology*). If not, and if it is not yet present in the clause set S, it is added to S, and is considered for further resolution inferences.
4. If after applying a resolution rule the empty clause is derived, the complete formula is unsatisfiable (or contradictory), and hence it can be concluded that the initial conjecture follows from the axioms.
5. If, on the other hand, the empty clause cannot be derived, and the resolution rule cannot be applied to derive any more new clauses, the conjecture is not a theorem of the original knowledge base.

In first order logic resolution condenses the traditional syllogisms of logical inference down to a single rule.

Fundamental Prolog concepts are *unification*, *tail recursion*, and *backtracking* (a strategy for finding solutions to *constraint satisfaction problems*). The concept of unification is one of the main ideas behind Prolog. It represents the mechanism of binding the contents of variables and can be viewed as a kind of one–time assignment. In Prolog, this operation is denoted by symbol '='. In traditional Prolog, a variable X which is uninstantiated, i.e., no previous unifications were performed on it, can be unified with an atom, a term, or another uninstantiated variable, thus effectively becoming its alias. In many modern Prolog dialects and in first–order logic calculi, a variable cannot be unified with a term that contains it; this is the so called 'occurs check'.

A *Prolog atom* can be unified only with the same atom. Similarly, a *Prolog term* can be unified with another term if the top function symbols and arities of the terms are identical and if the parameters can be unified simultaneously (note that this is a *recursive behaviour*). Due to its declarative nature, the order in a sequence of unifications is (usually) unimportant [BS01].

The tail recursion (or *tail–end recursion*) is a special case of recursion that can be easily transformed into an *iteration*. Such a transformation is possible if the recursive call is the last thing that happens in a function. Replacing recursion with iteration, either manually or automatically, can drastically decrease the amount of stack space used and improve efficiency. This technique is commonly used with functional programming languages, where the declarative approach and explicit handling of state promote the use of recursive functions that would otherwise rapidly fill the call stack.

Prolog has a built in mechanism for parsing *context–free grammar* (CFG), a formal grammar in which every *production rule* is of the form: $V \to w$, where V is a non–terminal symbol and w is a string consisting of terminals and/or non–terminals. The term 'context–free' comes from the fact that the non–terminal V can always be replaced by w, regardless of the context in which it occurs. A formal language is context–free if there is a context–free grammar that generates it.

Context–free grammars are powerful enough to describe the syntax of most programming languages; in fact, the syntax of most programming languages are specified using context–free grammars. On the other hand, context–free grammars are simple enough to allow the construction of efficient parsing algorithms which, for a given string, determine whether and how it can be generated from the grammar. The metasyntax called *Backus–Naur form* (BNF), is the most common notation used to express context–free grammars.

ACT–R: Combining Natural and Computational Intelligence

ACT–R (Adaptive Control of Thought–Rational) is a cognitive architecture mainly developed by John Anderson[134] at the Carnegie Mellon University (see [And83, And80, And90]). Like any cognitive architecture, ACT–R aims to define the basic and irreducible basic cognitive and perceptual operations that enable the human mind. In theory, each task that humans can perform should consist of a series of these discrete operations. Most of the ACT–R basic

[134] John Robert Anderson (born 1947 in Vancouver, British Columbia) is a professor of psychology and computer science at Carnegie Mellon University. He is widely known for his cognitive architecture ACT–R [And84]. He has published many papers on cognitive psychology, served as present of the Cognitive Science Society, and received many scientific awards, including one from the American Academy of Arts and Sciences. He is a fellow of the National Academy of Sciences. Anderson was an early leader in research on intelligent tutoring systems, and many of Anderson's former students, such as Kenneth Koedinger and Neil Heffernan, have become leaders in that area.

assumptions are also inspired by the progresses of cognitive neuroscience, and, in fact, ACT–R can be seen and described as way of specifying how the brain itself is organized in a way that enables individual processing modules to produce cognition.

Like other influential cognitive architectures (including Soar and EPIC), the ACT–R theory has a computational implementation as an interpreter of a special coding language. The interpreter itself is written in Lisp, and might be loaded into any of the most common distributions of the Lisp language. This enables researchers to specify models of human cognition in the form of a script in the ACT–R language. The language primitives and data–types are designed to reflect the theoretical assumptions about human cognition. These assumptions are based on numerous facts derived from experiments in cognitive psychology and brain imaging.

In recent years, ACT–R has also been extended to make quantitative predictions of patterns of activation in the brain, as detected in experiments with fMRI. In particular, ACT–R has been augmented to predict the exact shape and time–course of the BOLD response of several brain areas, including the hand and mouth areas in the motor cortex, the left prefrontal cortex, the anterior cingulate cortex, and the basal ganglia.

ACT–R's most important assumption is that human knowledge can be divided into two irreducible kinds of representations: declarative and procedural. Within the ACT–R code, declarative knowledge is represented in form of chunks, i.e., vector representations of individual properties, each of them accessible from a labelled slot. On the other hand, chunks are held and made accessible through buffers, which are the front–end of what are modules, i.e. specialized and largely independent brain structures.

There are two types of modules:

1. Perceptual–motor modules, which take care of the interface with the real world (i.e., with a simulation of the real world). The most well–developed perceptual–motor modules in ACT–R are the visual and the manual modules.
2. Memory modules. There are two kinds of memory modules in ACT–R: (i) Declarative memory, consisting of facts such as Washington, D.C. is the capital of United States, France is a country in Europe, or $2 + 3 = 5$; and (ii) Procedural memory, made of productions. Productions represent knowledge about how we do things: for instance, knowledge about how to type the letter 'Q' on a keyboard, about how to drive, or about how to perform addition.

Over the years, ACT–R models has been used in more than 500 different scientific publications, and has been cited in a huge amount of others. It has been applied in the following areas:

1. Learning and Memory
2. Higher level cognition, Problem solving and Decision making

3. Natural language, including syntactic parsing, semantic processing and language generation
4. Perception and Attention

More recently, more than two dozen papers made use of ACT–R for predicting brain activation patterns during imaging experiments, and it has also been tentatively used to model neuropsychological impairments and mental disorders.

Beside its scientific application in cognitive psychology, ACT–R used in other, more application–oriented domains.

1. Human–computer interaction to produce user models that can assess different computer interfaces,
2. Education, where ACT–R–based cognitive tutoring systems try to 'guess' the difficulties that students may have and provide focused help
3. Computer–generated forces to provide cognitive agents that inhabit training environments

Some of the most successful applications, the Cognitive Tutors for Mathematics, are used in thousands of schools across the United States. Such 'Cognitive Tutors' are being used as a platform for research on learning and cognitive modelling as part of the Pittsburgh Science of Learning Center.

After the publication of 'The Atomic Components of Thought' [And90], Anderson become more and more interested in the underlying neural plausibility of his life–time theory, and began to use brain imaging techniques pursuing his own goal of understanding the computational underpinnings of human mind. The necessity of accounting for brain localization pushed for a major revision of the theory. ACT–R 5.0, presented in 2002, introduced the concept of modules, specialized sets of procedural and declarative representations that could be mapped to known brain systems. In addition, the interaction between procedural and declarative knowledge was mediated by newly introduced buffers, specialized structures for holding temporarily active information (see the section above). Buffers were thought to reflect cortical activity, and a subsequent series of studies later confirmed that activations in cortical regions could be successfully related to computational operations over buffers. The theory was first described in the 2004 paper 'An Integrated Theory of Mind' [ABB04]. No major changes have occurred since then in the theory, but a new version of the code, completely rewritten, was presented in 2005 as ACT–R 6.0. It also included significant improvements in the ACT–R coding language.

Facial Recognition and Biometrics

A Facial Recognition (FR) system is a computer–driven application for automatically identifying a person from a digital image. It does that by comparing selected facial features in the live image and a facial database. It is typically used for security systems and can be compared to other biometrics such as fingerprint or eye iris recognition systems.

Popular FR algorithms include *eigenfaces*, the *Hidden Markov models*, and the neuronal motivated *dynamic link matching*. A newly emerging trend, claimed to achieve previously unseen accuracies, is 3D face recognition. Another emerging trend uses the visual details of the skin, as captured in standard digital or scanned images. Tests on the FERET database, the widely used industry benchmark, showed that this approach is substantially more reliable than previous algorithms.

FR is based on the computer identification of unknown face images by comparison with a single known image or database of known images. A FR may be used for access control (one–to–one) or for surveillance of crowds to locate people of interest (one–to–many or many–to–many). Access control FRs are often used in highly controlled environments, which means that the input data is of predictable quality, resulting in relatively high levels of performance. Surveillance applications (which are often covert), may call for a large number of faces to be compared with a large stored database of images to determine if there are any matches. This can result in a large number of false alarms. In addition, due to the nature of the surveillance application, the images obtained are often of poor quality, since it is often difficult to adequately control all the environmental conditions. This can reduce the ability of the FR to find a correct match with an enrolled image.

Modes of Operation

FR systems have two functional modes: enrolment and operation. Each mode used the same signal processing approach to extract salient information from the sensor data. In the enrolment phase, face data on known subjects is extracted and stored in a database of known persons (often called the 'gallery'). In general, each individual is sampled a number of times during enrolment, to ensure that the stored data is truly representative of that individual.

Once a database of known subjects is enrolled, the system may be used in the operational mode. In this mode, data from people who are not yet identified are processed in the same way as the enrolment data and the salient features are compared with the database to see if there is a match. When the degree of match is above some form of threshold, an action is generally required. A key to effective operation of an FRS is the image processing that extracts the salient features of faces for comparison with stored data.

Signal Processing Operations

The signal processing operations typically involved in FR include those listed below, either as discrete operations (an algorithmic approach) or in combination (e.g., a neural network approach):

(i) Face Capture: The first stage in the FR process is to identify objects that could be faces and then discard the rest of the scene. The face capture process could be as simple as a blob detector that sorts on size and shape, or it may include higher level processes that look for features such as eye/nose/mouth geometry, color information or motion and location to identify objects that are face–like.

(ii) Normalization: Once faces have been identified they must be presented to the classifier in a form that compensates for variability in brightness/colour due to lighting, camera and frame grabber characteristics, as well as geometric distortions due to distance, pose and viewing aspect angles. Typical intensity normalization may involve grey scale modification of regions of interest to provide fixed average levels and contrast. Scale errors may be minimized by re-sampling the faces to produce constant size inputs to following stages. In general, the distance between eye pupils is used as the baseline measure to re–scale images and it is critical that this parameter be accurately determined, either by the software or manually.

(iii) Feature Extraction: Feature extraction is the process that takes the normalized version of each real–world face image and generates a compact data vector that uniquely describes it for use by the classification/database engine.

(iv) Database Comparison: Unknown subjects and a target sample are compared with the known database. Face images are gathered using the same (or a similar) sensor as was used for enrolment and this data is processed in the same way as the enrolment data. Following salient feature extraction, the incoming data vector is compared with each template in the database to determine the goodness of match with known data and a match measure is generated for each comparison.

(v) Decision and Action: A decision making process follows the match measurement, whereby the outcome is declared to be either a true match or a non–match, based on the match measure. This decision may be made by comparing the match value to a threshold setting. Any match measure that is on higher side of the threshold is declared to be a true match and any on the other is a non-match. The process of facial recognition is complex and many of the processes outlined above are highly dependent upon external variables. This can lead to considerable difficulty in the evaluation of the technologies involved.

Evaluation Methods

Phillips *et al.* [PMW00] have given a general introduction to evaluating biometric systems. They focused on biometric applications that give the user some control over data acquisition. These applications recognize subjects from mug shots, passport photos and scanned fingerprints. They concentrated on two major kinds of biometric systems: identification and verification. In identification systems, a biometric signature of an unknown person is presented to a system. The system compares the new biometric signature with a database of biometric signatures of known individuals. On the basis of the comparison, the system reports (or estimates) the identity of the unknown person from this database. Systems that rely on identification include those that check for multiple applications by the same person for welfare benefits and driver's licences.

In verification systems, a user presents a biometric signature and a claim that a particular identity belonged to the biometric signature. The algorithm either accepts or rejects the claim. Alternatively, the algorithm could return a confidence measurement of the claim's validity. Verification applications include those that authenticate identity during point-of-sale transactions or that control access to computers or secure buildings.

Performance statistics for verification applications differ substantially from those for identification systems. The main performance measure for an identification system is that system's ability to identify the owner of a biometric signature. More specifically, the performance measure is equal to the percentage of queries in which the correct answer could be found in the top few matches.

Mansfield and Wayman [MW02] elaborated best practice in testing and reporting the performance of biometric devices. The purpose of their report, which is a revision of their original version [MW00], was to summarize the current understanding by the biometrics community of the best scientific practices for conducting technical performance testing toward the end of field performance estimation. The aims of the authors were as follows:

(1) To provide a framework for developing and fully describing test protocols.
(2) To help avoid systematic bias due to incorrect data collection or analytic procedures in evaluations.
(3) To help testers achieve the best possible estimate of field performance while expending minimum effort in conducting their evaluation.
(4) To improve understanding of the limits of applicability of test results and test methods.

The recommendations in this paper were extremely general in nature. It was noted that it might not be possible to follow best practice completely in any test. Compromises often need to be made. In such situations the experimenter has to decide on the best compromise to achieve the evaluation objectives, but should also report what has been done to enable a correct interpretation to be made of the results.

The FERET Program

The Face Recognition Technology (FERET) program, which was sponsored by the Department of Defense (DoD) Counterdrug Technology Program, commenced in September 1993. The primary mission of the FERET program was to develop automatic face recognition capabilities that could be employed to assist security, intelligence and law enforcement personnel in the performance of their duties.

The FERET program initially consisted of three one year phases. The objective of the first phase was to establish the viability of automatic face recognition algorithms, and to determine a performance baseline against which to measure future progress. The goals of the other two phases was to further develop face recognition technology. After the completion of phase 2 the FERET demonstration effort was commenced, with the goals to port FERET evaluated algorithms to real–time experimental/demonstration systems.

The program focused on three major areas:

1. Sponsoring Research: The goal of the sponsored research was to develop facial recognition algorithms. After a broad agency announcement for algorithm development proposals, twenty–four submissions were received and evaluated by DoD and law enforcement personnel. Five contracts were initially awarded, and three of these teams were selected to continue their development for phase 2.
2. Collecting the FERET database: The FERET database of facial images was a vital part of the overall FERET program and promised to be key to future work in face recognition, because it provided a standard database for algorithm development, test and evaluation, and most importantly, the images were gathered independently from the algorithm developers. The images were collected in a semi–controlled environment, with the same physical setup used in each photography session to maintain a degree of consistency throughout the database. However, because the equipment had to be reassembled for each session, there was some minor variation in images collected on different dates. The FERET database was collected in 15 sessions between August 1993 and July 1996. The database contains 1564 sets of images for a total of 14,126 images that includes 1199 individuals and 365 duplicate sets of images. A duplicate set is a second set of images of a person already in the database and was usually taken on a different day.
3. Performing the FERET evaluations: Before the FERET database was created, a large number of papers reported outstanding recognition results (usually >95% correct recognition) on limited–size databases (usually <50 individuals). Only a few of these algorithms reported results on images utilizing a common database – the FERET database made it possible for researchers to develop algorithms on a common database and to report results in the literature using this database. More importantly, the FERET database and evaluations clarified the state of the art in face recognition and pointed out general directions for future research. Three sets of evaluations were performed, with the last two evaluations being administered multiple times. The first FERET evaluation took place in August 1994, the Aug94 evaluation. This evaluation was designed to measure performance on algorithms that could automatically locate, normalize, and identify faces from a database. The test consisted of three subtests, each with a different gallery and probe set. The first subtest examined the ability of algorithms to recognize faces from a gallery of 316 individuals. The second subtest was the false-alarm test, which measured how well an algorithm

rejects faces not in the gallery. The third subtest baselined the effects of pose changes on performance. The second FERET evaluation took place in March of 1995, the Mar95 evaluation. The goal was to measure progress since the initial FERET evaluation, and to evaluate these algorithms on larger galleries (817 individuals). An added emphasis of this evaluation was on probe sets that contained duplicate images, where a duplicate image was defined as an image of a person whose corresponding gallery image was taken on a different date. The third FERET evaluations took place in September of 1996, the Sep96 evaluation. For the Sept96 evaluation, a new evaluation protocol was designed which required algorithms to match a set of 3323 images against a set of 3816 images. The new protocol design allowed the determination of performance scores for multiple galleries and probe sets, and perform a more detailed performance analysis. There were two versions of the September 1996 evaluation. The first tested partially automatic algorithms by providing the images with the coordinates of the center of the eyes. The second tested fully automatic algorithms by providing the images only.
Further details on the methodology of the FERET program can be found in [PMW00].

Eigenfaces

Recall that *eigenfaces* are a set of *eigenvectors*[135] used in the computer vision problem of human FR. These eigenvectors are derived from the *covariance matrix* of the *probability distribution* of the high–dimensional vector space of possible human faces, in a similar fashion as in *factor analysis* described above. Many authors prefer the term eigenimage rather than eigenface, as the technique has been used for handwriting, lip reading, voice recognition, and medical imaging.

In layman's terms, eigenfaces are a set of 'standardized face ingredients', derived from multivariate correlation analysis of many pictures of faces. Any human face can be considered to be a combination of these standard faces. One person's face might be made up of 10% from face 1, 24% from face 2

[135] Recall that an *eigenvector* of a transformation is a non–null vector whose direction is unchanged by that transformation. The factor by which the magnitude is scaled is called the *eigenvalue* of that vector. Often, a transformation is completely described by its eigenvalues and eigenvectors. An *eigenspace* is a set of eigenvectors with a common eigenvalue. These concepts play a major role in several branches of both pure and applied mathematics — appearing prominently in linear algebra, functional analysis, and even a variety of nonlinear situations.

It is common to prefix any natural name for the solution with eigen instead of saying eigenvector. For example, *eigenfunction* if the eigenvector is a function, eigenmode if the eigenvector is a harmonic mode, eigenstate if the eigenvector is a quantum state, and so on (e.g. the eigenface example below). Similarly for the eigenvalue, e.g., *eigenfrequency* if the eigenvalue is (or determines) a frequency.

and so on. This means that if we want to record someone's face for use by FR software, we can use far less space than would be taken up by a digitised photograph.

To generate a set of eigenfaces, a large set of digitized images of human faces, taken under the same lighting conditions, are normalized to line up the eyes and mouths. They are then all resampled at the same pixel resolution (say $m \times n$), and then treated as mnD vectors whose components are the values of their pixels. The eigenvectors of the covariance matrix of the statistical distribution of face image vectors are then extracted. It should be noted that these are the same as the eigenvectors from principal components analysis (PCA, see above), the statistical method from which eigenimaging is derived. Since the eigenvectors belong to the same vector space as face images, they can be viewed as if they were $m \times n$ pixel face images: hence the name eigenfaces. Viewed in this way, the principal eigenface looks like a bland androgynous average human face. Some subsequent eigenfaces can be seen to correspond to generalized features such as left–right and top–bottom asymmetry, or the presence or absence of a beard. Other eigenfaces are hard to categorize, and look rather strange.

When properly weighted, eigenfaces can be summed together to create an approximate gray–scale rendering of a human face. Remarkably few eigenvector terms are needed to give a fair likeness of most people's faces, so eigenfaces provide a means of applying data compression to faces for identification purposes (see, e.g., [Abd88]).

Dynamic Link Matching

The *dynamic link matching* (DLM) is a neural FR–system based on the Gabor–wavelet transform [Mal85, Mal88, KMM94, LVB93, Wis95]. The system is inherently invariant with respect to shift, and is robust against many other variations, most notably rotation in depth and deformation. The system consists of an image domain and a model domain, which is tentatively identified with primary visual cortex and infero–temporal cortex. Both domains have the form of neural sheets of hypercolumns, which are composed of simple feature detectors (modeled as Gabor wavelets). Each object is represented in memory by a separate model sheet, that is, a 2D array of features. The match of the image to the models is performed by network self–organization, in which rapid reversible synaptic plasticity of the connections ('dynamic links') between the two domains is controlled by signal correlations, which are shaped by fixed inter–columnar connections and by the dynamic links themselves. The system requires very little genetic or learned structure, relying essentially on the rules of rapid synaptic plasticity and the a priori constraint of preservation of topography to find matches. This constraint is encoded within the neural sheets with the help of lateral connections, which are excitatory over short range and inhibitory over long range.

Topographical relationships between nodes in the DLM–system are encoded by excitatory and inhibitory lateral connections (see Figure 1.4). The

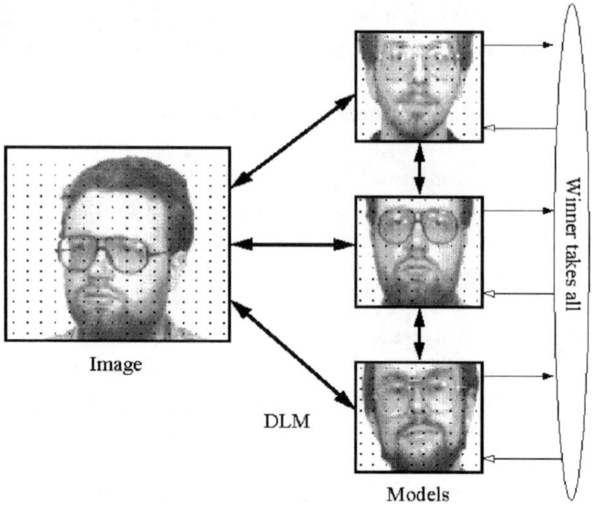

Fig. 1.4. Architecture of the DLM face recognition system. Several models are stored as neural layers of local features on a 1010 grid, as indicated by the black dots. A new image is represented by a 1617 layer of nodes. Initially, the image is connected all–to–all with the models. The task of DLM is to find the correct mapping between the image and the models, providing translational invariance and robustness against distortion. Once the correct mapping is found, a simple winner–take–all mechanism can detect the model that is most active and most similar to the image (adapted from [Mal85, Mal88, KMM94, LVB93, Wis95]).

model graphs are scaled horizontally and vertically and aligned manually, such that certain nodes of the graphs are placed on the eyes and the mouth. Model layers (1010 neurons) are smaller than the image layer (1617 neurons). Since the face in the image may be arbitrarily translated, the connectivity between model and image domain has to be all–to–all initially. The connectivity matrices are initialized using the similarities between the jets of the connected neurons. DLM serves as a process to restructure the connectivity matrices and to find the correct mapping between the models and the image. The models cooperate with the image depending on their similarity. A simple winner–take–all mechanism sequentially rules out the least active and least similar models, and the best–fitting one eventually survives.

Face Recognition Vendor Tests

Face Recognition Vendor Tests (FRVT) provide independent government evaluations of commercially available and mature prototype face recognition systems. During the FERET program face recognition technology was primarily found in prototype systems in universities and research labs. By 2000 systems were available on the commercial market, so FRVT 2000 was instigated to evaluate the capabilities of these commercial systems. Sponsored by the

Defense Advanced Research Projects Agency (DARPA), DoD Counterdrug Technology Development Program Office and National Institute of Justice (NIJ), and designed by the National Institute of Standards and Technology (NIST) the FRVT 2000 was based on the FERET evaluations and the evaluation methodology philosophy outlined by [PMW00].

The FRVT 2000 was a technology evaluation consisting of two components: the Recognition Performance Test and the Product Usability Test. The goal of the Recognition Performance Test was to compare competing techniques for performing facial recognition, with all systems tested on a standardized database. The product usability test examined system properties for performing access control. Five commercial products were evaluated, and the results of the tests can be found at get references from Lit Review Report. Under the USA Patriot Act, NIST is mandated to measure the accuracy of biometric technologies. In accordance with this legislation, NIST, in cooperation with other Government agencies, is conducting the Face Recognition Vendor Test 2002 FRVT 2002. Now sponsored or supported by 16 organisations, including some non–US agencies, the FRVT 2002 aims to assess the state–of–the–art in face recognition technology, and is conducting a technology evaluation of both mature prototype and commercially available systems face recognition systems.

Hidden Markov Models

A *hidden Markov model* (HMM) is a statistical model where the system being modelled is assumed to be a *Markov process*[136] with unknown parameters, and the challenge is to determine the hidden parameters from the observable

[136] Recall that a *Markov process* is a stochastic process that has a *Markov property*, or *Markov assumption*. Technically, there are three well–known special cases of the *Chapman–Kolmogorov equation*, describing a general Markov process (see [Gar85]):

1. When both $B_{ij}[x(t),t]$ and $W(t)$ are zero, i.e., in the case of pure deterministic motion, it reduces to the *Liouville equation*
$$\partial_t P(x',t'|x'',t'') = -\sum_i \frac{\partial}{\partial x^i}\left\{A_i[x(t),t]\,P(x',t'|x'',t'')\right\}.$$
2. When only $W(t)$ is zero, it reduces to the *Fokker–Planck diffusion equation*
$$\partial_t P(x',t'|x'',t'') = -\sum_i \frac{\partial}{\partial x^i}\left\{A_i[x(t),t]\,P(x',t'|x'',t'')\right\} + \frac{1}{2}\sum_{ij}\frac{\partial^2}{\partial x^i \partial x^j}\left\{B_{ij}[x(t),t]\,P(x',t'|x'',t'')\right\}.$$
3. When both $A_i[x(t),t]$ and $B_{ij}[x(t),t]$ are zero, i.e., the state–space consists of integers only, it reduces to the *Master equation* of discontinuous jumps

parameters.[137] The extracted model parameters can then be used to perform further analysis, for example for pattern recognition applications. A HMM can be considered as the simplest *dynamic Bayesian network*.

In a regular Markov model, the state is directly visible to the observer, and therefore the state transition probabilities are the only parameters. In a hidden Markov model, the state is not directly visible, but variables influenced by the state are visible. Each state has a probability distribution over the possible output tokens. Therefore the sequence of tokens generated by an HMM gives some information about the sequence of states.

The HMM–architecture is depicted in Figure 1.5. From this diagram, it is clear that the value of the *hidden variable* $x(t)$ (at time t) only depends on the value of the hidden variable $x(t-1)$ (at time $t-1$). Similarly, the value of the *observed variable* $y(t)$ only depends on the value of the hidden variable $x(t)$ (both at time t).

The probability of observing a sequence $Y = y(0), y(1), \dots, y(L-1)$ of length L in HMM is given by:

$$P(Y) = \sum_X P(Y \mid X) P(X),$$

$$\partial_t P(x', t'|x'', t'') = \int dx \left\{ W(x'|x'', t) \, P(x', t'|x'', t'') - W(x''|x', t) \, P(x', t'|x'', t'') \right\}.$$

The *Markov assumption* can now be formulated in terms of the conditional probabilities $P(x^i, t_i)$: if the times t_i increase from right to left, the conditional probability is determined entirely by the knowledge of the most recent condition. Markov process is generated by a set of conditional probabilities whose probability–density $P = P(x', t'|x'', t'')$ evolution obeys the general *Chapman–Kolmogorov integro–differential equation*

$$\begin{aligned} \partial_t P = &-\sum_i \frac{\partial}{\partial x^i} \{A_i[x(t), t] \, P\} \\ &+ \frac{1}{2} \sum_{ij} \frac{\partial^2}{\partial x^i} \partial x^j \{B_{ij}[x(t), t] \, P\} \\ &+ \int dx \left\{ W(x'|x'', t) \, P - W(x''|x', t) \, P \right\} \end{aligned}$$

including: *deterministic drift*, *diffusion fluctuations* and *discontinuous jumps* (given respectively in the first, second and third rows).

[137] Hidden Markov Models were first described in a series of statistical papers by Leonard Baum in the second half of the 1960s. One of the first applications of HMMs was speech recognition, starting in the mid–1970s. In the second half of the 1980s, HMMs began to be applied to the analysis of biological sequences, in particular DNA. Since then, they have become ubiquitous in the field of bioinformatics.

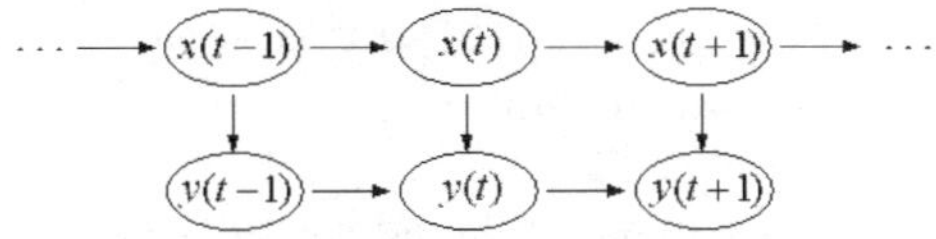

Fig. 1.5. Generic architecture of a Hidden Markov model. Each oval shape represents a random variable that can adopt a number of values. The random variable $x(t)$ is the value of the *hidden variable* at time t. The random variable $y(t)$ is the value of the *observed variable* at time t. The arrows in the diagram denote *conditional dependencies*.

where the sum runs over all possible hidden node sequences $X = x(0), x(1), \ldots, x(L-1)$. A *brute force* calculation of $P(Y)$ is intractable for realistic problems, as the number of possible hidden node sequences typically is extremely high. The calculation can however be speeded up enormously using a *dynamic programming* algorithm, called the *forward algorithm*.

Recall that dynamic programming, invented by Richard Bellman,[138] is a method for reducing the runtime of algorithms exhibiting the properties of:

1. Overlapping subproblems (the problem can be broken down into subproblems which are reused several times),[139]
2. Optimal substructure (optimal solution can be constructed efficiently from optimal solutions to its subproblems; used to determine the usefulness of dynamic programming and *greedy algorithms*[140] in a problem), and

[138] Richard Ernest Bellman (1920–1984) was an applied mathematician, celebrated for his invention of *dynamic programming* in 1953, and important contributions in other fields of mathematics, including the *Bellman equation* and *Hamilton–Jacobi–Bellman equation.*

A well–known term in computation coined by Bellman is *curse of dimensionality*: the problem caused by the rapid increase in volume associated with adding extra dimensions to a (mathematical) space (e.g., 'rules explosion' in fuzzy logic systems). Similarly, the curse of dimensionality is a significant obstacle in machine learning problems that involve learning from few data samples in a high–dimensional feature space.

[139] For example, the problem of computing the *Fibonacci sequence* exhibits *overlapping subproblems.* The problem of computing the nth Fibonacci number, $F(n)$, can be broken down into the subproblems of computing $F(n-1)$ and $F(n-2)$, and then adding the two. The subproblem of computing $F(n-1)$ can itself be broken down into a subproblem that involves computing $F(n-2)$. Therefore the computation of $F(n-2)$ is reused, and the *Fibonacci sequence* thus exhibits overlapping subproblems.

[140] A *greedy algorithm* is an algorithm that follows the problem solving metaheuristic of making the locally optimum choice at each stage with the hope of finding the global optimum. For instance, applying the greedy strategy to the *traveling salesman problem* yields the following algorithm: 'At each stage visit the unvisited city nearest to the current city'. In general, greedy algorithms have five pillars: (i) a candidate set, from which a solution is created; (ii) a selection function,

3. Memoization (speeding up programs by storing the results of functions for later reuse, rather than recomputing them).[141]

---

which chooses the best candidate to be added to the solution; (iii) a feasibility function, that is used to determine if a candidate can be used to contribute to a solution; (iv) an objective function, which assigns a value to a solution, or a partial solution; and (v) a solution function, which will indicate when we have discovered a complete solution.

There are two ingredients that are exhibited by most problems that lend themselves to a greedy strategy:

1. Greedy Choice Property: We can make whatever choice seems best at the moment and then solve the subproblems arising after the choice is made. The choice made by a greedy algorithm may depend on choices so far. But, it cannot depend on any future choices or all the solutions to the subproblem, it progresses in a fashion making one greedy choice after another iteratively reducing each given problem into a smaller one. This is the main difference between it and dynamic programming. Dynamic programming is exhaustive and is guaranteed to find the solution. After every algorithmic stage, dynamic programming makes decisions based on the all the decisions made in the previous stage, and may reconsider the previous stage's algorithmic path to solution. A greedy algorithm makes the decision early and changes the algorithmic path after decision, and will never reconsider the old decisions. It may not be accurate for some problems.
2. Optimal Sub structure: A problem exhibits optimal sub–structure, if an optimal solution to the sub–problem contains within its optimal solution to the problem.

For most problems, greedy algorithms mostly (but not always) fail to find the globally optimal solution, because they usually do not operate exhaustively on all the data. They can make commitments to certain choices too early which prevent them from finding the best overall solution later. For example, all known greedy algorithms for the graph coloring problem and all other NP-complete problems do not consistently find optimum solutions. Nevertheless, they are useful because they are quick to think up and often give good approximations to the optimum. If a greedy algorithm can be proven to yield the global optimum for a given problem class, it typically becomes the method of choice because it is faster than other optimisation methods like dynamic programming. Examples of such greedy algorithms are *Kruskal's algorithm*, *Dijkstra's algorithms* for finding single–source shortest paths and *Prim's algorithm* for finding minimum spanning trees and the algorithm for finding optimum *Huffman trees*. The theory of *matroids* provide whole classes of such algorithms.

[141] Functions can only be memoized if they are referentially transparent, that is, if they will always return the same result given the same arguments. Operations which are not referentially transparent, but whose results are not likely to change rapidly, can still be cached with methods more complicated than memoization. In general, memoized results are not expired or invalidated later, while caches generally are. In imperative languages, both memoization and more general caching are typically implemented using some form of associative array.

In a *functional programming language* it is possible to construct a higher–order function memoize which will create a memoized function for any

Dynamic programming usually takes one of two approaches:

- Top–down approach: The problem is broken into subproblems, and these subproblems are solved and the solutions remembered, in case they need to be solved again. This is recursion and memoization combined together.
- Bottom–up approach: All subproblems that might be needed are solved in advance and then used to build up solutions to larger problems. This approach is slightly better in stack space and number of function calls, but it is sometimes not intuitive to figure out all the subproblems needed for solving given problem.

There are 3 canonical problems associated with HMMs (see, e.g., [Rab89]):

1. Given the parameters of the model, compute the probability of a particular output sequence. This problem is solved by the *forward algorithm.*
2. Given the parameters of the model, find the most likely sequence of hidden states that could have generated a given output sequence. This problem is solved by the *Viterbi algorithm.*[142]

---

referentially transparent function. In languages without higher–order functions, memoization must be implemented separately in each function that is to benefit from it.

The term 'memoization' was coined by Donald Michie in his 1968 paper 'Memo functions and machine learning' in Nature.

[142] The Viterbi algorithm is a dynamic programming algorithm for finding the most likely sequence of hidden states, called the *Viterbi path*, that result in a sequence of observed events, especially in the HMM context. The *forward algorithm* is a closely related algorithm for computing the probability of a sequence of observed events. These algorithms form a subset of modern information theory.

The algorithm makes a number of assumptions. First, both the observed events and hidden events must be in a sequence. This sequence often corresponds to time. Second, these two sequences need to be aligned, and an observed event needs to correspond to exactly one hidden event. Third, computing the most likely hidden sequence up to a certain point t must depend only on the observed event at point t, and the most likely sequence at point t - 1. These assumptions are all satisfied in a first-order hidden Markov model.

The terms 'Viterbi path' and 'Viterbi algorithm' are also applied to related dynamic programming algorithms that discover the single most likely explanation for an observation. For example, in stochastic parsing a dynamic programming algorithm can be used to discover the single most likely context-free derivation (parse) of a string, which is sometimes called the 'Viterbi parse'.

The Viterbi algorithm was conceived by Andrew Viterbi as an error–correction scheme for noisy digital communication links, finding universal application in decoding the convolutional codes used in both CDMA and GSM digital cellular, dial–up modems, satellite, deep–space communications, and 802.11 wireless LANs. It is now also commonly used in speech recognition, keyword spotting, computational linguistics, and bioinformatics. For example, in *speech–to–text translation*, the acoustic signal is treated as the observed sequence of events, and a string of text is considered to be the 'hidden cause' of the

3. Given an output sequence or a set of such sequences, find the most likely set of state transition and output probabilities. In other words, train the parameters of the HMM given a dataset of sequences. This problem is solved by the *Baum–Welch algorithm*.[143]

Hidden Markov models are especially known for their applications in *speech recognition*, *machine translation* and *bioinformatics*.

Bayesian Belief Networks

A Bayesian belief network is a form of probabilistic graphical model developed by Judea Pearl.[144] Bayesian network represents joint probability distribution of a set of variables with explicit independency assumptions. It is a *directed acyclic graph* with nodes representing variables and arcs representing probabilistic dependency relations among the variables.

If there is an arc from node A to another node B, then variable B depends directly on variable A and A is called a *parent node* of B. If the variable represented by a node has a known value then the node is said to be an *evidence node*. A node can represent any kind of variable, be it a measured parameter, a latent variable or a hypothesis. Nodes are not restricted to representing random variables; this is what is 'Bayesian' about a Bayesian network.

Let the variables be $X_1, \ldots, X_n$. Let $parents(A)$ be the parents of the node A. Then the joint distribution for X_1 through X_n is represented as the product of the probability distributions for $i = 1$ to n. If X_i has no parents, its probability distribution is said to be unconditional, otherwise it is conditional.

Questions about incongruent dependence among variables can be answered by studying the graph alone. It can be shown that conditional independence

acoustic signal. The Viterbi algorithm finds the most likely string of text given the acoustic signal.

143 The *Baum–Welch algorithm* is an expectation–maximization (EM) algorithm (see [BPS70]). It can compute *maximum likelihood estimates* and *posterior–mode estimates* for the parameters (transition and emission probabilities) of an HMM, when given only emissions as training data. The algorithm has two steps: (i) Calculating the forward probability and the backward probability for each HMM state; and (ii) On the basis of this, determining the frequency of the *transition–emission pair* values and dividing it by the probability of the entire string. This amounts to calculating the expected count of the particular transition–emission pair. Each time a particular transition is found, the value of the quotient of the transition divided by the probability of the entire string goes up, and this value can then be made the new value of the transition.

144 Judea Pearl is a computer scientist and statistician, best known for his prominent work on the probabilistic approach to artificial intelligence, and in particular on Bayesian belief networks. His work is also intended as a *high–level cognitive model*. He is interested in the philosophy of causality, artificial intelligence and knowledge representation, probabilistic and causal reasoning, nonstandard logics, and learning strategies. Pearl is described as 'one of the giants in the field of artificial intelligence'.

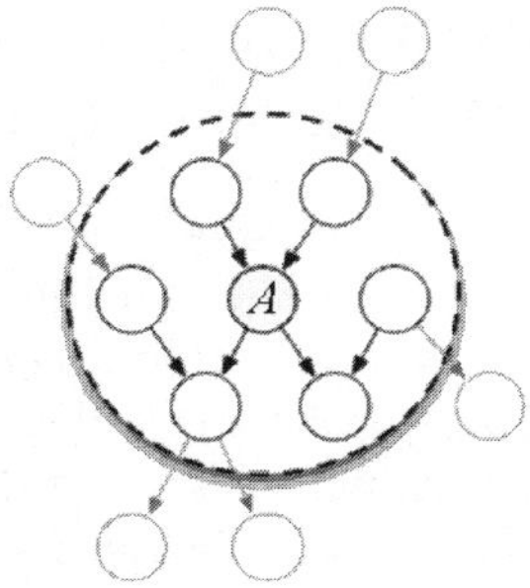

Fig. 1.6. A generic Markov blanket: the set of nodes $MB(A)$ composed of A's parents, its children, and its children's parents.

is represented in the graph by the graphical property of $d-$separation: nodes X and Y are $d-$separated in the graph, given specified evidence nodes, iff variables X and Y are independent given the corresponding evidence variables. The set of all other nodes on which node X can directly depend is given by X's *Markov blanket*.

The Markov blanket (see Figure 1.6) for a node A in a Bayesian network is the set of nodes $MB(A)$ composed of A's parents, its children, and its children's parents. In a *Markov network*, the Markov blanket of a node is its set of neighboring nodes. Every node in the network is conditionally independent of A when conditioned on the set $MB(A)$, that is, when conditioned on the Markov blanket of the node A. Formally, for distinct nodes A and B, we have

$$\Pr(A \mid MB(A), B) = \Pr(A \mid MB(A)).$$

The values of the parents and children of a node evidently give information about that node. However, its children's parents also have to be included, because they can be used to explain away the node in question. The Markov blanket of a node is interesting because it identifies all the variables that shield off the node from the rest of the network. This means that the Markov blanket of a node is the only knowledge that is needed to predict the behavior of that node.

A *causal Bayesian network* is a Bayesian network where the directed arcs of the graph are interpreted as representing *causal relations*[145] in some real domain. The directed arcs do not have to be interpreted as representing causal

[145] Recall that the philosophical *concept of causality*, the principles of causes, or causation, the working of causes, refers to the set of all particular 'causal' or 'cause–and–effect' relations. A neutral definition is notoriously hard to provide since every aspect of causation has been subject to much debate. Most generally, causation is a relationship that holds between events, properties, variables, or states of affairs. Causality always implies at least some relationship of dependency between the cause and the effect. For example, deeming something a cause may imply that, all other things being equal, if the cause occurs the effect does as well, or at least that the probability of the effect occurring increases. It is

relations; however in practice knowledge about causal relations is very often used as a guide in drawing Bayesian network graphs, thus resulting in causal Bayesian networks.

In the simplest case, a Bayesian network is specified by an expert and is then used to perform inference after some of the nodes are fixed to observed values. In some applications, such as finding *gene regulatory networks* (see [II06b]), a more complex problem of finding dependencies between variables arises. This can be solved by learning a Bayesian network that fits to the data.

Learning the structure of a Bayesian network (i.e., the graph) is a very important part of *machine learning*. Given the information that the data is being

---

also usually presumed that the cause chronologically precedes the effect. In natural languages, causal relationships can be expressed by the following causative expressions:

(i) a set of causative verbs (cause, make, create, do, effect, produce, occasion, perform, determine, influence; construct, compose, constitute; provoke, motivate, force, facilitate, induce, get, stimulate; begin, commence, initiate, institute, originate, start; prevent, keep, restrain, preclude, forbid, stop, cease);
(ii) a set of causative names (actor, agent, author, creator, designer, former, originator; antecedent, causality, causation, condition, fountain, occasion, origin, power, precedent, reason, source, spring; reason, grounds, motive, need, impulse);
(iii) a set of effective names (consequence, creation, development, effect, end, event, fruit, impact, influence, issue, outcome, outgrowth, product, result, upshot).

Causality is the centerpiece of the universe and so the main subject of human knowledge; for comprehending the nature, meaning, kinds, varieties, and ordering of cause and effect amounts to knowing the beginnings and endings of things, to uncovering the implicit mechanisms of world dynamics, or to having the fundamental scientific knowledge.

Ancient Hindu scriptures, the Upanishads (namely Chandogya Upanishad, Sarva Sara Upanishad and Mandukya Upanishad) and some other texts (namely Brahma Sutras, Yoga Vashishta, Avadhuta Gita and Astavakra Gita) mention causality. However, the mention is limited to the purpose of explaining creation of the universe: 'Cause is the effect concealed, effect is the cause revealed', which is also expressed as 'Cause is the effect unmanifested, effect is the cause manifested' (reference Complete Works of *Swami Vivekananda*, as well as *Yoga Vashishta*); 'Effect is same as cause only' (reference *Sankaracharya's* commentary on *Bhagavad Gita*).

In Metaphysics and Posterior Analytics, Aristotle stated: "All causes of things are beginnings; that we have scientific knowledge when we know the cause; that to know a thing's existence is to know the reason why it is." With this, he set the guidelines for all the subsequent causal theories by specifying the number, nature, principles, elements, varieties, order of causes as well as the modes of causation. Aristotle's account of the causes of things may be qualified as the most comprehensive model up to now.

The modern deterministic world–view is one in which the universe is nothing more than a chain of events following one after another according to the law of cause and effect.

generated by a Bayesian network and that all the variables are visible in every iteration, the following methods are used to learn the structure of the acyclic graph and the conditional probability table associated with it. The elements of a *structure–finding algorithm* are a *scoring function* and a *search strategy*. The time requirement of an exhaustive search returning back a structure that maximizes the score is superexponential in the number of variables. A local search algorithm makes incremental changes aimed at improving the score of the structure. A global search algorithm like *Markov–chain Monte–Carlo* (MCMC) can avoid getting trapped in local minima.

In order to fully specify the Bayesian network and thus fully represent the joint probability distribution, it is necessary to further specify for each node X the probability distribution for X conditional upon X's parents. The distribution of X conditional upon its parents may have any form. It is common to work with discrete or Gaussian distributions since that simplifies calculations. Sometimes only constraints on a distribution are known; one can then use the principle of maximum entropy to determine a single distribution, the one with the greatest entropy given the constraints. Analogously, in the specific context of a dynamic Bayesian network, one commonly specifies the conditional distribution for the hidden state's temporal evolution to maximize the entropy rate of the implied stochastic process. Often these conditional distributions include parameters which are unknown and must be estimated from data, sometimes using the maximum likelihood approach. Direct maximization of the likelihood (or of the posterior probability) is often complex when there are unobserved variables. A classical approach to this problem is the *expectation–maximization algorithm* which alternates computing expected values of the unobserved variables conditional on observed data, with maximizing the complete likelihood (or posterior) assuming that previously computed expected values are correct. Under mild regularity conditions this process converges on maximum likelihood (or maximum posterior) values for parameters. A more fully Bayesian approach to parameters is to treat parameters as additional unobserved variables and to compute a full posterior distribution over all nodes conditional upon observed data, then to integrate out the parameters. This approach can be expensive and lead to large dimension models, so in practise classical parameter–setting approaches are more common.

The goal of inference in Bayesian networks is typically to find the distribution of a subset of the variables, conditional upon some other subset of variables with known values (the evidence), with any remaining variables integrated out. This is known as the posterior distribution of the subset of the variables given the evidence. The posterior gives a universal sufficient statistic for detection applications, when one wants to choose values for the variable subset which minimize some expected loss function, for instance the probability of decision error.

A Bayesian network can thus be considered a mechanism for automatically constructing extensions of *Bayes' theorem* to more complex problems. Bayes' theorem relates the conditional and marginal probability distributions of random variables. In some interpretations of probability, Bayes' theorem tells how to update or revise beliefs in light of new evidence: a posteriori. The probability of an event A conditional on another event B is generally different from the probability of B conditional on A. However, there is a definite relationship between the two, and Bayes' theorem is the statement of that relationship.

As a formal theorem, Bayes' theorem is valid in all interpretations of probability. However, frequentist and Bayesian interpretations disagree about the kinds of things to which probabilities should be assigned in applications: frequentists assigned probabilities to random events according to their frequencies of occurrence or to subsets of populations as proportions of the whole; Bayesians assign probabilities to propositions that are uncertain. A consequence is that Bayesians have more frequent occasion to use Bayes' theorem. The articles on Bayesian probability and frequentist probability discuss these debates at greater length.

Formally, *Bayes' theorem* relates the conditional and marginal probabilities of stochastic events A and B as

$$\Pr(A|B) = \frac{\Pr(B|A)\Pr(A)}{\Pr(B)} \propto L(A|B)\Pr(A),$$

where $L(A|B)$ is the likelihood of A given fixed B. Each term in Bayes' theorem has a conventional name:

$Pr(A)$ is the prior probability or marginal probability of A. It is 'prior' in the sense that it does not take into account any information about B;

$Pr(A|B)$ is the conditional probability of A, given B. It is also called the posterior probability because it is derived from or depends upon the specified value of B.

$Pr(B|A)$ is the conditional probability of B given A.

$Pr(B)$ is the prior or marginal probability of B, and acts as a normalizing constant.

With this terminology, the theorem may be paraphrased as:

$$\text{posterior} = \frac{\text{likelihood} \times \text{prior}}{\text{normalizing constant}},$$

or, in words: *the posterior probability is proportional to the prior probability times the likelihood.* In addition, the ratio $Pr(B|A)/Pr(B)$ is sometimes called the *standardised likelihood*, so the theorem may also be paraphrased as:

$$\text{posterior} = \text{standardised likelihood} \times \text{prior}.$$

The most common exact inference methods are variable elimination which eliminates (by integration or summation) the non–observed non–query variables one by one by distributing the sum over the product, clique tree propagation which caches the computation so that the many variables can be queried at one time, and new evidence can be propagated quickly, recursive conditioning which allows for a space-time tradeoff but still allowing for the efficiency of variable elimination when enough space is used. All of these methods have complexity that is exponential in tree width. The most common approximate inference algorithms are stochastic MCMC simulation, mini–bucket elimination which generalizes loopy belief propagation, and variational methods.

Bayesian networks are used for modelling knowledge in gene regulatory networks, medicine, engineering, text analysis, image processing, data fusion, and decision support systems.

Support Vector Machines

Recall that *support vector machines* (SVMs, see Figure 1.7) are a set of related *supervised learning* methods used for *classification* and *regression* (see [Vap95, Vap98, SS01, CS00]). Their common factor is the use of a technique known as the '*kernel trick*' to apply *linear classification* techniques to *nonlinear classification* problems.

SVMs implement the statistical learning theory. They are a radically different type of classifier from artificial neural networks (ANNs, see below) that has attracted a great deal of attention lately due to the novelty of the concepts that they bring to pattern recognition, their strong mathematical foundation, and their excellent results in practical problems. SVM represents the coupling of the following two concepts: the idea that transforming the data into a high–dimensional space makes linear discriminant functions practical, and the idea of large margin classifiers to train the standard ANNs like MLP or RBF. It is another type of a kernel classifier: it places Gaussian kernels over the data and linearly weights their outputs to create the system output. To implement the SVM–methodology, we can use the Adatron–kernel algorithm, a sophisticated nonlinear generalization of the RBF networks, which maps inputs to a high–dimensional feature space, and then optimally separates data into their respective classes by isolating those inputs, which fall close to the data boundaries. Therefore, the Adatron–kernel is especially effective in separating sets of data, which share complex boundaries, as well as for the training for non-linearly separable patterns. The support vectors allow the network to rapidly converge on the data boundaries and consequently classify the inputs.

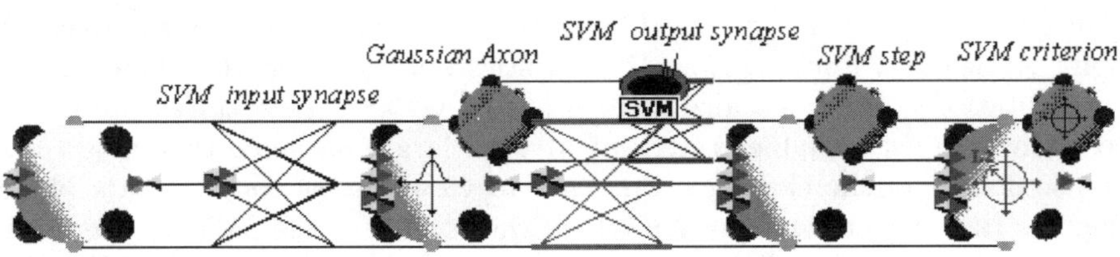

Fig. 1.7. Adatron–kernel based support vector machine (SVM) network, arranged using $NeuroSolutions^{TM}$.

The main advantage of SVMs over MLPs is that the learning task is a *convex optimization problem* which can be reliably solved even when the example data require the fitting of a very complicated function [Vap95, Vap98]. A common argument in computational learning theory suggests that it is dangerous to utilize the full flexibility of the SVM to learn the training data perfectly when these contain an amount of noise. By fitting more and more noisy data, the machine may implement a rapidly oscillating function rather than the smooth mapping which characterizes most practical learning tasks. Its prediction ability could be no better than random guessing in that case. Hence, modifications of SVM training [CS00] that allow for training errors were suggested to be necessary for realistic noisy scenarios. This has the drawback of introducing extra model parameters and spoils much of the original elegance of SVMs.

Mathematics of SVMs is based on real *Hilbert space* methods.

Linear Classification Problem

Suppose we want to classify some data points into two classes. Often we are interested in classifying data as part of a machine–learning process. These data points may not necessarily be points in $\mathbb{R}^2$ but may be multidimensional $\mathbb{R}^p$ (statistics notation) or $\mathbb{R}^n$ (computer science notation) points. We are interested in whether we can separate them by a *hyperplane*. As we examine a hyperplane, this form of classification is known as linear classification. We also want to choose a hyperplane that separates the data points 'neatly', with maximum distance to the closest data point from both classes – this distance is called the *margin*. We desire this property since if we add another data point to the points we already have, we can more accurately classify the new point since the separation between the two classes is greater. Now, if such a hyperplane exists, the hyperplane is clearly of interest and is known as the *maximum–margin hyperplane* or the *optimal hyperplane*, as are the vectors that are closest to this hyperplane, which are called the *support vectors*.

Formalization

Consider data points of the form

$$\{(\mathbf{x}_1, c_1), (\mathbf{x}_2, c_2), \dots, (\mathbf{x}_n, c_n)\},$$

where the c_i is either 1 or -1; this constant denotes the class to which the point $\mathbf{x}_i$ belongs. Each $\mathbf{x}_i$ is a pD (statistics notation), or nD (computer science notation) vector of scaled $[0, 1]$ or $[-1, 1]$ values. The scaling is important to guard against variables (attributes) with larger variance that might otherwise dominate the classification. We can view this as *training data*, which denotes the correct classification which we would like the SVM to eventually distinguish, by means of the dividing hyperplane, which takes the form:

$$\mathbf{w} \cdot \mathbf{x} - b = 0.$$

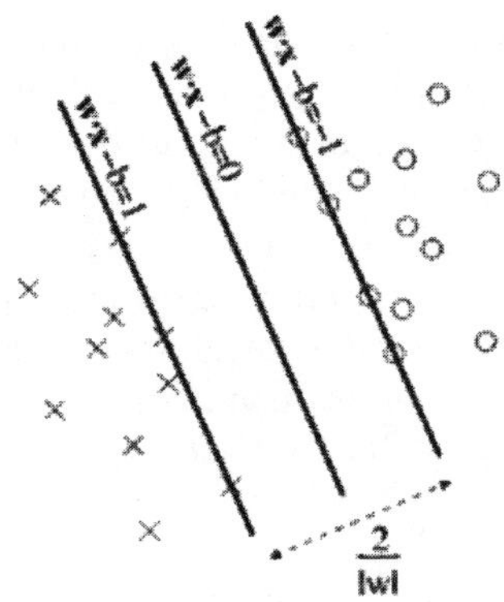

Fig. 1.8. Maximum–margin hyperplanes for a SVM trained with samples from two classes. Samples along the hyperplanes are called the support vectors.

As we are interested in the maximum margin, we are interested in the support vectors and the parallel hyperplanes (to the optimal hyperplane) closest to these support vectors in either class (see Figure 1.8). It can be shown that these parallel hyperplanes can be described by equations

$$\mathbf{w} \cdot \mathbf{x} - b = 1, \tag{1.6}$$

$$\mathbf{w} \cdot \mathbf{x} - b = -1. \tag{1.7}$$

We would like these hyperplanes to maximize the distance from the dividing hyperplane and to have no data points between them. By using geometry, we find the distance between the hyperplanes being $2/|\mathbf{w}|$, so we want to minimize $|\mathbf{w}|$. To exclude data points, we need to ensure that for all i either

$$\mathbf{w} \cdot \mathbf{x}_i - b \geq 1, \qquad \text{or}$$
$$\mathbf{w} \cdot \mathbf{x}_i - b \leq -1.$$

This can be rewritten as

$$c_i(\mathbf{w} \cdot \mathbf{x}_i - b) \geq 1, \qquad (1 \leq i \leq n). \tag{1.8}$$

The problem now is to minimize $|w|$ subject to the constraint (1.8). This is a *quadratic programming optimization* (QP) problem.

After the SVM has been trained, it can be used to classify unseen 'test' data. This is achieved using the following decision rule,

$$\hat{c} = \begin{cases} 1 & \text{if} \quad \mathbf{w} \cdot \mathbf{x} + b \geq 0, \\ -1 & \text{if} \quad \mathbf{w} \cdot \mathbf{x} + b \leq 0. \end{cases}$$

Writing the classification rule in its dual form reveals that classification is only a function of the support vectors, i.e., the training data that lie on the margin.

The use of the maximum–margin hyperplane is motivated by *Vapnik–Chervo-nenkis SVM theory*, which provides a probabilistic test *error bound* that is minimized when the margin is maximized. However the utility of this theoretical analysis is sometimes questioned given the large slack associated with these bounds: the bounds often predict more than 100% error rates.

The parameters of the maximum–margin hyperplane are derived by solving the optimization. There exist several specialized algorithms for quickly solving the QP problem that arises from SVMs. The most common method for solving the QP problem is Platt's *SMO algorithm*.

Nonlinear Classification

The original optimal hyperplane algorithm proposed by Vladimir Vapnik in 1963 was a *linear classifier*. However, in 1992, B. Boser, I. Guyon and Vapnik suggested a way to create nonlinear classifiers by applying the *kernel trick* (originally proposed by Aizerman) to maximum–margin hyperplanes. The resulting algorithm is formally similar, except that every *dot product* is replaced by a nonlinear *kernel* function. This allows the algorithm to fit the maximum–margin hyperplane in the transformed feature *space*. The transformation may be nonlinear and the transformed space high dimensional; thus though the classifier is a hyperplane in the high–dimensional feature space it may be nonlinear in the original input space.

If the kernel used is a Gaussian *radial basis function*, the corresponding feature space is a *Hilbert space* of infinite dimension. Maximum margin classifiers are well *regularized*, so the infinite dimension does not spoil the results. Some common kernels include:

1. *Polynomial (homogeneous)*:
$$k(\mathbf{x},\mathbf{x}') = (\mathbf{x}\cdot\mathbf{x}')^d;$$
2. *Polynomial (inhomogeneous)*:
$$k(\mathbf{x},\mathbf{x}') = (\mathbf{x}\cdot\mathbf{x}'+1)^d;$$
3. *Radial Basis Function*:
$$k(\mathbf{x},\mathbf{x}') = \exp(-\gamma\|\mathbf{x}-\mathbf{x}'\|^2), \qquad \text{for} \qquad \gamma>0;$$
4. *Gaussian radial basis function*:
$$k(\mathbf{x},\mathbf{x}') = \exp(-\frac{\|\mathbf{x}-\mathbf{x}'\|^2}{2\sigma^2}); \qquad \text{and}$$
5. *Sigmoid:*
$$k(\mathbf{x},\mathbf{x}') = \tanh(\kappa\mathbf{x}\cdot\mathbf{x}'+c),$$
for some (not every) $\kappa>0$ and $c<0$.

Soft Margin

In *1995*, *Corinna Cortes* and Vapnik suggested a modified maximum margin idea that allows for mislabeled examples. If there exists no hyperplane that can split the 'yes' and 'no' examples, the so-called *soft margin method* will choose a hyperplane that splits the examples as cleanly as possible, while still maximizing the distance to the nearest cleanly split examples. This work popularized the expression *Support Vector Machine* or *SVM*. This method introduces slack variables and the equation (1.8) now transforms to

$$c_i(\mathbf{w} \cdot \mathbf{x_i} - b) \geq 1 - \xi_i, \qquad (1 \leq i \leq n), \tag{1.9}$$

and the optimization problem becomes

$$\min ||w||^2 + C \sum_i \xi_i \qquad \text{such that} \quad c_i(\mathbf{w} \cdot \mathbf{x_i} - b) \geq 1 - \xi_i, \qquad (1 \leq i \leq n),$$

This constraint in (1.9) along with the objective of minimizing $|w|$ can be solved using *Lagrange multipliers* or setting up a dual optimization problem to eliminate the slack variable.

SV Regression

A version of a SVM for regression was proposed in *1995* by Vapnik, S. Golowich, and A. Smola (see [Vap98, SS01]). This method is called *support vector regression* (SVR). The model produced by support vector classification (as described above) only depends on a subset of the training data, because the *cost function* for building the model does not care about training points that lie beyond the margin. Analogously, the model produced by SVR only depends on a subset of the training data, because the cost function for building the model ignores any training data that is close (within a threshold ε) to the model prediction.

Intelligent Agents

Recall that the *agent theory* concerns the definition of the so-called *belief-desire-intention agents* (BDI-agents, for short), as well as multi-agent systems, properties, architectures, communication, cooperation and coordination capabilities (see [RG98]).

A common definition of an agent reads: An agent is a computer system that is situated in some environment, and that is capable of autonomous action in this environment in order to meet its design requirements [Woo00].

Practical side of the agent theory concerns the agent languages and platforms for programming and experimenting with agents. According to [Fer99], a BDI-agent is a physical or virtual entity which:

1. *is capable of limited perceiving its environment* (see Figure 1.9),
2. *has only a partial representation of its environment,*

Fig. 1.9. A basic agent–environment loop (modified from [Woo00]).

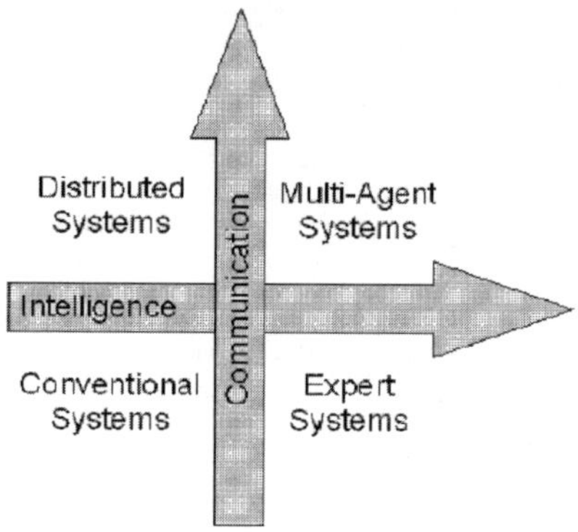

Fig. 1.10. Agent technology compared to relevant technologies.

3. *is capable of acting in an environment,*
4. *can communicate directly with other agents,*
5. *is driven by a set of tendencies,*[146]
6. *possesses resources of its own,*
7. *possesses some skills and can offer services,*
8. *may be able to reproduce itself,*
9. *whose behavior tends towards satisfying its objectives,*

– taking into account the resources and skills available to it and depending on its perception, its representation and the communications it receives. Agents' actions affect the environment which, in turn, affects future decisions of agents. The *multi–agent systems* have been successfully applied in numerous fields (see [Fer99] for the review).

Agents embody a new software development paradigm that attempts to merge some of the theories developed in artificial intelligence research with computer science. The power of agents comes from their intelligence and also their ability to communicate with each other. A simple mapping of agent technology compared to relevant technologies is illustrated in Figure 1.10. Agents can be considered as the successors of *object–oriented programming* techniques, applied to certain problem domains. However, the additional layer

[146] in the form of individual objectives or of a satisfaction/survival function which it tries to optimize

of implementation in agents provides some key functionalities and deliberately creates a separation between the implementation of an agent from the application being developed. This is done in order to achieve one of the core properties of agents, autonomy. Objects are able to assert a certain amount of control over themselves via private variables and methods, and other objects via public variables and methods. Consequently, a particular object is able to directly change public variables of other objects and also execute public methods of other objects. Hence, objects have no control over the values of public variables and who and when executes their public methods. Conversely, agents are explicitly separated, and can only request from each other to perform a particular task. Furthermore, it cannot be assumed that after a particular agent makes a request, another agent will do it. This is because performing a particular action may not be in the best interests of the other agent, in which case it would not comply [Woo00].

Types of Intelligent Agents

Here we give a general overview of different types of agents and groups them into several intuitive categories based on the method that they perform their reasoning [Woo00].

Deliberate Agents

Deliberate agents are agents that perform rational reasoning, take actions that are rational after deliberating using their *knowledge base* (KB), carefully considering the possible effects of different actions available to them. There are two subtypes of deliberate agents: *deductive reasoning agents* and *production-rule agents*.

1. Deductive reasoning agents are built using expert systems theory, they operate using an internal symbolic KB of the environment. Desired behavior is achieved by manipulating the environment and updating the KB accordingly. A utility function is implemented that provides an indication on how good a particular state is compared on what the agent should achieve. An example of the idea behind these type of agents is an agent that explores a building. It has the ability to move around and it uses a video camera, the video signal is processed and translated to some symbolic representation. As the agent explores the world it maintains a data structure of what it has explored. The internal structure of deductive reasoning agents is illustrated in Figure 1.11. There are two key problems encountered when trying to build deductive reasoning agents. Firstly, the transduction problem is the problem of translating the real world into an accurate, symbolic description in time for it to be useful. Secondly, the representation or reasoning problem is the problem of representing acquired information symbolically and getting agents to manipulate/reason with it [Woo00].

2. Production systems are also an extension of expert systems. However they place more emphasis how decisions are made based on the state of the

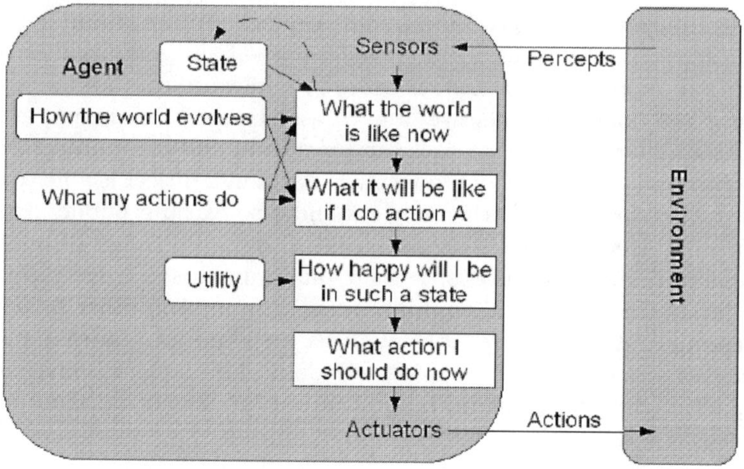

Fig. 1.11. A concept of deductive reasoning agents (modified from [RN03]).

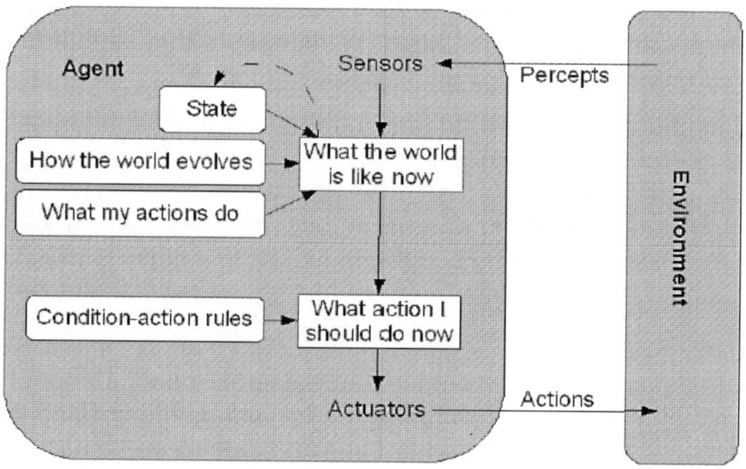

Fig. 1.12. A concept of production–rule agents (modified from [RN03]).

KB. The general structure of production system agents is illustrated in Figure 1.12. The KB is called working memory and is aimed to resemble short term memory. They also allow a designer to create a large set of condition-action rules called productions that resemble long term memory. When a production is executed it is able cause changes to the environment or directly change the working memory. This in turn possibly activates other productions. Production systems typically contain a small working memory, and a large number of rules that can be executed so fast that production systems are able to operate in real time with thousands of rules [RN03]. An example of a production–rule agent development environment is called SOAR (State,

Operator And Result). SOAR uses a KB as a problem space and production rules to look for solutions in a problem. IT has a powerful problem solving mechanism whereby every time that it is faced with more than one choice of productions (via a lack of knowledge about what is the best way to proceed) it creates an impasse that results in branching of the paths that it takes through the problem space. The impasse asserts subgoals that force the creation of sub–states of problem solving behavior with the aim to resolve the super–state impasse [Sio05].

Reactive Agents

Deliberate agents were originally developed using traditional software engineering techniques. Such techniques define pre–conditions required for operation and post–conditions that define the required output after operation. Some agents however, cannot be easily developed using this method because they maintain a constant interaction with a dynamic environment, hence they are called reactive agents. Reactive agents as especially suited for real–time applications where there are strict time constraints (i.e., milliseconds) on choosing actions.

Reactive systems are studied by behavioral means where researchers have tried to use entirely new approaches that reject any symbolic representation and decision making. Instead, they argue that intelligent and rational behavior emerges from the interaction of various simpler behaviors and is directly linked to the environment that the agent occupies [Woo00]. The general structure of reactive agents is illustrated in Figure 1.13 The main contributor of reactive agent research is Rod Brooks from MIT, with his *subsumption architecture*, where decision making is realized through a set of task–accomplishing

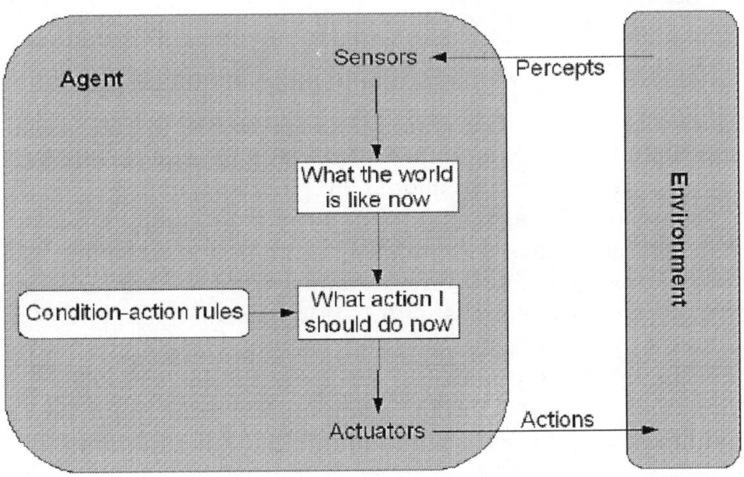

Fig. 1.13. A concept of reactive agents (modified from [RN03]).

behaviors. Behaviors are arranged into layers where lower layers have a higher priority and are able to inhibit higher layers that represent more abstract behaviors [Bro86]. A simple example of the subsumption architecture is a multi–agent system used to collect a specific type of rock scattered in a particular area on a distant planet. Agents are able to move around, collect rocks and return to the mother–ship. Due to obstacles on the surface of the planet, agents are not able to communicate directly, however they can carry special radioactive crumb that they drop on the ground for other agents to detect. The crumbs are used to leave a trail for other agents to follow. Additionally, a powerful locater signal is transmitted from the mother–ship, agents can find the ship by moving towards a stronger signal. A possible behavior architecture for this scenario are the following set of *heuristic IF–THEN rules*:

1. IF detect an obstacle THEN change direction (this rule ensures that the agent avoids obstacles when moving);
2. IF carrying samples and at the base THEN drop samples (this rule allows agent to drop samples in the mother–ship);
3. IF carrying samples and not at the base THEN drop 2 crumbs and travel up signal strength (this rule either reinforces a previous trail or creates a new one);
4. IF detect a sample THEN pick sample up (this rule collects samples);
5. IF sense crumbs THEN pick up 1 crumb and travel away from signal strength (this rule follows a crumb trail that should end at a mineral deposit; crumbs are picked up to weaken the trail such that it disappears when the mineral deposit has depleted);
6. IF true THEN move randomly (this rule explores the area until it stumbles upon a mineral deposit or a crumb trail).

Hybrid Agents

Hybrid agents are capable of expressing both reactive and pro-active behavior. They do this by breaking reactive and proactive behavior into different subsystems called layers. The lowest layer is the reactive layer and it provides immediate responses to changes for the environment, similarly to the subsumption architecture. The middle layer is the planning layer that is responsible for telling the agent what to do by reviewing internal plans, and selecting a particular plan that would be suitable for achieving a goal. The highest layer is the modelling layer that manages goals. A major issue encountered when developing solutions with hybrid reasoning agents is that agents must be able to balance the time spent between thinking and acting. This includes being able to stop planning at some point and commit to goal, even if that goal is not optimal [Woo00]. The general structure of hybrid agents is illustrated in Figure 1.14.

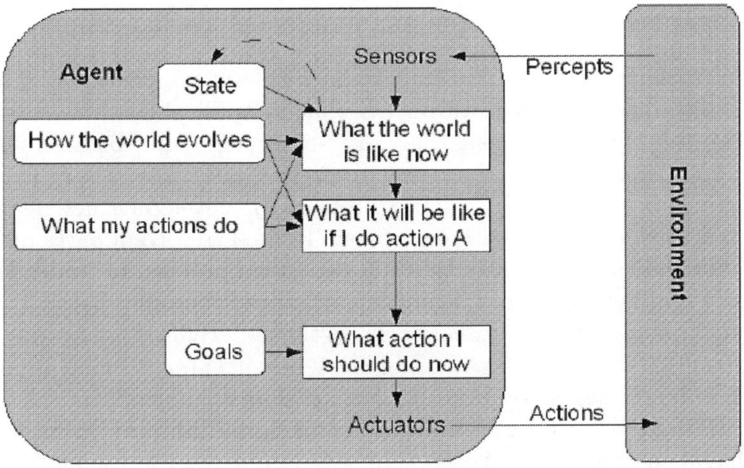

Fig. 1.14. A concept of hybrid, goal–directed agents (modified from [RN03]).

Agent–Oriented Software Development

Agent–oriented development is concerned with the techniques of software development that are specifically suited for developing agent systems. This is an important issue because existing software development techniques are unsuitable for agents as there exists a fundamental mismatch between traditional software engineering concepts and agents. Consequently, traditional techniques fail to adequately capture an agent's autonomous problem–solving behavior as well as the complex issues involved in multi–agent interaction [Sio05].

The first agent–oriented methodology was proposed by Wooldridge and is called Gaia. Gaia is deemed appropriate for agent systems with the following characteristics: (i) Agents are smart enough to require significant computational resources. (ii) Agents may be implemented using different programming languages, architectures or techniques. (iii) The system has a static organization structure such that inter–agent relationships do not change during operation. (iv) The abilities of agents and the services they provide do not change during operation. (v) The system requires only small amount of agents. Gaia splits the development process into three phases: Requirements, Analysis and Design. The requirements phase is treated in the same way as traditional systems. The analysis phase is concerned with the roles that agents play in the system as well as the interactions required between agents. The design phase is concerned with the agent types that will make up the system. The agent main services that are required to realize the agent's roles, and finally, the lines of communication between the different agents. The Gaia methodology was the inspiration for the more detailed methodology described in the next section (see [Woo00]).

Agents Environments

Agent technology has been applied to many different application areas, each focusing on a specific aspect of agents that is applicable to the domain at hand. The role that BDI–agents play in their environment distinctly depends on the application domain. The agent research community is very active and environments are mostly viewed as test–beds for developing new features in agents and showing how they are successfully used to solve a particular problem. Fortunately, in most cases this is a two–sided process, by understanding, developing and improving new agent technologies it becomes possible to solve similar real life problems. Consequently, as the underlying foundation of agent software matures, new publications describe how agents are being applied successfully in increasingly complex application domains [Sio05].

The BDI–agent is usually understood to be a *decision–maker* and anything that it interacts with, comprising everything outside the agent itself, is referred to as the *environment*. The environment has a number of features and generates *sensations* that contain some information about the features. A *situation* is commonly understood as a complete snapshot of the environment for a particular instance in time.[147] Hence, if an agent is able to get or deduce the situation of its environment it would know everything about the environment at that time. A *state* is here defined as a snapshot of the agent's beliefs corresponding to its limited understanding of the environment. This means that the state may or may not be a complete or accurate representation of the situation. This distinction supports research being conducted on improving the agent's *situation awareness* (SA), whereby SA measures how similar the state is as opposed to the situation.

The agent and the environment interact continually, the agent selects actions and the environment responds to the actions by presenting new sensations to the agent [SB98]. The interaction is normally segmented in a sequence of discrete time steps, whereby, at a particular time step the agent receives data from the environment and on that basis selects an action. In the next time step, the agent finds itself in a new state (see Figure 1.9).

Various properties of environments have been classified into six categories [RN03]:

1. *Fully observable or partially observable.* A fully observable environment provides the agent with complete, accurate and up–to–date information of the entire situation. However, as the complexity of environments increases, they become less and less observable. The physical world is considered a partially observable environment because it is not possible to know everything that happens in it [Woo00]. On the other hand, depending on the

[147] In a number of references, the term state is used with the same meaning. In this section a clear distinction is made between the two terms, a situation is defined as a complete snapshot of the real environment.

application, the environment should not be expected to be completely observable (e.g., if an agent is playing a card game it should not be expected to know the cards of every other player). Hence, in this case, even though there is hidden information in the environment and this information would be useful if the agent knew it, is not necessary for making rational decisions [SB98]. An extension of this property is when sensations received from the environment are able to summarize past sensations in a compact way such that all relevant information from the situation can be deduced. This requires that the agent maintains a history of all past sensations. When sensations succeeds in retaining all relevant information, they are said to have the Markov property. An example of a Markov sensation for a game of checkers is the current configuration of the pieces on the board, this is because it summarizes the complete sequence of sensations that led to it. Even though much of the information about the sequence is lost, all important information about the future of the game is retained. A difficulty encountered when dealing with partially observable environments is when the agent is fooled to perceiving two or more different situations as the same state, this problem is known as perceptual aliasing. If the same action is required for the different situations then aliasing is a desirable effect, and can be considered a core part of the agent's design, this technique is commonly called state generalization [SB98].

2. *Deterministic or stochastic.* Deterministic is the property when actions in the environment have a single guaranteed effect. In other words, if the same action is performed from the same situation, the result is always the same. A useful consequence of a deterministic environment is the ability to predict what will happen before an action is taken, giving rise to the possibility of evaluating multiple actions depending on their predicted effects. The physical world is classified as a stochastic environment as stated by [Woo00]. However, if an environment is partially observable it may appear to be stochastic because not all changes are observed and understood [RN03], if more detailed observations are made, including additional information, the environment becomes increasingly deterministic.
3. *Episodic or sequential.* Within an episodic environment, the situations generated are dependent on a number of distinct episodes, and there is no direct association between situations of different episodes. Episodic environments are simpler for agent development because the reasoning of the agent is based only on the current episode, there is no reason to consider future episodes [Woo00]. An important assumption made when designing agents for episodic environments, is that all episodes eventually terminate no matter what actions are selected [SB98]. This is particularly true when using learning techniques that only operate on the completion of an episode through using a captured history of situations that occurred within the episode. Actions made in sequential environments, on the other hand, affect all future decisions. Chess is an example of a sequential environment because short–term actions have long–term consequences.

4. *Static or dynamic.* A static environment is one that remains unchanged unless the agent explicitly causes changes through actions taken. A dynamic environment is one that contains other entities that cause changes in ways beyond the agents control. The physical world continuously changes with external means and is therefore considered a highly dynamic environment [Woo00]. An example of a static environment, is an agent finding its way though a 2D maze. In this case all changes are caused by the same agent. An advantage of static environments is that the agent does not need to continuously observe the environment while its deciding the next action. It can take as much time as it needs to make a decision and the environment will be the same as when previously observed [RN03].
5. *Discrete or continuous.* An environment is discrete if there is a fixed, finite number of actions and situations in it [Woo00]. Simulations and computer games are examples of discrete environments because they involve capturing actions performed by entities, processing the changes caused by the actions and providing an updated situation. Sometimes however, this process is so quick that the simulation appears to be running continuously. An example of a continuous environment is taxi driving, because the speed and location of the taxi and other cars changes smoothly over time [RN03].
6. *Single–agent or multi–agent.* Although the distinction between single and multi–agent environments may seem trivial, recent research has surfaced some interesting issues. These arise from the question of what in the environment may be viewed as another agent [RN03]. For example, does a taxi driver agent need to treat another car as an agent? What about a traffic light or a road sign? An extension to this question is when humans are included as part of the design of the system, giving rise to the new research area called human–agent teaming [Sio05].

Agents' Reasoning and Learning

The environments described above illustrate the need for *adaptation* when agent systems are required to interact with complex environments. Here we sill review how agents and humans are understood to perform reasoning and learning when they are faced with a particular environment.

Reasoning is understood as the thinking process that occurs within an agent that needs to make a particular decision. This topic has been tackled via two parallel directions with two different schools of thought. The first school of thought focuses on how agents can perform rational reasoning where the decisions made are a direct reflection of knowledge. The advantage of this approach is that decisions made by an agent can be understood simply by looking within its internal data structures, as the agent only makes decisions based on what it knows. This process includes maintaining the agent's knowledge base such that it contains accurate information about its environment, by performing operations in order to keep all knowledge consistent. Decisions are made through a collection of rules applied on the knowledge base that define what should occur as knowledge changes [Sio05].

Another school of thought is concerned with the way that humans perform reasoning and apply any concepts developed to agent technology. Humans are known to perform practical reasoning every day, their decisions are based on their desires and their understanding in regards to how to go about achieving them. The process that takes place between observing the world, considering desires and taking actions can be broken up into four main stages, each of which consists of a number of smaller components. Through learning, it also becomes possible to create agents that are able to change the way that they were originally programmed to behave. This can be advantageous when an agent is faced with a situation that it does not know how to proceed. Furthermore, it is useful when an agent is required to improve its performance with experience.

Reasoning and Behavior

Research on artificial reasoning and behavior has been tackled from different angles that can be categorized along two main dimensions (see Figure 1.15). The vertical dimension illustrates the opposing nature of reasoning and behavior that correspond to thinking versus acting respectively. This is an important feature concept in every application using AI techniques. Great emphasis is given to the balance between processing time for making better decisions, and the required speed of operation. Approaches falling to the left side are based on how humans reason and behave while approaches falling on the right side are concerned with building systems that are rational, meaning that they are required to think and act as best they can, given their limited knowledge [RN03].

Rational Reasoning

1. Representation and search. Recall that the way that information is represented and used for intelligent problem solving forms a number of important but difficult challenges that lie within the core of AI research. Knowledge representation is concerned with the principles of correct reasoning. This involves two parallel topics of research. One side is concerned with the development of formal representation languages with the ability to maintain consistent

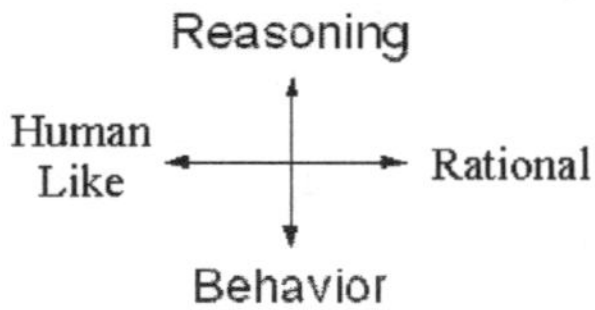

Fig. 1.15. Reasoning dimensions (modified from [RN03]).

knowledge about the world, the other side is concerned with the development of reasoning processes that bring the knowledge to life. The output of both of these areas results in a Knowledge Base (KB) system. KBs try to create a model of the real world via the collection of a number of sentences. An agent is normally able to add new sentences to the knowledge base as well as query the KB for information. Both of these tasks may require the KB to perform inference on its knowledge, where an inference is defined as the process of deriving new sentences from known information. An additional requirement of KBs is that when an agent queries the KB, the answer should be inferred from information previously added to the KB and not from unknown facts. The most important part of a KB is the logic in which the its sentences are represented. This is because all sentences in a KB are in fact expressed according to the syntax and semantics of the logic's representation language. The syntax of the logic is required for implementing well formed sentences while the semantics define the truth of each sentence with respect to a model of the environment being represented [RN03].

Problem solving using KBs involves the use of search algorithms that are able to search for solutions between different states of information within the KB. Searching involves starting from an initial state and expanding across different successor state possibilities until a solution is found. When a search algorithm is faced with a choice of possibilities to consider, each possibility is thoroughly searched before moving to the next possibility. Search however has a number of issues, including [Lug02]:
(i) Guarantee of a solution being available; (ii) Termination of the search algorithm; (iii) The optimality of a particular solution found; and (iv) The complexity of the search algorithm with respect to the time and memory usage.

State space analysis is done with the use of graphs. A graph is a set of nodes with arcs that connect them, each node can have a label to distinguish it from another node and arcs can have directions to indicate the direction of movement between the nodes. A path in the graph connects a sequence of nodes with arcs and the root is a node that has a path to all other nodes in the graph.

There are two ways to search a state space, the first way is to use data-driven search by which the search starts by a given set of facts and rules for changing states. The search proceeds until it generates a path that leads to the goal condition. Data driven search is more appropriate for problems in which the initial problem state is well defined, or there are a large number of potential goals and only a few facts to start with, or the goal state is unclear [Lug02].

The second way is to use goal-driven search by which the search starts by taking the goal state and determining what conditions must be true to move into the goal state. These conditions are then treated as subgoals to be searched. The search then continues backwards through the subgoals until it reaches the initial facts of the problem. Goal driven search is more appropriate for problems in which the goal state is well defined, or there are a large number of initial facts making it impractical to prefer data driven search, or the initial data is not given and must be acquired by the system [Lug02].

The choice of which of the options to expand first is defined by the algorithm's search strategy. Two well known search strategies are: Breadth-first, where all successors of a given depth are expanded first before any nodes at the next level. Depth-first search involves expanding the deepest node for a particular option before moving to the next option. There are also strategies that include both elements, for example defining a depth limit for searching in a tree. It is also possible to use heuristics to help with choosing branches that are more likely to lead to an acceptable solution. Heuristics are usually applied when a problem does not have an exact solution or the computational cost to find an exact solution is too big. They reduce the state space by following the more promising paths through the state space [RN03].

An additional layer of complexity in knowledge representation and search is due to the fact that agents almost never have access a truly observable environment. Which means that agents are required to act under *uncertainty*. There are two techniques that have been used for reasoning in uncertain situations. The first involves the use of probability theory in assigning a value that represents a degree of belief in facts in the KB. The second method involves the use of *fuzzy sets* (see below) for representing how well a particular object satisfies a vague description [RN03].

2. Expert systems. Recall that knowledge–based reasoning systems are commonly called *expert systems* because they work by accumulating knowledge extracted from different sources, and use different strategies on the knowledge in order to solve problems. Simply put, expert systems try to replicate what a human expert would do if faced with the same problem. They can be classified into different categories depending on the type of problem they are used to solve [Lug02]:

- *interpretation*: making conclusions or descriptions from collections of raw data;
- *prediction/forecasting*: predicting the consequences of given situations;
- *diagnosis*: finding the cause of malfunctions based on the symptoms observed;
- *design*: finding a configuration of components that best meets performance goals when considering several design constraints;
- *planning*: finding a sequence of actions to achieve some given goals using specific starting conditions and run–time constraints;
- *monitoring*: observing a system's behavior and comparing it to its expected behavior at run–time;
- *debugging*: finding problems and repairing caused malfunctions; and
- *control*: controlling how a complex system behaves.

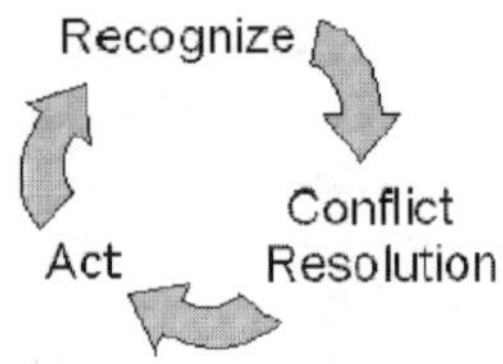

Fig. 1.16. A recognize–act operation cycle of production systems (modified from [Lug02]).

A common way to represent data in a expert systems is using first–order predicate calculus formulae. For example, the sentence 'If a bird is a crow then it is black' is represented as:

$$\forall X(crow(X) \Longrightarrow black(X)).$$

3. Production systems. They are based on a model of computation that uses search algorithms and models human problem solving. Production systems consist of production rules and a working memory. Production rules are pre-defined rules that describe a single segment of problem–solving knowledge. They are represented by a condition that determines when the production is applicable to be executed, and an action which defines what to do when executed. The working memory is an integrated KB that contains an ever–changing state of the world.

The operation of production systems generally follows a *recognize–act cycle* (see Figure 1.16). Working memory is initialized with data from the initial problem description and is subsequently updated with new information. At every step of operation, the state presented by the working memory is continuously captured as patterns and applied to conditions of productions. If a pattern is recognized against a condition, the associated production is added to a conflict set. A conflict resolution operation chooses between all enabled productions and the chosen production is fired by executing its associated action. The actions executed can have two effects. Firstly, they can cause changes to the agent's environment which indirectly changes the working memory. Secondly, they can explicitly cause changes in the working memory. The cycle then restarts using the modified working memory until a situation when no subsequent productions are enabled. Some production systems also contain the means to do backtracking when there are no further enabled productions but the goal of the system has still not been reached. Backtracking allows the system to work backwards and try some different options in order to achieve its goal [Lug02].

Human Reasoning

The so–called *practical reasoning* is concerned with studying the way that humans reason about what to do in everyday activities and applying this to

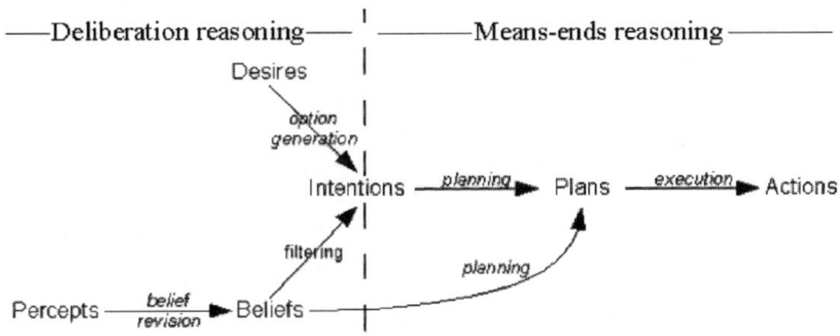

Fig. 1.17. BDI–reasoning process (modified from [Woo00]).

the design of intelligent agents. Practical reasoning is specifically geared to reasoning towards actions, it involves weighing conflicting considerations of different options that are available depending on what a person desires to do. Practical reasoning can be divided into two distinct activities (see Figure 1.17). The first activity is called *deliberation reasoning*, it involves deciding on what state to achieve. The second activity is called *means–ends reasoning* and it involves deciding on how to achieve this state of affairs [Woo00]. Recall that the central component of practical reasoning is the concept of *intention* because it is used to characterize both the action and thinking process of a person. For example 'intending to do something' characterizes a persons thinking while 'intentionally doing something' characterizes the action being taken.

The precursors of an intention are a persons's desires and beliefs and hence all of the beliefs, desires and intentions must be consistent. In other words, intending to do something must be associated with a relevant desire, as well as the belief that the intended action will help to achieve the desire. Maintaining this consistency is challenging due to the dynamic nature of desires and beliefs. Desires are always changing according to internal self-needs while beliefs are constantly updated using information obtained from senses through a process called belief revision, from the external environment.

Forming an intention involves performing two concurrent operations. Firstly, option generation uses the current desires to generate a set of possible alternatives. Secondly, filtering chooses between these alternatives based on the current intentions and beliefs. An intention also requires assigning a degree of commitment toward performing a particular action or set of actions in the future. There are four important characteristics emerging by this commitment are [Woo00]:

1. Intentions drive means-ends reasoning by forcing the agent to decide on how to achieve them.
2. Intentions persist by forcing a continuous strive to achieve them. Hence, after a particular action has failed, other alternative actions are attempted until it comes to be believed that it is not possible to achieve the intention, or the relevant desire is not longer present.

3. Intentions constrain future deliberation because it is not necessary to consider desires that are inconsistent with the current intentions.
4. Intentions influence beliefs by introducing future expectations. This is due the requirement of believing that a desired state is possible before and during execution the intention to satisfy it.

The process that occurs after forming an intention in order to take action is identified as planning, it involves selecting and advancing through a sequence of plans that dictate what actions to take. Plans are understood to consist of pre-condition that characterizes the state in which a plan is applicable for execution and a post-condition characterizes the resulting state after executing the plan. Finally, a body containing the recipe defining the actions to take [Woo00]. From the theory of practical reasoning, researchers have been able to develop intuitive agent development architectures. The transition between the theory and implementation has required the identification of equivalent software constructs for each of the BDI–components [Sio05].

Cognitive systems engineering takes into account, during the design and implementation of systems, that systems will be used by humans. It acknowledges that humans are dynamic entities that are part of the system itself but cannot be modelled as static components of a system. When humans use a system they adapt to the functional characteristics of the system. In addition, sometimes they can modify the system's functional characteristics in order to suit their own needs and preferences. This means that in order to understand the behavior of the system once the adaptation has happened is to abstract the structural elements into a purely functional level and identify and separate the functional relationships. This concept can best be understood using a simple example from [RPG94]:

> "When a novice is driving a car, it is based on an instruction manual identifying the controls of the car and explaining the use of instrument readings, that is, when to shift gears, what distance to maintain to the car ahead (depending on the speed), and how to use the steering wheel. In this way, the function of the car is controlled by discrete rules related to separate observations, and navigation depends on continuous observation of the heading error and correction by steering wheel movements. This aggregation of car characteristics and instructed input–output behavior makes it possible to drive; it initiates the novice by synchronizing them to the car functions. However, when driving skill evolves, the picture changes radically. Behavior changes from a sequence of separate acts to a complex, continuous behavioral pattern. Variables are no longer observed individually. Complex patterns of movements are synchronized with situational patterns and navigation depends on the perception of a field of safe driving. The drivers are perceiving the environment in terms of their driving goals. At this stage, the behavior of the system cannot be decomposed into

structural elements. A description must be based on abstraction into functional relationships."

A new design approach is introduced that shifts away from the traditional software engineering perspective to a functional perspective. There are two different ways to define functional characteristics. Firstly, relational representations are based on mathematical equations that relate physical, measurable environments. Secondly, casual representations are connections between different events. [RPG94] presented a framework that made it possible to relate conceptual characteristics. The framework takes into account that in order to bridge system behaviors into human profiles and preferences, several different perspectives of analysis and languages of representation are needed (see Figure 1.18).

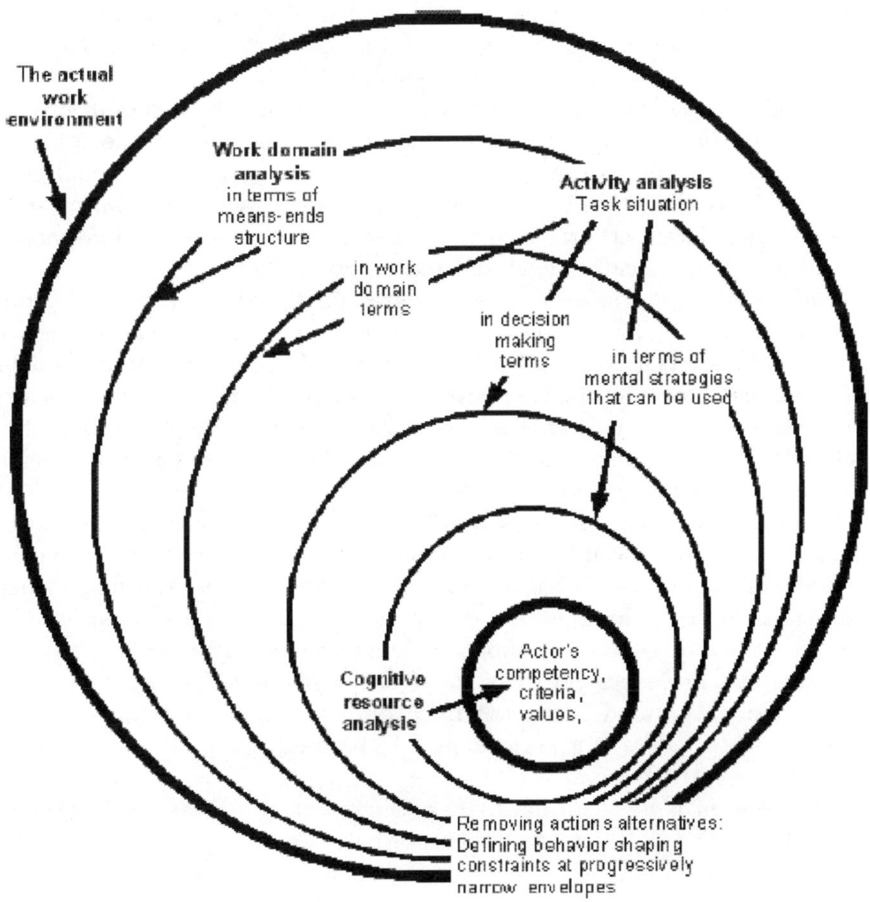

Fig. 1.18. Relating Work Environment to Cognitive Resource Profiles of Actors (adapted from [RPG94]).

In this framework, the *work domain analysis* is used to make explicit the goals, constraints and resources found in a work system. They are represented by a general inventory of system elements that are categorized by functional elements and their means-ends relations. The analysis identifies the structure and general content of the global knowledge of the work system. Activity analysis is divided into three different dimensions. Firstly, activity analysis in domain terms focuses on the freedom left for activities after the constraints posed by time and the functional space of the task. Generalizations are made in terms of objectives, functions and resources. Secondly, activity analysis in decision terms use functional languages to identify decision making functions within relevant tasks. This results of this analysis are used to identify prototype knowledge states that connect different decision functions together. Thirdly, mental strategies are used to compare task requirements with cognitive resource profiles of the individual actors and how they perform their work, thus supplies the designer with mental models, data formats and rule sets that can be incorporated into the interface of the system and used by actors of varying expertise and competence.

The *work organization analysis* is used to identify the actors involved in the decisions of different situations. This is done by finding the principles and criteria that govern the allocation of roles among the groups and group members. This allocation is dynamically dependent on circumstances and is governed by different criteria such as actor competency, access to information, minimizing communication load and sharing workload.

The *social organization analysis* focuses on the social aspect of groups working together. This is useful for understanding communication between team members, such communication may include complex information like intentions used for coordinating activities and resolving ambiguities or misinterpretations. Finally, User Analysis is used to help judge which strategy is likely to be chosen by an actor in a given situation focusing on the expertise and the performance criteria of each actor.

Rasmussen further proposes a framework for representing the various states of knowledge and information processes of human reasoning, it is called the *decision ladder* (see Figure 1.19). The ladder models the human decision making process through a set of generic operations and standardized key nodes or states of knowledge about the environment. The circles illustrated are states of knowledge and the squares are operations. The decision ladder was developed as a model for performing work domain analysis, however, the structure of the ladder is generic enough to be used as a guide in the context of describing agent reasoning.

The decision ladder can be further segmented into three levels of expertise [RPG94]. The skill (lowest) level represents very fast, automated sensory–motor performance and it is illustrated in the ladder via the heuristic shortcut links in the middle. The rule (medium) level represents the use of rules and/or procedures that have been pre–defined, or derived empirically using experience, or communicated by others, it traverses the bottom half of the ladder.

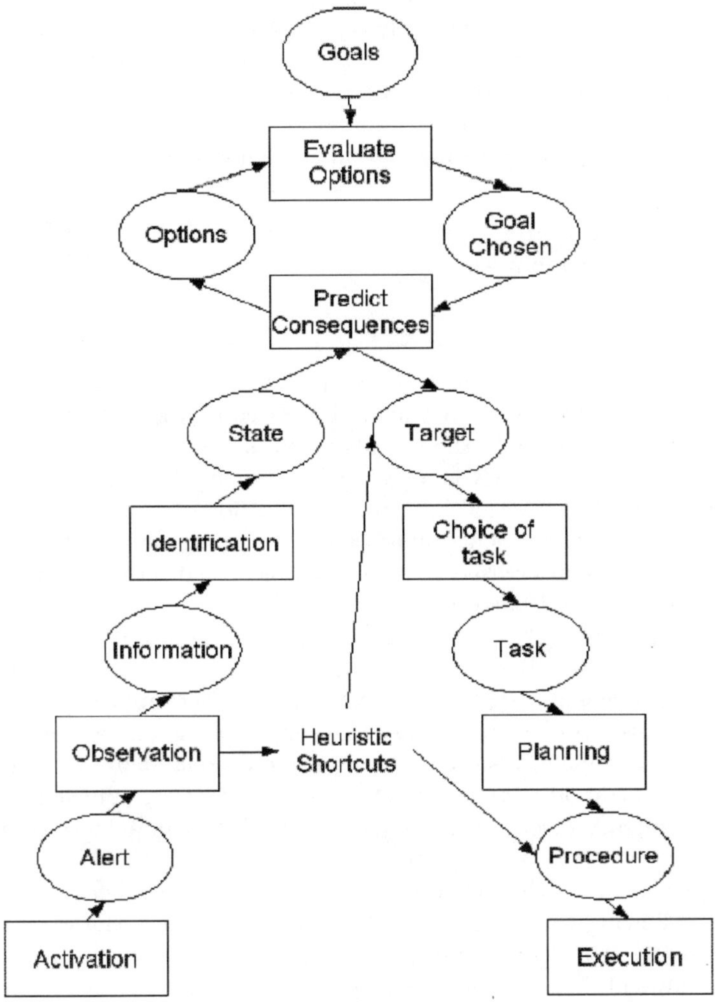

Fig. 1.19. Rassmusen's *decision ladder* (adapted from [RPG94]).

Finally, the knowledge (highest) level represents behaviors during less–familiar situations when someone is faced with an environment where there are no rules or skills available, in such cases a more detailed analysis of the environment is required with respect to the goals the agent is trying to achieve, the entire ladder is used for this case.

1.2.2 Computational Intelligence

Computational intelligence (CI) is a modern, more specifically defined AI branch. CI research aims to use learning, adaptive, or evolutionary computation to create programs that are, in some sense, intelligent. Computational intelligence research either explicitly rejects statistical methods (as is the case with fuzzy systems), or tacitly ignores statistics (as is the case with most neural network research). In contrast, machine learning research rejects non–statistical approaches to learning, adaptivity, and optimization. Main subjects in CI, as defined by IEEE Computational Intelligence Society, are:

1. Neural networks,
2. Fuzzy systems, and
3. Evolutionary computation.

Neural Networks

Recall that an *artificial neural network* (ANN) is an interconnected group of artificial neurons that uses a mathematical or computational model for information processing based on the so–called *connectionist approach* to computation. In most cases an ANN is an *adaptive system* that changes its structure based on external or internal information that flows through the network.

In more practical terms neural networks are nonlinear statistical data modelling tools. They can be used to model complex relationships between inputs and outputs or to find patterns in data.

Dynamically, the ANNs are *nonlinear dynamical systems* that act as *functional approximators* [Kos92]. The ANN builds *discriminant functions* from its processing elements (PE)s. The ANN topology determines the *number* and *shape* of the discriminant functions. The shapes of the discriminant functions change with the topology, so ANNs are considered *semi–parametric classifiers*. One of the central advantages of ANNs is that they are sufficiently powerful to create arbitrary discriminant functions so ANNs can achieve optimal classification.

The placement of the discriminant functions is controlled by the network weights. Following the ideas of non–parametric training, the weights are adjusted directly from the training data without any assumptions about the data's statistical distribution. Hence one of the central issues in neural network design is to utilize systematic procedures, the so–called *training algorithm*, to modify the weights so that as accurate a classification as possible is achieved. The accuracy is quantified by an error criterion [PEL00].

The training is usually performed in the following way. First, data is presented, and an output is computed. An error is obtained by comparing the output $\{y\}$ with a desired response $\{d\}$ and it is used to modify the weights with a training algorithm. This procedure is repeated using all the data in the training set until a convergence criterion is met. Thus, in ANNs (and in adaptive systems in general) the designer does not have to specify the *parameters* of the system. They are automatically extracted from the input data and the desired response by means of the training algorithm. The two central issues in neural network design (semi–parametric classifiers) are the selection of the shape and number of the discriminant functions and their placement in pattern space such that the classification error is minimized [PEL00].

Biological Versus Artificial Neural Nets

In biological neural networks, signals are transmitted between neurons by electrical pulses (action potentials or spike trains) travelling along the axon. These pulses impinge on the afferent neuron at terminals called synapses. These are found principally on a set of branching processes emerging from the cell body (soma) known as dendrites. Each pulse occurring at a synapse initiates the release of a small amount of chemical substance or neurotransmitter which travels across the synaptic cleft and which is then received at postsynaptic receptor sites on the dendritic side of the synapse. The neurotransmitter becomes bound to molecular sites here which, in turn, initiates a change in the dendritic membrane potential. This postsynaptic potential (PSP) change may serve to increase (hyperpolarize) or decrease (depolarize) the polarization of the postsynaptic membrane. In the former case, the PSP tends to inhibit generation of pulses in the afferent neuron, while in the latter, it tends to excite the generation of pulses. The size and type of PSP produced will depend on factors such as the geometry of the synapse and the type of neurotransmitter. Each PSP will travel along its dendrite and spread over the soma, eventually reaching the base of the axon (axonhillock). The afferent neuron sums or integrates the effects of thousands of such PSPs over its dendritic tree and over time. If the integrated potential at the axonhillock exceeds a threshold, the cell fires and generates an action potential or spike which starts to travel along its axon. This then initiates the whole sequence of events again in neurons contained in the efferent pathway.

ANNs are very loosely based on these ideas. In the most general terms, a ANN consists of large numbers of simple processors linked by weighted connections. By analogy, the processing nodes may be called artificial neurons. Each node output depends only on information that is locally available at the node, either stored internally or arriving via the weighted connections. Each unit receives inputs from many other nodes and transmits its output to yet other nodes. By itself, a single processing element is not very powerful; it generates a scalar output, a single numerical value, which is a simple nonlinear function of its inputs. The power of the system emerges from the combination of many units in an appropriate way [FS92].

ANN is specialized to implement different functions by varying the connection topology and the values of the connecting weights. Complex functions can be implemented by connecting units together with appropriate weights. In fact, it has been shown that a sufficiently large network with an appropriate structure and property chosen weights can approximate with arbitrary accuracy any function satisfying certain broad constraints. In ANNs, the design motivation is what distinguishes them from other mathematical techniques: an ANN is a processing device, either an algorithm, or actual hardware, whose design was motivated by the design and functioning of animal brains and components thereof.

There are many different types of ANNs, each of which has different strengths particular to their applications. The abilities of different networks can be related to their structure, dynamics and learning methods.

Multilayer Perceptrons

The most common ANN model is the *feedforward neural network* with one input layer, one output layer, and one or more hidden layers, called *multilayer perceptron* (MLP, see Figure 1.20). This type of neural network is known as a *supervised network* because it requires a desired output in order to learn. The goal of this type of network is to *create a model* $f : x \to y$ that correctly maps the input x to the output y using historical data so that the model can then be used to produce the output when the desired output is unknown [Kos92].

In MLP the inputs are fed into the input layer and get multiplied by interconnection weights as they are passed from the input layer to the first hidden layer. Within the first hidden layer, they get summed then processed by a nonlinear function (usually the hyperbolic tangent). As the processed data leaves the first hidden layer, again it gets multiplied by interconnection weights, then summed and processed by the second hidden layer. Finally the data is multiplied by interconnection weights then processed one last time within the output layer to produce the neural network output.

MLPs are typically trained with *static backpropagation*. These networks have found their way into countless applications requiring static pattern classification. Their main advantage is that they are easy to use, and that they can approximate any input/output map. The key disadvantages are that they train slowly, and require lots of training data (typically three times more training samples than the number of network weights).

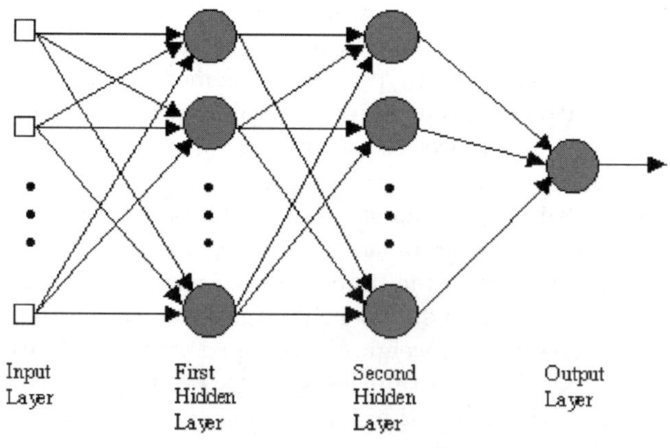

Fig. 1.20. Multilayer perceptron (MLP) with two hidden layers.

McCulloch–Pitts Processing Element

MLPs are typically composed of *McCulloch–Pitts neurons* (see [MP43]). This processing element (PE) is simply a sum–of–products followed by a threshold nonlinearity. Its input–output equation is

$$y = f(\text{net}) = f\left(w_i x^i + b\right), \qquad (i = 1, \dots, D),$$

where D is the number of inputs, x^i are the inputs to the PE, w_i are the weights and b is a bias term (see e.g., [MP69]). The activation function is a *hard threshold* defined by *signum* function,

$$f(\text{net}) = \begin{cases} 1, & \text{for} \quad \text{net} \geq 0, \\ -1, & \text{for} \quad \text{net} < 0. \end{cases}$$

Therefore, McCulloch–Pitts PE is composed of an adaptive linear element (*Adaline*, the weighted sum of inputs), followed by a signum nonlinearity [PEL00].

Sigmoidal Nonlinearities

Besides the hard threshold defined by signum function, other nonlinearities can be utilized in conjunction with the McCulloch–Pitts PE. Let us now smooth out the threshold, yielding a sigmoid shape for the nonlinearity. The most common nonlinearities are the *logistic* and the *hyperbolic tangent threshold activation functions*,

$$\begin{aligned} \text{hyperbolic}: \qquad & f(\text{net}) = \tanh(\alpha\,\text{net}), \\ \text{logistic}: \qquad & f(\text{net}) = \frac{1}{1 + \exp(-\alpha\,\text{net})}, \end{aligned}$$

where α is a *slope parameter* and normally is set to 1. The major difference between the two sigmoidal nonlinearities is the range of their output values. The logistic function produces values in the interval $[0, 1]$, while the hyperbolic tangent produces values in the interval $[-1, 1]$. An alternate interpretation of this PE substitution is to think that the discriminant function has been generalized to

$$g(x) = f(w_i x^i + b), \qquad (i = 1, \dots, D),$$

which is sometimes called a *ridge* function. The combination of the synapse and the tanh axon (or the sigmoid axon) is usually referred to as the modified McCulloch–Pitts PE, because they all respond to the full input space in basically the same functional form (a sum of products followed by a global nonlinearity). The output of the logistic function varies from 0 to 1. Under some conditions, the logistic function allows a very powerful interpretation of the output of the PE as a'posteriori probabilities for Gaussian–distributed input classes. The tanh is closely related to the logistic function by a linear transformation in the input and output spaces, so neural networks that use either of these can be made equivalent by changing weights and biases [PEL00].

Gradient Descent on the Net's Performance Surface

The *search* for the weights to meet a *desired response* or internal constraint is the essence of any *connectionist* computation. The central problem to be solved on the road to machine–based classifiers is how to automate the process of *minimizing the error* so that the machine can independently make these weight changes, without need for hidden agents, or external observers. The optimality criterion to be minimized is usually the *mean square error* (MSE)

$$J = \frac{1}{2N}\sum_{i=1}^{N}\varepsilon_i^2,$$

where ε_i is the instantaneous error that is added to the output y_i (the linearly fitted value), and N is the number of observations. The function $J(w)$ is called the *performance surface* (the total error surface plotted in the space of weights w).

The search for the minimum of a function can be done efficiently using a broad class of methods based on *gradient information*. The gradient has two main advantages for the search:

1. It can be computed locally, and
2. It always points in the direction of maximum change.

The *gradient of the performance surface*, $\nabla J = \nabla_w J$, is a vector (with the dimension of w) that always points toward the direction of maximum $J-$change and with a magnitude equal to the slope of the tangent of the performance surface. The minimum value of the error J_{min} depends on both the input signal x^i and the desired signal d_i,

$$J_{min} = \frac{1}{2N}\left[\sum_i d_i^2 - \frac{(d_i x^i)}{\sum_i x^i}\right], \qquad (i = 1, \dots, D).$$

The location in coefficient space where the minimum w^* occurs also depends on both x^i and d_i. The performance surface shape depends only on the input signal x^i [PEL00].

Now, if the goal is to reach the minimum, the search must be in the direction opposite to the gradient. The overall method of gradient searching can be stated in the following way: Start the search with an arbitrary initial weight $w(0)$, where the iteration number is denoted by the index in parentheses. Then compute the gradient of the performance surface at $w(0)$, and modify the initial weight proportionally to the negative of the gradient at $w(0)$. This changes the operating point to $w(1)$. Then compute the gradient at the new position $w(1)$, and apply the same procedure again, that is,

$$w(n+1) = w(n) - \eta\nabla J(n),$$

where η is a small constant and $\nabla J(n)$ denotes the gradient of the performance surface at the nth iteration. The constant η is used to maintain stability in the search by ensuring that the operating point does not move too far along the performance surface. This search procedure is called the *steepest descent method*.

In the late 1960s, Widrow proposed an extremely elegant algorithm to estimate the gradient that revolutionized the application of gradient descent procedures. His idea is very simple: Use the instantaneous value as the estimator for the true quantity:

$$\nabla J(n) = \frac{\partial}{\partial w(n)} J \approx \frac{1}{2}\frac{\partial}{\partial w(n)} \left(\varepsilon^2(n)\right) = -\varepsilon(n)\, x(n),$$

i.e., instantaneous estimate of the gradient at iteration n is simply the product of the current input $x(n)$ to the weight $w(n)$ times the current error $\varepsilon(n)$. The amazing thing is that the gradient can be estimated with one multiplication per weight. This is the gradient estimate that led to the celebrated *least means square algorithm* (LMS):

$$w(n+1) = w(n) + \eta\varepsilon(n)\, x(n), \tag{1.10}$$

where the small constant η is called the *step size*, or the *learning rate*. The estimate will be noisy, however, since the algorithm uses the error from a single sample instead of summing the error for each point in the data set (e.g., the MSE is estimated by the error for the current sample).

Now, for fast convergence to the neighborhood of the minimum a large step size is desired. However, the solution with a large step size suffers from rattling. One attractive solution is to use a large learning rate in the beginning of training to move quickly toward the location of the optimal weights, but then the learning rate should be decreased to get good accuracy on the final weight values. This is called *learning rate scheduling*. This simple idea can be implemented with a variable step size controlled by

$$\eta(n+1) = \eta(n) - \beta,$$

where $\eta(0) = \eta_0$ is the initial step size, and β is a small constant [PEL00].

Perceptron and Its Learning Algorithm

Rosenblatt perceptron (see [Ros58b, MP69]) is a *pattern–recognition machine* that was invented in the 1950s for optical character recognition. The perceptron has an input layer fully connected to an output layer with multiple McCulloch–Pitts PEs,

$$y_i = f(\underset{i}{\text{net}}) = f(w_i x^i + b_i), \qquad (i = 1, \dots, D),$$

where b_i is the bias for each PE. The number of outputs y_i is normally determined by the number of classes in the data. These PEs add the individual scaled contributions and respond to the entire input space.

F. Rosenblatt proposed the following procedure to directly minimize the error by changing the weights of the McCulloch–Pitts PE: Apply an input example to the network. If the output is correct do nothing. If the response is incorrect, tweak the weights and bias until the response becomes correct. Get the next example and repeat the procedure, until all the patterns are correctly classified. This procedure is called the *perceptron learning algorithm*, which can be put into the following form:

$$w(n+1) = w(n) + \eta(d(n) - y(n))\, x(n),$$

where η is the step size, y is the network output, and d is the desired response.

Clearly, the functional form is the same as in the LMS algorithm (1.10), that is, the old weights are incrementally modified proportionally to the product of the error and the input, but there is a significant difference. We cannot say that this corresponds to gradient descent since the system has a discontinuous nonlinearity. In the perceptron learning algorithm, $y(n)$ is the output of the nonlinear system. The algorithm is directly minimizing the difference between the response of the McCulloch–Pitts PE and the desired response, instead of minimizing the difference between the Adaline output and the desired response [PEL00].

This subtle modification has tremendous impact on the performance of the system. For one thing, the McCulloch–Pitts PE learns only when its output is wrong. In fact, when $y(n) = d(n)$, the weights remain the same. The net effect is that the final values of the weights are no longer equal to the linear regression result, because the nonlinearity is brought into the weight update rule. Another way of phrasing this is to say that the weight update became much more selective, effectively gated by the system performance. Notice that the LMS update is also a function of the error to a certain degree. Larger errors have more effect on the weight update than small errors, but all patterns affect the final weights implementing a 'smooth gate'. In the perceptron the net effect is that the placement of the discriminant function is no longer controlled smoothly by all the input samples as in the Adaline, only by the ones that are important for placing the discriminant function in a way that explicitly minimizes the output error.

The Delta Learning Rule

One can show that the LMS rule is equivalent to the chain rule in the computation of the *sensitivity* of the cost function J with respect to the unknowns. Interpreting the LMS equation (1.10) with respect to the sensitivity concept, we see that the gradient measures the sensitivity. LMS is therefore updating the weights proportionally to how much they affect the performance, i.e., proportionally to their sensitivity.

The LMS concept can be extended to the McCulloch–Pitts PE, which is a nonlinear system. The main question here is how can we compute the sensitivity through a nonlinearity? [PEL00] The so–called $\delta-$*rule* represents a direct extension of the LMS rule to nonlinear systems with smooth nonlinearities. In case of the McCulloch–Pitts PE, *delta–rule* reads:

$$w_i(n+1) = w_i(n) + \eta\, \varepsilon_p(n)\; x_p^i(n)\, f'(\underset{p}{\mathrm{net}}(n)),$$

where $f'(\mathrm{net})$ is the partial derivative of the static nonlinearity, such that the *chain rule* is applied to the network topology, i.e.,

$$f'(\mathrm{net})\, x^i = \frac{\partial y}{\partial w_i} = \frac{\partial y}{\partial\, \mathrm{net}} \frac{\partial}{\partial w_i}. \tag{1.11}$$

As long as the PE nonlinearity is smooth we can compute how much a change in the weight δw_i affects the output y, or from the point of view of the sensitivity, how sensitive the output y is to a change in a particular weight δw_i. Note that we compute this output sensitivity by a product of partial derivatives through intermediate points in the topology. For the nonlinear PE there is only one intermediate point, net, but we really do not care how many of these intermediate points there are. The chain rule can be applied as many times as necessary. In practice, we have an error at the output (the difference between the desired response and the actual output), and we want to adjust all the PE weights so that the error is minimized in a statistical sense. The obvious idea is to distribute the adjustments according to the sensitivity of the output to each weight.

To modify the weight, we actually *propagate back the output error* to intermediate points in the network topology and scale it along the way as prescribed by (1.11) according to the element transfer functions:

$$\begin{aligned}
\text{forward path}:&\quad x^i \longmapsto w_i \longmapsto \mathrm{net} \longmapsto y\\
\text{backward path 1}:&\quad w_i \overset{\partial\,\mathrm{net}/\partial w}{\hookleftarrow} \mathrm{net} \overset{\partial y/\partial\,\mathrm{net}}{\hookleftarrow} y\\
\text{backward path 2}:&\quad w_i \overset{\partial y/\partial w}{\hookleftarrow} y\,.
\end{aligned}$$

This methodology is very powerful, because we do not need to know explicitly the error at intermediate places, such as net. The chain rule automatically derives the error contribution for us. This observation is going to be crucial for adapting more complicated topologies and will result in the *backpropagation* algorithm, discovered in 1988 by Werbos [Wer89].

Now, several key aspects have changed in the performance surface (which describes how the cost changes with the weights) with the introduction of the nonlinearity. The nice, parabolic performance surface of the linear least squares problem is lost. The performance depends on the topology of the network through the output error, so when nonlinear processing elements are used to solve a given problem the 'performance – weights' relationship becomes nonlinear, and there is no guarantee of a single minimum. The performance surface may have several minima. The minimum that produces the smallest error in the search space is called the *global minimum*. The others are called *local* minima. Alternatively, we say that the performance surface is *nonconvex.* This affects the search scheme because gradient descent uses local information to search the performance surface. In the immediate neighborhood, local minima are indistinguishable from the global minimum, so the gradient search algorithm may be caught in these suboptimal performance points, 'thinking' it has reached the global minimum [PEL00].

$\delta-$rule extended to perceptron reads:

$$w_{ij}(n+1) = w_{ij}(n) - \eta \frac{\partial J}{\partial w_{ij}} = w_{ij}(n) + \eta\, \delta_{ip}\, x_p^j\,,$$

which are local quantities available at the weight, that is, the activation x_p^j that reaches the weight w_{ij} from the input and the local error δ_{ip} propagated from the cost function J. This algorithm is local to the weight. Only the local error δ_i and the local activation x^j are needed to update a particular weight. This means that it is immaterial how many PEs the net has and how complex their interconnection is. The training algorithm can concentrate on each PE individually and work only with the local error and local activation [PEL00].

Backpropagation

The multilayer perceptron constructs input–output mappings that are a nested composition of nonlinearities, that is, they are of the form

$$y = f\left(\sum f\left(\sum (\cdot)\right)\right),$$

where the number of function compositions is given by the number of network layers. The resulting map is very flexible and powerful, but it is also hard to analyze [PEL00].

MLPs are usually trained by generalized $\delta-$rule, the so–called *backpropagation* (BP). The weight update using backpropagation is

$$w_{ij}(n+1) = w_{ij}(n) + \eta f'(\underset{i}{\text{net}}(n)) \left(\varepsilon^k(n)\, f'(\underset{k}{\text{net}}(n))\, w_{ki}(n)\right) y_j(n). \quad (1.12)$$

The summation in (1.12) is a sum of local errors δ_k at each network output PE, scaled by the weights connecting the output PEs to the ith PE. Thus the term in parenthesis in (1.12) effectively computes the total error reaching the ith PE from the output layer (which can be thought of as the ith PE's contribution to the output error). When we pass it through the ith PE nonlinearity, we have its local error, which can be written as

$$\delta_i(n) = f'(\underset{i}{\text{net}}(n))\, \delta^k w_{ki}(n).$$

Thus there is a unifying link in all the gradient–descent algorithms. All the weights in gradient descent learning are updated by multiplying the local error

$\delta_i(n)$ by the local activation $x^j(n)$ according to Widrow's estimation of the instantaneous gradient first shown in the LMS rule:

$$\Delta w_{ij}(n) = \eta\, \delta_i(n)\, y_j(n).$$

What differs is the calculation of the local error, depending on whether the PE is linear or nonlinear and if the weight is attached to an output PE or a hidden–layer PE [PEL00].

Momentum Learning

Momentum learning is an improvement to the straight gradient–descent search in the sense that a memory term (the past increment to the weight) is used to speed up and stabilize convergence. In *momentum learning* the equation to update the weights becomes

$$w_{ij}(n+1) = w_{ij}(n) + \eta\, \delta_i(n)\, x_j(n) + \alpha\, (w_{ij}(n) - w_{ij}(n-1))\,,$$

where α is the momentum constant, usually set between 0.5 and 0.9. This is called momentum learning due to the form of the last term, which resembles the momentum in mechanics. Note that the weights are changed proportionally to how much they were updated in the last iteration. Thus if the search is going down the hill and finds a flat region, the weights are still changed, not because of the gradient (which is practically zero in a flat spot), but because of the rate of change in the weights. Likewise, in a narrow valley, where the gradient tends to bounce back and forth between hillsides, the momentum stabilizes the search because it tends to make the weights follow a smoother path. Imagine a ball (weight vector position) rolling down a hill (performance surface). If the ball reaches a small flat part of the hill, it will continue past this local minimum because of its momentum. A ball without momentum, however, will get stuck in this valley. Momentum learning is a robust method to speed up learning, and is usually recommended as the default search rule for networks with nonlinearities.

Advanced Search Methods

The popularity of *gradient descent method* is based more on its simplicity (it can be computed locally with two multiplications and one addition per weight) than on its search power. There are many other search procedures more powerful than backpropagation. For example, *Newtonian method* is a second–order method because it uses the information on the curvature to adapt the weights. However Newtonian method is computationally much more costly to implement and requires information not available at the PE, so it has been used little in neurocomputing. Although more powerful, Newtonian method is still a local search method and so may be caught in local minima or diverge due to the difficult neural network performance landscapes. Other techniques such as *simulated annealing*[148] and *genetic algorithms* (GA)[149] are global search procedures, that is, they can avoid local minima. The issue is that they are more costly to implement in a distributed system like a neural network, either because they are inherently slow or because they require nonlocal quantities [PEL00].

The problem of search with local information can be formulated as an approximation to the functional form of the *matrix cost function* $J(\mathbf{w})$ at the operating point $\mathbf{w}_0$. This immediately points to the Taylor series expansion of J around $\mathbf{w}_0$,

$$J(\mathbf{w}-\mathbf{w}_0) = J_0 + (\mathbf{w}-\mathbf{w}_0)\nabla J_0 + \frac{1}{2}(\mathbf{w}-\mathbf{w}_0)\mathbf{H}_0(\mathbf{w}-\mathbf{w}_0)^T + \cdots,$$

where ∇J is our familiar gradient, and $\mathbf{H}$ is the Hessian matrix, that is, the matrix of second derivatives with entries

$$H_{ij}(\mathbf{w}_0) = \left.\frac{\partial^2 J(w)}{\partial w_i \partial w_j}\right|_{w=w_0},$$

evaluated at the operating point. We can immediately see that the Hessian cannot be computed with the information available at a given PE, since it uses information from two different weights. If we differentiate J with respect to the weights, we get

$$\nabla J(\mathbf{w}) = \nabla J_0 + \mathbf{H}_0(\mathbf{w}-\mathbf{w}_0) + \cdots \tag{1.13}$$

so we can see that to compute the full gradient at $\mathbf{w}$ we need all the higher terms of the derivatives of J. This is impossible. Since the performance surface tends to be bowl shaped (quadratic) near the minimum, we are normally interested only in the first and second terms of the expansion [PEL00].

If the expansion of (1.13) is restricted to the first term, we get the gradient–search methods (hence they are called *first–order–search methods*), where the gradient is estimated with its value at $\mathbf{w}_0$. If we expand to use the second–order term, we get *Newton method* (hence the name second–order method). If we equate the truncated relation (1.13) to 0 we immediately get

$$w = w_0 - \mathbf{H}_0^{-1}\nabla J_0,$$

148 Simulated annealing is a global search criterion by which the space is searched with a random rule. In the beginning the variance of the random jumps is very large. Every so often the variance is decreased, and a more local search is undertaken. It has been shown that if the decrease of the variance is set appropriately, the global optimum can be found with probability one. The method is called simulated annealing because it is similar to the annealing process of creating crystals from a hot liquid.

149 Genetic algorithms are global search procedures proposed by J. Holland that search the performance surface, concentrating on the areas that provide better solutions. They use 'generations' of search points computed from the previous search points using the operators of crossover and mutation (hence the name).

which is the equation for the Newton method, which has the nice property of quadratic termination (it is guaranteed to find the exact minimum in a finite number of steps for quadratic performance surfaces). For most quadratic performance surfaces it can converge in one iteration.

The real difficulty is the memory and the computational cost (and precision) to estimate the Hessian. Neural networks can have thousands of weights, which means that the Hessian will have millions of entries. This is why methods of approximating the Hessian have been extensively researched. There are two basic classes of approximations [PEL00]:

1. Line search methods, and
2. Pseudo–Newton methods.

The information in the first type is restricted to the gradient, together with line searches along certain directions, while the second seeks approximations to the Hessian matrix. Among the line search methods probably the most effective is the *conjugate gradient method*. For quadratic performance surfaces the conjugate gradient algorithm preserves quadratic termination and can reach the minimum in D steps, where D is the dimension of the weight space. Among the Pseudo–Newton methods probably the most effective is the *Levenberg–Marquardt algorithm* (LM), which uses the Gauss–Newton method to approximate the Hessian. LM is the most interesting for neural networks, since it is formulated as a sum of quadratic terms just like the cost functions in neural networks.

The *extended Kalman filter* (EKF) forms the basis of a second–order neural network training method that is a practical and effective alternative to the batch–oriented, second–order methods mentioned above. The essence of the recursive EKF procedure is that, during training, in addition to evolving the weights of a network architecture in a sequential (as opposed to batch) fashion, an approximate error covariance matrix that encodes second–order information about the training problem is also maintained and evolved.

Homotopy Methods

The most popular method for solving nonlinear equations in general is the *Newton–Raphson method*. Unfortunately, this method sometimes fails, especially in cases when nonlinear equations possess multiple solutions (zeros). An emerging family of methods that can be used in such cases are homotopy (continuation) methods. These methods are robust and have good convergence properties.

Homotopy methods or *continuation methods* have increasingly been used for solving variety of nonlinear problems in fluid dynamics, structural mechanics, systems identifications, and integrated circuits (see [Wat90]). These methods, popular in mathematical programming, are globally convergent provided that certain coercivity and continuity conditions are satisfied by the equations that need to be solved [Wat90]. Moreover, they often yield all the solutions to the nonlinear system of equations.

The idea behind a homotopy or continuation method is to embed a parameter λ in the nonlinear equations to be solved. This is why they are sometimes referred to as *embedding methods*. Initially, parameter λ is set to zero, in which case the problem is reduced to an easy problem with a known or easily–found solution. The set of equations is then gradually deformed into the originally posed difficult problem by varying the parameter λ . The original problem is obtained for $\lambda = 1$. Homotopies are a class of continuation methods, in which parameter λ is a function of a path arc length and may actually increase or decrease as the path is traversed. Provided that certain coercivity conditions imposed on the nonlinear function to be solved are satisfied, the homotopy path does not branch (bifurcate) and passes through all the solutions of the nonlinear equations to be solved.

The zero curve of the homotopy map can be tracked by various techniques: an *ODE–algorithm*, a *normal flow algorithm*, and an *augmented Jacobian matrix algorithm*, among others [Wat90].

As a typical example, homotopy techniques can be applied to find the zeros of the gradient function $F:\mathbb{R}^N\to\mathbb{R}^N$, such that

$$F(\theta) = \frac{\partial E(\theta)}{\partial \theta_k}, \qquad 1\leq k\leq N,$$

where $E = (\theta)$ is the certain error function dependent on N parameters θ_k. In other words, we need to solve a system of nonlinear equations

$$F(\theta) = 0. \tag{1.14}$$

In order to solve equation (1.14), we can create a linear homotopy function

$$H(\theta,\lambda) = (1-\lambda)(\theta - a) + \lambda F(\theta),$$

where a is an arbitrary starting point. Function $H(\theta,\lambda)$ has properties that equation $H(\theta,0)=0$ is easy to solve, and that $H(\theta,1)\equiv F(\theta)$.

ANNs as Functional Approximators

The *universal approximation theorem* of Kolmogorov states [Hay94]:
Let $\phi(\cdot)$ be a nonconstant, bounded, and monotone–increasing continuous (C^0) function. Let I^N denote ND unit hypercube $[0,1]^N$. The space of C^0–functions on I^N is denoted by $C(I^N)$. Then, given any function $f\in C(I^N)$ and $\epsilon>0$, there exist an integer M and sets of real constants $\alpha_i,\theta_i,\omega_{ij}$, $i=1,\dots,M$; $j=1,\dots,N$ such that we may define

$$F(x_1,\ldots,x_N) = \alpha_i\phi(\omega_{ij}x_j - \theta_i),$$

as an approximate realization of the function $f(\cdot)$; that is

$$|F(x_1,\ldots,x_N) - f(x_1,\ldots,x_N)| < \epsilon \qquad \text{for all} \quad \{x_1,\ldots,x_N\} \in I^N.$$

This theorem is directly applicable to *multilayer perceptrons*. First, the logistic function $1/[1+\exp(-v)]$ used as the sigmoidal nonlinearity in a neuron model for the construction of a multilayer perceptron is indeed a nonconstant, bounded, and monotone–increasing function; it therefore satisfies the conditions imposed on the function $\phi(\cdot)$. Second, the upper equation represents the output of a multilayer perceptron described as follows:

1. The network has n input nodes and a single hidden layer consisting of M neurons; the inputs are denoted by $x_1,\ldots,x_N$.
2. ith hidden neuron has synaptic weights $\omega_{i1},\ldots,\omega_{iN}$ and threshold θ_i.
3. The network output y_j is a linear combination of the outputs of the hidden neurons, with $\alpha_i,\ldots,\alpha_M$ defining the coefficients of this combination.

The theorem actually states that a single hidden layer is sufficient for a multilayer perceptron to compute a uniform ϵ approximation to a given training set represented by the set of inputs $x_1,\ldots,x_N$ and desired (target) output $f(x_1,\ldots,x_N)$. However, the theorem does not say that a single layer is *optimum* in the sense of learning time or ease of implementation.

Recall that training of multilayer perceptrons is usually performed using a certain clone of the BP algorithm (1.2.2). In this forward–pass/backward–pass gradient–descending algorithm, the adjusting of synaptic weights is defined by the extended $\delta-$*rule*, given by equation

$$\Delta\omega_{ji}(N) = \eta\cdot\delta_j(N)\cdot y_i(N), \tag{1.15}$$

where $\Delta\omega_{ji}(N)$ corresponds to the *weight correction*, η is the *learning–rate parameter*, $\delta_j(N)$ denotes the *local gradient* and $y_i(N)$ – the *input signal of neuron* j; while the *cost function* E is defined as the instantaneous sum of squared errors e_j^2

$$E(n) = \frac{1}{2}\sum_j e_j^2(N) = \frac{1}{2}\sum_j [d_j(N) - y_j(N)]^2, \tag{1.16}$$

where $y_j(N)$ is the output of jth neuron, and $d_j(N)$ is the desired (target) response for that neuron. The slow BP convergence rate (1.15–1.16) can be accelerated using the faster LM algorithm (see subsection 1.2.2 above), while its robustness can be achieved using an appropriate fuzzy controller (see subsection (1.2.2) below).

Summary of Supervised Learning Methods

Gradient Descent Method

Given the $(D+1)$D weights vector $\mathbf{w}(n) = [w_0(n), \ldots, w_D(n)]^T$ (with $w_0 = bias$), and the correspondent MSE–gradient (including partials of MSE w.r.t. weights)

$$\nabla \mathbf{e} = \left[\frac{\partial e}{\partial w_0}, \ldots, \frac{\partial e}{\partial w_D}\right]^T,$$

and the learning rate (step size) η, we have the vector learning equation

$$\mathbf{w}(n+1) = \mathbf{w}(n) - \eta \nabla \mathbf{e}(n),$$

which in index form reads

$$w_i(n+1) = w_i(n) - \eta \nabla e_i(n).$$

LMS Algorithm

$$\mathbf{w}(n+1) = \mathbf{w}(n) + \eta \varepsilon(n)\, x(n),$$

where x is an input (measurement) vector, and ε is a zero–mean Gaussian noise vector uncorrelated with input, or

$$w_i(n+1) = w_i(n) + \eta \varepsilon(n)\, x^i(n).$$

Newton's Method

$$\mathbf{w}(n+1) = \mathbf{w}(n) - \eta\, \mathbf{R}^{-1} \mathbf{e}(n),$$

where $\mathbf{R}$ is input (auto)correlation matrix, or

$$\mathbf{w}(n+1) = \mathbf{w}(n) + \eta\, \mathbf{R}^{-1} \varepsilon(n)\, x(n),$$

Conjugate Gradient Method

$$\begin{aligned} \mathbf{w}(n+1) &= \mathbf{w}(n) + \eta\, \mathbf{p}(n), \\ \mathbf{p}(n) &= -\nabla \mathbf{e}(n) + \beta(n) \mathbf{p}(n-1), \\ \beta(n) &= \frac{\nabla \mathbf{e}(n)^T \nabla \mathbf{e}(n)}{\nabla \mathbf{e}(n-1)^T \nabla \mathbf{e}(n-1)}. \end{aligned}$$

Levenberg–Marquardt Algorithm

Putting

$$\nabla \mathbf{e} = \mathbf{J}^T \mathbf{e},$$

where $\mathbf{J}$ is the Jacobian matrix, which contains first derivatives of the network errors with respect to the weights and biases, and $\mathbf{e}$ is a vector of network errors, LM algorithm reads

$$\mathbf{w}(n+1) = \mathbf{w}(n) - [\mathbf{J}^T\mathbf{J} + \mu\mathbf{I}]^{-1}\mathbf{J}^T\mathbf{e}. \qquad (1.17)$$

Generalized Feedforward Nets

The *generalized feedforward network* (GFN, see Figure 1.21) is a generalization of MLP, such that connections can jump over one or more layers, which in practice, often solves the problem much more efficiently than standard MLPs. A classic example of this is the two–spiral problem, for which standard MLP requires hundreds of times more training epochs than the generalized feedforward network containing the same number of processing elements. Both MLPs and GFNs are usually trained using a variety of backpropagation techniques and their enhancements like the nonlinear LM algorithm (1.17). During training in the spatial processing, the weights of the GFN converge iteratively to the analytical solution of the 2D Laplace equation.

Modular Feedforward Nets

The *modular feedforward networks* are a special class of MLP. These networks process their input using several parallel MLPs, and then recombine the results. This tends to create some structure within the topology, which

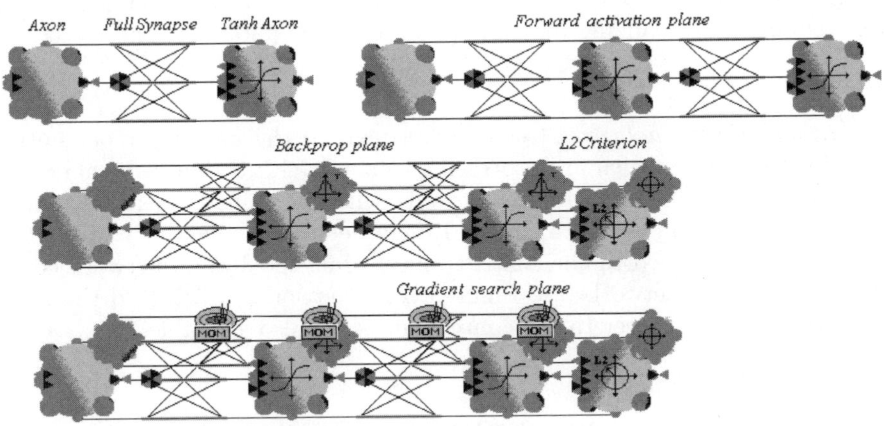

Fig. 1.21. Generalized feedforward network (GFN), arranged using *NeuroSolutions*^{TM}.

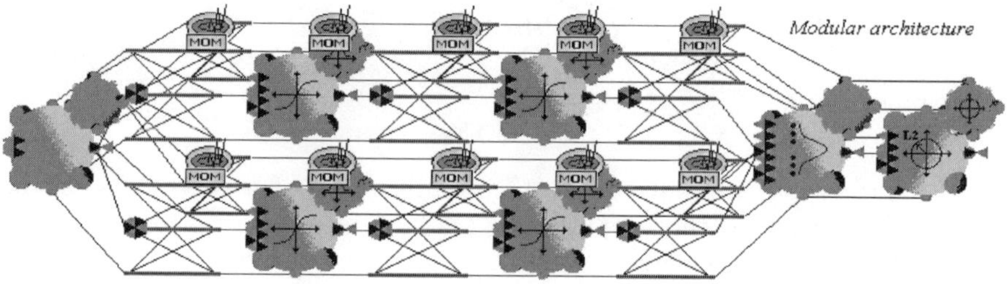

Fig. 1.22. Modular feedforward network, arranged using $NeuroSolutions^{TM}$.

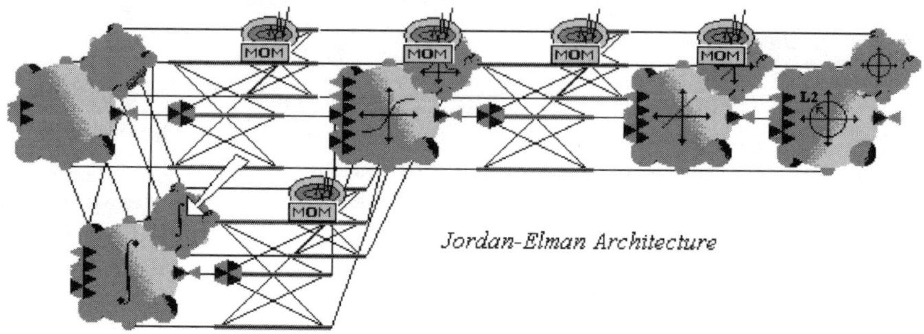

Fig. 1.23. Jordan and Elman network, arranged using $NeuroSolutions^{TM}$.

will foster specialization of function in each submodule (see Figure 1.22). In contrast to the MLP, modular networks do not have full inter–connectivity between their layers. Therefore, a smaller number of weights are required for the same size network (i.e., the same number of PEs). This tends to speed up training times and reduce the number of required training exemplars. There are many ways to segment a MLP into modules. It is unclear how to best design the modular topology based on the data. There are no guarantees that each module is specializing its training on a unique portion of the data.

Jordan and Elman Nets

Jordan and Elman networks (see [Elm90]) extend the multilayer perceptron with context units, which are processing elements (PEs) that remember past activity. Context units provide the network with the ability to extract temporal information from the data. In the Elman network, the activity of the first hidden PEs are copied to the context units, while the Jordan network copies the output of the network (see Figure 1.23). Networks which feed the input and the last hidden layer to the context units are also available.

Kohonen Self–Organizing Map

Kohonen self–organizing map (SOM, see Figure 1.24) is widely used for image pre–processing as well as a pre–processing unit for various hybrid architectures. SOM is a winner–take–all neural architecture that quantizes the input

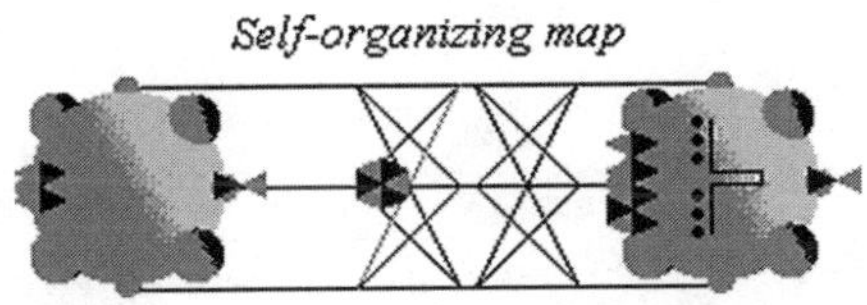

Fig. 1.24. Kohonen self–organizing map (SOM) network, arranged using $NeuroSolutions^{TM}$.

space, using a distance metric, into a discrete feature output space, where neighboring regions in the input space are neighbors in the discrete output space. SOM is usually applied to neighborhood clustering of random points along a circle using a variety of distance metrics: Euclidean, L^1, L^2, and L^n, Machalanobis, etc. The basic SOM architecture consists of a layer of Kohonen synapses of three basic forms: line, diamond and box, followed by a layer of winner–take–all axons. It usually uses added Gaussian and uniform noise, with control of both the mean and variance. Also, SOM usually requires choosing the proper initial neighborhood width as well as annealing of the neighborhood width during training to ensure that the map globally represents the input space.

The Kohonen SOM algorithm is defined as follows: Every stimulus $\mathbf{v}$ of an Euclidian input space V is mapped to the neuron with the position $\mathbf{s}$ in the neural layer R with the highest neural activity, the 'center of excitation' or 'winner', given by the condition

$$|\mathbf{w_s} - \mathbf{v}| = \min_{\mathbf{r}\in R} |\mathbf{w_r} - \mathbf{v}|,$$

where $|.|$ denotes the Euclidian distance in input space. In the Kohonen model the learning rule for each synaptic weight vector $\mathbf{w_r}$ is given by

$$\mathbf{w_r^{\mathsf{new}}} = \mathbf{w_r^{\mathsf{old}}} + \eta \cdot g_{\mathbf{rs}} \cdot (\mathbf{v} - \mathbf{w_r^{\mathsf{old}}}), \tag{1.18}$$

with $g_{\mathbf{rs}}$ as a gaussian function of Euclidian distance $|\mathbf{r} - \mathbf{s}|$ in the neural layer. Topology preservation is enforced by the common update of all weight vectors whose neuron $\mathbf{r}$ is adjacent to the center of excitation $\mathbf{s}$. The function $g_{\mathbf{rs}}$ describes the topology in the neural layer. The parameter η determines the speed of learning and can be adjusted during the learning process.

Radial Basis Function Nets

The *radial basis function network* (RBF, see Figure 1.25) provides a powerful alternative to MLP for function approximation or classification. It differs from MLP in that the overall input–output map is constructed from local contributions of a layer of Gaussian axons. It trains faster and requires fewer training samples than MLP, using the hybrid supervised/unsupervised method. The unsupervised part of an RBF network consists of a competitive synapse followed by a layer of Gaussian axons. The means of the Gaussian axons are

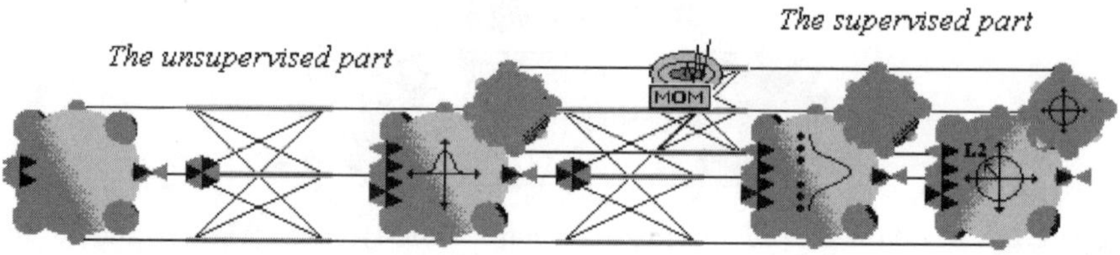

Fig. 1.25. Radial basis function network, arranged using $NeuroSolutions^{TM}$.

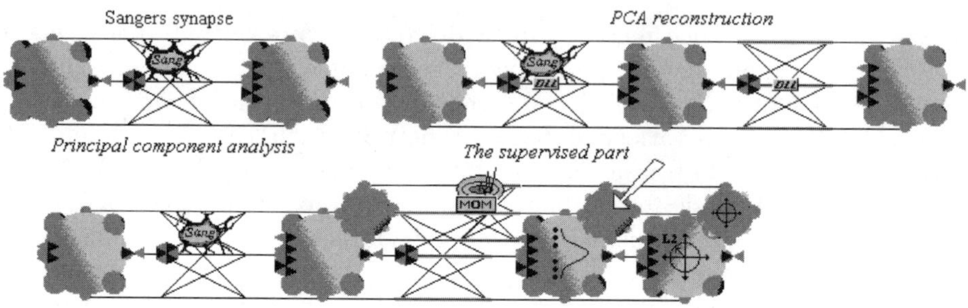

Fig. 1.26. Principal component analysis (PCA) network, arranged using $Neuro$-$Solutions^{TM}$.

found through competitive clustering and are, in fact, the weights of the Conscience synapse. Once the means converge the variances are calculated based on the separation of the means and are associated with the Gaussian layer. Having trained the unsupervised part, we now add the supervised part, which consists of a single–layer MLP with a soft–max output.

Principal Component Analysis Nets

The *principal component analysis networks* (PCAs, see Figure 1.26) combine unsupervised and supervised learning in the same topology. Principal component analysis is an unsupervised linear procedure that finds a set of uncorrelated features, principal components, from the input. A MLP is supervised to perform the nonlinear classification from these components. More sophisticated are the *independent component analysis networks* (ICAs).

Co–active Neuro–Fuzzy Inference Systems

The *co–active neuro–fuzzy inference system* (CANFIS, see Figure 1.27), which integrates adaptable fuzzy inputs with a modular neural network to rapidly and accurately approximate complex functions. Fuzzy–logic inference systems (see next section) are also valuable as they combine the explanatory nature of rules (membership functions) with the power of 'black box' neural networks.

Genetic ANN–Optimization

Genetic optimization, added to ensure and speed–up the convergence of all other ANN–components, is a powerful tool for enhancing the efficiency and

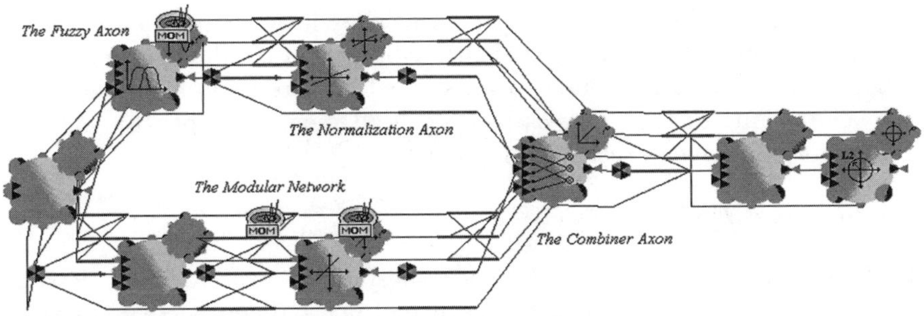

Fig. 1.27. Co–active neuro–fuzzy inference system (CANFIS) network, arranged using $NeuroSolutions^{TM}$.

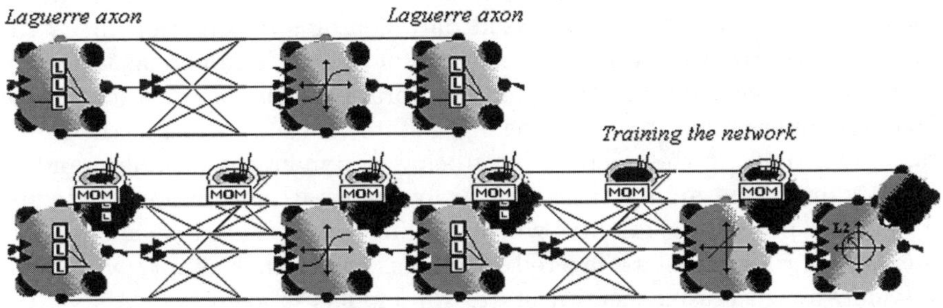

Fig. 1.28. Time–lagged recurrent network (TLRN), arranged using $Neuro$-$Solutions^{TM}$.

effectiveness of a neural network. Genetic optimization can fine–tune network parameters so that network performance is greatly enhanced. Genetic control applies a *genetic algorithm* (GA, see next section), a part of broader *evolutionary computation*, see MIT journal with the same name) to any network parameters that are specified. Also through the *genetic control*, GA parameters such as mutation probability, crossover type and probability, and selection type can be modified.

Time–Lagged Recurrent Nets

The *time–lagged recurrent networks* (TLRNs, see Figure 1.28) are MLPs extended with short term memory structures [Wer90]. Most real–world data contains information in its time structure, i.e., how the data changes with time. Yet, most neural networks are purely static classifiers. TLRNs are the state of the art in nonlinear time series prediction, system identification and temporal pattern classification. Time–lagged recurrent nets usually use memory Axons, consisting of IIR filters with local adaptable feedback that act as a variable memory depth. The time–delay neural network (TDNN) can be considered a special case of these networks, examples of which include the Gamma and Laguerre structures. The Laguerre axon uses locally recurrent all–pass IIR filters to store the recent past. They have a single adaptable parameter that controls the memory depth. Notice that in addition to providing memory for the input, we have also used a Laguerre axon after the hidden Tanh axon. This further increases the overall memory depth by providing memory for that layer's recent activations.

Fully Recurrent ANNs

The *fully recurrent networks* feed back the hidden layer to itself. Partially recurrent networks start with a fully recurrent net and add a feedforward connection that bypasses the recurrency, effectively treating the recurrent part as a state memory. These recurrent networks can have an infinite memory depth and thus find relationships through time as well as through the instantaneous input space. Most real–world data contains information in its time structure. Recurrent networks are the state of the art in nonlinear time series prediction, system identification, and temporal pattern classification. In case of large number of neurons, here the firing states of the neurons or their membrane potentials are the microscopic stochastic dynamical variables, and one is mostly interested in quantities such as average state correlations and global information processing quality, which are indeed measured by macroscopic observables. In contrast to layered networks, one cannot simply write down the values of successive neuron states for models of recurrent ANNs; here they must be solved from (mostly stochastic) coupled dynamic equations. For nonsymmetric networks, where the asymptotic (stationary) statistics are not known, dynamical techniques from non–equilibrium statistical mechanics are the only tools available for analysis. The natural set of macroscopic quantities (or order parameters) to be calculated can be defined in practice as the smallest set which will obey closed deterministic equations in the limit of an infinitely large network.

Being high–dimensional nonlinear systems with extensive feedback, the dynamics of recurrent ANNs are generally dominated by a wealth of attractors (fixed–point attractors, limit–cycles, or even more exotic types), and the practical use of recurrent ANNs (in both biology and engineering) lies in the potential for creation and manipulation of these attractors through adaptation of the network parameters (synapses and thresholds) (see [Hop82, Hop84]). Input fed into a recurrent ANN usually serves to induce a specific initial configuration (or firing pattern) of the neurons, which serves as a cue, and the output is given by the (static or dynamic) attractor which has been triggered by this cue. The most familiar types of recurrent ANN models, where the idea of creating and manipulating attractors has been worked out and applied explicitly, are the so–called *attractor associative memory ANNs*, designed to store and retrieve information in the form of neuronal firing patterns and/or sequences of neuronal firing patterns. Each pattern to be stored is represented as a microscopic state vector. One then constructs synapses and thresholds such that the dominant attractors of the network are precisely the pattern vectors (in the case of static recall), or where, alternatively, they are trajectories in which the patterns are successively generated microscopic system states. From an initial configuration (the cue, or input pattern to be recognized) the system is allowed to evolve in time autonomously, and the final state (or trajectory) reached can be interpreted as the pattern (or pattern sequence) recognized by network from the input. For such programmes to work one clearly needs recurrent ANNs with extensive ergodicity breaking: the state vector will during the course of the dynamics (at least on finite time–scales) have to be confined to a restricted region of state–space (an ergodic component), the location of which is to depend strongly on the initial conditions. Hence our interest will mainly be in systems with many attractors. This, in turn, has implications at a theoretical/mathematical level: solving models of recurrent ANNs with extensively many attractors requires advanced tools from disordered systems theory, such as replica theory (statics) and generating functional analysis (dynamics).

Complex–Valued ANNs

It is expected that *complex–valued ANNs*, whose parameters (weights and threshold values) are all complex numbers, will have applications in all the fields dealing with complex numbers (e.g., telecommunications, quantum physics). A complex–valued, feedforward, multi–layered, back–propagation neural network model was proposed independently by T. Nitta [NF91, Nit97, Nit00, Nit04], G. GK92 [GK92] and N. Benvenuto [BP92, BP92], and demonstrated its characteristics:

(a) the properties greatly different from those of the real–valued back–propagation network, including 2D motion structure of weights and the orthogonality of the decision boundary of a complex–valued neuron;
(b) the learning property superior to the real–valued back–propagation;
(c) the inherent 2D motion learning ability (an ability to transform geometric figures); and
(d) the ability to solve the XOR problem and detection of symmetry problem with a single complex–valued neuron.

Following [NF91, Nit97, Nit00, Nit04], we consider here the complex–valued neuron. Its input signals, weights, thresholds and output signals are all complex numbers. The net input U_n to a complex–valued neuron n is defined as

$$U_n = W_{mn} X_m + V_n,$$

where W_{mn} is the (complex–valued) weight connecting the complex–valued neurons m and n, V_n is the (complex–valued) threshold value of the complex–valued neuron n, and X_m is the (complex–valued) input signal from the complex–valued neuron m. To get the (complex–valued) output signal, convert the net input U_n into its real and imaginary parts as follows: $U_n = x + iy = z$, where $i = \sqrt{-1}$. The (complex–valued) output signal is defined to be

$$\sigma(z) = \tanh(x) + i\tanh(y),$$

where $\tanh(u) = (\exp(u) - \exp(-u)) = (\exp(u) + \exp(-u))$, $u \in \mathbb{R}$. Note that $-1 < \mathrm{Re}[\sigma]$, $\mathrm{Im}[\sigma] < 1$. Note also that σ is not regular as a complex function, because the Cauchy–Riemann equations do not hold.

A complex–valued ANN consists of such complex–valued neurons described above. A typical network has 3 layers: $m \to n \to 1$, with $w_{ij} \in \mathbb{C}$ – the weight between the input neuron i and the hidden neuron j, $w_{0j} \in \mathbb{C}$ – the threshold of the hidden neuron j, $c_j \in \mathbb{C}$ – the weight between the hidden neuron j and the output neuron $(1 \leq i \leq m; 1 \leq j \leq n)$, and $c_0 \in \mathbb{C}$ – the threshold of the output neuron. Let $y_j(z), h(z)$ denote the output values of the hidden neuron j, and the output neuron for the input pattern $z = [z_1, \dots, z_m]^t \in \mathbb{C}^m$, respectively. Let also $\nu_j(z)$ and $\mu(z)$ denote the net inputs to the hidden neuron j and the output neuron for the input pattern $z \in \mathbb{C}^m$, respectively. That is,

$$\nu_j(z) = w_{ij} z_i + w_{0j}, \qquad \mu(z) = c_j y_j(z) + c_0,$$
$$y_j(z) = \sigma(\nu_j(z)), \qquad h(z) = \sigma(\mu(z)).$$

The set of all $m \to n \to 1$ complex–valued ANNs described above is usually denoted by $N_{m,n}$. The Complex–BP learning rule [NF91, Nit97, Nit00, Nit04] has been obtained by using a steepest–descent method for such (multilayered) complex–valued ANNs.

Common Continuous ANNs

Virtually all computer–implemented ANNs (mainly listed above) are discrete dynamical systems, mainly using supervised training (except Kohonen SOM) in one of gradient–descent searching forms. They are good as problem–solving tools, but they fail as models of animal nervous system. The other category of ANNs are continuous neural systems that can be considered as models of animal nervous system. However, *as models of the human brain, all current ANNs are simply trivial.*

Neurons as Functions

According to B. Kosko, neurons behave as functions [Kos92]; they transduce an unbounded input *activation* $x(t)$ into output *signal* $S(x(t))$. Usually a sigmoidal (S–shaped, bounded, monotone-nondecreasing: $S' \geq 0$) function describes the transduction, as well as the input–output behavior of many operational amplifiers. For example, the *logistic signal* (or, the *maximum–entropy*) function

$$S(x) = \frac{1}{1 + \mathrm{e}^{-cx}}$$

is sigmoidal and strictly increases for positive scaling constant $c > 0$. Strict monotonicity implies that the *activation derivative* of S is positive:

$$S' = \frac{dS}{dx} = cS(1 - S) > 0.$$

An infinitely steep logistic signal function gives rise to a threshold signal function

$$S(x^{n+1}) = \begin{cases} 1, & \text{if} \quad x^{n+1} > T, \\ S(x^n), & \text{if} \quad x^{n+1} = T, \\ 0, & \text{if} \quad x^{n+1} < T, \end{cases}$$

for an arbitrary real–valued threshold T. The index n indicates the discrete time step.

In practice signal values are usually binary or bipolar. *Binary signals*, like logistic, take values in the unit interval $[0, 1]$. *Bipolar signals* are signed; they take values in the bipolar interval $[-1, 1]$. Binary and bipolar signals transform into each other by simple scaling and translation. For example, the bipolar logistic signal function takes the form

$$S(x) = \frac{2}{1 + \mathrm{e}^{-cx}} - 1.$$

Neurons with bipolar threshold signal functions are called *McCulloch–Pits neurons*.

A naturally occurring bipolar signal function is the *hyperbolic–tangent* signal function

$$S(x) = \tanh(cx) = \frac{\mathrm{e}^{cx} - \mathrm{e}^{-cx}}{\mathrm{e}^{cx} + \mathrm{e}^{-cx}},$$

with activation derivative

$$S' = c(1 - S^2) > 0.$$

The *threshold linear* function is a binary signal function often used to approximate neuronal firing behavior:

$$S(x) = \begin{cases} 1, & \text{if} \quad cx \geq 1, \\ 0, & \text{if} \quad cx < 0, \\ cx, & \text{else}, \end{cases}$$

which we can rewrite as

$$S(x) = \min(1, \max(0, cx)).$$

Between its upper and lower bounds the threshold linear signal function is trivially monotone increasing, since $S' = c > 0$.

Gaussian, or bell–shaped, signal function of the form $S(x) = \mathrm{e}^{-cx^2}$, for $c > 0$, represents an important exception to signal monotonicity. Its activation derivative $S' = -2cx\mathrm{e}^{-cx^2}$ has the sign opposite the sign of the activation x.

Generalized Gaussian signal functions define potential or radial basis functions $S_i(x^i)$ given by

$$S_i(x) = \exp[-\frac{1}{2\sigma_i^2}\sum_{j=1}^{n}(x_j - \mu_j^i)^2],$$

for input activation vector $x = (x^i) \in \mathbb{R}^n$, variance σ_i^2, and mean vector $\boldsymbol{\mu}_i = (\mu_j^i)$. Each radial basis function S_i defines a spherical *receptive field* in $\mathbb{R}^n$. The ith neuron emits unity, or near-unity, signals for sample activation vectors x that fall in its receptive field. The mean vector $\boldsymbol{\mu}$ centers the receptive field in $\mathbb{R}^n$. The variance σ_i^2 localizes it. The radius of the Gaussian spherical receptive field shrinks as the variance σ_i^2 decreases. The receptive field approaches $\mathbb{R}^n$ as σ_i^2 approaches ∞.

The *signal velocity* $\dot{S} \equiv dS/dt$ is the *signal time derivative*, related to the activation derivative by

$$\dot{S} = S'\dot{x},$$

so it depends explicitly on *activation velocity*. This is used in unsupervised learning laws that adapt with *locally available information*.

The signal $S(x)$ induced by the activation x represents the neuron's firing frequency of action potentials, or pulses, in a sampling interval. The firing frequency equals the average number of pulses emitted in a sampling interval.

Short–term memory is modelled by *activation dynamics*, and *long–term memory* is modelled by *learning dynamics*. The overall neural network behaves as an *adaptive filter* (see [Hay91]).

In the simplest and most common case, neurons are not topologically ordered. They are related only by the synaptic connections between them. Kohonen calls this *lack of topological structure* in a *field of neurons* the *zeroth–order topology*. This suggests that ANN–models are *abstractions*, not *descriptions* of the brain neural networks, in which order does matter.

Basic Activation and Learning Dynamics

One of the oldest continuous training methods, based on Hebb's biological synaptic learning [Heb49], is *Oja–Hebb learning rule* [Oja82], which calculates the weight update according to the ODE

$$\dot{\omega}_i(t) = O(t)\,[I_i(t) - O(t)\,\omega_i(t)],$$

where $O(t)$ is the output of a simple, linear processing element; $I_i(t)$ are the inputs; and $\omega_i(t)$ are the synaptic weights.

Related to the Oja–Hebb rule is a special matrix of synaptic weights called *Karhunen–Loeve covariance matrix* $\mathbf{W}$ (KL), with entries

$$W_{ij} = \frac{1}{N}\omega_i^\mu\,\omega_j^\mu, \qquad \text{(summing over } \mu)$$

where N is the number of vectors, and ω_i^μ is the ith component of the μth vector. The KL matrix extracts the principal components, or directions of maximum information (correlation) from a dataset.

In general, continuous ANNs are *temporal dynamical systems*. They have two coupled dynamics: activation and learning. First, a general system of coupled ODEs for the output of the ith *processing element* (PE) x^i, called the *activation dynamics*, can be written as

$$\dot{x}^i = g_i(x^i, \underset{i}{\text{net}}), \tag{1.19}$$

with the *net input* to the ith PE x^i given by $\quad \text{net}_i = \omega_{ij}x^j$.

For example,

$$\dot{x}^i = -x^i + f_i(\underset{i}{\text{net}}),$$

where f_i is called *output*, or *activation, function.* We apply some input values to the PE so that $\text{net}_i > 0$. If the inputs remain for a sufficiently long time, the output value will reach an equilibrium value, when $\dot{x}^i = 0$, given by $x^i = f_i(\text{net}_i)$. Once the unit has a nonzero output value, removal of the inputs will cause the output to return to zero. If $\text{net}_i = 0$, then $\quad \dot{x}^i = -x^i$, which means that $x \to 0$.

Second, a general system of coupled ODEs for the *update* of the synaptic weights ω_{ij}, i.e, *learning dynamics*, can be written as a generalization of the Oja–Hebb rule, i.e..

$$\dot{\omega}_{ij} = G_i(\omega_{ij}, x^i, x^i),$$

where G_i represents the *learning law*; the learning process consists of finding weights that encode the knowledge that we want the system to learn. For most realistic systems, it is not easy to determine a closed–form solution for this system of equations, so the approximative solutions are usually enough.

Standard Models of Continuous Nets

Hopfield Continuous Net

One of the first physically–based ANNs was developed by J. Hopfield. He first made a discrete, Ising–spin based network in [Hop82], and later generalized it to the continuous, graded–response network in [Hop84], which we briefly describe here. Later we will give full description of Hopfield models. Let $\text{net}_i = u_i$ – the net input to the ith PE, biologically representing the summed action potentials at the axon hillock of a neuron. The PE *output function* is

$$v_i = g_i(\lambda u_i) = \frac{1}{2}(1 + \tanh(\lambda u_i)),$$

where λ is a constant called the *gain parameter*. The network is described as a transient RC circuit

$$C_i\dot{u}_i = T_{ij}v_j - \frac{u_i}{R_i} + I_i, \tag{1.20}$$

where I_i, R_i and C_i are inputs (currents), resistances and capacitances, and T_{ij} are synaptic weights.

The Hamiltonian energy function corresponding to (1.20) is given as

$$H = -\frac{1}{2}T_{ij}v_iv_j + \frac{1}{\lambda}\frac{1}{R_i}\int_0^{v_i} g_i^{-1}(v)\,dv - I_iv_i, \qquad (j \neq i) \tag{1.21}$$

which is a generalization of a discrete, *Ising–spin Hopfield network* with energy function

$$E = -\frac{1}{2}\omega_{ij}x^ix^j, \qquad (j \neq i).$$

where $g_i^{-1}(v) = u$ is the inverse of the function $v = g(u)$. To show that (1.21) is an appropriate *Lyapunov function* for the system, we shall take its time derivative assuming T_{ij} are symmetric:

$$\dot{H} \;=\; -\dot{v}_i(T_{ij}v_j - \frac{u_i}{R_i} + I_i) \;=\; -C_i\dot{v}_i\dot{u}_i \;=\; -C_i\dot{v}_i^2\frac{\partial g_i^{-1}(v_i)}{\partial v_i}. \tag{1.22}$$

All the factors in the summation (1.22) are positive, so $\dot{H}$ must decrease as the system evolves, until it eventually reaches the stable configuration, where $\dot{H} = \dot{v}_i = 0$.

Hecht–Nielsen Counterpropagation Net

Hecht–Nielsen counterpropagation network (CPN) is a full–connectivity, graded–response generalization of the standard BP algorithm (see [Hec87, Hec90]). The outputs of the PEs in CPN are governed by the set of ODEs

$$\dot{x}^i = -Ax_i + (B - x^i)I_i - x^i\sum_{j\neq i} I_j,$$

where $0 < x^i(0) < B$, and $A, B > 0$. Each PE receives a net excitation (on–center) of $(B - x^i)I_i$ from its corresponding input value, I. The addition of inhibitory connections (off–surround), $-x^iI_j$, from other units is responsible for preventing the activity of the processing element from rising in proportion to the absolute pattern intensity, I_i. Once an input pattern is applied, the PEs quickly reach an equilibrium state ($\dot{x}^i = 0$) with

$$x^i = \Theta_i\frac{BI_i}{A + I_i},$$

with the normalized *reflectance pattern* $\Theta_i = I_i\left(\sum_i I_i\right)^{-1}$, such that $\sum_i \Theta_i = 1$.

Competitive Net

Activation dynamics is governed by the ODEs

$$\dot{x}^i = -Ax_i + (B - x^i)[f(x^i) + \underset{i}{\text{net}}] - x^i \left[\sum_{j \neq i} f(x_j) + \sum_{j \neq i} \text{net}_j \right],$$

where $A, B > 0$ and $f(x^i)$ is an output function.

Kohonen's Continuous SOM and Adaptive Robotics Control

Kohonen continuous self organizing map (SOM) is actually the original Kohonen model of the biological neural process (see [Koh88]). SOM activation dynamics is governed by

$$\dot{x}^i = -r_i(x^i) + \underset{i}{\text{net}} + z_{ij} x_j, \tag{1.23}$$

where the function $r_i(x^i)$ is a general form of a loss term, while the final term models the lateral interactions between units (the sum extends over all units in the system). If z_{ij} takes the form of the Mexican–hat function, then the network will exhibit a bubble of activity around the unit with the largest value of net input.

SOM learning dynamics is governed by

$$\dot{\omega}_{ij} = \alpha(t)(I_i - \omega_{ij})U(x^i),$$

where $\alpha(t)$ is the learning momentum, while the function $U(x^i) = 0$ unless $x^i > 0$ in which case $U(x^i) = 1$, ensuring that only those units with positive activity participate in the learning process.

Kohonen's continuous SOM (1.23–1.2.2) is widely used in adaptive robotics control. Having an $n-$segment robot arm with n chained $SO(2)-$joints, for a particular initial position x and desired velocity $\dot{x}^j_{desir}$ of the end–effector, the required torques T_i in the joints can be found as

$$T_i = a_{ij}\, \dot{x}^j_{desir},$$

where the inertia matrix $a_{ij} = a_{ij}(x)$ is learned using SOM.

Adaptive Resonance Theory

Principles derived from an analysis of experimental literatures in vision, speech, cortical development, and reinforcement learning, including attentional blocking and cognitive–emotional interactions, led to the introduction of S. Grossberg's *adaptive resonance theory* (ART) as a theory of human *cognitive information processing* (see [CG03]). The theory has evolved as a series of real–time neural network models that perform unsupervised and supervised learning, pattern recognition, and prediction. Models of unsupervised learning include ART1, for binary input patterns, and fuzzy–ART and ART2, for analog input patterns [Gro82, CG03]. ARTMAP models combine two unsupervised modules to carry out supervised learning. Many variations of the basic supervised and unsupervised networks have since been adapted for technological applications and biological analyzes.

A central feature of all ART systems is a *pattern matching process* that compares an external input with the internal memory of an active code. ART matching leads either to a resonant state, which persists long enough to permit learning, or to a parallel memory search. If the search ends at an established code, the memory representation may either remain the same or incorporate new information from matched portions of the current input. If the search ends at a new code, the memory representation learns the current input. This match–based learning process is the foundation of ART *code stability.* Match–based learning allows memories to change only when input from the external world is close enough to internal expectations, or when something completely new occurs. This feature makes ART systems well suited to problems that require on–line learning of large and evolving databases (see [CG03]).

Many ART applications use fast learning, whereby adaptive weights converge to equilibrium in response to each input pattern. Fast learning enables a system to adapt quickly to inputs that occur rarely but that may require immediate accurate recall. Remembering details of an exciting movie is a typical example of learning on one trial. Fast learning creates memories that depend upon the order of input presentation. Many ART applications exploit this feature to improve accuracy by voting across several trained networks, with voters providing a measure of confidence in each prediction.

Match–based learning is complementary to *error–based learning*, which responds to a mismatch by changing memories so as to reduce the difference between a target output and an actual output, rather than by searching for a better match. Error–based learning is naturally suited to problems such as adaptive control and the learning of *sensory–motor maps*, which require ongoing adaptation to present statistics. Neural networks that employ error–based learning include backpropagation and other multilayer perceptrons (MLPs).

Activation dynamics of ART2 is governed by the ODEs [Gro82, CG03]

$$\epsilon \dot{x}_i = -Ax_i + (1 - Bx_i)I_i^+ - (C + Dx_i)I_i^-,$$

where ϵ is the 'small parameter', I_i^+ and I_i^- are excitatory and inhibitory inputs to the ith unit, respectively, and $A, B, C, D > 0$ are parameters.

General *Cohen–Grossberg activation equations* have the form:

$$\dot{v}_j = -a_j(v_j)[b_j(v_j) - f_k(v_k)m_{jk}], \qquad (j = 1, \dots, N), \tag{1.24}$$

and the *Cohen–Grossberg theorem* ensures the global stability of the system (1.24). If

$$a_j = 1/C_j, b_j = v_j/R_j - I_j, f_j(v_j) = u_j,$$

and constant $m_{ij} = m_{ji} = T_{ji}$, the system (1.24) reduces to the Hopfield circuit model (1.20).

ART and distributed ART (dART) systems are part of a growing family of self–organizing network models that feature attentional feedback and stable code learning. Areas of technological application include industrial design and manufacturing, the control of mobile robots, face recognition, remote sensing land cover classification, target recognition, medical diagnosis, electrocardiogram analysis, signature verification, tool failure monitoring, chemical analysis, circuit design, protein/DNA analysis, 3D visual object recognition, musical analysis, and seismic, sonar, and radar recognition. ART principles have further helped explain parametric behavioral and brain data in the areas of visual perception, object recognition, auditory source identification, variable–rate speech and word recognition, and *adaptive sensory–motor control* (see [CG03]).

Spatiotemporal Networks

In *spatiotemporal networks*, activation dynamics is governed by the ODEs

$$\begin{aligned} \dot{x}^i &= A(-ax_i + b[I_i - \Gamma]^+), \\ \dot{\Gamma} &= \alpha(S - T) + \beta\dot{S}, \qquad \text{with} \\ [u]^+ &= \begin{cases} u & \text{if} \quad u > 0 \\ 0 & \text{if} \quad u \leq 0 \end{cases}, \\ A(u) &= \begin{cases} u & \text{if} \quad u > 0 \\ cu & \text{if} \quad u \leq 0 \end{cases}. \end{aligned}$$

where $a, b, \alpha, \beta > 0$ are parameters, $T > 0$ is the *power–level target*, $S = \sum_i x^i$, and $A(u)$ is called the *attack function.*

Learning dynamics is given by *differential Hebbian law*

$$\begin{aligned} \dot{\omega}_{ij} &= (-c\omega_{ij} + dx_i x_j)U(\dot{x}^i)U(-\dot{x}_j), \qquad \text{with} \\ U(s) &= \begin{cases} 1 & \text{if} \quad s > 0 \\ 0 & \text{if} \quad s \leq 0 \end{cases} \qquad \text{where } c, d > 0 \text{ are constants.} \end{aligned}$$

Fuzzy Systems

Recall that f*uzzy expert systems* are based on *fuzzy logic* (FL), which is itself derived from *fuzzy set theory* dealing with reasoning that is approximate rather than precisely deduced from classical *predicate logic*.[150] FL, introduced in 1965 by Prof. Lotfi Zadeh at the University of California, Berkeley, can be thought of as the application side of fuzzy set theory dealing with well thought out real world expert values for a complex problem.[151] FL allows for set membership values between and including 0 and 1, shades of gray as well as black and white, and in its linguistic form, imprecise concepts like 'slightly', 'quite' and 'very'. Specifically, it allows partial membership in a set. It is related to fuzzy sets and possibility theory.

---

[150] Recall that *predicate* or *propositional logic* (PL) is a system for evaluating the validity of arguments by encoding them into sentential variables and boolean operator and is part of the philosophy of *formal logic*. The actual truth of the *premises* is not particularly relevant in PL; it is dealing mostly with the structure of an argument so that if it so happens that the premises are true, the conclusion either must be true, or could perhaps be false. If it is demonstrable that the conclusion must be true then the original argument can be said to be valid. However, if it is possible for all of the premises to be true, and yet still have a false conclusion, the sequent is invalid. In an ordinary PL, there is one unitary operator, four binary operators and two quantifiers. The only unary operator in PL is the negation, usually denoted by $\neg P$, which is the opposite of the predicate (i.e., Boolean variable) P. The binary operators are: (i) *conjunction* $\wedge$, which is true iff both of the Boolean conjuncts are true; (ii) *disjunction* $\vee$, which is false iff both of the Boolean disjuncts are false; (iii) implication (or, conditional), meaning, *if* P *then* Q, and denoted $P \Longrightarrow Q$, where P is *antecedent* and Q is *consequent*; implication is false only iff from true P follows false Q; (iv) equivalence, or bi–conditional is a double–sided implication, $(P \Longrightarrow Q) \wedge (Q \Longrightarrow P)$; it is false iff from true P follows false Q and from true Q follows false P. Besides, PL also has the *universal quantifier* $\forall$, meaning 'for al', and the *existential quantifier* $\exists$, meaning 'there is'.

[151] Note that *degrees of truth* in fuzzy logic are often confused with probabilities. However, they are conceptually distinct; fuzzy truth represents membership in vaguely defined sets, not likelihood of some event or condition. To illustrate the difference, consider this scenario: Bob is in a house with two adjacent rooms: the kitchen and the dining room. In many cases, Bob's status within the set of things 'in the kitchen' is completely plain: he's either 'in the kitchen' or 'not in the kitchen'. What about when Bob stands in the doorway? He may be considered 'partially in the kitchen'. Quantifying this partial state yields a fuzzy set membership. With only his big toe in the dining room, we might say Bob is 99% 'in the kitchen' and 1% 'in the dining room', for instance. No event (like a coin toss) will resolve Bob to being completely 'in the kitchen' or 'not in the kitchen', as long as he's standing in that doorway. Fuzzy sets are based on vague definitions of sets, not randomness. Fuzzy logic is controversial in some circles, despite wide acceptance and a broad track record of successful applications. It is rejected by some control engineers for validation and other reasons, and by some statisticians who hold that probability is the only rigorous mathematical description of uncertainty. Critics also argue that it cannot be a superset of ordinary set theory since membership functions are defined in terms of conventional sets.

'Fuzzy Thinking'

'There is no logic in logic', pronounced the father of fuzzy logic, Lotfi Zadeh. His cryptic play–on–words, he explained, means that the kind of logic that people use to solve most real world problems rather than the artificial problems for which mathematical solutions are available is not the kind of logic that engineers are taught in school. 'An engineer can solve problems throughout his whole career without ever needing to resort to the brand of logic he was trained in', said Zadeh. 'Why? Because all people, even engineers, compute with words not the logical symbols taught in school', Zadeh maintained. 'In the future, computing will be done with words from natural languages, rather than with symbols that are far removed from daily life.'

In 1973, Zadeh proposed the concept of linguistic or fuzzy variables [Zad65, Zad78, Yag87]. Think of them as linguistic objects or words, rather than numbers. The sensor input is a noun, e.g., temperature, displacement, velocity, ow, pressure, etc. Since error is just the difference, it can be thought of the same way. The fuzzy variables themselves are adjectives that modify the variable (e.g., large positive error, small positive error, zero error, small negative error, and large negative error). As a minimum, one could simply have positive, zero, and negative variables for each of the parameters.

Additional ranges such as very large and very small could also be added to extend the responsiveness to exceptional or very nonlinear conditions, but are not necessary in a basic system. Normal logic is just not up to modelling the real world, claims Bart Kosko [Kos92, Kos93, Kos96, Kos99], perhaps the worlds most active proponent of fuzzy logic. According to Kosko, there is always *ambiguity* in our perceptions and measurements that is difficult to reflect in traditional logic. Probability attempts to reflect ambiguity by resorting to statistical averages over many events. But fuzzy theory describes the ambiguity of individual events. It measures the degree to which an event occurs, not whether it occurs.

Fuzzy Sets

Recall that a crisp (ordinary mathematical) set X is defined by a binary characteristic function $\chi_X(x)$ of its elements x

$$\chi_X(x) = \begin{cases} 1, & \text{if } x \in X, \\ 0, & \text{if } x \notin X, \end{cases}$$

while a fuzzy set is defined by a continuous characteristic function

$$\chi_X(x) = [0, 1] \, ,$$

including all (possible) real values between the two crisp extremes 1 and 0, and including them as special cases.

More precisely, a fuzzy set X is defined as a collection of ordered pairs

$$X = \{(x, \mu(x))\}, \tag{1.25}$$

where $\mu(x)$ is the *fuzzy membership function* representing the grade of membership of the element x in the set X. A single pair is called a fuzzy *singleton.*

Lotfi Zadeh claimed that many *sets* in the world that surrounds us are defined by a non-distinct boundary. Indeed, the *set of high mountains* is an example of such sets. Zadeh decided to extend two–valued logic, defined by the binary pair $\{0, 1\}$ to the whole continuous interval $[0, 1]$ thereby introducing a gradual transition from falsehood to truth. The original and pioneering papers on fuzzy sets by Zadeh [Zad65, Zad78, Yag87] explain the theory of fuzzy sets that result from the extension as well as a fuzzy logic based on the set theory.

Fuzzy sets are a further development of the mathematical concept of a set. Sets were first studied formally by German mathematician Georg Cantor (1845–1918). His theory of sets met much resistance during his lifetime, but nowadays most mathematicians believe it is possible to express most, if not all, of mathematics in the language of set theory. Many researchers are looking at the consequences of 'fuzzifying' set theory, and much mathematical literature is the result.

Conventional sets. A set is any collection of objects which can be treated as a whole. Cantor described a set by its members, such that an item from a given universe is either a member or not. Almost anything called a *set* in ordinary conversation is an acceptable set in the mathematical sense. A set can be specified by its members, they characterize a set completely. The list of members $A = \{0, 1, 2, 3\}$ specifies a finite set. Nobody can list all elements of an *infinite set*, we must instead state some property which characterizes the elements in the set, for instance the predicate $x > 10$. That set is defined by the elements of the *universe of discourse* which make the predicate true. So there are two ways to describe a set: explicitly in a list or implicitly with a predicate.

Fuzzy sets. Following Zadeh many sets have more than an *Either–Or* criterion for membership. Take for example the set of *young people*. A one year old baby will clearly be a member of the set, and a 100 years old person will not be a member of this set, but what about people at the age of 20, 30, or 40 years? Another example is a weather report regarding high temperatures, strong winds, or nice days. In other cases a criterion appears nonfuzzy, but is perceived as fuzzy: a speed limit of 60 kilometers per hour, a check–out time at 12 noon in a hotel, a 50 years old man. Zadeh proposed a *grade of membership*, such that the transition from membership to non–membership is gradual rather than abrupt.

The grade of membership for all its members thus describes a fuzzy set. An item's grade of membership is normally a real number between 0 and 1, often denoted by the Greek letter μ. The higher the number, the higher the membership. Zadeh regards Cantor's set as a special case where elements have full membership, i.e., $\mu = 1$. He nevertheless called Cantor's sets *nonfuzzy*; today the term *crisp* set is used, which avoids that little dilemma.

The membership for a 50 year old in the set *young* depends on one's own view. The grade of membership is a precise, but subjective measure that depends on the context.

A fuzzy membership function is different from a statistical probability distribution. A possible event does not imply that it is probable. However, if it is probable it must also be possible. We might view a fuzzy membership function as our personal distribution, in contrast with a statistical distribution based on observations.

Universe of discourse. Elements of a fuzzy set are taken from a *universe of discourse*. It contains all elements that can come into consideration. Even the universe of discourse depends on the context. An application of the universe is to suppress faulty measurement data. In case we are dealing with a non–numerical quantity, for instance *taste*, which cannot be measured against a numerical scale, we cannot use a numerical universe. The elements are then said to be taken from a *psychological continuum*.

Membership Functions. Every element in the universe of discourse is a member of the fuzzy set to some grade, maybe even zero. The set of elements that have a non–zero membership is called the *support* of the fuzzy set. The function that ties a number to each element $x$of the universe is called the *membership function*.

Continuous and discrete representations. There are two alternative ways to represent a membership function in a computer: continuous or discrete. In the continuous form the membership function is a mathematical function, possibly a program. A membership function is for example bell-shaped (also called a $\pi-curve$), $s-$shaped (called an $s-curve$), a reverse $s-curve$ (called $z-curve$), triangular, or trapezoidal. In the discrete form the membership function and the universe are discrete points in a list (vector). Sometimes it can be more convenient with a sampled (discrete) representation. As a very crude rule of thumb, the continuous form is more CPU intensive, but less storage demanding than the discrete form.

Normalization. A fuzzy set is *normalized* if its largest membership value equals 1. We normalize by dividing each membership value by the largest membership in the set, $a/max(a)$.

Singletons. Strictly speaking, a fuzzy set A is a collection of ordered pairs: $A = \{(x, \mu(x))\}$.

Item x belongs to the universe and $\mu(x)$ is its *grade of membership* in A. A single pair $(x, \mu(x))$ is called a fuzzy *singleton*; thus the whole set can be viewed as the union of its constituent singletons.

Linguistic variables. Just like an algebraic variable takes numbers as values, a *linguistic variable* takes words or sentences as values [Yag87, Kos92]. The set of values that it can take is called its *term set*. Each value in the term set is a *fuzzy variable* defined over a *base variable*. The base variable defines the universe of discourse for all the fuzzy variables in the term set. In short, the hierarchy is as follows:

$$\text{linguistic variable} \to \text{fuzzy variable} \to \text{base variable}.$$

Primary terms. A *primary term* is a term or a set that must be defined a priori, for example *Young* and *Old*, whereas the sets *Very Young* and *Not Young* are modified sets.

Fuzzy set operations. A fuzzy *set operation* creates a new set from one or several given sets.

Let A and B be fuzzy sets on a mutual universe of discourse X. If these were ordinary (crisp) sets, we would have the following definitions:

The *intersection* of A and B is: $A \cap B \equiv \min\{A, B\}$, where *min* is an item–by–item minimum operation.

The *union* of A and B is: $A \cup B \equiv \max\{A, B\}$, where *max* is an item–by–item maximum operation.

The *complement* of A is: $\neg A \equiv 1 - A$, where in a each membership value is subtracted from 1.

However, as A and B are fuzzy sets, the following definitions are more appropriate:

The *intersection* of A and B is: $A \cap B \equiv \min\{\mu_A(X), \mu_B(X)\}$, where *min* is an item–by–item minimum operation.

The *union* of A and B is: $A \cup B \equiv \max\{\mu_A(X), \mu_B(X)\}$, where *max* is an item–by–item maximum operation.

The *complement* of A is: $\neg A \equiv 1 - \mu_A(X)$, where in a each membership value is subtracted from 1.

Fuzzy Example

Using fuzzy membership functions $\mu(x)$, we can express both physical and non–physical quantities (e.g., temperature, see Figure 1.29) using *linguistic variables*.

Various logical combinations of such linguistic variables leads to the concept of fuzzy–logic control. Recall that basic logical operations AND, OR, NOT are defined as:

$$\begin{aligned} AND: &\quad C \cap W \quad - \quad \text{intersection of crisp sets} \quad C, W, \\ OR: &\quad C \cup W \quad - \quad \text{union of crisp sets} \quad C, W, \\ NOT: &\quad \neg C \quad - \quad \text{complement of a crisp set} \quad C. \end{aligned}$$

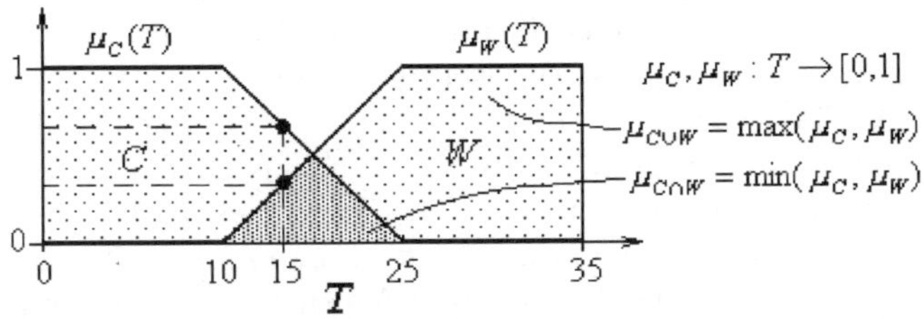

Fig. 1.29. Fuzzy–set description of *cold* (C) and *warm* (W) *temperature* (T), using the membership functions $\mu_C(T)$ and $\mu_W(T)$, respectively. For example, fuzzy answers to the questions "How cold is 15°?" and "How warm is 15°?" are given by: "15° *is quite cold* as $\mu_C(15) = 2/3$" and "15° *is not really warm* as $\mu_W(15) = 1/3$", respectively.

The corresponding fuzzy–logic operations are defined as:

$$\begin{aligned} AND: &\quad \mu_{C\cap W}(T) = \min\{\mu_C(T), \mu_W(T)\}, \\ OR: &\quad \mu_{C\cup W}(T) = \max\{\mu_C(T), \mu_W(T)\}, \\ NOT: &\quad \mu_{\neg C}(T) = 1 - \mu_C(T). \end{aligned}$$

Fuzziness of the Real World

The real world consists of all subsets of the universe and the only subsets that are not fuzzy are the constructs of classical mathematics.

From small errors to satisfied customers to safe investments to noisy signals to charged particles, each element of the real world is in some measure fuzzy. For instance, satisfied customers can be somewhat unsatisfied, safe investments somewhat unsafe and so on. What is worse, most events more or less smoothly transition into their opposites, making classification difficult near the midpoint of the transition. Unfortunately, textbook events and their opposites are crisp, unlike the real world. Take the proposition that there is a 50% chance that an apple is in the refrigerator. That is an assertion of crisp logic. But suppose upon investigation it is found that there is half an apple in the refrigerator, that is fuzzy.

But regardless of the realities, the crisp logic in vogue today assumes that the world is really unambiguous and that the only uncertainty is the result of random samples from large sets. As the facts about these large sets become better known, the randomness supposedly dissipates, so that if science had access to all the facts, it would disappear. Unfortunately, if all the facts were in, a platypus would remain only roughly an mammal.

On the other hand, fuzzy logic holds that uncertainty is deterministic and does not dissipate as more elements of a set are examined. Take an ellipse, for instance. It is approximately a circle, to whatever degree that it resembles a perfect circle. There is nothing random about it. No matter how precisely it is measured it remains only approximately a circle. All the facts are in and yet uncertainty remains.

Traditional crisp logic has a difficult time applying itself to very large sets, since probability fades to unity, as well as to individual events where probabilities cannot be defined at all. Nevertheless, crisp logic continues to rein supreme based on long standing western traditions that maintain that rationality would vanish if there were not crisp logical ideals to which we should aspire. These laws of (rational) thought were first characterized by Aristotle as the principle of non-contradiction and the principle of the excluded middle. The principle of non-contradiction, stated in words, says that nothing can be both A and $\neg A$. The law of the excluded middle says that anything must be either A or $\neg A$.

'Fuzziness is the denial of both these so–called laws', says E. Cox [Cox92, Cox94]). The classical example is of a platypus which both is and is not a mammal. In such individual cases, even appending probability theory to crisp logic cannot resolve the paradox. For instance, take the now classical paradox formulated by B. Russell: If a barber shaves everyone in a village who does not shave himself, then who shaves the barber? This paradox was devised to assault G. Cantor's set theory as the foundation for G. Boole's digital logic. It has been restated in many forms, such as the liar from Crete who said that all Creatans are liars. Russell solved it by merely disqualifying such self–referential statements in his set theory. Probability theory solves it by assuming a population of barbers 50% of whom do, and 50% of whom do not, shave themselves. But fuzzy logic solves it by assigning to this individual barber a 50% membership value in the set self-shaving barbers. Further, it shows that there is a whole spectrum of other situations that are less fuzzy and which correspond to other degrees of set membership. Such as, barbers who shave themselves 70% of the time.

Kosko illustrates these various degrees of ambiguity by geometrically plotting various degrees of set membership inside a *unit fuzzy hypercube* $[0,1]^n$ [Kos92, Kos93, Kos96, Kos99]. This sets–as–points approach holds that a fuzzy set is a point in a unit hypercube and a non–fuzzy set is a corner of the hypercube. Normal engineering practice often visualizes binary logical values as the corners of a hypercube, but only fuzzy theory uses the inside of the cube. Fuzzy logic is a natural filling–in of traditional set theory. Any engineer will recognize the 3D representation of all possible combinations three Boolean values: $\{0,0,0\},\{0,0,1\},\{0,1,0\},\{0,1,1\},\{1,0,0\},\{1,0,1\},\{1,1,0\},\{1,1,1\}$, which correspond to the corners of the unit hypercube. But fuzzy logic also allows any other fractional values inside the hypercube, such as $\{0.5,0.7,0.3\}$ corresponding to degrees of set membership.

Fuzzy logic holds that any point inside the unit hypercube is a fuzzy set with Russell's paradox located at the point of maximum ambiguity in the center of the hypercube.

Fuzzy Entropy

Degrees of fuzziness are referred to as entropy by Kosko. *Fuzzy mutual entropy* measures the *ambiguity of a situation*, information and entropy are inversely related – if you have a maximum–entropy solution, then you have a minimum–information solution, and visa versa, according to Kosko. But minimum–information does not mean that too little information is being used. On the contrary, the principle of maximum entropy ensures that only the relevant information is being used.

This idea of maximizing entropy, according to Kosko, is present throughout the sciences, although it is called by different names. 'From the quantum level up to astrophysics or anywhere in–between for pattern recognition, you want to use all and only the available information,' Kosko claims. This emergent model proposes that scientists and engineers estimate the uncertainty structure of a given environment and maximize the entropy relative to the known information, similar to the Lagrange technique in mathematics. The principle of maximum entropy states that any other technique has to be biased, because it has less entropy and thus uses more information than is really available.

Fuzzy theory provides a measure of this entropy factor. It measures ambiguity with operations of union $\cup$, intersection $\cap$ and complement $\neg$.

In traditional logic, these three operators are used to define a set of axioms that were proposed by Aristotle to be the immutable laws of (rational) thought, namely, the principle of *non–contradiction* and the principle of the *excluded middle.* The principle of non–contradiction, that nothing can be both A and $\neg A$, and the law of the excluded middle, that anything must be either A or $\neg A$, amounts to saying that the intersection of a set and its complement is always empty and that the union of a set and its complement always equals the whole *universe of discourse*, respectively. But if we do not know A with certainty, then we do not know $-A$ with certainty either, else by double negation we would know A with certainty. This produces non–degenerate *overlap* $(A \cap \neg A)$, which breaks the law of non–contradiction. Equivalently, it also produced non–degenerate *underlap* $(A \cup \neg A)$ which breaks the law of the excluded middle. In fuzzy logic both these so–called laws are denied. A set and its complement can both be overlap and underlap.

What is worse, there is usually ambiguity in more than one parameter or dimension of a problem. To represent multi–dimensional ambiguity, Kosko shows fuzzy entropy geometrically with a hypercube.

All these relationships are needed in fuzzy logic to express its basic structures for *addition*, *multiplication*, and most important, *implication* $IF \Rightarrow THEN$. They all follow from the subsethood relationships between fuzzy sets. The subset relation by itself, corresponds to the implication relation in crisp logic. For instance, $A \Rightarrow B$ is *false only* if the *antecedent* A is *true* and the *consequent* B is *false.* The same holds for subsets, A is a subset of B if there is no element that belongs to A but not to B.

But in fuzzy logic, degrees of subsethood permit some A to be somewhat of a subset of B even though some of its elements are not elements of B. The degree to which A is a subset of B can be measured as the distance from the origin to $(A \cap B)$ divided by the distance from the origin to A.

This structure is derived as a theorem of fuzzy logic, whereas for probability theory equivalent conditional probability theorem has to be assumed, making fuzzy logic a more fundamental.

The *fuzzy mutual entropy* measures how close a fuzzy description of the world is to its own opposite [Kos99]. It has no random analogue in general. The *fuzzy fluid* leads to a type of wave equation. The wave shows how the *extended Shannon entropy potential* $S : [0,1]^n \to \mathbb{R}$, defined on the *entire fuzzy cube* $[0,1]^n$, fluctuates in time. It has the form of a *reaction–diffusion* equation

$$\dot{S} = -c\,\nabla^2 S, \tag{1.26}$$

where c is the *fuzzy diffusion parameter*. The *fuzzy wave equation* (1.26) implies $\dot{S} > 0$, and thus resembles the entropy increase of the S−theorem of the *Second Law of thermodynamics.*

Similar equations occur in all branches of science and engineering. The Schrödinger wave equation (see [II06a, II06b]) has this form, as well as most models of diffusion. The fuzzy wave equation (1.26) assumes only that information is conserved. The total amount of information is fixed and we do not create or destroy information. Some form of the wave equation would still apply if information were conserved locally or in small regions of system space. The space itself is a fuzzy cube of high dimension. It has as many dimensions as there are objects of interest. The Shannon entropy S changes at each point in this cube and defines a *fuzzy wave.* The obvious result is that the entropy S can only grow in time in the spirit of the second law.

The entropy always grows but its *rate of growth* depends on the system's position in the *fuzzy parameter space*. A deeper result is that entropy changes slowest at the fuzzy cube *midpoint of maximum fuzz.* That is the only point in the cube where the fuzzy description equals its own opposite. The Shannon entropy wave grows faster and faster away from the cube midpoint and near its skin. The skin or surface of the fuzzy cube is the only place where a 0 or 1 appears in the system description. The fuzzy wave equation (1.26) shows that the entropy S changes infinitely fast iff it touches the cubes's skin. However, this is impossible in a universe with finite bounds on velocity like the speed of light. So, the result is never a *bit* – it is always a *fit* [Kos99].

Fuzzy Patches for System Modelling

Like ANNs, the fuzzy logic systems are generic *function approximators* [Kos92]. Namely, fuzzy system modelling is performed as a *nonlinear function approximation* using the so–called *fuzzy patches* (see Figure 1.30), which approximate the given function $y = f(x)$, i.e., the *system input–output relation.* The fuzzy patches R_i are given by a set of canonical fuzzy IF–THEN rules:

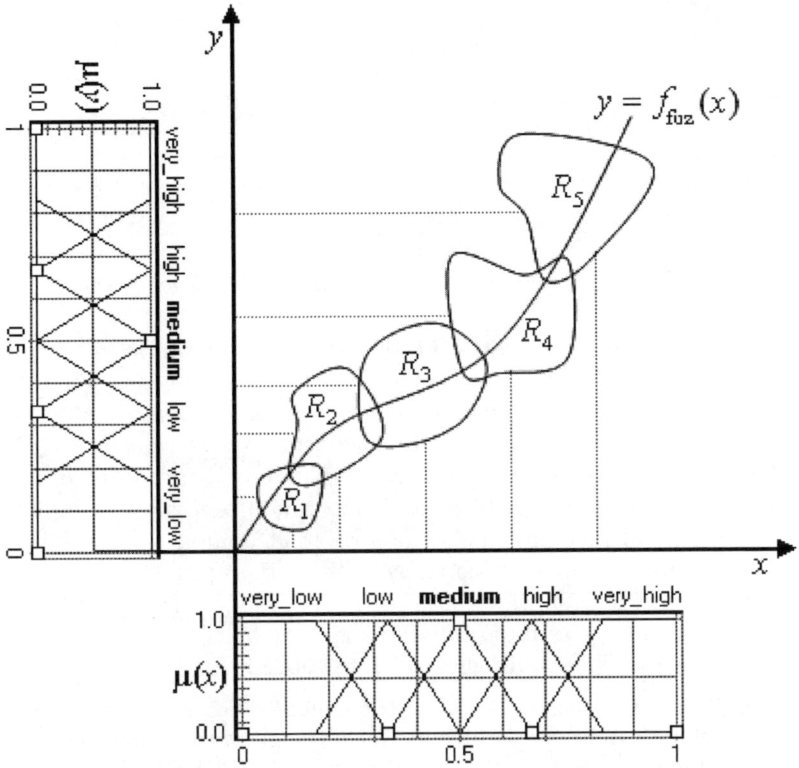

Fig. 1.30. Fuzzy–logic approximation $y = f_{\text{fuz}}(x)$ of an arbitrary function $y = f(x)$ using *fuzzy patches* R_i given by a set of canonical fuzzy IF–THEN rules.

$$
\begin{aligned}
&R_1 : IF \quad x \quad is \quad A_1 \quad THEN \quad y \quad is \quad R_1, \\
&R_2 : IF \quad x \quad is \quad A_2 \quad THEN \quad y \quad is \quad R_2, \\
&\quad\vdots \\
&R_n : IF \quad x \quad is \quad A_n \quad THEN \quad y \quad is \quad R_n.
\end{aligned}
$$

Fuzzy Inference Engine

In the realm of fuzzy logic the above generic nonlinear function approximation is performed by means of fuzzy inference engine. The *fuzzy inference engine* is an *input–output dynamical system* which *maps* a set of input linguistic variables (IF–part) into a set of output linguistic variables ($THEN$–part). It has three sequential modules (see Figure 1.31):

1. *Fuzzification*; in this module numerical crisp input variables are fuzzified; this is performed as an overlapping partition of their universes of discourse by means of fuzzy membership functions $\mu(x)$ (1.25), which can have

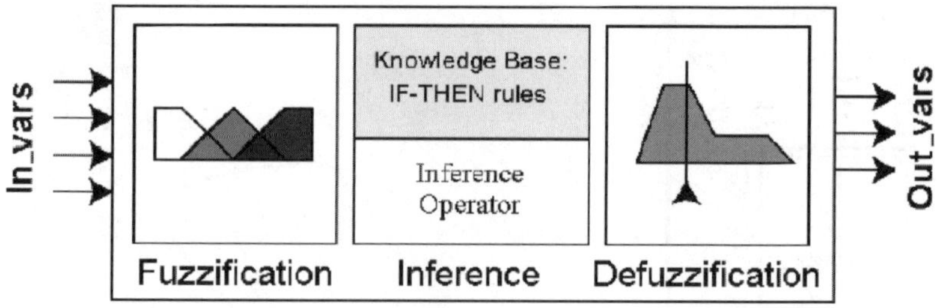

Fig. 1.31. Basic structure of the fuzzy inference engine.

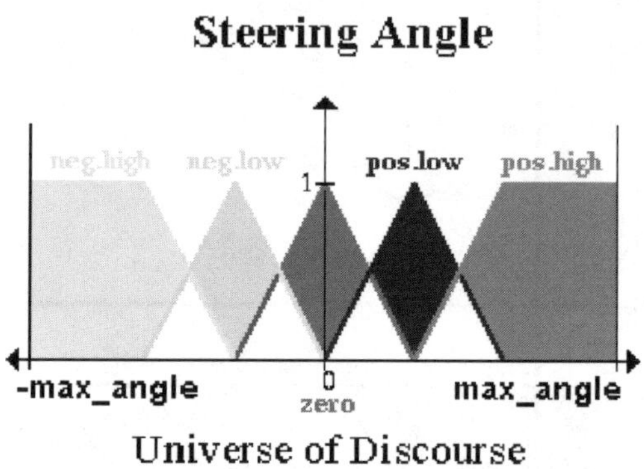

Fig. 1.32. Fuzzification example: set of triangular–trapezoidal membership functions partitioning the universe of discourse for the angle of the hypothetical steering wheel; notice the white overlapping triangles.

various shapes, like triangular–trapezoidal (see Figure 1.32), Gaussian–bell, $\mu(x) = \exp\left[\frac{-(x-m)^2}{2\sigma^2}\right]$ (with mean m and standard deviation σ), sigmoid $\mu(x) = \left[1 + \left(\frac{x-m}{\sigma}\right)^2\right]^{-1}$, or some other shapes.
B. Kosko and his students have done extensive computer simulations looking for the best shape of fuzzy sets to model a known test system as closely as possible. They let fuzzy sets of all shapes and sizes compete against each other. They also let neural systems tune the fuzzy–set curves to improve how well they model the test system. The main conclusion from these experiments is that 'triangles never do well' in such contests. Suppose we want an adaptive fuzzy system $F : \mathbb{R}^n \to \mathbb{R}$ to approximate a test function (or, approximand) $f : \mathbb{R}^n \to \mathbb{R}$ as closely as possible in the sense of minimizing the mean–squared error between them, $\left(\|f - F\|^2\right)$. Then the ith scalar 'sinc' function (as commonly used in signal processing),

$$\mu_i(x) = \frac{\sin\left(\frac{x-m_i}{d_i}\right)}{\frac{x-m_i}{d_i}}, \qquad (i = 1, \dots, n), \qquad (1.27)$$

with center m_i and dispersion (width) $d_i = \sigma_i^2 > 0$, often gives the best performance for IF–part mean–squared function approximation, even though this generalized function can take on negative values (see [Kos99]).

2. *Inference*; this module has two submodules:
(i) The expert–knowledge base consisting of a set of $IF - THEN$ rules relating input and output variables, and
(ii) The inference method, or implication operator, that actually combines the rules to give the fuzzy output; the most common is *Mamdani Min–Max inference*, in which the membership functions for input variables are first combined inside the $IF - THEN$ rules using AND ($\cap$, or Min) operator, and then the output fuzzy sets from different $IF - THEN$ rules are combined using OR ($\cup$, or Max) operator to get the common fuzzy output (see Figure 1.33).
3. *Defuzzification*; in this module fuzzy outputs from the inference module are converted to numerical crisp values; this is achieved by one of the several defuzzification algorithms; the most common is the Center of Gravity method, in which the crisp output value is calculated as the abscissa under the center of gravity of the output fuzzy set (see Figure 1.33).

In more complex technical applications of general function approximation (like in complex control systems, signal and image processing, etc.), two optional blocks are usually added to the fuzzy inference engine [Kos92, Kos96, Lee90]:

(0) *Preprocessor*, preceding the fuzzification module, performing various kinds of normalization, scaling, filtering, averaging, differentiation or integration of input data; and

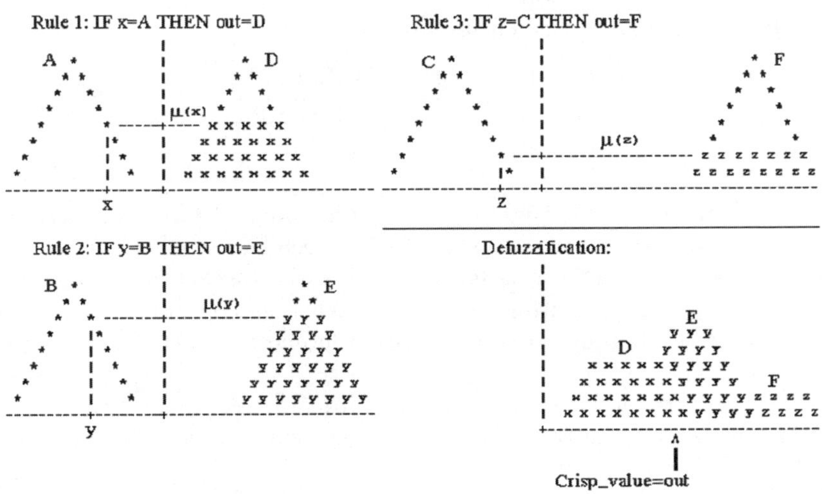

Fig. 1.33. Mamdani's Min–Max inference method and Center of Gravity defuzzification.

(4) *Postprocessor*, succeeding the defuzzification module, performing the analog operations on output data.

Common fuzzy systems have a simple feedforward mathematical structure, the so–called *Standard Additive Model* (SAM), which aids the spread of applications. Almost all applied fuzzy systems use some form of SAM, and some SAMs in turn resemble the ANN models (see [Kos99]).

In particular, an *additive fuzzy system* $F : \mathbb{R}^n \to \mathbb{R}^p$ stores m rules of the patch form $A_i \times B_i \subset \mathbb{R}^n \times \mathbb{R}^p$, or of the word form '**If** $X = A_i$ **Then** $Y = B_i$' and adds the 'fired' Then–parts $B_i^{'}(x)$ to give the output set $B(x)$, calculated as

$$B(x) = w_i B_i^{'}(x) = w_i \mu_i(x) B_i(x), \qquad (i = 1, \dots, n), \tag{1.28}$$

for a scalar rule weight $w_i > 0$. The factored form $B_i^{'}(x) = \mu_i(x) B_i(x)$ makes the additive system (1.28) a SAM system. The fuzzy system F computes its output $F(x)$ by taking the centroid of the output set $B(x)$: $F(x) = \text{Centroid}(B(x))$. The *SAM theorem* then gives the centroid as a simple ratio,

$$F(x) = p_i(x) c_i, \qquad (i = 1, \dots, n),$$

where the convex coefficients or discrete probability weights $p_i(x)$ depend on the input x through the ratios

$$p_i(x) = \frac{w_i \mu_i(x) V_i}{w_k \mu_k(x) V_k}, \qquad (i = 1, \dots, n). \tag{1.29}$$

V_i is the finite positive volume (or area if $p = 1$ in the codomain space $\mathbb{R}^p$) [Kos99],

$$V_i = \int_{\mathbb{R}^p} b_i(y_1, \dots, y_p) dy_1 \dots dy_p > 0,$$

and c_i is the centroid of the Then–part set $B_i(x)$,

$$c_i = \frac{\int_{\mathbb{R}^p} y \, b_i(y_1, \dots, y_p) dy_1 \dots dy_p}{\int_{\mathbb{R}^p} b_i(y_1, \dots, y_p) dy_1 \dots dy_p}.$$

Fuzzy Logic Control

The most common and straightforward applications of fuzzy logic are in the domain of nonlinear control [Kos92, Kos96, Lee90, DSS96]. Fuzzy control is a nonlinear control method based on fuzzy logic. Just as fuzzy logic can be described simply as computing with words rather than numbers, fuzzy control can be described simply as control with sentences rather than differential equations.

A fuzzy controller is based on the fuzzy inference engine, which acts either in the feedforward or in the feedback path, or as a supervisor for the conventional PID controller.

A fuzzy controller can work either directly with fuzzified dynamical variables, like direction, angle, speed, or with their fuzzified errors and rates of change of errors. In the second case we have rules of the form:

1. IF error is Neg AND change in error is Neg THEN output is NB.
2. IF error is Neg AND change in error is $Zero$ THEN output is NM.

The collection of rules is called a rule base. The rules are in $IF-THEN$ format, and formally the $IF-$side is called the condition and the $THEN-$side is called the conclusion (more often, perhaps, the pair is called antecedent – consequent). The input value Neg is a linguistic term short for the word Negative, the output value NB stands for $Negative\_Big$ and NM for $Negative\_Medium$. The computer is able to execute the rules and compute a control signal depending on the measured inputs error and change in error.

The rule–base can be also presented in a convenient form of one or several rule matrices, the so–called $FAM-$matrices, where FAM is a shortcut for Kosko's *fuzzy associative memory* [Kos92, Kos96]. For example, a 9×9 graded FAM matrix can be defined in a symmetrical weighted form:

$$\mathrm{FAM} = \begin{pmatrix} 0.6\mathrm{S}4 & 0.6\mathrm{S}4 & 0.7\mathrm{S}3 & ... & \mathrm{CE} \\ 0.6\mathrm{S}4 & 0.7\mathrm{S}3 & 0.7\mathrm{S}3 & ... & 0.9\mathrm{B}1 \\ 0.7\mathrm{S}3 & 0.7\mathrm{S}3 & 0.8\mathrm{S}2 & ... & 0.9\mathrm{B}1 \\ ... & ... & ... & ... & 0.6\mathrm{B}4 \\ \mathrm{CE} & 0.9\mathrm{B}1 & 0.9\mathrm{B}1 & ... & 0.6\mathrm{B}4 \end{pmatrix},$$

in which the vector of nine linguistic variables L^9 partitioning the *universes of discourse* of all three variables (with trapezoidal or Gaussian bell–shaped *membership functions*) has the form

$$L^9 = \{S4, S3, S2, S1, CE, B1, B2, B3, B4\}^T,$$

to be interpreted as: 'small 4', ... , 'small 1', 'center', 'big 1', ... , 'big 4'. For example, the left upper entry $(1,1)$ of the FAM matrix means: IF red is S4 and blue is S4, THEN result is 0.6S4; or, entry $(3,7)$ means: IF red is S2 and blue is B2, THEN result is center, etc.

Here we give three design examples for fuzzy controllers, the first one in detail, and the other two briefly.

Example: Mamdani Fuzzy Controller

The problem is to balance θ a pole of mass m and inertia moment I on a mobile platform of mass M that can be forced by F to move only (left/right) along $x-$axis (see Figure 1.34). This is quite an involved problem for conventional PID controller, based on differential equations of the pole and platform motion. Instead, we will apply fuzzy linguistic technique called *Mamdani inference* (see previous subsection).

Firstly, as a *fuzzification* part, we have to define (subjectively) what high speed, low speed etc. of the platform M is. This is done by specifying the membership functions for the fuzzy set partitions of the *platform speed* universe of discourse, using the following linguistic variables: (i) negative high

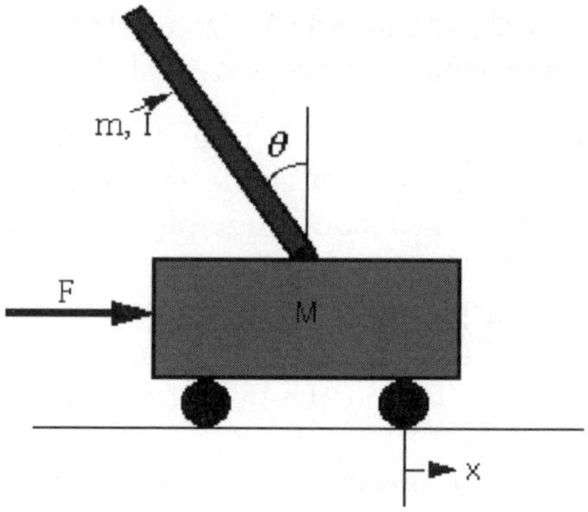

Fig. 1.34. Problem of balancing an inverted pendulum.

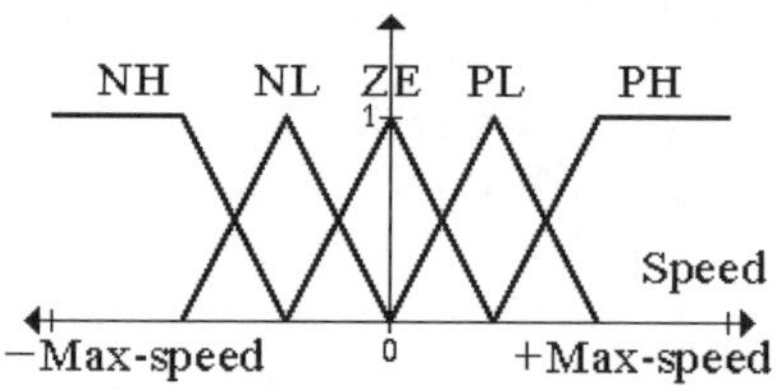

Fig. 1.35. Fuzzy membership functions for speed of the platform.

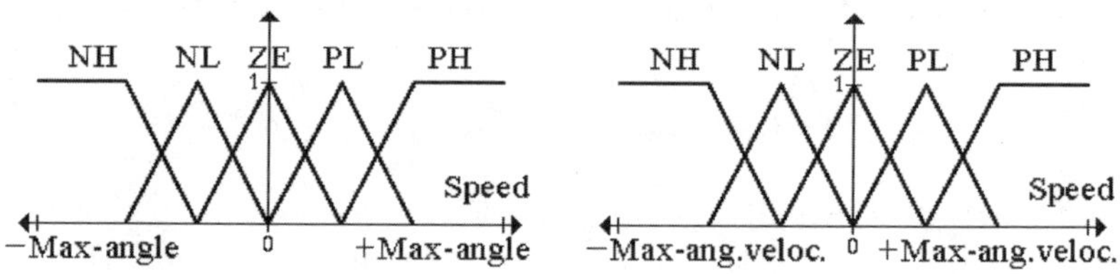

Fig. 1.36. Fuzzy membership functions for speed of the platform.

(NH), (ii) negative low (NL), (iii) zero (ZE), (iv) positive low (PL), and (v) positive high (PH) (see Figure 1.35).[152]

Also, we need to do the same for the angle θ between the platform and the pendulum and the angular velocity $\dot{\theta}$ of this angle (see Figure 1.36).

Secondly, as an *inference* part, we give several *fuzzy IF–THEN rules* that will tell us what to do in certain situations. Consider for example that the pole is in the upright position (angle θ is zero) and it does not move (angular velocity $\dot{\theta}$ is zero). Obviously this is the desired situation, and therefore we don't have to do anything (speed is zero). Let us consider also another case: the pole is in upright position as before but is in motion at *low* velocity in *positive*

[152] For simplicity, we assume that in the beginning the pole is in a nearly upright position so that an angle θ greater than, 45 degrees in any direction can never occur.

direction. Naturally we would have to compensate the pole's movement by moving the platform in the same direction at *low* speed.

So far we've made up two rules that can be put into a more formalized form like this:

IF angle is zero AND angular velocity is zero THEN speed shall be zero.

IF angle is zero AND angular velocity is positive low THEN speed shall be positive low.

We can summarize all applicable rules in the following FAM table (see previous subsection):

		Angle				
	Speed	NH	NL	ZE	PL	PH
V	NH			NH		
e	NL			NL	ZE	
l	ZE	NH	NL	ZE	PL	PH
o	PL		ZE	PL		
c	PH			PH		

Now, we are going to define two explicit values for angle and angular velocity to calculate with. Consider the situation given in Figure 1.37, and let us apply the following rule:

IF angle is zero AND angular velocity is zero THEN speed is zero

– to the values that we have previously selected (see Figure 1.38)

Only four rules yield a result (*rules fire*, see Figure 1.39), and we overlap them into one single result (see Figure 1.40).

Fan: the Temperature Control System

In this simple example, the input linguistic variable is:

$$temperature\_error = desired\_temperature - current\_temperature.$$

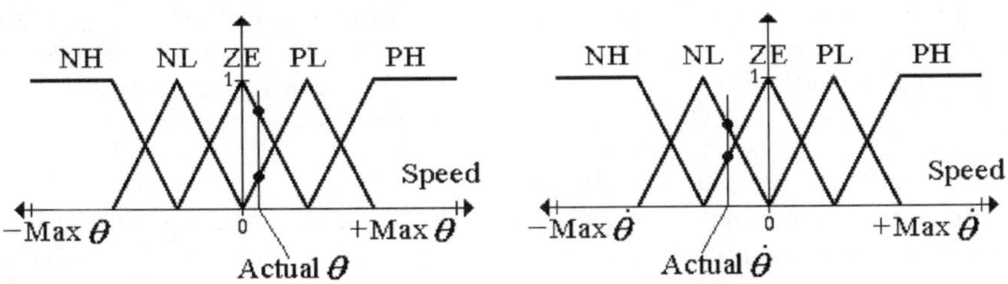

Fig. 1.37. Actual values for angle θ and angular velocity $\dot{\theta}$.

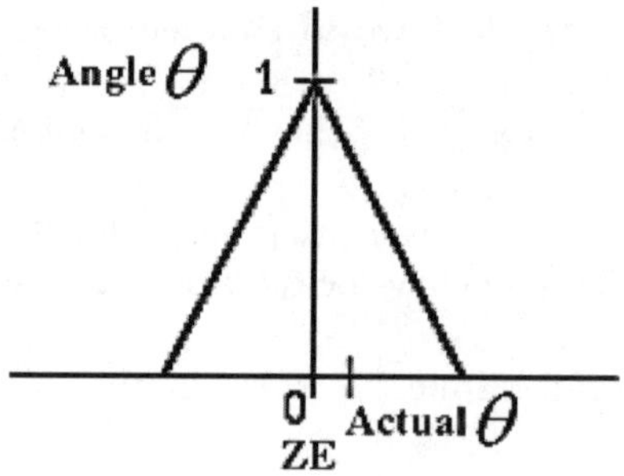

Fig. 1.38. Here is the linguistic variable *angle θ* where we zoom–in on the fuzzy set zero (ZE) and the actual angle.

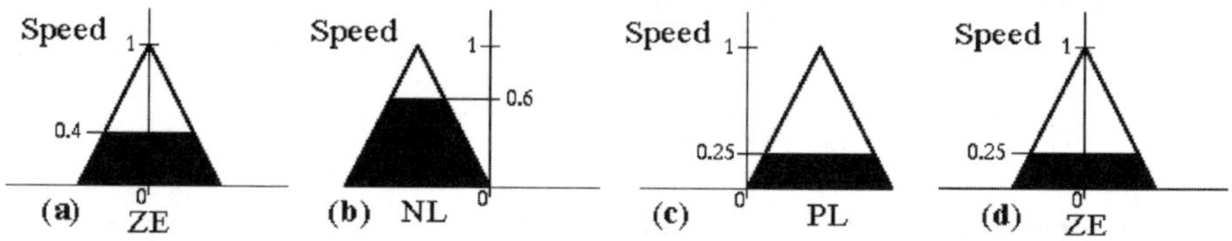

Fig. 1.39. Four fuzzy rules firing: (a) the result yielded by the rule: IF angle is zero AND angular velocity is zero THEN speed is zero; (b) the result yielded by the rule: IF angle is zero AND angular velocity is negative low THEN speed is negative low; (c) the result yielded by the rule: IF angle is positive low AND angular velocity is zero THEN speed is positive low; (d) the result yielded by the rule: IF angle is positive low AND angular velocity is negative low THEN speed is zero.

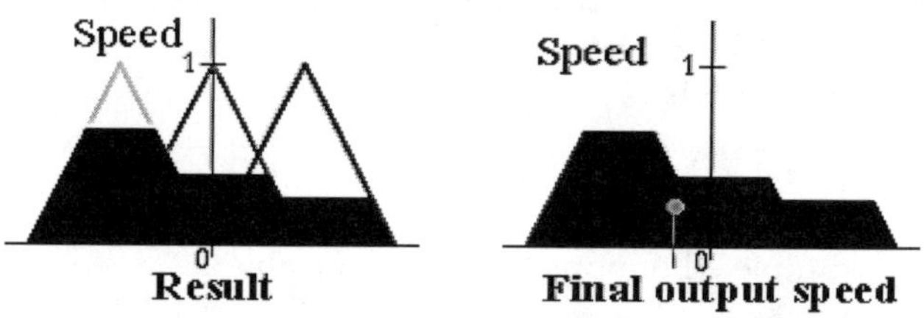

Fig. 1.40. Left: Overlapping single–rule results to yield the overall result. Right: The result of the fuzzy controller so far is a fuzzy set (of speed), so we have to choose one representative value as the final output; there are several heuristic *defuzzification* methods, one of them is to take the center of gravity of the fuzzy set. This is called *Mamdani fuzzy controller*.

The two output linguistic variables are:
hot_fan_speed, and $cool\_fan\_speed$. The universes of discourse, consisting of membership functions, i.e., overlapping triangular–trapezoidal shaped intervals, for all three variables are:

$invar$: $temperature\_error = \{Negative\_Big, Negative\_Medium,$ $Negative\_Small, Zero, Positive\_Small, Positive\_Medium, Positive\_Big\}$, with the range $[-110, 110]$ degrees;

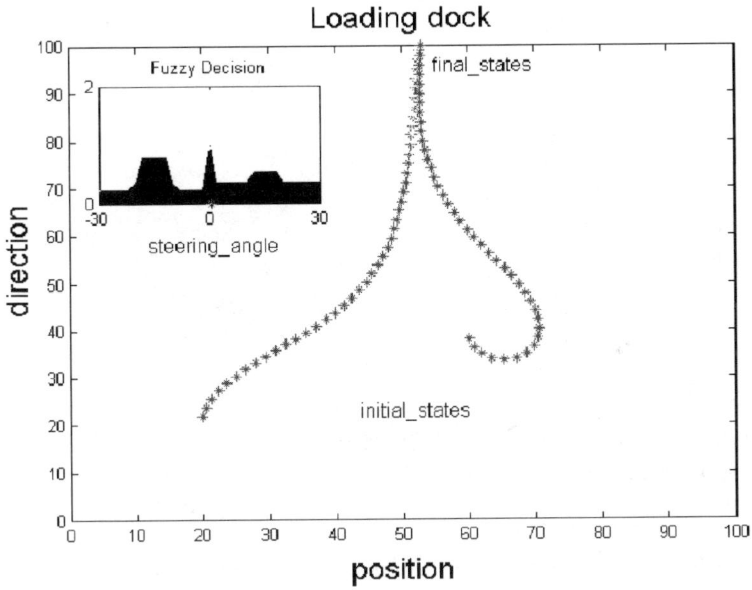

Fig. 1.41. Truck backer–upper steering control system.

outvars: *hot_fan_speed* and *cool_fan_speed* = {*zero*, *low*, *medium*, *high*, *very_high*}, with the range [0, 100] rounds–per–meter.

Truck Backer–Upper Steering Control System

In this example there are two input linguistic variables: position and direction of the truck, and one output linguistic variable: steering angle (see Figure 1.41). The universes of discourse, partitioned by overlapping triangular–trapezoidal shaped intervals, are defined as:

invars: *position* = $\{NL, NS, ZR, PS, PL\}$, and *direction* = $\{NL, NM, NS, ZR, PS, PM, PL\}$, where NL denotes Negative_Large, NM is Negative_Medium, NS is Negative_Small, etc.

outvar: *steering_angle* = $\{NL, NM, NS, ZR, PS, PM, PL\}$.

The rule–base is given as:

IF direction is NL, AND position is NL, THEN steering angle is NL;
IF direction is NL, AND position is NS, THEN steering angle is NL;
IF direction is NL, AND position is ZE, THEN steering angle is PL;
IF direction is NL, AND position is PS, THEN steering angle is PL;
IF direction is NL, AND position is PL, THEN steering angle is PL;
IF direction is NM, AND position is NL, THEN steering angle is ZE;
. . . .
IF direction is PL AND position is PL, THEN steering angle is PL.

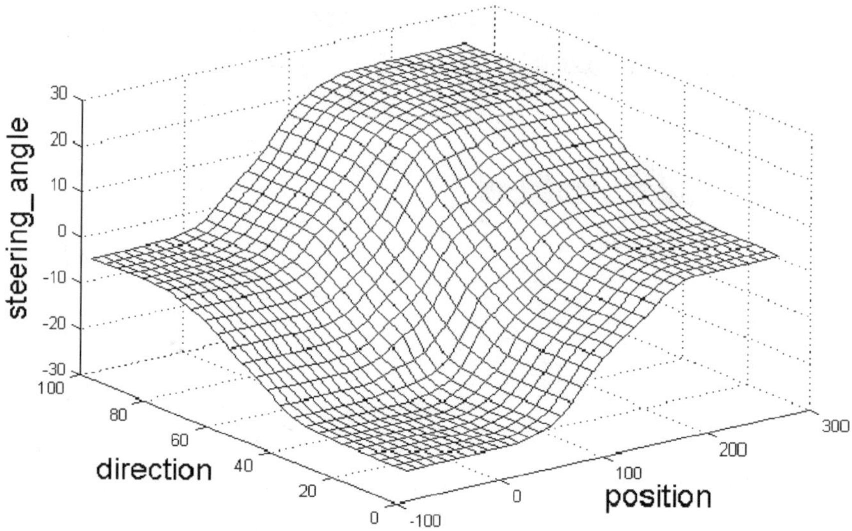

Fig. 1.42. Control surface for the truck backer–upper steering control system.

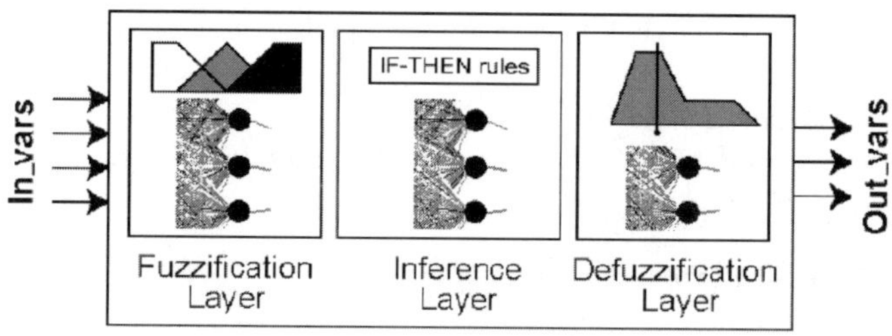

Fig. 1.43. Neuro–fuzzy inference engine.

The so–called *control surface* for the truck backer–upper steering control system is depicted in Figure 1.42.

To distinguish between more and less important rules in the knowledge base, we can put weights on them. Such weighted knowledge base can be then trained by means of artificial neural networks. In this way we get *hybrid neuro–fuzzy trainable expert systems.*

Another way of the hybrid neuro–fuzzy design is the fuzzy inference engine such that each module is performed by a layer of hidden artificial neurons, and ANN–learning capability is provided to enhance the system knowledge (see Figure 1.43).

Again, the fuzzy control of the BP learning (1.15–1.16) can be implemented as a set of heuristics in the form of fuzzy $IF - THEN$ rules, for the purpose of achieving a faster rate of convergence. The heuristics are driven by the behavior of the instantaneous sum of squared errors.

Finally, most *feedback fuzzy systems* are either discrete or continuous generalized SAMs [Kos99], given respectively by

$$x(k+1) = p_i(x(k))B_i(x(k)), \qquad \text{or} \qquad \dot{x}(t) = p_i(x(t))B_i(x(t)),$$

with coefficients p_i given by (1.29) above.

General Characteristics of Fuzzy Control

As demonstrated above, fuzzy logic offers several unique features that make it a particularly good choice for many control problems, among them [Lee90, DSS96]:

1. It is inherently robust since it does not require precise, noise–free inputs and can be programmed to fail safely if a feedback sensor quits or is destroyed. The output control is a smooth control function despite a wide range of input variations.
2. Since the fuzzy logic controller processes user–defined rules governing the target control system, it can be modified and tweaked easily to improve or drastically alter system performance. New sensors can easily be incorporated into the system simply by generating appropriate governing rules.
3. Fuzzy logic is not limited to a few feedback inputs and one or two control outputs, nor is it necessary to measure or compute rate–of–change parameters in order for it to be implemented. Any sensor data that provides some indication of a systems actions and reactions is sufficient. This allows the sensors to be inexpensive and imprecise thus keeping the overall system cost and complexity low.
4. Because of the rule-based operation, any reasonable number of inputs can be processed (1–8 or more) and numerous outputs (1–4 or more) generated, although defining the rule–base quickly becomes complex if too many inputs and outputs are chosen for a single implementation since rules defining their interrelations must also be defined. It would be better to break the control system into smaller chunks and use several smaller fuzzy logic controllers distributed on the system, each with more limited responsibilities.
5. Fuzzy logic can control nonlinear systems that would be difficult or impossible to model mathematically. This opens doors for control systems that would normally be deemed unfeasible for automation.

A *fuzzy logic controller* is usually designed using the following steps:

1. Define the control objectives and criteria: What am I trying to control? What do I have to do to control the system? What kind of response do I need? What are the possible (probable) system failure modes?
2. Determine the input and output relationships and choose a minimum number of variables for input to the fuzzy logic engine (typically error and rate–of–change of error).

3. Using the rule–based structure of fuzzy logic, break the control problem down into a series of *IF X AND Y THEN Z* rules that define the desired system output response for given system input conditions. The number and complexity of rules depends on the number of input parameters that are to be processed and the number fuzzy variables associated with each parameter. If possible, use at least one variable and its time derivative. Although it is possible to use a single, instantaneous error parameter without knowing its rate of change, this cripples the systems ability to minimize overshoot for a step inputs.
4. Create fuzzy logic membership functions that define the meaning (values) of Input/Output terms used in the rules.
5. Test the system, evaluate the results, tune the rules and membership functions, and re–test until satisfactory results are obtained.

Therefore, fuzzy logic does not require precise inputs, is inherently robust, and can process any reasonable number of inputs but system complexity increases rapidly with more inputs and outputs. Distributed processors would probably be easier to implement. Simple, plain–language rules of the form *IF X AND Y THEN Z* are used to describe the desired system response in terms of linguistic variables rather than mathematical formulas. The number of these is dependent on the number of inputs, outputs, and the designers control response goals. Obviously, for very complex systems, the rule–base can be enormous and this is actually the only drawback in applying fuzzy logic.

Evolving Fuzzy–Connectionist Systems

Recently, [Kas02] introduced a new type of fuzzy inference systems, denoted as dynamic evolving (see next subsection) neuro–fuzzy inference system (DENFIS), for adaptive online and off–line learning, and their application for dynamic time series prediction. *DENFIS system* evolves through incremental, hybrid (supervised/unsupervised), learning, and accommodates new input data, including new features, new classes, etc., through local element tuning. New fuzzy rules are created and updated during the operation of the system. At each time moment, the output of DENFIS is calculated through a fuzzy inference system based on $m-$most activated fuzzy rules which are dynamically chosen from a fuzzy rule set. Two approaches are proposed: (i) dynamic creation of a first–order Takagi–Sugeno–type (see, e.g., [Tan93]) fuzzy rule set for a DENFIS online model; and (ii) creation of a first–order Takagi–Sugeno–type fuzzy rule set, or an expanded high–order one, for a DENFIS offline model. A set of fuzzy rules can be inserted into DENFIS before or during its learning process. Fuzzy rules can also be extracted during or after the learning process. An evolving clustering method (ECM), which is employed in both online and off–line DENFIS models, is also introduced. It was demonstrated that DENFIS could effectively learn complex temporal sequences in an adaptive way and outperform some well–known, existing models.

Evolutionary Computation

In general, *evolutionary computation* (see [Fog98, ES03, BFM97]) is a CI–subfield involving *combinatorial optimization* problems.[153] It can be loosely recognized by the following criteria:

1. iterative progress, growth or development;
2. population based;
3. guided random search;
4. parallel processing; and
5. often biologically inspired.

This mostly involves the so–called *metaheuristic optimization algorithms*, such as *evolutionary algorithms* and *swarm intelligence*. In a lesser extent, evolutionary computation also involves *differential evolution*, *artificial life*, *artificial immune systems* and *learnable evolution model*.

Evolutionary Algorithms

In a narrow sense, evolutionary computation is represented by evolutionary algorithms (EAs), which are generic population–based *metaheuristic optimization algorithms* [Bac96]. The so–called *candidate solutions*[154] to the optimization problem play the role of individuals in a population, and the *cost function*[155] determines the environment within which the solutions 'live'. Evolution of the population then takes place after the repeated application of the above operators. Artificial evolution (AE) describes a process involving

[153] Recall that *combinatorial optimization* is a branch of optimization in applied mathematics and computer science, related to *operations research*, *algorithm theory* and *computational complexity theory*. Combinatorial optimization algorithms are often implemented in an efficient imperative programming language, in an expressive declarative programming language such as Prolog, or some compromise, perhaps a functional programming language such as Haskell, or a multi–paradigm language such as Lisp. A study of computational complexity theory helps to motivate combinatorial optimization. Combinatorial optimization algorithms are typically concerned with problems that are NP–hard. Such problems are not believed to be efficiently solvable in general. However, the various approximations of complexity theory suggest that some instances (e.g. 'small' instances) of these problems could be efficiently solved. This is indeed the case, and such instances often have important practical ramifications. The domain of combinatorial optimization is optimization problems where the set of *feasible solutions* is discrete or can be reduced to a discrete one, and the goal is to find the best possible solution.

[154] Recall that a *candidate solution* is a member of a set of possible solutions to a given problem. A candidate solution does not have to be a likely or reasonable solution to the problem. The space of all candidate solutions is called the *feasible region* or the feasible area.

[155] Recall that a generic optimization problem can be represented as:
Given: a function $f : A \to \mathbb{R}$ from some set A to the real numbers,

individual evolutionary algorithms; EAs are individual components that participate in an AE. EAs perform consistently well approximating solutions to all types of problems because they do not make any assumption about the underlying *fitness landscape*, evidenced by success in fields as diverse as engineering, art, biology, economics, genetics, operations research, robotics, social sciences, physics, and chemistry. Apart from their use as mathematical optimizers, EAs have also been used as an experimental framework within which to validate theories about biological evolution and natural selection, particularly through work in the field of artificial life. EAs involve biologically–inspired techniques implementing mechanisms such as:

1. *Reproduction*, which is the biological process by which new individual organisms are produced. Reproduction is a fundamental feature of all known life; each individual organism exists as the result of reproduction. The known methods of reproduction are broadly grouped into two main types: sexual and asexual. In asexual reproduction, an individual can reproduce without involvement with another individual of that species. The division of a bacterial cell into two daughter cells is an example of asexual reproduction. Asexual reproduction is not, however, limited to single–celled organisms. Most plants have the ability to reproduce asexually. On the other hand, sexual reproduction requires the involvement of

---

Sought: an element $x_0 \in A$ such that $f(x_0) \leq f(x)$ for all $x \in A$ ('minimization') or such that $f(x_0) \geq f(x)$ for all $x \in A$ ('maximization').

Typically, A is some subset of the *Euclidean space* $\mathbb{R}^n$, often specified by a set of constraints, equalities or inequalities that the members of A have to satisfy. The elements of A are called feasible solutions. The function f is called an *objective function*, or *cost function*. A feasible solution that minimizes (or maximizes, if that is the goal) the objective function is called an *optimal solution*. The domain A of f is called the *search space*, while the elements of A are called *candidate solutions* or *feasible solutions*.

Generally, when the feasible region or the objective function of the problem does not present *convexity*, there may be several local minima and maxima, where a local minimum x^* is defined as a point for which there exists some $\delta > 0$ so that for all x such that $\|\mathbf{x} - \mathbf{x}^*\| \leq \delta$, the expression $f(\mathbf{x}^*) \leq f(\mathbf{x})$ – holds; that is to say, on some region around x^* all of the function values are greater than or equal to the value at that point. Local maxima are defined similarly. For twice–differentiable functions, unconstrained problems can be solved by finding the points where the *gradient* of the objective function is zero (that is, the stationary points) and using the *Hessian matrix* to classify the type of each point. If the Hessian is positive definite, the point is a local minimum, if negative definite, a local maximum, and if indefinite it is some kind of saddle point. Constrained problems can often be transformed into unconstrained problems with the help of *Lagrange multipliers*. Note that a large number of algorithms proposed for solving non–convex problems, including the majority of commercially available solvers, are not capable of making a distinction between local optimal solutions and rigorous optimal solutions, and will treat the former as actual solutions to the original problem.

two individuals, typically one of each sex. Normal human reproduction is a common example of sexual reproduction. In general, more–complex organisms reproduce sexually while simpler, usually unicellular, organisms reproduce asexually.

2. *Mutation*, which is the biological change to the genetic material (usually DNA or RNA). Mutations can be caused by copying errors in the genetic material during cell division and by exposure to radiation, chemicals (mutagens), or viruses, or can occur deliberately under cellular control during processes such as meiosis or hypermutation. In multicellular organisms, mutations can be subdivided into germline mutations, which can be passed on to descendants, and somatic mutations. The somatic mutations cannot be transmitted to descendants in animals. Plants sometimes can transmit somatic mutations to their descendants asexually or sexually (in case when flower buds develop in somatically mutated part of plant). Mutations create variation in the gene pool, and then less favorable (or deleterious) mutations are removed from the gene pool by natural selection, while more favorable (beneficial or advantageous) ones tend to accumulate – this is evolution. Neutral mutations are defined as mutations whose effects do not influence the fitness of either the species or the individuals who make up the species. These can accumulate over time due to genetic drift. The overwhelming majority of mutations have no significant effect, since DNA repair is able to revert most changes before they become permanent mutations, and many organisms have mechanisms for eliminating otherwise permanently mutated somatic cells.
3. *Recombination*, which is the biological process of *genetic recombination* and *meiosis*, a genetic event that occurs during the formation of sperm and egg cells. It is also referred to as *crossover* or *phase change*.
4. *Natural selection*, which is the biological process by which individual organisms with favorable traits are more likely to survive and reproduce than those with unfavorable traits. Natural selection works on the whole individual, but only the heritable component of a trait will be passed on to the offspring, with the result that favorable, heritable traits become more common in the next generation. Given enough time, this passive process can result in adaptations and speciation. Natural selection is one of the cornerstones of modern biology. The term was introduced by Charles Darwin in his 1859 book 'The Origin of Species', by analogy with artificial selection, by which a farmer selects his breeding stock.
5. *Survival of the fittest*, a biological phrase, which is a shorthand for a concept relating to competition for survival or predominance. Originally applied by Herbert Spencer[156] in his 'Principles of Biology' of 1864, Spencer drew parallels to his ideas of economics with Charles Darwin's

[156] Herbert Spencer (27 April 1820 – 8 December 1903) was an English philosopher and prominent liberal political theorist. He is best known as the father of *Social Darwinism*, a school of thought that applied the evolutionist theory of survival of the fittest (a phrase coined by Spencer) to human societies. He also contributed to

theories of evolution by what Darwin termed natural selection. The phrase is a metaphor, not a scientific description; and it is not generally used by biologists, who almost exclusively prefer to use the phrase 'natural selection'.

Each evolutionary algorithm uses some mechanisms inspired by biological evolution: *reproduction*, *mutation*, *recombination*, *natural selection* and *survival of the fittest*. Candidate solutions to the optimization problem play the role of individuals in a population, and the cost function determines the environment within which the solutions 'live'. Evolution of the population then takes place after the repeated application of the above operators. The so–called *artificial evolution* (AE) describes a process involving individual evolutionary algorithms; EAs are individual components that participate in an AE.

Evolutionary algorithms perform consistently well approximating solutions to all types of problems because they do not make any assumption about the underlying fitness landscape, evidenced by success in fields as diverse as engineering, art, biology, economics, genetics, operations research, robotics, social sciences, physics, and chemistry. However, consider the no–free–lunch theorem.

Apart from their use as mathematical optimizers, EAs have also been used as an experimental framework within which to validate theories about biological evolution and natural selection, particularly through work in the field of artificial life. Techniques from evolutionary algorithms applied to the modelling of biological evolution are generally limited to explorations of microevolutionary processes, however some computer simulations, such as Tierra and Avida, attempt to model macroevolutionary dynamics.

In general, an evolutionary algorithm is based on three main statements:

1. It is a process that works at the chromosomic level. Each individual is codified as a set of chromosomes.
2. The process follows the Darwinian theory of evolution, say, the survival and reproduction of the fittest in a changing environment.
3. The evolutionary process takes place at the reproduction stage. It is in this stage when mutation and crossover occurs. As a result, the progeny chromosomes can differ from their parents ones.

Starting from a guess initial population, an evolutionary algorithm basically generates consecutive generations (offprints). These are formed by a set of chromosomes, or character (genes) chains, which represent possible solutions to the problem under consideration. At each algorithm step, a fitness function is applied to the whole set of chromosomes of the corresponding generation in order to check the goodness of the codified solution. Then, according

---

a wide range of subjects, including ethics, metaphysics, religion, politics, rhetoric, biology and psychology. Spencer is today widely criticized as a perfect example of scientism, while he had many followers and admirers in his time.

to their fitting capacity, couples of chromosomes, to which the crossover operator will be applied, are chosen. Also, at each step, a mutation operator is applied to a number of randomly chosen chromosomes.

The two most commonly used methods to randomly select the chromosomes are:

1. The *roulette wheel algorithm*. It consists in building a roulette, so that to each chromosome corresponds a circular sector proportional to its fitness.
2. The *tournament method*. After shuffling the population, their chromosomes are made to compete among them in groups of a given size (generally in pairs). The winners will be those chromosomes with highest fitness. If we consider a binary tournament, say the competition is between pairs, the population must be shuffled twice. This technique guarantees copies of the best individual among the parents of the next generation.

After this selection, we proceed with the sexual reproduction or crossing of the chosen individuals. In this stage, the survivors exchange chromosomic material and the resulting chromosomes will codify the individuals of the next generation. The forms of sexual reproduction most commonly used are:

(i) With one crossing point. This point is randomly chosen on the chain length, and all the chain portion between the crossing point and the chain end is exchanged.

(ii) With two crossing points. The portion to be exchanged is in between two randomly chosen points.

For the algorithm implementation, the crossover normally has an assigned percentage that determines the frequency of its occurrence. This means that not all of the chromosomes will exchange material but some of them will pass intact to the next generation. As a matter of fact, there is a technique, named elitism, in which the fittest individual along several generations does not cross with any of the other ones and keeps intact until an individual fitter than itself appears.

Besides the selection and crossover, there is another operation, mutation, that produces a change in one of the characters or genes of a randomly chosen chromosome. This operation allows to introduce new chromosomic material into the population. As for the crossover, the mutation is handled as a percentage that determines its occurrence frequency. This percentage is, generally, not greater than 5%, quite below the crossover percentage.

Once the selected chromosomes have been crossed and muted, we need some substitution method. Namely, we must choose, among those individuals, which ones will be substituted for the new progeny. Two main substitution ways are usually considered. In one of them, all modified parents are substituted for the generated new individuals. In this way an individual does never coexist with its parents. In the other one, only the worse fitted individuals of the whole population are substituted, thus allowing the coexistence among parents and progeny.

Since the answer to our problem is almost always unknown, we must establish some criterion to stop the algorithm. We can mention two such criteria [SRV04]:

(i) the algorithm is run along a maximum number of generations; and

(ii) the algorithm is ended when the population stabilization has been reached, i.e., when all, or most of, the individuals have the same fitness.

A limitation of EAs is their lack of a clear *genotype–phenotype distinction* [Bac96]. In nature, the fertilized egg cell undergoes a complex process known as embryogenesis to become a mature phenotype. This indirect encoding is believed to make the genetic search more robust (i.e., reduce the probability of fatal mutations), and also may improve the evolvability of the organism. Recent work in the field of artificial embryogeny, or artificial developmental systems, seeks to address these concerns.

Evolutionary algorithms usually comprise: *genetic algorithms, genetic programming, evolutionary programming, evolution strategy* and *learning classifier systems.*

Genetic Algorithms

The *genetic algorithm* (GA) is a search technique pioneered by John Holland[157] [Hol92] and used in computing to find true or approximate solutions to optimization and search problems (see [Gol89, Mit96, Vos99, Mic99]). GAs find application in computer science, engineering, economics, physics, mathematics and other fields. GAs are categorized as global search heuristics. GAs are implemented as a computer simulation in which a population of abstract representations (called chromosomes or the genotype) of candidate solutions (called individuals, creatures, or phenotypes) to an optimization problem evolves toward better solutions. Traditionally, solutions are represented in binary as strings of 0s and 1s, but other encodings are also possible. The evolution usually starts from a population of randomly generated individuals and happens in generations. In each generation, the fitness of every individual in the population is evaluated, multiple individuals are stochastically selected from the current population (based on their fitness), and modified (mutated or recombined) to form a new population. The new population is then used

[157] John Henry Holland (February 2, 1929) is a pioneer in complex system and nonlinear science. He is known as the father of genetic algorithms. Holland is Professor of Psychology and Professor of Electrical Engineering and Computer Science at the University of Michigan, Ann Arbor. He is also a member of The Center for the Study of Complex Systems (CSCS) at the University of Michigan, and a member of Board of Trustees and Science Board of the Santa Fe Institute. Holland is the author of a number of books about *complex adaptive systems* (CAS), including Hidden Order: How Adaptation Builds Complexity (1995), Emergence: From Chaos to Order (1998) and his ground–breaking book on genetic algorithms, 'Adaptation in Natural and Artificial Systems' (1975,1992). Holland also frequently lectures around the world on his own research, and on current research and open questions in CAS studies (see [Hol95, Hol95]).

in the next iteration of the algorithm. A typical GA requires two things to be defined: (i) a genetic representation of the solution domain, and (ii) a fitness function to evaluate the solution domain.

A standard representation of the solution is as an array of bits. Arrays of other types and structures can be used in essentially the same way. The main property that makes these genetic representations convenient is that their parts are easily aligned due to their fixed size, that facilitates simple crossover operation. Variable length representations were also used, but crossover implementation is more complex in this case. The *fitness function*[158] is defined over the genetic representation and measures the quality of the represented solution. The fitness function is always problem dependent. For instance, in the *knapsack problem*, we want to maximize the total value of objects that we can put in a knapsack of some fixed capacity. A representation of a solution might be an array of bits, where each bit represents a different object, and the value of the bit (0 or 1) represents whether or not the object is in the knapsack. Not every such representation is valid, as the size of objects may exceed the capacity of the knapsack. The fitness of the solution is the sum of values of all objects in the knapsack if the representation is valid, or 0 otherwise. In some problems, it is hard or even impossible to define the fitness expression; in these cases, *interactive genetic algorithms* are used. Once we have the genetic representation and the fitness function defined, GA proceeds to initialize a population of solutions randomly, then improve it through repetitive application of mutation, crossover, and selection operators. Another way of looking at fitness functions is in terms of a *fitness landscape*,[159] which shows the fitness for each possible chromosome (see [Mit96]).

---

[158] A *fitness function* is a particular type of *objective function* that quantifies the optimality of a solution (that is, a chromosome) in a genetic algorithm so that particular chromosome may be ranked against all the other chromosomes. Optimal chromosomes, or at least chromosomes which are more optimal, are allowed to breed and mix their datasets by any of several techniques, producing a new generation that will (hopefully) be even better. An ideal fitness function correlates closely with the algorithm's goal, and yet may be computed quickly. Speed of execution is very important, as a typical genetic algorithm must be iterated many, many times in order to produce a useable result for a non–trivial problem. Definition of the fitness function is not straightforward in many cases and often is performed iteratively if the fittest solutions produced by GA are not what is desired. In some cases, it is very hard or impossible to come up even with a guess of what fitness function definition might be. Interactive genetic algorithms address this difficulty by out–sourcing evaluation to external agents (normally humans).

[159] Fitness landscapes or adaptive landscapes are used to visualize the relationship between genotypes (or phenotypes) and reproductive success. It is assumed that every genotype has a well defined replication rate (often referred to as fitness). This fitness is the 'height' of the landscape. Genotypes which are very similar are said to be 'close' to each other, while those that are very different are 'far'

It is well–known in biology that any organism can be represented by its *phenotype*, which virtually determines *what* exactly the object is in the real world, and its *genotype* containing all the information about the object at the chromosome set level. Each gene, that is the genotype's information element, is reflected in the phenotype. Thus, to be able to solve problems we have to represent every attribute of an object in a form suitable for use in genetic algorithms. All further operation of genetic algorithm is done on the genotype level, making the information about the object's internal structure redundant. This is why this algorithm is widely used to solve all sorts of problems.

In the most frequently used variant of genetic algorithm, an object's genotype is represented by bit strings. Each attribute of an object in the phenotype has a single corresponding gene in the genotype. The gene is represented by a bit string, usually of a fixed length, which represents the value of the attribute.

The simplest variant can be used to encode such attributes that is the bit value of the attribute. Then it will be quite easy to use a gene of certain length, sufficient to represent all possible values of such an attribute. Unfortunately this encoding method is not perfect. Its main disadvantage is that neighboring numbers differ in several bits' values. Thus, for example, such numbers as 7 and 8 in the bit representation have four different bits, which complicates the gene algorithm functioning and increases time necessary for its convergence. To avoid this problem another encoding method should be used, in which neighboring numbers have less differences, ideally differing in only one bit.

---

from each other. The two concepts of height and distance are sufficient to form the concept of a 'landscape'. The set of all possible genotypes, their degree of similarity, and their related fitness values is then called a fitness landscape. In evolutionary optimization problems, fitness landscapes are evaluations of a fitness function for all candidate solutions.

Apart from the field of evolutionary biology, the concept of a fitness landscape has also gained importance in evolutionary optimization methods, in which one tries to solve real–world engineering or logistics problems by imitating the dynamics of biological evolution. For example, a delivery truck with a number of destination addresses can take a large variety of different routes, but only very few will result in a short driving time. In order to use evolutionary optimization, one has to define for every possible solution s to the problem of interest (i.e., every possible route in the case of the delivery truck) how 'good' it is. This is done by introducing a scalar–valued function $f(s)$ (scalar valued means that $f(s)$ is a simple number, such as 0.3, while s can be a more complicated object, for example a list of destination addresses in the case of the delivery truck), which is called the fitness function or fitness landscape. A high $f(s)$ implies that s is a good solution. In the case of the delivery truck, $f(s)$ could be the number of deliveries per hour on route s. The best, or at least a very good, solution is then found in the following way. Initially, a population of random solutions is created. Then, the solutions are mutated and selected for those with higher fitness, until a satisfying solution has been found.

Binary coding			**Coding using the Gray code**		
Dec.code	Bin.value	Hex.value	Dec.code	Bin.value	Hex.value
0	0000	0h	0	0000	0h
1	0001	1h	1	0001	1h
2	0010	2h	3	0011	3h
3	0011	3h	2	0010	2h
4	0100	4h	6	0110	6h
5	0101	5h	7	0111	7h
6	0110	6h	5	0101	5h
7	0111	7h	4	0100	4h
8	1000	8h	12	1100	Ch
9	1001	9h	13	1101	Dh
10	1010	Ah	15	1111	Fh
11	1011	Bh	14	1110	Eh
12	1100	Ch	10	1010	Ah
13	1101	Dh	11	1011	Bh
14	1110	Eh	9	1001	9h
15	1111	Fh	8	1000	8h

Table 1.2. Correspondence between decimal codes and the Gray codes.

One of such codes is the Gray code, which is appropriate to be used with genetic algorithms. The table below shows the Gray code values:
Accordingly, when encoding an integer–valued attribute, we break it into quadruples and then convert each quadruple according to *Gray code*. Usually, there is no need to convert attribute values into gene values in practical use of GAs. In practice, inverse problem occurs, when it is necessary to find the attribute value from the corresponding gene value. Thus, the problem of decoding gene values, which have corresponding integer–valued attributes, is trivial. The simplest coding method, which first comes to mind, is to use bit representation. However, this variant is equally imperfect as in the case of integers. For this reason, the following sequence is used in practice:

1. All the interval of the attribute's allowed values is split into segments with adequate accuracy.
2. The value of the gene is accepted as an integer defining the interval number (using the Gray code).
3. The midpoint number of the interval is taken as the parameter value.

Let us consider a specific example of the sequence of operations described above: Assume that the attribute values are located in the interval $[0, 1]$. During the encoding the segment is split into 256 intervals. Thus we will need 8 bits to code their numbers. Let us suppose the number of the gene is $00100101bG$ (the capital letter 'G' stands for 'Gray code'). For a start we shall find the corresponding interval number using the following Gray code: $25hG \to 36h \to 54d$. Now let us see what interval corresponds to it… Simple calculation gives us the interval: $[\ 0.20703125\ ,\ 0.2109375\ ]$.

Then, the value of the parameter is $(0.20703125 + 0.2109375)/2 = 0.208984375$.

To encode nonnumeric data, we have to convert it into numbers. More detailed description can be found on our web site in the articles dedicated to the use of neural nets.

Thus, to find an object's *phenotype* (i.e., values of the attributes describing the object) we only have to know the values of the genes corresponding to these attributes, i.e., the object's *genotype*. The aggregate of the genes describing the object's genotype represents the *chromosome*. In some implementations it is called an individual. Thus, when implementing genetic algorithm, a chromosome is a bit string of a fixed length. Each segment of a string has its corresponding gene. Genes inside a chromosome can have equal or different lengths. Genes of equal length are used most often. Let us consider an example of a chromosome and interpretation of its value. Let us assume that the object has five attributes, each encoded by a gene 4 elements long. Then, the length of the chromosome is $5 \cdot 4 = 20$ bits:

0010	1010	1001	0100	1101

Now we can define the values of the attributes:

Attribute	Gene value	Binary value of the attribute	Decimal value of the attribute
Attribute 1	0010	0011	3
Attribute 2	1010	1100	12
Attribute 3	1001	1110	14
Attribute 4	0100	0111	7
Attribute 5	1101	1001	9

As it is known in the evolution theory, the way the parents' attributes are inherited by their offsprings is of high importance. In genetic algorithms an operator called *crossing* (also known as crossover or crossing over) is in charge of passing the attributes from parents to their offsprings. It works in the following way:

1. Two individuals are selected from the population to become parents;
2. A break point is determined (usually at random); and
3. The offspring is determined as concatenation of the first and the second parents' parts.

Let us see how this operator works:
Now, if we put the break after the third bit of the chromosome, then we have:

Chromosome_1:	0000000000
Chromosome_2:	1111111111

Chromosome_1:	0000000000	>>	000	1111111	Resulting_chromosome_1
Chromosome_2:	1111111111	>>	111	0000000	Resulting_chromosome_2

After that, one of the resulting chromosomes is taken as an offspring with the 0.5 probability.

The next genetic operator is intended for maintaining the diversity of individuals in the population. It is called *mutation*. When it is used on a chromosome, each bit in it gets inverted with a certain probability.

Besides, one more operator is used, called *inversion*. Applying it makes a chromosome break in two parts, which then trade places. This can be shown schematically as follows:

000	1111111	>>	1111111	000

Theoretically, these two genetic operators are enough to make the genetic algorithm work. However, in practice some additional operators are used, as well as modifications of these two operators. For instance, in addition to the single-point crossover (described above) there can be a multipoint one, when several break points (usually two) are formed. Besides, in some implementations of the algorithm the mutation operator performs the inversion of only one randomly selected bit of a chromosome.

Having found out how to interpret the values of the genes, we proceed to describing the genetic algorithm operation. Let us consider the flow chart of genetic algorithm operation in its classic variant.

1. Initialize the start time $t = 0$. At random fashion form the initial population consisting of k individuals: $B_0 = \{A_1, A_2, \ldots, A_k)$.
2. Calculate the *fitness* of every individual: $F_{\mathrm{Ai}} = fit(A_i), (i = 1 \ldots k)$, and of the population as a whole: $F_t = fit(B_t)$. The value of this function determines how suitable for solving the problem the individual described by this chromosome is.
3. Select the individual A_c from the population: $A_c = Get(B_t)$.
4. With a certain crossover probability P_c select the second individual from the population: $A_{c1} = Get(B_t)$, and apply the crossover operator: $A_c = Crossing(A_c, A_{c1})$.
5. With a certain mutation probability P_m apply the mutation operator: $A_c = mutation(A_c)$.
6. With a certain inversion probability P_i apply the inversion operator: $A_c = inversion(A_c)$.
7. Place the resulting chromosome in the new population: $insert(B_{t+1}, A_c)$.

8. Repeat steps 3 to $7k$ times.
9. Increase the current epoch number $t = t + 1$.
10. If the stop condition is met, terminate the loop, else go to step 2.

Now let us examine in detail the individual steps of the algorithm. The steps 3 and 4 play the most important role in the successful operation of the algorithm when parent chromosomes are selected. Various alternatives are possible. The most frequently used *selection method* is called *roulette*. When using it, the probability of a chromosome selection is determined by its fitness, i.e.,

$$P_{\mathrm{Get(Ai)}} \sim Fit(A_{\mathrm{i}})/Fit(B_{\mathrm{t}}).$$

This method increases the probability of the attributes propagation that belong to the most adjusted individuals. Another frequently used method is the *tournament selection*. It means that several individuals (usually two) are selected in the population at random. The one wins which is more adjusted. Besides, in some implementations of the algorithm the so–called *elitism strategy* is used, which means that the best–adjusted individuals are guaranteed to enter the new population. Using the elitism method is usually helpful to accelerate the genetic algorithm convergence. The disadvantage of this strategy is increased probability of the algorithm getting in the local minimum.

Another important point is the algorithm stop criteria determination. Usually the highest limit of the algorithm functioning epochs is taken as such, or the algorithm is stopped upon stabilization of its convergence, normally measured by means of comparing the population's fitness on various epochs.

Genetic Programming

The *genetic programming* (GP) is an automated methodology inspired by biological evolution to find computer programs that best perform a user–defined task. It is therefore a particular machine learning technique that uses an evolutionary algorithm to optimize a population of computer programs according to a fitness landscape determined by a program's ability to perform a given computational task. The first experiments with GP were described in the book 'Genetic Programming' by John Koza (see [Koz92, Koz95, KBA99, KKS03]. Computer programs in GP can be written in a variety of programming languages. In the early (and traditional) implementations of GP, program instructions and data values were organized in tree–structures, thus favoring the use of languages that naturally embody such a structure (an important example pioneered by Koza is Lisp). Other forms of GP have been suggested and successfully implemented, such as the simpler linear representation which suits the more traditional imperative languages. The commercial GP software Discipulus, for example, uses linear genetic programming combined with machine code language to achieve better performance. Differently, the MicroGP uses an internal representation similar to linear genetic programming to generate programs that fully exploit the syntax of a given assembly language. GP is very computationally intensive and so in the 1990s it was mainly used to solve relatively simple problems. However, more recently, thanks to various improvements in GP technology and to the well known exponential growth in CPU power, GP has started delivering a number of outstanding results. At the time of writing, nearly 40 human–competitive results have been gathered, in areas such as quantum computing, electronic design, game playing, sorting, searching and many more. These results include the replication or infringement of several post–year–2000 inventions, and the production of two patentable new inventions. Developing a theory for GP has been very difficult and so in the 1990s genetic programming was considered a sort of pariah amongst the various techniques of search. However, after a series of breakthroughs in the early 2000s, the theory of GP has had a formidable and rapid development. So much so that it has been possible to build exact probabilistic models of GP (schema theories and Markov chain models) and to show that GP is more general than, and in fact includes, GAs. On the other hand, techniques have now been applied to evolvable hardware as well as computer programs. Finally, the so–called *meta–GP* is the technique of evolving a GP–system using GP itself; critics have argued that it is theoretically impossible, but more research is needed.

Evolutionary Programming

The *evolutionary programming* (EP) was first used by Lawrence Fogel [FOW66] in 1960 in order to use simulated evolution as a learning process aiming to generate artificial intelligence. Fogel used finite state machines as predictors and evolved them. Currently evolutionary programming is a wide evolutionary computing dialect with no fixed structure, (representation), in contrast with the other three dialects. It is becoming harder to distinguish from evolutionary strategies. Its main variation operator is *mutation*; members of the population are viewed as part of a specific species rather than members of the same species therefore each parent generates an offspring, using a $(\mu + \mu)$ *survivor selection.*

Selection is the stage of a EP or GA in which individual genomes are chosen from a population for later breeding (recombination or crossover). There are several generic selection algorithms. One of the common ones is the so–called roulette wheel selection, which can be implemented as follows:

1. The fitness function is evaluated for each individual, providing fitness values, which are then normalized. Normalization means multiplying the fitness value of each individual by a fixed number, so that the sum of all fitness values equals 1.
2. The population is sorted by descending fitness values.
3. Accumulated normalized fitness values are computed (the accumulated fitness value of an individual is the sum of its own fitness value plus the fitness values of all the previous individuals). The accumulated fitness of

the last individual should of course be 1 (otherwise something went wrong in the normalization step).

4. A random number R between 0 and 1 is chosen.
5. The selected individual is the first one whose accumulated normalized value is greater than R. There are other selection algorithms that do not consider all individuals for selection, but only those with a fitness value that is higher than a given (arbitrary) constant. Other algorithms select from a restricted pool where only a certain percentage of the individuals are allowed, based on *fitness value*.

Evolution Strategy

The *evolution strategy* (ES) is an optimization technique based on ideas of adaptation and evolution [Bey01, BS02]. ESs primarily use real–vector coding, and mutation, recombination, and environmental selection as its search operators. As common with EAs, the operators are applied in order:

1. mating selection,
2. recombination,
3. mutation,
4. fitness function evaluation, and
5. environmental selection.

Performing the loop one time is called a generation, and this is continued until a termination criterion is met. The first ES variants were not population based, but memorized only one search point (the parent) and one ($(1+1)-$ES) or more offspring ($(1+\lambda)-$ES) at a time. Contemporary versions usually employ a population ($(\mu+\lambda)-$ES) and are thus believed to be less prone to get stuck in local optima. Mutation is performed by adding a gaussian distributed random value simultaneously to each vector element. The step size or mutation strength (ie. the standard deviation of this distribution) is usually learned during the optimization. This process is called self–adaptation, and it should keep the evolutionary process within the *evolution window*.

It was observed in ES that during an evolutionary search the progress toward the fitness/objective function's optimum, generally, happens in a narrow band of mutation step size σ. That progress is called evolution window. So far, there is not an optimum tunning method for the mutation step size σ to keep the search inside the evolution window and how to fast achieve this window, although there are some investigations about that subject.

Learning Classifier Systems

The *learning classifier systems* (LCS) are machine learning systems with close links to reinforcement learning and genetic algorithms. First described by John Holland (see [Hol92, Hol95, Hol95]), an LCS consists of a population of binary rules on which a genetic algorithm altered and selected the best rules.

Instead of a using fitness function, rule utility is decided by a reinforcement learning technique. Learning classifier systems can be split into two types depending upon where the genetic algorithm acts. A Pittsburgh–type LCS has a population of separate rule sets, where the genetic algorithm recombines and reproduces the best of these rule sets. In a Michigan–style LCS there is only a single population and the algorithm's action focuses on selecting the best classifiers within that ruleset. Michigan–style LCSs have two main types of reinforcement learning, fitness sharing (ZCS) and accuracy–based (XCS). Initially the classifiers or rules were binary, but recent research has focused on improving this representation. This has been achieved by using populations of neural networks and other methods. Learning classifier systems are not well–defined mathematically and doing so remains an area of active research. Despite this, they have been successfully applied in many problem domains.

Swarm Intelligence

The *swarm intelligence* (SI) is based around the study of collective behavior in decentralized, self–organized systems (see, e.g., [Eng06]). The expression 'swarm intelligence' was introduced by Beni & Wang in 1989, in the context of *cellular automata*[160]. SI–systems are typically made up of a population of simple agents interacting locally with one another and with their environment. Although there is normally no centralized control structure dictating how individual agents should behave, local interactions between such agents often lead to the emergence of global behavior. Examples of systems like this can be found in nature, including ant colonies, bird flocking, animal herding, bacteria molding and fish schooling. Application of swarm principles to large numbers of robots is called as swarm robotics. SI–systems comprise:

1. The *ant colony optimization* (ACO), which is a *metaheuristic optimization algorithm* that can be used to find approximate solutions to difficult combinatorial optimization problems. In ACO artificial ants build solutions by moving on the problem graph and they, mimicking real ants, deposit artificial pheromone on the graph in such a way that future artificial ants can build better solutions. ACO has been successfully applied to an impressive number of optimization problems.

[160] Recall that a *cellular automaton* (plural: cellular automata, CA) is a discrete dynamical system invented by Stanislaw Ulam and John von Neumann. CA are studied in computability theory, mathematics, and theoretical biology. It consists of an infinite, regular grid of cells, each in one of a finite number of states. The grid can be in any finite number of dimensions. Time is also discrete, and the state of a cell at time t is a function of the states of a finite number of cells (called its neighborhood) at time $t-1$. These neighbors are a selection of cells relative to the specified cell, and do not change. Though the cell itself may be in its neighborhood, it is not usually considered a neighbor. Every cell has the same rule for updating, based on the values in this neighbourhood. Each time the rules are applied to the whole grid a new generation is created. See below for furhter details.

2. The *particle swarm optimization* (PSO), which is a global optimization algorithm for dealing with problems in which a best solution can be represented as a point or surface in an nD space. Hypotheses are plotted in this space and seeded with an initial velocity, as well as a communication channel between the particles. Particles then move through the solution space, and are evaluated according to some fitness criterion after each timestep. Over time, particles are accelerated towards those particles within their communication grouping which have better fitness values. The main advantage of such an approach over other global minimization strategies such as *simulated annealing* is that the large number of members that make up the particle swarm make the technique impressively resilient to the problem of local minima.
3. The *stochastic diffusion search* (SDS), which is an agent based probabilistic global search and optimization technique best suited to problems where the objective function can be decomposed into multiple independent partial–functions. Each agent maintains a hypothesis which is iteratively tested by evaluating a randomly selected partial objective function parameterised by the agent's current hypothesis. In the standard version of SDS such partial function evaluations are binary resulting in each agent becoming active or inactive. Information on hypotheses is diffused across the population via inter–agent communication. Unlike the stigmergetic communication used in ACO, in SDS agents communicate hypotheses via a $1-1$ communication strategy analogous to the tandem running procedure observed in some species of ant. A positive feedback mechanism ensures that, over time, a population of agents stabilise around the global–best solution. SDS is both an efficient and robust search and optimisation algorithm, which has been extensively mathematically described.

In a lesser extent, *evolutionary computation* also involves:

1. The *self–organization*,[161] comprising:
   a) The *self–organizing maps* (SOMs, or Kohonen[162] maps), which are a subtype of ANNs (see above), trained using unsupervised learning

[161] Recall that *self–organization* is a process in which the internal organization of a system, normally an open system, increases in complexity without being guided or managed by an outside source. Self–organizing systems usually display *emergent properties*. Self-organization usually relies on four basic ingredients: (i) positive feedback, (ii) negative feedback, (iii) balance of exploitation and exploration, and (iv) multiple interactions.

[162] Teuvo Kohonen, Dr. Ing (born July 11, 1934), is a Finnish academican and prominent researcher. He has made many contributions to the field of neural networks, including the Learning Vector Quantization algorithm, fundamental theories of distributed associative memory and optimal associative mappings, the learning subspace method and novel algorithms for symbol processing like

to produce low–dimensional representation of the training samples while preserving the topological properties of the input space; this makes SOMs especially good for visualizing high–dimensional data [Koh82, Koh88, Koh91]. SOM is a single layer feedforward network where the output neurons are arranged in low dimensional (usually 2D or 3D) grid. Each input is connected to all output neurons. Attached to every neuron there is a weight vector with the same dimensionality as the input vectors. The number of input dimensions is usually a lot higher than the output grid dimension. SOMs are mainly used for dimensionality reduction rather than expansion. The goal of SOM training is to associate different parts of the SOM lattice to respond similarly to certain input patterns. This is partly motivated by how visual, auditory or other sensory information is handled in separate parts of the cerebral cortex in the human brain. The weights of the neurons are initialized either to small random values or sampled evenly from the subspace spanned by the two largest principal component eigenvectors. The latter alternative will speed up the training significantly because the initial weights already give good approximation of SOM weights. The training utilizes competitive learning. Like most ANNs, SOM has two modes of operation:

i. During the training process a map is built, the neural network organises itself, using a competitive process. The network must be given a large number of input vectors, as much as possible representing the kind of vectors that are expected during the second phase (if any). Otherwise, all input vectors must be administered several times.
ii. During the mapping process a new input vector may quickly be given a location on the map, it is automatically classified or categorised. There will be one single winning neuron: the neuron whose weight vector lies closest to the input vector (this can be simply determined by calculating the Euclidean distance between input vector and weight vector).

b) The *growing neural gas* (GNG), which is a self–organized neural network proposed by B. Fritzke [Fri94]. It is based on the previously proposed *neural gas*, a biologically inspired adaptive algorithm, coined by Martinetz and Schulten in 1991, which sorts for the input signal according to how far away they are; a certain number of them are selected by distance in order, then the number of adaption units and strength are decreased according to a fixed schedule. On the other hand, GNG can add and delete nodes during algorithm execution.

redundant hash addressing. He has published several books and over 200 peer–reviewed papers. His most famous contribution is the self–organizing map (SOM) (also known as the Kohonen map, although Kohonen himself prefers SOM).

The growth mechanism is based on growing cell structures and competitive Hebbian learning.

c) The *competitive learning* (see, e.g., [Gro87]). In this area a large number of models exist which have a common goal to distribute a certain number of vectors in a possibly high–dimensional space. The distribution of these vectors reflects the probability distribution of the input signals which in general is not given explicitly but only through sample vectors. Two closely related concepts from computational geometry are the *Voronoi tessellation* and the *Delaunay triangulation* (see, e.g., [PS90]).

2. The *differential evolution* (DE), which grew out of K. Price's attempts to solve the *Chebyshev polynomial fitting problem* that had been posed to him by R. Storn. A breakthrough happened when Price came up with the idea of using vector differences for perturbing the vector population. Since this seminal idea a lively discussion between Price and Storn and endless ruminations and computer simulations on both parts yielded many substantial improvements which make DE the versatile and robust tool it is today. DE is a very simple population based, stochastic function minimizer which is very powerful at the same time. DE managed to finish 3rd at the First International Contest on Evolutionary Computation (Nagoya, 1996). DE turned out to be the best genetic type of algorithm for solving the real–valued test function suite of the 1st ICEO (the first two places were given to non–GA type algorithms which are not universally applicable but solved the test–problems faster than DE). The crucial idea behind DE is a scheme for generating trial parameter vectors. Basically, DE adds the weighted difference between two population vectors to a third vector. This way no separate probability distribution has to be used which makes the scheme completely self–organizing (see, e.g., [Lam02]).
3. The *artificial life* (*alife*), which is the study of life through the use of human–made analogs of living systems, evolving software that is more alive than a virus (see, e.g., [Lev92]). Theoretically, later it will become intelligent life. Computer scientist Christopher Langton coined the term in the late 1980s when he held the first Int. Conference on the Synthesis and Simulation of Living Systems (otherwise known as Artificial Life I) at the Los Alamos National Laboratory in 1987. Researchers of alife have focused on the 'bottom–up' nature of *emergent behaviors*. The alife field is characterized by the extensive use of computer programs and computer simulations which include evolutionary algorithms (EA), genetic algorithms (GA), genetic programming (GP), swarm intelligence (SI), ant colony optimization (ACO), artificial chemistries (AC), agent–based models, and cellular automata (CA). Often those techniques are seen as subfields of alife.The so–called *strong alife* position states that 'life is a process which can be abstracted away from any particular medium'.

Notably, Tom Ray declared that his program 'Tierra'[163] was not simulating life in a computer, but was synthesizing it. On the ther hand, the *weak alife* position denies the possibility of generating a 'living process' outside of a carbon–based chemical solution. Its researchers try instead to mimic life processes to understand the appearance of single phenomena. The usual way is through an agent based model, which usually gives a minimal possible solution. Closely related to alife is a *digital organism*, which is a self–replicating computer program that mutates and evolves. Digital organisms are used as a tool to study the dynamics of *Darwinian evolution*, and to test or verify specific hypotheses or mathematical models of evolution.

4. The *artificial immune system* (AIS), which is a type of optimisation algorithm inspired by the principles and processes of the vertebrate immune system (see [FPP86, Das99]). The algorithms typically exploit the immune system's characteristics of learning and memory to solve a problem. They are closely related to GAs. Processes simulated in AlS include pattern recognition, hypermutation and clonal selection for B cells,

[163] *Tierra* is a computer simulation developed by ecologist Thomas S. Ray in the early 1990s in which computer programs compete for central processing unit (CPU) time and access to main memory. The computer programs in Tierra are evolvable and can mutate, self–replicate and recombine. Tierra is a frequently cited example of an artificial life model; in the metaphor of the Tierra, the evolvable computer programs can be considered as digital organisms which compete for energy (CPU time) and resources (main memory). The basic Tierra model has been used to experimentally explore in silico the basic processes of evolutionary and ecological dynamics. Processes such as the dynamics of punctuated equilibrium, host–parasite co–evolution and density dependent natural selection are amenable to investigation within the Tierra framework. A notable difference to more conventional models of evolutionary computation, such as genetic algorithms is that there is no explicit, or exogenous fitness function built into the model. Often in such models there is the notion of a function being 'optimized'; in the case of Tierra, the fitness function is endogenous: there is simply survival and death. According to Ray and others this may allow for more 'open–ended' evolution, in which the dynamics of the feedback between evolutionary and ecological processes can itself change over time, although this promise has not been realized, like most other open–ended digital evolution systems, it eventually comes to a point where novelty ceases to be created, and the system at large begins either looping or evolving statically; some descendant systems like *Avida* try to avoid this pitfall. While the dynamics of Tierra are highly suggestive, the significance of the dynamics for real ecological and evolutionary behavior are still a subject of debate within the scientific community. Tierra is an abstract model, but any quantitative model is still subject to the same validation and verification techniques applied to more traditional mathematical models, and as such, has no special status. More detailed models in which more realistic dynamics of biological systems and organisms are incorporated is now an active research field.

negative selection of T cells, affinity maturation and immune network theory. In AIS, *antibody* and *antigen* representation is commonly implemented by strings of attributes. Attributes may be binary, integer or real–valued, although in principle any ordinal attribute could be used. Matching is done on the grounds of *Euclidean distance* $= \sum_{i=1}^{n}(x_i - y_i)^2$, *Manhattan distance*[164] or *Hamming distance*.[165] The so–called *clonal selection algorithms* are commonly used for *antibody hypermutation*. This allows the attribute string to be improved (as measured by a *fitness function*) using mutation alone.

5. The *learnable evolution model* (LEM), which is a novel, non–Darwinian methodology for evolutionary computation that employs machine learning

---

[164] The so–called *taxicab geometry*, considered by Hermann Minkowski in the 19th century, is a form of geometry in which the usual metric of Euclidean geometry is replaced by a new metric in which the distance between two points is the sum of the (absolute) differences of their coordinates. More formally, we can define the *Manhattan distance*, also known as the L^1–distance, between two points in an Euclidean space with fixed Cartesian coordinate system as the sum of the lengths of the projections of the line segment between the points onto the coordinate axes. Manhattan distance is also known as city block distance or taxi-cab distance. It is named so because it is the shortest distance a car would drive in a city laid out in square blocks, like Manhattan (discounting the facts that in Manhattan there are one–way and oblique streets and that real streets only exist at the edges of blocks, i.e., there is no 3.14th Avenue). Any route from a corner to another one that is 3 blocks East and 6 blocks North, will cover at least 9 blocks. All direct routes cover exactly 9. Taxicab geometry satisfies all of Hilbert's axioms except for the side–angle–side axiom, as one can generate two triangles with two sides and the angle between the same and have them not be congruent. A circle in taxicab geometry consists of those points that are a fixed Manhattan distance from the center. These circles are squares whose sides make a 45° angle with the coordinate axes.

In chess, the distance between squares on the chessboard for rooks is measured in Manhattan distance; kings and queens use Chebyshev distance, and bishops use the Manhattan distance (between squares of the same color) on the chessboard rotated 45 degrees, i.e., with its diagonals as coordinate axes. To reach from one square to another, only kings require the number of moves equal to the distance; rooks, queens and bishops require one or two moves (on an empty board, and assuming that the move is possible at all in the bishop's case).

[165] The Hamming distance between two strings of equal length is the number of positions for which the corresponding symbols are different. Put another way, it measures the number of substitutions required to change one into the other, or the number of errors that transformed one string into the other. For example: (i) The Hamming distance between 1011101 and 1001001 is 2; (ii) The Hamming distance between 2143896 and 2233796 is 3; (iii) The Hamming distance between 'toned' and 'roses' is 3. The *Hamming weight* of a string is its Hamming distance from the zero string (string consisting of all zeros) of the same length. That is, it is the number of elements in the string which are not zero: for a binary string this is just the number of 1's, so for instance the Hamming weight of 11101 is 4.

to guide the generation of new individuals (candidate problem solutions) [WM06]. Unlike standard, Darwinian–type evolutionary computation methods that use random or semi–random operators for generating new individuals (such as mutations and/or recombinations), LEM employs hypothesis generation and instantiation operators. The hypothesis generation operator applies a machine learning program to induce descriptions that distinguish between high–fitness and low–fitness individuals in each consecutive population. Such descriptions delineate areas in the search space that most likely contain the desirable solutions. Subsequently the instantiation operator samples these areas to create new individuals.

Cellular Automata

It is common in nature to find systems whose overall behavior is extremely complex, yet whose fundamental component parts are each very simple. The complexity is generated by the cooperative effect of many simple identical components. Much has been discovered about the nature of the components in physical and biological systems; little is known about the mechanisms by which these components act together to give the overall complexity observed. According to Steve Wolfram [Wol02, Wol84], what is needed is a general mathematical theory to describe the nature and generation of complexity.

Cellular automata (CA) are examples of mathematical systems constructed from many identical components, each simple, but together capable of complex behavior. From their analysis one may, on the one hand, develop specific models for particular systems, and, on the other hand, hope to abstract general principles applicable to a wide variety of complex systems.

1D Cellular Automata

Recall that a 1D CA consists of a line of sites, with each site carrying a value 0 or 1 (or in general $0, \dots, k-1$). The value α_i of the site at each position i is updated in discrete time steps according to an identical deterministic rule depending on a neighborhood of sites around it [Wol02, Wol84]:

$$\alpha_i^{t+1} = \varphi[\alpha_{i-r}^t, \alpha_{i-r+1}^t, \dots, \alpha_{i+r}^t]. \tag{1.30}$$

Even with $k = 2$ and $r = 1$ or 2, the overall behavior of CA constructed in this simple way can be extremely complex.

Consider first the patterns generated by CA evolving from simple 'seeds' consisting of a few non–zero sites. Some local rules φ give rise to simple behavior; others produce complicated patterns. An extensive empirical study suggests that the patterns take on four qualitative forms (see Figure 1.44):

1. Disappears with time;
2. Evolves to a fixed finite size;
3. Grows indefinitely at a fixed speed; and
4. Grows and contracts irregularly.

Fig. 1.44. Classes of patterns generated by the evolution of CA from simple 'seeds'. Successive rows correspond to successive time steps in the CA evolution. Each site is updated at each time step according to equation (1.30) by CA rules that depend on the values of a neighborhood of sites at the previous time step. Sites with values 0 and 1 are represented by white and black squares, respectively. Despite the simplicity of their construction, patterns of some complexity are seen to be generated. The rules shown exemplify the four classes of behavior found. In the third case, a self–similar pattern is formed (adapted from [Wol02, Wol84]).

Patterns of type 3 are often found to be self–similar or scale invariant. Parts of such patterns, when magnified, are indistinguishable from the whole. The patterns are characterized by a *fractal dimension*, with the most common value $log_2 3 \simeq 1.59$. Many of the self–similar patterns seen in natural systems may in fact, be generated by CA evolution.

Different initial states with a particular CA rule yield patterns that differ in detail, but are similar in form and statistical properties. Different CA rules yield very different patterns. An empirical study, nevertheless, suggests that four qualitative classes may be identified, yielding four characteristic limiting forms:

1. Spatially homogeneous state;
2. Sequence of simple stable or periodic structures;
3. Chaotic aperiodic behavior; and
4. Complicated localized structures, some propagating.

All CA within each class, regardless of the details of their construction and evolution rules, exhibit qualitatively similar behavior. Such universality should make general results on these classes applicable to a wide variety of systems modelled by CA.

CA Applications

Mathematical models of natural systems are usually based on differential equations which describe the smooth variation of one parameter as a function of a few others. Cellular automata provide alternative and in some respects complementary models, describing the discrete evolution of many (identical) components. Models based on CA are typically most appropriate in highly nonlinear regimes of physical systems, and in chemical and biological systems where discrete thresholds occur. Cellular automata are particularly suitable as models when *growth inhibition effects* are important [Wol02, Wol84].

As one example, CA provide global models for the growth of dendritic crystals (such as snowflakes). Starting from a simple seed, sites with values representing the solid phase are aggregated according to a 2D rule that accounts for the inhibition of growth near newly–aggregated sites, resulting in a fractal pattern of growth. Nonlinear chemical reaction–diffusion systems give another example: a simple CA rule with growth inhibition captures the essential features of the usual partial differential equations, and reproduces the spatial patterns seen. Turbulent fluids may also potentially be modelled as CA with local interactions between discrete vortices on lattice sites [Wol02, Wol84].

If probabilistic noise is added to the time evolution rule (1.30), then CA may be identified as generalized *Ising–spin* models. Phase transitions may occur if retains some deterministic components, or in more than one dimension.

Cellular automata may serve as suitable models for a wide variety of biological systems. In particular, they may suggest mechanisms for biological pattern formation. For example, the patterns of pigmentation found on many mollusc shells bear a striking resemblance to patterns generated by class 2 and 3 CA, and CA models for the growth of some pigmentation patterns have been constructed [Wol02, Wol84].

Two Approaches to CA Mathematics

Rather than describing specific applications of CA, here we concentrate on general mathematical features of their behavior. Two complementary approaches provide characterizations of the four classes of behavior [Wol02, Wol84].

In the first approach, CA are viewed as discrete dynamical systems (see, e.g., [GH83]), or discrete idealizations of partial differential equations. The set of possible (infinite) configurations of a CA forms a *Cantor set*. CA evolution may be viewed as a continuous mapping on this Cantor set. Quantities such as entropies, dimensions and Lyapunov exponents may then be considered for CA.

In the second approach, CA are instead considered as information–processing systems (see, e.g., [HU79]), or parallel–processing computers of simple construction. Information represented by the initial configuration is processed by the evolution of the CA. The results of this information processing may then be characterized in terms of the types of formal languages generated.[166]

---

[166] Note that the mechanisms for information processing in natural system appear to be much closer to those in CA than in conventional serial–processing computers: CA may, therefore, provide efficient media for practical simulations of many natural systems.

CA Entropies and Dimensions

Most CA rules have the important feature of irreversibility: several different configurations may evolve to a single configuration, and, with time, a contracting subset of all possible configurations appears. Starting from all possible initial configurations, the CA evolution may generate only special 'organized' configurations, and 'self–organization' may occur.

For class 1 CA, essentially all initial configurations evolve to a single final configuration, analogous to a limit point in a continuous dynamical system. Class 2 CA evolve to limit sets containing essentially only periodic configurations, analogous to limit cycles. Class 3 CA yield chaotic aperiodic limit sets, containing analogues of *strange attractors* [Wol02, Wol84].

Entropies and dimensions give a generalized measure of the density of the configurations generated by CA evolution. The (set) dimension or limiting (topological) entropy for a set of CA configurations is defined as (compare with [GH83]):

$$d^{(x)} = \lim_{X\to\infty} \frac{1}{X} \log_k N(X), \tag{1.31}$$

where $N(X)$ gives the number of distinct sequences of $X-$site values that appear. For the set of possible initial configurations, $d^{(x)} = 1$. For a limit set containing only a finite total number of configurations, $d^{(x)} = 0$. For most class 3 CA, $d^{(x)}$ decreases with time, giving, $0 < d^{(x)} < 1$, and suggesting that a fractal subset of all possible configurations occurs.

A dimension or limiting entropy $d^{(t)}$ corresponding to the time series of values of a single site may be defined in analogy with equation (1.31)[167] $d^{(t)} = 0$, for periodic sets of configurations.

Both $d^{(x)}$ and $d^{(t)}$ may be modified to account for the probabilities of configurations by defining

$$d_\mu^{(x)} = -\lim_{X\to\infty} \frac{1}{X} \sum_{i=1}^{k^\mu} p_i \log_k p_i, \tag{1.32}$$

and its $d^{(t)}-$analogue, where p_i are probabilities for possible length $X-$sequences. These measure dimensions may be used to delineate the large time behavior of the different classes of CA:[168]

1. $d_\mu^{(x)} = d_\mu^{(t)} = 0$;
2. $d_\mu^{(x)} > 0, d_\mu^{(t)} = 0$;
3. $d_\mu^{(x)} > 0, d_\mu^{(t)} > 0$.

---

[167] The analogue of equation (1.31) for a sufficiently wide patch of sites yields a topologically–invariant entropy for the CA mapping.

[168] Dimensions are usually undefined for class 4 CA.

CA Information Propagation

Cellular automata may also be characterized by the stability or predictability of their behavior under small perturbations in initial configurations, usually resulting from a change in a single initial site value (see Figure 1.45). Such perturbations have characteristic effects on the four classes of CA:

1. No change in final state;
2. Changes only in a finite region;
3. Changes over an ever-increasing region; and
4. Irregular changes.

In class 1 and 2 CA, information associated with site values in the initial state propagates only a finite distance; in class 3 CA, it propagates an infinite distance at a fixed speed, while in class 4 CA, it propagates irregularly but over an infinite range. The speed of information propagation is related to the Lyapunov exponent for the CA evolution, and measures the degree of sensitivity to initial conditions. It leads to different degrees of predictability for the outcome of CA evolution [Wol02, Wol84]:

1. Entirely predictable, independent of initial state;
2. Local behavior predictable from local initial state;
3. Behavior depends on an ever–increasing initial region; and
4. Behavior effectively unpredictable.

Information propagation is particularly simple for the special class of additive CA (whose local rule function φ is linear modulo k), in which patterns generated from arbitrary initial states may be obtained by superposition of patterns generated by evolution of simple initial states containing a single non–zero site. A rather complete algebraic analysis of such CA may be given. Most CA are not additive; however, with special initial configurations it is often possible for them to behave just like additive rules. Thus, for example, the evolution of an initial configuration consisting of a sequence of 00 and

Fig. 1.45. Evolution of small initial perturbations in CA, as shown by the difference (modulo two) between patterns generated from two disordered initial states differing in the value of a single site. The examples shown illustrate the four classes of behavior found. Information on changes in the initial state almost always propagates only a finite distance in the first two classes, but may propagate an arbitrary distance in the third and fourth classes (adapted from [Wol02, Wol84]).

01 diagrams under one rule may be identical to the evolution of the corresponding 'blocked' configuration consisting of 0 and 1 under another rule. In this way, one rule may simulate another under a blocking transformation (analogous to a renormalization group transformation). Evolution from an arbitrary initial state may be attracted to (or repelled from) the special set of configurations for which such a simulation occurs. Often several phases exist, corresponding to different blocking transformations: sometimes phase boundaries move at constant speed, and one phase rapidly takes over; in other cases, phase boundaries execute random walks, annihilating in pairs, and leading to a slow increase in the average domain size. Many rules appear to follow attractive simulation paths to additive rules, which correspond to fixed points of blocking transformations, and thus exhibit self similarity. The behavior of many rules at large times, and on large spatial scales, is therefore determined by the behavior of additive rules.

CA Thermodynamics

Decreases with time in the spatial entropies and dimensions of equations (1.31)–(1.32) signal irreversibility in CA evolution. Some CA rules are, however, reversible, so that each and every configuration has a unique predecessor in the evolution, and the spatial entropy and dimension of equations (1.31)–(1.32) remain constant with time.

Now, conventional *thermodynamics* gives a general description of systems whose microscopic evolution is reversible; it may, therefore, be applied to reversible CA. As usual, the 'fine–grained' entropy for sets (ensembles) of configurations, computed as in (1.32) with perfect knowledge of each site value, remains constant in time. The 'coarse–grained' entropy for configurations is, nevertheless, almost always non–decreasing with time, as required by the second law of thermodynamics. Coarse graining emulates the imprecision of practical measurements, and may be implemented by applying almost any contractive mapping to the configurations (a few iterations of an irreversible CA rule suffice). For example, coarse–grained entropy might be computed by applying (1.32) to every fifth site value. In an ensemble with low coarse–grained entropy, the values of every fifth site would be highly constrained, but arbitrary values for the intervening sites would be allowed. Then in the evolution of a class 3 or 4 CA the disorder of the intervening site values would 'mix' with the fifth–site values, and the coarse–grained entropy would tend towards its maximum value. Signs of self–organization in such systems must be sought in temporal correlations, often manifest in 'fluctuations' or meta–stable 'pockets' of order.

While all fundamental physical laws appear to be reversible, macroscopic systems often behave irreversibly, and are appropriately described by irreversible laws. Thus, for example, although the microscopic molecular dynamics of fluids is reversible, the relevant macroscopic velocity field obeys the irreversible *Navier–Stokes equations*. Conventional thermodynamics does not apply to such intrinsically irreversible systems; new general principles must be found. Thus, for CA with irreversible evolution rules, coarse–grained entropy typically increases for a short time, but then decreases to follow the fine grained entropy. Measures of the structure generated by self–organization in the large time limit are usually affected very little by coarse graining.

CA and Formal Language Theory

Quantities such as entropy and dimension, suggested by information theory, give only rough characterizations of CA behavior. Computation theory suggests more complete descriptions of self–organization in CA (and other systems). Sets of CA configurations may be viewed as formal languages, consisting of sequences of symbols (site values) forming words according to definite grammatical rules.

The set of all possible initial configurations corresponds to a trivial formal language. The set of configurations obtained after any finite number of time steps are found to form a regular language. The words in a regular language correspond to the possible paths through a finite graph representing a finite state machine. It can be shown that a unique smallest finite graph reproduces any given regular language (see [HU79]). Examples of such graphs are shown in Figure 1.46. These graphs give complete specifications for sets of CA configurations (ignoring probabilities). The number of nodes in the smallest graph corresponding to a particular set of configurations may be defined as the 'regular language complexity' of the set. It specifies the size of the minimal description of the set in terms of regular languages. Larger correspond to more complicated sets.

The regular language complexity Ξ for sets generated by CA evolution almost always seems to be nondecreasing with time. Increasing Ξ signals increasing self–organization. Ξ may thus represent a fundamental property of self–organizing systems, complementary to entropy. It may, in principle, be extracted from experimental data [Wol02, Wol84].

Cellular automata that exhibit only class 1 and 2 behavior always appear to yields sets that correspond to regular languages in the large time limit. Class 3 and 4 behavior typically gives rise, however, to a rapid increase of Ξ with time, presumably leading to limiting sets not described by regular languages.

Formal languages are recognized or generated by idealized computers with a 'central processing unit' containing a fixed finite number of internal states, together with a 'memory'. Four types of formal languages are conventionally identified, corresponding to four types of computer:

1. Regular languages: no memory required.
2. Context–free languages: memory arranged as a last–in, first–out stack.
3. Context–sensitive languages: memory as large as input word required.
4. Unrestricted languages: arbitrarily large memory required (general Turing machine).

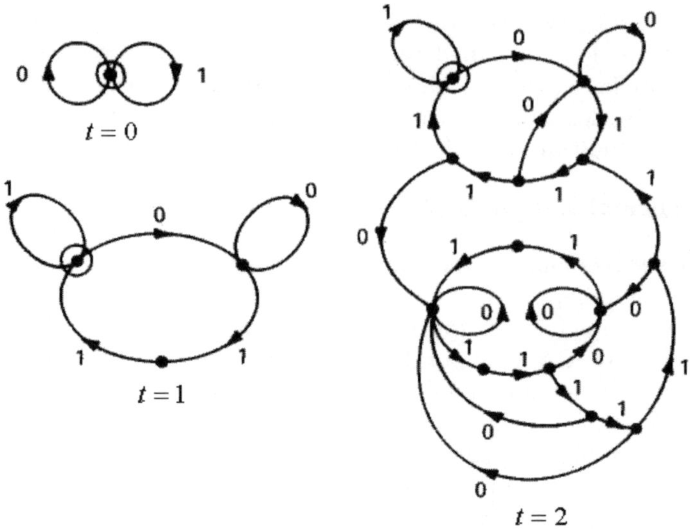

Fig. 1.46. Graphs representing the sets of configurations generated in the first few time steps of evolution according to a typical class 3 CA rule ($k = 2, r = 1$, rule number 126). Possible configurations correspond to possible paths through the graphs, beginning at the encircled node. At $t = 0$, all possible configurations are allowed. With time, a contracting subset of configurations are generated (e.g., after one time step no configuration containing the sequence of site value 101 can appear) At each time step, the complete set of possible configurations forms a regular formal language: the graph gives a minimal complete specification of it. The number of nodes in the graph gives a measure of the complexity Ξ of the set, viewed as a regular language. As for other class 3 CA, the complexity of the sets Ξ grows rapidly with time (modified and adapted from [Wol02, Wol84]).

Examples are known of CA whose limiting sets correspond to all four types of language. Arguments can be given that the limit sets for class 3 CA typically form context–sensitive languages, while those for class 4 CA correspond to unrestricted languages.[169]

CA and Computation Theory

While dynamical systems theory concepts suffice to define class 1, 2 and 3 CA, computation theory is apparently required for class 4 CA. Varied and complicated behavior, involving many different time scales is evident. Persistent structures are often generated. It seems that the structures supported by

[169] While a minimal specification for any regular language may always be found, there is no finite procedure to get a minimal form for more complicated formal languages; no generalization of the regular language complexity may thus be given.

this and other class 4 CA rule may be combined to implement arbitrary information processing operations. Class 4 CA would then be capable of universal computation: with particular initial states, their evolution could implement any finite algorithm. A few percent of CA rules with $k > 2$ or $r > 1$ are found to exhibit class 4 behavior: all these would then, in fact, be capable of arbitrarily complicated behavior. This capability precludes a smooth infinite size limit for entropy or other quantities: as the size of CA considered increases, more and more complicated phenomena may appear [Wol02, Wol84].

CA evolution may be viewed as a computation. Effective prediction of the outcome of CA evolution requires a short–cut that allows a more efficient computation than the evolution itself. For class 1 and 2 CA, such short cuts are clearly possible: simple computations suffice to predict their complete future. The computational capabilities of class 3 and 4 CA may, however, be sufficiently great that, in general, they allow no short-cuts. The only effective way to determine their evolution from a given initial state would then be by explicit observation or simulation: no finite formulae for their general behavior could be given.[170] Their infinite time limiting behavior could then not, in general, be determined by any finite computational process, and many of their limiting properties would be formally undecidable. Thus, for example, the 'halting problem' of determining whether a class 4 CA with a given finite initial configuration ever evolves to the null configuration would be undecidable. An explicit simulation could determine only whether halting occurred before some fixed time, and not whether it occurred after an arbitrarily long time.

For class 4 CA, the outcome of evolution from almost all initial configurations can probably be determined only by explicit simulation, while for class 3 CA this is the case for only a small fraction of initial states. Nevertheless, this possibility suggests that the occurrence of particular site value sequences in the infinite time limit is in general undecidable. The large time limit of the entropy for class 3 and 4 CA would then, in general, be non-computable: bounds on it could be given, but there could be no finite procedure to compute it to arbitrary precision.[171]

Undecidability and intractability are common in problems of mathematics and computation. They may well afflict all but the simplest CA. One may speculate that they are widespread in natural systems, perhaps occurring almost whenever nonlinearity is present. No simple formulae for the behavior of many natural systems could then be given; the consequences of their evolution could be found effectively only by direct simulation or observation.

For more details on CA, complexity and computation, see [Wol02].

---

[170] If class 4 CA are indeed capable of universal computation, then the variety of their possible behavior would preclude general prediction, and make explicit observation or simulation necessary.

[171] This would be the case if the limit sets for class 3 and 4 CA formed at least context–sensitive languages.

Adaptive Business Intelligence

Recall that businesses and government agencies are mostly interested in two fundamental things [Mic06]: (i) knowing what will happen next (prediction), and (ii) making the best decision under risk and uncertainty (optimization) (see Figure 1.47). Therefore, from CI–perspective, the goal is to provide CI–based solutions for modelling, simulation, and optimization to address these two fundamental needs.

Information technology applications that support decision–making processes and problem–solving activities have proliferated and evolved over the past few decades. In the 1970s, these applications were simple and based on spreadsheet software. During the 1980s, decision-support systems incorporated optimization models, which originated in the operations research and management science communities. In the 1990s, these systems were further enhanced with components from artificial intelligence and statistics [MSM05]. This evolution led to many different types of *decision–support systems* with somewhat confusing names, including *management information systems*, *intelligent information systems*, *expert systems*, *management–support systems*, and *knowledge–based systems*. Because businesses realized that data was a precious asset, they often based these 1intelligent' systems on data warehousing and online analytical processing technologies. They gathered and stored a lot of data, assuming valuable assets were implicitly coded in it. Raw data, however, is rarely beneficial. Its value depends on a user's ability to extract knowledge that is useful for decision support. Thousands of 'business intelligence' companies thus emerged to provide such services. After analyzing a corporation's operational data, for example, these companies might return intelligence (in the form of tables, graphs, charts, and so on) stating that, say,

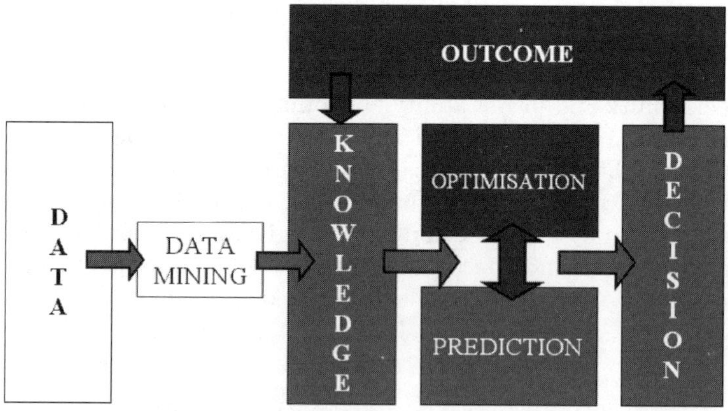

Fig. 1.47. Adaptive business intelligence: the diagram shows the flow from data acquisition to recommended action, including an adaptive feedback loop (adapted from [Mic06]).

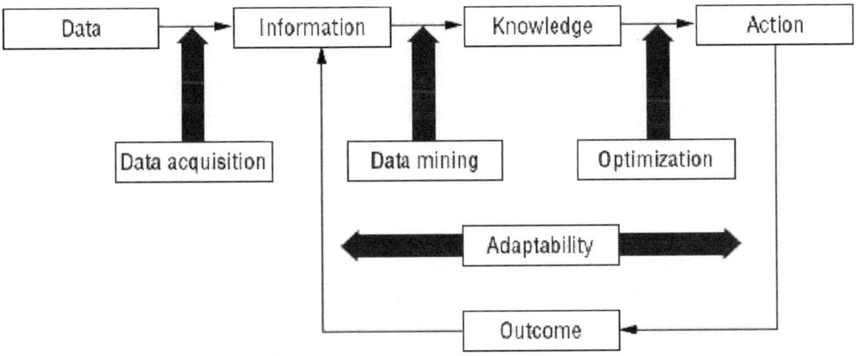

Fig. 1.48. Adaptive business intelligence: the diagram shows the flow from data acquisition to recommended action, including an adaptive feedback loop (adapted from [MSM05]).

57 percent of the corporation's customers are between 40 and 50, or product Q sells much better in Florida than in Georgia.

Many businesses have realized, however, that the return on investment for pure 'business intelligence' is much smaller than initially thought. The 'discovery' that 57 percent of our customers are between 40 and 50 doesn't directly lead to decisions that increase profit or market share. Moreover, we live in a dynamic environment where everything is in flux. Interest rates change, new fraud patterns emerge, weather conditions vary, the stock markets rise and fall, new regulations and policies surface, and so on. These economic and environmental changes render some data obsolete and make other data—which might have been useless just six weeks ago—suddenly meaningful.

Michalewicz *et al.* developed a software system (see Figure 1.48) to address these complexities and implemented it on a real distribution problem for a large car manufacturer. The system detects data trends in a dynamic environment, incorporates optimization modules to recommend a near–optimum decision, and includes self–learning modules to improve future recommendations. As Figure 1.48 shows, such a system lets enterprizes monitor business trends, evolve and adapt quickly as situations change, and make intelligent decisions based on uncertain and incomplete information. This intelligent system combines three modules: prediction, optimization and adaptation.

Research Issues in Dynamic Optimization

Most data–mining and optimization algorithms assume static data and a static objective. Typically, they search for a snapshot of 'knowledge' and a near-optimum solution with respect to some fixed measure (or set of measures), such as profit maximization or minimization of task–completion time. However, real–world applications operate in dynamic environments, where it's often necessary to modify the current solution due to changes in the problem setting, such as machine breakdown or employee illness; or the environment, such as consumer trends or changes in weather patterns or economic indicators. It's therefore important to investigate adaptive algorithms that don't require restart every time a change is recorded. In many commercial situations, such restarts are not an option.

Evolutionary Techniques

An obvious starting point here is evolutionary computation techniques [MF04], which are optimization algorithms inspired by the continuously changing natural environment. However, it is important to investigate which evolutionary algorithm extensions are actually useful in business scenarios. Unfortunately, most current approaches ignore dynamics and assume that re–optimization should occur at regular intervals. However, significant benefits can be realized when researchers explicitly address dynamism.

Many researchers have proposed various benchmarks for studying optimization in dynamic environments. Among the proposals are the moving peaks benchmark, the dynamic knapsack problem, dynamic bit–matching, scheduling with new jobs arriving over time, and the greenhouse control problem. Researchers have also proposed various measures, including off–line error, percentage of covered peaks, and diversity. Among the partial conclusions reached in this research [Bra01]:

- standard evolutionary algorithms get stuck on a single peak;
- diversity preservation slows down the convergence;
- random immigrants introduce high diversity from the beginning, but offer limited benefits;
- memory without diversity preservation is counterproductive; and
- nonadaptive memory suffers significantly if peaks move.

However, several essential points are seemingly missing in the key research on optimization in dynamic environments. Most researchers emphasize an ultimate goal of approximating real–world environments, but they fail to address several key issues for successful adaptive–system development. The following issues, which constitute the conceptual research framework, are essential for creating a methodology for building *intelligent systems* [MSM05].

Non–Stationary Constraints

Here, the task is to optimize a non–stationary *objective function* $f(x,t)$, subject to non–stationary constraints, $ci(x,t) \leq 0$, $(i = 1, 2, \ldots, k)$. This approach was applied successfully in the context of a collision situation at sea [SM00]. By accounting for particular maneuvering–region boundaries, along with information on navigation obstacles and other moving ships, the authors reduced the *collision–avoidance problem* to a *dynamic optimization task* with static and dynamic constraints. The proposed algorithm computed a safe and optimum ship path in both static and dynamic environments.

Prediction Component

Environmental changes are seldom random. In a typical real–world scenario, where constraints change over time, it's possible to calculate some failure probabilities by analyzing past data, and thus predict a possible environmental change. The above mentioned work on collisions at sea [SM00] offers a good example here as well. The authors based a ship's safe trajectory in a collision situation on predicted speeds and the other ships' directions. Studying dynamic environments where change is somewhat predictable is important, but so far, little work exists along these lines.

Parameter Adaptation

In nonstationary environments, researchers must study parameter control, particularly when the adaptive system includes predictive methods [EHM99].

Solution Robustness

Research into robustness concentrates on questions such as: What constitutes flexibility in the specific context? How can we integrate a flexibility goal into the algorithm? To answer these questions, we must take into account a predictive model (for environmental changes) and the prediction's estimated error. This has yet to occur [MSM05]. Many researchers have recognized the importance of solution robustness [Bra01]. Existing approaches vary, from techniques to 'disturb' individuals in the population to those using search history. Some researchers have considered an aspect of robustness, sometimes called flexibility, in which the problem requires sequential decision–making under an uncertain future, and the decision influences the system's future state. In such situations, the decision–making process should anticipate future needs. That is, rather than focusing exclusively on the primary objective function, it should try to move the system into a flexible state.

2

Chaotic Brain/Mind Dynamics

In this Chapter we present a chaos theory of the computational mind.

2.1 Chaos in Human EEG

During the last decade there has been a heated debate about whether chaos theory can be applied to the dynamics of the human brain and mind [LEA00]. While it is obvious that nonlinear mechanisms are crucial in neural systems, there has been strong criticism of attempts to identify at strange attractors in brain signals and to measure their fractal dimensions, Lyapunov exponents, etc. Conventional methods analyzing brain dynamics are largely based on linear models and on Fourier spectra. Regardless of the existence of strange attractors (see below) in brain activity, the neuro–sciences should benefit greatly from alternative methods that have been developed in recent years for the analysis of nonlinear and chaotic behavior.

Recall that *electroencephalography* is the neurophysiologic measurement of the electrical activity of the brain by recording from electrodes placed on the scalp or, in special cases, subdurally or in the cerebral cortex. The resulting traces are known as an *electroencephalogram* (EEG) and represent an electrical signal (postsynaptic potentials) from a large number of neurons. These are sometimes called brain–waves, though this use is discouraged [Cob83]. The EEG is a brain function test, but in clinical use it is a 'gross correlate of brain activity' [Ebe02]. Electrical currents are not measured, but rather voltage differences between different parts of the brain. It is well established that the electroencephalogram EEG! is directly proportional to the local field potential recorded by electrodes on the brain's surface [DLC01].

EEGs are frequently used in experimentation because the process is non–invasive to the research subject. The subject does not need to make a decision or behavioral action in order to log data, and it can detect covert responses to stimuli, such as reading. The EEG is capable of detecting changes in electrical

V.G. Ivancevic and T.T. Ivancevic: *Chaotic Brain/Mind Dynamics*, Studies in Computational Intelligence (SCI) **60**, 271–459 (2007)
www.springerlink.com

activity in the brain on a millisecond–level. It is one of the few techniques available that has such high temporal resolution.[1]

Now, frequently asked question of researchers in brain dynamics over the past decade has been [DBL02]: "Is there chaos in the brain?", with this very question the subject of serious experimental investigation over the past half–decade [PD95]. Alongside these experimental endeavors has been the immense corpus of research due to W. Freeman and colleagues over the past half–century [Fre92]. Based on both his experimental and theoretical studies of the mammalian olfactory system Freeman has suggested that chaos is the very property which allows perception to take place and gives brains the flexibility to rapidly respond in a coherent manner to perceptual stimuli [Fre91]. According to Freeman, brains create macroscopic order from microscopic disorder by neurodynamics in perception [ABL00]. Freeman's repeated exhortations of the existence of chaos in the brain (as reflected by the existence of chaos in its electrical dynamics) unfortunately has, up until recently, received scant theoretical support, with other existing theories of the electroencephalogram (EEG) either not showing chaos or being unable to do so because of the details of their mathematical construction [Zha84, Nun00, RPW97, RSE82].

The general theory of the electroencephalogram developed by D. Liley and collaborators [LCW99, LCD02, DLC01] leads to a mathematical model describing the behavior of two coupled populations of neurones: excitatory (being approximately representative of the pyramidal neurones of neocortex) and inhibitory (being approximately representative of the inter–neurones of neocortex). The scale of modelling here is approximately that of a cortical macro–column, which is a small volume of neocortex containing approximately 10^5 neurones. Each of the two modelled populations is connected to the other population and each population feeds back onto itself in either a mutually excitatory (for the excitatory population) or mutually inhibitory (for the inhibitory population) fashion. Both modelled populations have external excitatory and inhibitory inputs. Inputs to each population, whether from external sources, or from the other population, are modelled based on the dynamics of fast–acting synapses. The mean soma membrane potential of the excitatory population (he) is directly related to the local field potential of the neuronal mass, which overwhelmingly dominates the composition of the scalp–recorded electroencephalogram (EEG) [DLC01, Nun81].

Chaos in the brain would manifest itself as unpredictable and seemingly random electrical activity in a population of nerve cells, or neurons. Chaos may have an important neurological function: it could provide, as researchers have

[1] The other common technique is *magneto–encephalography* (MEG), which is an imaging technique used to measure the magnetic fields produced by electrical activity in the brain via extremely sensitive devices such as SQUIDs. These measurements are commonly used in both research and clinical settings. There are many uses for the MEG, including assisting surgeons in localizing a pathology, assisting researchers in determining the function of various parts of the brain, neuro–feedback, and others.

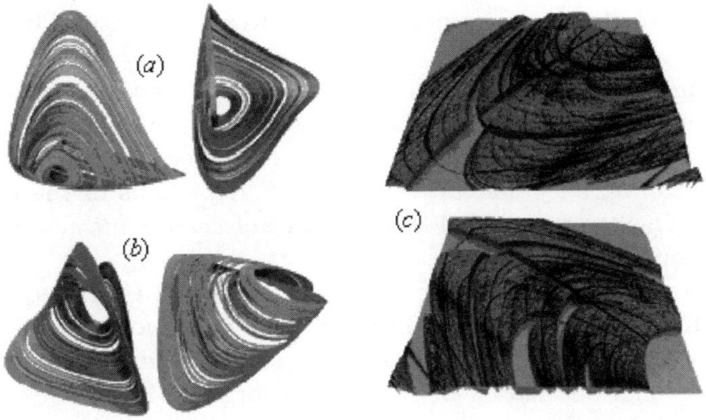

Fig. 2.1. Chaotic attractors in an EEG model (modified and adapted from [DBL02]). (*a*) Two complementary views of a chaotic attractor from the model shown from two different perspective. Note the shadowing and how it provides an enhanced sense of the depth of the attractor. (*b*) Two different views of another chaotic attractor. Note how the rendering provides a perspective on the interleaving of the attractor's sheets and folds. (*c*) Two different views of the parameter–space plane of the model with the largest Lyapunov exponent of the system as the dependent variable and the external excitatory input pulse density to the excitatory and inhibitory populations as the independent variables.

speculated, a flexible and rapid means for the brain to discriminate between different sounds, odors, and other perceptual stimuli.

EEGs record electrical activity in the cerebral cortex, but they, and all other current experimental techniques, may never be able to detect clear and unequivocal signs of chaos, since the cortex also emits a very large amount of obscuring 'noise' or random electrical activity.

Using realistic models of brain physiology, many researchers are trying to devise models which reproduce the output of EEGs yet also offer new insights into the brain's inner workings. However, previous models either do not allow for chaos to appear, or have been unable to demonstrate that chaos can occur under the conditions imposed by the structure of the brain.

In the present work, the researchers model the behavior of two large populations of neurons: excitatory (which bring other neurons closer to firing) and inhibitory (which make it more difficult for other neurons to fire). Specifically, they look at the 'mean soma membrane potential,' the electric potential between the outside and inside of the neuron's cell body (higher potential means more frequent firing).

Varying the rate of external electrical impulses to each neuron population, they found the mean electrical activity was irregular and noise–like (it looked like noise but really wasn't) for a wide range of external inputs. Quantitatively such behavior is associated with a *positive Lyapunov exponent*, a hallmark of chaos. The existence of chaos, the researchers say, would provide a means for the brain to change its response rapidly to even slightly different stimuli [DLC01].

It is a major source of contention in *brain dynamics* as to whether the electrical rhythms of the brain show signs of chaotic behavior. In [DBL02] the authors discussed the evidence for the existence of chaos in a theory of brain electrical activity and provided unique depictions of the dynamics of this model. They demonstrated the existence of *chaotic attractor* (see below) in human brain's EEG.

In a spontaneously bursting neuronal network *in vitro*, chaos can be demonstrated by the presence of unstable *fixed–point* behavior (see Figure 2.1 above). The techniques of *chaos control* have been used to increase the periodicity of neuronal population bursting behavior [Sch94].

2.2 Basics of Nonlinear Dynamics and Chaos Theory

Recall from [II06b] that the concept of *dynamical system* has its origins in *Newtonian mechanics*.[2] There, as in other natural sciences and engineering

[2] Recall that a *Newtonian deterministic system* is a system whose *present state is fully determined by its initial conditions* (at least, in principle), in contrast to a stochastic (or, random) system, for which the initial conditions determine the present state only partially, due to noise, or other external circumstances beyond our control. For a stochastic system, the present state reflects the past initial conditions plus the particular realization of the noise encountered along the way. So, in view of classical science, we have either deterministic or stochastic systems.

However, "Where chaos begins, classical science stops ... After relativity and quantum mechanics, chaos has become the 20th century's third great revolution in physical sciences." [Gle87].

For a long time, scientists avoided the irregular side of nature, such as disorder in a turbulent sea, in the atmosphere, and in the fluctuation of wild–life populations [Sha06]. Later, the study of this unusual results revealed that irregularity, nonlinearity, or chaos was the organizing principle of nature [Gle87]. Thus nonlinearity, most likely in its extreme form of chaos, was found to be ubiquitous [Hil94, CD98]. For example, in theoretical physics, chaos is a type of moderated randomness that, unlike true randomness, contains complex patterns that are mostly unknown [CD98]. Chaotic behavior appeared in the weather, the clustering of cars on an expressway, oil flowing in underground pipes [Gle87], convecting fluid, simple diode-circuits [Hil94], neural networks, digital filters, electronic devices, non-linear optics, lasers [CD98], and in complex systems like thrust-vectored fighter aircraft [Mos96]. No matter what the system, its behavior obeyed the same newly discovered law of nonlinearity and chaos [Gle87]. Thus, nonlinear dynamical system theory transcended the boundaries of different scientific disciplines, because it appeared to be a science of the global nature of systems [Sha06]. As a result, nonlinear dynamics found applications in physics, chemistry, meteorology, biology, medicine, physiology, psychology, fluid dynamics, engineering and various other disciplines. It has now become a common language used by scientists in various domains to study any system that obeys the same universal law [Gle87].

disciplines, the evolution rule of dynamical systems is given implicitly by a relation that gives the state of the system only a short time into the future. This relation is either a differential equation or difference equation. To determine the state for all future times requires iterating the relation many times–each advancing time a small step. The iteration procedure is referred to as solving the system or integrating the system. Once the system can be solved, given an initial point it is possible to determine all its future points, a collection known as a *trajectory* or *orbit*. All possible system trajectories comprise its *flow* in the phase–space.

In more simpler words, the *phase space* (also known as the *state space*) is the set of all possible states of a dynamical system. Solutions, such as a resting state or oscillations, correspond to geometric objects, such as points or closed curves, in phase space. Since it is usually impossible to derive an explicit formula for the solution of a nonlinear equation, the phase space provides an extremely useful way for visualizing and understanding qualitative features of solutions.

More precisely, a *dynamical system* geometrically represents a *vector–field* (or, more generally, a *tensor–field*) in the system's phase–space manifold M [II06b], which upon *integration* (governed by the celebrated *existence & uniqueness theorems for ordinary differential equations (ODEs)*) defines a *phase–flow* in M (see Figures 2.2 and 2.3). This phase–flow $f_t \in M$, describing the complete behavior of a dynamical system at every time instant, can be either linear, nonlinear or chaotic.

On the other hand, a modern scientific term *deterministic chaos* depicts an *irregular and unpredictable* time evolution of many (simple) deterministic dynamical systems, characterized by nonlinear coupling of its variables (see,

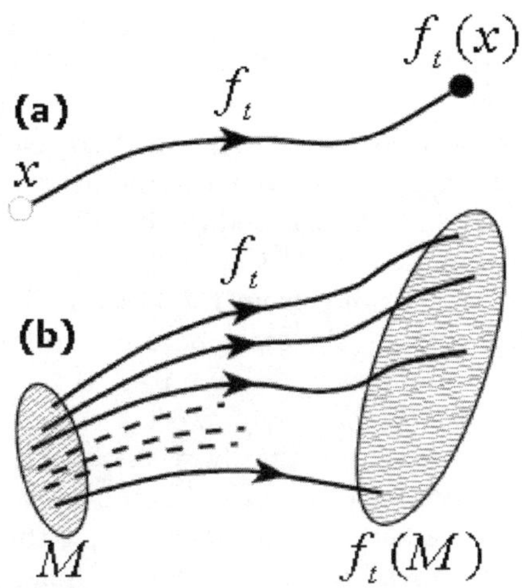

Fig. 2.2. Action of the *phase–flow* f_t in the phase–space manifold M: (a) Trajectory of a single initial point $x(t) \in M$, (b) Transporting the whole manifold M.

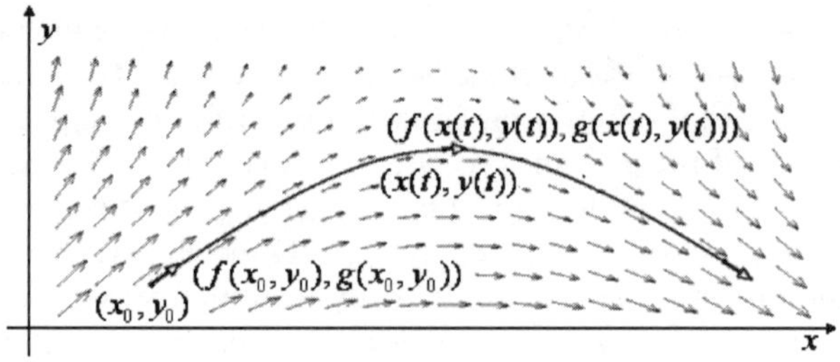

Fig. 2.3. The phase plane. The right hand side of the 2D dynamical system defined a vector–field. Solutions of the equations define curves or trajectories in the phase plane. The vector–field always points in the direction that the trajectories are flowing.

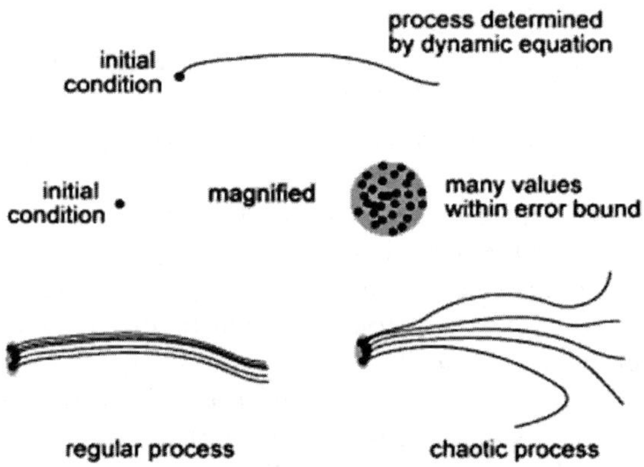

Fig. 2.4. Regular v.s. chaotic process.

e.g., [GOY87, YAS96, BG96, Str94]). Given an initial condition, the dynamic equation determines the dynamic process, i.e., every step in the evolution. However, the initial condition, when magnified, reveals a cluster of values within a certain error bound. For a regular dynamic system, processes issuing from the cluster are bundled together, and the bundle constitutes a predictable process with an error bound similar to that of the initial condition. In a chaotic dynamic system, processes issuing from the cluster diverge from each other exponentially, and after a while the error becomes so large that the dynamic equation losses its predictive power (see Figure 2.4).

For example, in a *pinball game*, any two trajectories that start out very close to each other separate exponentially with time, and in a finite (and in practice, a very small) number of bounces their separation $\delta x(t)$ attains the magnitude of L, the characteristic linear extent of the whole system. This property of sensitivity to initial conditions can be quantified as

$$|\delta x(t)| \approx \mathrm{e}^{\lambda t}|\delta x(0)|,$$

where λ, the *mean rate of separation of trajectories* of the system, is called the *Lyapunov exponent*. For any finite accuracy $|\delta x(0)| = \delta x$ of the initial data, the dynamics is predictable only up to a finite *Lyapunov time*

$$T_{Lyap} \approx -\frac{1}{\lambda} \ln |\delta x/L|,$$

despite the deterministic and infallible simple laws that rule the pinball motion.

However, a positive Lyapunov exponent does not in itself lead to chaos (see [CAM05]). One could try to play 1– or 2–disk pinball game, but it would not be much of a game; trajectories would only separate, never to meet again. What is also needed is mixing, the coming together again and again of trajectories. While locally the nearby trajectories separate, the interesting dynamics is confined to a globally finite region of the phase–space and thus the separated trajectories are necessarily folded back and can re–approach each other arbitrarily closely, infinitely many times. For the case at hand there are 2^n topologically distinct n bounce trajectories that originate from a given disk. More generally, the number of distinct trajectories with n bounces can be quantified as

$$N(n) \approx \mathrm{e}^{hn},$$

where the *topological entropy* h ($h = \ln 2$ in the case at hand) is the growth rate of the number of topologically distinct trajectories.

When a physicist says that a certain system "exhibits chaos", he means that the system obeys deterministic laws of evolution, but that the outcome is highly sensitive to small uncertainties in the specification of the initial state. The word "chaos" has in this context taken on a narrow technical meaning. If a deterministic system is locally unstable (positive Lyapunov exponent) and globally mixing (positive entropy), it is said to be *chaotic.*

While mathematically correct, the definition of chaos as "positive Lyapunov exponent + positive entropy" is useless in practice, as a measurement of these quantities is intrinsically asymptotic and beyond reach for systems observed in nature. More powerful is Poincaré's vision of chaos as the interplay of local instability (unstable periodic orbits) and global mixing (intertwining of their stable and unstable manifolds). In a chaotic system any open ball of initial conditions, no matter how small, will in finite time overlap with any other finite region and in this sense spread over the extent of the entire asymptotically accessible phase–space. Once this is grasped, the focus of theory shifts from attempting to predict individual trajectories (which is impossible) to a description of the geometry of the space of possible outcomes, and evaluation of averages over this space.

A definition of "turbulence" is even harder to come by. Intuitively, the word refers to *irregular behavior of an infinite–dimensional dynamical system described by deterministic equations of motion* – say, a bucket of boiling water – *described by the Navier–Stokes equations.* But in practice the word "turbulence" tends to refer to messy dynamics which we understand poorly. As soon as a phenomenon is understood better, it is reclaimed renamed as: “a route to chaos”, or “spatio–temporal chaos”, etc. (see [CAM05]).

Before the advent of fast computers, solving a dynamical system required sophisticated mathematical techniques and could only be accomplished for a small class of linear dynamical systems. Numerical methods executed on computers have simplified the task of determining the orbits of a dynamical system.

For simple dynamical systems, knowing the trajectory is often sufficient, but most dynamical systems are too complicated to be understood in terms of individual trajectories. The difficulties arise because:

1. The systems studied may only be known approximately–the parameters of the system may not be known precisely or terms may be missing from the equations. The approximations used bring into question the validity or relevance of numerical solutions. To address these questions several notions of stability have been introduced in the study of dynamical systems, such as *Lyapunov stability* or *structural stability*. The stability of the dynamical system implies that there is a class of models or initial conditions for which the trajectories would be equivalent. The operation for comparing orbits to establish their equivalence changes with the different notions of stability.
2. The type of trajectory may be more important than one particular trajectory. Some trajectories may be periodic, whereas others may wander through many different states of the system. Applications often require enumerating these classes or maintaining the system within one class. Classifying all possible trajectories has led to the qualitative study of dynamical systems, that is, properties that do not change under coordinate changes. Linear dynamical systems and systems that have two numbers describing a state are examples of dynamical systems where the possible classes of orbits are understood.
3. The behavior of trajectories as a function of a parameter may be what is needed for an application. As a parameter is varied, the dynamical systems may have *bifurcation points* where the qualitative behavior of the dynamical system changes. For example, it may go from having only periodic motions to apparently erratic behavior, as in the transition to *turbulence* of a fluid.
4. The trajectories of the system may appear erratic, as if random. In these cases it may be necessary to compute averages using one very long trajectory or many different trajectories. The averages are well defined for ergodic systems and a more detailed understanding has been worked out for *hyperbolic systems*. Understanding the probabilistic aspects of dynamical systems has helped establish the foundations of statistical mechanics and of chaos.

Dynamics Tradition and Chaos

Traditionally, a dynamicist would believe that to write down a system's equations is to understand the system. How better to capture the essential features? For example, for a *playground swing*, the equation of motion ties together the pendulum's angle, its velocity, its friction, and the force driving it. But because of the little bits of nonlinearity in this equation, a dynamicist would find himself helpless to answer the easiest practical questions about the future of the system. A computer can simulate the problem numerically calculating each cycle (i.e., integrating the pendulum equation), but simulation brings its own problem: the tiny imprecision built into each calculation rapidly takes over, because this is a system with sensitive dependence on initial condition. Before long, the signal disappears and all that remains is noise [Gle87].

For example, in 1960s, Ed Lorenz from MIT created a simple weather model in which small changes in starting conditions led to a marked ('catastrophic') changes in outcome, called *sensitive dependence on initial conditions*, or popularly, the *butterfly effect* (i.e., "the notion that a butterfly stirring the air today in Peking can transform storm systems next month in New York, or, even worse, can cause a hurricane in Texas"). Thus long–range prediction of imprecisely measured systems becomes an impossibility.

At about the same time, Steve Smale from Berkeley studied an oscillating system (the Van der Pol oscillator) and found that his initial conjecture "that all systems tend to a steady state" – was not valid for certain nonlinear dynamical systems. He represented behavior of these systems with topological foldings called *Smale's horseshoe* in the system's phase–space. These foldings allowed graphical display of why points close together could lead to quite different outcomes, which is again sensitive dependence on initial conditions.

The unique character of chaotic dynamics may be seen most clearly by imagining the system to be started twice, but from slightly different initial conditions (in case of human motion, these are initial joint angles and angular velocities). We can think of this small initial difference as resulting from measurement error. For non–chaotic systems, this uncertainty leads only to an error in prediction that *grows linearly* with time. For chaotic systems, on the other hand, the error *grows exponentially* in time, so that the state of the system is essentially unknown after very short time. This phenomenon, firstly recognized by H. Poincaré, the father of topology, in 1913, which occurs only when the governing equations are nonlinear, with nonlinearly coupled variables, is known as *sensitivity to initial conditions*. Another type of sensitivity of chaotic systems is *sensitivity to parameters*: a small variation of system parameters (e.g., mass, length and moment of inertia of human body segments) results in great change of system output (dynamics of human movement).

If prediction becomes impossible, it is evident that a chaotic system can resemble a stochastic system, say a Brownian motion. However, the source of the irregularity is quite different. For chaos, the irregularity is part of the intrinsic dynamics of the system, not random external influences (for example, random muscular contractions in human motion). Usually, though, chaotic systems are predictable in the short–term. This *short–term predictability* is useful in various domains ranging from weather forecasting to economic forecasting.

Recall that some aspects of chaos have been known for over a hundred years. Isaac Newton was said to get headaches thinking about the 3−body problem (Sun, Moon, and Earth). In 1887, King Oscar II of Sweden announced a prize for anyone who could solve the $n-$body problem and hence demonstrate stability of the solar system. The prize was awarded to Henri Poincaré, who showed that even the 3−body problem has no analytical solution [Pet93, BG79]. He went on to deduce many of the properties of chaotic systems including the sensitive dependence on initial conditions. With the successes of linear models in the sciences and the lack of powerful computers, the work of these early nonlinear dynamists went largely unnoticed and undeveloped for many decades. In 1963, Ed Lorenz from MIT published a seminal paper [Lor63, Spa82] in which he showed that chaos can occur in systems of autonomous (no explicit time dependence) ordinary differential equations (ODEs) with as few as three variables and two quadratic nonlinearities. For continuous flows, the *Poincaré–Bendixson theorem* [HS74] implies the necessity of three variables, and chaos requires at least one nonlinearity. More explicitly, the theorem states that the long–time limit of any 'smooth' two–dimensional flow is either a fixed–point or a periodic solution. With the growing availability of powerful computers, many other examples of chaos were subsequently discovered in algebraically simple ODEs. Yet the sufficient conditions for chaos in a system of ODEs remain unknown [SL00].

So, *necessary condition* for *existence of chaos* satisfies any autonomous continuous–time dynamical system (a vector–field) of dimension three or higher, with at least two nonlinearly coupled variables (e.g., a single human swivel joint like a shoulder or hip, determined by three joint angles and three angular momenta). In case of non–autonomous continuous–time systems, chaos can happen in dimension two, while in case of discrete–time systems – even in dimension one. Now, whether the behavior (a flow), of any such system will actually be chaotic or not depends upon the values of its parameters and/or initial conditions. Usually, for some values of involved parameters, the system behavior is oscillating in a stable regime, while for another values of the parameters the behavior becomes chaotic, showing a *bifurcation*, or a *phase transition* – from one regime/phase to a totally different one. If a change in the system's behavior at the bifurcation point is really sharp, we could probably be able to recognize one of the celebrated polynomial *catastrophes* of R. Thom (see [Tho75, Arn92]). A series of such bifurcations usually depicts a *route to chaos*.

Chaos theory has developed special mathematical procedures to *understand* irregularity and unpredictability of low–dimensional nonlinear systems, including Poincaré sections, bifurcation diagrams, power spectra, Lyapunov exponents, period doubling, fractal dimension, stretching and folding, special identification and estimation techniques, etc. (see e.g., [Arn89, Arn78, Arn88, Arn93, YAS96, BG96]). Understanding these phenomena has enabled science to *control* the chaos (see, e.g., [OGY90, CD98]).

There are many practical reasons for *controlling* or *ordering chaos*. For example, in case of a distributed artificial intelligence system, which is usually characterized by a massive collection of decision–making agents, the fact that an agent's decision also depends on decisions made by other agents – leads to extreme complexity and nonlinearity of the overall system. More often than not, the information received by agents about the 'state' of the system may be 'tainted'. When the system contains imperfect information, its agents tend to make poor decisions concerning choosing an optimal problem–solving strategy or cooperating with other agents. This can result in certain chaotic behavior of the agents, thereby downgrading the performance of the entire system. Naturally, chaos should be reduced as much as possible, or totally suppressed, in these situations [CD98].

In contrast, recent research has shown that chaos may actually be useful under certain circumstances, and there is growing interest in utilizing the richness of chaos [Gle87, Mos96, DGY97]. Since a chaotic, or *strange attractor*, usually has embedded within it a dense set of unstable limit cycles, if any of these limit cycles can be stabilized, it may be desirable to stabilize one that characterizes certain maximal system performance [OGY90]. The key is, in a situation where a system is meant for multiple purposes, switching among different limit cycles may be sufficient for achieving these goals. If, on the other hand the attractor is not chaotic, then changing the original system configuration may be necessary to accommodate different purposes. Thus, when designing a system intended for multiple uses, purposely building chaotic dynamics into the system may allow for the desired flexibilities [OGY90].

Within the context of *brain dynamics*, there are suggestions that 'the controlled chaos of the brain is more than an accidental by–product of the brain complexity, including its myriad connections' and that 'it may be the chief property that makes the brain different from an artificial–intelligence machine [FS92]. The so–called *anti–control of chaos* has been proposed for solving the problem of driving the system trajectories of a human brain model away from the stable direction and, hence, away from the stable equilibrium (in the case of a saddle type equilibrium), thereby preventing the periodic behavior of neuronal population bursting. Namely, in a spontaneously bursting neuronal network in vitro, chaos can be demonstrated by the presence of unstable fixed–point behavior. Chaos control techniques can increase the periodicity of such neuronal population bursting behavior. Periodic pacing is also effective in entraining such systems, although in a qualitatively different fashion. Using a strategy of anti–control such systems can be made less periodic. These techniques may be applicable to *in vivo* epileptic foci [SJD94].

2.2.1 Language of Nonlinear Dynamics

Recall that nonlinear dynamics is a language to talk about dynamical systems. Here, brief definitions are given for the basic terms of this language.

- *Dynamical system:* A part of the world which can be seen as a self–contained entity with some temporal behavior. In nonlinear dynamics, speaking about a dynamical system usually means to speak about an abstract mathematical system which is a model for such an entity. Mathematically, a dynamical system is defined by its *state* and by its *dynamics*. A pendulum is an example for a dynamical system.
- *State of a system:* A number or a vector (i.e., a list of numbers) defining the state of the dynamical system uniquely. For the free (un–driven) pendulum, the state is uniquely defined by the angle θ and the angular velocity $\dot{\theta} = d\theta/dt$. In the case of driving, the driving phase ϕ is also needed because the pendulum becomes a non–autonomous system. In spatially extended systems, the state is often a *field* (a scalar–field or a vector–field). Mathematically spoken, fields are functions with space coordinates as independent variables. The velocity vector–field of a fluid is a well–known example.
- *Phase space:* All possible states of the system. Each point in the phase–space corresponds to a unique state (see Figure 2.5). In the case of the free pendulum, the phase–space has 2D whereas for driven pendulum it has 3D. The dimension of the phase–space is infinite in cases where the system state is defined by a field.
- *Dynamics, or equation of motion:* The causal relation between the present state and the next state in the future. It is a deterministic rule which tells us what happens in the next time step. In the case of a continuous time, the time step is infinitesimally small. Thus, the equation of motion is an ordinary differential equation (ODE) (or a system of ODEs):

$$\dot{x} = f(x),$$

where x is the state and t is the time variable (overdot is the time derivative – as always). An example is the equation of motion of an un–driven and un–damped pendulum. In the case of a discrete time, the time steps are nonzero and the dynamics is a map:

$$x_{n+1} = f(x_n),$$

with the discrete time n. Note, that the corresponding physical time points t_n do not necessarily occur equidistantly. Only the order has to be the same. That is,

$$n < m \qquad \Longrightarrow \qquad t_n < t_m.$$

The dynamics is *linear* if the causal relation between the present state and the next state is linear. Otherwise it is *nonlinear*. If we have the case in which the next state is not uniquely defined by the present one, this is generally an indication that the *phase–space is not complete.* Thus, there are important variables determining the state which had been forgotten. This is a crucial point while modelling a real–life systems. Beside this, there are two important classes of systems where the phase–space is incomplete: the *non–autonomous and stochastic systems.* A non–autonomous system has an equation of motion which depends explicitly on time. Thus, the dynamical rule governing the next state not only depends on the present state but also at the time it applies. A driven pendulum is a classical example of a *non–autonomous system.* Fortunately, there is an easy way to make the phase–space complete: we simply include the time into the definition of the state. Mathematically, this is done by introducing a new state variable: t. Its dynamics reads

$$\dot{t} = 1, \qquad \text{or} \qquad t_{n+1} = t_n,$$

depending on whether time is continuous or discrete. For the periodically driven pendula, it is also natural to take the driving phase as the new state variable. Its equation of motion reads

$$\dot{\theta} = 2\pi w,$$

where w is the driving frequency (so that the angular driving frequency is $2\pi w$). On the other hand, in a *stochastic system*, the number and the nature of the variables necessary to complete the phase–space is usually unknown. Therefore, the next state can not be deduced from the present one. The deterministic rule is replaced by a stochastic one. Instead of the next state, it gives only the probabilities of all points in the phase–space to be the next state.

- *Orbit or trajectory:* A solution of the equation of motion. In the case of continuous time, it is a curve in phase–space parametrized by the time variable. For a discrete system it is an ordered set of points in the phase–space.
- *Phase Flow:* The mapping (or, map) of the whole phase–space of a continuous dynamical system onto itself for a given time step t. If t is an infinitesimal time step dt, the flow is just given by the right–hand side of the equation of motion (i.e., f). In general, the flow for a finite time step is not known analytically because this would be equivalent to have a solution of the equation of motion. For example, Figure 2.5 shows the *phase–flow* of a *damped pendulum* in the $(\theta, \dot{\theta})$–phase–plane.

Vector–Fields in the Phase Plane

Consider the following system of two first–order ODEs

$$\dot{x} = f(x, y), \qquad \dot{y} = g(x, y).$$

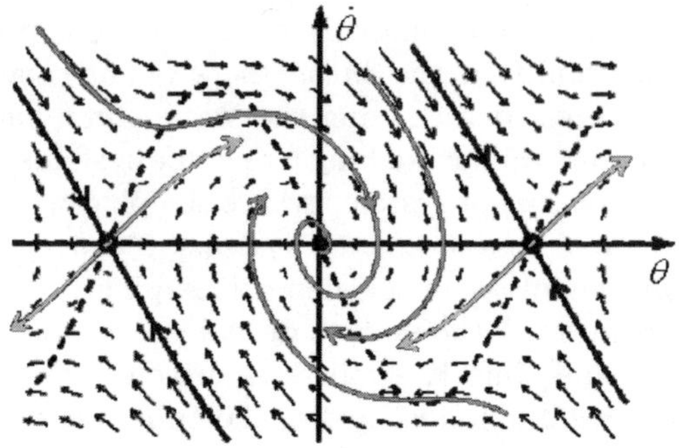

Fig. 2.5. Phase–portrait of a damped pendulum: Arrows denote the phase–flow, dashed line is a null–cline, filled dot is a stable fixed–point, open dot is an unstable fixed–point, dark gray curves are trajectories starting from sample initial points, dark lines with arrows are stable directions (manifolds), light lines with arrows are unstable directions (manifolds), the area between the stable manifolds is basin of attraction.

Here, f and g are given (smooth) functions. The phase space for this system is simply the (x, y)–plane; this is usually referred to as the *phase plane*. If $(x(t), y(t))$ is a solution of the system, then at each time t_0, the vector $(x(t_0), y(t_0))$ defines a point in the phase plane. The point changes with time, so the entire solution, $(x(t), y(t))$, traces out a curve, or *trajectory*, in the phase plane (see Figure 2.6).

Obviously, not every arbitrarily drawn curve in the phase plane represents a solution. What is special about solution trajectories is that the velocity vector at each point along the trajectory is given by the right hand side of the dynamical system above. That is, the velocity vector of the trajectory $(x(t), y(t))$ at a point (x_0, y_0) is given by

$$(\dot{x}(t), \dot{y}(t)) = (f(x_0, y_0), g(x_0, y_0)).$$

This geometrical property, that the vector $(f(x, y), g(x, y))$ always points in the direction that the solution is flowing, completely characterizes the solution trajectories. The set of all vectors (f, g) is called the *vector–field*.

Equilibria

Recall that *equilibrium points* (sometimes called *fixed points* or *rest points*) of the dynamical system are where both f and g vanish; that is, (x_0, y_0) is an equilibrium if

$$f(x_0, y_0) = g(x_0, y_0) = 0.$$

If (x_0, y_0) is an equilibrium, then $(x(t), y(t)) \equiv (x_0, y_0)$ for all time is a (constant) solution of the system. Equilibria can be either stable or unstable. One can usually determine whether an equilibrium is stable or unstable using

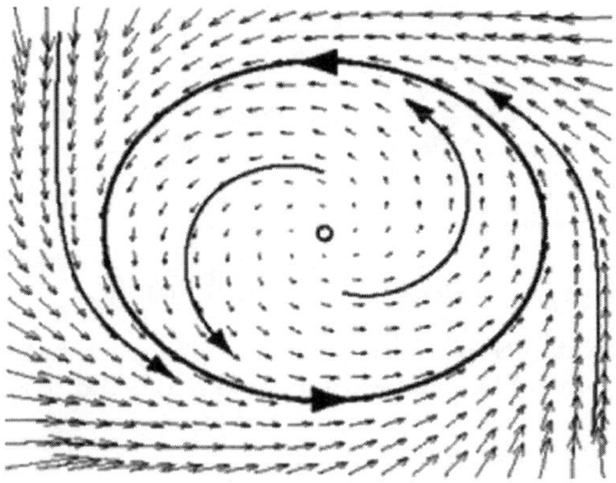

Fig. 2.6. Periodic solutions correspond to closed curves in the phase plane.

the *linearization* method. That is, suppose that (x_0, y_0) is an equilibrium and consider the following *Jacobian matrix*

$$M = \begin{bmatrix} \partial_x f(x_0, y_0) & \partial_y f(x_0, y_0) \\ \partial_x g(x_0, y_0) & \partial_y g(x_0, y_0) \end{bmatrix}.$$

If both eigenvalues of the matrix M have negative real part, then (x_0, y_0) is stable, while if at least one eigenvalue has positive real part, then the equilibrium is unstable. One can classify different types of equilibria on a phase plane in terms of properties of the eigenvalues, as follows:

(i) A *node*: all eigenvalues are real and have the same sign. Stable (unstable) nodes have negative (respectively positive) eigenvalues.
(ii) A *saddle*: all eigenvalues are real but have different signs. Saddles are always unstable.
(iii) A *focus*: there is a pair of complex–conjugate eigenvalues. Stable (unstable) foci have eigenvalues with negative (resp. positive) real part. Foci are often called spirals due to the shape of trajectories near them.

In higher dimensions, there could be other types of equilibria, such as saddle–focus, focus–focus, focus–node.

In 2D systems, every bounded solution must be an equilibrium, a closed orbit, or the solution must approach one of these in forwards and backwards time. This follows from the famous *Poincaré–Bendixson theorem*, which states:

If a trajectory enters and does not leave a closed and bounded region of phase space which contains no equilibria, then the trajectory must approach a periodic orbit as $t \to \infty$.

This theorem can sometimes be used to establish the existence of a (stable) periodic orbit for a planar vector–field. However, it does not hold in higher dimensions and much more complicated dynamics, including *chaos*, may arise.

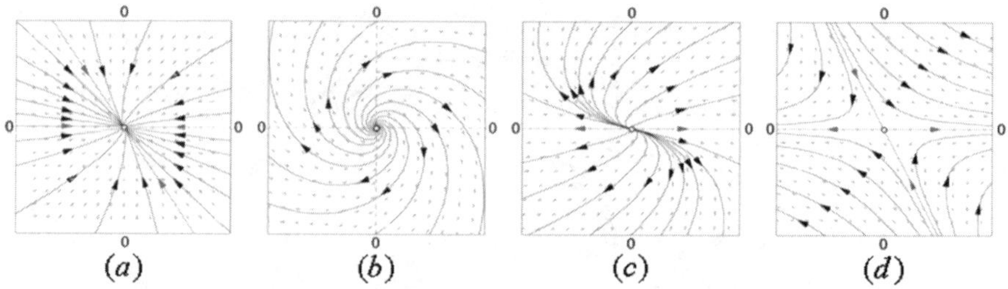

Fig. 2.7. Several examples of linear vector–fields, generated in Mathematica^{TM}: (a) stable node, (b) unstable focus, (c) unstable node, and (d) saddle (unstable).

2.2.2 Linearized Autonomous Dynamics

Recall that *linear dynamical systems* can be solved in terms of simple functions and the behavior of all orbits can be classified (see Figure 2.7). In a linear system the phase–space is the nD Euclidean space, so any point in phase–space can be represented by a vector with n numbers. The analysis of linear systems is possible because they satisfy a *superposition principle*: if $u(t)$ and $w(t)$ satisfy the differential equation for the vector–field (but not necessarily the initial condition), then so will $u(t) + w(t)$.

Flow of a Linear ODE

Fundamental theorem for linear autonomous ODEs states that if A is an $n \times n$ real matrix then the initial value problem

$$\dot{x} = Ax, \qquad x(0) = a \in \mathbb{R}^n \tag{2.1}$$

has the unique solution

$$x(t) = \mathrm{e}^{tA}a, \qquad \text{for all } t \in \mathbb{R}. \tag{2.2}$$

(Here a is the state at time $t = 0$ and e^{tA} is the state at time t). To prove the existence, let $x(t) = \mathrm{e}^{tA}a$ then

$$\frac{dx}{dt} = \frac{d(\mathrm{e}^{tA}a)}{dt} = A\mathrm{e}^{tA}a = Ax,$$
$$x(0) = \mathrm{e}^{0}a = Ia = a,$$

shows that $x(t)$ satisfies the initial value problem (2.1) (here I denotes the $n \times n$ identity matrix).

To prove the uniqueness, let $x(t)$ be any solution of (2.1). It follows that

$$\frac{d}{dt}\left[\mathrm{e}^{-tA}x(t)\right] = 0.$$

Thus $\mathrm{e}^{-tA}x(t) = C$, a constant. The initial condition implies that $C = a$ and hence $x(t) = \mathrm{e}^{tA}a$.

The unique solution of the ODE (2.1) is given by (2.2) for all t. Thus, for each $t \in \mathbb{R}$, the matrix $\mathrm{e}^{tA}a$ maps

$$a \mapsto \mathrm{e}^{tA}a.$$

The set $\{\mathrm{e}^{tA}\}_{t\in\mathbb{R}}$ is a 1−parameter family of linear maps of $\mathbb{R}^n$ into $\mathbb{R}^n$, and is called the *linear flow* of the ODE (for comparison with the general flow notion, see [II06b]).

We write

$$g^t = \mathrm{e}^{tA}$$

– to denote the flow. The flow describes the evolution in time of the physical system for all possible initial states. As the physical system evolves in time, one can think of the state vector x as a moving point in state space, its motion being determined by the flow $g^t = \mathrm{e}^{tA}$. The linear flow satisfies two important properties, which also hold for nonlinear flows.

The linear flow $g^t = \mathrm{e}^{tA}$ satisfies:

$$\begin{aligned} \mathrm{F1}: &\quad g^0 = I, \qquad \text{identity map}, \qquad \text{and} \\ \mathrm{F2}: &\quad g^{t_1+t_2} = g^{t_1} \circ g^{t_2}, \qquad \text{composition}. \end{aligned}$$

Note that properties F1 and F2 imply that the flow $\{g^t\}_{t\in\mathbb{R}}$ forms a *group* under composition of maps.

The flow g^t of the ODE (2.1) partitions the state–space $\mathbb{R}^n$ into subsets called *orbits*, defined by

$$\gamma(a) = \{g^t a : t \in \mathbb{R}\}.$$

The set $\gamma(a)$ is called the orbit of the ODE through a. It is the image in $\mathbb{R}^n$ of the solution curve $x(t) = \mathrm{e}^{tA}a$. It follows that for $a, b \in \mathbb{R}^n$, either $\gamma(a) = \gamma(b)$ or $\gamma(a) \cap \gamma(b) = \emptyset$, since otherwise the uniqueness of solutions would be violated.

For example, consider

$$\dot{x} = Ax, \qquad \text{for all} \quad x \in \mathbb{R}^2;$$

with

$$A = \begin{pmatrix} 0 & 1 \\ -1 & 0 \end{pmatrix},$$

the linear flow is

$$\mathrm{e}^{tA} = \begin{pmatrix} \cos t & \sin t \\ -\sin t & \cos t \end{pmatrix}.$$

The *action of the flow* on $\mathbb{R}^2$, $a \mapsto \mathrm{e}^{tA}a$ corresponds to a *clockwise rotation about the origin*. Thus if $a \neq 0$, the orbit $\gamma(a)$ is a circle centered at the origin passing through a. The origin is a *fixed–point* of the flow, since $\mathrm{e}^{tA}0 = 0$, for all $t \in \mathbb{R}$. The orbit $\gamma(0) = \{0\}$ is called a *point orbit*. All other orbits are called *periodic orbits* since $\mathrm{e}^{2\pi A}a = a$, i.e., the flow maps onto itself after a time $t = 2\pi$ has elapsed.

Classification of Orbits of an ODE

1. If $g^t a = a$ for all $t \in \mathbb{R}$, then $\gamma(a) = \{a\}$ and it is called a *point orbit*. Point orbits correspond to equilibrium points.
2. If there exists a $T > 0$ such that $g^T a = a$, then $\gamma(a)$ is called a *periodic orbit*. Periodic orbits describe a system that evolves periodically in time.
3. If $g^t a \neq a$ for all $t \neq 0$, then $\gamma(a)$ is called a *non–periodic orbit*.

Note that:

1. Non–periodic orbits can be of great complexity even for linear ODEs if $n > 3$ (for nonlinear ODEs if $n > 2$).

2. A *solution curve* of an ODE is a parameterized curve and hence contains information about the flow of time t. The *orbits* are paths in state–space (or subsets of state space). Orbits which are not point orbits are *directed paths* with the direction defined by increasing time. The orbits thus do not provide detailed information about the flow of time.

For an autonomous ODE, the slope of the solution curves depend only on x and hence the tangent vectors to the solution curves define a vector–field $f(x)$ in $x-$space. Infinitely many solution curves may correspond to a single orbit. On the other hand, a non–autonomous ODE does not define a flow or a family of orbits.

Canonical Linear Flows in $\mathbb{R}^2$

Jordan Canonical Forms

For any 2×2 real matrix A, there exists a non–singular matrix P such that

$$J = P^{-1}AP,$$

and J is one of the following matrices:

$$\begin{pmatrix} \lambda_1 & 0 \\ 0 & \lambda_2 \end{pmatrix}, \qquad \begin{pmatrix} \lambda & 1 \\ 0 & \lambda \end{pmatrix}, \qquad \begin{pmatrix} \alpha & \beta \\ -\beta & \alpha \end{pmatrix}.$$

Two linear ODEs, $\dot{x} = Ax$ and $\dot{x} = Bx$, are linearly equivalent iff there exists a non–singular matrix P and a positive constant k such that

$$A = kP^{-1}BP.$$

In other words, the linear ODEs, $\dot{x} = Ax$ and $\dot{x} = Bx$ are *linearly equivalent* iff there exists an invertible matrix P and a positive constant k such that

$$Pe^{tA} = \mathrm{e}^{ktB}P, \qquad \text{for all } t \in \mathbb{R}.$$

Case I: two eigen–directions

Jordan canonical form is

$$J = \begin{pmatrix} \lambda_1 & 0 \\ 0 & \lambda_2 \end{pmatrix}.$$

The flow is

$$e^{tJ} = \begin{pmatrix} e^{\lambda_1 t} & 0 \\ 0 & e^{\lambda_2 t} \end{pmatrix},$$

and the eigenvectors are $e_1 = \begin{pmatrix} 1 \\ 0 \end{pmatrix}$ and $e_2 = \begin{pmatrix} 0 \\ 1 \end{pmatrix}$. The solutions are $y(t) = e^{tJ}b$ for all $b \in \mathbb{R}^2$, i.e., $y_1 = e^{\lambda_1 t}b_1$ and $y_2 = e^{\lambda_2 t}b_2$.

Ia. $\lambda_1 = \lambda_2 < 0$: *attracting focus*;
Ib. $\lambda_1 < \lambda_2 < 0$: *attracting node*;
Ic. $\lambda_1 < \lambda_2 = 0$: *attracting line*;
Id. $\lambda_1 < 0 < \lambda_2$: *saddle*;
Ie. $\lambda_1 = 0 < \lambda_2$: *repelling line*;
If. $0 < \lambda_1 < \lambda_2$: *repelling node*;
Ig. $0 < \lambda_1 = \lambda_2$: *repelling focus*.

Case II: one eigen–direction

Jordan canonical form is

$$J = \begin{pmatrix} \lambda & 1 \\ 0 & \lambda \end{pmatrix}.$$

The flow is

$$e^{tJ} = e^{\lambda_t} \begin{pmatrix} 1 & t \\ 0 & 1 \end{pmatrix},$$

and the single eigenvector is $e = \begin{pmatrix} 1 \\ 0 \end{pmatrix}$.

IIa. $\lambda < 0$: *attracting Jordan node*;
IIb. $\lambda = 0$: *neutral line*;
IIc. $\lambda > 0$: *repelling Jordan node*.

Case III: no eigen–directions

Jordan canonical form is

$$J = \begin{pmatrix} \alpha & \beta \\ -\beta & \alpha \end{pmatrix}.$$

The given ODE is linearly equivalent to $\dot{y} = Jy$.

IIIa. $\alpha < 0$: *attracting spiral*;
IIIb. $\alpha = 0$: *center*;
IIIc. $\alpha > 0$: *repelling spiral*.

In terms of the Jordan canonical form of two matrices A and B, the corresponding ODEs are linearly equivalent iff:

1. A and B have the same number of eigen–directions, and
2. The eigenvalues of A are multiple (k) of the eigenvalues of B.

Topological Equivalence

Now, cases **Ia**, **Ib**, **IIa**, and **IIIa** have common characteristic that all orbits approach the origin (an equilibrium point) as $t \to \infty$. We would like these flows to be 'equivalent' in some sense. In fact, it can be shown, that for all flows of these types, *the orbits of one flow can be mapped onto the orbits of the simplest flow* **Ia**, using a (nonlinear) map $h : \mathbb{R}^2 \to \mathbb{R}^2$, which is a *homeomorphism* on $\mathbb{R}^2$.

Recall that map $h : \mathbb{R}^n \to \mathbb{R}^n$ is a *homeomorphism* on $\mathbb{R}^n$ iff (i) h is one–to–one and onto, (ii) h is continuous, and (iii) h^{-1} is continuous. Two linear flows e^{tA} and e^{tB} on $\mathbb{R}^n$ are said to be *topologically equivalent* if there exists a homeomorphism h on $\mathbb{R}^n$ and a positive constant k such that

$$h(\mathrm{e}^{tA}x) = \mathrm{e}^{ktB}h(x), \qquad \text{for all } x \in \mathbb{R}^n, \text{ and for all } t \in \mathbb{R}.$$

A *hyperbolic* linear flow in $\mathbb{R}^2$ is one in which the real parts of the eigenvalues are all non–zero (i.e., $\mathcal{R}e(\lambda_i) \neq 0$, for $i = 1, 2$.)

Any hyperbolic linear flow in $\mathbb{R}^2$ is topologically equivalent to the linear flow e^{tA}, where A is one of the following matrices:

1. $A = \begin{pmatrix} -1 & 0 \\ 0 & -1 \end{pmatrix}$, *standard sink.*
2. $A = \begin{pmatrix} 1 & 0 \\ 0 & 1 \end{pmatrix}$, *standard source.*
3. $A = \begin{pmatrix} -1 & 0 \\ 0 & 1 \end{pmatrix}$, *standard saddle.*

Any non–hyperbolic linear flow in $\mathbb{R}^2$ is linearly (and hence topologically) equivalent to the flow e^{tA}, where A is one of the following matrices:

$$\begin{pmatrix} 0 & 0 \\ 0 & 0 \end{pmatrix}, \quad \begin{pmatrix} 0 & -1 \\ 1 & 0 \end{pmatrix}, \quad \begin{pmatrix} 0 & 1 \\ 0 & 0 \end{pmatrix}, \quad \begin{pmatrix} -1 & 0 \\ 0 & 0 \end{pmatrix}, \quad \begin{pmatrix} 1 & 0 \\ 0 & 0 \end{pmatrix}.$$

These five flows are topologically equivalent.

2.2.3 Oscillations and Periodic Orbits

A non–constant solution $(x(t), y(t))$ of a dynamical system is *periodic solution* if $(x(0), y(0)) = (x(T), y(T))$ for some $T > 0$. The minimal T that satisfies this requirement is called the *period*. As $(x(t), y(t)) = (x(t + T), y(t + T))$ for all t, a periodic solution corresponds to a *closed curve in the phase plane*

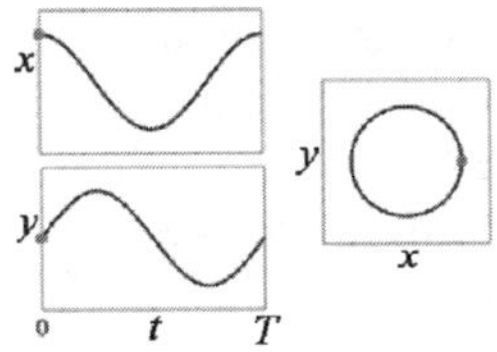

Fig. 2.8. Oscillation shown as a time–series for a vector–field (left), as well as a periodic orbit in phase space.

(see Figure 2.8). Periodic solutions can be either stable or unstable. Roughly speaking, a periodic solution is stable if solutions that begin near the closed curve remain near for all $t > 0$. An asymptotically stable periodic solution is often referred to as a *limit cycle*.

It is usually much more difficult to locate periodic solutions than it is to locate equilibria. An equilibrium point (x_0, y_0) satisfies the equations $f(x_0, y_0) = g(x_0, y_0) = 0$ and these equations can usually be solved with straightforward numerical methods. We also note that an equilibrium is a local object – it is simply a point in phase space. Oscillations or limit cycles are global objects; they correspond to an entire curve in phase space that retraces itself. This curve may be quite complicated.

Higher–Dimensional Dynamical Systems

More generally, consider a system of n first order ordinary differential equations (ODEs) of the form:

$$\dot{x} = F(x), \qquad x \in \mathbb{R}^n$$

The phase–space is simply nD *Euclidean space* and every solution, $(x(t))$, corresponds to a trajectory in phase space parameterized by the independent variable t. As before, $F(x)$ defines a vector–field in the phase space; at each point, $x(t_0)$, the vector $F(x(t_0))$ must be tangent to the solution curve $x(t)$. Moreover, equilibria are where $F(x) = 0$ and periodic solutions correspond to closed orbits.

Note that every system of ODEs is equivalent to a system of the form above. Hence, every solution of every ODE can be viewed geometrically as a trajectory in phase space. Clearly, the phase space may be quite complicated to analyze, especially if $n > 2$.

Periodic Orbit for a Vector Field

Consider a system of ODEs,

$$\dot{x} = f(x), \qquad x \in \mathbb{R}^n \qquad (n \geq 2), \qquad \text{or}$$
$$\dot{x} = f(x, t), \qquad x \in \mathbb{R}^n \qquad (n \geq 1),$$

corresponding to an autonomous or non–autonomous vector–field, respectively. A non–constant solution to such a system, $x(t)$, is periodic if there exists a constant $T > 0$, such that $x(t) = x(t+T)$ for all t. The period of this solution is defined to be the minimum such T. The image of the periodicity interval under in the state space is called the *periodic orbit* or *cycle*.

Limit Cycle

A periodic orbit Γ on a plane (or on a 2D manifold) is called a *limit cycle* if it is the $\alpha-$*limit set* or $\omega-$limit set of some point z not on the periodic orbit, that is, the set of accumulation points of either the forward or backward trajectory through z, respectively, is exactly Γ. Asymptotically stable and unstable periodic orbits are examples of limit cycles.

For example, consider the vector–field given by ODEs

$$\dot{x} = \alpha x - y - \alpha x(x^2 + y^2)$$
$$\dot{y} = x + \alpha y - \alpha y(x^2 + y^2),$$

where $\alpha > 0$ is a parameter. Transforming to radial coordinates, we see that the periodic orbit lies on a circle with unit radius for any $\alpha > 0$,

$$\dot{r} = \alpha r(1 - r^2), \qquad \dot{\theta} = 1.$$

This periodic orbit is a stable limit cycle for $\alpha > 0$ and unstable limit cycle for $\alpha < 0$. When $\alpha = 0$, the system above has infinite number of periodic orbits and no limit cycles.

Periodic Orbit for a Map

A periodic orbit with period k for a map

$$x_{i+1} = g(x_i), \qquad x \in \mathbb{R}^n \qquad (n \geq 1),$$

is the set of distinct points $\{p_j = g^j(p_0)|j = 0, \cdots, k-1\}$ with $g^k(p_0) = p_0$ [GH83]. Here g^k represents the composition of g with itself k times. The smallest positive value of k for which this equality holds is the period of the orbit. An example of a periodic orbit for a map is shown in the Figure 2.9.

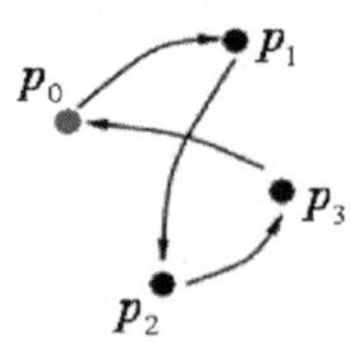

Fig. 2.9. A periodic orbit for a map $x_{i+1} = g(x_i)$, consisting of distinct points $\{p_j = g^j(p_0)|j = 0, \cdots, k-1\}$ with $g^k(p_0) = p_0$.

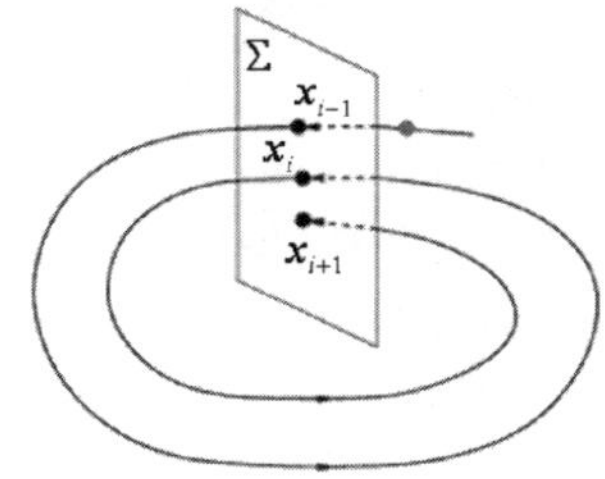

Fig. 2.10. Poincaré map for a vector–field.

Stability of a Periodic Orbit

The stability of a periodic orbit for an autonomous vector–field can be calculated by considering the *Poincaré map*, which replaces the flow of the nD continuous vector–field with an $(n-1)$D map [GH83, Str94]. Specifically, an $(n-1)$D surface of section Σ is chosen such that the flow is always transverse to Σ (see Figure 2.10). Let the successive intersections in a given direction of the solution $x(t)$ with Σ be denoted by x_i. The Poincaré map

$$x_{i+1} = g(x_i)$$

determines the $(i+1)$th intersection of the trajectory with Σ from the ith intersection. A periodic orbit of an autonomous vector–field corresponds to a fixed point x_f of this Poincaré map, characterized by . The linearization of the Poincaré map about x_f is usually written

$$\xi_{i+1} = \mathrm{D}g(x_f)\xi_i.$$

If all eigenvalues of $\mathrm{D}g$ have modulus less than unity, then x_f (and thus the corresponding periodic orbit) is asymptotically stable. If any eigenvalues of $\mathrm{D}g$ have modulus greater than unity, then x_f (and thus the corresponding periodic orbit) is unstable. The stability properties of a periodic orbit are independent of the cross section Σ [Wig90]. If x_f is stable then it is an *attractor of the Poincaré map*, and the corresponding periodic orbit is an attractor of the vector–field.

Periodic Orbits, Bifurcations and Chaos

Recall that a *bifurcation* represents a qualitative change in the behavior of a dynamical system as a system parameter is varied. This could involve a change in the stability properties of a periodic orbit, and/or the creation or destruction of one or more periodic orbits. Bifurcation analysis[3] can thus provide

[3] Bifurcation theory studies and classifies phenomena characterized by sudden shifts in behavior arising from small changes in circumstances, analyzing how the qualitative nature of equation solutions depends on the parameters that appear in

another (analytical or numerical) method for establishing the existence or non–existence of a periodic orbit.

Among codimension–1 bifurcations of periodic orbits for vector–fields, the most important are the following [GH83]:

1. The *Andronov–Hopf bifurcation*,[4] which results in the appearance of a small–amplitude periodic orbit.

---

the equation. This may lead to sudden and dramatic changes, for example the unpredictable timing and magnitude of a landslide. Closely related to the bifurcation theory is the *catastrophe theory*, which was originated with the work of the French mathematician Ren Thom in the 1960s, and became very popular not least due to the efforts of Christopher Zeeman in the 1970s, considers the special case where the long–run stable equilibrium can be identified with the minimum of a smooth, well–defined potential function (Lyapunov function). Small changes in certain parameters of a nonlinear system can cause equilibria to appear or disappear, or to change from attracting to repelling and vice versa, leading to large and sudden changes of the behavior of the system. However, examined in a larger parameter space, catastrophe theory reveals that such bifurcation points tend to occur as part of well–defined qualitative geometrical structures.

Catastrophe theory analyzes degenerate critical points of the *potential function* of a given dynamical system – points where not just the first derivative, but one or more higher derivatives of the potential function are also zero. These are called the germs of the catastrophe geometries. The degeneracy of these critical points can be unfolded by expanding the potential function as a Taylor series in small perturbations of the parameters. When the degenerate points are not merely accidental, but are structurally stable, the degenerate points exist as organizing centers for particular geometric structures of lower degeneracy, with critical features in the parameter space around them. If the potential function depends on three or fewer active variables, and five or fewer active parameters, then there are only seven generic structures for these bifurcation geometries, with corresponding standard forms into which the Taylor series around the catastrophe germs can be transformed by diffeomorphism (a smooth transformation whose inverse is also smooth). These seven fundamental polynomial types for 1D and 2D systems are given below, with the names that Thom gave them:
(i) Fold (1D): $V = x^3 + ax$;
(ii) Cusp (1D): $V = x^4 + ax^2 + bx$;
(iii) Swallowtail (1D): $V = x^5 + ax^3 + bx^2 + cx$;
(iv) Butterfly (1D): $V = x^6 + ax^4 + bx^3 + cx^2 + dx$;
(v) Hyperbolic umbilical (2D): $V = x^3 + y^3 + axy + bx + cy$;
(vi) Elliptic umbilical (2D): $V = x^3/3 - xy^2 + a(x^2 + y^2) + bx + cy$; and
(vii) Parabolic umbilical (2D): $x^2y + y^4 + ax^2 + by^2 + cx + dy$.

[4] *The Andronov–Hopf bifurcation* is the birth of a limit cycle from an equilibrium in dynamical systems generated by ODEs, when the equilibrium changes stability via a pair of purely imaginary eigenvalues. The bifurcation can be supercritical or subcritical, resulting in stable or unstable (within an invariant 2D manifold) limit cycle, respectively.

2. The *saddle–node bifurcation*[5] of periodic orbits, in which two periodic orbits coalesce and annihilate each other.
3. The saddle–node on invariant circle bifurcation (SNIC), in which a periodic orbit appears from a homoclinic orbit to a saddle-node equilibrium (along the central manifold).
4. The *homoclinic bifurcations*, in which periodic orbits appear from homoclinic orbits to a saddle, saddle–focus, or focus–focus equilibrium.
5. The *period–doubling bifurcation* (also known flip bifurcation), in which a periodic orbit of period appears near a periodic orbit of period.
6. The *Neimark–Sacker bifurcation*, in which an invariant torus appears near a periodic orbit.
7. The *blue–sky bifurcation*,[6] in which a periodic orbit of large period appears 'out of a blue sky' (actually, the orbit is homoclinic to a saddle–node periodic orbit).

These bifurcations result in the appearance or disappearance of periodic orbits, depending on the direction in which the bifurcation parameter is varied. The (dis)appearing orbits may be stable or unstable, depending, among other factors, on whether the bifurcations are subcritical or supercritical.

On the other hand, as a system parameter is varied, chaotic behavior can appear via an infinite sequence of period doubling bifurcations of periodic orbits. This is known as the *Feigenbaum phenomenon* or the *period doubling route to chaos* [Ott93]. Moreover, a chaotic attractor typically has a dense set of unstable periodic orbits embedded within it. Suitable averages over such periodic orbits can be used to approximate descriptive quantities for chaotic attractors such as *Lyapunov exponents* and *fractal dimensions*. Such periodic orbits can sometimes be stabilized (and the chaos thus suppressed) through small manipulations of a system parameter, an approach called *chaos control* [Ott93].

2.2.4 Conservative versus Dissipative Dynamics

Recall (see [II05, II06a, II06b]) that *conservative–reversible systems* are in classical dynamics described by Hamilton's equations

$$\dot{q}^i = \partial_{p_i} H, \qquad \dot{p}_i = \partial_{q^i} H, \qquad (i = 1, \dots, n), \tag{2.3}$$

[5] The *saddle–node bifurcation* or *tangential bifurcation* (in continuous dynamical systems) is a local bifurcation in which two fixed points (or, equilibria) of a dynamical system collide and annihilate each other. In discrete dynamical systems, the same bifurcation is often instead called a *fold bifurcation*. Saddle-node bifurcations may be associated with *hysteresis loops* and *catastrophes*. The normal form of a saddle–node bifurcation in a 1D phase–space is $\dot{x} = \mu + x^2$, where x is the state variable and μ is the bifurcation parameter.

[6] The *blue–sky bifurcation* is a codimension–1 bifurcation featuring a stable periodic orbit of infinite period and length, far from equilibrium states.

with a *constant Hamiltonian energy function*

$$H = H(q,p) = E_{kin}(p) + E_{pot}(q) = E = \text{const.} \tag{2.4}$$

Conservative dynamics visualizes the time evolution of a system in a *phase–space* P, in which the coordinates are q^i and p_i. The instantaneous state of the system is the *representative point* (q,p) in P. As time varies, the representative point (q,p) describes the *phase trajectory*. A particular case of a phase trajectory is the *position of equilibrium*, in which both $\dot{q}^i = 0$ and $\dot{p}_i = 0$.

Dissipative Systems

In addition to *conservative–reversible* systems, we must consider systems that give rise to *irreversible processes* and *dissipative structures* of *Nobel Laureate Ilya Prigogine* (see [GN90, II06a]).

A typical example is a chemical reaction in which a molecule of species A (say the hydroxyl radical OH) can combine with a molecule of species B say molecular hydrogen H_2) to produce one molecule of species C and one molecule of species D (respectively H_2O and atomic hydrogen H in our example). This process is symbolized

$$A + B \xrightarrow{k} C + D, \tag{2.5}$$

in which k is the rate constant, generally a function of temperature and pressure. On the l.h.s of (2.5), the *reactants* A and B combine and disappear in the course of time, whereas on the r.h.s the *products* C and D are formed and appear as the reaction advances. The rate at which particles of species A are consumed is proportional to the frequency of encounters of molecules of A and B – which, if the system is dilute, is merely proportional to the product of their concentrations, c,

$$\dot{c}_A = -k\, c_A\, c_B. \tag{2.6}$$

Clearly, if we reverse time, $t' = -t$, and denote by c'_A, c'_B the values of the concentrations as functions of t', (2.6) becomes

$$\dot{c}_A = k\, c_A\, c_B,$$

and describes a process in which c_A would be produced instead of being consumed. This is certainly not equivalent to the phenomenon described by (2.6).

Further examples are *heat conduction*, given by *Fourier equation*

$$\partial_t T = \kappa\, \nabla^2 T, \qquad \kappa > 0, \qquad \left(\partial_t \equiv \frac{\partial}{\partial_t}\right), \tag{2.7}$$

and *diffusion*, described by *Fick equation*

$$\partial_t T = D\, \nabla^2 c, \qquad D > 0. \tag{2.8}$$

Here T is the temperature, c is the concentration of a certain substance dissolved in the fluid, κ is the heat diffusivity coefficient and D is the mass diffusivity. Both experiments and these two equations show that when a slight temperature variation (respectively, inhomogeneity) is imposed in an isothermal (respectively, uniform) fluid, it will *spread out* and eventually disappear.

Again, if we reverse time, we get the completely different laws

$$\partial_t T = -\kappa \nabla^2 T, \qquad \partial_t c = -D \nabla^2 c,$$

describing a situation in which an initial temperature or concentration disturbance would be amplified rather than damped.

Both the *concentration* and the *temperature* variables are examples of so–called *even variables*, whose sign does not change upon time reversal. In contrast, the *momentum of a particle* and the *convection velocity of a fluid* are *odd variables*, since they are ultimately expressed as time derivatives of position–like variables and change their sign with time reversal.

This leads us to the following general property of the evolution equation of a dissipative system. Let $\{X_i\}$ denote a complete set of macroscopic variables of such a system. *Dissipative evolution laws* have the form

$$\partial_t X_i = F_i(\{X_j\}, \lambda), \tag{2.9}$$

where λ denote *control parameters*, and F_i are functions of $\{X_i\}$ and λ.

The basic feature of (2.9) is that, whatever the form of the functions F_i, in the absence of constraints they must reproduce the steady state of equilibrium

$$F_i\left(\{X_{j,eq}\}, \lambda_{eq}\right) = 0. \tag{2.10}$$

More generally, for a non–equilibrium steady state,

$$F_i(\{X_{j,s}\}, \lambda_s) = 0. \tag{2.11}$$

These relations impose certain restrictions. For instance, the evolution laws must ensure that positive values are attained for temperature or chemical concentrations that come up as solutions, or that detailed balance is attained. This is an important point, for it shows that the analysis of physical systems cannot be reduced to a mathematical game. In many respects physical systems may be regarded as highly atypical, specific, or nongeneric from the mathematical point of view. In these steady state relations, the *nonlinearity*, relating the control parameters λ to the steady state values $X_{j,s}$, begins to play the prominent role.

Thermodynamic Equilibrium

In mechanics, (static) equilibrium is a particular 'state of rest' in which both the velocities and the accelerations of all the material points of a system are equal to zero. By definition the net balance of forces acting on each point is zero at each moment. If this balance is disturbed, equilibrium will be broken. This is what happens when a piece of metal fractures under the effect of load (see [GN90, II06a]).

Now, the notion of *thermodynamic equilibrium* is sharply different. Contrary to mechanical equilibrium, the molecules constituting the system are subject to forces that are not balanced and move continuously in all possible directions unless the temperature becomes very low. 'Equilibrium' refers here to some collective properties $\{X_i\}$ characterizing the system as a whole, such as temperature, pressure, or the concentration of a chemical constituent.

Consider a system $\{X_i\}$ embedded in a certain environment $\{X_{ie}\}$. Dynamic role of the sets of properties $\{X_i\}$ and $\{X_{ie}\}$ resides primarily in their exchanges between the system and the environment. For instance, if the system is contained in a vessel whose walls are perfectly rigid, permeable to heat but impermeable to matter, one of these quantities will be identical to the temperature, T and will control the exchange of energy in the form of heat between the system and its environment.

We say that the system is in thermodynamic equilibrium if it is completely identified with its environment, that is, if the properties X_i and X_{ie} have identical values. In the previous example, thermodynamic equilibrium between the system and its surroundings is tantamount to $T = T_e$ at all times and at all points in space. But because the walls of the vessel are impermeable to matter, system and environment can remain highly differentiated in their chemical composition, c. If the walls become permeable to certain chemical substances i, thermodynamic equilibrium will prevail when the system and the environment become indistinguishable as far as those chemicals are concerned. In simple cases this means that the corresponding composition variables will satisfy the equality $c_i = c_{ie}$, but more generally equilibrium will be characterized by the equality for a quantity known as the chemical potential, $\mu_i = \mu_{ie}$. Similarly, if the walls of the vessel are not rigid, the system can exchange mechanical energy with its environment. Equilibrium will then also imply the equality of pressures, $p = p_e$.

According to the above definitions, equilibrium is automatically a stationary state, $\partial X_i/\partial t = 0$: the properties X_i do not vary with time. As they are identical in the properties X_i, the system and the environment have nothing to exchange. We express this by saying that there are no net *fluxes* across the system,

$$J_i^{eq} = 0. \tag{2.12}$$

Nonlinearity

Here is a simple example. Let X be the unique state variable, k a parameter, and let λ represent the applied constraint. We can easily imagine a mechanism such as $A \rightleftarrows X \rightleftarrows D$ in which X evolves according to

$$\dot{X} = \lambda - kX,$$

yielding a stationary state value given by $\lambda - kX_s = 0$, or $X_s = \lambda/k$. In the linear law linking the steady state value X_s to the control parameter λ the behavior is bound to be qualitatively similar to that in equilibrium, even in the presence of strongly correlated non–equilibrium constraints. In the nonlinear law linking the steady state value X_s to the control parameter λ there is an unlimited number of possible forms describing nonlinear dependencies. For the certain values of λ the system can present several distinct solutions.

Nonlinearity combined with *non–equilibrium constraints* allows for multiple solutions and hence for the diversification of the behaviors presented by a system (see [GN90, II06b]).

The Second Law of Thermodynamics

According to this law there exists a function of the state variables (usually chosen to be the *entropy*, S) of the system that varies monotonically during the approach to the unique final state of thermodynamic equilibrium:

$$\dot{S} \geq 0 \qquad \text{(for any isolated system).} \tag{2.13}$$

It is usually interpreted as a *tendency to increased disorder*, i.e., an irreversible trend to maximum disorder.

The above interpretation of entropy and a second law is fairly obvious for systems of *weakly interacting particles*, to which the arguments developed by Boltzmann referred.

Let us now turn to non–isolated systems, which exchange energy or matter with the environment. The entropy variation will now be the sum of two terms. One, entropy flux, d_eS, is due to these exchanges; the other, entropy production, d_iS, is due to the phenomena going on within the system. Thus the entropy variation is

$$\dot{S} = \frac{d_iS}{dt} + \frac{d_eS}{dt}. \tag{2.14}$$

For an isolated system $d_eS = 0$, and (2.14) together with (2.13) reduces to $dS = d_iS \geq 0$, the usual statement of the second law. But even if the system is non–isolated, d_iS will describe those (irreversible) processes that would still go on even in the absence of the flux term d_eS. We thus require the following extended form of the second law:

$$\dot{S} \geq 0 \qquad \text{(nonisolated system).} \tag{2.15}$$

As long as d_iS is strictly positive, irreversible processes will go on continuously within the system. Thus, $d_iS > 0$ is equivalent to the condition of dissipativity as time irreversibility. If, on the other hand, d_iS reduces to zero, the process will be reversible and will merely join neighboring states of equilibrium through a slow variation of the flux term d_eS.

Among the most common irreversible processes contributing to d_iS are chemical reactions, heat conduction, diffusion, viscous dissipation, and relaxation phenomena in electrically or magnetically polarized systems. For each of these phenomena two factors can be defined: an appropriate internal *flux*, J_i, denoting essentially its rate, and a driving *force*, X_i, related to the maintenance of the non–equilibrium constraint. A most remarkable feature is that d_iS becomes a *bilinear form* of J_i and X_i. The following table summarizes the fluxes and forces associated with some commonly observed irreversible phenomena (see [GN90, II06a])

Phenomenon	**Flux**	**Force**	**Rank**
Heat conduction	Heat flux, $\mathbf{J}_{th}$	$grad(1/T)$	Vector
Diffusion	Mass flux, $\mathbf{J}_d$	$-[grad(\mu/T) - \mathbf{F}]$	Vector
Viscous flow	Pressure tensor, $\mathbf{P}$	$(1/T)\ grad\,\mathbf{v}$	Tensor
Chemical reaction	Rate of reaction, ω	Affinity of reaction	Scalar

In general, the fluxes J_k are very complicated functions of the forces X_i. A particularly simple situation arises when their relation is linear, then we have the celebrated *Onsager relations* (named after *Nobel Laureate Lars Onsager*),

$$J_i = L_{ik}X_k, \tag{2.16}$$

in which L_{ik} denote the set of *phenomenological coefficients.* This is what happens near equilibrium where they are also symmetric, $L_{ik} = L_{ki}$. Note, however, that certain states far from equilibrium can still be characterized by a linear dependence of the form of (2.16) that occurs either accidentally or because of the presence of special types of regulatory processes.

Geometry of Phase Space

Now, we reduce (2.9) to the *temporal* systems, in which there is no space dependence in the operator F_i, so that $\partial \to d$, and we have

$$\dot{X}_i = F_i(\{X_j\}, \lambda), \qquad i = 1, \ldots, n. \tag{2.17}$$

Moreover, we restrict ourselves to autonomous systems, for which F_i does not depend explicitly on time, a consequence being that the trajectories in phase–space are invariant. Note that in a Hamiltonian system n must be even and F_i must reduce to the characteristic structure imposed by (2.3).

A first kind of phase–space trajectory compatible with (2.17) is given by

$$\dot{X}_i = 0. \tag{2.18}$$

It includes as particular cases the states of mechanical equilibrium encountered in conservative systems and the steady states encountered in dissipative systems. In phase–space such trajectories are quite degenerate, since they are given by the solutions of the n algebraic equations for n unknowns, $F_i = 0$. They are represented by *fixed–points*.

If (2.18) is not satisfied, the representative point will not be fixed but will move along a phase–space trajectory defining a curve. The line element along this trajectory for a displacement corresponding to $(dX_1, \dots, dX_n)$ along the individual axes is given by Euclidean metrics

$$ds^2 = dX_i dX_j = F_i F_j \, dt, \qquad (i, j = 1, \dots, n). \tag{2.19}$$

Thus, the projections of the tangent of the curve along the axes are given by

$$\frac{dX_\alpha}{ds} = \frac{F_\alpha}{\sqrt{F_i F_j}}, \tag{2.20}$$

and are well defined everywhere. The points belonging to such curves are called *regular points*. In contrast, the tangent on the fixed–points is *ill–defined* because of the simultaneous vanishing of all F_i's. Therefore, the fixed–points could be also referred to as the singular points of the flow generated by (2.17). The set of fixed–points and phase–space trajectories constitutes the *phase portrait* of a dynamical system.

One property that plays a decisive role in the structure of the phase portrait relates to the *existence–uniqueness theorem* of the solutions of ordinary differential equations. This important result of A. Cauchy asserts that *under quite mild conditions on the functions* F_i, *the solution corresponding to an initial condition not on a fixed–point exists and is unique for all times in a certain interval* $(0, \tau)$, *whose upper bound* τ *depends on the specific structure of the functions* F_i. In the phase–space representation, the theorem automatically rules out the intersection of two trajectories in any regular point (see [GN90, II06a]).

A second structure of great importance is the *existence and structure of invariant sets of the flow.* By this we mean objects embedded in the phase–space that are bounded and are mapped onto themselves during the evolution generated by (2.17). An obvious example of an invariant set is the ensemble of fixed–points. Another is a closed curve in phase–space representing a periodic motion.

The *impossibility of self–intersection of the trajectories* and the *existence of invariant sets* of a certain form (fixed–points, limit circles,...) determine, to a large extent, the structure of the phase portrait in 2D phase–spaces, and through it the type of behavior that may arise. *In three or more dimensions,* however, the constraints imposed by these properties are much less severe, since the trajectories have many more possibilities to avoid each other by 'gliding' within the 'gaps' left between invariant sets, thus implying the *possibility for chaos.*

In principle, the solution of (2.17) constitutes a *well–posed problem,* in the sense that *a complete specification of the state* $(X_1, \dots, X_n)$ *at any one time allows prediction of the state at all other times.* But in many cases such a complete specification may be operationally meaningless. For example, in a Hamiltonian system composed of particles whose number is of the order of Avogadro's number, or in a chaotic regime, it is no longer meaningful to argue in terms of individual trajectories. New modes of approach are needed, and one of the most important is a description of the system in terms of *probability concepts.* For this purpose we consider not the rather special case of a single system, but instead focus attention on the *Gibbs ensemble* of a very large number of identical systems, which are in general in different states, but all subject to exactly the same constraints. They can therefore be regarded as emanating from an initial ensemble of systems whose representative phase points were contained in a certain phase–space volume V_0 (see [GN90, II06a]).

2.2.5 Attractors

Roughly speaking, an attracting set, or more popularly an *attractor*, for a certain dynamical system is a closed subset A of its *phase space* such that for various *initial conditions* the system will evolve towards A. The word attractor is usually reserved for an attracting set which contains a dense orbit (this condition insures that it is not just the union of smaller attracting sets). In the case of an *iterated map*, with discrete time steps, the simplest attractors are *attracting fixed points.* Similarly, for solutions of an autonomous differential equation, with continuous time, the simplest examples are *attracting equilibrium points.* In both cases, the next simplest examples are attracting periodic orbits. The union of all orbits which converge towards A is called the *basin of attraction* (see Figure 2.11) and denoted $B(A)$.

A simple example is that of a point particle moving in a two–well potential $V(x)$ with friction, as in Figure 2.11(a). Due to the friction, all initial conditions, except those at $x = \dot{x} = 0$, or on its *stable manifold* eventually come to rest at either $x = x_0$ or $x = -x_0$, which are the two attractors of the system. A point initially placed on the unstable equilibrium point $x = 0$, will stay there forever; and this state has a 1D stable manifold. Figure 2.11(b) shows

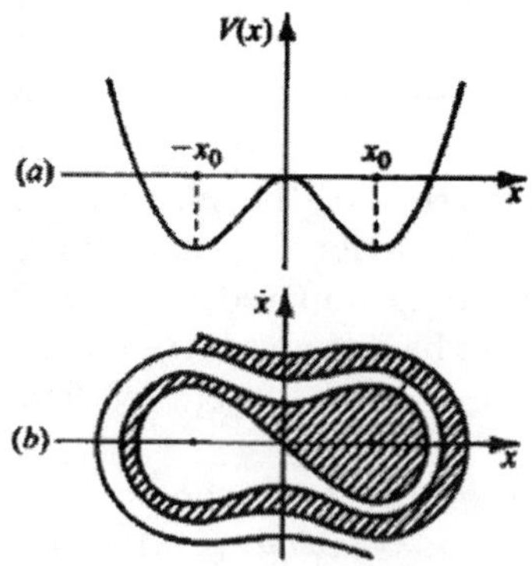

Fig. 2.11. (a) Double well potential $V(x)$, and (b) the resulting basins of attraction in the $x - \dot{x}$ phase–plane.

the basins of attraction of the two stable equilibrium points $x = \pm x_0$, where the crosshatched region is the basin for the attractor at $x = x_0$ and the blank region is the basin for the attractor at $x = -x_0$. The boundary separating these two basins is the stable manifold of the unstable equilibrium $x = 0$.

It is very common for dynamical systems to have more than one attractor. For each such attractor, its basin of attraction is the set of initial conditions leading to long–time behavior that approaches that attractor. Thus the qualitative behavior of the long–time motion of a given system can be fundamentally different depending on which basin of attraction the initial condition lies in (e.g., attractors can correspond to periodic, quasi–periodic or chaotic behaviors of different types). Regarding a basin of attraction as a region in the state space, it has been found that the basic topological structure of such regions can vary greatly from system to system. In what follows we give examples and discuss several qualitatively different kinds of basins of attraction and their practical implications.

Classical Examples of Attractors

Here we present numerical simulations of several popular chaotic systems (see, e.g., [Wig90, BCB92, Ach97]). Generally, to observe chaos in continuous time system, it is known that the dimension of the equation must be three or higher. That is, *there is no chaos in any phase plane* (see [Str94]), we need the third dimension for chaos in continuous dynamics. However, note that *all forced oscillators have actually dimension 3, although they are commonly written as second–order ODEs.*[7] On the other hand, in discrete–time systems like logistic map or Hénon map, we can see chaos even if the dimension is one.

Simple Pendulum

Recall (see [II05, II06a, II06b]) that a simple *undamped pendulum* (see Figure 2.12), given by equation

$$\ddot{\theta} + \frac{g}{l}\sin\theta = 0, \qquad (2.21)$$

swings forever; it has closed orbits in a 2D phase–space (see Figure 2.13).

The conservative (un–damped) pendulum equation (2.21) does not take into account the effects of friction and dissipation. On the other hand, a simple *damped pendulum* (see Figure 2.12) is given by modified equation, including a damping term proportional to the velocity,

$$\ddot{\theta} + \gamma\dot{\theta} + \frac{g}{l}\sin\theta = 0,$$

[7] Both Newtonian equation of motion and RLC circuit can generate chaos, provided they have a forcing term. This forcing (driving) term in second–order ODEs is the motivational reason for development of the jet–bundle formalism for non–autonomous dynamics (see [II06b]).

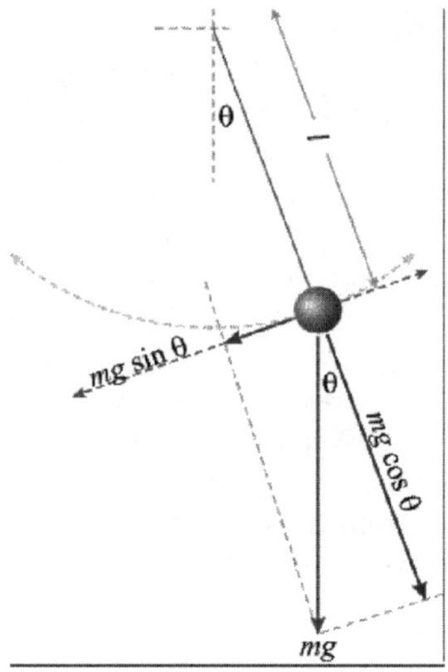

Fig. 2.12. Force diagram of a simple gravity pendulum.

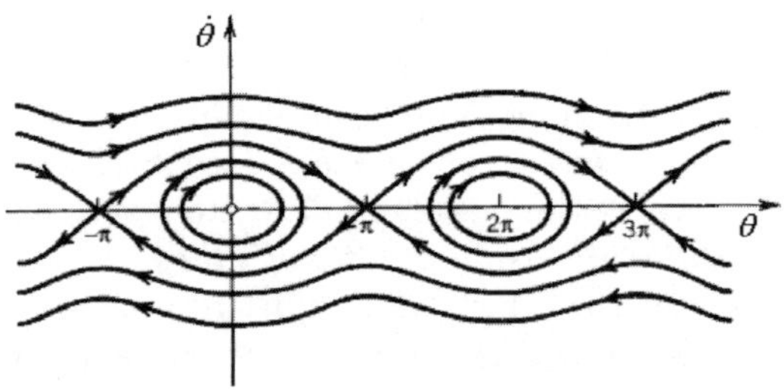

Fig. 2.13. Phase portrait of a simple gravity pendulum.

with the positive constant damping γ. This pendulum settles to rest (see Figure 2.14). Its spiralling orbits lead to a point attractor (focus) in a 2D phase–space. All closed trajectories for periodic solutions are destroyed, and the trajectories spiral around one of the critical points, corresponding to the vertical equilibrium of the pendulum. On the phase plane, these critical points are stable spiral points for the underdamped pendulum, and they are stable nodes for the overdamped pendulum. The unstable equilibrium at the inverted vertical position remains an unstable saddle point. It is clear physically that damping means loss of energy. The dynamical motion of the pendulum decays due to the friction and the pendulum relaxes to the equilibrium state in the vertical position.

Finally, a *driven pendulum*, periodically forced by a force term $F\cos(w_D t)$, is given by equation

$$\ddot{\theta} + \gamma\dot{\theta} + \frac{g}{l}\sin\theta = F\cos(w_D t). \tag{2.22}$$

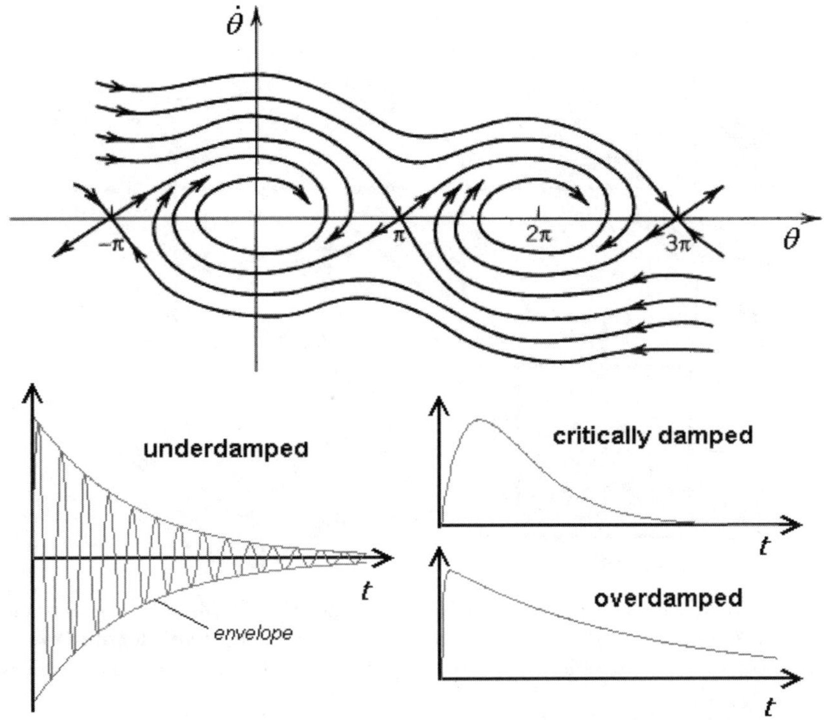

Fig. 2.14. A damped gravity pendulum settles to a rest: its phase portrait (up) shows spiralling orbits that lead to a focus attractor; its time plot (down) shows three common damping cases.

It has a 3D phase–space and can exhibit chaos (for certain values of its parameters, see Figure 2.15).

Van der Pol Oscillator

The unforced Van der Pol oscillator has the form of a second order ODE

$$\ddot{x} = \alpha\,(1 - x^2)\,\dot{x} - \omega^2\,x. \tag{2.23}$$

Its celebrated *limit cycle* is given in Figure 2.16. The simulation is performed with zero initial conditions and parameters α = random(0, 3), and $\omega = 1$. The Van der Pol oscillator was the first *relaxation oscillator*, used in 1928 as a model of human heartbeat (ω controls how much voltage is injected into the system, and α controls the way in which voltage flows through the system). The oscillator was also used as a model of an electronic circuit that appeared in very early radios in the days of vacuum tubes. The tube acts like a normal resistor when current is high, but acts like a negative resistor if the current is low. So this circuit pumps up small oscillations, but drags down large oscillations. α is a constant that affects how nonlinear the system is.

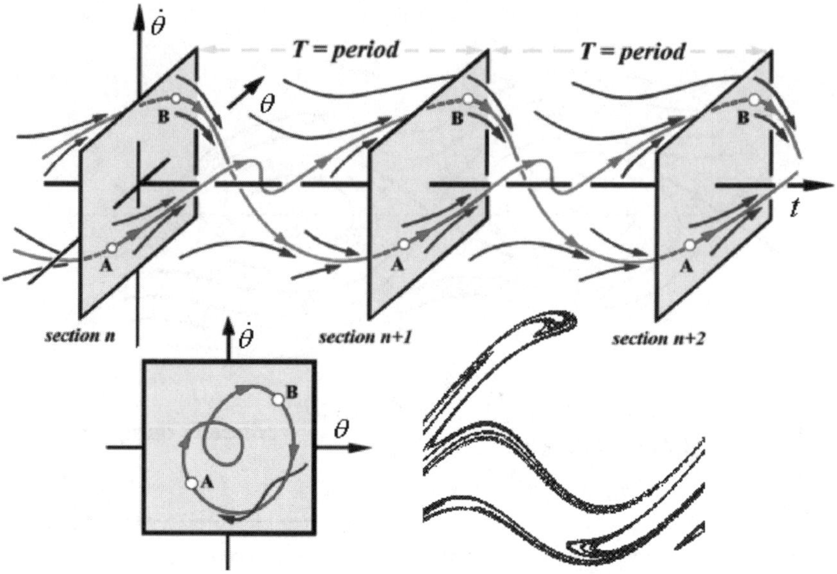

Fig. 2.15. A driven pendulum has a 3D phase–space with angle θ, angular velocity $\dot\theta$ and time t. Dashed lines denote steady states, while solid lines denote transients. Right–down we see a sample chaotic attractor (adapted and modified from [TS01]).

For α equal to zero, the system is actually just a linear oscillator. As α grows the nonlinearity of the system becomes considerable.

The *sinusoidally–forced Van der Pol oscillator* is given by equation

$$\ddot{x} - \alpha\,(1 - x^2)\,\dot{x} + \omega^2\;x = \gamma\cos(\phi t), \tag{2.24}$$

where ϕ is the forcing frequency and γ is the amplitude of the forcing sinusoid.

Nerve Impulse Propagation

The nerve impulse propagation along the axon of a neuron can be studied by combining the equations for an excitable membrane with the differential equations for an electrical core conductor cable, assuming the axon to be an infinitely long cylinder. A well known approximation of FitzHugh [Fit61] and Nagumo [NAY60] to describe the propagation of voltage pulses $V(x,t)$ along the membranes of nerve cells is the set of coupled PDEs[8]

[8] Note that the FitzHugh–Nagumo model is an approximation for the celebrated *Hodgkin–Huxley model* (HH) for neural action potential [HH52, Hod64], described by the nonlinear coupled ODEs for the four variables, V for the membrane potential, and m, h and n for the gating variables of sodium and potassium channels,

$$C\dot{V} \;=\; -g_{\mathrm{Na}}m^3h(V - V_{\mathrm{Na}}) - g_{\mathrm{K}}n^4(V - V_{\mathrm{K}}) - g_{\mathrm{L}}(V - V_{\mathrm{L}}) + I_j^{\mathrm{ext}},$$
$$\dot{m} = -(a_m + b_m)\,m + a_m, \quad \dot{h} = -(a_h + b_h)\,h + a_h, \quad \dot{n} = -(a_n + b_n)\,n + a_n,$$

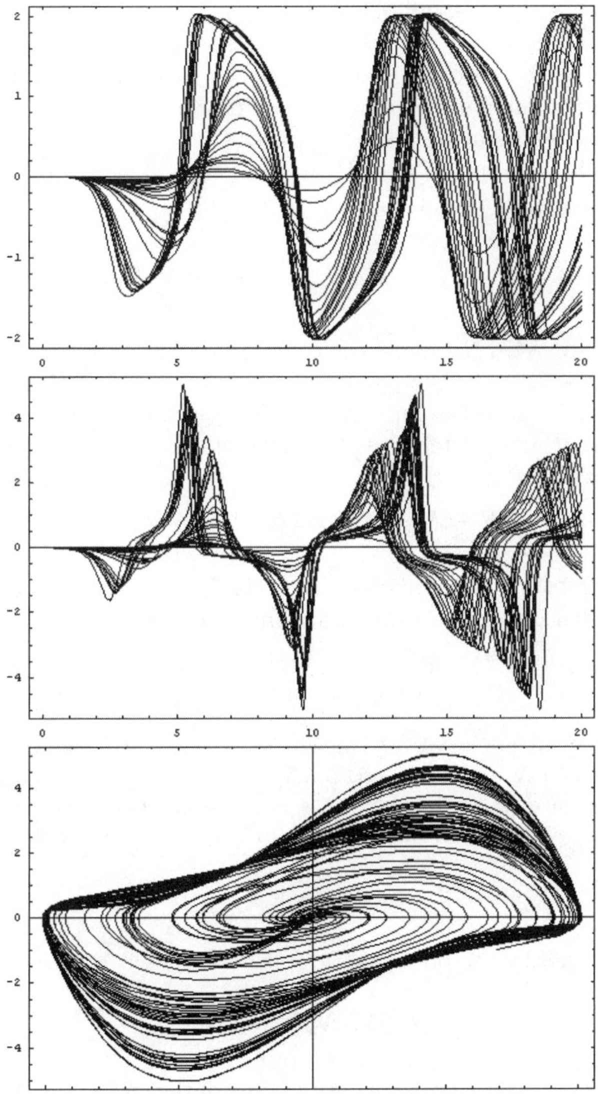

Fig. 2.16. Cascade of 30 unforced Van der Pol oscillators, simulated using *Mathematica*^{TM}; top–down: displacements, velocities and phase–plot (showing the celebrated limit cycle).

---

where

$$a_m = 0.1\,(V+40)/[1-\mathrm{e}^{-(V+40)/10}], \qquad b_m = 4\,\mathrm{e}^{-(V+65)/18},$$

$$a_h = 0.01\,(V+55)/[1-\mathrm{e}^{-(V+55)/10}], \qquad b_h = 0.125\,\mathrm{e}^{-(V+65)/80},$$

$$V_{xx} - V_t = F(V) + R - I, \qquad R_t = c(V + a - bR), \tag{2.27}$$

where $R(x,t)$ is the recovery variable, I the external stimulus and a, b, c are related to the membrane radius, specific resistivity of the fluid inside the membrane and temperature factor respectively.

When the spatial variation of V, namely V_{xx}, is negligible, (2.27) reduces to the Van der Pol oscillator,

$$\dot{V} = V - \frac{V^3}{3} - R + I, \qquad \dot{R} = c(V + a - bR),$$

with $F(V) = -V + \frac{V^3}{3}$. Normally the constants in (2.27) satisfy the inequalities $b < 1$ and $3a + 2b > 3$, though from a purely mathematical point of view this need not be insisted upon. Then with a periodic (ac) applied membrane current $A_1 \cos \omega t$ and a (dc) bias A_0, the Van der Pol equation becomes

$$\dot{V} = V - \frac{V^3}{3} - R + A_0 + A_1 \cos \omega t, \qquad \dot{R} = c(V + a - bR). \tag{2.28}$$

Further, (2.28) can be rewritten as a single second–order ODE by differentiating $\dot{V}$ with respect to time and using $\dot{R}$ for R,

---

$$a_n = 0.07\,\mathrm{e}^{-(V+65)/20}, \qquad b_n = 1/[1 + \mathrm{e}^{-(V+35)/10}].$$

Here the reversal potentials of Na, an K channels and leakage are $V_{\mathrm{Na}} = 50$ mV, $V_{\mathrm{K}} = -77$ mV and $V_{\mathrm{L}} = -54.5$ mV; the maximum values of corresponding conductivities are $g_{\mathrm{Na}} = 120$ mS/cm^2, $g_{\mathrm{K}} = 36$ mS/cm^2 and $g_{\mathrm{L}} = 0.3$ mS/cm^2; the capacity of the membrane is $C = 1$ μF/cm^2. The external, input current is given by

$$I_j^{\mathrm{ext}} = g_{syn}(V_a - V_c) \sum_n \alpha(t - t_{in}), \tag{2.25}$$

which is induced by the pre–synaptic spike–train input applied to the neuron i, given by

$$U_{\mathrm{i}}(t) = V_{\mathrm{a}} \sum_n \delta(t - t_{in}). \tag{2.26}$$

In (2.25) and (2.26), t_{in} is the nth firing time of the spike–train inputs, g_{syn} and V_c denote the conductance and the reversal potential, respectively, of the synapse, τ_{s} is the time constant relevant to the synapse conduction, and $\alpha(t)$ is the alpha function given by

$$\alpha(t) = (t/\tau_{\mathrm{s}})\,\mathrm{e}^{-t/\tau_{\mathrm{s}}}\Theta(t),$$

where $\Theta(t)$ is the Heaviside function. The HH model was originally proposed to account for the property of squid giant axons [HH52, Hod64] and it has been generalized with modifications of ion conductances [Arb98]. The HH–type models have been widely adopted for a study on activities of *transducer neurons* such as motor and thalamus relay neurons, which transform the amplitude–modulated input to spike–train outputs. In this section, we pay our attention to *data–processing neurons* which receive and emit the spike–train pulses.

$$\begin{aligned} &\ddot{V} - (1 - bc)\left\{1 - \frac{V^2}{1 - bc}\right\}\dot{V} - c(b-1)V + \frac{bc}{3}V^3 \\ &= c(A_0 b - a) + A_1 \cos(\omega t + \phi), \end{aligned} \tag{2.29}$$

where $\phi = \tan^{-1}\frac{\omega}{bc}$. Using the transformation $x = (1 - bc)^{-(1/2)}V$, $t \longrightarrow t' = t + \frac{\phi}{\omega}$, (2.29) can be rewritten as

$$\ddot{x} + p(x^2 - 1)\dot{x} + \omega_0^2 x + \beta x^3 = f_0 + f_1 \cos \omega t, \qquad \text{where} \tag{2.30}$$

$$p = (1 - bc), \qquad \omega_0^2 = c(1 - b), \qquad \beta = bc\frac{(1 - bc)}{3},$$

$$f_0 = c\frac{(A_0 b - a)}{\sqrt{1 - bc}}, \qquad f_1 = \frac{A_1}{\sqrt{1 - bc}}.$$

Note that (2.30), or its rescaled form

$$\ddot{x} + p(kx^2 + g)\dot{x} + \omega_0^2 x + \beta x^3 = f_0 + f_1 \cos \omega t, \tag{2.31}$$

is the *Duffing–Van der Pol equation*. In the limit $k = 0$, we have the Duffing equation discussed below (with $f_0 = 0$), and in the case $\beta = 0$ ($g = -1$, $k = 1$) we have the forced van der Pol equation. Equation (2.31) exhibits a very rich variety of bifurcations and chaos phenomena, including quasi–periodicity, phase lockings and so on, depending on whether the potential $V = \frac{1}{2}\omega_0^2 x^2 + \frac{\beta x^4}{4}$ is i) a double well, ii) a single well or iii) a double hump [Lak97, Lak03].

Duffing Oscillator

The forced *Duffing oscillator* [Duf18] has the form similar to (2.24),

$$\ddot{x} + b\,\dot{x} - a\,x\,(1 - x^2) = \gamma \cos(\phi t). \tag{2.32}$$

Stroboscopic *Poincaré sections* of a *strange attractor* can be seen (Figure 2.17), with the *stretch–and–fold* action at work. The simulation is performed with parameters: $a = 1$, $b = 0.2$, and $\gamma = 0.3$, $\phi = 1$. The Duffing equation is used to model a double well oscillator such as the magneto–elastic mechanical system. This system consists of a beam positioned vertically between two magnets, with the top end fixed, and the bottom end free to swing. The beam will be attracted to one of the two magnets, and given some velocity will oscillate about that magnet until friction stops it. Each of the magnets creates a fixed–point where the beam may come to rest above that magnet and remain there in equilibrium. However, when this whole system is shaken by a periodic forcing term, the beam may jump back and forth from one magnet to the other in a seemingly random manner. Depending on how big the shaking term is, there may be no stable fixed–points and no stable fixed cycles in the system.

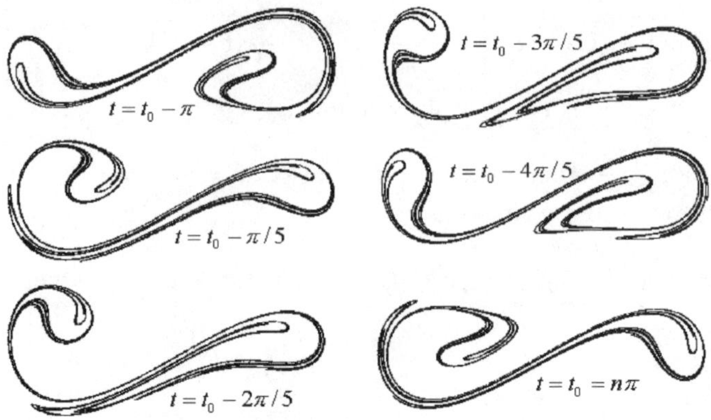

Fig. 2.17. Duffing strange attractor, showing stroboscopic Poincaré sections; simulated using *Dynamics Solver*^{TM}.

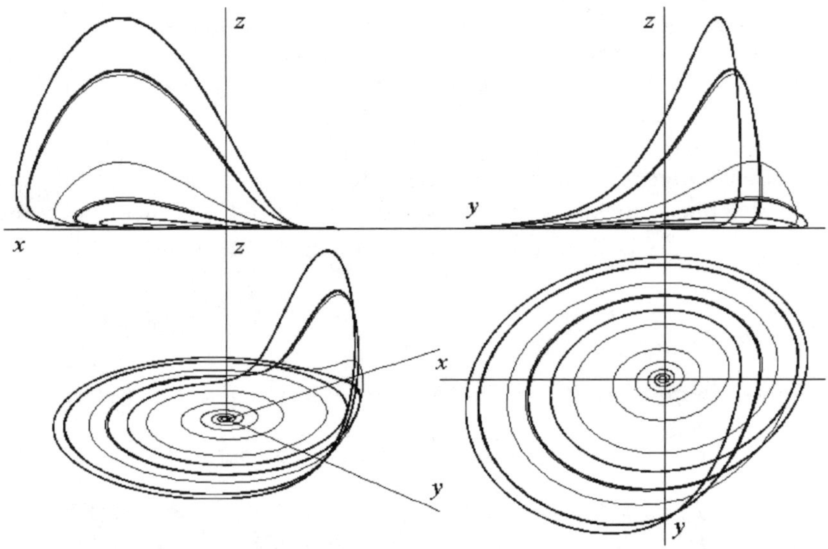

Fig. 2.18. The celebrated Rossler attractor, simulated using *Dynamics Solver*^{TM}.

Rossler System

Classical *Rossler system* is given by equations

$$\dot{x} = -y - z, \qquad \dot{y} = x + b\,y, \qquad \dot{z} = b + z\,(x - a). \tag{2.33}$$

Using the parameter values $a = 4$ and $b = 0.2$, the phase–portrait is produced (see Figure 2.18), showing the celebrated attractor. The system is credited to O. *Rossler* and arose from work in chemical kinetics.

Fig. 2.19. Ueda attractor in the $(x, \dot{x})$–plane.

Ueda Attractor

The Ueda attractor, discovered by Y. Ueda in 1961, appears to be a *trapped attractor* (see Figure 2.19). Here the plane is mapped into itself by following the trajectory of the *modified Duffing equation*

$$\ddot{x} + 0.05\,\dot{x} + x^3 = 7.5\,\cos(t),$$

for time $0 \leq t \leq 2\pi$.

Fractal Basin Boundaries

In the above example of a point particle moving in a two–well potential $V(x)$ with friction (Figure 2.11), the basin boundary was a smooth curve. However, other possibilities exist. An example of this occurs for the map

$$x_{n+1} = (3x_n) \mod 1, \qquad y_{n+1} = 1.5 + \cos 2\pi x_n$$

For almost any initial condition (except for those precisely on the boundary between the basins of attraction), $\lim_{n\to\infty} y_n$ is either $y = +\infty$ or $y = -\infty$, which we may regard as the two attractors of the system. Figure 2.20 shows the basin structure for this map, with the basin for the $y = -\infty$ attractor black and the basin of the $y = +\infty$ attractor blank. In contrast to the previous example, the basin boundary is no longer a smooth curve. In fact, it is a fractal curve with a box-counting dimension 1.62 .... We emphasize that, although fractal, this basin boundary is still a simple curve (it can be written as a

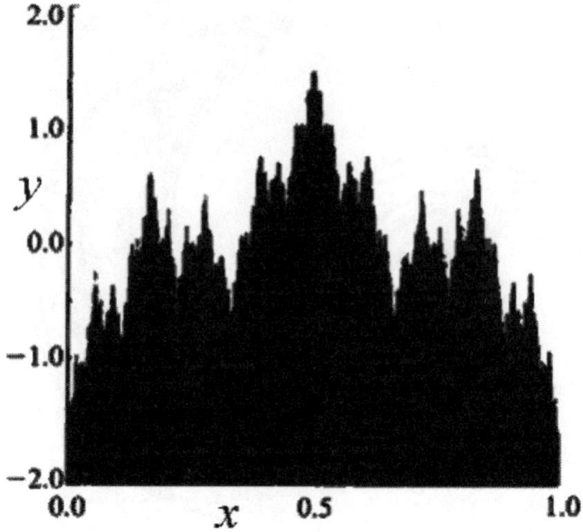

Fig. 2.20. A fractal curve as a basin boundary.

continuous parametric functional relationship $x = x(s), y = y(s)$ for $1 > s > 0$ such that

$$(x(s_1), y(s_1)) \neq (x(s_2), y(s_2))$$

if $s_1 \neq s_2$.

Another example of a system with a fractal basin boundary is the forced damped pendulum equation,

$$\ddot{\theta} + 0.1\,\dot{\theta} + \sin\theta = 2.1\cos t.$$

For these parameters, there are two attractors which are both periodic orbits [GOY87]. Figure 3 shows the basins of attraction of these two attractors with initial θ values plotted horizontally and initial values of $\dot{\theta}$ plotted vertically. The figure was made by initializing many initial conditions on a fine rectangular grid. Each initial condition was then integrated forward to see which attractor its orbit approached. If the orbit approached a particular one of the two attractors, a black dot was plotted on the grid. If it approached the other attractor, no dot was plotted. The dots are dense enough that they fill in a solid black region except near the basin boundary. The speckled appearance of much of this figure is a consequence of the intricate, fine–scaled structure of the basin boundary. In this case the basin boundary is again a fractal set (its box–counting dimension is about 1.8), but its topology is more complicated than that of the basin boundary of Figure 2.21 in that the Figure 2.21 basin boundary is not a simple curve. In both of the above examples in which fractal basin boundaries occur, the fractality is a result of chaotic motion (see transient chaos) of orbits on the boundary, and this is generally the case for fractal basin boundaries [MGO85].

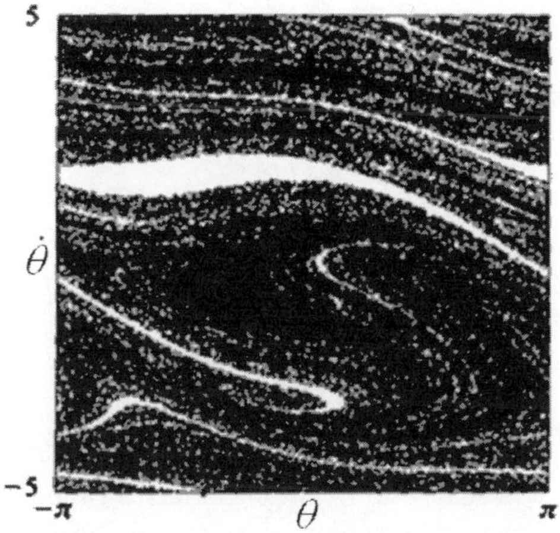

Fig. 2.21. Basins of attraction for a forced damped pendulum.

We have seen so far that there can be basin boundaries of qualitatively different types. As in the case of *attractors*, *bifurcations* can occur in which basin boundaries undergo *qualitative changes* as a system parameter passes through a *critical bifurcation value*. For example, for a system parameter $p < p_c$, the basin boundary might be a simple smooth curve, while for $p > p_c$ it might be fractal. Such basin boundary bifurcations have been called *metamorphoses* [GOY87].

The Uncertainty Exponent

Fractal basin boundaries, like those illustrated above, are extremely common and have potentially important practical consequences [Ott93]. In particular, they may make it more difficult to identify the attractor corresponding to a given initial condition, if that initial condition has some uncertainty. This aspect is already implied by the speckled appearance of Figure 2.21. A quantitative measure of this is provided by the *uncertainty exponent* [MGO85]. For definiteness, suppose we randomly choose an initial condition with uniform probability density in the area of initial condition space corresponding to the plot in Figure 2.21. Then, with probability one, that initial condition will lie in one of the basins of the two attractors (the basin boundary has zero *Lebesgue measure* (i.e., 'zero area') and so there is zero probability that a random initial condition is on the boundary). Now assume that we are also told that the initial condition has some given uncertainty, ϵ, and, for the sake of illustration, assume that this uncertainty can be represented by saying that the real initial condition lies within a circle of radius ϵ centered at the coordinates (x_0, y_0) that were randomly chosen. We ask what is the probability that the (x_0, y_0) could lie in a basin that is different from that of the true initial condition, i.e., what is the probability, $\rho(\epsilon)$, that the uncertainty ϵ could cause us to make a mistake in a determination of the attractor that the orbit goes to. Geometrically, this is the same as asking what fraction of the area of Figure 2.21 is within a distance ϵ of the basin boundary. This fraction scales as

$$\rho(\epsilon) \sim \epsilon^{\alpha},$$

where α is the *uncertainty exponent* and is given by $\alpha = D - D_0$, where D is the dimension of the initial condition space ($D = 2$ for Figure 2.21) and is the *box–counting dimension* of the basin boundary. For the example of Figure 2.21, since $D_0 \cong 1.8$, we have $\alpha \cong 0.2$. For small α it becomes very difficult to improve predictive capacity (i.e., to predict the attractor from the initial condition) by reducing the uncertainty. For example, if $\alpha = 0.2$, to reduce $\rho(\epsilon)$ by a factor of 10, the uncertainty ϵ would have to be reduced by a factor of 10^5. Thus, fractal basin boundaries (analogous to the *butterfly effect* of chaotic attractors, see next subsection) pose a *barrier to scientific prediction*, and this barrier is related to the presence of chaos [Ott93].

2.2.6 Chaotic Behavior

Lorenz Strange Attractor

Recall that an *attractor* is a set of system's states (i.e., points in the system's phase–space), invariant under the dynamics, towards which neighboring states in a given *basin of attraction* asymptotically approach in the course of dynamic evolution.[9] An attractor is defined as the smallest unit which cannot be itself decomposed into two or more attractors with distinct basins of attraction. This restriction is necessary since a dynamical system may have multiple attractors, each with its own basin of attraction.

Conservative systems do not have attractors, since the motion is periodic. For dissipative dynamical systems, however, volumes shrink exponentially, so attractors have 0 volume in nD phase–space.

In particular, a stable *fixed–point* surrounded by a dissipative region is an attractor known as a *map sink*.[10] Regular attractors (corresponding to 0 *Lyapunov exponents*) act as *limit cycles*, in which trajectories circle around a limiting trajectory which they asymptotically approach, but never reach. The so–called *strange attractors*[11] are bounded regions of phase–space (corresponding to positive Lyapunov characteristic exponents) having zero measure in the embedding phase–space and a *fractal dimension*. Trajectories within a strange attractor appear to skip around randomly.

[9] A *basin of attraction* is a set of points in the system's phase–space, such that initial conditions chosen in this set dynamically evolve to a particular attractor.

[10] A *map sink* is a stable fixed–point of a map which, in a dissipative dynamical system, is an attractor.

[11] A strange attractor is an attracting set that has zero measure in the embedding phase–space and has fractal dimension. Trajectories within a strange attractor appear to skip around randomly.

In 1963, Ed Lorenz from MIT was trying to improve weather forecasting. Using a primitive computer of those days, he discovered the first *chaotic attractor*. Lorenz used three Cartesian variables, (x, y, z), to define *atmospheric convection*. Changing in time, these variables gave him a trajectory in a (Euclidean) 3D–space. From all starts, trajectories settle onto a chaotic, or *strange attractor*.[12]

More precisely, Lorenz reduced the *Navier–Stokes equations* for *convective Bénard fluid flow* into three first order coupled nonlinear ODEs and

---

[12] Edward Lorenz is a professor of meteorology at MIT who wrote the first clear paper on *deterministic chaos*. The paper was called 'Deterministic Nonperiodic Flow' and it was published in the Journal of Atmospheric Sciences in 1963. Before that, in 1960, Lorenz began a project to simulate weather patterns on a computer system called the Royal McBee. Lacking much memory, the computer was unable to create complex patterns, but it was able to show the interaction between major meteorological events such as tornados, hurricanes, easterlies and westerlies. A variety of factors was represented by a number, and Lorenz could use computer printouts to analyze the results. After watching his systems develop on the computer, Lorenz began to see patterns emerge, and was able to predict with some degree of accuracy what would happen next. While carrying out an experiment, Lorenz made an accidental discovery. He had completed a run, and wanted to recreate the pattern. Using a printout, Lorenz entered some variables into the computer and expected the simulation to proceed the same as it had before. To his surprise, the pattern began to diverge from the previous run, and after a few 'months' of simulated time, the pattern was completely different. Lorenz eventually discovered why seemingly identical variables could produce such different results. When Lorenz entered the numbers to recreate the scenario, the printout provided him with numbers to the thousandth position (such as 0.617). However, the computer's internal memory held numbers up to the millionth position (such as 0.617395); these numbers were used to create the scenario for the initial run. This small deviation resulted in a completely divergent weather pattern in just a few months. This discovery creates the groundwork of chaos theory: In a system, small deviations can result in large changes. This concept is now known as a *butterfly effect*.

Lorenz definition of chaos is: "The property that characterizes a dynamical system in which most orbits exhibit sensitive dependence." Dynamical systems (like the weather) are all around us. They have recurrent behavior (it is always hotter in summer than winter) but are very difficult to pin down and predict apart from the very short term. 'What will the weather be tomorrow?' – can be anticipated, but 'What will the weather be in a months time?' is an impossible question to answer.

Lorenz showed that with a set of simple differential equations seemingly very complex turbulent behavior could be created that would previously have been considered as random. He further showed that accurate longer range forecasts in any chaotic system were impossible, thereby overturning the previous orthodoxy. It had been believed that the more equations you add to describe a system, the more accurate will be the eventual forecast.

demonstrated with these the idea of sensitive dependence upon initial conditions and chaos (see [Lor63, Spa82]).
We rewrite the celebrated Lorenz equations here as

$$\dot{x} = a(y - x), \qquad \dot{y} = bx - y - xz, \qquad \dot{z} = xy - cz, \tag{2.34}$$

where x, y and z are dynamical variables, constituting the 3D *phase–space* of the *Lorenz system*; and a, b and c are the parameters of the system. Originally, Lorenz used this model to describe the unpredictable behavior of the weather, where x is the rate of convective overturning (convection is the process by which heat is transferred by a moving fluid), y is the horizontal temperature overturning, and z is the vertical temperature overturning; the parameters are: $a \equiv P$–proportional to the *Prandtl number* (ratio of the fluid viscosity of a substance to its thermal conductivity, usually set at 10), $b \equiv R$–proportional to the Rayleigh number (difference in temperature between the top and bottom of the system, usually set at 28), and $c \equiv K$–a number proportional to the physical proportions of the region under consideration (width to height ratio of the box which holds the system, usually set at 8/3). The Lorenz system (2.34) has the properties:

1. *Symmetry*: $(x, y, z) \to (-x, -y, z)$ for all values of the parameters, and
2. The z–axis ($x = y = 0$) is *invariant* (i.e., all trajectories that start on it also end on it).

Nowadays it is well–known that the Lorenz model is a paradigm for low–dimensional chaos in dynamical systems in synergetics and this model or its modifications are widely investigated in connection with modelling purposes in meteorology, hydrodynamics, laser physics, superconductivity, electronics, oil industry, chemical and biological kinetics, etc.

The 3D *phase–portrait* of the Lorenz system (2.189) shows the celebrated '*Lorenz mask*', a special type of *fractal attractor* (see Figure 2.22). It depicts the famous '*butterfly effect*', (i.e., sensitive dependence on initial conditions) – the popular idea in meteorology that 'the flapping of a butterfly's wings in

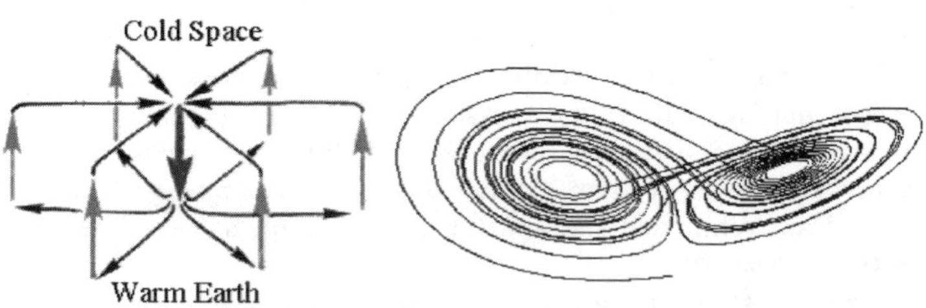

Fig. 2.22. Bénard cells, showing a typical vortex of a rolling air, with a warm air rising in a ring and a cool air descending in the center (left). A simple model of the Bénard cells provided by the celebrated 'Lorenz–butterfly' (or, 'Lorenz–mask') *strange attractor* (right).

Brazil can set off a tornado in Texas' (i.e., a tiny difference is amplified until two outcomes are totally different), so that the long term behavior becomes impossible to predict (e.g., long term weather forecasting). The Lorenz mask has the following characteristics:

1. Trajectory does not intersect itself in three dimensions;
2. Trajectory is not periodic or transient;
3. General form of the shape does not depend on initial conditions; and
4. Exact sequence of loops is very sensitive to the initial conditions.

Feigenbaum's Universality

Mitchell Jay Feigenbaum (born December 19, 1944; Philadelphia, USA) is a mathematical physicist whose pioneering studies in chaos theory led to the discovery of the *Feigenbaum constant*.

In 1964 he began graduate studies at the MIT. Enrolling to study electrical engineering, he changed to physics and was awarded a doctorate in 1970 for a thesis on dispersion relations under Francis Low. After short positions at Cornell University and Virginia Polytechnic Institute, he was offered a longer–term post at Los Alamos National Laboratory to study turbulence. Although the group was ultimately unable to unravel the intractable theory of turbulent fluids, his research led him to study chaotic maps.

Many mathematical maps involving a single linear parameter exhibit apparently random behavior known as chaos when the parameter lies in a certain range. As the parameter is increased towards this region, the map undergoes bifurcations at precise values of the parameter. At first there is one stable point, then bifurcating to oscillate between two points, then bifurcating again to oscillate between four points and so on. In 1975 Feigenbaum, using the HP-65 computer he was given, discovered that the ratio of the difference between the values at which such successive *period–doubling bifurcations* (called the *Feigenbaum cascade*) occur tends to a constant of around 4.6692. He was then able to provide a mathematical proof of the fact, and showed that the same behavior and the same constant would occur in a wide class of mathematical functions prior to the onset of chaos. For the first time this universal result enabled mathematicians to take their first huge step to unravelling the apparently intractable 'random' behavior of chaotic systems. This 'ratio of convergence' is now known as the Feigenbaum constant.

More precisely, the Feigenbaum constant δ is a universal constant for functions approaching chaos via successive period doubling bifurcations. It was discovered by Feigenbaum in 1975, while studying the fixed–points of the iterated function $f(x) = 1 - \mu|x|^r$, and characterizes the geometric approach of the bifurcation parameter to its limiting value (see Figure 2.23) as the parameter μ is increased for fixed x [Fei79].

The Logistic map is a well known example of the maps that Feigenbaum studied in his famous Universality paper [Fei78].

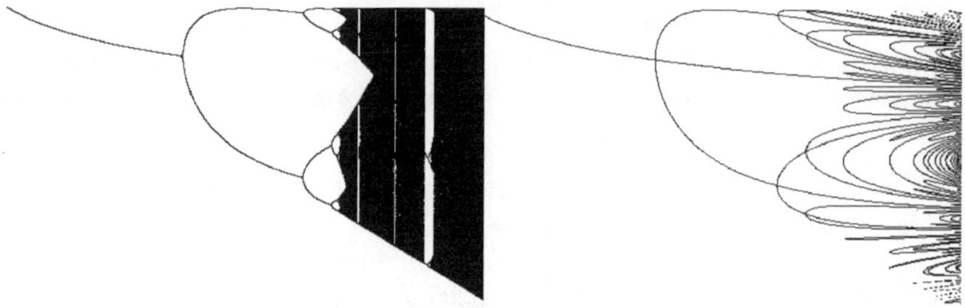

Fig. 2.23. Feigenbaum constant: approaching chaos via successive period doubling bifurcations. The plot on the left is made by iterating equation $f(x) = 1 - \mu|x|^r$ with $r = 2$ several hundred times for a series of discrete but closely spaced values of μ, discarding the first hundred or so points before the iteration has settled down to its fixed–points, and then plotting the points remaining. The plot on the right more directly shows the cycle may be constructed by plotting function $f^n(x) - x$ as a function of μ, showing the resulting curves for $n = 1, 2, 4$. Simulated in $Mathematica^{TM}$.

In 1986 Feigenbaum was awarded the Wolf Prize in Physics. He has been Toyota Professor at Rockefeller University since 1986.

For details on Feigenbaum universality, see [Gle87].

May's Logistic Map

Let $x(t)$ be the population of the species at time t; then the *conservation law* for the population is conceptually given by (see [Mur02])

$$\dot{x} = births - deaths + migration, \tag{2.35}$$

where $\dot{x} = dx/dt$. The above conceptual equation gave rise to a series of *population models*. The simplest continuous–time model, due to Thomas Malthus from 1798 [Mal798],[13] has no migration, while the birth and death terms are proportional to x,

[13] The Rev. Thomas Robert Malthus, FRS (February, 1766–December 23, 1834), was an English demographer and political economist best known for his pessimistic but highly influential views. Malthus's views were largely developed in reaction to the optimistic views of his father, Daniel Malthus and his associates, notably Jean-Jacques Rousseau and William Godwin. Malthus's essay was also in response to the views of the Marquis de Condorcet. In An Essay on the Principle of Population, first published in 1798, Malthus made the famous prediction that population would outrun food supply, leading to a decrease in food per person: "The power of population is so superior to the power of the earth to produce subsistence for man, that premature death must in some shape or other visit the human race. The vices of mankind are active and able ministers of depopulation. They are the precursors in the great army of destruction; and often finish the dreadful work themselves. But should they fail in this war of extermination, sickly seasons, epidemics, pestilence, and plague, advance in terrific array, and sweep

$$\dot{x} = bx - dx \qquad \Longrightarrow \qquad x(t) = x_0 \mathrm{e}^{(b-d)t}, \tag{2.36}$$

where b, d are positive constants and $x_0 = x(0)$ is the initial population. Thus, according to the *Malthus model* (2.36), if $b > d$, the population grows exponentially, while if $b < d$, it dies out. Clearly, this approach is fairly over-simplified and apparently fairly unrealistic. (However, if we consider the past and predicted growth estimates for the total world population from the 1900, we see that it has actually grown exponentially.)

This simple example shows that it is difficult to make long–term predictions (or, even relatively short–term ones), unless we know sufficient facts to incorporate in the model to make it a *reliable predictor*. In the long run, clearly, there must be some adjustment to such exponential growth. François Verhulst [Ver838, Ver845][14] proposed that a *self–limiting process* should operate when a population becomes too large. He proposed the so–called *logistic growth* population model,

$$\dot{x} = rx(1 - x/K), \tag{2.37}$$

off their thousands and tens of thousands. Should success be still incomplete, gigantic inevitable famine stalks in the rear, and with one mighty blow levels the population with the food of the world." This Principle of Population was based on the idea that population if unchecked increases at a geometric rate, whereas the food supply grows at an arithmetic rate. Only natural causes (eg. accidents and old age), misery (war, pestilence, and above all famine), moral restraint and vice (which for Malthus included infanticide, murder, contraception and homosexuality) could check excessive population growth. Thus, Malthus regarded his Principle of Population as an explanation of the past and the present situation of humanity, as well as a prediction of our future. The eight major points regarding evolution found in his 1798 *Essay* are: (i) Population level is severely limited by subsistence. (ii) When the means of subsistence increases, population increases. (iii) Population pressures stimulate increases in productivity. (iv) Increases in productivity stimulates further population growth. (v) Since this productivity can never keep up with the potential of population growth for long, there must be strong checks on population to keep it in line with carrying capacity. (vi) It is through individual cost/benefit decisions regarding sex, work, and children that population and production are expanded or contracted. (vii) Positive checks will come into operation as population exceeds subsistence level. (viii) The nature of these checks will have significant effect on the rest of the sociocultural system.

Evolutionists John Maynard Smith and Ronald Fisher were both critical of Malthus' theory, though it was Fisher who referred to the *growth rate* r (used in *logistic equation*) as the *Malthusian parameter*. Fisher referred to "... a relic of creationist philosophy ..." in observing the fecundity of nature and deducing (as Darwin did) that this therefore drove natural selection. Smith doubted that famine was the great leveller that Malthus insisted it was.

[14] François Verhulst (October 28, 1804–February 15, 1849, Brussels, Belgium) was a mathematician and a doctor in number theory from the University of Ghent in 1825. Verhulst published in 1838 the logistic demographic model (2.37).

where r, K are positive constants. In the Verhulst logistic model (2.37), the constant K is the *carrying capacity* of the environment (usually determined by the available sustaining resources), while the per capita birth rate $rx(1-x/K)$ is dependent on x. There are two steady states (where $\dot{x} = 0$) for (2.37): (i) $x = 0$ (unstable, since linearization about it gives $\dot{x} \approx rx$); and (ii) $x = K$ (stable, since linearization about it gives $\frac{d}{dt}(x - K) \approx -r(x - K)$, so $\lim_{t\to\infty} x = K$). The carrying capacity K determines the size of the stable steady state population, while r is a measure of the rate at which it is reached (i.e., the measure of the dynamics) – thus $1/r$ is a representative timescale of the response of the model to any change in the population. The solution of (2.37) is

$$x(t) = \frac{x_0 K \mathrm{e}^{rt}}{[K + x_0(\mathrm{e}^{rt} - 1)]} \qquad \Longrightarrow \qquad \lim_{t\to\infty} x(t) = K.$$

In general, if we consider a population to be governed by

$$\dot{x} = f(x), \tag{2.38}$$

where typically $f(x)$ is a nonlinear function of x, then the equilibrium solutions x^* are solutions of $f(x) = 0$, and are linearly stable to small perturbations if $\dot{f}(x^*) < 0$, and unstable if $\dot{f}(x^*) > 0$ [Mur02].

In the mid 20th century, ecologists realised that many species had no overlap between successive generations and so population growth happens in discrete–time steps x_t, rather than in continuous–time $x(t)$ as suggested by the conservative law (2.35) and its Maltus–Verhulst derivations. This leads to study *discrete–time models* given by *difference equations*, or, *maps*, of the form

$$x_{t+1} = f(x_t), \tag{2.39}$$

where $f(x_t)$ is some generic nonlinear function of x_t. Clearly, (2.39) is a discrete–time version of (2.38). However, instead of solving differential equations, if we know the particular form of $f(x_t)$, it is a straightforward matter to evaluate x_{t+1} and subsequent generations by simple recursion of (2.39). The skill in modelling a specific population's growth dynamics lies in determining the appropriate form of $f(x_t)$ to reflect known observations or facts about the species in question.

In 1970s, Robert May, a physicist by training, won the Crafoord Prize for 'pioneering ecological research in theoretical analysis of the dynamics of populations, communities and ecosystems', by proposing a simple *logistic map* model for the generic population growth (2.39).[15] May's model of population growth is the celebrated *logistic map* [May76, May73, May76],

[15] Lord Robert May received his Ph.D. in theoretical physics from University of Sydney in 1959. He then worked at Harvard University and the University of Sydney before developing an interest in animal population dynamics and the relationship between complexity and stability in natural communities. He moved to Princeton University in 1973 and to Oxford and the Imperial College in 1988. May was able to make major advances in the field of population biology through

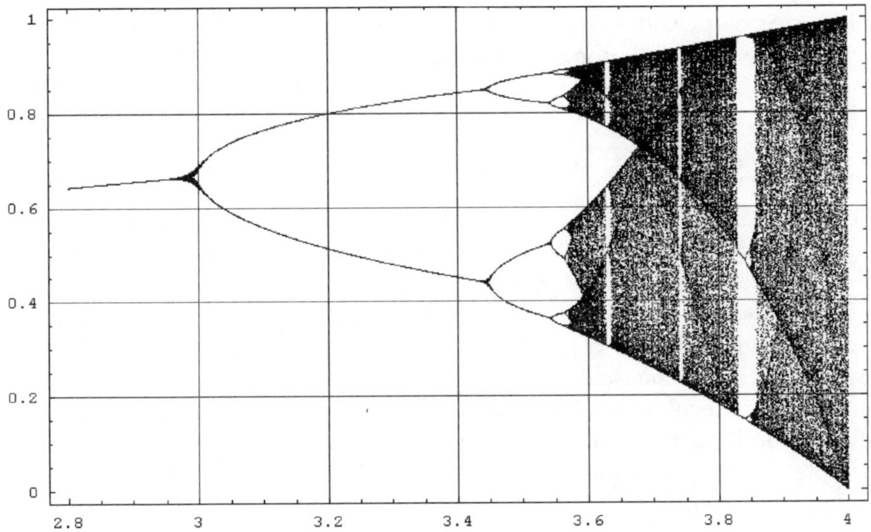

Fig. 2.24. Bifurcation diagram for the logistic map, simulated using *Mathematica*^{TM}.

$$x_{t+1} = r\, x_t\, (1 - x_t), \tag{2.40}$$

where r is the *Malthusian parameter* that varies between 0 and 4, and the initial value of the population $x_0 = x(0)$ is restricted to be between 0 and 1. Therefore, in May's logistic map (2.40), the generic function $f(x_t)$ gets a specific quadratic form

$$f(x_t) = r\, x_t\, (1 - x_t).$$

For $r < 3$, the x_t have a single value. For $3 < r < 3.4$, the x_t oscillate between two values (see *bifurcation diagram*[16] on Figure 2.24). As r increases, bifurcations occur where the number of iterates doubles. These *period–doubling bifurcations* continue to a limit point at $r_{lim} = 3.569944$ at which the period is 2^∞ and the dynamics become chaotic. The r values for the first two bifurcations can be found analytically, they are $r_1 = 3$ and $r_2 = 1 + \sqrt{6}$. We can label the successive values of r at which bifurcations occur as r_1, r_2, ... The universal number associated with such period doubling sequences is called the *Feigenbaum number*,

$$\delta = \lim_{k \to \infty} \frac{r_k - r_{k-1}}{r_{k+1} - r_k} \approx 4.669.$$

the application of mathematics. His work played a key role in the development of *theoretical ecology* through the 1970s and 1980s. He also applied these tools to the study of disease and to the study of *bio–diversity*.

[16] A bifurcation diagram shows the possible long–term values a variable of a system can get in function of a parameter of the system.

This series of period–doubling bifurcations says that close enough to r_{lim} the distance between bifurcation points decreases by a factor of δ for each bifurcation. The complex *fractal pattern* got in this way shrinks indefinitely.

Hénon's Map and Strange Attractor

Michel Hénon (born 1931 in Paris, France) is a mathematician and astronomer. He is currently at the Nice Observatory. In astronomy, Hénon is well known for his contributions to stellar dynamics, most notably the problem of *globular cluster* (see [Gle87]). In late 1960s and early 1970s he was involved in dynamical evolution of star clusters, in particular the globular clusters. He developed a numerical technique using *Monte Carlo methods*, to follow the dynamical evolution of a spherical star cluster much faster than the so–called n−body methods. In mathematics, he is well known for the Hénon map, a simple discrete dynamical system that exhibits chaotic behavior. Lately he has been involved in the restricted 3−body problem.

His celebrated *Hénon map* [Hen69] is a discrete–time dynamical system that is an extension of the *logistic map* (2.40) and exhibits a chaotic behavior. The map was introduced by Michel Hénon as a simplified model of the *Poincaré section* of the *Lorenz system* (2.34). This 2D–map takes a point (x, y) in the plane and maps it to a new point defined by equations

$$x_{n+1} = y_n + 1 - a x_n^2, \qquad y_{n+1} = b x_n,$$

The map depends on two parameters, a and b, which for the canonical Hénon map have values of $a = 1.4$ and $b = 0.3$ (see Figure 2.25). For the canonical values the Hénon map is chaotic. For other values of a and b the map may be chaotic, intermittent, or converge to a periodic orbit. An overview of the type of behavior of the map at different parameter values may be obtained

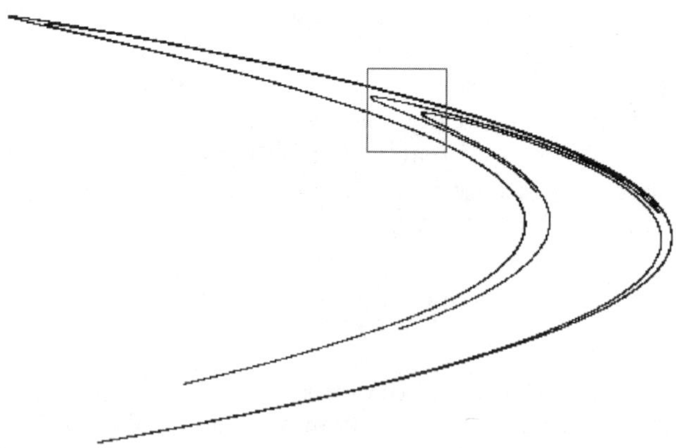

Fig. 2.25. *Hénon strange attractor* (see text for explanation), simulated using *Dynamics Solver*^{TM}.

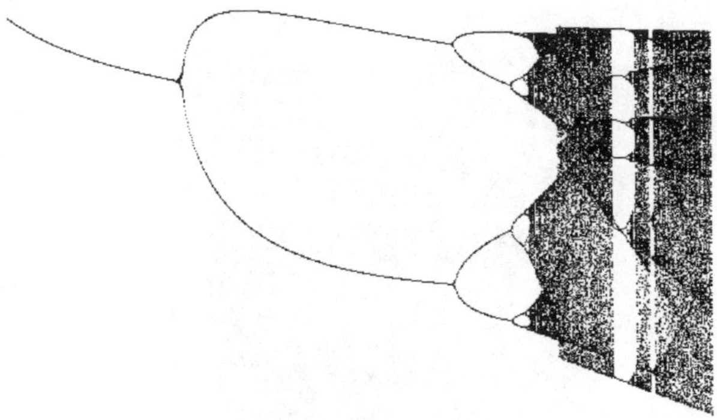

Fig. 2.26. Bifurcation diagram of the *Hénon strange attractor*, simulated using *Dynamics Solver*^{TM}.

from its orbit (or, bifurcation) diagram (see Figure 2.26). For the canonical map, an initial point of the plane will either approach a set of points known as the *Hénon strange attractor*, or diverge to infinity. The Hénon attractor is a fractal, smooth in one direction and a Cantor set in another. Numerical estimates yield a correlation dimension of 1.42 ± 0.02 (Grassberger, 1983) and a Hausdorff dimension of 1.261 ± 0.003 (Russel 1980) for the Hénon attractor. As a dynamical system, the canonical Hénon map is interesting because, unlike the logistic map, its orbits defy a simple description. The Hénon map maps two points into themselves: these are the invariant points. For the canonical values of a and b, one of these points is on the attractor: $x = 0.631354477...$ and $y = 0.189406343...$ This point is unstable. Points close to this fixed–point and along the slope 1.924 will approach the fixed–point and points along the slope -0.156 will move away from the fixed–point. These slopes arise from the linearizations of the *stable manifold* and *unstable manifold* of the fixed–point. The unstable manifold of the fixed–point in the attractor is contained in the strange attractor of the Hénon map. The Hénon map does not have a strange attractor for all values of the parameters a and b. For example, by keeping b fixed at 0.3 the bifurcation diagram shows that for a = 1.25 the Hénon map has a stable periodic orbit as an attractor. Cvitanovic *et al.* [CGP88] showed how the structure of the Hénon strange attractor could be understood in terms of unstable periodic orbits within the attractor.

For the (slightly modified) Hénon map: $x_{n+1} = ay_n + 1 - x_n^2$, $y_{n+1} = bx_n$, there are three *basins of attraction* (see Figure 2.27).

The *generalized Hénon map* is a 3D–system (see Figure 2.28)

$$x_{n+1} = a\,x_n - z\,(y_n - x_n^2)), \qquad y_{n+1} = z\,x_n + a\,(y_n - x_n^2)), \qquad z_{n+1} = z_n,$$

where $a = 0.24$ is a parameter. It is an *area–preserving map*, and simulates the *Poincaré map* of period orbits in *Hamiltonian systems*. Repeated random initial conditions are used in the simulation and their gray–scale color is selected at random.

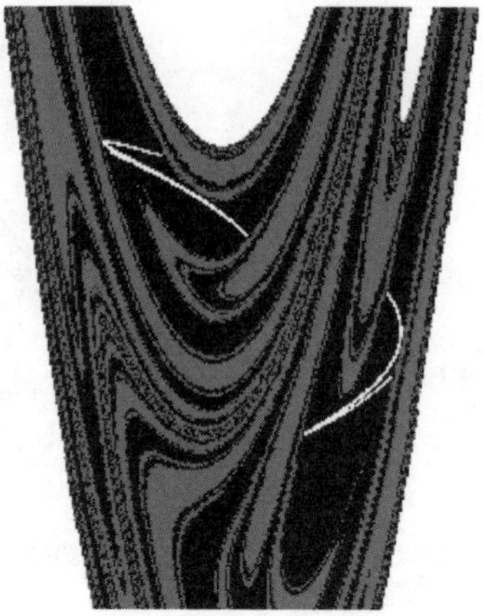

Fig. 2.27. Three basins of attraction for the Hénon map $x_{n+1} = ay_n + 1 - x_n^2$, $y_{n+1} = bx_n$, with $a = 0.475$.

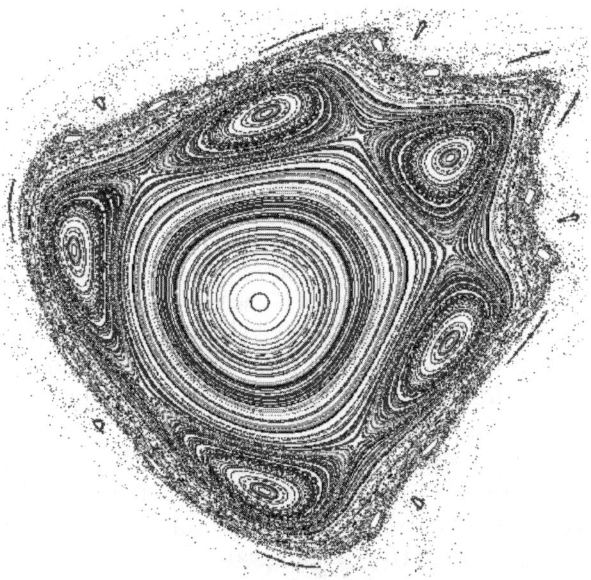

Fig. 2.28. Phase–plot of the *area–preserving generalized Hénon map*, simulated using *Dynamics Solver*^{TM}.

Other Famous 2D Chaotic Maps

1. The *standard map*:

$$x_{n+1} = x_n + y_{n+1}/2\pi, \qquad y_{n+1} = y_n + a\sin(2\pi x_n).$$

2. The *circle map*:

$$x_{n+1} = x_n + c + y_{n+1}/2\pi, \qquad y_{n+1} = by_n - a\sin(2\pi x_n).$$

3. The *Duffing map*:

$$x_{n+1} = y_n, \qquad y_{n+1} = -bx_n + ay_n - y_n^3.$$

4. The *Baker map*:

$$x_{n+1} = bx_n, \qquad y_{n+1} = y_n/a \qquad \text{if} \quad y_n \leq a,$$
$$x_{n+1} = (1-c) + cx_n, \qquad y_{n+1} = (y_n - a)/(1-a) \qquad \text{if} \quad y_n > a.$$

5. The *Kaplan–Yorke map*:

$$x_{n+1} = ax_n \bmod 1, \qquad y_{n+1} = -by_n + \cos(2\pi x_n).$$

6. The *Ott–Grebogi–Yorke map*:

$$x_{n+1} = x_n + w_1 + aP_1(x_n, y_n) \bmod 1,$$
$$y_{n+1} = y_n + w_2 + aP_2(x_n, y_n) \bmod 1,$$

where the nonlinear functions P_1, P_2 are sums of sinusoidal functions $A_{rs}^{(i)} \sin[2\pi(rx + sy + B_{rs}^{(i)})]$, with $(r,s) = (0,1), (1,0), (1,1), (1,-1)$, while $A_{rs}^{(i)}, B_{rs}^{(i)}$ were selected randomly in the range $[0,1]$.

Mandelbrot and Julia Sets

Recall that *Mandelbrot and Julia sets* (see Figure 2.29) are celebrated *fractals*. Recall that fractals are sets with *fractional dimension* (see Figure 2.30). The Mandelbrot and Julia fractals are defined either by a quadratic *conformal z–map* [Man80a, Man80b]

$$z_{n+1} = z_n^2 + c,$$

or by a real (x,y)–map

$$x_{n+1} = \sqrt{x_n} - \sqrt{y_n} + c_1, \qquad y_{n+1} = 2\,x_n\,y_n + c_2,$$

where c, c_1 and c_2 are parameters. For almost every c, this conformal transformation generates a fractal (probably, only for $c = -2$ it is not a fractal). Julia set J_c with $c \ll 1$, the *capacity dimension* is

$$d_{cap} = 1 + \frac{|c|^2}{4\ln 2} + O(|c|^3).$$

The set of all points for which J_c is connected is the Mandelbrot set.[17]

[17] The Mandelbrot set has its place in complex–valued dynamics, a field first investigated by the French mathematicians Pierre Fatou [Fat19, Fat22] and Gaston Julia [Jul18] at the beginning of the 20th century. For general families of holomorphic functions, the boundary of the Mandelbrot set generalizes to the bifurcation locus, which is a natural object to study even when the connectedness locus is not useful. A related *Mandelbar set* was encountered by mathematician John Milnor in his study of parameter slices of real cubic polynomials; it is not locally connected; this property is inherited by the connectedness locus of real cubic polynomials.

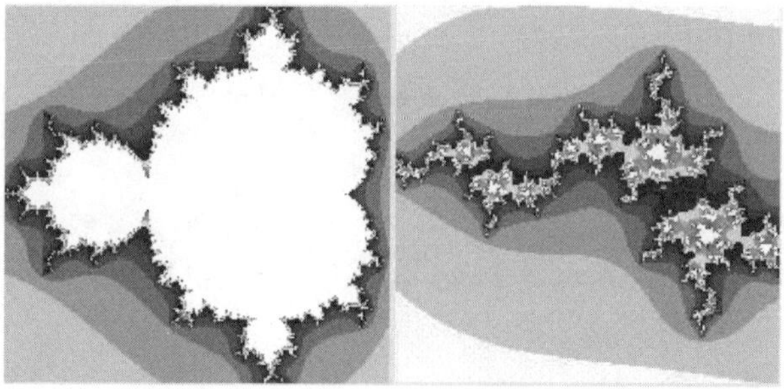

Fig. 2.29. The celebrated conformal Mandelbrot (left) and Julia (right) sets in the complex plane, simulated using *Dynamics Solver*^{TM}.

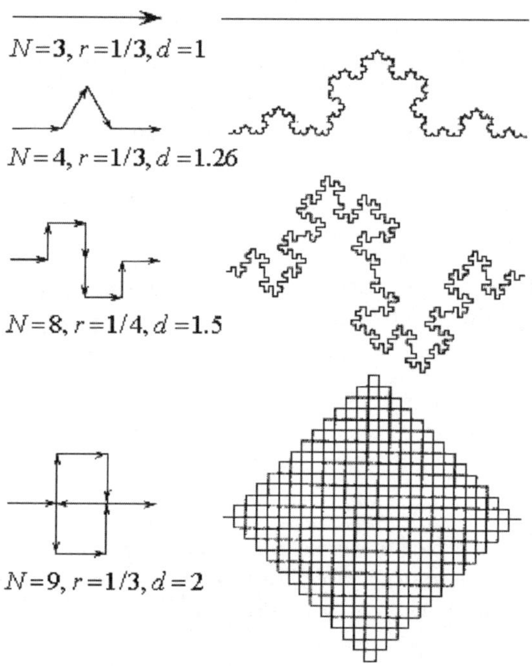

Fig. 2.30. Fractal dimension of curves in $\mathbb{R}^2$: $d = \frac{logN}{log1/r}$.

Let $K = K(f_c)$ be the filled Julia set, that is the union of all bounded orbits, for the quadratic map

$$f(z) = f_c(z) = z^2 + c .$$

Here both the parameter c and the dynamic variable z range over the complex numbers. The Mandelbrot set M can be defined as the compact subset of the parameter plane (or c–plane) consisting of all complex numbers c for which

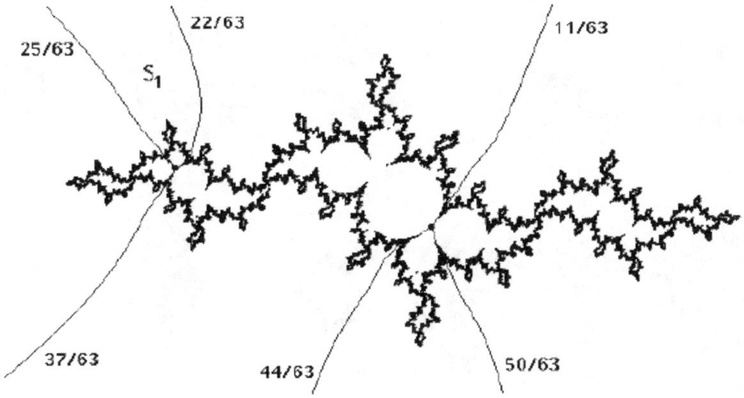

Fig. 2.31. Julia set showing the six rays landing on a period–2 parabolic orbit (adapted from [Mil99]).

$K(f_c)$ is connected. We can also identify the complex number c with one particular point in the dynamic plane (or $z-$plane), namely the *critical value* $f_c(0) = c$ for the map f_c. The parameter c belongs to M if and only if the orbit $f_c : 0 \mapsto c \mapsto c^2 + c \mapsto \cdots$ is bounded, or in other words if and only if $0, c \in K(f_c)$. Associated with each of the compact sets $K = K(f_c)$ in the dynamic plane there is a *potential function* or *Green's function* $G^K : \mathbb{C} \to [0, \infty)$ which vanishes precisely on K, is harmonic off K, and is asymptotic to $\log|z|$ near infinity. The family of *external rays* of K can be described as the orthogonal trajectories of the level curves $G^K =$ constant. Each such ray which extends to infinity can be specified by its angle at infinity $t \in \mathbb{R}/\mathbb{Z}$, and will be denoted by $\mathcal{R}_t^K$. Here c may be either in or outside of the Mandelbrot set. Similarly, we can consider the potential function G^M and the external rays $\mathcal{R}_t^M$ associated with the Mandelbrot set. We will use the term *dynamic ray* (or briefly $K-$ray) for an external ray of the filled Julia set, and *parameter ray* (or briefly $M-$ ray) for an external ray of the Mandelbrot set M [Mil99].

There is a theorem due Douady and Hubbard [DH85] related to a Mandelbrot set M saying that every parabolic point $c \neq 1/4$ in M is the *landing point* for exactly two external rays with angles which are periodic under doubling.[18]

Figure 2.31 shows the six rays landing on a period–2 parabolic orbit for the Julia set given by $z \longmapsto z^2 + (\frac{1}{4}e^{2\pi i/3} - 1)$.

Biomorphic Systems

Closely related to the Mandelbrot and Julia sets are *biomorphic systems*, which look like one–celled organisms. The term '*biomorph*' was proposed by C.

[18] By definition, a parameter point is parabolic iff the corresponding quadratic map has a periodic orbit with some root of unity as multiplier.

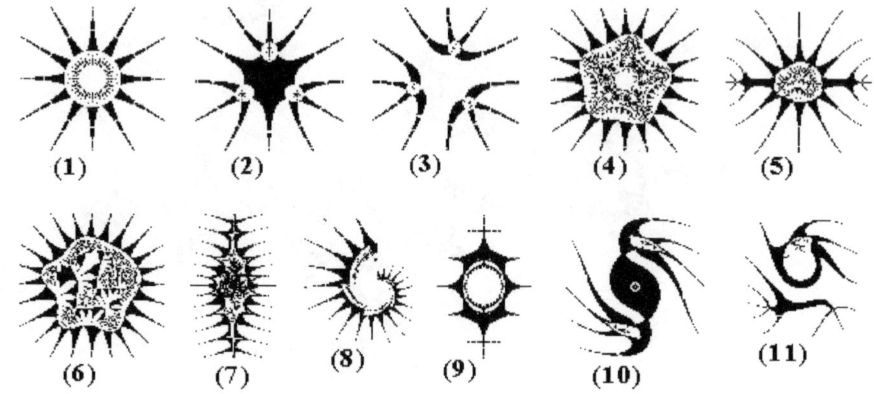

Fig. 2.32. *Pickover's biomorphs* (see text for details).

Pickover from IBM [Pic86, Pic87]. Pickover's biomorphs inhabit the complex plane like the Mandelbrot and Julia sets and exhibit a *protozoan morphology*. Biomorphs began for Pickover as a 'bug' in a program intended to probe the fractal properties of various formulas. He accidentally used an OR logical operator instead of an AND operator in the conditional test for the size of z's real and imaginary parts. The cilia that project from the biomorphs are a consequence of this 'error'. Each biomorph is generated by multiple iterations of a particular conformal map,

$$z_{n+1} = f(z_n, c),$$

where c is a parameter. Each iteration takes the output of the previous operations as the input of the next iteration. To generate a biomorph, one first needs to lay out a grid of points on a rectangle in the complex plane [And01]. The coordinate of each point constitutes the real and imaginary parts of an initial value, z_0, for the iterative process. Each point is also assigned a pixel on the computer screen. Depending on the outcome of a simple test on the 'size' of the real and imaginary parts of the final value, the pixel is colored either black or white. The biomorphs presented in Figure 2.32 are generated using the following conformal functions:

1. $f(z,c) = z^3$,
2. $f(z,c) = z^3 + c, \qquad c = 10$,
3. $f(z,c) = z^3 + c, \qquad c = 10 - 10i$,
4. $f(z,c) = z^5 + c, \qquad c = 0.77 - 0.77i$,
5. $f(z,c) = z^3 + \sin z + c, \qquad c = 1 - i$,
6. $f(z,c) = z^6 + \sin z + c, \qquad c = 0.5 - 0.5i$,
7. $f(z,c) = z^2 \sin z + c, \qquad c = 0.78 - 0.78i$,
8. $f(z,c) = z^c, \qquad c = 5 - i$,

9. $f(z,c) = |z|^c \sin z, \qquad c = 4,$
10. $f(z,c) = |z|^c \cos z + c, \qquad c = 3 + 3i,$
11. $f(z,c) = |z|^c(\cos z + z) + c, \qquad c = 3 + 2i.$

Lyapunov Exponents

The sensitive dependence on the initial conditions can be formalized in order to give it a quantitative characterization. The main growth rate of trajectory separation is measured by the first (or maximum) *Lyapunov exponent*, defined as (see, e.g., [BLV01])

$$\lambda_1 = \lim_{t\to\infty} \lim_{\Delta(0)\to 0} \frac{1}{t} \ln \frac{\Delta(t)}{\Delta(0)}, \tag{2.41}$$

As long as $\Delta(t)$ remains sufficiently small (i.e., infinitesimal, strictly speaking), one can regard the separation as a tangent vector $\mathbf{z}(t)$ whose time evolution is

$$\dot{z}_i = \frac{\partial f_i}{\partial x_j}|_{\mathbf{x}(t)} \cdot z_j, \tag{2.42}$$

and, therefore,

$$\lambda_1 = \lim_{t\to\infty} \frac{1}{t} \ln \frac{||\mathbf{z}(t)||}{||\mathbf{z}(0)||}. \tag{2.43}$$

In principle, λ_1 may depend on the initial condition $\mathbf{x}(0)$, but this dependence disappears for ergodic systems. In general there exist as many Lyapunov exponents, conventionally written in decreasing order $\lambda_1 \geq \lambda_2 \geq \lambda_3 \geq ...$, as the independent coordinates of the phase–space [BGG80]. Without entering the details, one can define the sum of the first k Lyapunov exponents as the growth rate of an infinitesimal kD volume in the phase–space. In particular, λ_1 is the growth rate of material lines, $\lambda_1 + \lambda_2$ is the growth rate of $2D$ surfaces, and so on. A numerical widely used efficient method is due to Benettin *et al.* [BGG80].

It must be observed that, after a transient, the growth rate of any generic small perturbation (i.e., distance between two initially close trajectories) is measured by the first (maximum) Lyapunov exponent λ_1, and $\lambda_1 > 0$ means chaos. In such a case, the state of the system is unpredictable on long times. Indeed, if we want to predict the state with a certain tolerance Δ then our forecast cannot be pushed over a certain time interval T_P, called *predictability time*, given by [BLV01]:

$$T_P \sim \frac{1}{\lambda_1} \ln \frac{\Delta}{\Delta(0)}. \tag{2.44}$$

The above relation shows that T_P is basically determined by $1/\lambda_1$, seen its weak dependence on the ratio $\Delta/\Delta(0)$. To be precise one must state that, for a series of reasons, relation (2.44) is too simple to be of actual relevance [BCF02].

Kolmogorov–Sinai Entropy

Deterministic chaotic systems, because of their irregular behavior, have many aspects in common with stochastic processes. The idea of using stochastic processes to mimic chaotic behavior, therefore, is rather natural [Chi79, Ben84]. One of the most relevant and successful approaches is symbolic dynamics [BS93]. For the sake of simplicity let us consider a discrete time dynamical system. One can introduce a partition $\mathcal{A}$ of the phase–space formed by N disjoint sets $A_1, \dots, A_N$. From any initial condition one has a trajectory

$$\mathbf{x}(0) \to \mathbf{x}(1), \mathbf{x}(2), \dots, \mathbf{x}(n), \dots \tag{2.45}$$

dependently on the partition element visited, the trajectory (2.45), is associated to a symbolic sequence

$$\mathbf{x}(0) \to i_1, i_2, \dots, i_n, \dots \tag{2.46}$$

where i_n $(n = 1, 2, \dots, N)$ means that $\mathbf{x}(n) \in A_{i_n}$ at the step n, for $n = 1, 2, \dots$. The coarse-grained properties of chaotic trajectories are therefore studied through the discrete time process (2.46).

An important characterization of symbolic dynamics is given by the *Kolmogorov–Sinai entropy* (KS), defined as follows. Let $C_n = (i_1, i_2, \dots, i_n)$ be a generic 'word' of size n and $P(C_n)$ its occurrence probability, the quantity [BLV01]

$$H_n = \sup_A [-\sum_{C_n} P(C_n) \ln P(C_n)], \tag{2.47}$$

is called *block entropy* of the $n-$sequences, and it is computed by taking the largest value over all possible partitions. In the limit of infinitely long sequences, the asymptotic entropy increment

$$h_{KS} = \lim_{n \to \infty} H_{n+1} - H_n, \tag{2.48}$$

is the Kolmogorov–Sinai entropy. The difference $H_{n+1} - H_n$ has the intuitive meaning of average information gain supplied by the $(n+1)-$th symbol, provided that the previous n symbols are known. KS–entropy has an important connection with the positive Lyapunov exponents of the system [Ott93]:

$$h_{KS} = \sum_{\lambda_i > 0} \lambda_i. \tag{2.49}$$

In particular, for low–dimensional chaotic systems for which only one Lyapunov exponent is positive, one has $h_{KS} = \lambda_1$.

We observe that in (2.47) there is a technical difficulty, i.e., taking the sup over all the possible partitions. However, sometimes there exits a special partition, called generating partition, for which one finds that H_n coincides with its superior bound. Unfortunately the generating partition is often hard to find, even admitting that it exist. Nevertheless, given a certain partition, chosen by physical intuition, the statistical properties of the related symbol sequences can give information on the dynamical system beneath. For example, if the probability of observing a symbol (state) depends only by the knowledge of the immediately preceding symbol, the symbolic process becomes a *Markov chain* (see [II06b]) and all the statistical properties are determined by the transition matrix elements W_{ij} giving the probability of observing a transition $i \to j$ in one time step. If the memory of the system extends far beyond the time step between two consecutive symbols, and the occurrence probability of a symbol depends on k preceding steps, the process is called *Markov process* of order k and, in principle, a k rank tensor would be required to describe the dynamical system with good accuracy. It is possible to demonstrate that if $H_{n+1} - H_n = h_{KS}$ for $n \geq k+1$, k is the (minimum) order of the required Markov process [Khi57]. It has to be pointed out, however, that to know the order of the suitable Markov process we need is of no practical utility if $k \gg 1$.

Pinball Game and Periodic Orbits

Confronted with a potentially chaotic dynamical system, we analyze it through a sequence of three distinct stages: (i) diagnose, (ii) count, (iii) measure. First we determine the intrinsic dimension of the system – the minimum number of coordinates necessary to capture its essential dynamics. If the system is very turbulent we are, at present, out of luck. We know only how to deal with the transitional regime between regular motions and chaotic dynamics in a few dimensions. That is still something; even an infinite–dimensional system such as a burning flame front can turn out to have a very few chaotic degrees of freedom. In this regime the chaotic dynamics is restricted to a space of low dimension, the number of relevant parameters is small, and we can proceed to step (ii); we count and classify all possible topologically distinct trajectories of the system into a hierarchy whose successive layers require increased precision and patience on the part of the observer. If successful, we can proceed with step (iii): investigate the weights of the different pieces of the system [CAM05].

With the game of pinball we are lucky: it is only a 2D system, free motion in a plane. The motion of a point particle is such that after a collision with one disk it either continues to another disk or it escapes. If we label the three disks by 1, 2 and 3, we can associate every trajectory with an itinerary, a sequence of labels indicating the order in which the disks are visited; for example, the two trajectories in Figure 1.2 have itineraries 2313, 23132321 respectively. The itinerary is finite for a scattering trajectory, coming in from infinity and escaping after a finite number of collisions, infinite for a trapped trajectory, and infinitely repeating for a periodic orbit.[19] Such labelling is the simplest example of *symbolic dynamics*. As the particle cannot collide two times in succession with the same disk, any two consecutive symbols must differ. This is an example of *pruning*, a rule that forbids certain subsequences of symbols. Deriving pruning rules is in general a difficult problem, but with the game of pinball we are lucky, as there are no further pruning rules.[20]

Suppose you wanted to play a good game of pinball, that is, get the pinball to bounce as many times as you possibly can – what would be a winning strategy? The simplest thing would be to try to aim the pinball so it bounces many times between a pair of disks – if you managed to shoot it so it starts out in the periodic orbit bouncing along the line connecting two disk centers, it would stay there forever. Your game would be just as good if you managed to get it to keep bouncing between the three disks forever, or place it on any periodic orbit. The only rub is that any such orbit is unstable, so you have to aim very accurately in order to stay close to it for a while. So it is pretty clear that if one is interested in playing well, unstable periodic orbits are important – they form the skeleton onto which all trajectories trapped for long times cling.

Now, recall that a trajectory is *periodic* if it returns to its starting position and momentum. It is custom to refer to the set of periodic points that belong to a given periodic orbit as a *cycle*.

Short periodic orbits are easily drawn and enumerated, but it is rather hard to perceive the systematics of orbits from their shapes. In mechanics a trajectory is fully and uniquely specified by its position and momentum at a given instant, and no two distinct phase–space trajectories can intersect. Their projections on arbitrary subspaces, however, can and do intersect, in rather unilluminating ways. In the pinball example, the problem is that we are looking at the projections of a 4D phase–space trajectories onto its 2D subspace, the configuration space. A clearer picture of the dynamics is obtained by constructing a phase–space Poincaré section.

The position of the ball is described by a pair of numbers (the spatial coordinates on the plane), and the angle of its velocity vector. As far as a classical dynamist is concerned, this is a complete description. Now, suppose that the pinball has just bounced off disk 1. Depending on its position and outgoing angle, it could proceed to either disk 2 or 3. Not much happens in between the bounces – the ball just travels at constant velocity along a straight line – so we can reduce the 4D flow to a 2D map f that takes the coordinates of the pinball from one disk edge to another disk edge. Let us state this more precisely: the trajectory just after the moment of impact is defined by marking s_n, the arc–length position of the nth bounce along the

[19] The words *orbit* and *trajectory* here are synonymous.

[20] The choice of symbols is in no sense unique. For example, as at each bounce we can either proceed to the next disk or return to the previous disk, the above 3–letter alphabet can be replaced by a binary $\{0, 1\}$ alphabet. A clever choice of an alphabet will incorporate important features of the dynamics, such as its symmetries.

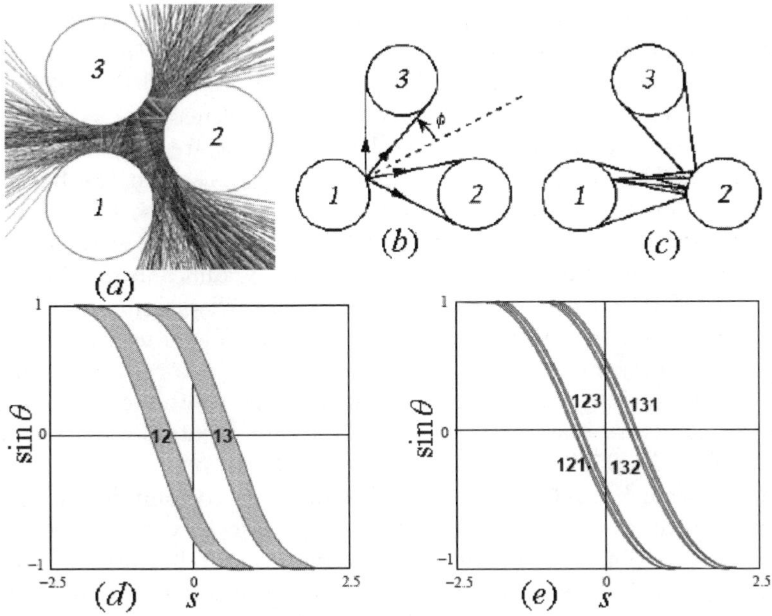

Fig. 2.33. A 3–disk pinball game. Up: (a) Elastic scattering around three hard disks (simulated in *Dynamics Solver*^{TM}); (b) A trajectory starting out from disk 1 can either hit another disk or escape; (c) Hitting two disks in a sequence requires a much sharper aim; the cones of initial conditions that hit more and more consecutive disks are nested within each other. Down: Poincaré section for the 3–disk pinball game, with trajectories emanating from the disk 1 with $x_0 = (arc - length, parallel\, momentum) = (s_0, p_0)$, disk radius: center separation ratio $a : R = 1 : 2.5$; (d) Strips of initial points M_{12}, M_{13} which reach disks 2, 3 in one bounce, respectively. (e) Strips of initial points M121, M131 M132 and M123 which reach disks 1, 2, 3 in two bounces, respectively; the Poincaré sections for trajectories originating on the other two disks are obtained by the appropriate relabelling of the strips (modified and adapted from [CAM05]).

billiard wall, and $p_n = p \sin \phi_n$ is the momentum component parallel to the billiard wall at the point of impact (see Figure 2.33). Such a section of a flow is called a *Poincaré section*, and the particular choice of coordinates (due to Birkhoff) is particularly smart, as it conserves the phase–space volume. In terms of the Poincaré section, the dynamics is reduced to the *return map*

$$P : (s_n, p_n) \to (s_{n+1}, p_{n+1}),$$

from the boundary of a disk to the boundary of the next disk.

Next, we mark in the Poincaré section those initial conditions which do not escape in one bounce. There are two strips of survivors, as the trajectories originating from one disk can hit either of the other two disks, or escape without further ado. We label the two strips M_0, M_1. Embedded within them there are four strips, $M_{00}, M_{10}, M_{01}, M_{11}$ of initial conditions that survive for two bounces, and so forth (see Figure 2.33). Provided that the disks are sufficiently separated, after n bounces the survivors are divided into 2^n distinct strips: the M_ith strip consists of all points with itinerary $i = s_1 s_2 s_3 ... s_n$, $s = \{0, 1\}$. The unstable cycles as a skeleton of chaos are almost visible here: each such patch contains a periodic point $\overline{s_1 s_2 s_3 ... s_n}$ with the basic block infinitely repeated. Periodic points are skeletal in the sense that as we look further and further, the strips shrink but the periodic points stay put forever.

We see now why it pays to utilize a symbolic dynamics; it provides a navigation chart through chaotic phase–space. There exists a unique trajectory for every admissible infinite length itinerary, and a unique itinerary labels every trapped trajectory. For example, the only trajectory labelled by 12 is the 2–cycle bouncing along the line connecting the centers of disks 1 and 2; any other trajectory starting out as 12 . . . either eventually escapes or hits the 3rd disk [CAM05].

Now we can ask what is a good physical quantity to compute for the game of pinball? Such system, for which almost any trajectory eventually leaves a finite region (the pinball table) never to return, is said to be open, or a *repeller*. The repeller escape rate is an eminently measurable quantity. An example of such a measurement would be an unstable molecular or nuclear state which can be well approximated by a classical potential with the possibility of escape in certain directions. In an experiment many projectiles are injected into such a non–confining potential and their mean escape rate is measured. The numerical experiment might consist of injecting the pinball between the disks in some random direction and asking how many times the pinball bounces on the average before it escapes the region between the disks. On the other hand, for a theorist a good game of pinball consists in predicting accurately the asymptotic lifetime (or the escape rate) of the pinball.

Here we briefly show how Cvitanovic's *periodic orbit theory* [Cvi91] accomplishes this for us. Each step will be so simple that you can follow even at the cursory pace of this overview, and still the result is surprisingly elegant. Let us consider Figure 2.33 again. In each bounce, the initial conditions get thinned out, yielding twice as many thin strips as at the previous bounce. The total area that remains at a given time is the sum of the areas of the strips, so that the fraction of survivors after n bounces, or the *survival probability* is given by

$$\begin{aligned}
\hat{\Gamma}_1 &= \frac{|M_0|}{|M|} + \frac{|M_1|}{|M|}, \qquad (2.50)\\
\hat{\Gamma}_2 &= \frac{|M_{00}|}{|M|} + \frac{|M_{10}|}{|M|} + \frac{|M_{01}|}{|M|} + \frac{|M_{11}|}{|M|},\\
&\dots\\
\hat{\Gamma}_n &= \frac{1}{|M|}\sum_{i=1}^{(n)} |M_i|,
\end{aligned}$$

where $i = 01, 10, 11, \ldots$ is a label of the ith strip (not a binary number), $|M|$ is the initial area, and $|M_i|$ is the area of the ith strip of survivors. Since at each bounce one routinely loses about the same fraction of trajectories, one expects the sum (2.50) to fall off exponentially with n and tend to the limit

$$\Gamma_{n+1}/\hat{\Gamma}_n = \mathrm{e}^{-\gamma n} \to \mathrm{e}^{-\gamma},$$

where the quantity γ is called the *escape rate* from the repeller. In [Cvi91] and subsequent papers, Cvitanovic has showed that the escape rate γ can be extracted from a highly convergent exact expansion by reformulating the sum (2.50) in terms of *unstable periodic orbits*.

2.2.7 Chaotic Repellors and Their Fractal Dimension

In addition to chaotic attractors, nonattracting chaotic sets (also called chaotic saddles or chaotic repellors) are also of great practical importance. In particular, such sets arise in the consideration of chaotic scattering, boundaries between basins of attraction, and chaotic transients. If a cloud of initial conditions is sprinkled in a bounded region including a nonattracting chaotic set, the orbits originating at these points eventually leave the vicinity of the set, and there is a characteristic escape time, τ, such that, at late time, the fraction of the cloud still in the region decays exponentially at the rate τ^{-1}.

In this subsection, mainly following [SO00], we will study the *fractal dimension* of nonattracting chaotic sets and their stable and unstable manifolds. Fractal dimension is of basic interest as a means of characterizing the geometric complexity of chaotic sets. In addition, a knowledge of the fractal dimension can, in some situations, provide quantitative information that is of potential practical use. For example, in the case of boundaries between different basins, the basin boundary is typically the stable manifold of a nonattracting chaotic set, and knowledge of the stable manifold's box–counting dimension (also called its capacity) quantifies the degree to which uncertainties in initial conditions result in errors in predicting the type of long–term motion that results (e.g., which attractor is approached; see [MGO85]). Our focus is on obtaining the *information dimension* of a suitable 'natural measure' μ lying on the chaotic set. The information dimension is a member of a one parameter (q) class of dimension definitions given by [SO00]

$$D_q = \lim_{\epsilon \to 0} \frac{1}{1-q} \frac{\ln \sum \mu_i^q}{\ln(1/\epsilon)}, \tag{2.51}$$

where q is a real index, ϵ is the grid spacing for a dD rectangular grid dividing the dD state space of the system, and μ_i is the natural measure of the ith grid cube. The *box–counting dimension* is given by (2.51) with $q = 0$, and the information dimension is given by taking the limit $q \to 1$ in (2.51),

$$D_1 = \lim_{\epsilon \to 0} \frac{I(\epsilon)}{\ln(1/\epsilon)}, \quad I(\epsilon) = \sum_i \mu_i \ln(1/\mu_i). \tag{2.52}$$

In general, the information dimension is a lower bound on the box–counting dimension, $D_1 \leq D_0$. In practice, in cases where D_1 and D_0 have been determined for chaotic sets, it is often found that their values are very close.

Following [SO00], we are specifically concerned with investigating formulae conjectured in [HOY96] that give the information dimensions for the nonattracting chaotic set and its stable and unstable manifolds in terms of Lyapunov exponents and the decay time, τ. These formulae generalize previous results for nonattracting chaotic sets of 2D maps with one positive and one negative Lyapunov exponent [KG85, HOG88], and for Hamiltonian systems of arbitrary dimensionality [Do95]. In turn, these past results for nonattracting chaotic sets were motivated by the *Kaplan–Yorke conjecture* which gives the information dimension of a chaotic attractor in terms of its *Lyapunov exponents* [KY79]. A rigorous result for the information dimension of an ergodic invariant chaotic set of a 2D diffeomorphism has been given by [LY85a, LY85b], and this result supports the Kaplan–Yorke conjecture for attractors and the 2D map results of [KG85] and [HOG88] for nonattracting chaotic sets.

Dimension Formulae

A *chaotic saddle*, Λ, is a *nonattracting*, *ergodic*, *invariant* set. By *invariant* we mean that all forward and reverse time evolutions of points in Λ are also in Λ. The *stable manifold* of Λ is the set of all initial conditions which converge to Λ upon forward time evolution. The *unstable manifold* of Λ is the set of all initial conditions which converge to Λ upon reverse time evolution. We say Λ is *nonattracting* if it does not completely contain its unstable manifold. In such a case there are points not in Λ that converge to it on backwards iteration.

To define the characteristic escape time, τ, first define a bounded region, R, which contains Λ and no other chaotic saddle. Uniformly sprinkle a large number, $N(0)$, of initial conditions in R. (In this section we take the dynamical system to be a discrete time system, i.e., a map.) Iterate the sprinkled initial conditions forward $n \gg 1$ times and discard all orbits which are no longer in R. Denote the remaining number of orbits $N(n)$. We define τ as [SO00]

$$\mathrm{e}^{-n/\tau} \sim \frac{N(n)}{N(0)}, \tag{2.53}$$

or, more formally, $\tau = \lim_{n\to\infty} \lim_{N(0)\to\infty} \ln[N(0)/N(n)]/n$. The Lyapunov exponents are defined with respect to the *natural transient measure* of the chaotic saddle [Ott93]. This measure is defined on an open set $C \subset R$ as

$$\mu(C) = \lim_{n\to\infty} \lim_{N(0)\to\infty} \frac{N(\xi n, n, C)}{N(n)}, \tag{2.54}$$

where $0 < \xi < 1$, and $N(m, n, C)$ is the number of sprinkled orbits still in R at time n that are also in C at the earlier time $m < n$. The above definition of $\mu(C)$ is presumed to be independent of the choice of ξ as long as $0 < \xi < 1$ (e.g., $\xi = 1/2$ will do).

We take the system to be MD with U positive and S negative Lyapunov exponents measured with respect to μ (where $U + S = M$) which we label according to the convention,

$$h_U^+ \geq h_{U-1}^+ \geq \cdots \geq h_1^+ > 0 > -h_1^- \geq \cdots \geq -h_{S-1}^- \geq -h_S^-.$$

Following [HOY96] we define a forward entropy,

$$H = \sum_{i=1}^{U} h_i^+ - \tau^{-1}.$$

We now define a natural transient measure μ_S on the stable manifold and a natural transient measure μ_U on the unstable manifold. Using the notation of (2.54),

$$\mu_S(C) = \lim_{n\to\infty} \lim_{N(0)\to\infty} \frac{N(0,n,C)}{N(n)}, \tag{2.55}$$

$$\mu_U(C) = \lim_{n\to\infty} \lim_{N(0)\to\infty} \frac{N(n,n,C)}{N(n)}. \tag{2.56}$$

Thus, considering the $N(n)$ orbits that remain in R up to time n, the fraction of those orbits that initially started in C gives $\mu_S(C)$, and the fraction that end up in C at the final time n gives $\mu_U(C)$. We use the measure (2.54), (2.55), (2.56) to define the information dimensions of the invariant set, the stable manifold, and the unstable manifold, respectively.

According to [HOY96], the dimension of the unstable manifold is then [SO00]

$$D_U = U + I + \frac{H - (h_1^- + \cdots + h_I^-)}{h_{I+1}^-}, \tag{2.57}$$

where I is defined by

$$h_1^- + \cdots + h_I^- + h_{I+1}^- \geq H \geq h_1^- + \cdots + h_I^-.$$

The dimension of the stable manifold is [HOY96]

$$D_S = S + J + \frac{H - (h_1^+ + \cdots + h_J^+)}{h_{J+1}^+}, \tag{2.58}$$

where J is defined by

$$h_1^+ + \cdots + h_J^+ + h_{J+1}^+ \geq H \geq h_1^+ + \cdots + h_J^+.$$

Considering the chaotic saddle to be the (generic) intersection of its stable and unstable manifolds, the generic intersection formula gives the dimension of the saddle,

$$D_\Lambda = D_U + D_S - M. \tag{2.59}$$

It is of interest to discuss some special cases of (2.57)–(2.59). In the case of a chaotic attractor, the invariant set is the attractor itself, the stable manifold is the basin of attraction, and we identify the unstable manifold with the attractor. Thus $D_S = M$ and $D_\Lambda = D_U$. Since points near the attractor never leave, we have $\tau = \infty$. Equation (2.57) then yields the *Kaplan–Yorke formula* [KY79],

$$D_\Lambda = U + I + \frac{(h_1^+ + \dots + h_U^+) - (h_1^- + \dots + h_I^-)}{h_{I+1}^-}, \tag{2.60}$$

where I is the largest integer for which $(h_1^+ + \dots + h_U^+) - (h_1^- + \dots + h_I^-)$ is positive.

In the case of a 2D map with one positive Lyapunov exponent h_1^+ and one negative Lyapunov exponent h_1^- with the exponents satisfying $h_1^+ - h_1^- - 1/\tau \le 0$, equations (2.57) and (2.58) give the result of [KG85] and [HOG88],

$$D_U = 1 + \frac{h_1^+ - 1/\tau}{h_1^-} \qquad \text{and} \qquad D_S = 1 + \frac{h_1^+ - 1/\tau}{h_1^+}.$$

Another case is that of a nonattracting chaotic invariant set of a 1D map. In this case $S = 0$ and $U = 1$. The unstable manifold of the invariant set has dimension one, $D_U = 1$. Recalling the definition of the stable manifold as the set of points that approach the invariant set as time increases, we can identify the stable manifold with the invariant set itself. This is because points in the neighborhood of the invariant set are repelled by it unless they lie precisely on the invariant set. Thus, $D_S = D_\Lambda$, and from (2.58) and (2.59) we have

$$D_S = D_\Lambda = H/h_1^+, \qquad \text{where} \qquad H = h_1^+ - 1/\tau.$$

Still another simple situation is the case of a 2D map with two positive Lyapunov exponents. This case is particularly interesting because we will be able to use it to gain understanding of the nature of the natural measure whose dimension we are calculating. In this case $U = 2$ and $S = 0$. Thus $D_U = 2$ and $D_S = D_\Lambda$. There are two cases (corresponding to $J = 0$ and $J = 1$ in (2.57)). For $h_2^+\tau \le 1$, we have that $D_S = D_\Lambda$ is between zero and one,

$$D_S = D_\Lambda = 1 + \frac{h_2^+}{h_1^+} - \frac{1}{h_1^+\tau}. \tag{2.61}$$

For $h_2^+\tau \ge 1$, we have that $D_S = D_\Lambda$ is between one and two,

$$D_S = D_\Lambda = 2 - \frac{1}{h_2^+\tau}. \tag{2.62}$$

In the next section we will be concerned with testing and illustrating (2.61) and (2.62) by use of a simple model.

Illustrative Expanding 2D–Map Model

We consider the following example [SO00],

$$x_{n+1} = 2x_n \,\mathrm{modulo}\, 1, \tag{2.63}$$

$$y_{n+1} = \lambda(x_n)y_n + \frac{\eta}{2\pi}\sin(2\pi x_n), \tag{2.64}$$

where $\lambda(x) > 1$, and the map is defined on the cylinder $-\infty \leq y \leq +\infty$, $1 \geq x \geq 0$, with x regarded as angle-like. We take $\lambda(x)$ to be the piecewise constant function,

$$\lambda(x) = \begin{cases} \lambda_1 \; 0 < x < 1/2, \\ \lambda_2 \; 1/2 < x < 1, \end{cases} \tag{2.65}$$

and, without loss of generality, we assume $\lambda_1 \leq \lambda_2$.

For this map, almost every initial condition generates an orbit that either tends toward $y = +\infty$ or toward $y = -\infty$. Initial conditions on the border of these two regions stay on the border forever. Thus, the border is an invariant set. It is also ergodic by virtue of the ergodicity of the map

$$x_{n+1} = 2x_n \, mod\, 1.$$

We wish to apply (2.61) and (2.62) to this invariant set and its natural measure.

The Jacobian matrix for this model is [SO00]

$$\mathcal{J}(x) = \begin{bmatrix} 2 & 0 \\ \eta\cos 2\pi x & \lambda(x) \end{bmatrix}.$$

Thus, for an ergodic invariant measure of the map, the two Lyapunov exponents are

$$h_a = p\ln\lambda_1 + (1-p)\ln\lambda_2 \qquad \text{and} \tag{2.66}$$
$$h_b = \ln 2,$$

where p is the measure of the region $x < 1/2$. To find h_a we thus need to know the measure of the invariant set.

The Decay Time and the Natural Measure

Consider a vertical line segment of length ℓ_0 whose x coordinate is x_0 and whose center is at $y = y_0$. After one iterate of the map (2.63)–(2.65), this line segment will have length $\ell_1 = \lambda(x_0)\ell_0$ and be located at $x = x_1$ with its center at $y = y_1$, where (x_1, y_1) are the iterates of (x_0, y_0) using the map (2.63)–(2.65). Thus we see that vertical line segments are expanded by the multiplicative factor $\lambda(x) \geq \lambda_1 > 1$. Now consider the strip, $-K \leq y \leq K$, and sprinkle many initial conditions uniformly in this region with density ρ_0. A vertical line segment, $x = x_0$, $-K \leq y \leq K$, iterates to $x = x_1$ and with its center at $y_1 = (\eta/2\pi)\sin 2\pi x_0$. We choose $K > (\eta/2\pi)(\lambda_1 - 1)^{-1}$ so that the iterated line segment spans the strip $-K \leq y \leq K$. After one iterate, the density will still be uniform in the strip: The region $x < 1/2$ ($x > 1/2$), $-K \leq y \leq K$, is expanded uniformly vertically by λ_1 (λ_2) and horizontally by 2. Thus, after one iterate, the new density in the strip is $\rho_1 = [(\lambda_1^{-1} + \lambda_2^{-1})/2]\rho_0$, and, after n iterates, we have [SO00]

$$\rho_n = [(\lambda_1^{-1} + \lambda_2^{-1})/2]^n \rho_0.$$

Hence the exponential decay time for the number of orbits remaining in the strip is

$$\frac{1}{\tau} = \ln \left[\frac{1}{2} \left(\frac{1}{\lambda_1} + \frac{1}{\lambda_2} \right) \right]^{-1}. \qquad (2.67)$$

To find the natural stable manifold transient measure of any x–interval $s_m^{(n)} = [m/2^n, (m+1)/2^n]$, where $m = 0, 1, \ldots, 2^n - 1$, we ask what fraction of the orbits that were originally sprinkled in the strip and are still in the strip at time n started in this interval. Let $s_m^{(n)}$ experience $n_1(m)$ vertical stretches by λ_1 and $n_2(m) = n - n_1(m)$ vertical stretches by λ_2. Then the initial subregion of the $s_m^{(n)}$ still in the strip after n iterates has vertical height $K\lambda_1^{-n_1(m)}\lambda_2^{-n_2(m)}$. Hence the natural measure of $s_m^{(n)}$ is

$$\mu(s_m^{(n)}) = \frac{2^{-n}\lambda_1^{-n_1(m)}\lambda_2^{-n_2(m)}}{\left[\frac{1}{2}(\lambda_1^{-1} + \lambda_2^{-1})\right]^n} = \frac{\lambda_1^{n_2(m)}\lambda_2^{n_1(m)}}{(\lambda_1 + \lambda_2)^n}. \qquad (2.68)$$

(Note that this is consistent with $\mu([0,1]) = \sum_{m=0}^{n-1} \mu(s_m^{(n)}) = 1$.) Thus the measures of the intervals $[0, 1/2]$ and $[1/2, 1]$ are

$$p = \mu(s_0^{(1)}) = \frac{\lambda_2}{\lambda_1 + \lambda_2}$$

and

$$1 - p = \mu(s_1^{(1)}) = \frac{\lambda_1}{\lambda_1 + \lambda_2}.$$

It is important to note that our natural transient measures p and $(1-p)$ for the 2D map are different from the natural measures of the same x–intervals for the 1D map,

$$x_{n+1} = 2x_n \, mod \, 1,$$

alone. In that case, with probability one, a random choice of x_0 produces an orbit which spends half its time in $[0, 1/2]$ and half its time in $[1/2, 1]$, so that in this case the natural measures of these regions are $p = (1-p) = 1/2$. The addition of the y–dynamics changes the natural measure of x-intervals.

From (2.66) we get

$$h_a = \frac{\lambda_2}{\lambda_1 + \lambda_2} \ln \lambda_1 + \frac{\lambda_1}{\lambda_1 + \lambda_2} \ln \lambda_2. \qquad (2.69)$$

For a general function $f(Z)$ with $d^2 f/dZ^2 < 0$, averaging over different values of Z gives the well-known inequality $\langle f(Z) \rangle \leq f(\langle Z \rangle)$ where $\langle (\cdots) \rangle$ denotes the average of the quantity $(\cdots)$. Using $f(Z) = \ln Z$ with $Z = \lambda_1$ with probability $p = \lambda_2/(\lambda_1 + \lambda_2)$ and $Z = \lambda_2$ with probability $(1-p)$, this inequality and (2.67) and (2.69) yield the result that

$$h_a \leq \frac{1}{\tau}. \tag{2.70}$$

Application of the Dimension Formulae

Let $\lambda_2 = r\lambda_1$, $r > 1$, and imagine that we fix r and vary λ_1. Applying (2.61) and (2.62) to our example we get three cases,

(a) $h_b > 1/\tau > h_a$ (λ_1 small),
(b) $1/\tau > h_b > h_a$ (λ_1 moderate), and
(c) $1/\tau > h_a > h_b$ (λ_1 large).

Corresponding to these three cases (2.61) and (2.62) yield the following values for D_Λ, the dimension of the invariant set [SO00],

$$D_a = 1 + \frac{\ln(1+r^{-1}) - \ln\lambda_1}{\ln 2}, \quad \text{for } \lambda_1 \leq \lambda_a, \tag{2.71}$$

$$D_b = \frac{\ln(1+r^{-1}) + (1+r)^{-1}\ln r}{\ln\lambda_1 + (1+r)^{-1}\ln r}, \quad \text{for } \lambda_a \leq \lambda_1 \leq \lambda_b, \tag{2.72}$$

$$D_c = \frac{\ln(1+r^{-1}) + (1+r)^{-1}\ln r}{\ln 2}, \quad \text{for } \lambda_b \leq \lambda_1, \tag{2.73}$$

where $\ln\lambda_a = \ln(1+r^{-1})$ and $\ln\lambda_b = \ln 2 - (1+r)^{-1}\ln r$. Note that for large λ_1, $D_\Lambda = D_c$ is independent of λ_1.

It is also instructive to consider the case of uniform stretching ($r = 1$) for which $\lambda_1 = \lambda_2$. In that case, $h_a = 1/\tau$, and there is a rigorous known result for the dimension [KMY84]. For $\lambda_1 = \lambda_2$, (2.71)–(2.73) yield

$$D_\Lambda = \begin{cases} 2 - \frac{\ln\lambda_1}{\ln 2} & \text{for } 1 \leq \lambda_1 \leq 2, \\ 1 & \text{for } \lambda_1 \geq 2. \end{cases} \tag{2.74}$$

(For $r \to 1$ region (b), where $D_\Lambda = D_b$, shrinks to zero width in λ_1). For $r = 1$ the natural transient measure is uniform; from (2.68) we have $\mu(s_m^{(n)}) = 2^{-n}$ independent of the interval (i.e, independent of m). In this case there is no difference between the capacity dimension of the invariant set and the information dimension of its measure. Equation (2.74) agrees with the rigorous known result, thus lending support to the original conjecture.

Numerical Tests

The formulae (2.71)–(2.73) were verified by numerical measurements of the information dimension, D_1, of Λ at various values of λ_1 with $r = \lambda_2/\lambda_1$ fixed at $r = 3$. Shown for comparison is the box–counting dimension, D_0. The values of the box–counting dimension are numerically indistinguishable from the values of the information dimension when D_0, $D_1 > 1$, or $\lambda_1 < \lambda_a$, the region corresponding to formula (2.71). For $\lambda_1 > \lambda_a$, Λ is a smooth curve and so has a box–counting dimension of $D_0 = 1$. No points for D_1 are shown near $\lambda_1 = \lambda_b$. It can be argued [SO00] that numerical convergence is too slow here to yield accurate measurements of the dimension.

To numerically determine the information dimension of Λ, we place a square $x - y$ grid with a spacing ε between grid points over a region containing Λ. Using the method dscribed in the next paragraph we compute the natural measure in each grid box and repeat for various ε. The information dimension is then given by

$$D_1 = \lim_{\varepsilon \to 0} \frac{I(\varepsilon)}{\ln(1/\varepsilon)}, \qquad \text{where}$$

$$I(\varepsilon) = \sum_{i=1}^{N(\varepsilon)} \mu_i \ln(1/\mu_i)$$

is a sum over the $N(\varepsilon)$ grid boxes which intersect Λ and μ_i is the natural measure in the ith box. The slope of a plot of $I(\varepsilon)$ versus $\ln \varepsilon$ gives D_1. The box–counting dimension is given by:

$$D_0 = \lim_{\varepsilon \to 0} \frac{\ln N(\varepsilon)}{\ln 1/\varepsilon}$$

and calculated in an analogous way.

To determine which boxes intersect Λ and what measure is contained in each of them we take advantage of the fact that Λ is a function [Ott93]. That is, for each value of x there is only one corresponding value of y in Λ, which we denote $y = y_\Lambda(x)$. We divide the interval $0 \leq x < 1$ into 2^n intervals of width $\delta \equiv 2^{-n}$. We wish to approximate $y_\Lambda(x_0)$ for x_0 in the center of the $x-$interval. To do this we iterate x_0 forward using (2.63) m times until the condition

$$\frac{\delta}{2} \lambda_1^{m_1} \lambda_2^{m_2} \geq 1 \tag{2.75}$$

is first met, where m_1 (m_2) is the number of times the orbit lands in $0 \leq x < 1/2$ ($1/2 \leq x < 1$), and $m_1 + m_2 = m$ (we will see the reason for this condition below). All of the values, x_i, of the iterates are saved. Starting now from x_m and taking $y_m = 0$ we iterate backward m times. For η small enough, Λ is contained in $-1 \leq y \leq 1$, so the point (x_m, $y_m = 0$) is within a distance 1 in the $y-$direction of Λ. The m reverse iterations shrink the segment $y_m \leq y \leq y_\Lambda(x_m)$ by a factor $\lambda_1^{m_1} \lambda_2^{m_2}$ so that

$$|y_0 - y_\Lambda(x_0)| \leq \delta/2$$

by condition (2.75). Thus, we have found a point y_0 that approximates $y_\Lambda(x_0)$ to within δ. Since we will be using ϵ boxes with $\epsilon \gg \delta$, we may regard y_0 as being essentially equal to $y_\Lambda(x_0)$. The measure in the δ width interval containing x_0 is found by iterating x_0 forward n times and using equation (2.68),

$$\mu = \frac{\lambda_1^{n_2} \lambda_2^{n_1}}{(\lambda_1 + \lambda_2)^n},$$

where n_1 (n_2) is the number of times the orbit lands in $0 \leq x < 1/2$ ($1/2 \leq x < 1$), and $n_1 + n_2 = n$. We associate this measure with the point (x_0, y_0) (with y_0 found by the above procedure).

Note that for fractal $y_\Lambda(x)$ the y interval occupied by the curve $y = y_\Lambda(x)$ in an x interval of width $\delta \ll 1$ is of order $\delta^{D_0 - 1}$ which is large compared to δ. We now cover the region with new grids having successively larger spacing, $\varepsilon_i = 2^i \delta = 2^{i-n}$, and calculate $I(\varepsilon_i)$ and $N(\varepsilon_i)$ based on the data taken from the first $\delta-$grid. For i large enough, such that the y extent of the curve $y_\Lambda(x)$ in a typical δ width interval is less than ϵ_i (i.e., $\epsilon_i \gtrsim \delta^{2-D_0}$ or $i \gtrsim (D_0 - 1)n$) we observe linear scaling of $\log I(\epsilon_i)$ and $\log N(\epsilon_i)$ with $\log \epsilon_i$, and we use the slope of such plots to determine D_1 and D_0. The dimensions D_1 and D_0 are then determined as described above.

Atypical Case

The conjecture of [HOY96] is that the above dimension formulae apply for 'typical' systems. To see the need for this restriction consider (2.64) for the case where $\eta = 0$. It is easily shown that the dimension formulae can be violated in this case. The claim, however, is that $\eta = 0$ is special, or 'atypical', in that, as soon as we give η any nonzero value, the validity of the dimension formulae is restored. In this connection it is important to note that as long as $\eta \neq 0$, the dimension of the invariant set is independent of the value of η. This follows since if $\eta \neq 0$ we can always rescale the value of η to one by the change of variables $\tilde{y} = y/\eta$. To see the violation of the dimension formulae for $\eta = 0$, we note that in this case, by virtue of (2.64), the line $y = 0$ is invariant. Thus the measure is distributed on a 1D subspace, the $x-$axis. Using the definition of the information dimension and dividing the $x-$axis into intervals of width 2^{-n}, the information dimension of the natural measure is [SO00]

$$D_\Lambda = \lim_{n \to \infty} \frac{\sum_{m=0}^{n-1} \mu(s_m^{(n)}) \ln[1/\mu(s_m^{(n)})]}{\ln(2^n)}. \tag{2.76}$$

The quantity whose limit is taken in (2.76) is in fact independent of n. Thus, taking $n = 1$ we get for D_Λ the result that, for $\eta = 0$,

$$D_\Lambda = D_c,$$

for all λ_1 and $\lambda_2 > 1$, where D_c is given by (2.73). Thus, for $h_a < h_b$, D_Λ is greater when $\eta \neq 0$ than when $\eta = 0$, and, thus, the above conjectured stable manifold dimension formula is violated. For $h_a > h_b$, D_Λ is the same in both cases.

General Considerations

The previous considerations readily generalize to the case of an arbitrary smooth function $\lambda(x) > 1$ and a general chaotic map, $x_{n+1} = M(x_n)$, which replaces (3.1). Consider the finite time vertical Lyapunov exponent [SO00],

$$\tilde{h}(x,n) = \frac{1}{n}\sum_{m=1}^{n} \ln \lambda(M^{m-1}(x))$$

computed for the initial condition x. Choosing x randomly with uniform probability distribution in the relevant basin for chaotic motion [e.g., x in $[0,1]$ for (2.63)], $\tilde{h}(x,n)$ can be regarded as a random variable. Let $\tilde{P}(h,n)$ denote its probability distribution function. For large n, we invoke large deviation theory to write $\tilde{P}(h,n)$ as [Ott93]

$$\ln \tilde{P}(h,n) = -nG(h) + o(n),$$

or, more informally,

$$\tilde{P}(h,n) \sim \mathrm{e}^{-nG(h)}, \tag{2.77}$$

where the specific form of $G(h)$ depends on $M(x)$ and the specific $\lambda(x)$, and $G(h)$ is convex, $d^2G(h)/dh^2 \geq 0$. For the normalization, $\int \tilde{P}(h,n)dh = 1$, to hold for $n \to \infty$, we have that

$$\min_h G(h) = 0,$$

where $\bar{h}$ denotes the value of h for which the above minimum is attained. As $n \to \infty$ we see that $\tilde{P}$ approaches a delta function, $\delta(h - \bar{h})$. Thus, $\bar{h}$ is the usual infinite time Lyapunov exponent for almost all initial conditions with respect to Lebesgue measure in $0 \leq x \leq 1$.

As described above $\tilde{P}(h,n)$ is the probability distribution of $h(x,n)$ for x chosen randomly with respect to a uniform distribution in $[0,1]$. We now ask what the probability distribution of $h(x,n)$ is for x chosen randomly with respect to the natural transient measure for our expanding map, $x_{n+1} = M(x_n)$ and (2.64). To answer this question we proceed as before and consider an initial vertical line segment $|y| \leq K$ starting at x (with $K > (\eta/2\pi)(\lambda_{\min} - 1)^{-1}$, $\lambda_{\min} = \min_x \lambda(x) > 1$). After n iterations, this line segment lengthens by the factor $\exp[n\tilde{h}(x,n)]$. Thus, the fraction of the line still remaining in the strip $|y| < K$ is $\exp[-n\tilde{h}(x,n)]$. Hence, the fraction of points sprinkled uniformly in the strip that still remains after n iterates is

$$\mathrm{e}^{-n/\tau} \sim \int \mathrm{e}^{-nG(h)-nh} dh, \tag{2.78}$$

and the probability distribution of finite time vertical Lyapunov exponents for x chosen randomly with respect to the natural transient measure is [SO00]

$$P(h,n) \sim \frac{\mathrm{e}^{-nG(h)-nh}}{\int \mathrm{e}^{-nG(h)-nh}dh}. \tag{2.79}$$

Evaluating (2.78) for large n we have $\int \mathrm{e}^{-n[G(h)+h]}dh \sim \mathrm{e}^{-n[G(h_*)+h_*]}$, where $\min[G(h)+h] = G(h_*)+h_*$ and h_* is the solution of $dG(h_*)/dh_* = -1$. Thus,

$$1/\tau = G(h_*) + h_*. \tag{2.80}$$

The infinite time vertical Lyapunov exponent for the transient natural measure is

$$h_a = \int hP(h,n)dh. \tag{2.81}$$

Using (2.79) and again letting n be large (2.81) yields $h_a = h_*$. We have

$$h_a \leq 1/\tau,$$

that is, (2.70) is valid for general $M(x)$ and $\lambda(x)$ and not just for $M(x)$ and $\lambda(x)$ given by (2.63) and (2.65).

A 3D Billiard Chaotic Scatterer

We consider a 3 DOF billiard. The billiard is formed by a hard ellipsoid of revolution, placed in a hard, infinitely long tube. The center of the ellipsoid is placed at the center of the tube. Following [SO00], we consider two cases: (a) the major axis of the ellipsoid coincides with the $z-$axis, (b) the major axis of the ellipsoid lies in the $y-z$ plane and makes an angle ξ with the z-axis. The ratio of the minor radius of the ellipsoid ($r_{\|}$) to the width of a side of the tube is 1/4. This leaves the major radius, $r_{\perp}$, and the tilt angle, ξ, as parameters. A point particle injected into the system experiences specular reflection from the ellipsoid and the walls (i.e., the angle of reflection is equal to the angle of incidence, where both are taken with respect to the normal to the surface off of which the particle bounces). When the orbit has passed the top (bottom) of the ellipsoid, with positive (negative) $z-$velocity, we say that it has exited upward (downward). We fix the conserved energy so that $|\mathbf{v}| = 1$.

Pictures of the Stable Manifold

By the symmetry of the geometry of the billiard with the ellipsoid axis along z, the chaotic saddle, Λ, of this system is the collection of initial conditions satisfying $z = v_z = 0$. Started with these initial conditions, an orbit will have $z = v_z = 0$ for all forward and reverse time. The surface normals of the ellipsoid and walls at $z = 0$ lie in the $z = 0$ plane, and thus the particle cannot acquire a non–zero v_z. The $z = 0$ slice through the 3D billiard is a 2D billiard with concave walls. It is known [Ott93] that a typical orbit in this billiard will fill the phase space ergodically. Near $z = 0$, we can picture a typical point on the stable manifold (denoted SM) as having, for example,

z slightly less than zero and v_z slightly greater than zero. The particle will hit the ellipsoid below its equator and, thus, v_z will be decreased with each bounce, yet remain positive. With successive bounces, the orbit on SM slowly approaches $z = v_z = 0$, the chaotic saddle.

To visualize SM, we note that it forms the boundary between initial conditions which escape upward and those which escape downward. We say that points which, when iterated, eventually escape upward (downward) are in the *basin* of upward (downward) escape. Points which are on the boundary between the two basins never escape at all, i.e. they are in SM. We initiate a (2D) grid of orbits (500×500) on the plane, $-3 < x < 3$, $y = 5.1$, $-2.5 < z < 0$, $v_x = 0$, $v_z = 0.1$, and v_y is given by the condition $|\mathbf{v}| = 1$. We iterate each of these initial conditions forward until it escapes. The boundary between the white and black regions is then the intersection of SM lying in the phase space of the 5D billiard (x, y, z, v_x, v_y, v_z constrained by $|\mathbf{v}| = 1$) with the specified 2D x, z–plane. SM appears to take the form of a nowhere–differentiable curve. This is true in various 2D slices, none of which are chosen specially, which suggests that SM has this form in a typical slice. A similar procedure can be followed for the case of the tilted ellipsoid.

Lyapunov Exponents, Decay Times, and Approximate Formulae for the Stable Manifold

Again, we begin with the untilted case. To construct a map from this system we record the cylindrical coordinates (z, ϕ) and their corresponding z and ϕ velocity components, which we denote (v, ω), each time the particle hits the ellipsoid. The coordinate r is constrained, for a given z, by the shape of the ellipsoid surface, and v_r is given by the energy conservation condition $|\mathbf{v}| = 1$. The four components $(z,\ v,\ \phi,\ \omega)_n$ give the state of the system at discrete time n, where n labels the number of bounces from the ellipsoid. Let $\mathbf{z} \equiv \begin{bmatrix} z \\ v \end{bmatrix}$ and $\boldsymbol{\phi} \equiv \begin{bmatrix} \phi \\ \omega \end{bmatrix}$. We express the map using the following notation [SO00],

$$\begin{aligned} \mathbf{z}_{n+1} &= M_z(\mathbf{z}_n, \boldsymbol{\phi}_n), \\ \boldsymbol{\phi}_{n+1} &= M_\phi(\mathbf{z}_n, \boldsymbol{\phi}_n). \end{aligned} \tag{2.82}$$

In what follows, when we refer to an orbit, saddle, invariant set, stable manifold, etc., we are referring to these quantities for the discrete time map (rather than the original continuous time system).

In the case of the untilted ellipsoid, linearizing about an orbit on Λ, (i.e., $\mathbf{z}_n = 0,\ \boldsymbol{\phi}_n$), we get, for the evolution of differential orbit perturbations $\delta\mathbf{z}$ and $\delta\boldsymbol{\phi}$,

$$\begin{aligned} \delta\mathbf{z}_{n+1} &= DM_z(0, \boldsymbol{\phi}_n)\delta\mathbf{z}_n, \\ \delta\boldsymbol{\phi}_{n+1} &= DM_\phi(0, \boldsymbol{\phi}_n)\delta\boldsymbol{\phi}_n, \end{aligned}$$

where $\boldsymbol{\phi}_{n+1} = M_\phi(0, \boldsymbol{\phi}_n)$ is the map for the 2D billiard, $DM_z(0, \boldsymbol{\phi})$ is the tangent map for differential orbit perturbations in $\mathbf{z}$ evaluated at $\mathbf{z} = 0$, and

$DM_\phi(0, \boldsymbol{\phi})$ is the tangent map for differential perturbations lying in Λ. Let $\pm h_z$ and $\pm h_\phi$ denote the Lyapunov exponents with respect to the *natural transient measure* for perturbations in $\mathbf{z}$ and in $\boldsymbol{\phi}$, respectively (these exponents occur in positive–negative pairs due to the Hamiltonian nature of the problem).

In principle, one could numerically evaluate h_z and h_ϕ by sprinkling a large number, N, of initial conditions in the vicinity of Λ, iterating $n \gg 1$ times, evaluating h_z and h_ϕ over those orbits still near Λ, and averaging h_z and h_ϕ over those orbits. We could also find τ by this procedure; it is the exponential rate of decay of the orbits from the vicinity of Λ. For cases where the escape time τ is not long, this procedure, however, becomes problematic. For finite N the number of retained orbits can be small or zero if n is too large. Thus, we adopt an alternate procedure which we found to be less numerically demanding.

In particular, we define the *uniform measure* as the measure generated by uniformly sprinkling many initial conditions in Λ (the hyperplane $z = v_z = 0$). An average over these orbits of the tangent space stretching exponents would yield uniform measure Lyapunov exponents. We denote the distribution of finite-time Lyapunov exponents with respect to this measure by

$$P(\tilde{h}_\phi, \tilde{h}_z, n) \sim \mathrm{e}^{-nG(\tilde{h}_\phi, \tilde{h}_z)}$$

(as in (2.77)), where the tilde indicates finite time exponents for initial conditions distributed according to the uniform measure.

To compute the decay time and the Lyapunov exponents with respect to the natural transient measure we note that orbits near a point $\boldsymbol{\phi}$ in Λ iterate away from Λ as $\exp[n\tilde{h}_z(\boldsymbol{\phi}, n)]$. Thus, the fraction of a large number of initial conditions sprinkled near Λ which remain near Λ after n iterates is

$$\int P(\tilde{h}_\phi, \tilde{h}_z, n)\mathrm{e}^{-n\tilde{h}_z} d\tilde{h}_z,$$

and h_ϕ (the infinite time Lyapunov exponent with respect to the natural transient measure) is

$$h_\phi = \lim_{n\to\infty} \frac{\int \tilde{h}_\phi P(\tilde{h}_\phi, \tilde{h}_z, n)\mathrm{e}^{-n\tilde{h}_z} dh}{\int P(\tilde{h}_\phi, \tilde{h}_z, n)\mathrm{e}^{-n\tilde{h}_z} dh}.$$

This expression can be approximated numerically by choosing N initial conditions, $\boldsymbol{\phi}_i$, uniformly in Λ and calculating

$$\left\langle n\tilde{h}_\phi \right\rangle_n \equiv \frac{n \sum_{i=1}^N \tilde{h}_\phi(\boldsymbol{\phi}_i, n)\mathrm{e}^{-n\tilde{h}_z(\boldsymbol{\phi}_i, n)}}{N \sum_{i=1}^N \mathrm{e}^{-n\tilde{h}_z(\boldsymbol{\phi}_i, n)}}.$$

We calculate the finite-time Lyapunov exponents $\tilde{h}_\phi(\boldsymbol{\phi}_i, n)$ and $\tilde{h}_z(\boldsymbol{\phi}_i, n)$ using the QR decomposition method [Aba96]. Since we chose the $\boldsymbol{\phi}_i$ uniformly in

Λ these exponents are distributed according to $P(n, \tilde{h}_\phi, \tilde{h}_z)$. N is taken to be large enough that there are at least 100 terms contributing to 90% of each sum. The range of n is from 10 to about 40. We find that $\left\langle n\tilde{h}_\phi \right\rangle_n$ versus n is well-fitted by a straight line and we take its slope as our estimate of h_ϕ.

To find τ, first note that we numerically find that h_ϕ does not vary much from $\bar{h}_\phi$ ($.91 \leq h_\phi/\bar{h}_\phi \leq 1$) as we change the system parameter (the height of the ellipsoid), where the overbar denotes an infinite–time Lyapunov exponent with respect to the uniform measure on Λ. Thus, we make the approximation $P(\tilde{h}_\phi, \tilde{h}_z, n) \approx P(\tilde{h}_z, n)$, $G(\tilde{h}_\phi, \tilde{h}_z) \approx G(\tilde{h}_z)$ and plot $G(\tilde{h}_z)$ versus $\tilde{h}_z$. The value of $1/\tau$ is given by (2.80), and h_z is given by $dG(n, h_z)/d\tilde{h}_z = -1$. (Note that $h_z \leq 1/\tau \leq \bar{h}_z$.) A third order polynomial is fit to the data for $G(\tilde{h}_z)$ and used to find τ and h_z. This calculation is performed at a value of n which allows a significant number of points in $G(\tilde{h}_z)$ versus $\tilde{h}_z$ to be collected near in the range $h_z < \tilde{h}_z < \bar{h}_z$.

For the case of the tilted ellipsoid, we will consider a very small tilt angle, $\xi = 2\pi/100$. With this small tilt angle, the Lyapunov exponents and decay times for the tilted and the untilted cases are approximately the same. Thus, for the tilted case, we will use the same Lyapunov exponent values, $\pm h_z$ and $\pm h_\phi$ and decay time τ, that we numerically calculated for the untilted case. The above *dimension formula* for SM becomes [SO00]

$$\begin{aligned} D_S &= 4 - (h_\phi \tau)^{-1} \quad \text{for } h_\phi \tau \geq 1, \\ D_S &< 3 \qquad\qquad\quad \text{for } h_\phi \tau < 1. \end{aligned} \tag{2.83}$$

If the tilt angle is made to be zero ($\xi = 0$), we find that D_S is not given by (2.83), but by the following formula

$$\begin{aligned} D_S &= 4 - \tfrac{h_z + 1/\tau}{h_\phi} \quad \text{for } h_\phi \geq h_z + 1/\tau, \\ D_S &< 3 \qquad\qquad\quad \text{for } h_\phi < h_z + 1/\tau. \end{aligned} \tag{2.84}$$

Since D_S is given by (2.84) only if the tilt angle ξ is precisely zero, we say that the untilted ellipsoid scattering system is atypical. As conjectured in [HOY96], D_S from (2.83) is greater than or equal to D_S from (2.84). Note that D_S from the first line of (2.83) is larger that D_S from the first line of (2.84) by the factor h_z/h_ϕ. Although the transition of D_S from $\xi = 0$ to $\xi \neq 0$ is strictly discontinuous, there is also a continuous aspect: In numerically calculating the dimension of a measure one typically plots $\ln I(\epsilon)$ versus $\ln(1/\epsilon)$, where

$$I(\epsilon) = \sum \mu_i \ln[1/\mu_i]$$

and μ_i is the measure of the ith cube in an ϵ grid. One then estimates the dimension as the slope of a line fitted to small ϵ values in such a plot. In the case of very small tilt, such a plot is expected to yield a slope given by (2.84) for $\epsilon \gtrsim \epsilon_*$ and subsequently, for $\epsilon \lesssim \epsilon_*$, to yield a slope given by (2.83). Here ϵ_* is a small tilt–dependent cross–over value, where $\epsilon_* \to 0$ as $\xi \to 0$. In such a case, the dimension, which is defined by the $\epsilon \to 0$ limit, is given by (2.83).

Numerical Computations for the Three-Dimensional Billiard Scatterer

To verify that the untilted system is atypical we numerically calculated the box–counting dimension of SM for various values of the parameter $r_\perp$, then introduced a small tilt perturbation in the form of a $-2\pi/100$ radian tilt of the ellipsoid about the $x-$axis and repeated the dimension calculations. The results confirm (2.83) and (2.84).

We compare measured values of the box–counting dimension to the predicted values of the information dimension [SO00]. The box–counting dimension gives an upper bound on the information dimension, but often the values of the two dimensions are very close. In particular, we can compare the result for the tilted ellipsoid system to the 2D map. In the regime $h_\phi \geq 1/\tau$, our system is similar to case (a) studied above. Changing $r_\perp$ while leaving $r_\parallel$ fixed changes τ while h_ϕ changes only slightly. This is similar to varying λ_1 of the 2D map while keeping λ_2 fixed, i.e., varying r.

For $h_\phi\tau < 1$ ($h_\phi < h_z + 1/\tau$) the information dimension of SM is predicted to be less than three for the tilted (untilted) ellipsoid system.

The box–counting dimension of SM was computed using the *uncertainty dimension method* [MGO85, Ott93]. This method gives the box–counting dimension of the basin boundary which, as discussed above, coincides with SM.

The uncertainty dimension method was carried out as follows [SO00]:

1. Choose a point, $\mathbf{x}$, at random in a region of a 2D plane intersecting the basin boundary and determine by iteration in which basin it lies.
2. Determine in which basins the perturbed initial points $\mathbf{x} \pm \boldsymbol{\delta}$ lie ($\boldsymbol{\delta}$ is some small vector).
3. If the three points examined in (1) and (2) do not all lie in the same basin, then $\mathbf{x}$ is called 'uncertain'.
4. Repeat 1 to 3 for many points $\mathbf{x}$ randomly chosen in the 2D plane, and get the fraction of these that are uncertain.
5. The fraction of points which is uncertain for a given $\boldsymbol{\delta}$, denoted f, scales like [Pel85] $f \sim |\boldsymbol{\delta}|^{2-d_0}$, where d_0 is the box–counting dimension of the intersection of SM with the 2D plane. The box–counting dimension of SM in the full 4D state space of the map is $D_0 = 2 + d_0$. (The dimension of a generic intersection of a 2D plane with a set having dimension D_0 in a 4D space is $d_0 = 2 + D_0 - 4$, which gives $D_0 = 2 + d_0$.) Thus, plot $\ln f$ versus $\ln|\boldsymbol{\delta}|$, fit a straight line to the plot, and estimate D_0 as 4 minus the slope of this line.

Structure of the Stable Manifold

Untilted Ellipsoid

Due to the symmetry to the untilted ellipsoid billiard, the chaotic saddle of the untilted ellipsoid system has a special geometry (i.e., it lies in $z = v = 0$), which, as we show, accounts for the dimension being lower than the predicted value for a typical system. Similarly, the symmetry induces a special geometry on the stable manifold (SM). The slice is at a fixed value of ω. The axes are z, v, and ϕ, but one could have chosen an arbitrary line through (ϕ,ω) as the third axes and seen a plot which was qualitatively the same. The stable manifold is organized into rays emanating from the $\phi-$axis with oscillations along the $\phi-$direction. The magnitude of the oscillations decreases to zero as the $\phi-$axis is approached.

To understand the structure of SM in more detail, assume that $|\mathbf{z}|$ is small. Then, since $|\mathbf{z}| = 0$ is invariant, we can approximate the dynamics by expanding to first order in $\mathbf{z}$ [SO00],

$$\mathbf{z}_{n+1} \cong DM_z(0, \boldsymbol{\phi}_n)\mathbf{z}_n, \tag{2.85}$$

$$\boldsymbol{\phi}_{n+1} \cong M_\phi(0, \boldsymbol{\phi}_n). \tag{2.86}$$

Say $(\mathbf{z}_{SM}, \boldsymbol{\phi}_{SM})$ is a point on SM. As this point is iterated we have that $|\mathbf{z}| \to 0$ with increasing n. However, since (2.85) is linear in $\mathbf{z}_n$, for any constant α, and the initial condition, $(\alpha\mathbf{z}_{SM}, \boldsymbol{\phi}_{SM})$, the subsequent orbit must also have $|\mathbf{z}| \to 0$. Consequently, if $(\mathbf{z}_{SM}, \boldsymbol{\phi}_{SM})$ lies in SM, so does $(\alpha\mathbf{z}_{SM}, \boldsymbol{\phi}_{SM})$. Thus, for the system (2.85), (2.86), the stable manifold at any point $\boldsymbol{\phi}$ lies on a straight line through the origin of the 2D $\mathbf{z}-$space. Put another way, in the approximation (2.85), (2.86), the stable manifold can be specified by an equation giving the *angle* of $\mathbf{z}$ as a function of $\boldsymbol{\phi}$. Thus, decomposing into polar coordinates (ρ, χ), the stable manifold for $|\mathbf{z}| \to 0$ approaches the form

$$\chi = \chi(\boldsymbol{\phi}). \tag{2.87}$$

For $|\mathbf{z}|$ finite the linearity of (2.85) is not exact, and we expect that the behavior of the stable manifold at constant $\boldsymbol{\phi}$ is not a straight line through the origin of the $\mathbf{z}$ plane. Rather, as $|\mathbf{z}|$ becomes larger we expect (and numerically observe) the straight line for small $|\mathbf{z}|$ to appear as a smooth curve through $\mathbf{z} = 0$.

Since for fixed $\boldsymbol{\phi}$ SM varies smoothly with increasing $\rho = |\mathbf{z}|$, the dimension of SM is not affected by the approximation (2.85) and (2.86). That is, to find the dimension of SM, we can attempt to find it in the region of small $|\mathbf{z}|$ where (2.85) and (2.86) are valid, and that determination will apply to the whole of SM.

The task of analytically determining D_S is too hard for us to accomplish in a rigorous way for the system, (2.85), (2.86), applying to our billiard. Thus, to make progress, we adopt a model system with the same structure as (2.85), (2.86). In particular, we wish to replace the 2D billiard map (2.86) by a simpler map, $M_\phi \to \bar{M}_\phi$, that, like the original 2D billiard map, is chaotic and describes a Hamiltonian system. For this purpose we choose the cat map [SO00],

$$\boldsymbol{\phi}_{n+1} = \bar{M}_\phi(\boldsymbol{\phi}_n) \equiv C\boldsymbol{\phi}_n \,\mathrm{modulo}\, 1, \tag{2.88}$$

where C is the *cat map matrix*,

$$C = \begin{bmatrix} 2 & 1 \\ 1 & 1 \end{bmatrix}.$$

Similarly, we replace $DM_z(0, \phi)$ in (2.85) by a simple symplectic map depending on ϕ,

$$\mathbf{z}_{n+1} = \bar{M}_z(\mathbf{z}_n) = \begin{bmatrix} \lambda & f(\phi_n) \\ 0 & \lambda^{-1} \end{bmatrix} \mathbf{z}_n, \qquad (2.89)$$

where $\lambda > 1$, and $f(\phi)$ is a smooth periodic function with period one in ϕ and π. Note that vertical (i.e., parallel to z) line segments are uniformly expanded by the factor λ, and thus, by the same argument as above, we have $1/\tau = \ln \lambda$ and $h_z = 1/\tau$.

Since only the angle of $\mathbf{z}$ is needed to specify the stable manifold, we introduce the variable $\nu = z/v = \tan \chi$. We can then derive a map for ν: From (2.89) we have

$$\nu_{n+1} v_{n+1} = \lambda \nu_n v_n + v_n f(\phi_n) \qquad \text{and} \qquad v_{n+1} = \lambda^{-1} v_n.$$

Dividing the first equation by the second, v_{n+1} and v_n cancel and we get [SO00]

$$\nu_{n+1} = \lambda^2 \nu_n + \lambda f(\phi). \qquad (2.90)$$

We now consider the dynamical system consisting of (2.88) and (2.90). Note that the system (2.88), (2.90) is a 3D map, unlike the system (2.88), (2.89), which is a four D map.

For (2.88), (2.90), the stable manifold is given by

$$\nu = \nu_S(\phi) = -\lambda^{-1} \sum_{i=0}^{\infty} \lambda^{-2i} f(C^i \phi_0), \qquad (2.91)$$

where C is the cat map matrix.

To verify that this is SM we note that points above SM, $\nu > \nu_S(\phi)$ (below SM, $\nu < \nu_S(\phi)$) are repelled toward $\nu \to \infty$ ($\nu \to -\infty$). Thus, on backwards iteration, points go toward SM. We take advantage of this behavior to determine SM. Imagine that we iterate ϕ forward n iterates to $\phi_n = C^n \phi$ modulo 1, then choose a value of ν_n, and iterate it backwards using (2.90). We find that the initial value of ν at time zero giving the chosen value ν_n at time n is

$$\nu_0 = (\nu_n / \lambda^{2n}) - \lambda^{-1} \sum_{i=0}^{n-1} \lambda^{-2i} f(C^i \phi).$$

Keeping ν_n fixed and letting $n \to +\infty$, the value of ν_0 approaches $\nu_S(\phi)$, given by (2.91).

Results proven in [KMY84] show that the box–counting dimension (the capacity) of the graph of the function $\nu = \nu_S(\phi)$ given by (2.91) is

$$\hat{D}_S = \begin{cases} 3 - 2\frac{\ln \lambda}{\ln B}, & \text{for } \lambda \leq B, \\ 2, & \text{for } \lambda > B, \end{cases} \tag{2.92}$$

where $B = \frac{3+\sqrt{5}}{2} > 1$ is the larger eigenvalue of the matrix C.The formula for $\lambda < B$ holds for almost all (with respect to Lebesgue measure) values of λ. Since $\lambda > 1$, the sum in (2.91) converges absolutely, implying that $\nu_S(\phi)$ is a continuous function of ϕ. Thus, when the first result in (2.92) applies (i.e., the surface is fractal with $\hat{D}_S > 2$), the stable manifold is a continuous non–differentiable surface.

For example, evaluating (2.91) on the surface $\phi = s\hat{u}_+$ where $\hat{u}_+$ is the unit vector in the eigen–direction of C corresponding to the expanding eigenvalue B, we have that ν versus s is of the form [SO00]

$$\nu = -\sum_{i=0}^{\infty} \lambda^{-2i} g(B^i s).$$

Thus, the graph of ν versus s for $B > \lambda$ has the form of Weierstrass' famous example of a continuous, nowhere–differentiable curve.

To get (2.92) in another way, we again consider the map (2.88), (2.90). We claim that (2.88), (2.90) can be regarded as a *typical* system, and that the above dimension formulae should apply to it. That is, in contrast to the existence of a symmetry for (2.89) [namely, $\mathbf{z} \to -\mathbf{z}$ leaves (2.89) invariant], (2.90) has no special symmetry. The Lyapunov exponent corresponding to (2.90) is $h_\nu = 2\ln\lambda$. Note that for the system (2.88), (2.90) [and also for the system (2.88), (2.89)] there are *no* fluctuations in the finite time Lyapunov exponents and thus the decay time for the system (2.88), (2.90) is given by $\tau_\nu^{-1} = h_\nu$. Noting that the Lyapunov exponents for the cat map are $\pm\ln B$ and applying (2.53) to the three D map (2.88), (2.90), we immediately recover (2.92). As discussed in Appendix B, this point of view can also be exploited for the original ellipsoid system [rather than just for the model system (2.88) and (2.89)].

Returning now to the full 4D system, (2.88), (2.89), and noting that SM is smooth along the direction that we eliminated when we went from (2.88), (2.89) to (2.88), (2.90), we have that the dimension D_S of the stable manifold of the invariant set ($\mathbf{z} = 0$) for (2.88), (2.89) is $D_S = \hat{D}_S + 1$. The Lyapunov exponents for $\mathbf{z}$ motion in the four-coordinate system are $\pm h_z = \pm\ln\lambda$ and $\pm h_\phi = \pm\ln B$ for ϕ motion. In terms of the Lyapunov exponents, D_S is

$$D_S = \begin{cases} 4 - 2\frac{h_z}{h_\phi}, & \text{for } h_z/h_\phi \leq 1/2, \\ 3, & \text{for } h_z/h_\phi > 1/2. \end{cases} \tag{2.93}$$

Since $h_z = 1/\tau$ for (2.88) and (2.89) we see that, for $h_z/h_\phi \leq 1/2$, (2.93) is the same as (2.84). Also, the system (2.88), (2.89) has no finite time Lyapunov exponent fluctuations, and, thus, the information dimension and the box–counting dimensions are the same. Hence, $D_S = 3$ when SM is smooth ($h_z/h_\phi > 1/2$). This is analogous to the situation $r = 1$ and (2.74) above.

Basin Boundary for a Map Modelling the Tilted Ellipsoid Billiard

We now wish to investigate the structure of the stable manifold when we give the ellipsoid a small tilt. Again, following [SO00], we adopt the above approach: we get a rigorous result by utilizing a simpler map model that preserves the basic features of the tilted ellipsoid case. Here we again use (2.88) but we now modify (2.89) to incorporate the main effect of a small tilt. The effect of this modification is to destroy the invariance of $\mathbf{z} = 0$. Thinking of the first non-zero term in a power series expansion for small $|\mathbf{z}|$, this invariance results because the first expansion term is linear in $\mathbf{z}$; i.e., the $\mathbf{z}$–independent term in the expansion is exactly zero. When there is tilt this is not so. Thus we replace (2.89) by

$$\begin{bmatrix} z_{n+1} \\ v_{n+1} \end{bmatrix} = \begin{bmatrix} \lambda & f(\phi_n) \\ 0 & \lambda^{-1} \end{bmatrix} \begin{bmatrix} z_n \\ v_n \end{bmatrix} + \begin{bmatrix} f_z(\phi_n) \\ f_v(\phi_n) \end{bmatrix}. \tag{2.94}$$

The simplest version of (2.94) which still has the essential breaking of $\mathbf{z} \to -\mathbf{z}$ symmetry is the case where $f = f_v = 0$. Because setting $f = f_v = 0$ greatly simplifies the analysis, we consider this case in what follows (we do not expect our conclusion to change if $f, f_v \neq 0$). Thus, we have [SO00]

$$\begin{bmatrix} z_{n+1} \\ v_{n+1} \end{bmatrix} = \begin{bmatrix} \lambda & 0 \\ 0 & \lambda^{-1} \end{bmatrix} \begin{bmatrix} z_n \\ v_n \end{bmatrix} + \begin{bmatrix} f_z(\phi_n) \\ 0 \end{bmatrix}. \tag{2.95}$$

The problem of finding the stable manifold for the invariant set of the map, (2.88) and (2.95), is now the same as for the previously considered case of (2.88) and (2.90) (compare the equation

$$z_{n+1} = \lambda z_n + f_z(\phi_n)$$

with (2.90)). Thus, making use of this equivalence we can immediately write down the equation for the stable manifold in the 4D state space (z, v, ϕ, ω) as

$$z = -v \sum_{i=0}^{\infty} \lambda^{-i} f_z(C^i \phi), \tag{2.96}$$

which is obtained from (2.91) using the replacements $\nu \to z/v$, $\lambda f \to f_z$, and $\lambda^2 \to \lambda$. The rigorous results of [KMY84] again show that this is a continuous, nowhere–differentiable surface for almost all λ in $1 < \lambda < B$, and, furthermore, when this is so ($\ln \lambda / \ln B = h_z/h_\phi \leq 1$) we have

$$D_S = 4 - \frac{h_z}{h_\phi}. \tag{2.97}$$

Also, $D_S = 3$ when $h_z/h_\phi > 1$. Since $h_z = 1/\tau$, this is the same as (2.83).

Derivation of Dimension Formulae

D_S for Typical Systems

Let R be the portion of the state space which contains all points within ϵ of a nonattracting, ergodic, invariant set, Λ of an MD map, P. Let sm be the portion of the stable manifold of Λ which is contained in R. The points in R are within ϵ of sm since Λ is a subset of sm. If we sprinkle a large number, N_0, of orbit initial conditions in R, then the number of orbits left in R after $n \gg 1$ iterates is assumed to scale like (2.53) $N_n/N_0 \sim \mathrm{e}^{-n/\tau}$. Let the map, P, have Lyapunov exponents [SO00]

$$h_U^+ \geq h_{U-1}^+ \geq \cdots \geq h_1^+ > 0 > -h_1^- \geq \cdots \geq -h_{S-1}^- \geq -h_S^-,$$

where $U + S = M$.

We will use the box–counting definition of dimension

$$N(\epsilon) \sim \epsilon^{-D}, \tag{2.98}$$

where $N(\epsilon)$ is the minimum number of MD hypercubes ('boxes') needed to cover sm and D is its box–counting dimension.

We wish to develop a covering of sm using small boxes and determine how the number of boxes in that covering scales as the size of the boxes is decreased. We will look at how the linearized system dynamics distorts a typical small box. This will help us determine how the Lyapunov exponents are related to the dimension of sm. Since we assume a smooth map, the dimension of the stable manifold of the map is equal to that of sm.

Cover sm with boxes which are of length ϵ on each of their M sides. Call this set of boxes C. The number of boxes in C is denoted N_0. Iterate each box forward n steps, where n is large, but not so large that the linearized dynamics do not apply to the boxes. Call the set of iterated boxes $P^n(C)$. A typical iterated box in $P^n(C)$ is a distorted MD parallelopiped and has dimensions

$$\epsilon \mathrm{e}^{nh_U^+} \times \cdots \times \epsilon \mathrm{e}^{nh_1^+} \times \epsilon \mathrm{e}^{-nh_1^-} \times \cdots \times \epsilon \mathrm{e}^{-nh_S^-}.$$

Each parallelopiped intersects sm since its preimage (a box) did.

To construct a refined covering of sm we begin by covering each parallelopiped of $P^n(C)$ with slabs of size

$$\overbrace{\epsilon \times \cdots \times \epsilon}^{U \text{ factors}} \times \epsilon \mathrm{e}^{-nh_1^-} \times \cdots \times \epsilon \mathrm{e}^{-nh_S^-}.$$

There are roughly $\exp[n(h_U^+ + \cdots + h_1^+)]$ such slabs. Only $\sim \mathrm{e}^{-n/\tau}$ of these slab are within ϵ of sm. Let C' denote the set of $N' \sim N_0 \exp[n(h_U^+ + \cdots + h_1^+ - 1/\tau)]$ slabs needed to cover the part of sm lying in $P^n(C)$. Let us define

$$H = h_U^+ + \cdots + h_1^+ - 1/\tau,$$

so that $N' \sim N_0 \mathrm{e}^{nH}$.

Iterate each of the N' slabs backward n steps. We now have the set $P^{-n}(C')$. It contains N' parallelopipeds of size

$$\epsilon e^{-nh_U^+} \times \cdots \times \epsilon e^{-nh_1^+} \times \overbrace{\epsilon \times \cdots \times \epsilon}^{\text{S factors}}.$$

The set $P^{-n}(C')$ forms a covering of sm. To calculate the dimension (which is defined in terms of boxes) we cover the N' parallelopipeds in $P^{-n}(C')$ with boxes which are $\epsilon_j = \epsilon e^{-nh_{j+1}^+}$ on each side. (We will choose the value of the index j below.) The number of boxes needed to cover sm, for boxes of size ϵ_j, scales as

$$N'' \sim N' \frac{\epsilon e^{-nh_j^+}}{\epsilon_j} \times \frac{\epsilon e^{-nh_{j-1}^+}}{\epsilon_j} \times \cdots \times \frac{\epsilon e^{-nh_1^+}}{\epsilon_j} \times \left(\frac{1}{\epsilon_j}\right)^S.$$

To cover the slabs with these boxes we need a factor of 1 boxes along each direction of a slab which is shorter than the edge length of a box and a factor of $\epsilon e^{-nh_k^+}/\epsilon_j$ boxes along each direction (here, the kth direction) which is longer than the edge length of a box. In terms of N_0,

$$N'' \sim N_0 \exp\left\{n[(S+j)h_{j+1} + H - h_1^+ - h_2^+ \cdots - h_j^+ - 1/\tau]\right\}.$$

To compute the dimension of sm, we compare $N(\epsilon) \equiv N_0$ to $N(\epsilon_j) \equiv N''$ using (2.98). This gives $N(\epsilon_j)/N(\epsilon) \sim (\epsilon/\epsilon_j)^D \sim \exp(nDh_{j+1}^+)$ which yields the following j-dependent dimension estimate [SO00],

$$D(j) = S + j + \frac{H - h_1^+ + h_2^+ + \cdots h_j^+}{h_{j+1}^+}. \qquad (2.99)$$

Our definition of box–counting dimension, (2.98), requires us to find the minimum number of boxes needed to cover the set. Since we are certain that the set is covered, (2.99) yields an upper bound for the dimension for any j. Thus, to find the best estimate of those given by (2.99), we minimize $D(j)$ over the index j. Comparing $D(j)$ to $D(j+1)$ yields the condition

$$h_1^+ + \cdots + h_J^+ + h_J^+ \geq H \geq h_1^+ + \cdots + h_J^+,$$

where J is the best choice for j (i.e., the choice giving the minimum upper bound). The conjecture is that this minimum upper bound $D(J)$ from (2.99) is the actual dimension of sm for typical systems. $D(J)$ is the same as the result (2.58) presented above for D_S. The derivation of the dimension formula for the unstable manifold, D_U, is similar to that just presented for the stable manifold.

[The derivation just presented gives the *information dimension* of the stable manifold, not the box–counting dimension. We were considering the sizes of typical boxes in the system and covered only those. This leaves boxes that have atypical stretching rates for large n (a box containing a periodic point, for example, will, in general, not stretch at the rates given by the Lyapunov exponents) unaccounted for. What we have actually computed is the box–counting dimension of most of the measure, which is the information dimension [Ott93], of the stable manifold. See [Ott93] for discussion of this point.]

As derived, $D(J)$ gives an upper bound on the dimension. We saw in the ellipsoid example above that the $z \to -z$ symmetry of the system lead to a special geometry for its stable manifold and the formula just derived did not apply (although it was an upper bound). It is the conjecture of [HOY96] that this formula gives not the upper bound, but the exact dimension of the stable manifold for typical systems. This is supported by the results for the tilted ellipsoid example.

D_S for the Untilted (Atypical) Case

Recall that our untilted ellipsoid map (which we call P here) has Lyapunov exponents $\pm h_z$ and $\pm h_\phi$ and decay time τ. Following [SO00], we consider the case

$$h_\phi > h_z + \tau > 0$$

Let us cover the region of state space which is within $\epsilon/2$ of Λ with N_0 boxes with edge lengths $\epsilon \times \epsilon \times \epsilon \times \epsilon$. Call this set C. The map, P, is linear in $\mathbf{z}$ (in particular, the $\mathbf{z}$ portion is of the form $\mathbf{z}_{n+1} = DM(\phi)\mathbf{z}_n$) so that the graph of SM in $\mathbf{z}$ space for a given value of ϕ is a straight line through $\mathbf{z} = 0$, i.e. $z = vg(\phi)$. We denote by sm the portion of the stable manifold which is contained in C. Since SM contains Λ, each box in C intersects sm.

Iterate these boxes forward $n \gg 1$ steps. They become a set distorted of parallelepipeds which we call $P^n(C)$. Each parallelepiped has dimensions

$$\epsilon e^{nh_\phi} \times \epsilon e^{nh_z} \times \epsilon e^{-nhz} \times \epsilon e^{-nh_\phi}$$

and intersects sm. We cover $P^n(C)$ by $N_0 \exp[n(h_z + h_\phi)]$ parallelepipeds with dimensions

$$\epsilon \times \epsilon \times \epsilon e^{-nh_z} \times \epsilon e^{-nh_\phi},$$

and discard all of these parallelepipeds which do not intersect sm. There are $N' \sim N_0 \exp[n(h_z + h_\phi - 1/\tau)]$ parallelepipeds remaining which cover sm. The portions of these parallelepipeds which are in $\epsilon > |z| > \epsilon e^{-nh_z}$ do not contain sm. Suppose some portion of sm did fall in $\epsilon > |z| > \epsilon e^{-nh_z}$. Since P is linear in $\mathbf{z}$, this outlying portion of sm would, upon n reverse iterations, map to the region $\epsilon e^{nh_z} > |z| > \epsilon$, which contradicts the definition of sm given above (i.e. sm is within $\epsilon/2$ of Λ). Therefore we can discard the portion of each parallelepiped which lies is $\epsilon > |z| > \epsilon e^{-nh_z}$. These N' parallelepipeds now have dimensions

$$\epsilon \times \epsilon e^{-nh_z} \times \epsilon e^{-nhz} \times \epsilon e^{-nh_\phi},$$

and are denoted C'. We now iterate the parallelepipeds in C' backward n times and call the resulting set $P^{-n}(C')$. This set contains N' parallelepipeds with dimensions

$$\epsilon e^{-nh_\phi} \times \epsilon e^{-2nh_z} \times \epsilon \times \epsilon.$$

We cover $P^{-n}(C')$ (which covers sm) with hypercubes which have as their edge length ϵe^{-nh_ϕ}. The number of hypercubes needed is $N'' \sim N_0 \exp[n(h_z + h_\phi - 1/\tau - 2h_z + 3h_\phi)]$. The (information) dimension is again found by comparing the number of ϵ–sized hypercubes needed to cover sm to the number of ϵe^{-nh_ϕ} sized hypercubes needed to cover sm,

$$\frac{N''}{N_0} \sim \left(\frac{\epsilon e^{-nh_z}}{\epsilon}\right)^{-D}, \qquad \text{which yields} \qquad D = 4 - \frac{h_z + 1/\tau}{h_\phi}.$$

This expression is between three and four for $(h_z + 1/\tau)/h_\phi < 1$. For more technical details, see [SO00].

2.2.8 Fractal Basin Boundaries and Saddle–Node Bifurcations

Recall from the subsection Attractors that it is common for dynamical systems to have two or more coexisting attractors. In predicting the long–term behavior of a such a system, it is important to determine sets of initial conditions of orbits that approach each attractor (i.e., the basins of attraction). The boundaries of such sets are often fractal (see [MGO85], as well as [Ott93] and references therein). The fine–scale fractal structure of such a boundary implies increased sensitivity to errors in the initial conditions: Even a considerable decrease in the uncertainty of initial conditions may yield only a relatively small decrease in the probability of making an error in determining in which basin such an initial condition belongs [MGO85, Ott93]. For discussion of fractal basin boundaries in experiments, see [Vir00].

Thompson and Soliman [TS91] showed that another source of uncertainty induced by fractal basin boundaries may arise in situations in which there is slow (adiabatic) variation of the system. For example, consider a fixed point attractor of a map (a node). As a system parameter varies slowly, an orbit initially placed on the node attractor moves with time, closely following the location of the solution for the fixed point in the absence of the temporal parameter variation. As the parameter varies, the node attractor may suffer a saddle–node bifurcation. For definiteness, say that the node attractor exists for values of the parameter μ in the range $\mu < \mu_*$, and that the saddle–node bifurcation of the node occurs at $\mu = \mu_*$. Now assume that, for a parameter interval $[\mu_L, \mu_R]$ with $\mu_L < \mu_* < \mu_R$, in addition to the node, there are also two other attractors A and B, and that the boundary of the basin of attractor A, attractor B and the node is a fractal basin boundary. We are interested in the typical case where, before the bifurcation, the saddle lies on the fractal basin boundary, and thus, at the bifurcation, the merged saddle–node orbit is on the basin boundary. In such a case an arbitrarily small ball about the saddle–node at $\mu = \mu_*$ contains pieces of the basins of both A and B. Thus, as μ slowly increases through μ_*, it is unclear whether the orbit following the node will go to A or to B after the node attractor is destroyed by the bifurcation. In practice, noise or round–off error may lead the orbit to go to one attractor or the other, and the result can often depend very sensitively on the specific value of the slow rate at which the system parameter varies.

We note that the study of orbits swept through an indeterminate saddle–node bifurcation belongs to the theory of dynamical bifurcations. Many authors have analyzed orbits swept through other bifurcations, like the period doubling bifurcation [Bae91], the pitchfork bifurcation [BG02, LP95], and the transcritical bifurcation [LP95]. In all these studies of the bifurcations listed above, the local structure before *and* after the bifurcation includes stable invariant manifolds varying smoothly with the bifurcation parameter (i.e., a stable fixed point that exists before or after the bifurcation, and whose location varies smoothly with the bifurcation parameter). This particular feature of the local bifurcation structure, not shared by the saddle–node bifurcation, allows for well–posed, locally defined, problems of dynamical bifurcations. The static saddle–node bifurcation has received much attention in theory and experiments [Kuz95, PM80, CKP02], but so far, no dynamical bifurcation problems have been defined for the saddle–node bifurcation. In this work, we demonstrate that, in certain common situations, global structure (i.e., an invariant Cantor set or a fractal basin boundary) adds to the local properties of the saddle–node bifurcation and allows for well–posed problems of dynamical bifurcations.

Situations where a saddle–node bifurcation occurs on a fractal basin boundary have been studied in 2D Poincaré maps of damped forced oscillators [TS91, NOY95, BNO03]. Several examples of such systems are known [TS91, BNO03], and it seems that this is a common occurence in dynamical systems. In this subsection, following [BNO03], we first focus on saddle–node bifurcations that occur for one parameter families of smooth 1D maps having multiple critical points (a critical point is a point at which the derivative of the map vanishes). Since 1D dynamics is simpler than 2D dynamics, indeterminate bifurcations can be more simply studied, without the distraction of extra mathematical structure. Taking advantage of this, we are able to efficiently investigate several scaling properties of these bifurcations. For 1D maps, a situation dynamically similar to that in which there is indeterminacy in which attractor captures the orbit can also occur in cases where there are two rather than three (or more) attractors. In particular, we can have the situation where one attractor persists for all values of the parameters we consider, and the other attractor is a node which is destroyed via a saddle–node bifurcation on the basin boundary separating the basins of the two attractors. In such a situation, an orbit starting on the node, and swept through the saddle–node bifurcation, will go to the remaining attractor. It is possible to distinguish different ways that the orbit initially on the node approaches the remaining attractor. We find that the way in which this attractor is approached can be indeterminate.

Indeterminacy in Which Attractor is Approached

We consider the general situation of a 1D real map $f_\mu(x)$ depending on a parameter μ. We assume the following [BNO03]:

(i) the map is twice differentiable with respect to x, and once differentiable with respect to μ (the derivatives are continuous);

(ii) f_μ has at least two attractors sharing a fractal basin boundary for parameter values in the vicinity of μ_*; and

(iii) an attracting fixed point x_* of the map $f_\mu(x)$ is destroyed by a saddle–node bifurcation as the parameter μ increases through a critical value μ_*, and this saddle–node bifurcation occurs on the common boundary of the basins of the two attractors.

We first recall the *saddle–node bifurcation theorem* (see, e.g., [Kuz95]). If the map $f_\mu(x)$ satisfies:

(a) $f_{\mu_*}(x_*) = x_*$,

(b) $\partial_x f_{\mu_*}(x_*) = 1$,

(c) $\partial_x^2 f_{\mu_*}(x_*) > 0$, and

(d) $\partial_\mu f(x_*; \mu_*) > 0$,

– then the map f_μ undergoes a backward saddle–node bifurcation (i.e., the node attractor is destroyed at x_* as μ increases through μ_*). If the inequality in either (c) or (d) is reversed, then the map undergoes a forward saddle–node bifurcation, while, if both these inequalities are reversed, the bifurcation remains backward. A saddle–node bifurcation in a 1D map is also called a tangent or a fold bifurcation.

Example: a 1D Map

As an illustration of an indeterminate saddle–node bifurcation in a 1D map, we construct an example in the following way. We consider the logistic map for a parameter value where there is a stable period three orbit. We denote this map $g(x)$ and its third iterate $g^{[3]}(x)$. The map $g^{[3]}(x)$ has three stable fixed points. We perturb the map $g^{[3]}(x)$ by adding a function (which depends on a parameter μ) that will cause a saddle–node bifurcation of one of the attracting fixed points but not of the other two. We investigate [BNO03]

$$f_\mu(x) = g^{[3]}(x) + \mu \sin(3\pi x), \quad \text{where} \quad g(x) = 3.832\, x(1-x).$$

Numerical calculations show that the function $f_\mu(x)$ satisfies all the conditions of the saddle–node bifurcation theorem for having a backward saddle–node bifurcation at $x_* \approx 0.15970$ and $\mu_* \approx 0.00279$.

For $\mu < \mu_*$, each of these colored sets has infinitely many disjoint intervals and a fractal boundary. As μ increases, the leftmost stable fixed point B_μ

is destroyed via a saddle–node bifurcation on the fractal basin boundary. In fact, in this case, for $\mu < \mu_*$, every boundary point of one basin is a boundary point for all three basins.[21] The basins are so–called *Wada basins* [KY91]. This phenomenon of a saddle–node bifurcation on the fractal boundary of Wada basins also occurs for the damped forced oscillators studied in [NOY95, BNO03]. Alternatively, if we look at the saddle–node bifurcation as μ decreases through the value μ_*, then the basin $B[\mu]$ of the newly created stable fixed point immediately has infinitely many disjoint intervals and its boundary displays fractal structure. According to the terminology of [RAO00], we may consider this bifurcation an example of an 'explosion'.

Dimension of the Fractal Basin Boundary

We can compute dimension D of the fractal basin boundary versus the parameter μ. For $\mu < \mu_*$, we observe that D appears to be a continuous function of μ. Park et al. [PGO89] argue that the fractal dimension of the basin boundary near μ_*, for $\mu < \mu_*$, scales as

$$D(\mu) \approx D_* - k(\mu_* - \mu)^{1/2},$$

with D_* the dimension at $\mu = \mu_*$ (D_* is less than the dimension of the phase space), and k a positive constant.

The existence of a fractal basin boundary has important practical consequences. In particular, for the purpose of determining which attractor eventually captures a given orbit, the arbitrarily fine-scaled structure of fractal basin boundaries implies considerable sensitivity to small errors in initial conditions. If we assume that initial points cannot be located more precisely than some $\epsilon > 0$, then we cannot determine which basin a point is in, if it is within ϵ of the basin boundary. Such points are called $\epsilon-$uncertain. The Lebesgue measure of the set of $\epsilon-$uncertain points (in a bounded region of interest) scales like $\epsilon^{D_0 - D}$, where D_0 is the dimension of the phase space ($D_0 = 1$ for 1D maps) and D is the *box–counting dimension* of the basin boundary [MGO85]. For the case of a fractal basin boundary $(D_0 - D) < 1$. When $D_0 - D$ is small, a large decrease in ϵ results in a relatively small decrease in $\epsilon^{D_0 - D}$. This is discussed in [MGO85] which defines the uncertainty dimension, D_u, as follows. Say we randomly pick an initial condition x with uniform probability density in a state–space region S. Then we randomly pick another initial condition y in S, such that $|y - x| < \epsilon$. Let $p(\epsilon, S)$ be the probability that x and y are in different basins. (We can think of $p(\epsilon, S)$ as the probability that an error will be made in determining the basin of an initial condition if the initial condition has uncertainty of size ϵ.) The uncertainty dimension of the basin boundary D_u is defined as the limit [MGO85]

[21] That is, an arbitrarily small $x-$interval centered about any point on the boundary of any one of the basins contains pieces of the other two basins.

$$\lim_{\epsilon\to 0} \ln p(\epsilon, S)/\ln(\epsilon).$$

Thus, the probability of error scales as $p(\epsilon, S) \sim \epsilon^{D_0 - D_u}$, where for fractal basin boundaries $D_0 - D_u < 1$. This indicates enhanced sensitivity to small uncertainty in initial conditions. For example, if $D_0 - D_u = 0.2$, then a decrease of the initial condition uncertainty ϵ by a factor of 10 leads to only a relative small decrease in the final state uncertainty $p(\epsilon, S)$, since p decreases by a factor of about $10^{0.2} \approx 1.6$. Thus, in practical terms, it may be essentially impossible to significantly reduce the final state uncertainty. In [MGO85] it was conjectured that the box–counting dimension equals the uncertainty dimension for basin boundaries in typical dynamical systems. In [NY92] it is proven that the box–counting dimension, the uncertainty dimension and the Hausdorff dimension are all equal for the basin boundaries of one and 2D systems that are uniformly hyperbolic on their basin boundary.

Now, from [PV88] it follows that the box–counting dimension and the Hausdorff dimension coincide for all intervals of μ for which the map f_μ is hyperbolic on the basin boundary, and that the dimension depends continuously on the parameter μ in these intervals. For $\mu > \mu_*$, there are many parameter values for which the map has a saddle–node bifurcation of a periodic orbit on the fractal basin boundary. At such parameter values, which we refer to as saddle–node bifurcation parameter values, the dimension is expected to be discontinuous (as it is at the saddle–node bifurcation of the fixed point, $\mu = \mu_*$). In fact, there exist sequences of saddle–node bifurcation parameter values converging to μ_* [BNO03]. Furthermore, for each parameter value $\mu > \mu_*$ for which the map undergoes a saddle–node bifurcation, there exists a sequence of saddle–node bifurcation parameter values converging to that parameter value. The basins of attraction of the periodic orbits created by saddle–node bifurcations of high period exist only for very small intervals of the parameter μ. We did not encounter them numerically by iterating initial conditions for a discrete set of values of the parameter μ, as we did for the basin of our fixed point attractor.

Scaling of the Fractal Basin Boundary

Just past μ_*, the remaining green and red basins display an alternating stripe structure. The red and green stripes are interlaced in a fractal structure. As we approach the bifurcation point, the interlacing becomes finer and finer scaled, with the scale approaching zero as μ approaches μ_*. Similar fine scaled structure is present in the neighborhood of all preiterates of x_*.

Now, consider the second order expansion of f_μ in the vicinity of x_* and μ_* [BNO03]

$$\hat{f}_{\hat{\mu}}(\hat{x}) = \hat{\mu} + \hat{x} + a\hat{x}^2, \quad \text{where} \begin{cases} \hat{x} = x - x_*, \\ \hat{\mu} = \mu - \mu_*, \end{cases} \tag{2.100}$$

and $a \approx 89.4315$. The trajectories of $\hat{f}_{\hat{\mu}}$ in the neighborhood of $\hat{x} = 0$, for $\hat{\mu}$ close to zero, are good approximations to trajectories of f_μ in the neighborhood of $x = x_*$, for μ close to μ_*. Assume that we start with a certain initial condition for $\hat{f}_{\hat{\mu}}$, $\hat{x}_0 = \hat{x}_s$, and we ask the following question: What are all the positive values of the parameter $\hat{\mu}$ such that a trajectory passes through a fixed position $\hat{x}_f > 0$ at some iterate n? For any given x_f which is not on the fractal basin boundary, there exists a range of μ such that iterates of x_f under f_μ evolve to the same final attractor, for all values of μ in that range. In particular, once $a\hat{x}^2$ appreciably exceeds $\hat{\mu}$, the subsequent evolution is approximately independent of $\hat{\mu}$. Thus, we can choose $\hat{x}_f \gg \sqrt{\hat{\mu}/a}$, but still small enough so that it lies in the region of validity of the canonical form (2.100). There exists a range of such $\hat{x}_f$ values satisfying these requirements provided that $|\hat{\mu}|$ is small enough.

Since consecutive iterates of $\hat{f}_{\hat{\mu}}$ in the neighborhood of $\hat{x} = 0$ for $\hat{\mu}$ close to zero differ only slightly, we approximate the 1D map, [BNO03]

$$\hat{x}_{n+1} = \hat{f}_{\hat{\mu}}(\hat{x}_n) = \hat{\mu} + \hat{x}_n + a\hat{x}_n^2,$$

by the differential equation [PM80],

$$\frac{d\hat{x}}{dn} = \hat{\mu} + a\hat{x}^2, \tag{2.101}$$

where in (2.101) n is considered as a continuous, rather than a discrete, variable. Integrating (2.101) from $\hat{x}_s$ to $\hat{x}_f$ yields

$$n\sqrt{a\hat{\mu}} = \arctan\left(\sqrt{\frac{a}{\hat{\mu}}}\hat{x}_f\right) - \arctan\left(\sqrt{\frac{a}{\hat{\mu}}}\hat{x}_s\right). \tag{2.102}$$

Close to the saddle–node bifurcation (i.e., $0 < \hat{\mu} \ll 1$, and $\hat{x}_{s,f}$ close to zero), $\hat{f}_{\hat{\mu}}$ is a good approximation to f_μ. For $|\hat{x}_{s,f}|\sqrt{(a/\hat{\mu})} \gg 1$ (2.102) becomes

$$n\sqrt{a\hat{\mu}} \approx \pi. \tag{2.103}$$

The values of $\hat{\mu}_n^{-1/2}$ satisfying (2.103) increase with n in step of $\sqrt{a}/\pi$. For our example we have $a \approx 89.4315$, thus $\sqrt{a}/\pi \approx 3.010$.

In order to investigate the structure of the fractal basin boundary in the vicinity of the saddle–node bifurcation (i.e., $\hat{x}_s$ close to $\hat{x}_* = 0$), we consider (2.102) in the case where we demand only $|\hat{x}_f|\sqrt{(a/\hat{\mu})} \gg 1$. Thus, (2.102) becomes

$$n\sqrt{a\hat{\mu}} \approx \frac{\pi}{2} - \arctan\left(\sqrt{\frac{a}{\hat{\mu}}}\hat{x}_s\right). \tag{2.104}$$

Let $\hat{\mu}_n^{-1/2}(\hat{x}_s)$ denote the solution of (2.104) for $\hat{\mu}$. Equation (2.104) implies the behavior of $\hat{\mu}_n^{-1/2}(\hat{x}_s)$ as function of $\hat{x}_s$ and n. For a fixed n, $\hat{\mu}_n^{-1/2}$ has a horizontal asymptote at the value $n\sqrt{a}/\pi$ as $\hat{x}_s \to -\infty$, and a vertical asymptote to infinity at $\hat{x}_s = 1/(an)$. For $\hat{x}_s < 0$, we have an infinite number of values of the parameter $\hat{\mu}$, for which an orbit of $\hat{f}_{\hat{\mu}}$ starting at $\hat{x}_s$ passes through the same position $\hat{x}_f$, after some number of iterations. For $\hat{x}_s = 0$ (i.e., $x_s = x_*$), we also have an infinite number of $\hat{\mu}_n^{-1/2}(0)$, but with constant step $2\sqrt{a}/\pi$ rather than $\sqrt{a}/\pi$. This is hard to verify from numerics, since ∂

$$_{\hat{x}_s}\hat{\mu}_n^{-1/2}(0) = a^{3/2}(2n/\pi)^2$$

increases with n^2, and the stripes become very tilted in the neighborhood of $\hat{x}_s = \hat{x}_* = 0$. For $\hat{x}_s > 0$, $\hat{\mu}_n^{-1/2}$ has only a limited number of values with $n_{\max} < 1/(a\hat{x}_0)$.

Sweeping Through an Indeterminate Saddle–Node Bifurcation

In order to understand the consequences of a saddle–node bifurcation on a fractal basin boundary for systems experiencing slow drift, we imagine the following experiment. We start with the dynamical system f_μ at parameter $\mu_s < \mu_*$, with x_0 on the attractor to be destroyed at $\mu = \mu_*$ by a saddle–node bifurcation (i.e., B_μ). Then, as we iterate, we slowly change μ by a small constant amount $\delta\mu$ per iterate, thus increasing μ from μ_s to $\mu_f > \mu_*$,[BNO03]

$$x_{n+1} = f_{\mu_n}(x_n), \qquad \mu_n = \mu_s + n\,\delta\mu. \tag{2.105}$$

When $\mu \geq \mu_f$ we stop sweeping the parameter μ, and, by iterating further, we determine to which of the remaining attractors of f_{μ_f} the orbit goes. Numerically, we observe that, if $(\mu_f - \mu_*)$ is not too small, then, by the time μ_f is reached, the orbit is close to the attractor of f_{μ_f} to which it goes. (From the subsequent analysis, 'not too small $|\mu_{s,f} - \mu_*|$' translates to choices of $\delta\mu$ that satisfy $(\delta\mu)^{2/3} \ll |\mu_{s,f} - \mu_*|$.) We repeat this for different values of $\delta\mu$ and we graph the final attractor position for the orbit versus $\delta\mu$.

Once $\mu = \mu_f$, the orbit typically lands in the green or the red basin of attraction and goes to the corresponding attractor. Due to sweeping, it is possible for the orbit to switch from being in one basin of attraction of the *time-independent* map f_μ to the other, since the basin boundary between $G[\mu]$ and $R[\mu]$ changes with μ. However, the sweeping of μ is slow (i.e., $\delta\mu$ is small), and, once $(\mu - \mu_*)$ is large enough, the orbit is far enough from the fractal basin boundary, and the fractal basin boundary changes too little to switch the orbit between $G[\mu]$ and $R[\mu]$.

In order to explain this result, we again consider the map $\hat{f}_{\hat{\mu}}$, the local approximation of f_μ in the region of the saddle–node bifurcation. Equations (2.105) can be approximated by [BNO03]

$$\hat{x}_{n+1} = \hat{f}_{\hat{\mu}_n}(\hat{x}_n) = \hat{\mu}_n + \hat{x}_n + a\hat{x}_n^2, \qquad \hat{\mu}_n = \hat{\mu}_s + n\,\delta\mu. \tag{2.106}$$

We perform the following numerical experiment. We consider orbits of our approximate 2D map given by (2.106) starting at $\hat{x}_s = -\sqrt{-\hat{\mu}_s/a}$. We define a final state function of an orbit swept with parameter $\delta\mu$ in the following way. It is 0 if the orbit has at least one iterate in a specified fixed interval far from the saddle–node bifurcation, and is 1, otherwise. In particular, we take the final state of a swept orbit to be 0 if there exists n such that $100 < \hat{x}_n < 250$, and to be 1 otherwise.

We are now in a position to give a theoretical analysis explaining the observed periodicity in $1/\delta\mu$. In particular, we now know that this can be explained using the canonical map (2.106), and that the periodicity result is thus universal (i.e., independent of the details of our particular example). For slow sweeping (i.e., $\delta\mu$ small), consecutive iterates of (2.106) in the vicinity of $\hat{x} = 0$ and $\hat{\mu} = 0$ differ only slightly, and we further approximate the system by the following *Ricatti ODE*, [BNO03]

$$\frac{d\hat{x}}{dn} = \hat{\mu}_s + n\delta\mu + a\hat{x}^2. \tag{2.107}$$

The solution of (2.107) can be expressed in terms of the *Airy functions Ai* and Bi and their derivatives, denoted by Ai' and Bi',

$$\hat{x}(n) = \frac{\eta Ai'(\xi) + Bi'(\xi)}{\eta Ai(\xi) + Bi(\xi)} \left(\frac{\delta\mu}{a^2}\right)^{1/3}, \qquad \text{where} \tag{2.108}$$

$$\xi(n) = -a^{1/3}\,\frac{\hat{\mu}_s + n\,\delta\mu}{\delta\mu^{2/3}},$$

and η is a constant to be determined from the initial condition. We are only interested in the case of slow sweeping,

$$\delta\mu \ll 1, \qquad \text{and} \qquad \hat{x}(0) \equiv \hat{x}_s = -\sqrt{-\hat{\mu}_s/a},$$

which is the stable fixed point of $\hat{f}_{\hat{\mu}}$ destroyed by the saddle–node bifurcation at $\hat{\mu} = 0$. In particular, we will consider the case where $\hat{\mu}_s < 0$ and $|\hat{\mu}_s| \gg \delta\mu^{2/3}$ (i.e., $|\xi(0)| \gg 1$). Using $\hat{x}(0) = -\sqrt{-\hat{\mu}_s/a}$ to solve for η yields

$$\eta \sim \mathcal{O}[\xi(0)e^{2\xi(0)}] \gg 1.$$

For positive large values of $\xi(n)$ (i.e., for n small enough), using the corresponding asymptotic expansions of the Airy functions [AS72], the lowest order in $\delta\mu$ approximation to (2.108) is

$$\hat{x}(n) \approx -\sqrt{-\frac{\hat{\mu}_s + n\,\delta\mu}{a}},$$

with the correction term of higher order in $\delta\mu$ being negative. Thus, for n sufficiently smaller than $-\hat{\mu}_s/\delta\mu$, the swept orbit lags closely behind the fixed point for $\hat{f}_{\hat{\mu}}$ with $\hat{\mu}$ constant. For $\xi \leq 0$, we use the fact that η is large to approximate (2.108) as

$$\hat{x}(n) \approx \frac{Ai'(\xi)}{Ai(\xi)} \left(\frac{\delta\mu}{a^2}\right)^{1/3}. \qquad (2.109)$$

Note that

$$\hat{x}(-\hat{\mu}_s/\delta\mu) \approx \frac{Ai'(0)}{Ai(0)} \left(\frac{\delta\mu}{a^2}\right)^{1/3} = (-0.7290...) \left(\frac{\delta\mu}{a^2}\right)^{1/3}$$

gives the lag of the swept orbit relative to the fixed point attractor evaluated at the saddle–node bifurcation. Equation (2.109) does not apply for $n > n_{\max}$, where $n_{\max}$ is the value of n for which $\xi(n_{\max}) = \tilde{\xi}$, the largest root of $Ai(\tilde{\xi}) = 0$ (i.e., $\tilde{\xi} = -2.3381...$). At $n = n_{\max}$, the normal form approximation predicts that the orbit diverges to $+\infty$. Thus, for n near $n_{\max}$, the normal form approximation of the dynamical system ceases to be valid. Note, however, that (2.109) can be valid even for $\xi(n)$ close to $\xi(n_{\max})$. This is possible because $\delta\mu$ is small. In particular, we can consider times up to the time n' where n' is determined by

$$\xi' \equiv \xi(n') = \tilde{\xi} + \delta\xi, \qquad (\delta\xi > 0 \quad \text{is small}),$$

provided $|\hat{x}(n')| \ll 1$ so that the normal form applies. That is, we require

$$[Ai'(\xi')/Ai(\xi')](\delta\mu/a^2)^{1/3} \ll 1,$$

which can be satisfied even if $[Ai'(\xi')/Ai(\xi')]$ is large. Furthermore, we will take the small quantity $\delta\xi$ to be not too small (i.e., $\delta\xi/(a\,\delta\mu)^{1/3} \gg 1$), so that $(n_{\max} - n') \gg 1$. We then consider (2.109) in the range, $-(\hat{\mu}_s/\delta\mu) \leq n < n'$, where the normal form is still valid.

We use (2.109) for answering the following question: What are all the values of the parameter $\delta\mu$ ($\delta\mu$ small) for which an orbit passes exactly through the same position $\hat{x}_f > 0$, at some iterate n_f? All such orbits would further evolve to the same final attractor, independent of $\delta\mu$, provided $a\hat{x}_f^2 \gg \hat{\mu}_s + n_f\,\delta\mu$; i.e., $\hat{x}_f$ is large enough that

$$\hat{\mu}_f = \hat{\mu}_s + n_f\,\delta\mu$$

does not much influence the orbit after $\hat{x}$ reaches $\hat{x}_f$. Let us denote $\xi(n_f)$ as $\xi(n_f) \equiv \xi_f$. Using (2.109) we can estimate when this occurs,

$$a\hat{x}_f^2 = [Ai'(\xi_f)/Ai(\xi_f)]^2(\delta\mu^2/a)^{1/3} \gg (\hat{\mu}_s + n_f\,\delta\mu), \qquad \text{or}$$

$$[Ai'(\xi_f)/Ai(\xi_f)]^2 \gg \xi_f.$$

This inequality is satisfied when ξ_f gets near $\tilde{\xi}$, which is the largest zero of Ai (i.e., $\xi_f = \tilde{\xi} + \delta\xi$, where $\delta\xi$ is a small positive quantity). We now rewrite (2.109) in the following way [BNO03]

$$\frac{1}{\delta\mu} = -\frac{n_f}{\hat{\mu}_s - \left[\frac{(\delta\mu)^2}{a}\right]^{1/3} K\left[\left(\frac{a^2}{\delta\mu}\right)^{1/3} \hat{x}_f\right]}, \tag{2.110}$$

representing a transcedental equation in $\delta\mu$ where $\hat{\mu}_s$ and $\hat{x}_f$ are fixed, n_f is a large positive integer (i.e., $n_f - 1$ is the integer part of $(\hat{\mu}_f - \hat{\mu}_s)/\delta\mu$), and $K(\zeta)$ is the inverse function of $Ai'(\xi)/Ai(\xi)$ in the neighborhood of

$$\zeta = (a^2/\delta\mu)^{1/3}\,\hat{x}_f \gg 1, \qquad \text{thus} \qquad |K[(a^2/\delta\mu)^{1/3}\,\hat{x}_f]| \lesssim |K(\infty)| = |\tilde{\xi}|.$$

The difference $[1/\delta\mu(x_f, n_f + 1) - 1/\delta\mu(x_f, n_f)]$, where $\delta\mu(x_f, n_f)$ is the solution of (2.110), yields the limit period of the attracting state versus $1/\delta\mu$ graph. We denote this limit period by $\Delta\,(1/\delta\mu)$. For small $\delta\mu$, the term involving $K[(a^2/\delta\mu)^{1/3}\,\hat{x}_f]$ in (2.110) can be neglected, and we get

$$\Delta\,(1/\delta\mu) = -\hat{\mu}_s^{-1} = (-\mu_s + \mu_*)^{-1}.$$

An alternate point of view on this scaling property is as follows. For $\hat{\mu} < 0$ (i.e., $\mu < \mu_*$) and slow sweeping (i.e., $\delta\mu$ small), the orbit closely follows the stable fixed point attractor of $\hat{f}_{\hat{\mu}}$, until $\hat{\mu} \geq 0$, and the saddle–node bifurcation takes place. However, due to the discreteness of n, the first nonnegative value of $\hat{\mu}$ depends on $\hat{\mu}_s$ and $\delta\mu$. Now consider two values of $\delta\mu$, one $\delta\mu_m$ satisfying $\hat{\mu}_s + m\,\delta\mu_m = 0$, and another $\delta\mu_{m+1}$ satisfying

$$\hat{\mu}_s + (m+1)\,\delta\mu_{m+1} = 0.$$

Because $\delta\mu_m$ and $\delta\mu_{m+1}$ are very close (for large m) *and* both lead $\hat{\mu}(n)$ to pass through $\hat{\mu} = \hat{\mu}_* = 0$ (one at time $n = m$, and the other at time $n = m+1$), it is reasonable to assume that their orbits for $\hat{\mu}_s/\delta\mu < n < n'$ are similar (except for a time shift $n \to n+1$); i.e., they go to the same attractor. Thus, the period of $1/\delta\mu$ is approximately

$$\Delta\,(1/\delta\mu) = 1/\delta\mu_{m+1} - 1/\delta\mu_m = -\hat{\mu}_s^{-1}.$$

We now discuss a possible experimental application of our analysis. The conceptually most straightforward method of measuring a fractal basin boundary would be to repeat many experiments each with precisely chosen initial conditions. By determining the final attractor corresponding to each initial condition, basins of attraction could conceivably be mapped out [Vir00]. However, it is commonly the case that accurate control of initial conditions is not feasible for experiments. Thus, the application of this direct method is limited, and, as a consequence, fractal basin boundaries have received little experimental study, in spite of their fundamental importance. If a saddle–node bifurcation occurs on the fractal basin boundary, an experiment can be arranged to take advantage of this. In this case, the purpose of the experiment would be to measure the dimension D' as an estimate of the fractal dimension of the basin boundary D. The measurements would determine the final attractor of orbits starting at the attractor to be destroyed by the saddle–node bifurcation, and swept through the saddle–node bifurcation at different velocities. This does not require precise control of the initial conditions of the orbits. It is sufficient for the initial condition to be in the basin of the attractor to be destroyed by the saddle–node bifurcation; after enough time, the orbit will be as close to the attractor as the noise level allows. Then, the orbit may be swept through the saddle–node bifurcation. The final states of the orbits are attractors; in their final states, orbits are robust to noise and to measurement perturbations. The only parameters which require rigorous control are the sweeping velocity (i.e., $\delta\mu$) and the initial value of the parameter to be swept (i.e., μ_s); precise knowledge of the parameter value where the saddle–node bifurcation takes place (i.e., μ_*) is not needed. It is also required that the noise level be sufficiently low.

Capture Time

A question of interest is how much time it takes for a swept orbit to reach the final attracting state. Namely, we ask how many iterations with $\mu > \mu_*$ are needed for the orbit to reach a neighborhood of the attractor having the green basin. Due to slow sweeping, the location of the attractor changes slightly on every iterate. If x_μ is a fixed point attractor of f_μ (with μ constant), then a small change $\delta\mu$ in the parameter μ, yields a change in the position of the fixed point attractor, [BNO03]

$$(x_{\mu+\delta\mu} - x_\mu) \equiv \delta x = \delta\mu \, \frac{\partial_\mu f(x_\mu; \mu)}{1 - \partial_x f_\mu(x_\mu)}.$$

We consider the swept orbit to have reached its final attractor if consecutive iterates differ by about δx (which is proportional to $\delta\mu$). For numerical purposes, we consider that the orbit has reached its final state if $|x_{n+1} - x_n| < 10\,\delta\mu$. In our numerical experiments, this condition is satisfied by every orbit before μ reaches its final value μ_f. We refer to the number of iterations with $\mu > \mu_*$ needed to reach the final state as the *capture time* of the corresponding orbit. Orbits swept with $\delta\mu$ at the centers of these intervals spend only a small number of iterations close to the common fractal boundary of $R[\mu]$ and $G[\mu]$. Thus, the capture time of such similar orbits does not depend on the structure of the fractal basin boundary. We use (2.109) as an approximate description of these orbits. A swept orbit reaches its final attracting state as $\hat{x}(n)$ becomes large. Then, the orbit is rapidly trapped in the neighborhood of one of the swept attractors of f_μ. Thus, we equate the argument of the Airy function in the denominator to its first root [see (2.109)], solve for n, and substract $-\hat{\mu}_s/\delta\mu$ (the time for $\hat{\mu}$ to reach the bifurcation value). This yields the following approximate formula for the capture time

$$n_C \approx |\tilde{\xi}|(a\,\delta\mu)^{-1/3},$$

where $\tilde{\xi} = -2.3381...$ is the largest root of the Airy function Ai. Thus, we predict that for small $\delta\mu$, a log-log plot of the capture time of the selected orbits versus $\delta\mu$ is a straight line with slope $-1/3$.

Indeterminate Saddle–Node Bifurcation in the Presence of Noise

We now consider the addition of noise. Thus, we change our swept dynamical system to [BNO03]

$$x_{n+1} = f_{\mu_n}(x_n) + A\,\epsilon_n, \qquad (2.111)$$
$$\mu_n = \mu_s + n\,\delta\mu,$$

where ϵ_n is random with uniform probability density in the interval $[-1, 1]$, and A is a parameter which we call the noise amplitude.

Now we take advantage of the asymptotically periodic structure of the noiseless final destination graph versus $1/\delta\mu$. We consider centers of the largest intervals of $1/\delta\mu$ for which an orbit reaches the middle attractor in the absence of noise. We chose five such values of $\delta\mu$, spread over two decades, where the ratio of consecutive values is approximately 3. We notice that all the curves have qualitatively similar shape. For a range from zero to small A, the probability is 1, and as A increases, the probability decreases to a horizontal asymptote. The rightmost curve in the family corresponds to the largest value of $\delta\mu$ ($\delta\mu \approx 3.445974 \times 10^{-5}$), and the leftmost curve corresponds to the smallest value of $\delta\mu$ ($\delta\mu \approx 4.243522 \times 10^{-7}$). All data collapse to a single curve, indicating that the probability that a swept orbit reaches the attractor G_{μ_f} depends only on the reduced variable $A/(\delta\mu)^{5/6}$. Later, we provide a theoretical argument for this scaling.

In order to gain some understanding of this result, we follow the above idea and use the canonical form $\hat{f}_{\hat{\mu}}$ to propose a simplified setup of our problem. We modify (2.106) by the addition of a noise term $A\,\epsilon_n$ in the right hand side of the first equation of (2.106). We are interested in the probability that a swept orbit has at least one iterate, $\hat{x}_n$, in a specified fixed interval far from the vicinity of the saddle–node bifurcation. More precisely, we analyze how this probability changes versus A and $\delta\mu$. Depending on the choice of interval and the choice of $\delta\mu$, the probability versus A graph (not shown) has various shapes. For numerical purposes, we choose our fixed interval to be the same as above, $100 \leq \hat{x} \leq 250$. We then select values of $\delta\mu$ for which a noiseless swept orbit, starting at $\hat{x}_s = -\sqrt{-\hat{\mu}_s/a}$, reaches exactly the center of our fixed interval. The inverse of these values of $\delta\mu$ are centers of intervals where the final state of the swept orbits is 0. We consider five such values of $\delta\mu$, where the ratio of consecutive values is approximately 3.

We now present a theoretical argument for why the probability of reaching an attractor depends on $\delta\mu$ and A only through the scaled variable $A/(\delta\mu)^{5/6}$ when $\delta\mu$ and A are small. We know that the scaling we wish to demonstrate should be obtainable by use of the canonical form $\hat{f}_{\hat{\mu}}$. Accordingly, we again use the differential equation approximation (2.107), but with a noise term added, [BNO03]

$$\frac{d\hat{x}}{dn} = n\,\delta\mu + a\hat{x}^2 + A\hat{\epsilon}(n), \qquad (2.112)$$

where $\hat{\epsilon}(n)$ is white noise,

$$\langle\hat{\epsilon}(n)\rangle = 0, \qquad \langle\hat{\epsilon}(n+n')\hat{\epsilon}(n)\rangle = \delta(n'),$$

and we have redefined the origin of the time variable n so that the parameter $\hat{\mu}$ sweeps through zero at $n = 0$ (i.e., we replaced n by $n - |\hat{\mu}_s|/\delta\mu$). Because we are only concerned with scaling, and not with the exact solution of (2.112), a fairly crude analysis will be sufficient.

First we consider the solution of (2.112) with the noise term omitted, and the initial condition

$$\hat{x}(0) = (-0.7290...)\left(\delta\mu/a^2\right)^{1/3}.$$

We define a characteristic point of the orbit, $\hat{x}_{\mathrm{nl}}(n_{\mathrm{nl}})$, where $a\hat{x}_{\mathrm{nl}}^2 \approx n_{\mathrm{nl}}\,\delta\mu$. For $n < n_{\mathrm{nl}}$, $n\,\delta\mu \leq d\hat{x}/dn < 2n\,\delta\mu$, and we can approximate the noiseless orbit as

$$\hat{x}(n) \approx \hat{x}(0) + \alpha(n)(n^2\,\delta\mu),$$

where $\alpha(n)$ is a slowly varying function of n of order 1 ($1/2 \leq \alpha(n) < 1$ for $n < n_{\mathrm{nl}}$). Setting $a\hat{x}^2 \approx n\,\delta\mu$, we find that n_{nl} is given by

$$n_{\mathrm{nl}} \sim (a\,\delta\mu)^{-1/3},$$

corresponding to

$$\hat{x}_{\mathrm{nl}} \sim (\delta\mu/a)^{1/3}.$$

For $n > n_{\mathrm{nl}}$ (i.e., $\hat{x}(n) > \hat{x}_{\mathrm{nl}}$), (2.112) can be approximated as $d\hat{x}/dn \approx a\hat{x}^2$. Starting at $\hat{x}(n) \sim \hat{x}_{\mathrm{nl}}$, integration of this equation leads to explosive growth of $\hat{x}$ to infinity in a time of order $(a\,\delta\mu)^{-1/3}$, which is of the same order as n_{nl}. Thus, the relevant time scale is $(a\,\delta\mu)^{-1/3}$.

Now consider the action of noise. For $n < n_{\mathrm{nl}}$, we neglect the nonlinear term $a\hat{x}^2$, so that (2.112) becomes

$$d\hat{x}/dn = n\,\delta\mu + A\hat{\epsilon}(n).$$

The solution of this equation is the linear superposition of the solutions of

$$d\hat{x}_a/dn = n\,\delta\mu, \qquad d\hat{x}_b/dn = A\hat{\epsilon}(n),$$

$$\text{or} \qquad \hat{x}(n) = \hat{x}_a(n) + \hat{x}_b(n);$$

$\hat{x}_a(n)$ is given by

$$\hat{x}_a(n) = \hat{x}(0) + n^2\delta\mu/2,$$

and $\hat{x}_b(n)$ is a random walk. Thus, for $n < n_{\rm nl}$, there is diffusive spreading of the probability density of $\hat{x}$,

$$\Delta_{\rm diff}(n) \equiv \sqrt{\langle \hat{x}_b^2(n) \rangle} \sim n^{1/2}A.$$

This diffusive spreading can blur out the natural pattern. How large does the noise amplitude A have to be to do this? We can estimate A by noting that the periodic structure results from orbits that take different integer times to reach $\hat{x} \sim \hat{x}_{\rm nl}$. Thus, for $n \approx n_{\rm nl}$ we define a scale $\Delta_{\rm nl}$ in $\hat{x}$ corresponding to the periodicity in $1/\delta\mu$ by

$$\hat{x}_{\rm nl} \pm \Delta_{\rm nl} \approx \hat{x}(0) + (n_{\rm nl} \pm 1)^2\delta\mu$$

which yields

$$\Delta_{\rm nl} \sim n_{\rm nl}\delta\mu.$$

If by the time $n \approx n_{\rm nl}$, the diffusive spread of the probability density of $\hat{x}$ becomes as large as $\Delta_{\rm nl}$, then the noise starts to wash out the periodic variations with $1/\delta\mu$. Setting $\Delta_{\rm diff}(n_{\rm nl})$ to be of the order of $\Delta_{\rm nl}$, we get $n_{\rm nl}^{1/2}A \sim n_{\rm nl}\delta\mu$, which yields

$$A \sim (\delta\mu)^{5/6}.$$

Thus, we expect a collapse of the two parameter $(A, \delta\mu)$ data by means of a rescaling of A by $\delta\mu$ raised to an exponent 5/6 (i.e., $A/(\delta\mu)^{5/6}$).

Scaling of Indeterminate Saddle–Node Bifurcations for a Periodically Forced 2nd Order ODE

In this section we demonstrate the scaling properties of sweeping through an indeterminate saddle–node bifurcation in the case of the *periodically forced Duffing oscillator* [BNO03],

$$\ddot{x} - 0.15\,\dot{x} - x + x^3 = \mu\cos t.$$

The unforced Duffing system (i.e., $\mu = 0$) is an example of an oscillator in a double well potential. It has two coexisting fixed point attractors corresponding to the two minima of the potential energy. For small μ, the forced Duffing oscillator has two attracting periodic orbits with the period of the forcing (i.e., 2π), one in each well of the potential. At $\mu = \mu_* \approx 0.2446$, a new attracting periodic orbit of period 6π arises through a saddle–node bifurcation. In [AS02], it is argued numerically that for a certain range of $\mu > \mu_*$ the basin of attraction of the 6π periodic orbit and the basins of attraction of the 2π periodic orbits have the Wada property. Thus, as μ decreases through the critical value μ_*, the period 6π attractor is destroyed via a saddle–node bifurcation on the fractal boundary of the basins of the other two attractors. This is an example of an indeterminate saddle–node bifurcation of the Duffing system which we study by considering the 2D map in the $(\dot{x}, x)$ plane resulting from a Poincaré section at constant phase of the forcing signal. We consider orbits starting in the vicinity of the period three fixed point attractor, and, as we integrate the Duffing system, we decrease μ from $\mu_s > \mu_*$ to $\mu_f < \mu_*$ at a small rate of $\delta\mu$ per one period of the forcing signal. As μ approaches μ_*, (with $\mu > \mu_*$,) the period three fixed point attractor of the unswept Duffing system approaches its basin boundary, and the slowly swept orbit closely follows its location. For $\mu - \mu_* < 0$ small, the orbit will approximately follow the 1D unstable manifold of the $\mu = \mu_*$ period three saddle–node pair. Thus, we can describe the sweeping through the indeterminate bifurcation of the Duffing oscillator by the theory we developed for 1D discrete maps. We believe that the scaling properties of the indeterminate saddle–node bifurcation we found in 1D discrete maps are also shared by higher dimensional flows [BNO03].

Indeterminacy in How an Attractor is Approached

In this section we consider the case of a 1D map f_μ having two attractors A and B, one of which (i.e., A) exists for all $\mu \in [\mu_s, \mu_f]$. The other (i.e., B) is a node which is destroyed by a saddle–node bifurcation on the boundary between the basins of A and B, as μ increases through μ_* ($\mu_* \in [\mu_s, \mu_f]$). When an orbit is initially on B, and μ is slowly increased through μ_*, the orbit will always go to A (which is the only attractor for $\mu > \mu_*$). However, it is possible to distinguish between two (or more) different ways of approaching A. (In particular, we are interested in ways of approach that can be distinguished in a coordinate–free (i.e., invariant) manner.) As we show in this section, the way in which A is approached can be indeterminate. In this case, the indeterminacy is connected with the existence of an invariant nonattracting Cantor set embedded in the basin of A for $\mu > \mu_*$.

As an illustration, we construct the following model [BNO03]

$$f_\mu(x) = -\mu + x - 3x^2 - x^4 + 3.6x^6 - x^8.$$

Calculations show that f_μ satisfies all the requirements of the saddle–node bifurcation theorem for undergoing a backward saddle–node bifurcation at $x_* = 0$ and $\mu_* = 0$. For every value of μ we consider, the map f_μ has invariant Cantor sets. The trajectories of points which are located on an invariant Cantor set, do not diverge to infinity. One way to display such Cantor sets, is to select uniquely defined intervals whose end points are on the Cantor set. For every fixed parameter value μ, the collection of points that are boundary points of the red and green regions, constitutes an invariant Cantor set. In order to describe these green and red regions, we introduce the following notations. For each parameter value μ, let p_μ be the leftmost fixed point of f_μ.

For every $x_0 < p_\mu$, the sequence of iterates $\{x_n = f_\mu^{[n]}(x_0)\}$ is decreasing and diverges to minus infinity. For each value of μ, let q_μ be the fixed point of f_μ to the right of $x = 0$ at which $\partial_x f_\mu(q_\mu) > 1$. A point $(x;\mu)$ is colored green if its trajectory diverges to minus infinity and it passes through the interval (q_μ, ∞), and it is colored red if its trajectory diverges to minus infinity and it does not pass through the interval (q_μ, ∞). Denote the collection of points $(x;\mu)$ that are colored green by $G[\mu]$, and the collection of points $(x;\mu)$ that are colored red by $R[\mu]$. Using the methods and techniques of [Nus87], it can be shown that the collection of points $(x;\mu)$ which are common boundary points of $G[\mu]$ and $R[\mu]$ is a Cantor set $C[\mu]$.[22] In particular, the results of [Nus87] imply that for $\mu = \mu_* = 0$ the point $x_* = 0$ belongs to the invariant Cantor set $C[\mu_*]$.

As discussed above, past the saddle–node bifurcation of f_μ at μ_*, infinitely many other saddle–node bifurcations of periodic orbits take place on the invariant Cantor set $C[\mu]$. We believe that μ_{**} is an approximate value of μ where such a saddle–node of a periodic orbit takes place.

- scaling of the fractal basin boundary of the static (i.e., unswept) system near the saddle–node bifurcation,
- the scaling dependence of the orbit's final destination with the inverse of the sweeping rate,
- the dependence of the time it takes for an attractor to capture a swept orbit with the $-1/3$ power of the sweeping rate,
- scaling of the effect of noise on the final attractor capture probability with the $5/6$ power of the sweeping rate.

2.2.9 Chaos Field Theory

In [Cvi00], Cvitanovic re–examined the path–integral formulation and the role that the classical solutions play in quantization of strongly nonlinear fields. In the path integral formulation of a field theory the dominant contributions come from saddle–points, the classical solutions of equations of

[22] For every μ $(-0.3 < \mu < 0.3)$, write $p_\mu^* = \max\{x \in \mathbb{R} : f_\mu(x) = p_\mu\}$ and $q_\mu^* = \max\{x \in \mathbb{R} : f_\mu(x) = q_\mu\}$. The interval $[q_\mu, q_\mu^*]$ also contains a Cantor set. By coloring this whole segment green, this information is lost. Therefore, the coloring scheme should be adapted if one wants to have the whole invariant Cantor set represented for every μ. For example, if a trajectory that diverges to minus infinity contains a point that is greater than p_μ^* then the initial point is colored green, if a trajectory that diverges to minus infinity contains a point that is greater than q_μ^* but not greater than p_μ^* then the initial point is colored yellow. A point is colored red, if its trajectory diverges to minus infinity and does not have a point that is greater than q_μ^*. Then the collection of boundary points (a point x is a boundary point if every open neighborhood of x contains points of at least two different colors) is a Cantor set that contains the Cantor set described above.

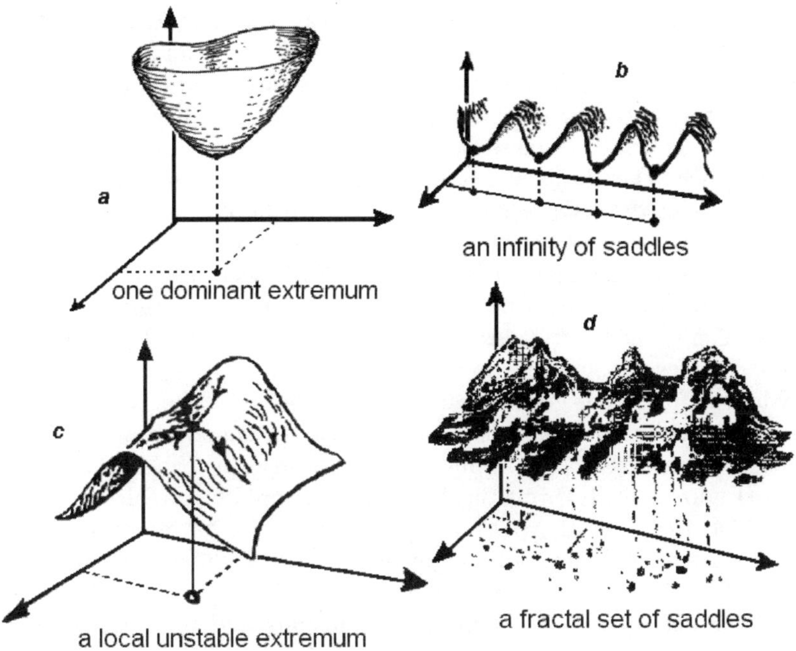

Fig. 2.34. Path integrals and chaos field theory (see text for explanation).

motion. Usually one imagines *one dominant saddle point*, the 'vacuum' (see Figure 2.34, (a)).

The *Feynman diagrams* of quantum electrodynamics (QED) and quantum chromodynamics (QCD), associated to their *path–integrals* (see next Chapter), give us a visual and intuitive scheme to calculate the correction terms to this starting semiclassical, *Gaussian saddlepoint approximation*. But there might be other saddles (Figure 2.34, (b)). That field theories might have a rich repertoire of classical solutions became apparent with the discovery of *instantons* [BPS75], analytic solutions of the classical $SU(2)$ *Yang–Mills relation*, and the realization that the associated *instanton vacua* receive contributions from countable ∞'s of saddles. What is not clear is whether these are the important classical saddles. Cvitanovic asks the question: could it be that the strongly nonlinear theories are dominated by altogether different classical solutions?

The search for the classical solutions of nonlinear field theories such as the *Yang–Mills* and *gravity* has so far been neither very successful nor very systematic. In modern field theories the main emphasis has been on symmetries (compactly collected in action functionals that define the theories) as guiding principles in writing down the actions. But writing down a differential equation is only the start of the story; even for systems as simple as 3 coupled ordinary differential equations one in general has no clue what the nature of the long time solutions might be.

These are hard problems, and in explorations of modern field theories the dynamics tends to be is neglected, and understandably so, because the wealth of the classical solutions of nonlinear systems can be truly bewildering. If the classical behavior of these theories is anything like that of the field theories that describe the classical world – the hydrodynamics, the magneto–hydrodynamics, the *Burgers dynamical system* , *Ginzburg–Landau equation*, or *Kuramoto–Sivashinsky equation* (see, e.g., [II06b]), there should be very many solutions, with very few of the important ones analytical in form; the strongly nonlinear classical field theories are turbulent, after all. Furthermore, there is not a dimmest hope that such solutions are either beautiful or analytic, and there is not much enthusiasm for grinding out numerical solutions as long as one lacks ideas as what to do with them.

By late 1970's it was generally understood that even the simplest nonlinear systems exhibit chaos. Chaos is the norm also for generic Hamiltonian flows, and for path integrals that implies that instead of a few, or countably few saddles (Figure 2.34, (c)), classical solutions populate fractal sets of saddles (Figure 2.34, (d)). For the path–integral formulation of quantum mechanics such solutions were discovered and accounted for by Gutzwiller, in late 1960's (see [Gut90]). In this framework the spectrum of the theory is computed from a set of its unstable classical periodic solutions and quantum corrections. The new aspect is that the individual saddles for classically chaotic systems are nothing like the harmonic oscillator degrees of freedom, the quarks and gluons of QCD – they are all unstable and highly nontrivial, accessible only by numerical techniques.

So, if one is to develop a semiclassical field theory of systems that are *classically chaotic* or *turbulent*, the problem one faces is twofold [Cvi00]

1. Determine, classify, and order by relative importance the classical solutions of nonlinear field theories.
2. Develop methods for calculating perturbative corrections to the corresponding classical saddles.

2.3 Chaos Control

2.3.1 Feedback versus Non–Feedback Algorithms

Although the presence of chaotic behavior is generic and robust for suitable nonlinearities, ranges of parameters and external forces, there are practical situations where one wishes to avoid or control chaos so as to improve the performance of the dynamical system. Also, although chaos is sometimes useful as in a mixing process or in heat transfer, it is often unwanted or undesirable. For example, increased drag in flow systems, erratic fibrillations of heart beats, extreme weather patterns and complicated circuit oscillations are situations where chaos is harmful. Clearly, the ability to control chaos, that is to convert chaotic oscillations into desired regular ones with a periodic time dependence would be beneficial in working with a particular system. The possibility of purposeful selection and stabilization of particular orbits in a normally chaotic system, using minimal, predetermined efforts, provides a unique opportunity to maximize the output of a dynamical system. It is thus of great practical importance to develop suitable control methods and to analyze their efficacy.

Let us consider a general nD nonlinear dynamical system,

$$\dot{x} = F(x, p, t), \tag{2.113}$$

where $x = (x_1, x_2, x_3, \ldots, x_n)$ represents the n state variables and p is a control or external parameter. Let $x(t)$ be a chaotic solution of (2.113). Different control algorithms are essentially based on the fact that one would like to effect the most minimal changes to the original system so that it will not be grossly deformed. From this point of view, controlling methods or algorithms can be broadly classified into two categories:

(i) feedback methods, and
(ii) non–feedback algorithms.

Feedback methods essentially make use of the intrinsic properties of chaotic systems, including their sensitivity to initial conditions, to stabilize orbits already existing in the systems. Some of the prominent methods are the following (see, [Lak97, Lak03, Sch88, II06b]):

1. Adaptive control algorithm;
2. Nonlinear control algorithm;
3. Ott–Grebogi–Yorke (OGY) method of stabilizing unstable periodic orbits;
4. Singer's method of stabilizing unstable periodic orbits; and
5. Various control engineering approaches.

In contrast to feedback control techniques, non–feedback methods make use of a small perturbing external force such as a small driving force, a small noise term, a small constant bias or a weak modulation to some system parameter. These methods modify the underlying chaotic dynamical system weakly so that stable solutions appear. Some of the important controlling methods of this type are the following.

1. Parametric perturbation method
2. Addition of a weak periodic signal, constant bias or noise
3. Entrainment–open loop control
4. Oscillator absorber method.

Here is a typical example of adaptive control algorithm. We can control the chaotic orbit $X_s = (x_s, y_s)$ of the *Van der Pol oscillator* (2.30) by introducing the following dynamics on the parameter A_1:

$$\dot{x} = x - \frac{x^3}{3} - y + A_0 + A_1 \cos\omega t, \qquad \dot{y} = c(x + a - by),$$
$$\dot{A}_1 = -\epsilon[(x - x_s) - (y - y_s)], \qquad \epsilon << 1.$$

On the other hand, recall from [II06b] that a generic SISO nonlinear system

$$\dot{x} = f(x) + g(x)\,u \qquad y = h(x) \tag{2.114}$$

is said to have *relative degree* r at a point x^o if

(i) $L_g L_f^k h(x) = 0$ for all x in a neighborhood of x^o and all $k < r - 1$

(ii) $L_g L_f^{r-1} h(x^o) \neq 0$, where L_g denotes the *Lie derivative* in the direction of the vector–field g.

Now, the Van der Pol oscillator (2.23) has the state space form

$$\dot{x} = f(x) + g(x)\,u = \begin{bmatrix} x_2 \\ 2\omega\zeta\,(1 - \mu x_1^2)\,x_2 - \omega^2 x_1 \end{bmatrix} + \begin{bmatrix} 0 \\ 1 \end{bmatrix} u. \tag{2.115}$$

Suppose the output function is chosen as

$$y = h(x) = x_1. \tag{2.116}$$

In this case we have

$$L_g h(x) = \frac{\partial h}{\partial x} g(x) = \begin{bmatrix} 1 & 0 \end{bmatrix} \begin{bmatrix} 0 \\ 1 \end{bmatrix} = 0, \qquad \text{and} \tag{2.117}$$

$$L_f h(x) = \frac{\partial h}{\partial x} f(x) = \begin{bmatrix} 1 & 0 \end{bmatrix} \begin{bmatrix} x_2 \\ 2\omega\zeta\,(1 - \mu x_1^2)\,x_2 - \omega^2 x_1 \end{bmatrix} = x_2. \tag{2.118}$$

Moreover

$$L_g L_f h(x) = \frac{\partial (L_f h)}{\partial x} g(x) = \begin{bmatrix} 0 & 1 \end{bmatrix} \begin{bmatrix} 0 \\ 1 \end{bmatrix} = 1 \tag{2.119}$$

and thus we see that the Van der Pol oscillator system has relative degree 2 at any point x^o.

However, if the output function is, for instance

$$y = h(x) = \sin x_2 \tag{2.120}$$

then $L_g h(x) = \cos x_2$. The system has relative degree 1 at any point x^o, provided that $(x^o)_2 \neq (2k+1)\,\pi/2$. If the point x^o is such that this condition is violated, no relative degree can be defined.

Both adaptive and nonlinear control methods can be naturally extended to other chaotic systems, e.g., *Lorenz attractor* (see Figure 2.35).

Hybrid Systems and Homotopy ODEs

Consider a *hybrid dynamical system of variable structure*, given by an nD ODE–system (see [MWH01])

$$\dot{x} = f(t, x), \tag{2.121}$$

where $x = x(t) \in \mathbb{R}^n$ and $f = f(t,x) : \mathbb{R}^+ \times \mathbb{R}^n \to \mathbb{R}^n$. Let the domain $G \subset \mathbb{R}^+ \times \mathbb{R}^n$, on which the vector–field $f(t,x)$ is defined, be divided into

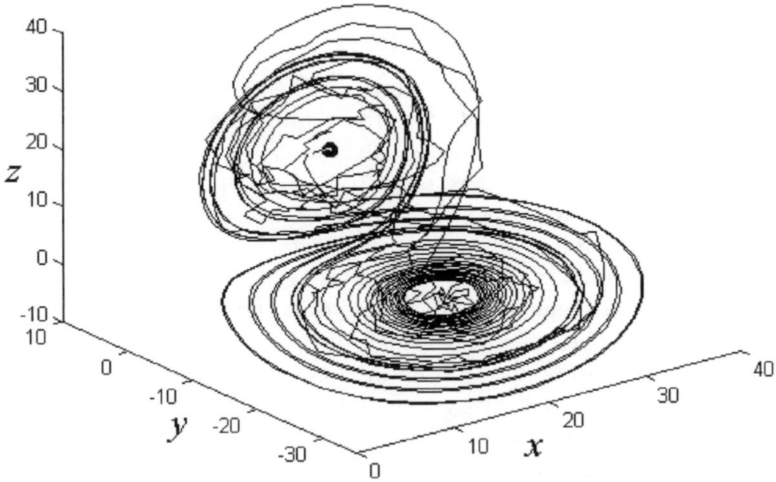

Fig. 2.35. Nonlinear control of the Lorenz system: targeting of unstable upper and lower states in the Lorenz attractor (after applying random perturbations, see [Pet96]), using a MIMO nonlinear controller (see [II06b]); simulated using $Matlab^{TM}$.

two subdomains, G^+ and G^-, by means of a smooth $(n-1)$–manifold M. In $G^+ \cup M$, let there be given a vector–field $f^+(t,x)$, and in $G^- \cup M$, let there be given a vector–field $f^-(t,x)$. Assume that both $f^+ = f^+(t,x)$ and $f^- = f^-(t,x)$ are continuous in t and smooth in x. For the system (2.121), let

$$f = \begin{cases} f^+ & \text{when } x \in G^+ \\ f^- & \text{when } x \in G^- \end{cases}.$$

Under these conditions, a solution $x(t)$ of ODE (2.121) is well–defined while passing through G until the manifold M is reached.

Upon reaching the manifold M, in physical systems with inertia, the transition

$$\text{from} \quad \dot{x} = f^-(t,x) \quad \text{to} \quad \dot{x} = f^+(t,x)$$

does not take place instantly on reaching M, but after some delay. Due to this delay, the solution $x(t)$ oscillates about M, $x(t)$ being displaced along M with some mean velocity.

As the delay tends to zero, the limiting motion and velocity along M are determined by the *linear homotopy ODE*

$$\dot{x} = f^0(t,x) \equiv (1-\alpha)\, f^-(t,x) + \alpha\, f^+(t,x), \tag{2.122}$$

where $x \in M$ and $\alpha \in [0,1]$ is such that the *linear homotopy segment* $f^0(t,x)$ is tangential to M at the point x, i.e., $f^0(t,x) \in T_xM$, where T_xM is the tangent space to the manifold M at the point x.

The vector–field $f^0(t,x)$ of the system (2.122) can be constructed as follows: at the point $x \in M$, $f^-(t,x)$ and $f^+(t,x)$ are given and their ends are joined by the linear homotopy segment. The point of intersection between this segment and T_xM is the end of the required vector–field $f^0(t,x)$. The vector function $x(t)$ which satisfies (2.121) in G^- and G^+, and (2.122) when $x \in M$, can be considered as a *solution* of (2.121) in a *general sense*.

However, there are cases in which the solution $x(t)$ cannot consist of a finite or even countable number of arcs, each of which passes through G^- or G^+ satisfying (2.121), or moves along the manifold M and satisfies the homotopic ODE (2.122). To cover such cases, assume that the vector–field $f = f(t,x)$ in ODE (2.121) is a Lebesgue–measurable function in a domain $G \subset \mathbb{R}^+ \times \mathbb{R}^n$, and that for any closed bounded domain $D \subset G$ there exists a summable function $K(t)$ such that almost everywhere in D we have $|f(t,x)| \leq K(t)$. Then the absolutely continuous vector function $x(t)$ is called the *generalized solution* of the ODE (2.121) *in the sense of Filippov* (see [MWH01]) if for almost all t, the vector $\dot{x} = \dot{x}(t)$ belongs to the least convex closed set containing all the limiting values of the vector–field $f(t,x^*)$, where x^* tends towards x in an arbitrary manner, and the values of the function $f(t,x^*)$ on a set of measure zero in $\mathbb{R}^n$ are ignored.

Such *hybrid systems* of variable structure occur in the study of nonlinear electric networks (endowed with electronic switches, relays, diodes, rectifiers, etc.), in models of both natural and artificial neural networks, as well as in feedback control systems (usually with continuous–time plants and digital controllers/filters).

2.3.2 Exploiting Critical Sensitivity

The fact that some dynamical systems showing the necessary conditions for chaotic behavior possess such a critical dependence on the initial conditions was known since the end of the last century. However, only in the last thirty years, experimental observations have pointed out that, in fact, chaotic systems are common in nature. They can be found, e.g., in chemistry (*Belouzov–Zhabotinski reaction*), in nonlinear optics (lasers), in electronics (*Chua–Matsumoto circuit*), in fluid dynamics (*Rayleigh–Bénard convection*), etc. Many natural phenomena can also be characterized as being chaotic. They can be found in meteorology, solar system, heart and brain of living organisms and so on.

Due to their critical dependence on the initial conditions, and due to the fact that, in general, experimental initial conditions are never known perfectly, these systems are intrinsically unpredictable. Indeed, the prediction trajectory emerging from an initial condition and the real trajectory emerging from the real initial condition diverge exponentially in course of time, so that the error in the prediction (the distance between prediction and real trajectories) grows exponentially in time, until making the system's real trajectory completely different from the predicted one at long times.

For many years, this feature made chaos undesirable, and most experimentalists considered such characteristic as something to be strongly avoided. Besides their critical sensitivity to initial conditions, chaotic systems exhibit two other important properties. Firstly, there is an infinite number of unstable periodic orbits embedded in the underlying chaotic set. In other words, the skeleton of a chaotic attractor is a collection of an infinite number of periodic orbits, each one being unstable. Secondly, the dynamics in the chaotic attractor is *ergodic*, which implies that during its temporal evolution the system ergodically visits small neighborhood of every point in each one of the unstable periodic orbits embedded within the chaotic attractor.

A relevant consequence of these properties is that a chaotic dynamics can be seen as shadowing some periodic behavior at a given time, and erratically jumping from one to another periodic orbit. The idea of controlling chaos is then when a trajectory approaches ergodically a desired periodic orbit embedded in the attractor, one applies small perturbations to stabilize such an orbit. If one switches on the stabilizing perturbations, the trajectory moves to the neighborhood of the desired periodic orbit that can now be stabilized. This fact has suggested the idea that the critical sensitivity of a chaotic system to changes (perturbations) in its initial conditions may be, in fact, very desirable in practical experimental situations. Indeed, if it is true that a small perturbation can give rise to a very large response in the course of time, it is also true that a judicious choice of such a perturbation can direct the trajectory to wherever one wants in the attractor, and to produce a series of desired dynamical states. This is exactly the idea of *targeting* [BGL00].

The important point here is that, because of chaos, one is able to produce an infinite number of desired dynamical behaviors (either periodic and not periodic) using the same chaotic system, with the only help of tiny perturbations chosen properly. We stress that this is not the case for a non–chaotic dynamics, wherein the perturbations to be done for producing a desired behavior must, in general, be of the same order of magnitude as the un–perturbed evolution of the dynamical variables.

The *idea* of *chaos control* was enunciated in 1990 at the University of Maryland, by E. Ott, C. Grebogi and J.A. Yorke [OGY90], widely referred to as Ott–Grebogi–Yorke (OGY, for short). In OGY–paper [OGY90], the ideas for controlling chaos were outlined and a method for stabilizing an unstable periodic orbit was suggested, as a proof of principle. The main idea consisted in waiting for a natural passage of the chaotic orbit close to the desired periodic behavior, and then applying a small judiciously chosen perturbation, in order to stabilize such periodic dynamics (which would be, in fact, unstable for the un–perturbed system). Through this mechanism, one can use a given laboratory system for producing an infinite number of different periodic behavior (the infinite number of its unstable periodic orbits), with a great flexibility in switching from one to another behavior. Much more, by constructing appropriate goal dynamics, compatible with the chaotic attractor, an operator may apply small perturbations to produce any kind of desired dynamics, even not periodic, with practical application in the coding process of signals.

A branch of the theory of dynamical systems has been developed with the aim of formalizing and quantitatively characterizing the sensitivity to initial conditions. The *largest Lyapunov exponent* λ (together with the related *Kaplan–Yorke dimension* d_{Kaplan}) and the *Kolmogorov–Sinai entropy* h_{KS} are the two indicators for measuring the *rate of error growth* and *information* produced by the dynamical system [ER85].

2.3.3 Lyapunov Exponents and Kaplan–Yorke Dimension

The characteristic Lyapunov exponents are somehow an extension of the linear stability analysis to the case of aperiodic motions. Roughly speaking, they measure the typical rate of exponential divergence of nearby trajectories. In this sense they give information on the rate of growth of a very small error on the initial state of a system [BCF02].

Consider an nD dynamical system given by the set of ODEs of the form

$$\dot{x} = f(x), \tag{2.123}$$

where $x = (x_1, \dots, x_n) \in \mathbb{R}^n$ and $f : \mathbb{R}^n \to \mathbb{R}^n$. Recall that since the r.h.s of equation (2.123) does not depend on t explicitly, the system is called *autonomous*. We assume that f is smooth enough that the evolution is well–defined for time intervals of arbitrary extension, and that the motion occurs in a bounded region R of the system phase–space M. We intend to study the separation between two trajectories in M, $x(t)$ and $x'(t)$, starting from two close initial conditions, $x(0)$ and $x'(0) = x(0)+\delta x(0)$ in $R_0 \subset M$, respectively.

As long as the difference between the trajectories, $\delta x(t) = x'(t) - x(t)$, remains infinitesimal, it can be regarded as a vector, $z(t)$, in the tangent space $T_x M$ of M. The time evolution of $z(t)$ is given by the linearized differential equations:

$$\dot{z}_i(t) = \left.\frac{\partial f_i}{\partial x_j}\right|_{x(t)} z_j(t).$$

Under rather general hypothesis, Oseledets [Ose68] proved that for almost all initial conditions $x(0) \in R$, there exists an orthonormal basis $\{e_i\}$ in the tangent space $T_x M$ such that, for large times,

$$z(t) = c_i e_i \exp(\lambda_i t), \tag{2.124}$$

where the coefficients $\{c_i\}$ depend on $z(0)$. The exponents $\lambda_1 \geq \lambda_2 \geq \cdots \geq \lambda_d$ are called *characteristic Lyapunov exponents*. If the dynamical system has an ergodic invariant measure on M, the spectrum of LEs $\{\lambda_i\}$ does not depend on the initial conditions, except for a set of measure zero with respect to the natural invariant measure.

Equation (2.124) describes how a dD spherical region $R = S^n \subset M$, with radius ϵ centered in $x(0)$, deforms, with time, into an ellipsoid of semi–axes

$\epsilon_i(t) = \epsilon \exp(\lambda_i t)$, directed along the e_i vectors. Furthermore, for a generic small perturbation $\delta x(0)$, the distance between the reference and the perturbed trajectory behaves as

$$|\delta x(t)| \sim |\delta x(0)| \exp(\lambda_1 t) \left[1 + O\left(\exp -(\lambda_1 - \lambda_2)t\right)\right].$$

If $\lambda_1 > 0$ we have a rapid (exponential) amplification of an error on the initial condition. In such a case, the system is chaotic and, unpredictable on the long times. Indeed, if the initial error amounts to $\delta_0 = |\delta x(0)|$, and we purpose to predict the states of the system with a certain tolerance Δ, then the prediction is reliable just up to a *predictability time* given by

$$T_p \sim \frac{1}{\lambda_1} \ln\left(\frac{\Delta}{\delta_0}\right).$$

This equation shows that T_p is basically determined by the *positive leading Lyapunov exponent*, since its dependence on δ_0 and Δ is logarithmically weak. Because of its preeminent role, λ_1 is often referred as 'the leading positive Lyapunov exponent', and denoted by λ.

Therefore, Lyapunov exponents are average rates of expansion or contraction along the principal axes. For the ith principal axis, the corresponding Lyapunov exponent is defined as

$$\lambda_i = \lim_{t\to\infty} \{(1/t) \ln[L_i(t)/L_i(0)]\}, \qquad (2.125)$$

where $L_i(t)$ is the radius of the ellipsoid along the ith principal axis at time t. For technical details on calculating Lyapunov exponents from any time series data, see [WSS85].

An initial volume V_0 of the phase–space region R_0 evolves on average as

$$V(t) = V_0 \mathrm{e}^{(\lambda_1+\lambda_2+\cdots+\lambda_{2n})t}, \qquad (2.126)$$

and therefore the rate of change of $V(t)$ is simply

$$\dot{V}(t) = \sum_{i=1}^{2n} \lambda_i V(t).$$

In the case of a 2D phase area A, evolving as $A(t) = A_0 \mathrm{e}^{(\lambda_1+\lambda_2)t}$, a *Lyapunov dimension* d_L is defined as

$$d_L = \lim_{\epsilon\to 0} \left[\frac{d(\ln(N(\epsilon)))}{d(\ln(1/\epsilon))}\right],$$

where $N(\epsilon)$ is the number of squares with sides of length ϵ required to cover $A(t)$, and d represents an ordinary *capacity dimension*,

$$d_c = \lim_{\epsilon\to 0} \left(\frac{\ln N}{\ln(1/\epsilon)}\right).$$

Lyapunov dimension can be extended to the case of nD phase–space by means of the *Kaplan–Yorke dimension* [KY79, YAS96, OGY90]) as

$$d_{Kaplan} = j + \frac{\lambda_1 + \lambda_2 + \cdots + \lambda_j}{|\lambda_{j+1}|},$$

where the λ_i are ordered (λ_1 being the largest) and j is the index of the smallest nonnegative Lyapunov exponent.

2.3.4 Kolmogorov–Sinai Entropy

The LE, λ, gives a first quantitative information on how rapidly we loose the ability of predicting the evolution of a system [BCF02]. A state, initially determined with an error $\delta x(0)$, after a time enough larger than $1/\lambda$, may be found almost everywhere in the region of motion $R \in M$. In this respect, the *Kolmogorov–Sinai* (KS) *entropy*, h_{KS}, supplies a more refined information. The error on the initial state is due to the maximal resolution we use for observing the system. For simplicity, let us assume the same resolution ϵ for each degree of freedom. We build a partition of the phase–space M with cells of volume ϵ^d, so that the state of the system at $t = t_0$ is found in a region R_0 of volume $V_0 = \epsilon^d$ around $x(t_0)$. Now we consider the trajectories starting from V_0 at t_0 and sampled at discrete times $t_j = j\,\tau$ ($j = 1, 2, 3, \dots, t$). Since we are considering motions that evolve in a bounded region $R \subset M$, all the trajectories visit a finite number of different cells, each one identified by a symbol. In this way a unique sequence of symbols $\{s(0), s(1), s(2), \dots\}$ is associated with a given trajectory $x(t)$. In a chaotic system, although each evolution $x(t)$ is univocally determined by $x(t_0)$, a great number of different symbolic sequences originates by the same initial cell, because of the divergence of nearby trajectories. The total number of the admissible symbolic sequences, $\widetilde{N}(\epsilon, t)$, increases exponentially with a rate given by the topological entropy

$$h_T = \lim_{\epsilon \to 0} \lim_{t \to \infty} \frac{1}{t} \ln \widetilde{N}(\epsilon, t)\,.$$

However, if we consider only the number of sequences $N_{eff}(\epsilon, t) \le \widetilde{N}(\epsilon, t)$ which appear with very high probability in the long time limit – those that can be numerically or experimentally detected and that are associated with the natural measure – we arrive at a more physical quantity, namely the *Kolmogorov–Sinai entropy* [ER85]:

$$h_{KS} = \lim_{\epsilon \to 0} \lim_{t \to \infty} \frac{1}{t} \ln N_{eff}(\epsilon, t) \le h_T. \tag{2.127}$$

h_{KS} quantifies the long time exponential rate of growth of the number of the effective coarse-grained trajectories of a system. This suggests a link with information theory where the Shannon entropy measures the mean asymptotic growth of the number of the typical sequences – the ensemble of which has probability almost one – emitted by a source.

We may wonder what is the number of cells where, at a time $t > t_0$, the points that evolved from R_0 can be found, i.e., we wish to know how big is the coarse–grained volume $V(\epsilon, t)$, occupied by the states evolved from the volume V_0 of the region R_0, if the minimum volume we can observe is $V_{min} = \epsilon^d$. As stated above (2.126), we have

$$V(t) \sim V_0 \exp(t \sum_{i=1}^{d} \lambda_i).$$

However, this is true only in the limit $\epsilon \to 0$. In this (unrealistic) limit, $V(t) = V_0$ for a conservative system (where $\sum_{i=1}^{d} \lambda_i = 0$) and $V(t) < V_0$ for a dissipative system (where $\sum_{i=1}^{d} \lambda_i < 0$). As a consequence of limited resolution power, in the evolution of the volume $V_0 = \epsilon^d$ the effect of the contracting directions (associated with the negative Lyapunov exponents) is completely lost. We can experience only the effect of the expanding directions, associated with the positive Lyapunov exponents. As a consequence, in the typical case, the coarse grained volume behaves as

$$V(\epsilon, t) \sim V_0 \,\mathrm{e}^{\left(\sum_{\lambda_i > 0} \lambda_i\right) t},$$

when V_0 is small enough. Since $N_{eff}(\epsilon, t) \propto V(\epsilon, t)/V_0$, one has

$$h_{KS} = \sum_{\lambda_i > 0} \lambda_i.$$

This argument can be made more rigorous with a proper mathematical definition of the metric entropy. In this case one derives the Pesin relation [Pes77, ER85]

$$h_{KS} \leq \sum_{\lambda_i > 0} \lambda_i. \tag{2.128}$$

Because of its relation with the Lyapunov exponents – or by the definition (2.127) – it is clear that also h_{KS} is a fine-grained and global characterization of a dynamical system.

The metric entropy is an invariant characteristic quantity of a dynamical system, i.e., given two systems with invariant measures, their KS–entropies exist and they are equal iff the systems are isomorphic [Bil65].

2.3.5 Chaos Control by Ott, Grebogi and Yorke)

Besides the occurrence of chaos in a large variety of natural processes, chaos may also occur because one may wish to design a physical, biological or chemical experiment, or to project an industrial plant to behave in a chaotic manner. The Ott–Grebogi–Yorke (OGY) idea is that chaos may indeed be desirable since it can be controlled by using small perturbation to some accessible parameter.

The major key ingredient for the OGY–control of chaos is the observation that a chaotic set, on which the trajectory of the chaotic process lives, has embedded within it a large number of unstable low–period periodic orbits. In addition, because of ergodicity, the trajectory visits or accesses the neighborhood of each one of these periodic orbits. Some of these periodic orbits may correspond to a desired system's performance according to some criterion. The second ingredient is the realization that chaos, while signifying sensitive dependence on small changes to the current state and henceforth rendering unpredictable the system state in the long time, also implies that the system's behavior can be altered by using small perturbations. Then, the accessibility of the chaotic systems to many different periodic orbits combined with its sensitivity to small perturbations allows for the control and the manipulation of the chaotic process. Specifically, the OGY approach is then as follows. One first determines some of the unstable low–period periodic orbits that are embedded in the chaotic set. One then examines the location and the stability of these orbits and chooses one which yields the desired system performance. Finally, one applies small control to stabilize this desired periodic orbit. However, all this can be done from data by using nonlinear time series analysis for the observation, understanding and control of the system. This is particularly important since chaotic systems are rather complicated and the detailed knowledge of the equations of the process is often unknown [BGL00].

Simple Example of Chaos Control: a 1D Map. The basic idea of controlling chaos can be understood [Lai94] by considering May's classical *logistic map* [May76] (2.40)

$$x_{n+1} = f(x_n, r) = r x_n (1 - x_n),$$

where x is restricted to the unit interval $[0, 1]$, and r is a control parameter. It is known that this map develops chaos via the *period–doubling bifurcation* route. For $0 < r < 1$, the asymptotic state of the map (or the attractor of the map) is $x = 0$; for $1 < r < 3$, the attractor is a nonzero fixed–point $x_F = 1 - 1/r$; for $3 < r < 1 + \sqrt{6}$, this fixed–point is unstable and the attractor is a stable period-2 orbit. As r is increased further, a sequence of period–doubling bifurcations occurs in which successive period–doubled orbits become stable. The period–doubling cascade accumulates at $r = r_\infty \approx 3.57$, after which chaos can arise.

Consider the case $r = 3.8$ for which the system is apparently chaotic. An important characteristic of a chaotic attractor is that there exists an infinite number of unstable periodic orbits embedded within it. For example, there are a fixed–point $x_F \approx 0.7368$ and a period-2 orbit with components $x(1) \approx 0.3737$ and $x(2) = 0.8894$, where $x(1) = f(x(2))$ and $x(2) = f(x(1))$.

Now suppose we want to avoid chaos at $r = 3.8$. In particular, we want trajectories resulting from a randomly chosen initial condition x_0 to be as close as possible to the period−2 orbit, assuming that this period−2 orbit gives the best system performance. Of course, we can choose the desired asymptotic state of the map to be any of the infinite number of unstable periodic orbits.

Suppose that the parameter r can be finely tuned in a small range around the value $r_0 = 3.8$, i.e., r is allowed to vary in the range $[r_0 - \delta, r_0 + \delta]$, where $\delta << 1$. Due to the nature of the chaotic attractor, a trajectory that begins from an arbitrary value of x_0 will fall, with probability one, into the neighborhood of the desired period−2 orbit at some later time. The trajectory would diverge quickly from the period−2 orbit if we do not intervene. Our task is to program the variation of the control parameter so that the trajectory stays in the neighborhood of the period−2 orbit as long as the control is present. In general, the small parameter perturbations will be time dependent [BGL00].

The logistic map in the neighborhood of a periodic orbit can be approximated by a linear equation expanded around the periodic orbit. Denote the target period−m orbit to be controlled as $x(i)$, $i = 1, \ldots, m$, where $x(i+1) = f(x(i))$ and $x(m+1) = x(1)$. Assume that at time n, the trajectory falls into the neighborhood of component i of the period−m orbit. The linearized dynamics in the neighborhood of component $i+1$ is then

$$\begin{aligned} x_{n+1} - x(i+1) &= \frac{\partial f}{\partial x}[x_n - x(i)] + \frac{\partial f}{\partial r}\Delta r_n \\ &= r_0[1 - 2x(i)][x_n - x(i)] + x(i)[1 - x(i)]\Delta r_n, \end{aligned}$$

where the partial derivatives are evaluated at $x = x(i)$ and $r = r_0$. We require x_{n+1} to stay in the neighborhood of m. Hence, we set $x_{n+1} - x(i+1) = 0$, which gives

$$\Delta r_n = r_0 \frac{[2x(i) - 1][x_n - x(i)]}{x(i)[1 - x(i)]}. \qquad (2.129)$$

Equation (2.129) holds only when the trajectory x_n enters a small neighborhood of the period-m orbit, i.e., when $|x_n - x(i)| << 1$, and hence the required parameter perturbation Δr_n is small. Let the length of a small interval defining the neighborhood around each component of the period−m orbit be 2ε. In general, the required maximum parameter perturbation δ is proportional to ε. Since ε can be chosen to be arbitrarily small, δ also can be made arbitrarily small. The average transient time before a trajectory enters the neighborhood of the target periodic orbit depends on ε (or δ). When the trajectory is outside the neighborhood of the target periodic orbit, we do not apply any parameter perturbation, so the system evolves at its nominal parameter value r_0. Hence we set $\Delta r_n = 0$ when $\Delta r_n > \delta$. The parameter perturbation Δr_n depends on x_n and is time–dependent.

The above strategy for controlling the orbit is very flexible for stabilizing different periodic orbits at different times. Suppose we first stabilize a chaotic trajectory around a period−2 orbit. Then we might wish to stabilize the fixed–point of the logistic map, assuming that the fixed–point would correspond to a better system performance at a later time. To achieve this change of control, we simply turn off the parameter control with respect to the period-2 orbit. Without control, the trajectory will diverge from the period-2 orbit exponentially. We let the system evolve at the parameter value r_0. Due to the nature of chaos, there comes a time when the chaotic trajectory enters a small neighborhood of the fixed–point. At this time we turn on a new set of parameter perturbations calculated with respect to the fixed–point. The trajectory can then be stabilized around the fixed–point [Lai94].

In the presence of external noise, a controlled trajectory will occasionally be 'kicked' out of the neighborhood of the periodic orbit. If this behavior occurs, we turn off the parameter perturbation and let the system evolve by itself. With probability one the chaotic trajectory will enter the neighborhood of the target periodic orbit and be controlled again. The effect of the noise is to turn a controlled periodic trajectory into an intermittent one in which chaotic phases (uncontrolled trajectories) are interspersed with laminar phases (controlled periodic trajectories). It is easy to verify that the averaged length of the laminar phase increases as the noise amplitude decreases [Lai94].

2.3.6 Floquet Stability Analysis and OGY Control

Controlling chaos, or stabilization of unstable periodic orbits of chaotic systems, has established to a field of large interest since the seed paper of Ott, Grebogi, Yorke [OGY90]. The idea is to stabilize by a feedback calculated at each *Poincaré section*, which reduces the control problem to stabilization of an unstable fixed–point of an iterated map. The feedback can, as in OGY scheme, be chosen proportional to the distance to the desired fixed–point, or proportional to the difference in phase–space position between actual and last but one Poincaré section. This difference control scheme [BDG93], being a time–discrete counterpart of the Pyragas approach [Pyr92, Pyr95], allows for stabilization of inaccurately known fixed–points, and can be extended by a memory term to overcome stability restrictions and to allow for tracking of drifting fixed–points [CMP98a].

In this section the stability of perturbations $x(t)$ around an unstable periodic orbit being subject to a Poincaré–based control scheme is analyzed by means of *Floquet theory* [HL93]. This approach allows to investigate viewpoints that have not been accessible by considering only the iteration dynamics between the Poincaré sections. Among these are primary the discussion of small measurement delays and variable impulse lengths. The impulse length is for both OGY and difference control usually a fixed parameter; and the iterated dynamics is uniquely defined only as long as this impulse length is not varied. The influence of the impulse length has not been point of consideration before; if reported at all, usually for both OGY and difference control a relative length of approximately $1/3$ is chosen without any reported sensitivity [Cla02b].

The linearized ODEs of both schemes are invariant under translation in time, $t \to t+T$. Therefore we can expand the solutions after periodic solutions $u(t+T) = u(t)$ according to

$$x(t) = \mathrm{e}^{\gamma T} u_{\gamma}(t).$$

The necessary condition for stability of the solution is $\mathrm{Re}(\gamma) < 0$; and $x(t) \equiv 0$ refers to motion along the orbit.

Whereas for the *Pyragas control* method (in which the delayed state feedback enforces a time–continuous description) a *Floquet stability analysis* is known [JBO97], here the focus is on the time–discrete control schemes.

Time–Continuous Stability Analysis of OGY Control

Due to the mathematically elegant and practical convenient description and application of OGY control in the Poincaré section up to now there seems to have been no need to calculate explicitly the *Floquet multiplicator* for a stability analysis. However, this allows a novel viewpoint on the differences between the local dynamics around an instable periodic orbit of a dynamical system being subject to Pyragas and OGY control.

For the 1D case, one has the dynamical system [Cla02b]

$$\dot{x}(t) = \lambda x(t) + \mu\varepsilon x(t - (t \mod T)).$$

In the first time interval between $t = 0$ and $t = T$ the differential equation reads

$$\dot{x}(t) = \lambda x(t) - \mu\varepsilon x(0), \qquad \text{for} \qquad 0 < t < T.$$

Integration of this differential equation yields

$$x(t) = \left((1 - \frac{\mu\varepsilon}{\lambda})\mathrm{e}^{\lambda t} + \frac{\mu\varepsilon}{\lambda}\right) x(0).$$

This gives us an iterated dynamics (here we label the beginning of the time period again with t)

$$x(t+T) = \left((1 - \frac{\mu\varepsilon}{\lambda})\mathrm{e}^{\lambda T} + \frac{\mu\varepsilon}{\lambda}\right) x(t).$$

The Floquet multiplier of an orbit therefore is

$$\mathrm{e}^{\gamma T} = (1 - \frac{\mu\varepsilon}{\lambda})\mathrm{e}^{\lambda T} + \frac{\mu\varepsilon}{\lambda}.$$

Influence of the Duration of the Control Impulse on OGY Control

The time–discrete viewpoint now allows to investigate the influence of timing questions on control. First we consider the case that the control impulse is applied timely in the Poincaré section, but only for a finite period $T \cdot p$ within the orbit period $(0 < p < 1)$.

This situation is described by the differential equation [Cla02b]

$$\dot{x}(t) = \lambda x(t)\mu\varepsilon x(t - (t \mod T)) \cdot \Theta((t \mod T) - p).$$

Here Θ is a step function ($\Theta(x) = 1$ for $x > 0$ and $\Theta(x) = 0$ elsewhere). In the first time interval between $t = 0$ and $t = T \cdot p$ the differential equation reads

$$\dot{x}(t) = \lambda x(t) + \mu\varepsilon x(0), \qquad \text{for} \qquad 0 < t < T \cdot p.$$

Integration of this differential equation yields

$$x(t) = \left((1 + \frac{\mu\varepsilon}{\lambda})\mathrm{e}^{\lambda t} - \frac{\mu\varepsilon}{\lambda}\right) x(0), \qquad x(T \cdot p) = \left((1 + \frac{\mu\varepsilon}{\lambda})\mathrm{e}^{\lambda T \cdot p} - \frac{\mu\varepsilon}{\lambda}\right) x(0).$$

In the second interval between $t = T \cdot p$ and $t = T$ the differential equation is the same as without control,

$$\dot{x}(t) = \lambda x(t), \qquad \text{for} \qquad T \cdot p < t < T.$$

From this one has immediately

$$x(t) = \mathrm{e}^{\lambda(t - T \cdot p)} x(T \cdot p).$$

If the beginning of the integration period again is denoted by t, this defines an iteration dynamics,

$$\begin{aligned} x(t+T) &= \mathrm{e}^{\lambda(1-p)T}\left(\left(1 + \frac{\mu\varepsilon}{\lambda}\right)\mathrm{e}^{\lambda T \cdot p} - \frac{\mu\varepsilon}{\lambda}\right) x(t) \\ &= \left(\left(1 + \frac{\mu\varepsilon}{\lambda}\right)\mathrm{e}^{\lambda T} - \frac{\mu\varepsilon}{\lambda}\mathrm{e}^{\lambda(1-p)T}\right), \end{aligned}$$

and the Floquet multiplier of an orbit is given by

$$\mathrm{e}^{\gamma T} = \left(1 - \frac{\mu\varepsilon}{\lambda}\right)\mathrm{e}^{\lambda T} + \frac{\mu\varepsilon}{\lambda}\mathrm{e}^{\lambda(1-p)T} = \mathrm{e}^{\lambda T}\left(1 - \frac{\mu\varepsilon}{\lambda}\left(1 - \mathrm{e}^{-\lambda pT}\right)\right). \qquad (2.130)$$

One finds that in zero order the 'strength' of control is given by the product $p \cdot \mu\varepsilon$; in fact there is a weak linear correction in p. For $\lambda pT \leq 1$ one has

$$\begin{aligned} \mathrm{e}^{\gamma T} &= \mathrm{e}^{\lambda T}\left(1 + \mu\varepsilon pT + \frac{1}{2}\mu\varepsilon\lambda p^2 T^2 + o\left(p^3\right)\right) \\ &= \mathrm{e}^{\lambda T}\left(1 + \mu\varepsilon pT\left(1 - \frac{1}{2}\lambda pT + o\left(p^2\right)\right)\right), \end{aligned}$$

i.e., to get a constant strength of control, one has to fulfill the condition

$$\mu\varepsilon pT = \frac{1}{1 - \frac{\lambda T}{2}p} = 1 + \frac{\lambda T}{2}p + o(p^2).$$

The result is, apart from a weak linear correction for OGY control the length of the impulse can be chosen arbitrarily, and the 'strength' of control in zero order is given by the time integral over the control impulse.

Floquet Stability Analysis of Difference Control

Again the starting point is the linearized equation of motion around the periodic orbit when control is applied. For difference control now there is a dependency on two past time steps,

$$\dot{x}(t) = \lambda x(t) + \mu\varepsilon x(t - (t \mod T)) - \mu\varepsilon x(t - T - (t \mod T)). \qquad (2.131)$$

Although the r.h.s of (2.131) depends on x at three different times, it can be nevertheless integrated exactly, which is mainly due to the fact that the two past times (of the two last Poincaré crossings) have a fixed time difference being equal to the orbit length. This allows not only for an exact solution, but also offers a correspondence to the time–discrete dynamics and the matrix picture used in time–delayed coordinates [CMP98a, CS98, CMP98b].

Stability Analysis of Difference Control

Now also for difference control the experimentally more common situation of a finite but small measurement delay $T{\cdot}s$ is considered, together with a finite impulse length $T{\cdot}p$ (here $0 < p < 1$ and $0 < (s + p) < 1$) [Cla02b].

In the first time interval between $t = 0$ and $t = T{\cdot}s$ the ODE reads

$$\dot{x}(t) = \lambda x(t), \qquad \text{for} \qquad 0 < t < T \cdot s.$$

The integration gives $x(T \cdot s) = \mathrm{e}^{\lambda T \cdot s} x(0)$.

For the second interval between $t = T \cdot s$ and $t = T \cdot (s + p)$ we have

$$\dot{x}(t) = \lambda x(t) - \mu\varepsilon x(0) = \lambda x(t) + \mu\varepsilon (x(0) - x(-T)), \quad \text{for} \quad T{\cdot}s < t < T{\cdot}(s+p).$$

Integration of this ODE yields

$$x(t) = -\frac{\mu\varepsilon}{\lambda}(x(0) - x(-T)) + \frac{\mu\varepsilon}{\lambda}(x(0) - x(-T)) + \mathrm{e}^{\lambda sT} x(0) \mathrm{e}^{\lambda(t-sT)}$$

$$x(T(s+p)) = -\frac{\mu\varepsilon}{\lambda}(x(0) - x(-T)) \frac{\mu\varepsilon}{\lambda}(x(0) - x(-T)) + \mathrm{e}^{\lambda pT} + \mathrm{e}^{\lambda(s+p)T} x(0).$$

For the third interval, the ODE is homogeneous again and one has

$$x(t) = \mathrm{e}^{\lambda(t-(s+p)T)} x(T \cdot (s+p)), \qquad \text{for} \qquad T \cdot (s+p) < t < T.$$

Insertion gives

$$x(T) = x(0)\mathrm{e}^{\lambda T}\left(1 + \frac{\mu\varepsilon}{\lambda}\mathrm{e}^{-\lambda sT}(1 - \mathrm{e}^{-\lambda pT})\right) - x(-T)\mathrm{e}^{\lambda T}\frac{\mu\varepsilon}{\lambda}\mathrm{e}^{-\lambda sT}(1 - \mathrm{e}^{-\lambda pT})$$

or, in time–delayed coordinates of the last and last but one Poincaré crossing [Cla02b]

$$\begin{pmatrix} x_{n+1} \\ x_n \end{pmatrix} = \begin{pmatrix} \mathrm{e}^{\lambda T}\left(1 + \frac{\mu\varepsilon\left(1-\mathrm{e}^{-\lambda pT}\right)}{\lambda \mathrm{e}^{\lambda sT}}\right) & -\mathrm{e}^{\lambda T}\frac{\mu\varepsilon\left(1-\mathrm{e}^{-\lambda pT}\right)}{\lambda \mathrm{e}^{\lambda sT}} \\ 1 & 0 \end{pmatrix} \begin{pmatrix} x_n \\ x_{n-1} \end{pmatrix}.$$

If we identify with the coefficients of the time–discrete case, $\lambda_{\mathsf{d}} = \mathrm{e}^{\lambda T}$ and $\mu_{\mathsf{d}}\varepsilon_{\mathsf{d}} = \mathrm{e}^{-\lambda sT}(1-\mathrm{e}^{\lambda pT})\frac{\mu\varepsilon}{\lambda}$, the dynamics in the Poincaré iteration $t = nT$ becomes identical with the pure discrete description; this again illustrates the power of the concept of the Poincaré map. Due to the low degree of the characteristic polynomial, one in principle can explicitly diagonalize the iteration matrix, allowing for a closed expression for the nth power of the iteration matrix. As for the stability analysis only the eigenvalues are needed, this straightforward calculation is excluded here.

For the Floquet multiplier one has [Cla02b]

$$\mathrm{e}^{2\gamma T} = \mathrm{e}^{\gamma T}\mathrm{e}^{\lambda T}\left(1+\frac{\mu\varepsilon}{\lambda}\mathrm{e}^{-\lambda sT}\left(1-\mathrm{e}^{-\lambda pT}\right)\right) - \mathrm{e}^{\lambda T}\frac{\mu\varepsilon}{\lambda}\mathrm{e}^{-\lambda sT}\left(1-\mathrm{e}^{-\lambda pT}\right).$$

This quadratic equation yields two Floquet multipliers,

$$\begin{aligned}\mathrm{e}^{\gamma T} &= \frac{1}{2}\mathrm{e}^{\lambda T}\left(1+\frac{\mu\varepsilon}{\lambda}\mathrm{e}^{-\lambda sT}\left(1-\mathrm{e}^{-\lambda pT}\right)\right)\\ &\pm \frac{1}{2}\sqrt{\left(\mathrm{e}^{\lambda T}\left(1+\frac{\mu\varepsilon}{\lambda}\mathrm{e}^{-\lambda sT}\left(1-\mathrm{e}^{-\lambda pT}\right)\right)\right)^2 + 4\mathrm{e}^{\lambda T}\frac{\mu\varepsilon}{\lambda}\mathrm{e}^{-\lambda sT}\left(1-\mathrm{e}^{-\lambda pT}\right)}.\end{aligned}$$

For $s = 0$ one gets the special cases discussed above.

2.3.7 Blind Chaos Control

One of the most surprising successes of chaos theory has been in biology: the experimentally demonstrated ability to control the timing of spikes of electrical activity in complex and apparently chaotic systems such as heart tissue [GSD92] and brain tissue [SJD94]. In these experiments, PPF control, a modified formulation of OGY control [OGY90], was applied to set the timing of external stimuli; the controlled system showed stable periodic trajectories instead of the irregular inter–spike intervals seen in the uncontrolled system. The mechanism of control in these experiments was interpreted originally as analogous to that of OGY control: unstable periodic orbits riddle the chaotic attractor and the electrical stimuli place the system's state on the stable manifold of one of these periodic orbits [KY79].

Alternative possible mechanisms for the experimental observations have been described by Zeng and Glass [GZ94] and Christini and Collins [CC95]. These authors point out that the controlling external stimuli serve to truncate the inter–spike interval to a maximum value. When applied, the control stimulus sets the next interval s_{n+1} to be on the line

$$s_{n+1} = As_n + C. \tag{2.132}$$

We will call this relationship the 'control line.' Zeng and Glass showed that if the uncontrolled relationship between inter–spike intervals is a chaotic 1D function, $s_{n+1} = f(s_n)$, then the control system effectively flattens the top of this map and the controlled dynamics may have fixed points or other periodic orbits [GZ94]. Christini and Collins showed that behavior analogous to the fixed–point control seen in the biological experiments can be accomplished even in completely random systems [CC95]. Since neither chaotic 1D systems nor random systems have a stable manifold, the interval–truncation interpretation of the biological experiments is different than the OGY interpretation. The interval–truncation method differs also from OGY and related control methods in that the perturbing control input is a fixed–size stimulus whose timing can be treated as a continuous parameter. This type of input is conventional in cardiology (e.g., [HCT97]).

Kaplan demonstrated in [KY79] that the state–truncation interpretation was applicable in cases where there was a stable manifold of a periodic orbit as well as in cases where there were only unstable manifolds. He found that superior control could be achieved by intentionally placing the system's state off of any stable manifold. That suggested a powerful scheme for the rapid experimental identification of fixed points and other periodic orbits in systems where inter–spike intervals were of interest.

The chaos control in [GSD92] and [SJD94] was implemented in two stages. First, inter–spike intervals s_n from the uncontrolled, 'natural' system were observed. Modelling the system as a function of two variables

$$s_{n+1} = f(s_n, s_{n-1}),$$

the location $s^\star$ of a putative unstable flip-saddle type fixed–point and the corresponding stable eigenvalue λ_s were estimated from the data[23] [CK97]. The linear approximation to the stable manifold lies on a line given by (2.132) with

$$A = \lambda_s \qquad \text{and} \qquad C = (1 - \lambda_s)s^\star.$$

Second, using estimated values of A and C, the control system was turned on. Following each observed interval s_n, the maximum allowed value of the next inter–spike interval was computed as

$$S_{n+1} = As_n + C.$$

If the next interval naturally was shorter than S_{n+1} no control stimulus was applied to the system. Otherwise, an external stimulus was provided to truncate the inter–spike interval at $s_{n+1} = S_{n+1}$.

In practice, the values of $s^\star$ and λ_s for a real fixed–point of the natural system are known only imperfectly from the data. Insofar as the estimates are inaccurate, the control system does not place the state on the true stable manifold. Therefore, we will analyze the controlled system without presuming that A and C in (2.132) correspond to the stable manifold.

If the natural dynamics of the system is modelled by

$$s_{n+1} = f(s_n, s_{n-1}),$$

[23] Since the fixed–point is unstable, there is also an unstable eigenvalue λ_u.

then the dynamics of the controlled system is given by [KY79]

$$s_{n+1} = \min \begin{cases} f(s_n, s_{n-1}) : \text{ Natural Dynamics}, \\ As_n + C \quad : \text{ Control Line}. \end{cases} \tag{2.133}$$

We can study the dynamics of the controlled system close to a natural fixed–point, $s^\star$, by approximating the natural dynamics linearly as[24]

$$s_{n+1} = f(s_n, s_{n-1}) = (\lambda_s + \lambda_u)s_n - \lambda_s\lambda_u s_{n-1} + s^\star(1 + \lambda_s\lambda_u - \lambda_s - \lambda_u).$$

Since the controlled system (2.133) is nonlinear even when $f()$ is linear, it is difficult to analyze its behavior by algebraic iteration. Nonetheless, the controlled system can be studied in terms of 1D maps.

Following any inter–spike interval when the controlling stimulus has been applied, the system's state (s_n, s_{n-1}) will lie somewhere on the control line. From this time onward the state will lie on an image of the control line even if additional stimuli are applied during future inter–spike intervals.

The stability of the controlled dynamics fixed–point and the size of its basin of attraction can be analyzed in terms of the control line and its image. When the previous inter–spike interval has been terminated by a control stimulus, the state lies somewhere on the control line. If the controlled dynamics are to have a stable fixed–point, this must be at the controller fixed–point $x^\star$ where the control line intersects the line of identity. However, the controller fixed–point need not be a fixed–point of the controlled dynamics. For example, if the image of the controller fixed–point is below the controller fixed–point, then the inter–spike interval following a stimulus will be terminated naturally.

For the controller fixed–point to be a fixed–point of the controlled dynamics, we require that the natural image of the controller fixed–point be at or above the controller fixed–point. Thus the dynamics of the controlled system, close to $x^\star$, are given simply by

$$s_{n+1} = As_n + C \tag{2.134}$$

The fixed–point of these dynamics is stable so long as $-1 < A < 1$. In the case of a flip saddle, we therefore have a simple recipe for successful state-truncation control: position $x^\star$ below the natural fixed–point $s^\star$ and set $-1 < A < 1$.

Fixed points of the controlled dynamics can exist for natural dynamics other than flip saddles. This can be seen using the following reasoning: Let ξ be the difference between the controller fixed–point and the natural fixed–point: $s^\star = x^\star + \xi$. Then the natural image of the controller fixed–point can be found from (2.134) to be [KY79]

$$s_{n+1} = (\lambda_s + \lambda_u)x^\star - \lambda_s\lambda_u x^\star + (1 + \lambda_s\lambda_u - \lambda_s - \lambda_u)(x^\star + \xi). \tag{2.135}$$

[24] Equation (2.134) is simply the linear equation $s_{n+1} = as_n + bs_{n-1} + c$ with a, b, and c set to give eigenvalues λ_s and λ_u and fixed–point $s^\star$.

Table 2.1. Cases which lead to a stable fixed–point for the controlled dynamics. In all cases, it is assumed that $|\mathcal{A}| < 1$. (For the cases where $\lambda_s < -1$, the subscript s in λ_s is misleading in that the corresponding manifold is unstable. For the spiral, there is no stable manifold (adapted from [KY79]).)

Type of FP	λ_u	λ_s	$x^\star$ Locat.
Flip saddle	$\lambda_u < -1$	$-1 < \lambda_s < 1$	$x^\star < s^\star$
Saddle	$\lambda_u > 1$	$-1 < \lambda_s < 1$	$x^\star > s^\star$
Single–flip repeller	$\lambda_u > 1$	$\lambda_s < -1$	$x^\star > s^\star$
Double–flip repeller	$\lambda_u < -1$	$\lambda_s < -1$	$x^\star < s^\star$
Spiral (complex λ)	$\lvert\lambda_u\rvert > 1$	$\lvert\lambda_s\rvert > 1$	$x^\star < s^\star$

The condition that

$$s_{n+1} \geq x^\star \tag{2.136}$$

will be satisfied depending only on λ_s, λ_u, and $\xi = s^\star - x^\star$. This means that for any flip saddle, so long as $x^\star < s^\star$, the point $x^\star$ will be a fixed–point of the controlled dynamics and will be stable so long as $-1 < A < 1$.

Equations (2.135) and (2.136) imply that control can lead to a stable fixed–point for any type of fixed–point except those for which both λ_u and λ_s are greater than 1 (so long as $-1 < A < 1$). Since the required relationship between $x^\star$ and $s^\star$ for a stable fixed–point of the controlled dynamics depends on the eigenvalues, it is convenient to divide the fixed points into four classes, as given in Table 2.1.

Beyond the issue of the stability of the fixed–point of the controlled dynamics, there is the question of the size of the fixed–point's basin of attraction. Although the local stability of the fixed–point is guaranteed for the cases in Table 2.1 for $-1 < A < 1$, the basin of attraction of this fixed–point may be small or large depending on A, C, $s^\star$, λ_u and λ_s.

The endpoints of the basin of attraction can be derived analytically [KY79]. The size of the basin of attraction will often be zero when A and C are chosen to match the stable manifold of the natural system. Therefore, in order to make the basin large, it is advantageous intentionally to misplace the control line and to put $x^\star$ in the direction indicated in Table 2.1. In addition, control may be enhanced by setting $A \neq \lambda_s$, for instance $A = 0$.

If the relationship between $x^\star$ and $s^\star$ is reversed from that given in Table 2.1, the controlled dynamics will not have a stable fixed points. To some extent, these can also be studied using 1D maps. The flip saddle and double–flip repeller can display stable period–2 orbits and chaos. For the non–flip saddle and single–flip repeller, control is unstable when $x^\star < s^\star$.

The fact that control may be successful or even enhanced when A and C are not matched to λ_s and $s^\star$ suggests that it may be useful to reverse the experimental procedure often followed in chaos control. Rather than first identifying the parameters of the natural unstable fixed points and then applying the control, one can blindly attempt control and then deduce the natural dynamics from the behavior of the controlled system. This use of PPF control is reminiscent of pioneering studies that used periodic stimulation to demonstrate the complex dynamics of biological preparations [GGS81].

As an example, consider the *Hénon map*:

$$s_{n+1} = 1.4 + 0.3 s_{n-1} - s_n^2.$$

This system has two distinct fixed points. There is a flip–saddle at $s^\star = 0.884$ with $\lambda_u = -1.924$ and $\lambda_s = 0.156$ and a non–flip saddle at $s^\star = -1.584$ with $\lambda_u = 3.26$ and $\lambda_s = -0.092$. In addition, there is an unstable flip–saddle orbit of period 2 following the sequence $1.366 \to -0.666 \to 1.366$. There are no real orbits of period 3, but there is an unstable orbit of period 4 following the sequence $.893 \to .305 \to 1.575 \to -.989 \to .893$. These facts can be deduced by algebraic analysis of the equations.

In an experiment using the controlled system, the control parameter $x^\star = C/(1-A)$ can be varied. The theory presented above indicates that the controlled system should undergo a bifurcation as $x^\star$ passes through $s^\star$. For each value of $x^\star$, the controlled system was iterated from a random initial condition and the values of s_n plotted after allowing a transient to decay. A bifurcation from a stable fixed–point to a stable period 2 as $x^\star$ passes through the flip–saddle value of $s^\star = 0.884$. A different type bifurcation occurs at the non–flip saddle fixed–point at $s^\star = -1.584$. To the left of the bifurcation point, the iterates are diverging to $-\infty$ and are not plotted.

Adding gaussian dynamical noise (of standard deviation 0.05) does not substantially alter the bifurcation diagram, suggesting that examination of the truncation control bifurcation diagram may be a practical way to read off the location of the unstable fixed points in an experimental preparation.

Unstable periodic orbits can be difficult to find in uncontrolled dynamics because there is typically little data near such orbits. Application of PPF control, even blindly, can stabilize such orbits and dramatically improve the ability to locate them. This, and the robustness of the control, may prove particularly useful in biological experiments where orbits may drift in time as the properties of the system change [KY79].

2.4 Synchronization in Chaotic Systems

Recall that *synchronization* phenomena occur abundantly in nature and in day to day life. A few well known examples are the observations in coupled systems such as pendulum clocks, radio circuits, swarms of light–emitting fireflies, groups of neurons and neuronal ensembles in sensory systems, chemical systems, Josephson junctions, cardiorespiratory interactions, etc. Starting from the observation of pendulum clocks by Huygens, a vast literature already exists which studies synchronization in coupled nonlinear systems – in systems of coupled maps as well as in oscillators and networks (see [PRK01] and references therein). In recent times, different kinds of synchronization have been classified – mutual synchronization, lag synchronization, phase synchronization and complete synchronization (see [RPK96, Bal06]).

2.4.1 Lyapunov Vectors and Lyapunov Exponents

Here, following [PGY06], we discuss a method to determine *Lyapunov exponents* (LEs) from suitable ensemble averages. It is easy to write down a formal meaningful definition, but the problem lies in translating it into a workable procedure. With reference to an ND discrete–time system, given by the *map*

$$\mathbf{x}_{t+1} = \mathbf{f}_d(\mathbf{x}_t), \qquad (\mathbf{x} \in \mathbb{R}^N), \tag{2.137}$$

one can express the ith LE (as usual, LE are supposed to be ordered from the largest to the smallest one) as

$$\lambda^{(i)} = \frac{1}{2}\int d\mathbf{x}P(\mathbf{x}) \ln\left[\frac{||\partial_x \mathbf{f}_d \mathbf{V}^{(i)}(\mathbf{x})||^2}{||\mathbf{V}^{(i)}(\mathbf{x})||^2}\right] \tag{2.138}$$

where $P(\mathbf{x})$ is the corresponding invariant measure, $\partial_x \mathbf{f}_d$ is the Jacobian of the transformation, and the Lyapunov vector $\mathbf{V}^{(i)}(\mathbf{x})$ identifies the ith most expanding direction in $\mathbf{x}$.

With reference to a continuous–time system, ruled by the ODE

$$\dot{\mathbf{x}} = \mathbf{f}_c(\mathbf{x}), \qquad (\mathbf{x} \in \mathbb{R}^N). \tag{2.139}$$

the ith LE is defined by

$$\lambda^{(i)} = \int d\mathbf{x}P(\mathbf{x}) \frac{[\partial_x \mathbf{f}_c \mathbf{V}^{(i)}(\mathbf{x})] \cdot \mathbf{V}^{(i)}(\mathbf{x})}{||\mathbf{V}^{(i)}(\mathbf{x})||^2}, \tag{2.140}$$

where $\cdot$ denotes the scalar product.

Unless a clear procedure to determine the LV is given, (2.138, 2.140) are nothing but formal statements. As anticipated in the introduction, $\mathbf{V}^{(i)}(\mathbf{x})$ can be obtained by following a two–step procedure. We start with a generic set of i linearly independent vectors lying in the tangent space and let them evolve in time. This is the standard procedure to determine LEs, and it is well known that the hyper–volume $\mathbf{Y}^{(i)}$ identified by such vectors contains for, large enough times, the i most expanding directions. Furthermore, with reference to the set of orthogonal cordinates got by implementing the *Gram–Schmidt procedure*, the component v_k of a generic vector $\mathbf{v}$ evolves according to the following ODE [EP98]

$$\dot{v}_k = \sum_{j=k}^{i} \sigma_{k,j}(\mathbf{x})v_j, \qquad (1 \le k \le i), \tag{2.141}$$

where $\sigma_{k,j}$ does not explicitly depend on time, but only through the position $\mathbf{x}$ in the phase–space. As a result, the ith Lyapunov exponent can be formally expressed as the ensemble average of the local expansion rate $\sigma_{i,i}$, i.e.,

$$\lambda^{(i)} = \int d\mathbf{x} P(\mathbf{x}) \sigma_{i,i}(\mathbf{x}). \qquad (2.142)$$

By comparing with (2.140), one finds the obvious equality

$$\sigma_{i,i} = \frac{[\partial_x \mathbf{f}_c \mathbf{V}^{(i)}(\mathbf{x})] \cdot \mathbf{V}^{(i)}(\mathbf{x})}{||\mathbf{V}^{(i)}(\mathbf{x})||^2}. \qquad (2.143)$$

In subsection 2.4.1 below, we will apply this formalism to a phase–synchronization problem, and we will find that the only workable way to get an analytic expression for $\sigma_{i,i}$ passes through the determination of the direction of the corresponding LV vector $\mathbf{V}^{(i)}(\mathbf{x})$.

Let us now consider the *backward evolution* of a generic vector $\mathbf{V}^{(i)} \in \mathbf{Y}^{(i)}$. Its direction is identified by the $(i-1)$D vector

$$\mathbf{u} \equiv (u_1, u_2, \dots, u_{i-1}), \qquad (2.144)$$

where $u_k = v_k/v_i$. From (2.141) and the definition of $\mathbf{u}$, it follows that the backward evolution follows the equation

$$\dot{u}_k = (\sigma_{i,i} - \sigma_{k,k}) u_k - \sum_{j=k+1}^{i-1} \sigma_{k,j}(t) u_j - \sigma_{k,i}, \qquad (1 \le k < i). \qquad (2.145)$$

This is a cascade of skew–product linear stable equations (they are stable because the Lyapunov exponents are organized in descending order). The overall stability is basically determined by the smallest $(\sigma_{k,k} - \sigma_{i,i})$ that is got for $k = i-1$. It is, therefore, sufficient to turn our attention to the last $(i-1)$ component of the vector $\mathbf{V}$. Its equation has the following structure

$$\dot{u}(t) = \gamma u + \sigma(t), \qquad (2.146)$$

where $\gamma = \lambda_i - \lambda_{i-1} < 0$ and we have dropped the subscript i for simplicity. The value of the direction u is got by integrating this equation. By neglecting the temporal fluctuations of γ (it is not difficult to include them, but this is not important for our final goal), the formal solution of (2.146) reads

$$u(\mathbf{x}(t)) = \int_{-\infty}^{t} \mathrm{e}^{\gamma(t-\tau)} \sigma(\mathbf{x})\, d\tau. \qquad (2.147)$$

This equation does not simply tell us the value of u at time t, but the value of u when the trajectory sits in $\mathbf{x}(\mathbf{t})$. It is in fact important to investigate the dependence of u on $\mathbf{x}$. We proceed by determining the deviation $\delta_j u$ induced by a perturbation δx_j of $\mathbf{x}$ along the jth direction,

$$\delta_j u = \int_{-\infty}^{t} \mathrm{e}^{\gamma(t-\tau)} \delta_j \sigma(\tau)\, d\tau, \qquad (2.148)$$

where, assuming a smooth dependence of σ on $\mathbf{x}$,

$$\delta_j\sigma(\tau) \approx \sigma_x(\tau)\delta x_j(\tau) = \sigma_x(\tau)\delta x_j(t)\mathrm{e}^{\lambda_j(t-\tau)} \tag{2.149}$$

(notice that the dynamics is flowing backward). If the Lyapunov exponent λ_j is negative, $\delta_j\sigma(\tau)$ decreases for $\tau \to -\infty$ and the integral over τ in (2.148) converges. As a result, $\delta_j u$ is proportional to δx_j, indicating that the direction of the LV is smooth along the jth direction. If λ_j is positive, $\delta_j\sigma(\tau)$ diverges, and below time t_0, where

$$\delta x_j(t)\mathrm{e}^{\lambda_j(t-t_0)} = 1, \tag{2.150}$$

linearization breaks down. In this case, $\delta\sigma(\tau)$ for $\tau < t_0$ is basically uncorrelated with its 'initial value' $\delta_j\sigma(t)$ and one can estimate $\delta_j u$, by limiting the integral to the range $[t_0, t]$

$$\delta_j u(t) = \delta x_j(t)\int_{t_0}^{t} d\tau \mathrm{e}^{(\lambda_j+\gamma)(t-\tau)}\sigma_x(\tau), \tag{2.151}$$

where t_0 is given by (2.150). By bounding σ_x with constant functions and thereby performing the integral in (2.151), we finally get

$$\delta_j u(t) \approx \delta x_j(t) + \delta x_j(t)^{-\gamma/\lambda_j}. \tag{2.152}$$

The scaling behavior is finally got as the smallest number between 1 and $-\gamma/\lambda_j$. If we now introduce the exponent η_j to identify the scaling behavior of the deviation of the LV direction when the point of reference is moved along the jth direction in phase–space, the results are summarized in the following way

$$\eta_j = \begin{cases} 1, & \text{for } \lambda_j \le -\gamma, \\ -\gamma/\lambda_j, & \text{for } \lambda_j > -\gamma. \end{cases} \tag{2.153}$$

The former case corresponds to a smooth behavior (the derivative is finite), while the latter one reveals a singular behavior that is the signature of a generalized synchronization.

Forced Rössler Oscillator

The first model where phase synchronization has been explored is the periodically *forced Rössler oscillator* [RPK96]. In this section we derive a discrete–time map describing a forced Rössler system in the limit of weak coupling [PZR97]. We start with the ODE,

$$\begin{aligned} \dot{x} &= -y - z + \varepsilon y\cos(\Omega t + \psi_0), \\ \dot{y} &= x + a_0 y - \varepsilon x\sin(\Omega t + \psi_0), \\ \dot{z} &= a_1 + z(x - a_2), \end{aligned} \tag{2.154}$$

where ψ_0 fixes the phase of the forcing term at time 0. It is convenient to introduce cylindrical coordinates, namely $\mathbf{u} = (\varphi, r, z)$, $(x = r\cos\phi,\ y = r\sin\phi)$. For the future sake of clarity, let us denote with $\mathbf{S_c}$ the 3D space parametrized by such coördinates, so that (2.154) reads

$$\dot{\mathbf{u}} = \mathbf{F}(\mathbf{u}) + \varepsilon\mathbf{G}(\mathbf{u}, \Omega t + \psi_0), \qquad \text{where} \tag{2.155}$$

$$\mathbf{F} = \left[1 + \frac{z}{r}\sin\phi + \frac{a_0}{2}\sin 2\phi,\ a_0 r\sin^2\phi - z\cos\phi,\ a_1 + z(r\cos\phi - a_2),\right]$$
$$\mathbf{G} = \Big[-\sin^2\phi\cos(\Omega t + \psi_0) - \cos^2\phi\sin(\Omega t + \psi_0),$$
$$\frac{r}{\sqrt{2}}\sin 2\phi\cos(\Omega t + \psi_0 + \pi/4),\ 0\Big].$$

Note that system (2.155) can be written in the equivalent autonomous form

$$\dot{\mathbf{u}} = \mathbf{F}(\mathbf{u}) + \varepsilon\mathbf{G}(\mathbf{u}, \psi), \qquad \dot{\psi} = \Omega,$$

where ψ denotes the phase of the forcing term.

We pass to a discrete–time description, by monitoring the system each time the phase ϕ is a multiple of 2π. In the new framework, the relevant variables are r, z, and ψ, all measured when the Poincaré section is crossed. The task is to determine the transformation map the state (r, z, ψ) onto (r', z', ψ').

In order to get the expression of the map, it is necessary to formally integrate the equations of motion from one to the next section. This can be done, by expanding around the un–perturbed solution for $\varepsilon = 0$ (which must nevertheless be obtained numerically). The task is anyhow worth, because it allows determining the structure of the resulting map, which turns out to be [PGY06]

$$\begin{aligned}
\psi' &= \psi + \langle T^{(0)}\rangle\,\Omega + A_1 + \varepsilon\left(B_1^c\cos\psi + B_1^s\sin\psi\right),\\
r' &= A_2 + \varepsilon\left(B_2^c\cos\psi + B_2^s\sin\psi\right),\\
z' &= A_3 + \varepsilon\left(B_3^c\cos\psi + B_3^s\sin\psi\right),
\end{aligned} \tag{2.156}$$

where $\langle T^{(0)}\rangle$ is the average period of the un–perturbed Rössler oscillator and A_m's and B_m's are functions of z and r. They can be numerically determined by integrating the appropriate set of equations. Up to first order in ε, the structure of the model is fairly general as it is got for a generic periodically forced oscillator represented in cylindrical coordinates (as long the phase of the attractor can be unambiguously identified).

For the usual parameter values, the Rössler attractor is characterized by a strong contraction along one direction [YML00]. As a result, one can neglect the z dependence since this variable is basically a function of r, and thus write

$$\begin{aligned}
\psi' &= \psi + \langle T^{(0)}\rangle\,\Omega + A_1(r) + \varepsilon\left(B_1^c(r)\cos\psi + B_1^s(r)\sin\psi\right),\\
r' &= A_2(r) + \varepsilon\left(B_2^c(r)\cos\psi + B_2^s(r)\sin\psi\right),
\end{aligned} \tag{2.157}$$

where all the functions can be obtained by integrating numerically the equations of motion of the single Rössler oscillator.

To simplify further manipulations, we finally recast equation (2.157) in the form

$$\begin{aligned} \psi' &= \psi + K + A_1(r) + \varepsilon g_1(r)\cos\left(\psi + \beta_1(r)\right), \\ r' &= A_2(r) + \varepsilon\, g_2(r)\cos\left(\psi + \beta_2(r)\right), \qquad \text{where} \\ B_i^c(r) &= g_i(r)\cos\beta_i(r), \qquad B_i^s(r) = -g_i(r)\sin\beta_i(r) \end{aligned} \tag{2.158}$$

for $i = 1, 2$. The parameter $K = \langle T^{(0)}\rangle\,\Omega - 2\pi$ represents the detuning between the original Rössler–system average frequency and the forcing frequency Ω.

The GSF (2.158) generalizes the model introduced in [PZR97], where the effect of the phase on the r dynamics was not included. This implies that the GSF looses the skew–product structure. This has important consequences on the orientation of the second Lyapunov vector that we determine in the next sections. Notice also that the GSF (2.158) generalizes and justifies the model invoked in [POR97].

For the sake of simplicity, we have analyzes the following model,

$$\begin{aligned} r' &= f(r) + 2\varepsilon c g(r)\cos(\psi + \alpha) \\ \psi' &= \psi + K + \Delta r + \varepsilon b\cos\psi, \qquad \text{where} \\ f(r) &= 1 - 2|r|, \quad g(r) = r^2 - |r|, \end{aligned} \tag{2.159}$$

with $r \in [-1, 1]$. The tent–map choice for r ensures that $[-1, 0]$ and $[0, 1]$ are the two atoms of a Markov partion. Moreover, since $g(r)$ is equal to 0 for $r = 0$ and $r = \pm 1$, this remains true also when the perturbation is switched on.

In this 2D setup, the formal expression of the ith LE (2.138) reads

$$\lambda^{(i)} = \frac{1}{2}\int_{-1}^{1} dr \int_{0}^{2\pi} d\psi P(r,\psi)\ln\left[\frac{||\mathbf{J}(r,\psi)\mathbf{V}^{(i)}(r,\psi)||^2}{||\mathbf{V}^{(i)}(r,\psi)||^2}\right], \tag{2.160}$$

and the Jacobian is given by

$$\mathbf{J}(r,\psi) = \begin{pmatrix} f_r(r) + 2\varepsilon c g_r(r)\cos(\psi+\alpha) & -2\varepsilon c g(r)\sin(\psi+\alpha) \\ \Delta & 1 - \varepsilon b\sin\psi \end{pmatrix},$$

where the subscript r denotes the derivative with respect to r. The computation of the Lyapunov exponent therefore, requires determining both the invariant measure $P(r, \psi)$ and the local direction of the Lyapunov vector $\mathbf{V}^{(i)}$.

Second Lyapunov Exponent: Perturbative Calculation

Here we derive a perturbative expression for the second LE of the GSF (2.159), by expanding (2.160). One of the key ingredients is the second LV, whose direction can be identified by writing $\mathbf{V} = (V, 1)$ (for the sake of clarity, from now on, we omit the superscript $i = 2$ in $\mathbf{V}$ and λ, as we shall refer only to the second direction). Due to the skew-product structure of the un–perturbed map (2.159), the second LV is, for $\varepsilon = 0$, aligned along the ψ direction (i.e., $V = 0$). It is therefore natural to expand V in powers of ε

$$V \approx \varepsilon v_1(r, \psi) + \varepsilon^2 v_2(r, \psi). \tag{2.161}$$

Accordingly, the logarithm of the norm of $\mathbf{V}$ is

$$\ln ||\mathbf{V}||^2 = \ln(1 + \varepsilon^2 v_1^2) = \varepsilon^2 v_1^2,$$

while its forward iterate writes as (including only those terms that contribute up to second order in the norm),

$$\mathbf{JV} = \begin{pmatrix} \varepsilon f_r(r) v_1 - 2c\varepsilon g(r) \sin(\psi + \alpha) \\ 1 + \varepsilon(\Delta v_1 - b \sin \psi) + \varepsilon^2 \Delta v_2) \end{pmatrix}, \tag{2.162}$$

Notice that we have omitted the (r, ψ) dependence of v_1 and v_2 to keep the notation compact [PGY06].

The Euclidean norm of the forward iterate is

$$\begin{aligned} ||\mathbf{JV}||^2 = 1 + 2\varepsilon(\Delta v_1 - b \sin \psi) + \varepsilon^2 \{(\Delta v_1 - b \sin \psi)^2 \\ + 2\Delta v_2 + [f_r(r) v_1 - 2cg(r) \sin(\psi + \alpha)]^2\}, \end{aligned}$$

and its logarithm is

$$\begin{aligned} \ln ||\mathbf{JV}||^2 = 2\varepsilon(\Delta v_1 - b \sin \psi) - \varepsilon^2 \{(\Delta v_1 - b \sin \psi)^2 \\ - 2\Delta v_2 - [f_r(r) v_1 - 2cg(r) \sin(\psi + \alpha)]^2\}. \end{aligned}$$

We now proceed by formally expanding the invariant measure in powers of ε

$$P(r, \psi) \approx p_0(\psi) + \varepsilon p_1(r, \psi) + \varepsilon^2 p_2(r, \psi). \tag{2.163}$$

The determination of the p_i coefficients is presented in the next section, but here we anticipate that, as a consequence of the skew-product structure for $\varepsilon = 0$, the zeroth-order component of the invariant measure does not depend on the phase ψ. Moreover, because of the structure of the tent-map, p_0 is also independent of r, i.e., $p_0 = 1/4\pi$. The second Lyapunov exponent can thus be written as

$$\begin{aligned} \lambda = \int_{-1}^{1} dr \int_0^{2\pi} d\psi \left(\frac{1}{4\pi} + \varepsilon p_1(r, \psi) \right) \{2\varepsilon(\Delta v_1(r, \psi) \\ - b \sin \psi) - \varepsilon^2 [(\Delta v_1(r, \psi) - b \sin \psi)^2 - 2\Delta v_2(r, \psi) \\ + [f_r(r) v_1(r, \psi) + 2cg(r) \sin(\psi + \alpha)]^2 + v_1^2(r, \psi)]\} + o(\varepsilon^2). \end{aligned} \tag{2.164}$$

As the variable ψ is a phase, it is not a surprise that some simplifications can be found by expanding the relevant functions into Fourier components. We start writing the first component of the invariant measure as

$$p_1(r,\psi) = \frac{1}{2\pi}\sum_n q_i(r)\mathrm{e}^{in\psi}. \tag{2.165}$$

We then consider the first order component $v_1(r,\psi)$ of the second LV (2.161). Due to the sinusoidal character of the forcing term in the GSF (2.159), it is easy to verify (see the next section) that $v_1(r,\psi)$ contains just the first Fourier component,

$$v_1(r,\psi) = c\left[L(r)\sin(\psi+\alpha) + R(r)\cos(\psi+\alpha)\right]. \tag{2.166}$$

By now, inserting (2.165–2.166) into (2.164) and performing the integration over ψ, we get

$$\begin{aligned}\lambda = \varepsilon^2 \int_{-1}^{1} dr \Big\{ &\Delta c[q_1^r[L(r)\sin\alpha + R(r)\cos\alpha] - q_1^i[L(r)\cos\alpha - R(r)\sin\alpha]] \\ &+ bq_1^i - \frac{b^2}{8} + \Delta\frac{bc}{4}\left[L(r)\cos\alpha - R(r)\sin\alpha\right] + \frac{c^2}{8}(3-\Delta^2)\left[L^2(r)+R^2(r)\right] \\ &+ \frac{c^2}{2}g^2(r) + c^2\frac{|r|}{r}g(r)L(r)\Big\} + \frac{\Delta I_2}{4\pi},\end{aligned} \tag{2.167}$$

where we have further decomposed $q_1(r)$ in its real and imaginary parts

$$q_1(r) = q_1^r(r) + iq_1^i(r),$$

and we have defined

$$I_2 = \int_{-1}^{1} dr \int_0^{2\pi} d\psi\, v_2(r,\psi), \tag{2.168}$$

which accounts for the contribution arising from the second order correction to the LV. This expansion shows that the highest–order contribution to the second Lyapunov exponent of the GSF scales quadratically with the perturbation amplitude. This is indeed a general result that does not depend on the particular choice of the functions used to define the GSF, but only on the skew–product structure of the un–perturbed time evolution and on the validity of the expansion assumed in (2.163).

According to [PGY06], we finally get the perturbative expression for the second LE,

$$\begin{aligned}\lambda = \varepsilon^2 \Bigg\{ &\frac{c^2}{30} - \frac{b^2}{4} + \int_{-1}^{1} dr [bq_1^i(r) + \frac{c^2}{16}(6-\Delta^2)[L^2(r)+R^2(r)] \\ &+ \Delta c q_1^r(r)[L(r)\sin\alpha + R(r)\cos\alpha] + \Delta c\left(\frac{b}{4} - q_1^i(r)\right)[L(r)\cos\alpha - R(r)\sin\alpha] \\ &+ c^2\frac{|r|}{r}g(r)L(r) + \frac{\Delta c^2}{4} r\sin\left(\frac{\Delta(1-r)}{2}\right)[L(r)\cos K - R(r)\sin K]]\Bigg\}.\end{aligned} \tag{2.169}$$

Accordingly, the numerical value of the second LE can be obtained by performing integrals which involve the four functions $q_1^r(r)$, $q_1^i(r)$, $L(r)$, and $R(r)$.

2.4.2 Phase Synchronization in Coupled Chaotic Oscillators

Over the past decade or so, *synchronization in chaotic oscillators* [FY83, PC90] has received much attention because of its fundamental importance in nonlinear dynamics and potential applications to laser dynamics [DBO01], electronic circuits [KYR98], chemical and biological systems [ESH98], and secure communications [KP95]. Synchronization in chaotic oscillators is characterized by the loss of exponential instability in the transverse direction through interaction. In coupled chaotic oscillators, it is known, various types of synchronization are possible to observe, among which are *complete synchronization* (CS) [FY83, PC90], *phase synchronization* (PS) [RPK96, ROH98], *lag synchronization* (LS) [RPK97] and *generalized synchronization* (GS) [KP96].

One of the noteworthy synchronization phenomena in this regard is PS which is defined by the phase locking between nonidentical chaotic oscillators whose amplitudes remain chaotic and uncorrelated with each other: $|\theta_1 - \theta_2| \leq$ const. Since the first observation of PS in mutually coupled chaotic oscillators [RPK96], there have been extensive studies in theory [ROH98] and experiments [DBO01]. The most interesting recent development in this regard is the report that the interdependence between physiological systems is represented by PS and *temporary phase–locking* (TPL) states, e.g., (a) *human heart beat and respiration* [SRK98], (b) a certain brain area and the tremor activity [TRW98, RGL99]. Application of the concept of PS in these areas sheds light on the analysis of non–stationary bivariate data coming from biological systems which was thought to be impossible in the conventional statistical approach. And this calls new attention to the PS phenomenon [KK00, KLR03].

Accordingly, it is quite important to elucidate a detailed transition route to PS in consideration of the recent observation of a TPL state in biological systems. What is known at present is that TPL[ROH98] transits to PS and then transits to LS as the coupling strength increases. On the other hand, it is noticeable that the phenomenon from non–synchronization to PS have hardly been studied, in contrast to the wide observations of the TPL states in the biological systems.

Here, following [KK00, KLR03], we study the characteristics of TPL states observed in the regime from non–synchronization to PS in coupled chaotic oscillators. We report that there exists a special locking regime in which a TPL state shows maximal periodicity, which phenomenon we would call *periodic phase synchronization* (PPS). We show this PPS state leads to local negativeness in one of the vanishing Lyapunov exponents, taking the measure by which we can identify the maximal periodicity in a TPL state. We present a qualitative explanation of the phenomenon with a nonuniform oscillator model in the presence of noise.

We consider here the unidirectionally coupled non–identical Roššler oscillators for first example:

$$\begin{aligned}
&\dot{x}_1 = -\omega_1 y_1 - z_1, \quad \dot{y}_1 = \omega_1 x_1 + 0.15 y_1, \quad \dot{z}_1 = 0.2 + z_1(x_1 - 10.0),\\
&\dot{x}_2 = -\omega_2 y_2 - z_2, \quad \dot{y}_2 = \omega_2 x_2 + 0.165 y_2 + \epsilon(y_1 - y_2),\\
&\dot{z}_2 = 0.2 + z_2(x_2 - 10.0),
\end{aligned} \tag{2.170}$$

where the subscripts imply the oscillators 1 and 2, respectively, $\omega_{1,2}(= 1.0 \pm 0.015)$ is the overall frequency of each oscillator, and ϵ is the coupling strength. It is known that PS appears in the regime $\epsilon \geq \epsilon_c$ and that 2π phase jumps arise when $\epsilon < \epsilon_c$. Lyapunov exponents play an essential role in the investigation of the transition phenomenon with coupled chaotic oscillators and as generally understood that PS transition is closely related to the transition to the negative value in one of the vanishing Lyapunov exponents [PC90].

A vanishing Lyapunov exponent corresponds to a phase variable of an oscillator and it exhibits the neutrality of an oscillator in the phase direction. Accordingly, the local negativeness of an exponent indicates this neutrality is locally broken [RPK96]. It is important to define an appropriate phase variable in order to study the TPL state more thoroughly. In this regard, several methods have been proposed methods of using linear interpolation at a Poincaré section [RPK96], phase–space projection [RPK96, ROH98], tracing of the center of rotation in phase–space [YL97], Hilbert transformation [RPK96], or wavelet transformation [KK00, KLR03]. Among these we take the method of phase–space projection onto the $x_1 - y_1$ and $x_2 - y_2$ planes with the geometrical relation

$$\theta_{1,2} = \arctan(y_{1,2}/x_{1,2}),$$

and get *phase difference* $\varphi = \theta_1 - \theta_2$.

The system of coupled oscillators is said to be in a TPL state (or laminar state) when $\langle\varphi\rangle < \Lambda_c$ where $\langle ... \rangle$ is the running average over appropriate short time scale and Λ_c is the cutoff value to define a TPL state. The locking length of the TPL state, τ, is defined by time interval between two adjacent peaks of $\langle\varphi\rangle$.

In order to study the characteristics of the locking length τ, we introduce a measure [KK00, KLR03]: $P(\epsilon) = \sqrt{\mathrm{var}(\tau)}/\langle\tau\rangle$, which is the ratio between the average value of time lengths of TPL states and their standard deviation. In terminology of stochastic resonance, it can be interpreted as noise–to–signal ratio [PK97, Jun93]. The measure would be minimized where the periodicity is maximized in TPL states.

To validate the argument, we explain the phenomenon in simplified dynamics. From (2.170), we get the equation of motion in terms of phase difference:

$$\dot{\varphi} = \Delta\omega + A(\theta_1, \theta_2, \epsilon)\sin\varphi + \xi(\theta_1, \theta_2, \epsilon), \quad \text{where} \tag{2.171}$$

$$A(\theta_1, \theta_2, \epsilon) = (\epsilon + 0.15)\cos(\theta_1 + \theta_2) - \frac{\epsilon}{2}\left(\frac{R_1}{R_2}\right),$$

$$\xi(\theta_1, \theta_2, \epsilon) = \frac{\epsilon}{2}\frac{R_1}{R_2}\sin(\theta_1 + \theta_2) + \frac{z_1}{R_1}\sin(\theta_1) - \frac{z_2}{R_2}\sin(\theta_2)$$
$$+(\epsilon + 0.015)\cos(\theta_2)\sin(\theta_2).$$

$$\text{Here,} \qquad \Delta\omega = \omega_1 - \omega_2, \qquad R_{1,2} = \sqrt{x_{1,2}^2 + y_{1,2}^2}.$$

And from (2.171) we get the simplified equation to describe the phase dynamics:

$$\dot{\varphi} = \Delta\omega + \langle A\rangle\sin(\varphi) + \xi,$$

where $\langle A\rangle$ is the time average of $A(\theta_1, \theta_2, \epsilon)$. This is a nonuniform oscillator in the presence of noise where ξ plays a role of effective noise [Str94] and the value of $\langle A\rangle$ controls the width of bottleneck (i.e, non–uniformity of the flow). If the bottleneck is wide enough, (i.e., faraway from the saddle–node bifurcation point: $\Delta\omega \gg -\langle A\rangle$), the effective noise hardly contributes to the phase dynamics of the system. So the passage time is wholly governed by the width of the bottleneck as follows:

$$\langle\tau\rangle \sim 1/\sqrt{\Delta\omega^2 - \langle A\rangle^2} \sim 1/\sqrt{\Delta\omega^2 - \epsilon^2/4},$$

which is a slowly increasing function of ϵ. In this region while the standard deviation of TPL states is nearly constant (because the widely opened bottlenecks periodically appears and those lead to small standard deviation), the average value of locking length of TPL states is relatively short and the ratio between them is still large.

On the contrary as the bottleneck becomes narrower (i.e., near the saddle–node bifurcation point: $\Delta\omega \geq -\langle A\rangle$) the effective noise begins to perturb the process of bottleneck passage and regular TPL states develop into intermittent ones [ROH98, KK00]. It makes the standard deviation increase very rapidly and this trend overpowers that of the average value of locking lengths of the TPL states. Thus we understand that the competition between width of bottleneck and amplitude of effective noise produces the crossover at the minimum point of $P(\epsilon)$ which shows the maximal periodicity of TPL states.

Rosenblum *et al.* firstly observed the dip in mutually coupled chaotic oscillators [RPK96]. However the origin and the dynamical characteristics of the dip have been left unclarified. We argue that the dip observed in mutually coupled chaotic oscillators has the same origin as observed above in unidirectionally coupled systems.

Common apprehension is that near the border of synchronization the phase difference in coupled regular oscillators is periodic [RPK96] whereas in coupled chaotic oscillators it is irregular [ROH98]. On the contrary, we report that the special locking regime exhibiting the maximal periodicity of a TPL state also exists in the case of coupled chaotic oscillators. In general, the phase difference of coupled chaotic oscillators is described by the 1D Langevin equation,

$$\dot{\varphi} = F(\varphi) + \xi,$$

where ξ is the effective noise with finite amplitude. The investigation with regard to PS transition is the study of scaling of the laminar length around the virtual fixed–point φ^* where $F(\varphi^*) = 0$ [KK00, KT01] and PS transition is established when

$$|\int_{\varphi}^{\varphi^*} F(\varphi) d\phi| > \max |\xi|.$$

Consequently, the crossover region, from which the value of P grows exponentially, exists because intermittent series of TPL states with longer locking length τ appears as PS transition is nearer. Eventually it leads to an exponential growth of the standard deviation of the locking length. Thus we argue that PPS is the generic phenomenon mostly observed in coupled chaotic oscillators prior to PS transition.

In conclusion, analyzing the dynamic behaviors in coupled chaotic oscillators with slight parameter mismatch we have completed the whole transition route to PS. We find that there exists a special locking regime called PPS in which a TPL state shows maximal periodicity and that the periodicity leads to local negativeness in one of the vanishing Lyapunov exponents. We have also made a qualitative description of this phenomenon with the nonuniform oscillator model in the presence of noise. Investigating the characteristics of TPL states between non–synchronization and PS, we have clarified the transition route before PS. Since PPS appears in the intermediate regime between non–synchronization and PS, we expect that the concept of PPS can be used as a tool for analyzing weak interdependences, i.e., those not strong enough to develop to PS, between non–stationary bivariate data coming from biological systems, for instance [KK00, KLR03]. Moreover PPS could be a possible mechanism of the chaos regularization phenomenon [Har92, Rul01] observed in neurobiological experiments.

2.4.3 The Onset of Synchronization in Chaotic Systems

Recall that systems of many coupled dynamical units are of great interest in a wide variety of scientific fields including physics, chemistry and biology. In particular, in [OSB02] Ott *et al.* were interested in the case of *global coupling* in which each element was coupled to all others. Beginning with the work of Kuramoto [Kur84] and Winfree [Win80], there has been much research on synchrony in systems of globally coupled limit cycle oscillators. Here, mainly following [OSB02], we present and apply a formal analysis of the stability of the unsynchronized state (or 'incoherent state') of a general system of globally coupled heterogeneous, continuous–time dynamical systems. In this treatment, no *a priori* assumption about the dynamics of the individual coupled elements is made; thus the systems can consist of elements whose natural uncoupled dynamics is chaotic or periodic, including the case where both types of elements are present.

We consider dynamical systems of the form

$$\dot{\mathbf{x}}_i = \mathbf{G}(\mathbf{x}_i(t), \boldsymbol{\Omega}_i) + \mathbf{K}(\langle\langle \mathbf{x} \rangle\rangle_* - \langle\langle \mathbf{x}(t) \rangle\rangle), \tag{2.172}$$

where $\mathbf{x}_i = (x_i^{(1)}, x_i^{(2)}, \dots, x_i^{(q)})^T$ is a qD vector; $\mathbf{G}$ is a qD vector function; $\mathbf{K}$ is a constant $q \times q$ coupling matrix; $i = 1, 2, \cdots, N$ is an index labeling components in the ensemble of coupled systems (in our analytical work we take the limit $N \to \infty$, while in our numerical work $N >> 1$ is finite); $\langle\langle \mathbf{x}(t) \rangle\rangle$ is the instantaneous average component state (referred to as the *order parameter* by H. Haken in his *synergetics* [Hak83, Hak93]),

$$\langle\langle \mathbf{x}(t) \rangle\rangle = \lim_{N \to \infty} N^{-1} \sum_i \langle \mathbf{x}_i(t) \rangle, \tag{2.173}$$

and, for each i, $\langle \mathbf{x}_i \rangle$ is the average of $\mathbf{x}_i$ over an infinite number of initial conditions $\mathbf{x}_i(0)$, distributed according to some chosen initial distribution on the attractor of the ith uncoupled system

$$\dot{\mathbf{x}}_i = \mathbf{G}(\mathbf{x}_i, \boldsymbol{\Omega}_i). \tag{2.174}$$

$\boldsymbol{\Omega}_i$ is a parameter vector specifying the uncoupled ($\mathbf{K} = 0$) dynamics, and $\langle\langle \mathbf{x} \rangle\rangle_*$ is the *natural measure* [Ott93] and i average of the state of the uncoupled system. That is, to compute $\langle\langle \mathbf{x} \rangle\rangle_*$, we set $\mathbf{K} = 0$, compute the solutions to (2.174), and get $\langle\langle \mathbf{x} \rangle\rangle_*$ from

$$\langle\langle \mathbf{x} \rangle\rangle_* = \lim_{N \to \infty} N^{-1} \sum_i \left[\lim_{\tau_0 \to \infty} \tau_0^{-1} \int_0^{\tau_0} \mathbf{x}_i(t)\, dt \right].$$

In what follows we assume that the $\boldsymbol{\Omega}_i$ are randomly chosen from a smooth probability density function $\rho(\boldsymbol{\Omega})$. Thus, we have

$$\langle\langle \mathbf{x} \rangle\rangle_* = \int \mathbf{x} \rho(\boldsymbol{\Omega})\, d\mu_{\boldsymbol{\Omega}}\, d\boldsymbol{\Omega},$$

where $\mu_{\boldsymbol{\Omega}}$ is the natural invariant measure for the system $\dot{\mathbf{x}} = \mathbf{G}(\mathbf{x}, \boldsymbol{\Omega})$. By construction, $\langle\langle \mathbf{x} \rangle\rangle = \langle\langle \mathbf{x} \rangle\rangle_*$ is a solution of the globally coupled system (2.172). This solution is called the 'incoherent state' [OSB02] because the coupling term cancels and the individual oscillators do not affect each other. The question we address is whether the incoherent state is stable. In particular, as a system parameter such as the coupling strength varies, the onset of instability of the incoherent state signals the start of coherent, synchronous behavior of the ensemble.

Stability Analysis

To perform the stability analysis, we assume that the system is in the incoherent state, so that at any fixed time t, and for each i, $\mathbf{x}_i(t)$ is distributed according to the natural measure. We then perturb the orbits

$$\mathbf{x}_i(t) \to \mathbf{x}_i(t) + \delta\mathbf{x}_i(t),$$

where $\delta\mathbf{x}_i(t)$ is an infinitesimal perturbation [OSB02]:

$$\frac{d\delta\mathbf{x}_i}{dt} = \mathbf{DG}(\mathbf{x}_i(t), \boldsymbol{\Omega}_i)\,\delta\mathbf{x}_i - \mathbf{K}\langle\langle\delta\mathbf{x}_i\rangle\rangle, \qquad \text{where} \tag{2.177}$$
$$\mathbf{DG}(\mathbf{x}_i(t), \boldsymbol{\Omega}_i)\,\delta\mathbf{x}_i = \delta\mathbf{x}_i \cdot \partial_{\mathbf{x}_i}\mathbf{G}(\mathbf{x}_i(t), \boldsymbol{\Omega}_i).$$

Introducing the fundamental matrix $\mathbf{M}_i(t)$ for system (2.177),

$$\dot{\mathbf{M}}_i = \mathbf{DG}\cdot\mathbf{M}_i, \tag{2.178}$$

where $\mathbf{M}_i(0) \equiv \mathbf{1}$, we can write the solution of (2.177) as

$$\delta\mathbf{x}_i(t) = -\int_{-\infty}^{t} \mathbf{M}_i(t)\mathbf{M}_i^{-1}(\tau)\,\mathbf{K}\langle\langle\delta\mathbf{x}\rangle\rangle_\tau\,d\tau, \tag{2.179}$$

where we use the notation $\langle\langle\delta\mathbf{x}\rangle\rangle_\tau$ to signify that $\langle\langle\delta\mathbf{x}\rangle\rangle$ is evaluated at time τ. Note that, through (2.178), $\mathbf{M}_i$ depends on the unperturbed orbits $\mathbf{x}_i(t)$ of the uncoupled nonlinear system (2.174), which are determined by their initial conditions $\mathbf{x}_i(0)$ (distributed according to the natural measure).

Assuming that the perturbed order parameter evolves exponentially in time (i.e., $\langle\langle\delta\mathbf{x}\rangle\rangle = \boldsymbol{\Delta}\mathrm{e}^{st}$), (2.179) yields

$$\{\mathbf{1} + \tilde{\mathbf{M}}(s)\mathbf{K}\}\,\boldsymbol{\Delta} = 0, \tag{2.180}$$

where s is complex, and

$$\tilde{\mathbf{M}}(s) = \left\langle\left\langle \int_{-\infty}^{t} \mathrm{e}^{-s(t-\tau)}\mathbf{M}_i(t)\mathbf{M}_i^{-1}(\tau)\,d\tau \right\rangle\right\rangle_* . \tag{2.181}$$

Thus the dispersion function determining s is

$$D(s) = \det\{\mathbf{1} + \tilde{\mathbf{M}}(s)\mathbf{K}\} = 0. \tag{2.182}$$

In order for equations (2.180) and (2.182) to make sense, the right side of (2.181) must be independent of time. As written, it may not be clear that this is so. We now demonstrate this, and express $\tilde{\mathbf{M}}(s)$ in a more convenient form. To do this, we make the dependence of $\mathbf{M}_i$ in (2.181) on the initial condition explicit,

$$\mathbf{M}_i(t)\mathbf{M}_i^{-1}(\tau) = \mathbf{M}_i(t, \mathbf{x}_i(0))\mathbf{M}_i^{-1}(\tau, \mathbf{x}_i(0)).$$

From the definition of $\mathbf{M}_i$, we have

$$\mathbf{M}_i(t,\mathbf{x}_i(0))\mathbf{M}_i^{-1}(\tau,\mathbf{x}_i(0)) = \mathbf{M}_i(t-\tau,\mathbf{x}_i(\tau)) = \mathbf{M}_i(T,\mathbf{x}_i(t-T)), \quad (2.183)$$

where we have introduced $T = t - \tau$. Using (2.183) in (2.181) we have

$$\tilde{\mathbf{M}}(s) = \left\langle\left\langle \int_0^\infty \mathrm{e}^{-sT}\mathbf{M}_i(T,\mathbf{x}_i(t-T)\,dT \right\rangle\right\rangle_* .$$

Note that our solution requires that the integral in the above converge. Since the growth of $\mathbf{M}_i$ with increasing T is dominated by h_i, the largest Lyapunov exponent for the orbit $\mathbf{x}_i$, we require

$$\mathrm{Re}(s) > \Gamma , \qquad \Gamma = \max_{\mathbf{x}_i,\boldsymbol{\Omega}_i} h_i.$$

In contrast with the chaotic case where $\Gamma > 0$, an ensemble of periodic attractors has $\Gamma = 0$ (for an attracting periodic orbit $h_i = 0$ corresponds to orbit perturbations along the flow). With the condition $\mathrm{Re}(s) > \Gamma$, the integral converges exponentially and uniformly in the quantities over which we average. Thus we can interchange the integration and the average,

$$\tilde{\mathbf{M}}(s) = \int_0^\infty \mathrm{e}^{-sT}\langle\langle \mathbf{M}_i(T,\mathbf{x}_i(t-T))\rangle\rangle_* dT. \quad (2.184)$$

In (2.184) the only dependence on t is through the initial condition $\mathbf{x}_i(t-T)$. However, since the quantity within angle brackets includes not only an average over i, but also an average over initial conditions with respect to the natural measure of each uncoupled attractor i, the time invariance of the natural measure ensures that (2.184) is independent of t. In particular, invariance of a measure means that if an infinite cloud of initial conditions $\mathbf{x}_i(0)$ is distributed on uncoupled attractor i at $t = 0$ according to its natural invariant measure, then the distribution of the orbits, as they evolve to any time t via the uncoupled dynamics (2.174), continues to give the same distribution as at time $t = 0$. Hence, although $\mathbf{M}_i(T,\mathbf{x}_i(t-T))$ depends on t, when we average over initial conditions, the result $\langle \mathbf{M}_i(T,\mathbf{x}_i(t-T))\rangle\rangle_*$ is independent of t for each i. Thus we drop the dependence of $\langle\langle \mathbf{M}_i\rangle\rangle_*$ on the initial values of the $\mathbf{x}_i$ and write

$$\tilde{\mathbf{M}}(s) = \int_0^\infty \mathrm{e}^{-sT}\langle\langle \mathbf{M}(T)\rangle\rangle_* dT, \quad (2.185)$$

where, for convenience we have also dropped the subscript i. Thus $\tilde{\mathbf{M}}$ is the Laplace transform of $\langle\langle \mathbf{M}\rangle\rangle_*$. This result for $\tilde{\mathbf{M}}(s)$ can be analytically continued into $\mathrm{Re}(s) < \Gamma$, as explained below.[25]

[25] Note that $\tilde{\mathbf{M}}(s)$ depends only on the solution of the linearized *uncoupled* system (2.178). Hence the utility of the dispersion function $D(s)$ given by (2.182) is that it determines the linearized dynamics of the globally coupled system in terms of those of the individual uncoupled systems.

Analytic Continuation of $\tilde{\mathbf{M}}(s)$

Consider the kth column of $\langle\langle\mathbf{M}(t)\rangle\rangle_*$, which we denote $[\langle\langle\mathbf{M}(t)\rangle\rangle_*]_k$. According to our definition of $\mathbf{M}_i$ given by (2.178), we can interpret $[\langle\langle\mathbf{M}(t)\rangle\rangle_*]_k$ as follows. Assume that for each of the uncoupled systems i in (2.174), we have a cloud of an infinite number of initial conditions sprinkled randomly according to the natural measure on the uncoupled attractor. Then, at $t = 0$, we apply an equal infinitesimal displacement δ_k in the direction k to each orbit in the cloud. That is, we replace $\mathbf{x}_i(0)$ by $\mathbf{x}_i(0) + \delta_k\mathbf{a}_k$, where $\mathbf{a}_k$ is a unit vector in $\mathbf{x}-$space in the direction k. Since the particle cloud is displaced from the attractor, it relaxes back to the attractor as time evolves. The quantity $[\langle\langle\mathbf{M}\rangle\rangle_*]_k\delta_k$ gives the time evolution of the i-averaged perturbation of the centroid of the cloud as it evolves back to the attractor and redistributes itself on the attractor.

We now argue that $\langle\langle\mathbf{M}\rangle\rangle_*$ decays to zero exponentially with increasing time. We consider the general case where the support of the smooth density $\rho(\mathbf{\Omega})$ contains open regions of $\mathbf{\Omega}$ for which the dynamical system (2.174) has attracting periodic orbits as well as a positive measure of $\mathbf{\Omega}$ on which (2.174) has chaotic orbits. Numerical experiments on chaotic attractors (including structurally unstable attractors) generally show that they are strongly mixing; i.e., a cloud of many particles rapidly arranges itself on the attractor according to the natural measure. Thus, for each $\mathbf{\Omega}_i$ giving a chaotic attractor, it is reasonable to assume that the average of $\mathbf{M}_i$ over initial conditions $\mathbf{x}_i(0)$, denoted $\langle\mathbf{M}_i\rangle_*$, decays exponentially. For a periodic attractor, however, $\langle\mathbf{M}_i\rangle_*$ does not decay: a distribution of orbits along a limit cycle comes to the same distribution after one period, and this repeats forever. Thus, if the distribution on the limit cycle was noninvariant, it remains noninvariant and oscillates forever at the period of the periodic orbit. On the other hand, periodic orbits exist in open regions of $\mathbf{\Omega}$, and, when we average over $\mathbf{\Omega}$, there is the possibility that with increasing time cancellation causing decay occurs via the process of 'phase mixing' [KT73]. For this case we appeal to an example. In particular, the explicit computation of $\langle\mathbf{M}_i\rangle_*$ for a simple model limit cycle ensemble results in

$$\langle\mathbf{M}_i\rangle_* = \frac{1}{2}\begin{bmatrix}\cos\Omega_i t & -\sin\Omega_i t\\ \sin\Omega_i t & \cos\Omega_i t\end{bmatrix},$$

and indeed this oscillates and does not decay to zero. However, if we average over the oscillator distribution $\rho(\Omega)$ we get [OSB02]

$$\langle\langle\tilde{\mathbf{M}}\rangle\rangle_* = \frac{1}{2}\begin{bmatrix}c(t) & -s(t)\\ s(t) & c(t)\end{bmatrix}, \qquad \text{where}$$

$$c(t) = \int \rho(\Omega)\cos\Omega t\, d\Omega, \qquad s(t) = \int \rho(\Omega)\sin\Omega t\, d\Omega.$$

For any analytic $\rho(\Omega)$ these integrals decay exponentially with time. Thus, based on these considerations of chaotic and periodic attractors, we see that for sufficiently smooth $\rho(\mathbf{\Omega})$, there is reason to believe that $\langle\langle\mathbf{M}\rangle\rangle_*$, the average of $\mathbf{M}_i$ over $\mathbf{x}_i(0)$ and over $\mathbf{\Omega}_i$, decays exponentially to zero with increasing time. Conjecturing this decay to be exponential [OSB02], we see that the integral in (2.185) converges for $\mathrm{Re}(s) > -\xi$. Thus, while (2.185) was derived under the assumption $\mathrm{Re}(s) > \Gamma > 0$, using analytic continuation, we can regard (2.185) as valid for $\mathrm{Re}(s) > -\xi$. Note that, for our purposes, it suffices to require only that $\|\langle\langle\mathbf{M}(t)\rangle\rangle_*\|$ be bounded, rather than that it decay exponentially. Boundedness corresponds to $\xi = 0$, which is enough for us, since, as soon as instability occurs, the relevant root of $D(s)$ has $\mathrm{Re}(s) > 0$.

The Distribution Function Approach

Much previous work has treated the Kuramoto problem and its various generalizations using a kinetic equation approach. Ott *et al.* [OSB02] have also obtained the main result (2.182) for $D(s)$ by this more traditional method. We briefly outline their procedure below.

Let $F(\mathbf{x}, \mathbf{\Omega}, t)$ be the distribution function (actually a generalized function) such that $F(\mathbf{x}, \mathbf{\Omega}, t)\, d\mathbf{x}d\mathbf{\Omega}$ is the fraction of oscillators at time t whose state and parameter vectors lie in the infinitesimal volume $d\mathbf{x}d\mathbf{\Omega}$ centered at $(\mathbf{x}, \mathbf{\Omega})$. Note that $\int F d\mathbf{x}$ is time independent, since it is equal to the distribution function $\rho(\mathbf{\Omega})$ of the oscillator parameter vector. The time evolution of F is simply obtained from the conservation of probability following the system evolution,

$$\partial_t F + \partial_{\mathbf{x}} \cdot [(\mathbf{G}(\mathbf{x}, \mathbf{\Omega}) + \mathbf{K} \cdot (\langle\langle\mathbf{x}\rangle\rangle_* - \langle\langle\mathbf{x}\rangle\rangle))F] = 0, \qquad \text{where} \quad (2.186)$$

$$\langle\langle\mathbf{x}\rangle\rangle = \int\int F\, d\mathbf{x}d\mathbf{\Omega}\mathbf{x}, \qquad \langle\langle\mathbf{x}\rangle\rangle_* = \int\int F_0\, d\mathbf{x}d\mathbf{\Omega}\mathbf{x},$$

and $F_0 = F_0(\mathbf{x}, \mathbf{\Omega}) = f(\mathbf{x}, \mathbf{\Omega})\rho(\mathbf{\Omega})$, in which $f(\mathbf{x}, \mathbf{\Omega})$ is the density corresponding to the natural invariant measure of the uncoupled attractor whose parameter vector is $\mathbf{\Omega}$. Thus $f(\mathbf{x}, \mathbf{\Omega})$, which is a generalized function, formally satisfies

$$\partial_{\mathbf{x}} \cdot [\mathbf{G}(\mathbf{x}, \mathbf{\Omega}) f(\mathbf{x}, \mathbf{\Omega})] = 0.$$

Hence, $F = F_0$ is a time–independent solution of (2.186) (the 'incoherent solution'). We examine the stability of the incoherent solution by linearly perturbing F, $F = F_0 + \delta F$, to get

$$\partial_t \delta F + \partial_{\mathbf{x}} \cdot [\mathbf{G}(\mathbf{x}, \mathbf{\Omega})\delta F - \mathbf{K}\langle\langle\delta\mathbf{x}\rangle\rangle F_0] = 0, \tag{2.187}$$

$$\langle\langle\delta\mathbf{x}\rangle\rangle = \int\int \delta F\, d\mathbf{x}d\mathbf{\Omega}\mathbf{x}. \tag{2.188}$$

We can then introduce the Laplace transform, solve the transformed version of (2.187), and substitute into (2.188) to get the same dispersion function $D(s)$ as in the stability analysis above. The calculation is somewhat lengthy, involving the formal solution of (2.187) by integration along the orbits of the uncoupled system.

We note that the computation outlined above is formal in that we treat the distribution functions as if they were ordinary, as opposed to generalized, functions. In this regard, we note that $f(\mathbf{x}, \boldsymbol{\Omega})$ is often extremely singular both in its dependence on $\mathbf{x}$ (because the measure on a chaotic attractor is typically a multi–fractal) and on $\boldsymbol{\Omega}$ (because chaotic attractors are often structurally unstable). We believe that both these sources of singularity are sufficiently mitigated by the regularizing effect of the averaging process over $(\mathbf{x}, \boldsymbol{\Omega})$, and that the above stability results are still valid. This remains a problem for future study. We note, however, that for structurally unstable attractors, a smooth distribution of system parameters $\rho(\boldsymbol{\Omega})$ is likely to be much less problematic than the case of identical ensemble components, $\rho(\boldsymbol{\Omega}) = \delta(\boldsymbol{\Omega} - \bar{\boldsymbol{\Omega}})$. In the case of identical structurally unstable chaotic components, an arbitrarily small change of $\bar{\boldsymbol{\Omega}}$ can change the character of the base state whose stability is being examined. In contrast, a small change of a smooth distribution $\rho(\boldsymbol{\Omega})$ results in a small change in the weighting of the ensemble members, but would seem not to cause any qualitative change.

Bifurcations

It is natural to ask what happens as a parameter of the system passes from values corresponding to stability to values corresponding to instability. Noting that the incoherent state represents a time independent solution of (2.172), we can seek intuition from standard results on the generic bifurcations of a fixed point of a system of ODEs [GH83]. There are two linear means by which such a fixed point can become unstable: (i) a real solution of $D(s) = 0$ can pass from negative to positive s values, and (ii) two complex conjugate solutions, s and s^*, can cross the imaginary $s-$axis, moving from $\mathrm{Re}(s) < 0$ to $\mathrm{Re}(s) > 0$.

In reference to case (i), we note that the incoherent steady state always exists for our above formulation. In this situation, in the absence of a system symmetry, the generic bifurcation of the system is a *transcritical bifurcation*.

In the presence of symmetry, the existence of a fixed point solution with $\langle\langle \mathbf{x} \rangle\rangle_* - \langle\langle \mathbf{x} \rangle\rangle$ nonzero may imply the simultaneous existence of a second fixed point solution with $\langle\langle \mathbf{x} \rangle\rangle_* - \langle\langle \mathbf{x} \rangle\rangle$ nonzero, where these solutions map to each other under the symmetry transformation of the system. In this case the transcritical bifurcation is ruled out, and the generic bifurcation is the pitchfork bifurcation, which can be either subcritical or supercritical .

In case (ii), where two complex conjugate solutions cross the $\mathrm{Im}(s)$ axis, the generic bifurcations are the subcritical and supercritical Hopf bifurcations. (In this case we note that although the individual oscillators may be behaving chaotically, their average coherent behavior is periodic.)

In the numerical experiments in [OSB02] the authors found cases of apparent subcritical and supercritical Hopf bifurcations, as well as a case of a subcritical pitchfork bifurcation. As their globally coupled system was a collection of coupled Lorenz equations [OSB02]

$$\begin{aligned} \dot{x}^{(1)} &= \sigma(x^{(2)} - x^{(1)}) \\ \dot{x}^{(2)} &= rx^{(1)} - x^{(2)} - x^{(1)}x^{(3)}, \\ \dot{x}^{(3)} &= -bx^{(3)} + x^{(1)}x^{(2)} \end{aligned} \tag{2.189}$$

with the symmetry $(x^{(1)}, x^{(2)}, x^{(3)}) \to (-x^{(1)}, -x^{(2)}, x^{(3)})$, and since the form of the coupling used respects this symmetry, the transcritical bifurcation is ruled out, which leaves only the pitchfork bifurcation.

Generalizations

One generalization is to consider a general nonlinear form of the coupling such that we replace system (2.172) by

$$\begin{aligned} \dot{\mathbf{x}}_i &= \hat{\mathbf{G}}(\mathbf{x}_i, \boldsymbol{\Omega}_i, \mathbf{y}), \\ \mathbf{y} &= \langle\langle \mathbf{x} \rangle\rangle_* - \langle\langle \mathbf{x} \rangle\rangle, \end{aligned} \tag{2.190}$$

and the role of the uncoupled system (analogous to (2.174)) is played by the equation

$$\dot{\mathbf{x}}_i = \tilde{\mathbf{G}}(\mathbf{x}_i, \boldsymbol{\Omega}_i, \mathbf{0}).$$

In this more general setting, following the steps of the above stability analysis yields [OSB02]

$$D(s) = \det\{\mathbf{1} + \tilde{\mathbf{Q}}(s)\}, \qquad \text{where} \tag{2.191}$$

$$\tilde{\mathbf{Q}}(s) = \int_0^\infty dT e^{-st} \langle\langle \mathbf{M}(T) \mathbf{D}_{\mathbf{y}} \hat{\mathbf{G}}(\mathbf{x}, \boldsymbol{\Omega}, \mathbf{0}) \rangle\rangle_*.$$

A still more general form of the coupling is

$$\dot{\mathbf{x}}_i = \hat{\hat{\mathbf{G}}}(\mathbf{x}_i, \boldsymbol{\Omega}_i, \langle\langle \mathbf{x} \rangle\rangle). \tag{2.192}$$

For (2.190) and (2.172), a unique incoherent solution $\langle\langle \mathbf{x} \rangle\rangle_*$ always exists and can be obtained by solving the nonlinear equations for each $\mathbf{x}_i(0)$ with $\mathbf{y} = (\langle\langle \mathbf{x} \rangle\rangle_* - \langle\langle \mathbf{x} \rangle\rangle)$ set equal to zero. In the case of (2.192), the existence of a unique incoherent state is not assured. By definition, $\langle\langle \mathbf{x} \rangle\rangle$ is time independent in an incoherent state. Thus replacing $\langle\langle \mathbf{x} \rangle\rangle$ in (2.192) by a constant vector $\mathbf{u}$, imagine that we solve (2.192) for an infinite number of initial conditions distributed for each i on the natural invariant measure of the system,

$$\dot{\mathbf{x}}_i = \hat{\hat{\mathbf{G}}}(\mathbf{x}_i, \boldsymbol{\Omega}_i, \mathbf{u}),$$

and then get the average $\langle\langle \mathbf{x} \rangle\rangle_{\mathbf{u}}$. This average depends on $\mathbf{u}$, so that $\langle\langle \mathbf{x} \rangle\rangle_{\mathbf{u}} = \mathbf{F}(\mathbf{u})$. We then define an incoherent solution $\langle\langle \mathbf{x} \rangle\rangle_*$ for (2.192) by setting $\langle\langle \mathbf{x} \rangle\rangle_{\mathbf{u}} = \mathbf{u} = \langle\langle \mathbf{x} \rangle\rangle_*$, so that $\langle\langle \mathbf{x} \rangle\rangle_*$ is the solution of the nonlinear equation

$$\langle\langle \mathbf{x} \rangle\rangle_* = \mathbf{F}(\langle\langle \mathbf{x} \rangle\rangle_*).$$

Generically, such a nonlinear equation may have multiple solutions or no solution. In this setting, if a stable solution of this equation exists for some paramter $k < k_c$, then the solution of the nonlinear system (2.192) (with appropriate initial conditions) will approach it for large t. If now, as k approaches k_c from below, a real eigenvalue approaches zero, then $k = k_c$ generically corresponds to a saddle-node bifurcation. That is, an unstable incoherent solution merges with the stable incoherent solution, and, for $k > k_c$, neither exist. In this case, loss of stability by the Hopf bifurcation is, of course, still generic, and the incoherent solution continues to exist before and after the Hopf bifurcation. $D(s)$ for (2.192) is given by (2.191) with $\mathbf{D_y}\hat{\mathbf{G}}$ replaced by $-\mathbf{D}_{\langle\langle\mathbf{x}\rangle\rangle}\hat{\hat{\mathbf{G}}}$ evaluated at the incoherent state ($\langle\langle\mathbf{x}\rangle\rangle = \langle\langle\mathbf{x}\rangle\rangle_*$) whose stability is being investigated.

Another interesting case is when the coupling is delayed by some linear deterministic process. That is, the ith oscillator does not sense $\langle\langle\mathbf{x}\rangle\rangle$ immediately, but rather responds to the time history of $\langle\langle\mathbf{x}\rangle\rangle$. Thus, using (2.190) as an example, the coupling term $\mathbf{y}$ is replaced by a convolution,

$$\mathbf{y}(t) = \int_{-\infty}^{t} \mathbf{\Lambda}(t-t')\cdot(\langle\langle\mathbf{x}\rangle\rangle_* - \langle\langle\mathbf{x}\rangle\rangle_{t'})\,dt'.$$

In this case a simple analysis shows that (2.191) is replaced by

$$D(\mathbf{s}) = \det\{\mathbf{1} + \tilde{\mathbf{Q}}(s)\cdot\mathbf{\Lambda}(s)\}, \qquad \text{where}$$
$$\tilde{\mathbf{\Lambda}}(s) = \int_0^\infty \mathrm{e}^{-st}\mathbf{\Lambda}(t')\,dt.$$

The simplest form of this would be a discrete delay

$$\mathbf{\Lambda}(t) = \mathbf{K}\delta(t-\eta),$$

in which case $\tilde{\mathbf{\Lambda}}(s) = \mathbf{1}\mathrm{e}^{-\eta s}$.

The Kuramoto Problem

As an example, we now consider a case that reduces to the well–studied Kuramoto problem. We consider the ensemble members to be 2D,

$$\mathbf{x}_i = (x_i(t), y_i(t))^T,$$

and characterized by a scalar parameter Ω_i. For the coupling matrix $\mathbf{K}$ we choose $k\mathbf{1}$. Thus (2.172) becomes

$$\dot{x}_i = G^{(x)}(x_i, y_i, \Omega_i) + k(\langle\langle x\rangle\rangle_* - \langle\langle x\rangle\rangle),$$
$$\dot{y}_i = G^{(y)}(x_i, y_i, \Omega_i) + k(\langle\langle y\rangle\rangle_* - \langle\langle y\rangle\rangle).$$

We assume that in polar coordinates ($x = r\cos\theta, y = r\sin\theta$), the uncoupled ($k = 0$) dynamical system is given by [OSB02]

$$\dot{\theta}_i = \Omega_i, \tag{2.193}$$
$$\dot{r}_i = (r_0 - r_i)/\tau, \tag{2.194}$$

where $\Omega_i\tau \ll 1$. That is, the attractor is the circle $r_i = r_0$, and it attracts orbits on a time scale τ that is very short compared to the limit cycle period. For $\Omega_i\tau \ll 1$ it will suffice to calculate $\mathbf{M}_i(t)$ for $t \gg \tau$. To do this, we consider an initial infinitesimal orbit displacement

$$\mathbf{\Delta}_{oi} = \mathbf{a}_x dx_{oi} + \mathbf{a}_y dy_{oi},$$

where $\mathbf{a}_{x,y}$ are unit vectors.

In a short time this displacement relaxes back to the circle, so that for $(2\pi/\Omega) \gg t \gg \tau$ we have $r = r_0$, $\theta = \theta_{oi}$, $\mathbf{\Delta}_i(t) \simeq \Delta^+_{oi}\mathbf{a}_\theta$, where θ_{oi} is the initial value $\theta_i(0)$, $\mathbf{a}_\theta$ is evaluated at $\theta_i(0)$, and $\Delta^+_{oi} = -\sin\theta_{oi}dx_{oi} + \cos\theta_{oi}dy_{oi}$. For later time $t \gg \tau$, we have $r = r_0$, $\theta_i(t) = \theta_{oi} + \Omega_i t$ and $\mathbf{\Delta}_i(t) = \Delta^+_{oi}\mathbf{a}_\theta$, with $\mathbf{a}_\theta$ evaluated at $\theta_i(t)$. In rectangular coordinates this is

$$\begin{bmatrix} dx_i(t) \\ dy_i(t) \end{bmatrix} = \begin{bmatrix} \sin(\theta_{oi} + \Omega_i t)\sin\theta_{io} & -\sin(\theta_{oi} + \Omega_i t)\cos\theta_{io} \\ -\cos(\theta_{oi} + \Omega_i t)\sin\theta_{oi} & \cos(\theta_{oi} + \Omega_i t)\cos\theta_{oi} \end{bmatrix} \begin{bmatrix} dx_{oi} \\ dy_{oi} \end{bmatrix}.$$

By definition, the above matrix is $\mathbf{M}_i$ appearing in the stability analysis above. Averaging (2.195) over the invariant measure on the attractor of (2.193) and (2.194) implies averaging over θ_{oi}. This yields

$$\langle \mathbf{M}_i \rangle_\theta = \frac{1}{2}\begin{bmatrix} \cos\Omega_i t & -\sin\Omega_i t \\ \sin\Omega_i t & \cos\Omega_i t \end{bmatrix}.$$

Averaging the rotation frequencies Ω_i over the distribution function $\rho(\Omega)$ and taking the Laplace transform gives $\tilde{\mathbf{M}}(s)$,

$$\tilde{\mathbf{M}}(s) = \begin{bmatrix} (q_+ + q_-) & i(q_+ - q_-) \\ -i(q_+ - q_-) & (q_+ + q_-) \end{bmatrix}, \qquad \text{where} \tag{2.195}$$

$$q_\pm(s) = \frac{1}{4}\left\langle \frac{1}{s \mp i\Omega} \right\rangle_\Omega \equiv \frac{1}{4}\int_{-\infty}^{+\infty} \frac{\rho(\Omega)\,d\Omega}{s \mp i\Omega}, \tag{2.196}$$

and, in doing the Laplace transform, we have neglected the contribution to the Laplace integral from the short time interval $0 \le t \le 0(\tau)$ (this contribution approaches zero as $\Omega\tau \to 0$). Using (2.195) and (2.196) in (2.182) then gives $D(s) = D_+(s)D_-(s)$, where $D_\pm(s)$ is the well–known result for the Kuramoto model (e.g., [Str00]),

$$D_\pm(s) = 1 + \frac{k}{2}\int_{-\infty}^{+\infty} \frac{\rho(\Omega)\,d\Omega}{s \pm i\Omega} = 0, \qquad \mathrm{Re}(s) > 0,$$

and $D_{\pm}(s)$ for $\mathrm{Re}(s) \leq 0$ is obtained by analytic continuation [Str00]. Note that the property $D^{\dagger}_{\pm}(s) = D_{\mp}(s^{\dagger})$, where $\dagger$ denotes complex conjugation, insures that complex roots of $D(s) = D_{+}(s)D_{-}(s) = 0$ come in conjugate pairs.

2.4.4 Neural Bursting and Consciousness

A neuron is said to *fire a burst of spikes* when it fires two or more action potentials followed by a period of quiescence. A burst of two spikes is called a doublet, of three spikes is called a triplet, four – quadruplet, etc. [Izh00] Almost every neuron can burst if stimulated or manipulated pharmacologically. Many burst autonomously due to the interplay of fast ionic currents responsible for spiking activity and slower currents that modulate the activity. Below is the list of the more popular bursting neurons [Izh07]:

1. Neocortex
   a) IB: Intrinsically bursting neurons, if stimulated with a long pulse of dc current, fire an initial burst of spikes followed by shorter bursts, and then tonic spikes [CG90]. These are predominantly pyramidal neurons in layer 5.
   b) CH: Chattering neurons can fire high–frequency bursts of 3–5 spikes with a relatively short interburst period [GM96]. Some call them fast rhythmic bursting (FRB) cells. These are pyramidal neurons in layer 2–4, mainly layer 3.
   c) Interneurons: Some cortical interneurons exhibit bursting activity in response to pulses of dc current [MTW04].
2. Hippocampus
   a) LTB: Low–threshold bursters fire high–frequency bursts in response to injected pulses of current. Some of these neurons burst spontaneously [SUK01]. These are pyramidal neurons in CA1 region.
   b) HTB: High–threshold bursters fire bursters only in response to strong long pulses of current.
3. Thalamus
   a) TC: Thalamocortical neurons can fire bursts if inhibited and then released from inhibition. This rebound burst is often called a low–threshold spike. Some fire bursts spontaneously in response to tonic inhibition.
   b) RTN: Reticular thalamic nucleus inhibitory neurons have bursting properties similar to those of TC cells.
4. Cerebellum
   a) PC: Purkinje cells in cerebellar slices usually fire tonically but when synaptic input is blocked they can switch to a trimodal pattern which includes a bursting phase [WK02].

It is relatively easy to identify bursts in response to simple stimuli, such as dc steps or sine waves, especially if recording intracellularly from a quiet in vitro slice. The bursts fully evolve and the hallmarks of burst responses are clear. However, responses to sensory stimuli are often comprised of doublets or triplets embedded in spike trains. Furthermore, these responses are usually recorded extracellularly so the experimenter does not have access to the membrane potential fluctuations that are indicative of bursting. Thus, it is difficult to distinguish burst responses from random multispike events. The statistical analysis of spike trains addresses this problem. Bimodal inter–spike interval (ISI) histograms can be indicative of burst responses. The rationale is that short ISIs occur more frequently when induced by burst dynamics than would occur if predicted by Poisson firing. Burst spikes with short ISIs form the first mode while quiescent periods correspond to the longer ISIs of the second mode. This is true for intrinsic or forced (stimulus driven and network–induced) bursting. Furthermore, the trough between the two modes may correspond to the refractory period of an intrinsic burst or the timescale of the network–induced bursting [DLL02, DCM03]. This method defines a criterion for burst identification so that further analysis and experimentation can determine the mechanism and function of the bursts. See [BN01] for a deeper analysis into burst detection from stochastic spike data.

Spiking versus Bursting Neural Networks

Recently, Izhikevich [HI97] discussed biological plausibility and computational efficiency of some of the most useful models of *spiking and bursting neurons* (see Figure 2.36). He compared their applicability to large–scale simulations of cortical neural networks.

Following [HI97], we present some widely used models of spiking and bursting neurons that can be expressed in the form of ODEs. Throughout this subsection, v denotes the membrane potential. All the parameters in the models are chosen so that v has mV scale and the time has ms scale. To compare computational cost, we assume that each model, written as a dynamical system $\dot{x} = f(x)$, is implemented using the simplest, fixed–step first–order Euler method, with the integration time step chosen to achieve a reasonable numerical accuracy.

Integrate–and–Fire Neuron

One of the most widely used models in computational neuroscience is the *leaky integrate–and–fire neuron*, (I&F neuron, for short) given by

$$\dot{v} = I + a - bv, \qquad \text{If} \quad v \geq v_{trsh} \quad \text{Then} \quad v \leftarrow c,$$

where v is the membrane potential, I is the input current, and a, b, c, and v_{trsh} are the parameters. When the membrane potential v reaches the threshold value v_{trsh}, the neuron is said to fire a *spike*, and v is reset to c. The I&F

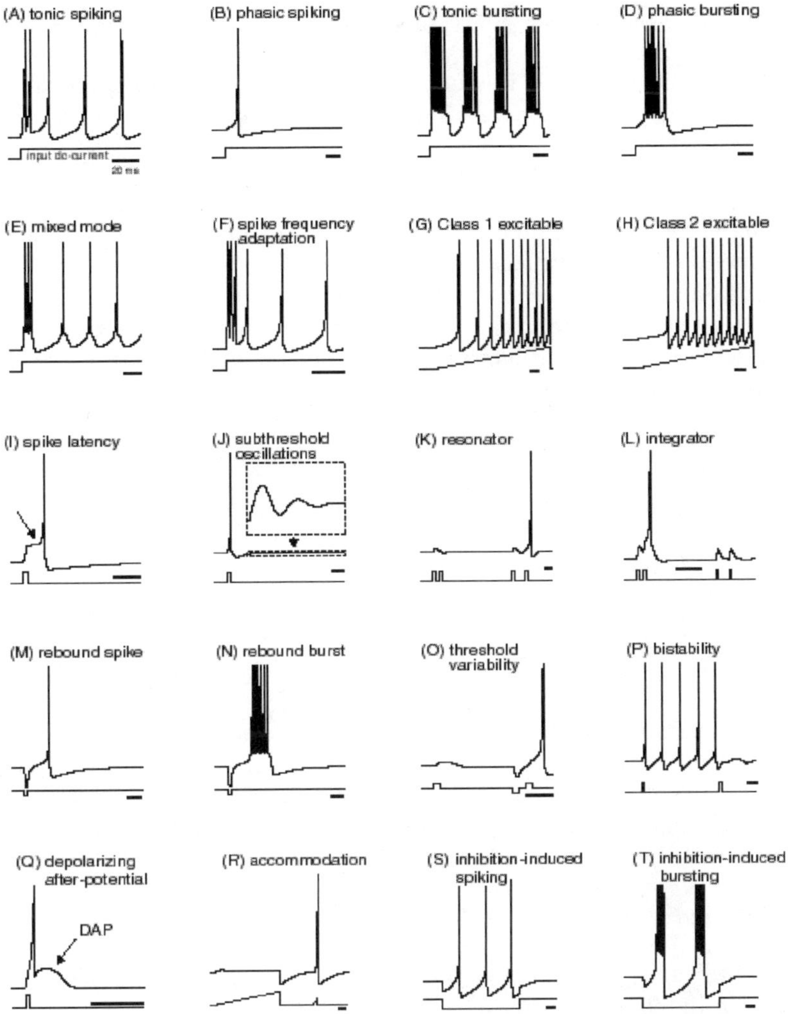

Fig. 2.36. Neuro–computational features of biological neurons (with permission from E. Izhikevich).

neuron can fire tonic spikes with constant frequency, and it is an integrator. The I&F neuron is *Class 1 excitable system* [Izh99a]; it can fire tonic spikes with constant frequency, and it is an integrator. It is the simplest model to implement when the integration time step τ is 1 ms. Because I&F has only one variable, it cannot have phasic spiking, bursting of any kind, rebound responses, threshold variability, bistability of attractors, or autonomous chaotic dynamics. Because of the fixed threshold, the spikes do not have latencies. In summary, despite its simplicity, I&F is one of the worst models to use in simulations, unless one wants to prove analytical results [HI97].

Integrate–and–Fire Neuron with Adaptation

The I&F model is $1D$, hence it cannot burst or have other properties of cortical neurons. One may think that having a second linear equation

$$\dot{v} = I + a - bv + g(d - v), \qquad \dot{g} = (e\delta(t) - g)/\tau,$$

describing activation dynamics of a high–threshold $K-$current, can make an improvement, e.g., endow the model with spike–frequency adaptation. Indeed, each firing increases the $K-$activation gate via Dirac $\delta-$function and produces an outward current that slows down the frequency of tonic spiking. This model is fast, yet still lacks many important properties of cortical spiking neurons.

Integrate–and–Fire–or–Burst Neuron

The *integrate–and–fire–or–burst neuron* model is given by

$$\dot{v} = I + a - bv + gH(v - v_h)h(v_T - v),$$

$$\text{If} \quad v \geq v_{trsh} \quad \text{Then} \quad v \leftarrow c, \qquad \dot{h} = \begin{cases} \frac{-h}{\tau^-}, & \text{if} \quad v > v_h, \\ \frac{1-h}{\tau^+}, & \text{if} \quad v < v_h \end{cases}$$

to model thalamo–cortical neurons. Here h describes the inactivation of the calcium $T-$current, g, v_h, v_T, τ^+ and τ^- are parameters describing dynamics of the $T-$current, and H is the Heaviside step function. Having this kind of a second variable creates the possibility for bursting and other interesting regimes [HI97], but is already a much slower (depending on the value of v).

Complex–Valued Resonate–and–Fire Neuron

The *resonate–and–fire neuron* is a complex–valued (i.e., $2D$) analogue of the I&F neuron [Izh01], given by

$$\dot{z} = I + (b + iw)z, \qquad \text{if} \quad \operatorname{Im} z = a_{trsh} \quad \text{then} \quad z \longleftarrow z_0(z), \tag{2.197}$$

where $z = x + iy \in \mathbb{C}$ is a complex–valued variable that describes oscillatory activity of the neuron. Here b, w, and a_{trsh} are parameters, $i = \sqrt{-1}$, and $z_0(z)$ is an arbitrary function describing activity–dependent after–spike reset. (2.197) is equivalent to the linear system

$$\dot{x} = bx - wy, \qquad \dot{y} = wx + by,$$

where the real part x is the current–like variable, while the imaginary part y is the voltage–like variable. The resonate–and–fire model is simple and efficient. When the frequency of oscillation $w = 0$, it becomes an integrator.

Quadratic Integrate–and–Fire Neuron

An alternative to the leaky I&F neuron is the *quadratic I&F neuron*, also known as the *theta–neuron*, or the Ermentrout–Kopell canonical model [Erm81, Gut98]. It can be presented as

$$\dot{v} = I + a(v - v_{rest})(v - v_{trsh}), \qquad \text{If} \quad v = v_{trsh} \ \text{ Then } \quad v \leftarrow v_{rest},$$

where v_{rest} and v_{trsh} are the resting and threshold values of the membrane potential. This model is canonical in the sense that any *Class 1 excitable system* [Izh99a] described by smooth ODEs can be transformed into this form by a continuous change of variables. It takes only seven operations to simulate 1 ms of the model, and this should be the model of choice when one simulates large–scale networks of integrators. Unlike its linear analogue, the quadratic I&F neuron has spike latencies, activity dependent threshold (which is v_{trsh} only when $I = 0$), and bistability of resting and tonic spiking modes.

FitzHugh–Nagumo Neuron

The parameters in the *FitzHugh–Nagumo neuron* model

$$\dot{v} = a + bv + cv^2 + dv^3 - u, \qquad \dot{u} = \varepsilon(ev - u),$$

can be tuned so that the model describes spiking dynamics of many resonator neurons. Since one needs to simulate the shape of each spike, the time step in the model must be relatively small, e.g., $\tau = 0.25\, ms$. Since the model is a $2D$ system of ODEs, without a reset, it cannot exhibit autonomous chaotic dynamics or bursting. Adding noise to this, or some other $2D$ models, allows for stochastic bursting.

Hindmarsh–Rose Neuron

The *Hindmarsh–Rose thalamic neuron* model [RH89] can be written as a $3D$ ODE system

$$\dot{v} = I + u - F(v) - w, \qquad \dot{u} = G(v) - u, \qquad \dot{w} = (H(v) - w)/\tau,$$

where $F, G,$ and H are some functions. This model is quite expensive to implement as a large–scale spike simulator [HI97].

Morris–Lecar Neuron

Morris and Lecar [ML81] suggested a simple $2D$ model to describe oscillations in barnacle giant muscle fiber. Because it has biophysically meaningful and measurable parameters, the *Morris–Lecar neuron* model became quite popular in computational neuroscience community. It consists of a membrane potential equation with instantaneous activation of Ca current and an additional equation describing slower activation of K current,

$$C\dot{V} = I - g_L(V - V_L) - g_{Ca}m_\infty(V)(V - V_{Ca}) - g_K n(V - V_K),$$
$$\dot{n} = \lambda(V)(n_\infty(V) - n), \qquad \text{where}$$
$$m_\infty(V) = \frac{1}{2}\left(1 + \tanh\left[\frac{V - V_1}{V_2}\right]\right), \qquad \text{and}$$
$$n_\infty(V) = \frac{1}{2}\left(1 + \tanh\left[\frac{V - V_3}{V_4}\right]\right), \qquad \lambda(V) = \bar{\lambda}\cosh\left[\frac{V - V_3}{2V_4}\right],$$

with parameters: $C = 20\,\mu F/cm^2$, $g_L = 2\,mmho/cm^2$, $V_L = -50\,mV$, $g_{Ca} = 4\,mmho/cm^2$, $V_{Ca} = 10\,mV$, $g_K = 8\,mmho/cm^2$, $V_K = -70\,mV$, $V_1 = 0\,mV$, $V_2 = 15\,mV$, $V_3 = 10\,mV$, $V_4 = 10\,mV$, $\bar{\lambda} = 0.1\,s^{-1}$, and applied current $I(\mu A/cm^2)$. The model can exhibit various types of spiking, but could exhibit tonic bursting only when an additional equation is added, e.g., slow inactivation of Ca current. In this case, the model becomes equivalent to the *Hodgkin–Huxley neuron model* [HH52, Hod64], which is extremely expensive to implement.

Burst as a Unit of Neuronal Information

There are many hypotheses on the importance of bursting activity in neural computation [Izh07]:

1. Bursts are more reliable than single spikes in evoking responses in post-synaptic cells. Indeed, excitatory post-synaptic potentials (EPSP) from each spike in a burst add up and may result in a superthreshold EPSP.
2. Bursts overcome synaptic transmission failure. Indeed, postsynaptic responses to a single presynaptic spike may fail (release does not occur), however in response to a bombardment of spikes, i.e., a burst, synaptic release is more likely [Lis97].
3. Bursts facilitate transmitter release whereas single spikes do not [Lis97]. Indeed, a synapse with strong short-term facilitation would be insensitive to single spikes or even short bursts, but not to longer bursts. Each spike in the longer burst facilitates the synapse so the effect of the last few spikes may be quite strong.
4. Bursts evoke long–term potentiation and hence affect synaptic plasticity much greater, or differently than single spikes [Lis97].
5. Bursts have higher signal–to–noise ratio than single spikes [She01]. Indeed, burst threshold is higher than spike threshold, i.e., generation of bursts requires stronger inputs.
6. Bursts can be used for selective communication if the postsynaptic cells have subthreshold oscillations of membrane potential. Such cells are sensitive to the frequency content of the input. Some bursts resonate with oscillations and elicit a response, others do not, depending on the inter-burst frequency [IDW03].
7. Bursts can resonate with short–term synaptic plasticity making a synapse a band–pass filter [IDW03]. A synapse having short–term facilitation and

depression is most sensitive to a burst having certain resonant interspike frequency. Such a burst evokes just enough facilitation, but not too much depression, so its effect on the postsynaptic target is maximal.

8. Bursts encode different features of sensory input than single spikes [OCD04]. For example, neurons in the electro–sensory lateral–line lobe (ELL) of weakly electric fish fire network induced–bursts in response to communication signals and single spikes in response to prey signals [DLL02, DCM03]. In the thalamus of the visual system bursts from pyramidal neurons encode stimuli that inhibit the neuron for a period of time and then rapidly excite the neuron [LS04]. Natural scenes are often composed of such events.
9. Bursts have more informational content than single spikes when analyzed as unitary events [RGS99]. This information may be encoded into the burst duration or in the fine temporal structure of interspike intervals within a burst.

In summary, burst input is more likely to have a stronger impact on the postsynaptic cell than single spike input, so some believe that bursts are all–or–none events, whereas single spikes may be noise.

Most spiking neurons can burst if stimulated with a current that slowly drives the neuron above and below the firing threshold. Such a current could be injected via an electrode or generated by the synaptic input (see Figure 2.37). Below are some examples of forced bursters:

1. RA neurons in the songbird burst in response to drive from HVC neurons. The bursting arises as a result of either network dynamics within RA or is inherited from HVC [FKH04].
2. Network induced bursts of electric fish [DCM03]. Here bursting arises because periodic inhibitory inputs reduce firing and create intervals of quiescence.
3. Electroreceptor afferents in paddlefish. Bursting occurs because of a pre–filtering of broadband stochastic stimuli that drives the receptors. The receptor dynamics can be modeled as a simple excitable system and the slow noise (filtered) pushes the neuron into periods of rapid firing and into periods of quiescence [NR02].

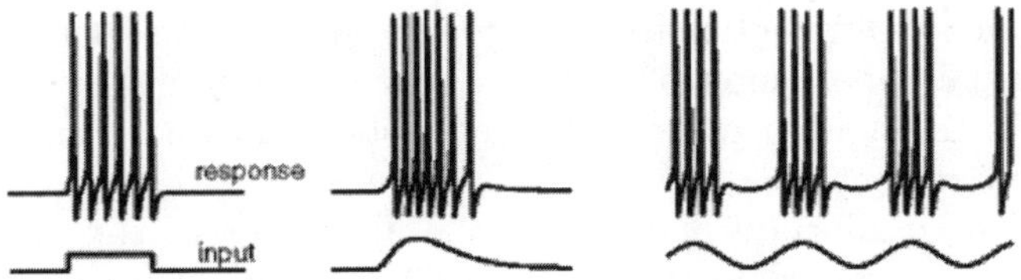

Fig. 2.37. Forced bursting in response to injected input.

Many neurons have slow intrinsic membrane currents that can modulate fast spiking activity. Typically, the currents build up during continuous spiking, hyperpolarize the cell and result in the termination of the spike train. While the cell is quiescent, the currents slowly decay, the cell recovers, and it is ready to fire another burst.

Different ionic mechanisms of bursting may result in different mathematical mechanisms, which in turn determine the neuro-computational properties of bursters, i.e., how they respond to the input [Izh00, Izh07]. Most mathematical models of bursters can be written in the fast–slow form:

$$\begin{aligned} \dot{x} &= f(x, y), && \text{fast spiking} \\ \dot{y} &= \mu g(x, y), && \text{slow modulation} \end{aligned}$$

where vector x describes the state of the fast subsystem responsible for spiking activity, vector y describes the state of the slow subsystem that modulates spiking, f and g are some *Hodgkin–Huxley–type functions*, and $\mu << 1$ is the ratio of time scales.

A standard method of analysis of fast–slow bursters, as well as of any singularly perturbed system, is to set and consider the fast and the slow subsystems separately. This is known as dissection of neuronal bursting [Rin85], since it allows us to study the fast subsystem

$$\dot{x} = f(x, y),$$

and treat y as a vector of slowly changing bifurcation parameters. Typically, the fast subsystem has a *limit cycle* (spiking) attractor for some values of y and an equilibrium (resting) attractor for other values of y. As the slow variable oscillates between the two values, the fast subsystem, and hence the whole system, burst.

Now, what makes the slow variable oscillate? In the simplest case, the slow subsystem

$$\dot{y} = \mu g(x, y)$$

may have a limit cycle attractor, which is relatively insensitive to the value of the fast variable. In this case, the slow variable exhibits an autonomous oscillation that periodically drives the fast subsystem over the threshold. Such a bursting is called *slow–wave bursting.* The slow subsystem must be at least 2D to exhibit slow–wave bursting. Slow–wave bursting in conductance-based models is usually more interesting than the simplest case described above. In such models, the slow subsystem often consists of activation and inactivation gates of slow currents.

When the equilibrium and limit cycle attractors of the fast subsystem co–exits for the same value of y, there is a bi–stability of resting and spiking states. This creates a hysteresis loop for the slow variable and such a bursting is called hysteresis–loop bursting. The slow variable y may be 1D in this case, oscillating between resting and spiking values via the hysteresis loop.

When the fast variable x is in the spiking state, the slow variable, governed by the equation, is pushed toward the region of quiescence (resting, rightward in the figure), and spiking abruptly stops. When the fast variable is quiescent, the slow variable is pushed toward the region of spiking (leftward in the figure) and after a while spiking abruptly starts. These transitions from spiking to resting and back correspond to bifurcations of the fast subsystem.

Bursters are distinguished qualitatively according to their topological type. There are two important bifurcations of the fast subsystem that determine the topological type:

1. Resting to spiking: Bifurcation of a stable equilibrium (resting) that results in the transition to limit cycle attractor (spiking).
2. Spiking to resting: Bifurcation of a limit cycle attractor that results in the transition to the equilibrium (resting).

Mathematical studies of bursters revealed that different topological types have different neuro–computational properties [Izh00, Izh07]:

1. Bursters that involve Andronov–Hopf bifurcation act as resonators, i.e., they are sensitive to the frequency content of the synaptic input. In contrast, the other types (fold and circle) act as integrators.
2. Bursters that involve fold, subcritical Andronov–Hopf, saddle homoclinic orbit, and fold limit cycle bifurcations have co–existence of resting and spiking states, and hence have a bistable or multistable dynamics. An appropriately timed input can switch bursting activity from spiking to quiescence and back. The input does not even have to be excitatory.
3. Different topological types of bursters have different synchronization properties. Some tend to synchronize in–phase, others tend to de–synchronize.

2.5 Complexity of Humanoid Robots

2.5.1 General Complexity

The ability of science and technology to augment human performance depends on an understanding of systems, not just components. The convergence of technologies is an essential aspect of the effort to enable functioning systems that include human beings and technology; and serve the human beings to enhance their well–being directly and indirectly through what they do, and what they do for other human beings. The recognition today that human beings function in teams, rather than as individuals, implies that technological efforts that integrate human beings across scales of tools, communication, biological and cognitive function are essential. Understanding the role of complex systems concepts in technology integration requires a perspective on how the concept of complexity is affecting science, engineering, and finally, technology integration [BY04].

The structure of scientific inquiry is being challenged by the broad relevance of *complexity* to the understanding of physical, biological and social systems [BY00, GA99]. Cross–disciplinary interactions are giving way to trans–disciplinary and unified efforts to address the relevance of large amounts of information to description, understanding and control of complex systems. From the study of biomolecular interactions [Ser99, Nor99, WBI] to the 21st Century Information Age, complexity has arisen as a unifying feature of challenges to understanding and action. In this arena of complex systems, information and action, structure and function are entangled. New approaches that recognize the importance of patterns of behavior, the multi–scale space of possibilities, and evolutionary or adaptive processes that select systems or behaviors that can be effective in a complex world are central to advancing our understanding and capabilities [BY97].

The failure of design and implementation of a new air–traffic control system, failures of Intel processors, medical errors, failures of medical drugs, even the failure of the Soviet Union, can be attributed to large system complexities [BY04]. Systematic studies of large scale engineering projects have revealed a remarkable proportion of failures in major high investment projects [CH94]. The precursors of such failures: multi–system integration, high performance constraints, many functional demands, high rates of response, and large context specific protocols, are symptomatic of complex engineering projects. The methods for addressing and executing major engineering challenges must begin from the recognition of the central role of complexity and the modern tools that can guide the design, or self–organize, highly complex systems. Central to effective engineering is the evaluation of the complexity of function of a system, and the recognition of fundamental engineering tradeoffs of structure, function, complexity and scale in system capabilities, and the application of indirection to specification, design and control of system development and the system itself.

One way to identify a complex task is as a problem where the number of distinct possibilities that must be considered, anticipated or dealt with is substantially larger than can be reasonably named or enumerated. Intuitively, the complexity of a task is the number of wrong choices for every right choice. We can casually consider in an explicit way tens of possibilities, a professional will readily deal with hundreds of possibilities, and a major project will deal with thousands, the largest projects deal with tens of thousands. For larger numbers of possibilities we must develop new strategies. Simplifying a complex task by ignoring the need for different responses is what leads to errors or failures that affect the success of the entire effort, leaving it as a gamble with progressively higher risks.

The source of complex tasks is complex systems. Complex systems are systems with interdependent parts. Interdependence means that we cannot identify the system behavior by just considering each of the parts and combining them. Instead we must consider how the relationships between the parts affect the behavior of the whole. Thus a complex task is also one for which many factors must be considered to determine the outcome of an action. While complex systems give rise to complex tasks, reliable responses to complex tasks can only be achieved by complex systems. Thus, the complex challenges that we face in the world can be met only by the development of complex systems that can address them [BY97].

The rapid development of nanotechnology and the convergence of biological, information, and cognitive sciences is creating a context in which complex systems concepts that enable effective organizations to meet complex challenges can be realized through technological implementation. At the same time, complex systems concepts and methods are an essential part of the framework in which this convergence is taking place. From the fine scale control of systems based upon nanotechnology to understanding the system properties of the integrated socio–technical system consisting of human beings and computer information networks, the synergy of complex systems and converging technologies is apparent as soon as we consider the transition between components and functions.

Human civilization, its various parts, including its technology, and its environmental context, are all complex. The most reliable prediction possible is that this complexity will continue to increase. The great opportunity of the convergence of nanotechnology, biomedical, information, and cognitive sciences is an explosive increase in what is possible through combining advances in all areas. This is, by definition, an increase in the complexity of the systems that will be formed out of technology and of the resulting behaviors of people who use them directly, or are affected by them. The increasing complexity suggests that there will be a growing need for widespread understanding of complex systems as a counter point to the increasing specialization of professions and professional knowledge. The insights of complex systems research and its methodologies may become pervasive in guiding what we build, how we build it, and how we use and live with it. Possibly the most visible outcome of these developments will be an improved ability of human beings aided by technology to address complex global social and environmental problems, third world development, poverty in developed countries, war and natural disasters. At an intermediate scale, the key advances will dramatically change how individuals work together in forming functional teams that are more directly suited to the specific tasks they are performing. In the context of individual human performance, the key to major advances is recognizing that the convergence of technology will lead to the possibility of designing (more correctly adapting) the environment of each individual for his or her individual needs and capabilities in play and work [BY00].

Fundamental Research in Complex Systems

Fundamental research in complex systems is designed to get characterizations of complex systems and relationships between quantities that characterize them. When there are well defined relationships, these are formalized as theorems or principles, more general characterizations and classifications of complex systems are described below in major directions of inquiry. These are only a sample of the ongoing research areas.

A theorem or principle of complex systems should apply to physical, biological, social and engineered systems. Similar to laws in physics, a law in complex systems should relate various quantities that characterize the system and its context. An example is Newton's 2nd Law that relates force, mass and acceleration. Laws in complex systems relate qualities of system, action, environment, function and information. Three examples follow.

(i) Functional complexity. Given a system whose function we want to specify, for which the environmental (input) variables have a complexity of $C(e)$, and the actions of the system have a complexity of $C(a)$, then the complexity of specification of the function of the system is [BY97]: $C(f) = C(a) \cdot 2^{C(e)}$, where complexity is defined as the logarithm (base 2) of the number of possibilities or, equivalently, the length of a description in bits. The proof follows from recognizing that a complete specification of the function is given by a table whose rows are the actions ($C(a)$ bits) for each possible input, of which there are $2^{C(e)}$. Since no restriction has been assumed on the actions, all actions are possible and this is the minimal length description of the function. Note that this theorem applies to the complexity of description as defined by the observer, so that each of the quantities can be defined by the desires of the observer for descriptive accuracy. This theorem is known in the study of *Boolean functions* (binary functions of binary variables) but is not widely understood as a basic theorem in complex systems. The implications of this theorem are widespread and significant to science and engineering. The exponential relationship between the complexity of function and the complexity of environmental variables implies that systems that have environmental variables (inputs) with more than a few bits (i.e. 100 bits or more of relevant input) have functional complexities that are greater than the number of atoms in a human being, and thus cannot be reasonably specified. Since this is true about most systems that we characterize as 'complex' the limitation is quite general. The implications are that fully phenomenological approaches to describing complex systems, such as the behaviorist approach to human psychology, cannot be successful. Similarly, the testing of response or behavioral descriptions of complex systems cannot be performed. This is relevant to various contexts from the testing of computer chips, today with over 100 bits of input, to testing of the effects of medical drugs in double blind population studies, today used in various combinations with various quantities for synergistic effects, with a need to avoid harmful drug interactions. In each case the number of environmental variables (inputs) is large enough that all cases cannot be tested.

(ii) Requisite variety. The Law of Requisite Variety states: The larger the variety of actions available to a control system, the larger the variety of perturbations it is able to compensate [Ash56]. Quantitatively, it specifies that the probability of success of a well adapted system in the context of its environment can be bounded: $-\log_2(P) < C(e) - C(a)$. Qualitatively, this theorem specifies the conditions in which success is possible: a matching between the environmental complexity and the system complexity, where success implies regulation of the impact of the environment on the system. The implications of this theorem are widespread in relating the complexity of desired function to the complexity of the system that can succeed in the desired function. This is relevant to discussions of the limitations of specific engineered control system structures, to the limitations of human beings and of human organizational structures.[26]

(iii) Non averaging. The Central Limit Theorem specifies that collective/aggregate properties of independent components with bounded probability distributions are Gaussian distributed with a standard deviation that diminishes as the square root of the number of components. This simple solution to the collective behavior of non–interacting systems does not extend to the study of interacting/interdependent systems. The lack of averaging of properties of complex systems is a statement that can be used to guide the study of complex systems more generally. It also is related to a variety of other formal results, including Simpson's paradox [Sim51] which describes the inability of averaged quantities to characterize the behavior of systems, and Arrow's Dictator Theorem which describes the generic dynamics of voting systems [Arr63, MB98]. The lack of validity of the Central Limit Theorem has many implications that affect experimental and theoretical treatments of complex systems. Many studies rely upon unjustified assumptions in averaging observations that lead to misleading if not false conclusions. The development of approaches that can identify the domain of validity of averaging and use more sophisticated approaches (like clustering) when they do not apply, are essential to progress in the study of complex systems. Another class of implications of the lack of validity of the Central Limit Theorem is the recognition of the importance of individual variations between different complex systems even when they appear to be within a single class. An example mentioned above is the importance of individual differences and the lack of validity of averaging in cognitive science studies. While snowflakes are often acknowledged as individual, research on human beings often is based on assuming their homogeneity. More generally, we see that the study of complex systems is concerned with their universal properties, and one of their universal properties is individual differences. This apparent paradox, one of many in complex systems (see below), reflects the importance of identifying when universality and common properties apply and when they do not, a key part of the universal study of complex systems [BY04].

[26] Note that this theorem, as formulated, does not take into account the possibility of avoidance (actions that compensate for multiple perturbations because they anticipate and thus avoid the direct impact of the perturbations), or the relative measure of the space of success to that of the space of possibilities. These limitations can be compensated for.

2.5.2 Humanoid Robotics

Anthropomorphism of Humanoid Robots

Current development of robotics indicates that the spectrum of robotic activities will expand significantly in the near future. Rapid development of humanoid robots brings about new shifts of the boundaries of robotics as a scientific and technological discipline.

New technologies of components, sensors, microcomputers, as well as new materials, have recently shifted the barriers to real–time integrated control of some very complex dynamic systems such as humanoid robots are, which already today possess about fifty degrees of freedom, and are updated in microseconds [VBB05].

For a long time already, robots have not been present only in industrial plants, at the time their traditional workspace, but have been increasingly more engaged in the close living and working environment of humans. This fact inevitably leads to the need of a *working coexistence* of man and robot and sharing their common working environment. The fact that no significant rearrangement of the humans' environment because of the presence of robots could be expected, robots will have to further 'adapt' to the environment previously dedicated only to man. However, in the time to come it will be inevitable to accept the necessity of cooperative activities of man and robot, and make a step in the direction of increasing comfort of their joint action. Besides, it is expected that the robots cooperating with humans will have operation efficiency as close as possible to that of humans. The working and living environment, adapted to humans, imposes on robots with their *mechanical–control structure* at least two classes of tasks: manipulating various objects from the human environment and motion in a specific environment with the obstacles of the type of staircases, thresholds, multi–level floors, etc. For fulfillment of diverse tasks in the environment highly adapted to humans the most promising is *human–like design*. The first step that would enable robots to realize tasks in the manner and with the efficiency similar to those of humans is to make robot's structure close to that of humans, i.e., anthropomorphic. Hence, the necessary degree of the robot's anthropomorphism may be more concretely conceived as the degree of similarity of its motion and global behavior, whereby the similarity should not be only visual, but some other aspects of anthropomorphism have to be also satisfied.[27]

In relation to this, the work raises also some new fundamental questions. One of them is surely to what extent 'human design' should be 'copied', or to what extent robot design ought to be similar to human's? This question could also be formulated in the following, more practical, way: How complex should be the robotic structure (i.e., how many degrees–of–freedom (DOFs)

[27] Activities in the common working and living environment of man and robot imply also some other similarities such as, for example, the interaction and man–robot communication (including also emotional aspects).

should the robot possess and which they are) in order it would be capable of attaining the desired (high enough) degree of anthropomorphism? It is clear that the mechanical complexity of the human skeleton is practically impossible, and perhaps senseless, to mimic, either from the viewpoint of mechanics or control. Besides, it is not a priori clear what are the DOFs that predominantly influence the degree of anthropomorphism. Hence, a thought–out and factuality–based answer to this delicate question is needed.

Another question is related to the anthropomorphism of the gait itself that is to be performed by the humanoid robot mechanism under real conditions. There are two aspects that should be borne in mind. The first is, how to synthesize a gait with the highest possible degree of anthropomorphism, and second, how to preserve the synthesized gait anthropomorphism in the course of its realization in the presence of disturbances, i.e., how to realize 'the most anthropomorphic' compensation of disturbances? [VBB05]

It should also be emphasized that in the control of legged locomotion, and especially that of biped robots, in view of the possibility of occurrence of unpowered (passive) DOFs between the foot and ground caused by larger disturbances, apart from the complete conventional dynamic control (tracking, i.e., maintaining the state of internal coordinates), it is essential to check all the time the fulfillment of the conditions of dynamic balance of the humanoid robot as a whole. In the case of an abrupt compensational movement, however, there may appear such inertial forces that represent a real threat of robot's rotation about the foot edge. Hence, it is necessary to have as natural (moderate) as possible compensation of disturbances, which will bring the robot again to the previous state of dynamic balance.

A fundamental question is how to more precisely define the anthropomorphism of an artificial gait and how to quantify it. Instead of giving a definite answer to this delicate question we will define some relevant attributes of anthropomorphism that are, in our opinion, dominant, so that we will focus our attention on them:

The amplitudes of particular DOFs of humanoid robots should be kept within the possible moderate range, whereby a decisive influence has the robot's trunk, both in the frontal and sagittal plane. Lower consumption of driving energy is therefore in correlation with smaller movements at robot's joints, namely of those realizing the compensational motion in the stage of forming nominal dynamics, i.e., the dynamic balance under ideal conditions of the synthesized artificial gait. Of course, it is necessary to mention that one can also speak about the relation between the anthropomorphism and compensational motion in the cases of real gait too, when the control mechanism is to solve the problem of maintaining dynamic balance of the humanoid robot in the circumstances of the ever-present disturbances of various types.

When speaking about the relationship between the magnitude of compensational movements and energy consumption in both above cases (the forming and maintaining of dynamic balance of humanoid robots) we should notice that our initial investigations of the model of gait dynamics with the imposed flat–foot contact [VHC73] showed somewhat lower energy consumption in comparison with the 'natural' gait, where the foot–ground contact is realized in three phases (heel strike, flat foot and deploy phase). We should also mention that, for example, in some types of parade marching step, the major part of the half–step has a flat–foot contact with the ground, whereas a smaller part of contact takes place in the form of deploy phase. In the case of walk on stairs (ascending or descending) the contact with the support is usually realized in the form of flat–foot and deploys phases. Finally, let us notice only one more data that the Honda [HHH99] robot realizes its gait via flat–foot contact with the ground [VBB05].

The number of prescribed *Zero–Moment Points* (ZMP) [VJ69, VBS90, VB04] and their distribution within the support polygon, either in the single–support or double–support gait phase influences the robot's anthropomorphism. Namely, simulations have confirmed the intuitive expectations that the increase in the number of ZMPs yields an increase in both the anthropomorphism of dynamic balance nominal model and control model, with the aim of maintaining dynamic balance in the real perturbation regimes, in which artificial gait of the humanoid robot takes place.

And the last, but not least important, attribute concerning the functional anthropomorphism of humanoid robots is related to the importance of the choice of mechanical DOFs, such as active segmentation of the foot and trunk, as well as the robot's active rotation about the vertical axis.

The above remarks concerning the anthropomorphism of humanoid robots testify to its significant complexity. The possibility to determine the degree of this integral performance as a solution of the high–complexity optimization problem involving numerous constraints seems to be rather unlikely. Hence we think it more practical to use the approach in which, instead of attempting to find an integral criterion of anthropomorphism, one considers a set of its particular attributes (for example, those mentioned above). Then, taking into account the maximal possible particular attributes of humanoid robots one should arrive at the maximum of its possible overall anthropomorphism, even when it has not been explicitly defined.

Basic Characteristics of Bipedal Systems

All of the biped mechanism joints are powered and directly controllable except for the contact of the foot and the ground (it can be considered as an additional DOF), which is the only site at which the mechanism interacts with the environment. This contact is essential for the walk realization because the mechanism's position with respect to the environment depends on the relative position of the foot with respect to the ground. The foot cannot be controlled directly but only in an indirect way – by ensuring appropriate dynamics of the mechanism above the foot. Thus, the overall indicator of the mechanism's behavior is the point where the influence of all the forces acting on the mechanism can be replaced by one single force. As mentioned above, this point was termed *Zero–Moment Point* (ZMP). Recognition of the significance and role of ZMP in biped artificial walk was a turning point in gait planning and control. Thus, irrespective of their structure and number of DOFs involved, a basic characteristic of all biped locomotion systems is the possibility of the appearance of unpowered DOFs, formed by the contact of the foot with the ground surface. In the case the motion takes place under conditions of small perturbations the basic task of control is to minimize the deviation of ZMP from its prescribed (nominal) position, which simultaneously ensures dynamic balance and prevents the loss of the regular contact of the foot with the ground. If tracking of the internal trajectories of all joints of the humanoid robot is thus ensured, we can speak of its overall dynamic control. However, if in the case of intensive disturbances the ZMP comes out of the support polygon or its zone from which it must not step out, the biped system may face the loss of the regular contact with the ground. When the regular contact with the ground is lost passive DOFs appear and the foot becomes partly deployed from the ground, losing thus the feedback involving dynamic reaction force, and the possibility of further maintaining dynamic balance is essentially endangered. In such a situation, the main task of the control system is to re-establish the broken foot-ground contact and reduce large disturbances to small ones, i.e., to bring the system to the state in which all feedback loops are operative, so that the usual procedure can be applied to control the bipedal gait of the humanoid robot under conditions of small perturbations.

The motion of a humanoid robot should be as anthropomorphic as possible. Hence, it is necessary to synthesize the most anthropomorphic motion under ideal conditions (in the absence of disturbances), which we call nominal. Then, such motion should be realized by the real system, so that the deviations from the nominal should be as small as possible, and corrections made in the most anthropomorphic way. In this work, to our knowledge the first one intending to call attention to the problem of anthropomorphism of humanoid robots, we will confine ourselves to the analysis of the synthesized nominal motion [VBB05].

For the gait synthesis (defining trajectories of all the mechanism joints) of crucial importance is the semi–inverse method [VJ69, VBS90, VB04], in which, upon prescribing the ZMP and trajectories for a part of mechanism joints, trajectories of the remaining joints are calculated and thus the dynamic balance of the overall humanoid robot is ensured. The motion of the mechanism was synthesized by the semi-inverse method in the following way:

The legs' motion was copied from a human subject's motion and adopted as the motion of the mechanism legs;

The trunk's motion was determined in the way ensuring dynamic equilibrium of the mechanism as a whole during the half-step, i.e., in the period considered, the point within the support polygon that at the given moment represents the ZMP is characterized by the equalities $\boldsymbol{M}_X = \boldsymbol{M}_Y = 0$[28]

Special attention should be paid to the role of the hands during the gait. There are three ways in which the hands in relation to the trunk may be treated and, consequently, participate in the process of gait synthesis. They can freely hang on the shoulders as physical pendulums and move only under the influence of inertial forces formed during the trunk motion. Further, the hands' joints can be powered and the hands can perform certain motion due to the action of the moments at their own joints, and finally, they can be immobile with respect to the trunk. In the first case, when the hands are freely hanging (passively swinging) as physical pendulums, the motion of the hands can also be synthesized along with the trunk motion, by prescribing additional conditions at the suspension points at which the moments are naturally equal to zero. In the second case, since their joints are powered, the hands can perform certain predefined motion with respect to the trunk. Therefore, in this case the motions of both the legs and hands are prescribed in advance and compensational motion of the trunk is determined in the process of synthesis in the way to satisfy the conditions of repeatability and dynamic balance. In the third case, when the hands are fixed to the trunk [VBS90], it can be assumed that they represent its constitutive part, augmenting only the mass and changing thus the inertia moments. Compensational motion of the trunk is calculated in the usual way.

If we want to consider the entire locomotion system of humanoid robot, we ought to take care of the anthropomorphism of its two basic subsystems that are strongly coupled: the legs' subsystem and the subsystem of the upper part (trunk). Evidently, different motions of the legs can cause different compensational motion of the trunk. Hence, the variation in the motion of the legs can influence the form of the synthesized trunk motion. Since the legs' motion has been copied from a human, the requirement for anthropomorphism is inherently satisfied. However, since the copied motion can never be faithfully reproduced by a humanoid system the question arises as to how the simplification of legs' motion can influence the trunk motion, i.e., how much abandoning (blocking) of the motion at particular DOFs at the main leg joints (the hip, knee, and ankle) can influence the anthropomorphism of the upper part of the system. Besides, there is an essential difference in the complexity of the human foot and the foots of humanoid robots that have been realized up to now. Another very interesting question is how much the anthropomorphism of the trunk motion is influenced by the complexity of construction of the foot of humanoid robot.

To answer the above questions it is necessary to find out the way how to estimate the level of anthropomorphism. Since the motion of the mechanism's legs is based on the motion of the legs of humans we think that the

---

[28] It can be also required that $\boldsymbol{M}_X = \boldsymbol{M}_Y = \boldsymbol{M}_Z = 0$ (all three components of the moment at the ZMP).

most essential attribute of anthropomorphism is the trunk swinging, i.e., its relevant parameters of the amplitudes and mean value of trunk inclination in the frontal and the sagittal plane, so that we will consider just these quantities to compare different types of gait of humanoid robots. The trunk motion synthesized on the basis of the legs motion copied from the human can be taken as the reference one. All other motions of the legs are derived from this pattern by 'excluding' the motions of particular DOFs, which means that the corresponding coordinates have been immobilized, i.e., kept constant. Compensational motion of the trunk synthesized using thus obtained 'new' motion of the legs, was compared with the reference one. In this way we could observe how the absence of motion at some of DOFs influences the trunk motion in the sense of its anthropomorphism.

We have also investigated the influence of the active motion of the hands on the synthesized trunk motion. For the reference motion of the legs, hands motions of different amplitudes were prescribed and the effect on the trunk motion was followed. In addition to the motion of the hands and legs we also investigated the effect of variation of the ZMP trajectory and change of the gait rate (the gait was accelerated for the same trajectories of the joint angles) on the synthesized motion of the trunk. Besides, in all the above cases, the compensation of each of the ZMP moment components ($\boldsymbol{M}_X$ and $\boldsymbol{M}_Y$) was realized with the aid of only one joint located just below the trunk link. In view of the fact that the compensation of disturbances by humans is performed using several DOFs, we have investigated how the gait anthropomorphism is influenced by the distribution of the task of compensation of one moment component ($\boldsymbol{M}_X$ or $\boldsymbol{M}_Y$) on more joints (we called it 'distributed' compensation), whereby the hip DOFs were included in compensation in one case, while the two–link trunk was modelled in the other [VBB05].

Basic Definitions Related to Humanoid Robots Locomotion

Let us consider first the definitions of some basic notions that appear in the area of biped locomotion. These notions have been tacitly accepted, probably because they represent basic notions, so that everybody has thought for himself (and from this stemmed also the collective acceptance) that they are understandable by themselves. Hence these basic notions have never been formally defined in the robotic literature. We think that these notions have still to be defined and, although the lack of definitions caused no serious confusion, it can be noticed that a number of very important notions have been defined by various authors in different ways, so that it would be desirable to have a unified terminology [VBP06].

Walk. According to [HCL74], under walk is understood the 'move by putting forward each foot in turn, not having both feet off the ground at once'. From this definition it comes out that walk is characterized by such displacement of legs in which both feet are not separated from the ground at the same time, and which ensures that the body motion in space - usually forward, though it is possible to consider a backward walk too. We think that this definition, though not originate from technical literature, satisfies the needs of humanoid robotics.

Gait. It is known from experience that the walk of every individual is specific and that a man walks differently in different situations. Each of these particular ways of walking represents a particular gait. Therefore, it can be said that gait represents the *manner of walking or running* [HCL74]. Hence, any walk is realized by a certain gait. By recording time changes of the angles at legs' joints during one step, one is recording in fact a particular gait. The basic notions that are related to gait and that should be considered are: step and repeatability conditions, periodicity and symmetry.

Step. When speaking of gait it should be pointed out the fact that has been indirectly pronounced by the formulation '... by putting forward each foot in turn ... ' [HCL74]. It suggests that in leg locomotion, even in the most general case, there exists a certain kind of repeatability: 'in the direction of motion, during the contact with the ground, the leg from the front position with respect to the trunk comes to the rear position, then it is deployed from the ground and in the transfer phase moves to the front position, to make again contact with the ground, and the cycle is repeated'. The described sequence of actions represents a basic cycle of walk, and it is called a *step*. It should be noticed that the instant from which we observe a step within this cycle can be arbitrarily selected (we need not to start as in the above example with the contact of the 'front' leg with the ground). The described repetition of movements is the basis of locomotion activity. Still, we should emphasize that each step can be generally different, an example being the staggering of a drunk man. Although a step can be divided into a large number of phases, we think that each step consists of at least two phases: 'single-support phase, when only one foot is in contact with the ground, and double-support phase in which both feet are simultaneously on the ground'. The two phases alternate regularly.

Periodic gait. If the gait is realized by repeating the same step in an identical way then we speak of a periodic gait. In that case the relative position of legs' links is repeated periodically. This does not mean that the other parts of the body (e.g., the arms or the head) behave obligatorily in a periodic manner, but it still should be pointed out that it is the most common case. Mathematically, the periodicity condition is expressed via the change of the internal coordinates q_j (joint angles) [VBP06]:

$$q_j(t+T) = q_j(t), \ \forall t, \ j = \ 1, \cdots$$

where j changes per each legs' joint of the locomotion mechanism, while T represents the step duration. 'Periodicity of the motion of legs' joints is a necessary and sufficient condition for a periodic gait t'. If all the body joints move also periodically then we speak of a periodic gait, and it has actually been considered in the majority of papers in the area of biped locomotion.

Let us add that periodic gait can be realized only if the motion is performed on the ground surface of appropriate characteristics that allows periodicity.

Repeatability conditions. In some literature sources, the attributes *periodic* and *repeatable* are considered as being synonymous. However, the term *periodic* has its firm footing in the mathematical definition of a periodic function, whereas *repeatability* and *repeatability conditions* require certain explanation. Namely, 'to attain periodicity a necessary condition is the equality of the system state (more precisely, the state of lower extremities) at the beginning of each step'. However, this condition is not sufficient because a periodic gait (in the sense of the above definition) will be realized only under the conditions that the humanoid performs each step on an identical ground, obeying the same control law, and in the absence of disturbances. Mathematically, repeatability conditions can be expressed in the following way [VBP06]:

$$q_j(t_i) = q_j(t_{i-1}), \quad \dot{q}_j(t_i) = \dot{q}_j(t_{i-1}), \quad j = 1, \cdots,$$

where j changes per each joint of the locomotion mechanism legs, and t_i represents the instant of the beginning of the i-th step. Therefore, the identity of the state at the beginning of each step offers the possibility to repeat the preceding step and thus realize a periodic gait. In the literature [Vuk75], we can find a somewhat different definition of repeatability conditions, requiring that 'the state at the end of a step is equal to the state at its beginning'. Such formulation is identical to the previous one, with an additional explanation that the end of a step coincides with the beginning of the next one, i.e., $(t_{i+1} = t_i + T)$. Here, we should comment the fact that the walk segment that represents a step can be chosen so that it ends by the foot touching the ground. At this instant, the impact occurs that might cause discontinuity in the system state (instantaneous change in velocities). The above definition assumes that the impact and potential change in the state are an integral part of a step.

A natural question is posed as to whether the notions of periodicity is synonymous with repeatability. Thus, why should a human/humanoid abandon repeating the previous step provided it could be realized? Is this a purely academic issue or something that can happen in reality? We think that it is a real possibility. For example, there may arise such situation in which (because of a certain disturbance) the robot performing periodic gait has to change the motion of one leg in some step phase (e.g., the foot has to be lifted somewhat higher because of the presence of an obstacle on the ground) and, after this 'intervention', return again to the previous trajectory, to complete the step with a state identical to the one at its beginning.

It should also be mentioned that repeatability conditions are unavoidable in gait synthesis, where the walk is formed by synthesizing one step that is then repeated [JV72, Vuk73, Vuk75].

Symmetric step and gait. Symmetry is a characteristic of a step, but a gait, being a sequence of symmetric steps, can also be called symmetric. A prerequisite for a symmetric step and gait is the symmetry of the extremities, i.e., of the left and right legs (which is almost always considered as being fulfilled). If a step can be divided in two equal time periods, and if the left leg in one period behaves as the right leg in the other, then we speak of a symmetric gait. The half–period and the motion realized in it are termed *half-step*. The symmetry condition can be mathematically expressed as [VBP06]

$$q_j^{right}(t + T/2) = q_j^{left}(t), \qquad \forall t,\ j = 1, \dots, n,$$

where j denotes the symmetric joints of the right and left leg, and $T/2$ is the half-step duration. For a symmetric gait, this necessary and sufficient condition should be fulfilled during the gait. If all the body joints move symmetrically[29]

It should be pointed out that a periodic and repeatable gait need not be symmetric, and that symmetry does not necessarily assume either periodicity of repeatability.

It is important to note that all above definitions assume implicitly the gait continuation, i.e., the human/humanoid is not going to fall. Hence, let the gait that is realized with two legs[30] and for which there are no any additionally preset conditions (symmetry, repeatability, etc.) be called – *sustained gait*.

Regular gait. Under the notion *regular gait* is understood a periodic gait in which the leg in the single-support phase is in contact with the ground by the whole foot area or with only its front part (the toes link with the two-link foot), and in the case of double-support phase the requirement applies to at least one foot. It should be noticed that regular gait can, but not necessarily, be symmetric (e.g., when the robot performs a turn and the 'internal' leg passes the shorter way). The gait consisting of the parts that are all regular is also regular. For example, climbing the staircases, straight–line gait forward, turning, etc., considered as a whole, represent also regular gaits.

Ideal gait. Ideal gait is a purely academic notion, and it represents a regular gait for which the repeatability and symmetry conditions can be mathematically checked. In view of the fact that there is always some difference between the data used in mathematical treatment (mechanism parameters, time changes of joints angles, characteristics of the ground on which the

---

[29] For the joints that have no their 'symmetric pair' (waist and neck), the condition of gait symmetry is somewhat different

$$q_j(t + T/2) = -q_j(t),$$

we speak of a symmetric motion of human/humanoid and the above mathematical expression expands to hold for the entire body. Symmetry assumes a straight–line gait, but it is also important to emphasize the need for the 'symmetry' of the support (ground), i.e., the equality of support conditions for the left and the right leg. With a symmetric gait, all kinematic and dynamic analyzes can be carried out on one half–step.

[30] The motion by crawling or staggering while using hands to hold on to something, cannot be considered a gait.

humanoid is walking, etc.) and real data, the data used in mathematical treatment are called ideal. Ideal gait is often used as a reference motion that the system is attempting to realize. Although the ideal gait coincides greatly (almost in full) with regular gait – the differences being in the level of 'refinement', the authors consider this notion necessary, and the term as being appropriate, because such gait is most often used in all theoretical investigations in this area.

Support area.[31] This is the surface determined by the contact of the foot and the ground. With regular gait, there is always support area of a finite size: 'in the single–support phase the support area coincides with the area of the foot in contact with the ground, whereas in the double–support phase, the support area is a convex area determined by the areas of the feet and the ground and common tangents, so that the encompassed area is maximized'. Support area does not exist only in the case when both feet are off the ground (ruling or jumping) or the contact area degenerated to a point or a line (this, however, means that the rigid foot rotates about an axis or point and that the mechanism as whole is overturning). In the case of the occurrence of any of the two instances, the gait of the humanoid cannot be considered regular [VBP06].

Honda Humanoid Series

Corresponding to Honda's Slogan 'The Power of Dreams', Honda set itself the ambitious goal to create a two–legged walking robot by developing revolutionary new technology. Research began by envisioning the ideal robot form for use in human society. The robot would need to be able to maneuver between objects in a room, be able to go up and down stairs and need to be able to walk on uneven ground. For this reason it had to have two legs, just like a person.

The first Honda robot, E0, was made in 1986. A two legged robot was made to walk. Walking by putting one leg before the other was successfully achieved. However, taking nearly five seconds between steps, it walked very slowly in a straight line. To increase walking speed, or to allow walking on uneven surfaces or slopes, fast walking must be realized [HHH99].

In the period 1987–1991, Honda made the next three robots in E–series: E1, E2, and E3. Human walking was thoroughly researched and analyzed. Based on this data a fast walking program was created, input into the robot and experiments were begun. The E2 robot achieved fast walking at a speed of 1.2 km/h on a flat surface. The next step was realized fast, stable walking

[31] Commonly used term is support polygon. This came out from the fact that all realized walking robots had feet of a rectangular shape. However, the future robots need not have such feet, which might be even of a shape close to that of human, and thus far from a rectangle or any polygon, so that the support area will not be of a polygonal shape.

in the human living environment, especially on uneven surfaces, slopes and steps, without falling down.

In the period 1991–1993, Honda made the next three robots in E–series: E4, E5, and E6. Honda investigated techniques for stabilizing walking, and developed three control techniques: (i) floor reaction control, (ii) target ZMP control, and (iii) foot planting location control. In particular, E5 robot achieved stable, two legged walking, even on steps or slopping surfaces. The next step was to attach the legs to a body and create a *humanoid robot*.

In the period 1994–1997, Honda made three humanoid robots in the new P–series: P1, P2, and P3. The first humanoid, P1, can turn external electrical and computer switches on and off, grab doorknobs, and pick up and carry things. Its height is 1.91 m, and its weight is 175 kg.

P2, the world's first self–regulating, two–legged humanoid walking robot debuted in December, 1996. Using wireless techniques, the torso contained a computer, motor drives, battery, wireless radio, all of which were build in. It is 1.82 m tall and weights 210 kg.

In September 1997 the two–legged humanoid walking robot P3 was completed. Size and weight were reduced by changing component materials and by decentralizing the control system. Its smaller size is better suited for use in the human environment. It is 1.60 m tall, and weights 130 kg.

Finally, in 2000 Honda released a humanoid robot Asimo. Using the know–how gained from the prototypes P2 and P3, research and development began on new technology for actual use. Asimo represents the fruition of this pursuit. Weight was reduced to 43 kg and height to 1.20 m.

Cerebellar Robotics

Now, a new trend in robotics research is the so–called *cerebellar robotics*. In a series of papers published in prestigious journals, M. Kawato and his collaborators [AHP00, Kaw99, SKG93, GK96, BOF01, IMT00, IKM03, WK98] investigated the information processing of the brain with the long–term goal that machines, either computer programs or robots, could solve the same computational problems as those that the human brain solves, while using essentially the same principles. With these general approaches, they made progresses in elucidating visual information processing, optimal control principles for arm trajectory planning, internal models in the cerebellum, teaching by demonstration for robots, human interfaces based on electromyogram, and applications in rehabilitation medicine.

They developed a 30 DOF humanoid robot 'DB' for computational neuroscience research. DB is quick in movements, very compliant, with the same dimension and weight with humans. It has four cameras, artificial vestibular sensor, joint angle sensors and force sensors for all the actuators. DB can demonstrate 24 different behaviors, classified into 3 main classes: (i) learning

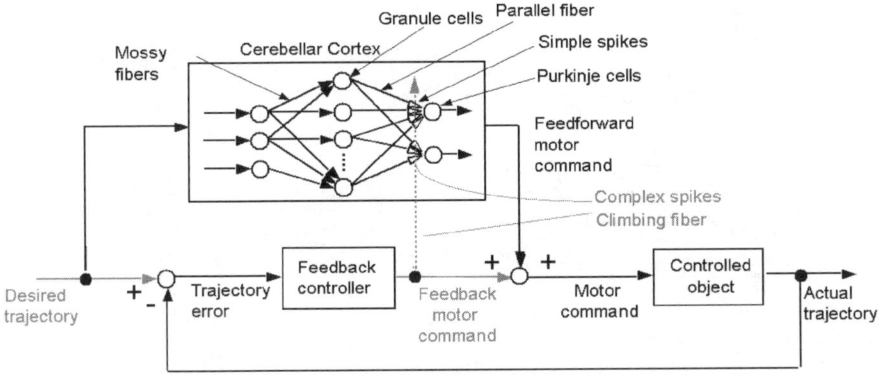

Fig. 2.38. Cerebellar feedback–error learning (see text for explanation).

from demonstration, (ii) eye movements, and (iii) behavior depending on task dynamics, physical interaction, and learning.

Essential computational principles of some of these demonstrations are: (i) cerebellar internal models, (ii) reinforcement learning in the basal ganglia, and (iii) cerebral stochastic internal model.

Their feedback error learning for cerebellum (see Figure 2.38) includes the following data: (1) Simple spike represents feedforward motor command; (2) Parallel–fibre inputs represent desired trajectory; (3) Cerebellar cortex constitutes inverse model; and (4) Complex spike represents error in motor–command coordinate.

Theories of motor control postulate that the brain uses internal models of the body to control movements accurately. Internal models are neural representations of how, for instance, the arm would respond to a neural command, given its current position and velocity. The cerebellar cortex can acquire internal models through motor learning. Because the human cerebellum is involved in higher cognitive function as well as in motor control, they proposed a coherent computational theory in which the phylogenetically newer part of the cerebellum similarly acquires internal models of objects in the external world (see Figure 2.39). While human subjects learned to use a new tool (a computer mouse with a novel rotational transformation), cerebellar activity was measured by functional magnetic resonance imaging. As predicted by their theory, two types of activity were observed. One was spread over wide areas of the cerebellum and was precisely proportional to the error signal that guides the acquisition of internal models during learning. The other was confined to the area near the posterior superior fissure and remained even after learning, when the error levels had been equalized, thus probably reflecting an acquired internal model of the new tool.

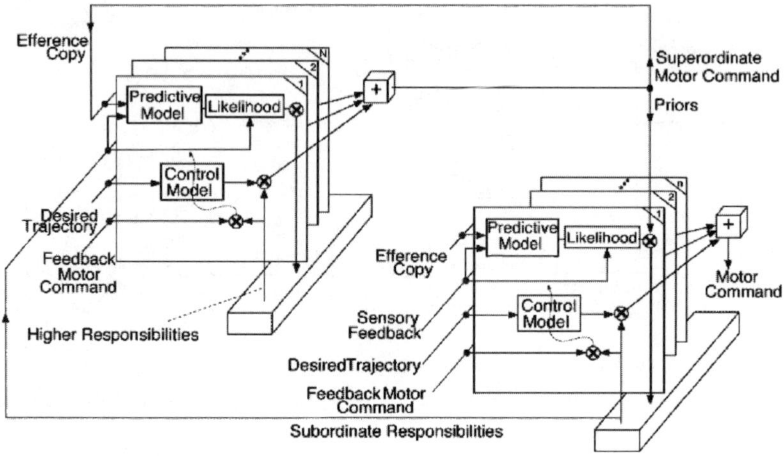

Fig. 2.39. Cerebellar modular selection and identification control

2.5.3 Humanoid Complexity

Now, recall that human (humanoid) bio–dynamics is a science of human (humanoid) motion. It is governed by both *Newtonian dynamics* and *biological control laws* [IB05, II05, II06a, II06b]. In its modern computational form, it also obeys *computational rules*. Thus, the human/humanoid bio–dynamics includes dynamical, control and computational complexities. This study shows that these three sources of complexity do not cancel each other. Instead, we have either their superposition or a kind of 'macro–entanglement' (see below in this subsection, as well as in the next Chapter) at work.

The mechanical part of a human bio–dynamical system determines the *lower limit of complexity*, which is simply defined by the *number of mechanical degrees–of–freedom*. The biological, in this case neuro–muscular, part of the combined system efficiently controls the complex dynamics of the mechanical skeleton. Such *biological complexity* cannot be explained by common complexity models such as cellular automata (CA), for the following reasons:[32]

1. Human bones neither die nor grow during the simulation period, so there is an absence of any cancellation of the physical degrees–of–freedom like in CA.
2. Averaging of these degrees–of–freedom does not work in general either, as explained below.
3. Low–dimensional linear physical systems can be successfully modelled using CA (e.g, modelling a single linear 1D wave equation using a

[32] The work presented in this subsection has been developed in collaboration with Dr. Sanjeev Sharma, Lead Human Factors, BAE Systems Australia, e-mail: Sanjeev.Sharma@baesystems.com

Margolus rule [BY97]). However, we are dealing with a system of 500 nonlinearly–coupled differential equations (see below), which has a completely different complexity level [33].

4. Human neural control (as well as humanoid–robotic control) has a natural hierarchical (multi–level) structure: spinal (reflex) level, cerebellar (synergetic) level, and cortical (planning) level. A system of this kind of complexity cannot be efficiently controlled using a single control level.

There are over 200 bones in the human skeleton driven by over 600 muscular actuators. It is sufficient to have a glimpse at the structure and function of a single skeletal muscle to get an impression of the natural complexity at work in bio–dynamics. The efficient '*orchestration*' of the whole musculo–skeletal dynamics is naturally performed by several levels of neural motor control:
(i) Spinal level of autogenetic reflexes;
(ii) Cerebellar level of muscular synergy; and
(iii) Cortical level of motion planing.

Here we need to emphasize that human joints are significantly more flexible than current robot joints, which implies their more general kinematics, dynamics and control. Bio–dynamically speaking, in each human synovial joint besides gross Eulerian rotational movements (roll, pitch and yaw), we also have some hidden and restricted translations along $(X, Y, Z)-$axes. For example, in the knee joint (see Figure 2.40), patella (knee cap) moves for about 7–10 cm from maximal extension to maximal flexion). It is well–known that even greater are translational amplitudes in the shoulder joint. In other words, within the realm of rigid body mechanics, a segment of a human arm or leg is not properly represented as a rigid body fixed at a certain point, but rather as a rigid body hanging on rope–like ligaments. More generally, the whole skeleton mechanically represents a system of flexibly coupled rigid bodies.

We can immediately foresee here the increased problems of gait balance, stability and control [VBB05], but we still cannot neglect reality.

Modern unified geometrical basis for both human biomechanics and humanoid robotics represents the *constrained* $SE(3)-$*group*, i.e., the so–called *special Euclidean group of rigid–body motions in 3D space* (see [PC05, Iva06a, II05, II06a, II06b]). In other words, during human movement, in each movable human joint there is an action of a constrained $SE(3)-$group. In other words, constrained $SE(3)-$group represents *general kinematics* of human–like joints. The corresponding nonlinear dynamics problem (resolved mainly for aircraft and spacecraft dynamics) is called the *dynamics on* $SE(3)-$*group*, while the associated nonlinear control problem (resolved mainly for general helicopter control) is called the *control on* $SE(3)-$*group*.

[33] When solving partial differential equations using CA, in a way we emulate the classical finite element method (FEM). However, FEM, even in its most recent (and most expensive software) versions is simply an unsuitable tool for any kind of serious robotics.

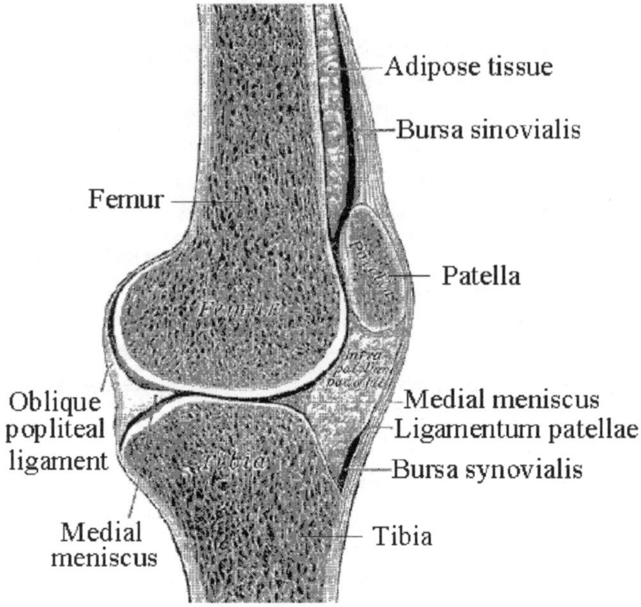

Fig. 2.40. Sagittal section through the knee joint.

The Euclidean $SE(3)$–group is defined as a semidirect (noncommutative) product of 3D rotations and 3D translations, $SE(3) := SO(3) \rhd \mathbb{R}^3$. Its most important subgroups are the following:

Subgroup	*Definition*
$SO(3)$, group of rotations in 3D (a spherical joint)	Set of all proper orthogonal 3×3 – rotational matrices
$SE(2)$, special Euclidean group in 2D (all planar motions)	Set of all 3×3 – matrices: $\begin{bmatrix} \cos\theta & \sin\theta & r_x \\ -\sin\theta & \cos\theta & r_y \\ 0 & 0 & 1 \end{bmatrix}$
$SO(2)$, group of rotations in 2D subgroup of $SE(2)$ – group (a revolute joint)	Set of all proper orthogonal 2×2 – rotational matrices included in $SE(2)$ – group
$\mathbb{R}^3$, group of translations in 3D (all spatial displacements)	Euclidean 3D vector space

Using a 'realistic model' of human bio–dynamics comprising all above complexities (see [IB05, II05, II06a]), as a well–defined example of both a general bio–physical system and a general human behavior, we propose the following conjecture: In a combined bio–physical system, where the action of the physical laws (or engineering rules) cannot be neglected, it is the physical part that determines the lower limit of the total complexity. This complexity is commonly defined as the *number of physical degrees–of–freedom*. The biological part of the combined system, as being 'more intelligent', naturally serves as a 'controller' for the physical 'plant'. Although, in some special cases, the behavior of the combined system might appear 'simple' externally (i.e., have a low–dimensional output space), the realistic internal state–space analysis shows that the complexity of the total system equals the sum of the complexities of the two parts. Neither 'mutual cancelling' nor 'averaging' of the physical degrees–of–freedom generally occurs in such bio–physical system. We demonstrate the validity of the above conjecture using the example of the human bio–dynamical system and its realistic computer model. We further discuss simplicity versus predictability (and controllability) in a complex combined system. Then we identify self–organization with training in human motion as a simple and well–defined example of general human behavior, and finally propose a new measure of complexity: the *observational resolution*.

Finally, we argue that there is a possible route to bio–dynamical simplicity in the form of oscillatory synchronization at the cost of long–term training.

Humanoid Bio–Dynamics Complexity

A physiologically realistic model of the human/humanoid bio–dynamics was developed in [IB05, IS01, Iva02, IP01a, IP01b, Iva04, Iva06b] and implemented in a software package called Human Biodynamics Engine (HBE) (for the preliminary, Lagrangian version of the spinal only HBE–simulator, see [Iva06a]). The model was developed using generalized Hamiltonian mechanics and nonlinear control on Lie groups. It includes 264 active degrees–of–freedom, driven by 132 equivalent muscular actuators[34] (each with its own excitation and contraction dynamics), as well as two levels of neural–like control (stretch–reflex and cerebellum–like Lie–derivative stabilizer and target tracker). The cortical level of motion planning is currently under development, using adaptive fuzzy logic.

In this bio–dynamical $SE(3)$–based model (see Figure 2.41), rotational joint dynamics is considered 'active', driven by Newton–Euler type forces and torques, combined with neuro–muscular stretch–reflex and higher cerebellum control. Translational dynamics is considered 'passive', representing intervertebral discs, joint tendons and ligaments as a nonlinear spring–damper system. The model was initially applied for prediction of spinal injuries [Iva06a], representing the total motion of the human spine as a dynamical chain of 25 constrained $SE(3)$ groups (i.e., special Euclidean groups of rigid body motion).

[34] An equivalent muscular actuator is a flexor–extensor pair of muscles, rotating a body segment (with all the masses attached to it) around a certain Euler axis. Each equivalent muscular actuator has its excitation dynamics, coming from the neural stimulus, as well as contraction dynamics, which generate the muscular torque in that joint. The muscular torque is the driving torque that counteracts inertial and gravity torques as well as joint elasticity and viscosity.

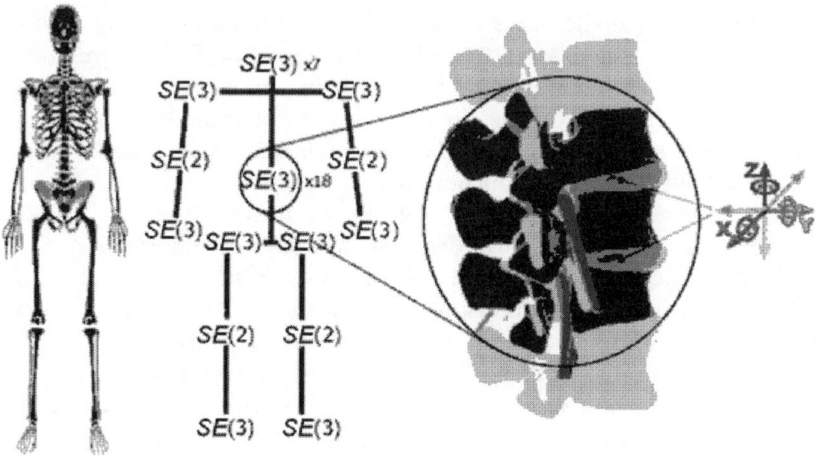

Fig. 2.41. Configuration manifold of human/humanoid skeleton, as modelled in the Human Biodynamics Engine.

Once the constrained $SE(3)-$based configuration manifold M^N is properly defined, we can define the full neuro–musculo–skeletal dynamics on its *momentum phase–space manifold* T^*M^N. The generalized Hamiltonian HBE–system is given, in a local canonical chart on T^*M^N, by (we skip here the symplectic geometry derivations, see [IS01, Iva02, IP01a, IP01b, Iva04, Iva06b] for technical details)

$$\dot{q}^i = \frac{\partial H_0}{\partial p_i} + \frac{\partial R}{\partial p_i}, \tag{2.198}$$

$$\dot{p}_i = T_i - \frac{\partial H_0}{\partial q^i} + \frac{\partial R}{\partial q^i}, \tag{2.199}$$

$$q^i(0) = q_0^i, \qquad p_i(0) = p_i^0, \tag{2.200}$$

$$(i = 1, \dots, N)$$

including the *contravariant velocity equation* (2.198) and the *covariant force equation* (2.199), with initial joint coordinates q_0^i and momenta p_i^0. Here the *physical Hamiltonian function* $H_0 : T^*M^N \to \mathbb{R}$ represents the total mechanical energy of the human motion

$$H_0(q,p) = \frac{1}{2} g^{ij}\, p_i\, p_j + V(q), \qquad (i,j = 1, \dots, N),$$

where $g^{ij} = g^{ij}(q,m)$ denotes the contravariant *material metric tensor* (associated with Riemannian metrics $g : TM^N \to \mathbb{R}$ on M^N), relating internal joint coordinates q^i and external Cartesian coordinates x^r, and including n segmental masses m_μ

$$g^{ij}(q,m) = \sum_{\mu=1}^{n} m_\mu \delta_{rs} \frac{\partial q^i}{\partial x^r} \frac{\partial q^j}{\partial x^s},$$
$$(i,j = 1,\dots,N), \qquad (r,s = 1,\dots,3n).$$

$R = R(q,p)$ denotes the Rayleigh nonlinear (usually biquadratic) dissipation function.

The driving covariant vector fields (i.e., one–forms), $T_i = T_i(t, q^i_{ang}, p_i^{ang}, u_i)$, are generalized muscular torques, depending on joint angles and angular momenta (not on translational coordinates and momenta), as well as on $u_i = u_i(t,q,p)$–corrections from the two neural control levels. Physiologically speaking, the torques T_i in the force equation (2.199) resemble neuro–muscular excitation dynamics, T_i^{EXC}, and contraction dynamics T_i^{CON}, of equivalent antagonistic muscular pairs in the ith joint, i.e., $T_i = T_i^{MUS} = T_i^{EXC} \cdot T_i^{CON}$ (see [IS01, IP01b, Iva04] for technical details).

Now, to make the highly nonlinear and high–dimensional system (2.198–2.200) even closer to bio–physical reality, namely to account for ever–present external noise as well as imprecision of anthropometric and physiological measurements, we had to add to it [IS01, Iva02]:

1. *Stochastic forces*, in the form of diffusion fluctuations $B_{ij}[q^i(t),t]$ and discontinuous jumps as N–dimensional Wiener process $W^j(t)$; and
2. *Fuzzification* of the system parameters (segmental lengths, masses, inertia moments, joint dampings, tendon elasticities, etc.) and initial conditions (body configurations),

to get the *fuzzy–stochastic HBE system:*

$$dq^i = \left(\frac{\partial H_0(q,p,\sigma_\mu)}{\partial p_i} + \frac{\partial R}{\partial p_i} \right) dt, \tag{2.201}$$

$$dp_i = B_{ij}[q^i(t),t]\, dW^j(t) + \left(\bar{T}_i - \frac{\partial H_0(q,p,\sigma_\mu)}{\partial q^i} + \frac{\partial R}{\partial q^i} \right) dt, \tag{2.202}$$

$$q^i(0) = \bar{q}^i_0, \qquad p_i(0) = \bar{p}^0_i,$$

where $\{\sigma\}_\mu$ (with $\mu \geq 1$) denote fuzzy sets of conservative parameters (segment lengths, masses and moments of inertia), dissipative joint dampings and actuator parameters (amplitudes and frequencies), while the bar $\bar{(.)}$ over a variable (.) denotes the corresponding fuzzified variable.

It is clear that the fuzzy–stochastic HBE system (2.201–2.202) is even more complex and nonlinear and therefore harder to predict/control, compared to the crisp–deterministic system (2.198–2.199). However, it is much closer to the reality of human motion.

Humanoid Control Complexity

As already stated, control of human motion is naturally and necessarily hierarchical, including three control levels: spinal, cerebellar and cortical. The first two levels have already been implemented in the software package HBE, while the cortical level is currently under the development. In this section, we briefly describe these three levels, so that the reader can get a 'feeling' for the control complexity involved.

Spinal–Like Reflex Force Control

The *force HBE servo-controller* is formulated as an affine control HBE–system. Introducing the coupling Hamiltonians $H^j = H^j(q,p)$, $j = 1, \dots, M \leq N$, corresponding to the system's active joints, we define an *affine Hamiltonian function* $H_a : T^*M^N \to \mathbb{R}$, in local canonical coordinates on T^*M^N given as

$$H_a(q,p,u) = H_0(q,p) - H^j(q,p)\, u_j, \tag{2.203}$$

where $u_i = u_i(t,q,p)$ are feedback–controls. Using (2.203) we come to the affine Hamiltonian control HBE–system, in deterministic form

$$\begin{aligned}
\dot{q}^i &= \frac{\partial H_0(q,p)}{\partial p_i} - \frac{\partial H^j(q,p)}{\partial p_i} u_j + \frac{\partial R}{\partial p_i}, \\
\dot{p}_i &= \bar{T}_i - \frac{\partial H_0(q,p)}{\partial q^i} + \frac{\partial H^j(q,p)}{\partial q^i} u_j + \frac{\partial R}{\partial q^i}, \\
o^i &= -\frac{\partial H_a(q,p,u)}{\partial u_i} = H^j(q,p), \\
q^i(0) &= q_0^i, \qquad p_i(0) = p_i^0, \\
&(i = 1, \dots, N; \qquad j = 1, \dots, M \leq N).
\end{aligned} \tag{2.204}$$

and in fuzzy–stochastic form

$$\begin{aligned}
dq^i &= \left(\frac{\partial H_0(q,p,\sigma_\mu)}{\partial p_i} - \frac{\partial H^j(q,p,\sigma_\mu)}{\partial p_i} u_j + \frac{\partial R(q,p)}{\partial p_i} \right) dt, \\
dp_i &= B_{ij}[q^i(t),t]\, dW^j(t) \qquad + \\
&\left(\bar{T}_i - \frac{\partial H_0(q,p,\sigma_\mu)}{\partial q^i} + \frac{\partial H^j(q,p,\sigma_\mu)}{\partial q^i} u_j + \frac{\partial R(q,p)}{\partial q^i} \right) dt, \\
d\bar{o}^i &= -\frac{\partial H_a(q,p,u,\sigma_\mu)}{\partial u_i} dt = H^j(q,p,\sigma_\mu)\, dt, \\
q^i(0) &= \bar{q}_0^i, \qquad p_i(0) = \bar{p}_i^0.
\end{aligned} \tag{2.205}$$

Both affine control HBE–systems (2.204–2.205) resemble an *autogenetic motor servo* [Hou79], acting on the spinal–reflex level of the human locomotion control. A voluntary contraction force F of human skeletal muscle is reflexly excited (positive feedback $+F^{-1}$) by the responses of its *spindle receptors* to stretch and is reflexly inhibited (negative feedback $-F^{-1}$) by the responses of its *Golgi tendon organs* to contraction. Stretch and unloading reflexes are mediated by combined actions of several autogenetic neural pathways, forming the so–called 'motor servo.' The term 'autogenetic' means that the stimulus excites receptors located in the same muscle that is the target of the reflex response. The most important of these muscle receptors are the primary and secondary endings in the muscle–spindles, which are sensitive to length change – positive length feedback $+F^{-1}$, and the Golgi tendon organs, which are sensitive to contractile force – negative force feedback $-F^{-1}$.

The gain G of the length feedback $+F^{-1}$ can be expressed as the *positional stiffness* (the ratio $G \approx S = dF/dx$ of the force–F change to the length–x change) of the muscle system. The greater the stiffness S, the less the muscle will be disturbed by a change in load. The autogenetic circuits $+F^{-1}$ and $-F^{-1}$ appear to function as *servoregulatory loops* that convey continuously graded amounts of excitation and inhibition to the large (*alpha*) skeletomotor neurons. Small (*gamma*) fusimotor neurons innervate the contractile poles of muscle spindles and function to modulate spindle–receptor discharge.

Cerebellum–Like Velocity and Jerk Control

Nonlinear *velocity and jerk* (time derivative of acceleration) *servo–controllers*, developed in HBE using the Lie–derivative formalism, resemble self–stabilizing and adaptive tracking action of the cerebellum [HBB96]. By introducing the vector–fields f and g, given respectively by

$$f = \left(\frac{\partial H_0}{\partial p_i}, -\frac{\partial H_0}{\partial q^i}\right), \qquad g = \left(-\frac{\partial H^j}{\partial p_i}, \frac{\partial H^j}{\partial q^i}\right)$$

we get the affine controller in the standard nonlinear MIMO–system form (see [Isi89, NS90])

$$\dot{x}_i = f(x) + g(x)\, u_j. \tag{2.206}$$

Finally, using the *Lie derivative formalism* [Iva04][35] and applying the *constant relative degree* r to all HB joints, the *control law* for asymptotic tracking

[35] Let $F(M)$ denote the set of all smooth (i.e., C^∞) real valued functions $f : M \to \mathbb{R}$ on a smooth manifold M, $V(M)$ – the set of all smooth vector–fields on M, and $V^*(M)$ – the set of all differential one–forms on M. Also, let the vector–field $\zeta \in V(M)$ be given with its local flow $\phi_t : M \to M$ such that at a point $x \in M$, $\frac{d}{dt}|_{t=0}\, \phi_t x = \zeta(x)$, and ϕ_t^* representing the *pull–back* by ϕ_t. The *Lie derivative* differential operator $\mathcal{L}_\zeta$ is defined:

(i) on a function $f \in F(M)$ as

$$\mathcal{L}_\zeta : F(M) \to F(M), \qquad \mathcal{L}_\zeta f = \frac{d}{dt}(\phi_t^* f)|_{t=0},$$

of the reference outputs $o_R^j = o_R^j(t)$ could be formulated as (generalized from [Isi89])

$$u_j = \frac{\dot{o}_R^{(r)j} - L_f^{(r)} H^j + \sum_{s=1}^r c_{s-1}(o_R^{(s-1)j} - L_f^{(s-1)} H^j)}{L_g L_f^{(r-1)} H^j}, \tag{2.207}$$

where c_{s-1} are the coefficients of the linear differential equation of order r for the *error function* $e(t) = x^j(t) - o_R^j(t)$

$$\mathrm{e}^{(r)} + c_{r-1}\mathrm{e}^{(r-1)} + \cdots + c_1\mathrm{e}^{(1)} + c_0 e = 0.$$

The affine nonlinear MIMO control system (2.206) with the Lie–derivative control law (2.207) resembles the self–stabilizing and synergistic output tracking action of the human cerebellum. To make it adaptive (and thus more realistic), instead of the 'rigid' controller (2.207), we can use the *adaptive Lie–derivative controller*, as explained in the seminal paper on geometrical nonlinear control [Si89].

Cortical–Like Fuzzy–Topological Control

For the purpose of our cortical control, the dominant, rotational part of the human configuration manifold M^N, could be first, reduced to an *N–torus*, and second, transformed to an *N–cube* ('hyper–joystick'), using the following topological techniques (see [IS01, Iva02, IP01a]).

Let S^1 denote the constrained unit circle in the complex plane, which is an Abelian Lie group. Firstly, we propose two reduction homeomorphisms, using the semidirect product $\ltimes$ of the constrained $SO(2)$–groups:

$$SO(3) \approx SO(2) \ltimes SO(2) \ltimes SO(2) \qquad \text{and} \qquad SO(2) \approx S^1.$$

Next, let I^N be the unit cube $[0,1]^N$ in $\mathbb{R}^N$ and '$\sim$' an equivalence relation on $\mathbb{R}^N$ obtained by 'gluing' together the opposite sides of I^N, preserving their orientation. Therefore, M^N can be represented as the quotient space of $\mathbb{R}^N$

(ii) on a vector–field $\eta \in V(M)$ as

$$\mathcal{L}_\zeta : V(M) \to V(M), \qquad \mathcal{L}_\zeta \eta = \frac{d}{dt}(\phi_t^* \eta)|_{t=0} \equiv [\zeta, \eta]$$

– the *Lie bracket*, and

(iii) on a one–form $\alpha \in V^*(M)$ as

$$\mathcal{L}_\zeta : V^*(M) \to V^*(M), \qquad \mathcal{L}_\zeta \alpha = \frac{d}{dt}(\phi_t^* \alpha)|_{t=0}.$$

In general, for any smooth tensor field $\mathbf{T}$ on M, the Lie derivative $\mathcal{L}_\zeta \mathbf{T}$ geometrically represents a directional derivative of $\mathbf{T}$ along the flow ϕ_t.

by the space of the integral lattice points in $\mathbb{R}^N$, that is an oriented and constrained N–dimensional torus T^N:

$$\begin{aligned} \mathbb{R}^N/Z^N &= I^N/\sim\approx \prod_{i=1}^{N} S_i^1 \equiv \{(q^i,\, i=1,\dots,N) : \mathrm{mod}2\pi\} \\ &= T^N. \end{aligned} \tag{2.208}$$

Its *Euler–Poincaré characteristic* is (by the *De Rham theorem*) both for the configuration manifold T^N and its *momentum phase–space* T^*T^N given by (see [IS01])

$$\chi(T^N, T^*T^N) = \sum_{p=1}^{N}(-1)^p b_p\,,$$

where b_p are the *Betti numbers* defined as

$$\begin{aligned} &b^0 = 1, \\ &b^1 = N, \dots b^p = \binom{N}{p}, \dots b^{N-1} = N, \\ &b^N = 1, \qquad\qquad (0 \le p \le N). \end{aligned}$$

Conversely by 'ungluing' the configuration space we get the primary unit cube. Let '$\sim^*$' denote an equivalent decomposition or 'ungluing' relation. According to Tychonoff's *product–topology theorem* (see, e.g., [AM78], [II05]), for every such quotient space there exists a 'selector' such that their quotient models are homeomorphic, that is, $T^N/\sim^*\approx A^N/\sim^*$. Therefore I_q^N represents a 'selector' for the configuration torus T^N and can be used as an N–directional '$\hat{q}$–command–space' for the *feedback control* (FC). Any subset of degrees–of–freedom on the configuration torus T^N representing the joints included in HB has its simple, rectangular image in the rectified $\hat{q}$–command space – selector I_q^N, and any joint angle q^i has its rectified image $\hat{q}^i$.

In the case of an end–effector, $\hat{q}^i$ reduces to the position vector in external–Cartesian coordinates z^r $(r = 1,\dots,3)$. If orientation of the end–effector can be neglected, this gives a topological solution to the standard inverse kinematics problem.

Analogously, all momenta $\hat{p}_i$ have their images as rectified momenta $\hat{p}_i$ in the $\hat{p}$–command space – selector I_p^N. Therefore, the total momentum phase–space manifold T^*T^N obtains its 'cortical image' as the $\widehat{(q,p)}$–command space, a trivial $2N$–dimensional bundle $I_q^N \times I_p^N$.

Now, the simplest way to perform the feedback FC on the cortical $\widehat{(q,p)}$–command space $I_q^N \times I_p^N$, and also to mimic the cortical–like behavior, is to use the $2N$– dimensional fuzzy–logic controller, in much the same way as in the popular 'inverted pendulum' examples (see [Kos92]).

We propose the fuzzy feedback–control map Ξ that maps all the rectified joint angles and momenta into the feedback–control one–forms

$$\Xi : (\hat{q}^i(t),\, \hat{p}_i(t)) \mapsto u_i(t, q, p), \tag{2.209}$$

so that their corresponding universes of discourse, $\hat{Q}^i = (\hat{q}^i_{max} - \hat{q}^i_{min})$, $\hat{P}_i = (\hat{p}_i^{max} - \hat{p}_i^{min})$ and $\hat{U}_i = (u_i^{max} - u_i^{min})$, respectively, are mapped as

$$\Xi : \prod_{i=1}^{N} \hat{Q}^i \times \prod_{i=1}^{N} \hat{P}_i \to \prod_{i=1}^{N} \hat{U}_i. \tag{2.210}$$

The $2N$–dimensional map Ξ (2.209,2.210) represents a *fuzzy inference system*, defined by (adapted from [IJB99a]):

1. *Fuzzification* of the crisp *rectified–and–discretized* angles, momenta and controls using Gaussian–bell membership functions

$$\mu_k(\chi) = exp[-\frac{(\chi - m_k)^2}{2\sigma_k}], \qquad (k = 1, 2, \ldots, 9),$$

   where $\chi \in D$ is the common symbol for $\hat{q}^i$, $\hat{p}_i$ and $u_i(q,p)$ and D is the common symbol for $\hat{Q}^i, \hat{P}_i$ and $\hat{U}_i$; the mean values m_k of the nine partitions of each universe of discourse D are defined as $m_k = \lambda_k D + \chi_{min}$, with partition coefficients λ_k uniformly spanning the range of D, corresponding to the set of nine linguistic variables $L = \{NL, NB, NM, NS, ZE, PS, PM, PB, PL\}$; standard deviations are kept constant $\sigma_k = D/9$. Using the linguistic vector L, the 9×9 FAM (fuzzy associative memory) matrix (a 'linguistic phase–plane'), is heuristically defined for each human joint, in a symmetrical weighted form

$$\mu_{kl} = \varpi_{kl}\, exp\{-50[\lambda_k + u(q,p)]^2\}, \qquad (k, l = 1, \ldots, 9)$$

   with weights: $\varpi_{kl} \in \{0.6, 0.6, 0.7, 0.7, 0.8, 0.8, 0.9, 0.9, 1.0\}$.
2. *Mamdani inference* is used on each FAM–matrix μ_{kl} for all human joints: (i) $\mu(\hat{q}^i)$ and $\mu(\hat{p}_i)$ are combined inside the fuzzy IF–THEN rules using AND (Intersection, or Minimum) operator,

$$\mu_k[\bar{u}_i(q,p)] = \min_l\{\mu_{kl}(\hat{q}^i),\, \mu_{kl}(\hat{p}_i)\}.$$

   (ii) the output sets from different IF–THEN rules are then combined using OR (Union, or Maximum) operator, to get the final output, fuzzy–covariant torques,

$$\mu[u_i(q,p)] = \max_k\{\mu_k[\bar{u}_i(q,p)]\}.$$

3. *Defuzzification* of the fuzzy controls $\mu[u_i(q,p)]$ with the 'center of gravity' method
$$u_i(q,p) = \frac{\int \mu[u_i(q,p)]\, du_i}{\int du_i},$$
to update the crisp feedback–control one–forms $u_i = u_i(t,q,p)$.

Now, it is easy to make this top–level controller *adaptive*, simply by *weighting* both the above fuzzy–rules and membership functions, by the use of any standard competitive neural–network (see, e.g., [Kos92]). Operationally, the construction of the cortical $\widehat{(q,p)}$–command space $I_q^N \times I_p^N$ and the $2N$–dimensional feedback map Ξ (2.209,2.210), mimic the regulation of the *motor conditioned reflexes* by the motor cortex [HBB96].

Humanoid Computational Complexity

A simplified version of the HBE system (2.201,2.202,2.205,2.206,2.207), with crisp parameters derived from the user anthropometry and physiology data, and simple random forces added to the crisp dynamics (2.198–2.200), has been developed at DSTO, Australia (together with a neural–like control described below), for the purpose of predicting the risk of musculo–skeletal injuries (see [Iva06a]). The system considered had 264 DOF (fingers and toes are not modelled), in the form of the set of 528 generalized Hamiltonian equations, with 132 Lie–derivative controllers. This huge set of nonlinearly–coupled nonlinear differential equations, were derived in Mathematica and then implemented in 'Delphi' compiler for MS Windows, using the specially developed *matrix–symplectic explicit integrator* of the 6th order.

It is practically *impossible to integrate* such a complex system of differential equations, even for 1 second, even with the best possible integrator, like Mathematica integrator NDSolve, the standard trick from modern mechanics and nonlinear control was adopted: *dynamical decoupling with simultaneous inertial* (static) *coupling* (see, e.g., [Sch98]).[36] Once Hamiltonian equations are decoupled, they can be both numerically solved (using a matrix symplectic integrator) and efficiently controlled (using a linear or polynomial controller derived by Lie–derivative formalism described above).

A sample HBE output is given in Figures 2.42–2.45, which simulate running with the speed of 5 m/s.

---

36 The basic idea of geometrical decoupling is to 'free' the angular momentum (resp. angular velocity) and torque variables from the inertia matrix (i.e., metric tensor) g_{ij}, by putting it on the other side of Hamiltonian (resp. Lagrangian) equations [Isi89, NS90].

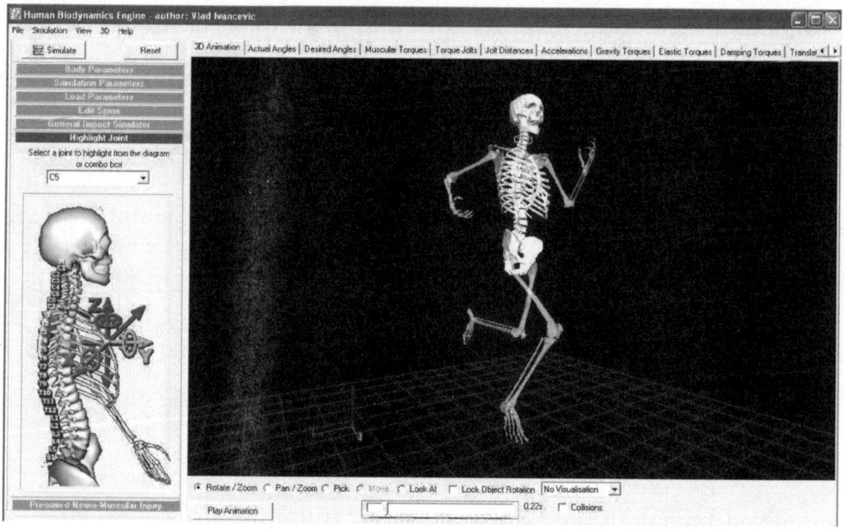

Fig. 2.42. Sample output from the Human Biodynamics Engine: running with the speed of 5 m/s – 3D animation view–port.

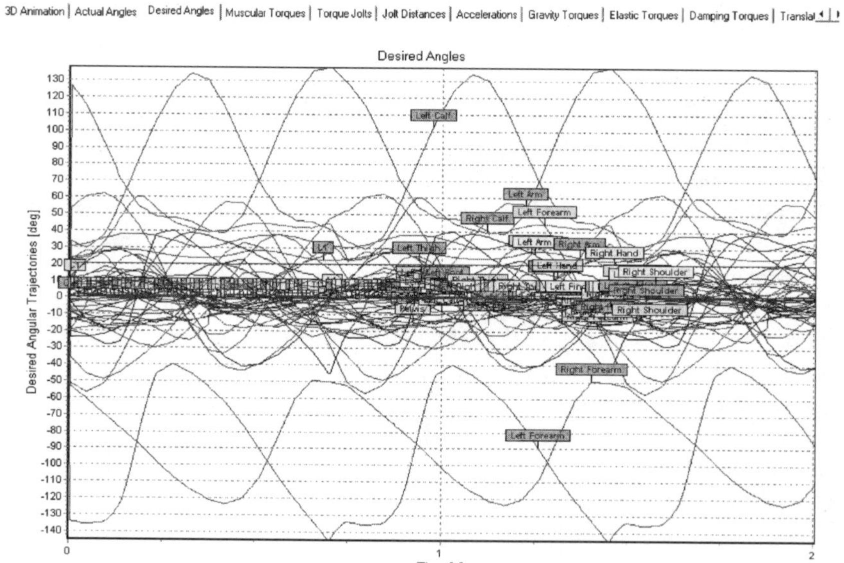

Fig. 2.43. Running HBE output: desired angular joint trajectories.

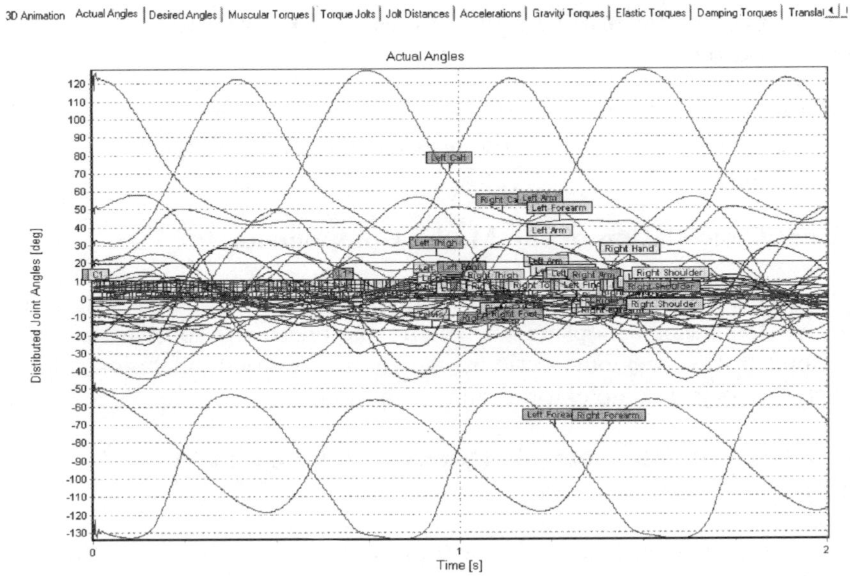

Fig. 2.44. Running HBE output: actual angular joint trajectories.

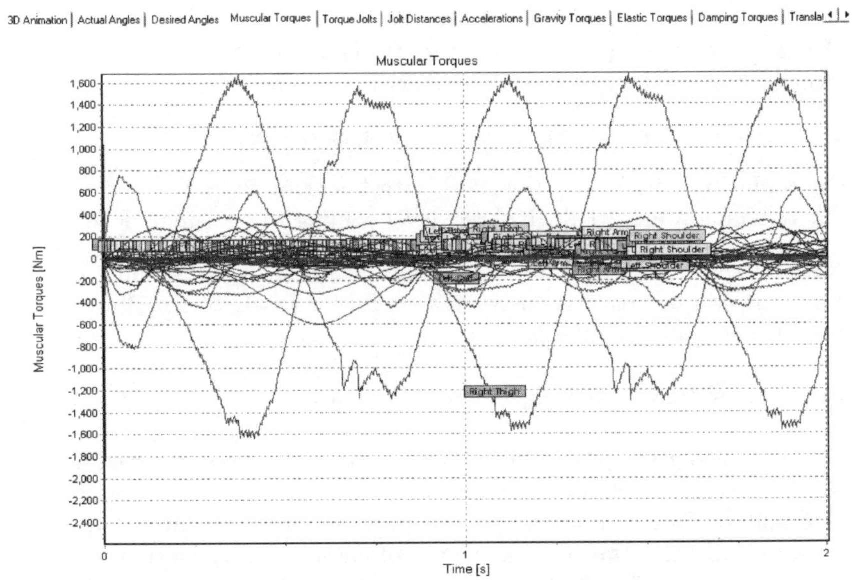

Fig. 2.45. Running HBE output: muscular torques in the joints calculated using half–inverse dynamics (forward dynamics starts with the test input torques; after the user–defined reaction time is reached, it switches to the controller–determined inverse dynamics).

Simplicity, Predictability and 'Macro–Entanglement' in Humanoid Bio–Dynamics

Here we argue that the simplification of the complex and realistic bio–dynamical model described in the previous section results in inaccurate prediction and control.

Mutual Cancellation of the Model Components

Cancellation of the skeletal components is technically called 'amputation'. Clearly, this is not an option for solving the enormous complexity problem described in the previous sections. We cannot just cut–off human limbs to reduce the overall complexity of human motion.

Reduction of Mechanical Degrees–of–Freedom and Associated Controllers

It is possible to reduce the number of mechanical degrees–of–freedom, and therefore the bio–dynamical configuration manifold, by the total factor of six:

1. By replacing three–axial joints with uniaxial ones, which reduces the system's dimension by a factor of three; and
2. By neglecting all (restricted) joint translations, as is done in robotics, which reduces the system's dimension by a factor of two.

It is also possible to simplify the control subsystem:

1. by replacing nonlinear controllers with linear ones; and
2. by reducing a hierarchical, three–level control to the single level.

The overall result of these two simplifications is commonly known as 'dummy'. It can be very expensive and useful for crash–testing, but it cannot be used for any human–like performance.

Averaging of Physical Degrees–of–Freedom

Let us consider the possibility of averaging the degrees–of–freedom in bio–dynamics, using a technique similar to Maxwell's techniques in thermodynamics and statistical physics. In the past, it has been an old practice in bio–dynamics to use the body's 'center–of–mass' (CoM) motion as a simple representative of the full human musculo–skeletal dynamics, which is a systematic kinematical procedure of averaging the segmental trajectories. However, at present, it is used only for the low–resolution global positioning system (GPS) tracking of soldiers. It simply fails in simulating/predicting any realistic human movement, which is well–known to the researchers in robotics. For example, if we use the Cartesian vector trajectory of the CoM to simulate the motion of an athlete in a successful 'high–jump' event, we will see that the CoM trajectory passes under the bar while at the same time his whole body passes over the bar, which is represented by all segmental trajectories. This simple example shows that the averaging of physical degrees–of–freedom simply does not work if realistic representation of human motion is needed.

Superposition of Complexities or 'Macro–Entanglement'

From the standard engineering viewpoint, having two systems (biological and mechanical) combined as a single 'working machine', we can expect that the total 'machine' complexity equals the sum of the two partial ones. For example, electrical circuitry has been a standard modelling framework in neurophysiology (A. Hodkgin and A. Huxley won a Nobel Prize for their circuit model of a single neuron, the celebrated HH–neuron [HH52]). Using the HH–approach for modelling human neuro–muscular circuitry as electrical circuitry, we get an electro–mechanical model for our bio–dynamical system, in which the superposition of complexities is clearly valid.

On the other hand, in a recent research on dissipative quantum brain modelling, one of the most popular issues has been *quantum entanglement*[37] between the *brain* and its *environment* (see [PV03, PV04]) where the brain–environment system has an entangled 'memory' state, identified with the ground (vacuum) state $|0>_{\mathcal{N}}$, that cannot be factorized into two single–mode states.[38] Similar to this microscopic brain–environment entanglement,

[37] The *quantum entanglement* (see next Chapter) is a quantum–mechanical phenomenon in which the quantum states of two or more objects have to be described with reference to each other, even though the individual objects may be spatially separated. This leads to correlations between observable physical properties of the systems. For example, it is possible to prepare two particles in a single quantum state such that when one is observed to be spin–up, the other one will always be observed to be spin–down and vice versa, this despite the fact that it is impossible to predict, according to quantum mechanics, which set of measurements will be observed. As a result, measurements performed on one system seem to be instantaneously influencing other systems entangled with it. Quantum entanglement does not enable the transmission of classical information faster than the speed of light.

Quantum entanglement is closely concerned with the emerging technologies of quantum computing and quantum cryptography, and has been used to experimentally realize *quantum teleportation*. At the same time, it prompts some of the more philosophically oriented discussions concerning quantum theory. The correlations predicted by quantum mechanics, and observed in experiment, reject the principle of local realism, which is that information about the state of a system should only be mediated by interactions in its immediate surroundings. Different views of what is actually occurring in the process of quantum entanglement can be related to different interpretations of quantum mechanics.

[38] In the Vitiello–Pessa dissipative quantum brain model [PV03, PV04] (see next Chapter), the evolution of the $\mathcal{N}$–coded memory system was represented as a trajectory of given initial condition running over time–dependent states $|0(t)>_{\mathcal{N}}$, each one minimizing the free energy functional.

we propose a kind of *macroscopic entanglement* between the operating modes of our neuro–muscular controller and mechanical skeleton.

In other words, we suggest that the *diffeomorphism* between the *brain motion manifold* ($N-$cube) and the *body motion manifold* M^N (which can be reduced to the constrained $N-$torus), described as the *cortical motion control* (subsection 3.3), can be considered a 'long–range correlation'.

Therefore, if the complexity of the two subsystems is not the 'expected' superposition of their partial complexities, then we have a kind of macro–entanglement at work.

Self–Organization, Synchronization and Resolution in Bio–Mechanics

Self–Organization versus Training

In the framework of human motion dynamics, *self–organization* represents *adaptive motor control*, i.e., physiological motor training performed by *iteration of conditioned reflexes*. For this, a combination of *supervised* and *reinforcement learning* is commonly used, in which a number of (nonlinear) *control parameters are iteratively adjusted* similar to the weights in neural networks, using either backpropagation–type or Hebbian–type learning, respectively. Every human motor skill is mastered using this general method. Once it is mastered, it appears as 'self–organized', and its output space appears 'low–dimensional'.

Therefore, bio–dynamical self–organization clearly represents an 'evolution' in the parameter-space of human motion control. One might argue that such an evolution can be modelled using CA. However, this parameter–space, though being a dynamical and possibly even a contractible structure, is not an independent set of parameters – it is necessarily coupled to the mechanical skeleton configuration space, the plant to be controlled.

The system of 200 bones and 600 muscles can an produce infinite number of different movements. In other words, the *output–space dimension* of a skilled human motion dynamics equals *infinity* – there is no upper limit to the number of possible different human movements, starting with simple walk, run, jump, throw, play, etc. Even for the simplest motions, like walking, a child needs about 12 months to master it (and Honda robots took a decade to achieve this).

Furthermore, as human motion represents a simplest and yet well–defined example of a general human behavior, it is possible that other human behavioral and performance skills are mastered (i.e., self–organized) in a similar way.

Observational Resolution: a True Measure for Humanoid Bio–Dynamics Complexity

Similar to a GPS tracking of soldiers' motion being reduced to the CoM motion, the *observational resolution* represents a true criterion underlying the apparent external complexity. For instance, if we 'zoom–out' sufficiently enough to get to the 'satellite–level' observation, then the collective motion of a crowd of 100,000 people looks like a single 'soliton'. On the other hand, if we 'zoom–in' deep to get to the 'Earth–level', then the full bio–dynamical system complexity and possibly an infinite–dimensional output space of a single human member within the same crowd is seen. There is a significant difference in the resolution of human motion while watching 'subtle' hand movements playing a piano, or 'coarse' movements of the crowd (on a football stadium) from an orbital satellite. CA can be a good model for the crowd motion, but certainly not for hierarchical neural control of the dynamics of human hands playing a piano. Thus, the eventual criterion that determines apparent complexity is the observational resolution. In other words, the bio–dynamical complexity is a resolution–dependent variable.

Synchronization: A Route to Simplicity in Humanoid Bio–Dynamics

Finally, there *is* also a possible route to simplicity in bio–dynamics. Namely, *synchronization* and *phase-locking* are ubiquitous in nature as well as in human brain (see [HI97, HI99, Izh01, HI01]). Synchronization can occur in *cyclic forms of human motion* (e.g., walking, running, cycling, swimming), both externally, in the form of *oscillatory dynamics*, and internally, in the form of *oscillatory cortical–control*. This oscillatory synchronization, e.g., in walking dynamics, has three possible forms: in–phase, anti–phase, and out–of–phase. The underlying phase–locking properties determined by type of oscillator (e.g., periodic/chaotic, relaxation, bursting[39], pulse-coupled, slowly

[39] Periodic bursting behavior in neurons is a recurrent transition between a quiescent state and a state of repetitive firing. Three main types of neural bursters are: (i) parabolic bursting ('circle/circle'), (ii) square–wave bursting ('fold/homoclinic'), and (iii) elliptic bursting ('subHopf/fold cycle'). Most burster models can be written in the singularly perturbed form:

$$x = f(x, y), \qquad y = \mu g(x, y),$$

where $x \in \mathbb{R}^m$ is a vector of fast variables responsible for repetitive firing (e.g., the membrane voltage and fast currents). The vector $y \in \mathbb{R}^k$ is a vector of slow variables that modulates the firing (e.g., slow (in)activation dynamics and changes in intracellular Ca^{2+} concentration). The small parameter $\mu << 1$ is a ratio of fast/slow time scales. The synchronization dynamics between bursters depends crucially on their spiking frequencies, i.e., the interactions are most effective when the presynaptic interspike frequency matches the frequency of postsynaptic

connected, or connections with time delay) involved in the cortical control system (motion planner). According to Izhikevich–Hoppensteadt work (ibid), phase–locking is prominent in the brain: it frequently results in coherent activity of neurons and neuronal groups, as seen in recordings of local field potentials and EEG. In essence, the purpose of brain control of human motion is reduction of mechanical configuration space; brain achieves this through synchronization.

While cyclic movements indeed present a natural route to oscillatory bio–dynamical synchronization, both on the dynamical and cortical–control level, the various forms of synchronized group behavior in sport (such as synchronized swimming, diving, acrobatics) or in military performance represent the imperfect products of hard training. The synchronized team performance is achievable, but the cost is a difficult long–term training and sacrifice of one's natural characteristics.

Summary on Humanoid Complexity

In this subsection we examined the complexity issues of a combined bio–dynamical system. Using a physiologically realistic model of humanoid bio–dynamics, the study demonstrated that in a combined bio–dynamical system, where the action of the Newtonian laws cannot be neglected, the mechanical part determines the lower limit of complexity of the combined system, defined by the number of physical degrees–of–freedom. The biological part of such system, being 'more intelligent' serves as a controller, of the mechanical plant. Although, in some special cases, the behavior of the combined system might appear 'simple', the analysis shows that the complexity of the total system equals the sum of the complexities of its parts, unless we have a kind of 'macro–entanglement' at work. The simplicity versus complexity issues are related to the system's predictability and controllability: a simple model can be useful for explanation, but not for prediction and control. In human motion, which represents a simple and well–defined example of general human behavior, self–organization means training using iterative conditioned reflexes. The true measure of complexity here is observational resolution.

The total complexity of the two subsystems, neuro–muscular and mechanical, is either the superposition of their partial complexities, or a kind of macro–entanglement.

Since human/humanoid motion is a simplest well–defined paradigm of *general human/humanoid behavior*, it is possible that the observational resolution underpins the apparent complexity of human behavior. This resolution–dependent complexity of human motion/behavior is totally different from

---

oscillations. The synchronization dynamics between bursters in the cortical motion planner induces synchronization dynamics between upper and lower limbs in oscillatory motions.

those proposed by CA. The number of physical degrees–of–freedom necessary to perform the motion task (e.g., high–jump or playing the piano) determines the lower limit of complexity. When this physical complexity is high, the controller's complexity needs to match it. And cortical motion planner does match this complexity with its command–space: the motor area in human cortex is not infinite – it is just high–dimensional. More important, it is not an average (statistical) value – as it precisely controls motion of every human bone (e.g., fingers playing the piano). Interestingly, the same human body that can play the piano can perform the high–jump or participate in the human crowd. Only the crowd movement can be modelled by CA.

Finally, a possible route to simplicity in bio–dynamics is represented by oscillatory synchronization, which appears in the external dynamics of oscillatory motion as a result of the synchronization in the brain control.

3

Quantum Computational Mind

In this Chapter we present a quantum theory of the computational mind.

3.1 Dirac–Feynman Quantum Dynamics

The most important discoveries in natural sciences are in some or other way connected to quantum mechanics. There is also a bias that biological phenomena will be explained by quantum theory in the future, since quantum theory already contains all basic principles of particle interactions and these principles had success in molecular dynamics, the basis of life. Recall that a little book entitled *What is Life?*, written by Noble Laureate Erwin Schrödinger, represents one of the great science classics of the twentieth century. A distinguished physicist's exploration of the question which lies at the heart of biology, it was written for the layman, but proved one of the spurs to the birth of molecular biology and the subsequent discovery of the structure of DNA, by Schrödinger's student, another Nobel laureate, Francis Crick.

If quantum phenomena really exist in the basis of *life*, then we can further argue that they are also essential for understanding both *the human and computational mind.* Therefore, before embarking on the journey into the quantum mind, we need to briefly review the basis of quantum theory.

3.1.1 Non–Relativistic Quantum Mechanics

Recall that *Heisenberg*, with his discovery of quantum mechanics (1925; see [Cas92]), introduced a *new outlook* on the nature of physical theory. Previously, it was always considered essential that there should be a detailed description of what is taking place in natural phenomena, and one used this description to calculate results comparable with experiment. Heisenberg put forward the view that it is sufficient to have a mathematical scheme from which one can calculate in a consistent manner the results of all experiments.

V.G. Ivancevic and T.T. Ivancevic: *Quantum Computational Mind*, Studies in Computational Intelligence (SCI) **60**, 461–638 (2007)
www.springerlink.com

That is, a detailed description in the traditional sense is unnecessary and may very well be impossible to establish [Dir28a, Dir28b, Dir26e].

Heisenberg's method focuses attention on the quantities which enter into experimental results. It was first applied to the *spectral theory*, for which these quantities are the energy levels of the atomic system and certain probability coefficients, which determine the probability of a radiative transition taking place from one level to another. The method sets up equations connecting these quantities and allows one to calculate them, but does not go beyond this. It does not provide any description of radiative transition processes. It does not even allow one to deduce how the results of a calculation are to be used, but requires one to assume *Einstein's laws of radiation* (the laws which tell how the probability of a radiative transition process depends on the intensity of the incident radiation), and to assume that certain quantities determined by the calculation are the coefficients appearing in the laws.

Shortly after Heisenberg's discovery, *Schrödinger* set up independently another form of quantum mechanics (1926; see [Moo89]), which also enables one to calculate energy levels and probability coefficients and gives results agreeing with those of Heisenberg, but which introduces an important new feature. It connects together, in one calculation, a *set of probability coefficients* that act together under certain conditions in Nature; e.g., the set of probability coefficients referring to *transitions* from one particular initial state to any final state. In this respect, *Schrödinger's method* is to be contrasted with Heisenberg's method, which connects together in one calculation all the probability coefficients for a dynamical system, i.e., *the probability coefficients from all initial states to all final states.*

This feature of Schrödinger's method gives it two important advantages [Dir25, Dir26e]. First, as a consequence of its enabling one to get fewer results at a time, it makes the computation much simpler. Secondly, it supplies, in a certain sense, a description of what is taking place in Nature, since a calculation leading to results that come into play together under certain conditions in Nature will be in close correspondence with the physical process that is taking place under those conditions, various points in the calculation having their counterparts in the physical process. A description in this limited sense seems to be the most that is possible for atomic processes. It implies a much less complete connection between the mathematics and the physics than one has in classical mechanics, and one might be disinclined to call it a description at all, but one may at least consider it as an appropriate generalization of what one usually means by a description. On account of Schrödinger's method allowing a description in this new sense while Heisenberg's allows none, Schrödinger's method introduces an outlook on the nature of physical theory intermediate between Heisenberg's and the old classical (Newton–Maxwellian) one.

When Heisenberg's and Schrödinger's theories were developed it was soon found by *Dirac* that they both rested on the same mathematical formalism and differed only with regard to the method of physical interpretation (see [Dir82]). *Dirac's formalism* is a generalization of the Hamiltonian form of classical

Newtonian dynamics, involving linear operators instead of ordinary algebraic variables, and is so natural and beautiful as to make one feel sure of its correctness as the foundation of the theory. The question of its interpretation, however, which involved unifying Heisenberg's and Schrödinger's ideas into a satisfactory comprehensive scheme, was not so easily settled.

The situation of a formalism (in this case, Dirac's) becoming established before one is clear about its interpretation should not be considered as surprising, but rather as a natural consequence of the drastic alterations which the development of physics had required in some of the basic physical concepts. This made it an easier matter to discover the mathematical formalism needed for a fundamental physical theory than its interpretation, since the number of things one had to choose between in discovering the formalism was very limited, the number of fundamental ideas in pure mathematics being not very great, while with the interpretation most unexpected things might turn up.

The best way of seeking the interpretation in such cases is probably from a discussion of simple examples. This way was used for the theory of quantum mechanics and led eventually to a satisfactory interpretation applicable to all phenomena for which relativistic effects are negligible. This interpretation is more closely connected with Schrödinger's method than Heisenberg's, as one would expect on account of the former affording in some sense a description of Nature, and is centered round a *Schrödinger's wave ψ–function*, which is one of the things that can be operated on by the linear operators which the dynamical variables have become. The correspondence which the existence of a description implies between the mathematics and the physics makes a wave ψ–function correspond to a state of motion of the atomic system, in such a way that, for example, a calculation which gives the transition probabilities from a particular initial state to any final state would be based on that wave ψ–function which represents the motion ensuing from this initial state. A wave ψ–function is a complex function $\psi = \psi(q_1, q_2, \dots, q_n, t)$ of all the coordinates $q_1, q_2, \dots, q_n, t$ of the system and of the time t, and it receives the interpretation that the square of its modulus, $|\psi(q_1, q_2, \dots, q_n, t)|^2$, is the *probability*, for the state of motion it corresponds to, of the coordinates having values in the neighborhood of $q_1, q_2, \dots, q_n$, per unit volume of coordinate space (or, configuration space), at the time t.

A wave ψ–function can be transformed so as to refer to other dynamical variables, for example, the momenta $p_1, p_2, \dots, p_n$, when it is said to be in another representation. The square of its modulus $|\psi(p_1, p_2, \dots, p_n, t)|^2$ is then the *probability*, per unit volume of momentum space (or, phase–space) , of the momenta having values in the neighborhood of $p_1, p_2, \dots, p_n$ at the time t. A wave ψ–function itself never has an interpretation, but only the square of its modulus, and the need for distinguishing between two wave functions having the same squares of their moduli arises only because, if they are transformed to a different representation, the squares of their moduli will in general become different. This brings out the *incompleteness of description*, which is possible with quantum mechanics [Dir28a, Dir28b, Dir26e, Dir82].

One may make a slight modification in the wave functions in any representation by introducing a weight factor λ and arranging for the probability to be $\lambda|\psi|^2$ instead of $|\psi|^2$. The weight factor may be any positive function of the variables occurring in the wave $\psi-$function.

Wave functions have to satisfy a certain *wave equation*, namely, the equation

$$i\hbar\,\partial_t\psi = H\psi, \tag{3.1}$$

where $\partial_t \equiv \partial/\partial_t$, $i = \sqrt{-1}$, $\hbar$ is the *Planck's constant*, and H is a *Hermitian (self–adjoint) linear operator* representing the Hamiltonian of the system (expressed in the representation concerned). The wave equation (3.1) is a generalization of the *Hamilton–Jacobi equation* of classical mechanics. If S is a solution of the latter equation, then

$$\psi = \mathrm{e}^{iS/\hbar} \tag{3.2}$$

will give a first approximation to a solution of the former.

An important property of the wave equation (3.1) is that it yields the *probability conservation law*: the total probability of the variables occurring in the wave $\psi-$function having any value is constant. The wave $\psi-$function should be *normalized* so as to make this probability initially unity and then it always remains unity. This conservation law is a mathematical consequence of the wave equation being linear in the operator ∂_t and of H being a self–adjoint operator.

The wave equation is linear and homogeneous in the wave $\psi-$function and so are the transformation equations. In consequence, one can add together two ψ's and get a third. The correspondence between ψ's and states of motion now allows one to infer that there is a relationship between the states of motion, such that one can add or superpose two states to get a third. This relationship constitutes the *Principle of superposition of states*, one of the general principles governing the interpretation of quantum mechanics.

Another of these principles is *Heisenberg's Principle of indeterminacy*. This is a consequence of the transformation laws connecting $\psi(q)$ and $\psi(p)$, which show that each of these functions is the *Fourier transform* of the other, apart from numerical coefficients, so that one meets the same limitations in giving values to a q and p as in giving values to the position and frequency of a train of waves [Dir26e, Dir82]. These general principles serve to bring out the departures needed from ordinary classical (Newton–Maxwellian) ideas. They are of so drastic and unexpected a nature that it is not to be wondered at that they were discovered only indirectly, as consequences of a previously established mathematical scheme, instead of being built up directly from experimental facts.

Dirac's Canonical Quantization

To make a leap into the quantum realm, recall that classical state–space for the biodynamic system of n point–particles is its $6N$D phase–space $\mathcal{P}$, including all position and momentum vectors, $\mathbf{r}_i = (x, y, z)_i$ and $\mathbf{p}_i = (p_x, p_y, p_z)_i$, respectively, for $i = 1, \dots, n$.

The *quantization* is performed as a *linear representation* of the real Lie algebra $\mathcal{L}_P$ of the phase–space $\mathcal{P}$, defined by the Poisson bracket $\{f, g\}$ of classical variables f, g – into the corresponding real Lie algebra $\mathcal{L}_H$ of the Hilbert space $\mathcal{H}$, defined by the commutator $[\hat{f}, \hat{g}]$ of skew–Hermitian operators $\hat{f}, \hat{g}$. This sounds like a functor, however it is not; as J. Baez says, 'First quantization is a mystery, but second quantization is a functor'. Mathematically, if quantization were *natural* it would be a functor from the category **Symplec**, whose objects are symplectic manifolds (i.e., phase–spaces) and whose morphisms are symplectic maps (i.e., canonical transformations) to the category **Hilbert**, whose objects are Hilbert spaces and whose morphisms are unitary operators.

Historically first, the so–called *canonical quantization* is based on the so–called *Dirac rules for quantization*. It is applied to 'simple' systems: finite number of degrees–of–freedom and 'flat' classical phase–spaces (an open set of $\mathbb{R}^{2n}$). Canonical quantization includes the following data [Dir82]:

1. *Classical description.* The system is described by the *Hamiltonian* or *canonical formalism*: its classical phase–space is locally coordinated by a set of *canonical coordinates* (q^j, p_j), the *position* and *momentum* coordinates. Classical observables are real functions $f(q^j, p_j)$. Eventually, a Lie group G of symmetries acts on the system.
2. *Quantum description.* The quantum phase–space is a complex Hilbert space $\mathcal{H}$. Quantum observables are Hermitian (i.e., self–adjoint) operators acting on $\mathcal{H}$. (The Hilbert space is complex in order to take into account the interference phenomena of wave functions representing the quantum states. The operators are self–adjoint in order to assure their eigenvalues are real.) The symmetries of the system are realized by a group of unitary operators $U_G(\mathcal{H})$.
3. *Quantization method.* As a Hilbert space we take the space of square integrable complex functions of the configuration space; that is, functions depending only on the position coordinates, $\psi(q^j)$. The quantum operator associated with $f(q^j, p_j)$ is obtained by replacing p_j by $-i\hbar\frac{\partial}{\partial q^j}$, and hence we have the correspondence $f(q^j, p_j) \mapsto \hat{f}(q^j, -i\hbar\frac{\partial}{\partial q^j})$. In this way, the classical commutation rules between the canonical coordinates are assured to have a quantum counterpart: the commutation rules between the quantum operators of position and momentum (which are related to the 'uncertainty principle' of quantum mechanics).

Quantum States and Operators

Quantum systems have two modes of evolution in time. The first, governed by standard, *time–dependent Schrödinger equation*:

$$i\hbar\, \partial_t \left|\psi\right\rangle = \hat{H} \left|\psi\right\rangle , \tag{3.3}$$

describes the time evolution of quantum systems when they are undisturbed by measurements. 'Measurements' are defined as *interactions* of the quantum system with its classical environment. As long as the system is sufficiently isolated from the environment, it follows Schrödinger equation. If an interaction with the environment takes place, i.e., a measurement is performed, the system abruptly *decoheres* i.e., collapses or reduces to one of its classically allowed states.

A *time–dependent state of a quantum system* is determined by a normalized, complex, *wave psi–function* $\psi = \psi(t)$. In Dirac's words, this is a unit *ket* vector $|\psi\rangle$, which is an element of the *Hilbert space* $L^2(\psi)$ with a coordinate basis (q^i). The state ket–vector $|\psi(t)\rangle$ is subject to action of the Hermitian operators, obtained by the procedure of *quantization* of classical biodynamic quantities, and whose real eigenvalues are being measured.

Quantum superposition is a generalization of the algebraic principle of linear combination of vectors. The Hilbert space has a set of states $|\varphi_i\rangle$ (where the index i runs over the degrees–of–freedom of the system) that form a basis and the most general state of such a system can be written as $|\psi\rangle = \sum_i c_i$ $|\varphi_i\rangle$. The system is said to be in a state $|\psi(t)\rangle$, describing the motion of the *de Broglie waves* (named after *Nobel Laureate, Prince Louis V.P.R. de Broglie*), which is a linear superposition of the basis states $|\varphi_i\rangle$ with weighting coefficients c_i that can in general be complex. At the microscopic or quantum level, the state of the system is described by the wave function $|\psi\rangle$, which in general appears as a linear superposition of all basis states. This can be interpreted as the system being in all these states at once. The coefficients c_i are called the *probability amplitudes* and $|c_i|^2$ gives the probability that $|\psi\rangle$ will collapse into state $|\varphi\rangle$ when it decoheres (interacts with the environment). By simple normalization we have the constraint that $\sum_i |c_i|^2 = 1$. This emphasizes the fact that the wavefunction describes a *real, physical system*, which must be in one of its allowable classical states and therefore by summing over all the possibilities, weighted by their corresponding probabilities, one must get unity. In other words, we have the *normalization condition* for the psi–function, determining the unit length of the state ket–vector

$$\langle\psi(t)|\psi(t)\rangle = \int \psi^*\psi \, dV = \int |\psi|^2 \, dV = 1,$$

where $\psi^* = \langle\psi(t)|$ denotes the *bra* vector, the complex–conjugate to the ket $\psi = |\psi(t)\rangle$, and $\langle\psi(t)|\psi(t)\rangle$ is their scalar product, i.e., Dirac *bracket*. For this reason the scene of quantum mechanics is the functional space of square–integrable complex psi–functions, i.e., the Hilbert space $L^2(\psi)$.

When the system is in the state $|\psi(t)\rangle$, the average value $\langle f\rangle$ of any physical observable f is equal to

$$\langle f\rangle = \langle\psi(t)| \hat{f} |\psi(t)\rangle,$$

where $\hat{f}$ is the Hermitian operator corresponding to f.

A quantum system is *coherent* if it is in a linear superposition of its basis states. If a measurement is performed on the system and this means that the system must somehow interact with its environment, the superposition is destroyed and the system is observed to be in only one basis state, as required classically. This process is called *reduction* or *collapse* of the wavefunction or simply *decoherence* and is governed by the form of the wavefunction $|\psi\rangle$.

Entanglement on the other hand, is a purely quantum phenomenon and has no classical analogue. It accounts for the ability of quantum systems to exhibit correlations in counterintuitive 'action–at–a–distance' ways. Entanglement is what makes all the difference in the operation of quantum computers versus classical ones. Entanglement gives 'special powers' to quantum computers because it gives quantum states the potential to exhibit and maintain correlations that cannot be accounted for classically. Correlations between bits are what make information encoding possible in classical computers. For instance, we can require two bits to have the same value thus encoding a relationship. If we are to subsequently change the encoded information, we must change the correlated bits in tandem by explicitly accessing each bit. Since quantum bits exist as superpositions, *correlations* between them also exist in superposition. When the superposition is destroyed (e.g., one qubit is measured), the correct correlations are *instantaneously* 'communicated' between the qubits and this communication allows *many qubits* to be accessed *at once*, preserving their correlations, something that is absolutely impossible classically.

More precisely, the *first quantization* is a *linear representation* of all classical dynamical variables (like coordinate, momentum, energy, or angular momentum) by linear *Hermitian* operators acting on the associated Hilbert state–space $L^2(\psi)$, which has the following properties [Dir82]:

1. Linearity:
$$\alpha f + \beta g \to \alpha \hat{f} + \beta \hat{g},$$
for all constants $\alpha, \beta \in \mathbb{C}$;
2. A 'dynamical' variable, equal to unity everywhere in the phase–space, corresponds to unit operator: $1 \to \hat{I}$; and
3. *Classical Poisson brackets*
$$\{f, g\} = \frac{\partial f}{\partial q^i}\frac{\partial g}{\partial p_i} - \frac{\partial f}{\partial p_i}\frac{\partial g}{\partial q^i}$$
quantize to the corresponding *commutators*
$$\{f, g\} \to -i\hbar[\hat{f}, \hat{g}], \qquad [\hat{f}, \hat{g}] = \hat{f}\hat{g} - \hat{g}\hat{f}.$$

Like Poisson bracket, commutator is bilinear and skew–symmetric operation, satisfying Jacobi identity. For Hermitian operators $\hat{f}, \hat{g}$ their commutator $[\hat{f}, \hat{g}]$ is anti–Hermitian; for this reason i is required in $\{f, g\} \to -i\hbar[\hat{f}, \hat{g}]$.

Property (2) is introduced for the following reason. In Hamiltonian mechanics each dynamical variable f generates some transformations in the phase–space via Poisson brackets. In quantum mechanics it generates transformations in the state–space by direct application to a state, i.e.,

$$\dot{u} = \{u, f\}, \qquad \partial_t|\psi\rangle = \frac{i}{\hbar}\hat{f}|\psi\rangle. \tag{3.4}$$

Exponent of anti–Hermitian operator is unitary. Due to this fact, transformations, generated by Hermitian operators

$$\hat{U} = \exp\frac{i\hat{f}t}{\hbar},$$

are unitary. They are *motions* – scalar product preserving transformations in the Hilbert state–space $L^2(\psi)$. For this property i is needed in (3.4).

Due to property (2), the transformations, generated by classical variables and quantum operators, have the same algebra.

For example, the quantization of energy E gives:

$$E \to \hat{E} = i\hbar\,\partial_t.$$

The relations between operators must be similar to the relations between the relevant physical quantities observed in classical mechanics.

For example, the quantization of the classical equation $E = H$, where

$$H = H(p_i, q^i) = T + U$$

denotes the Hamilton's function of the total system energy (the sum of the kinetic energy T and potential energy U), gives the Schrödinger equation of motion of the state ket–vector $|\psi(t)\rangle$ in the Hilbert state–space $L^2(\psi)$

$$i\hbar\,\partial_t|\psi(t)\rangle = \hat{H}\,|\psi(t)\rangle.$$

In the simplest case of a single particle in the potential field U, the operator of the total system energy – Hamiltonian is given by:

$$\hat{H} = -\frac{\hbar^2}{2m}\nabla^2 + U,$$

where m denotes the mass of the particle and ∇ is the classical gradient operator. So the first term on the r.h.s denotes the kinetic energy of the system, and therefore the momentum operator must be given by:

$$\hat{p} = -i\hbar\nabla.$$

Now, for each pair of states $|\varphi\rangle, |\psi\rangle$ their scalar product $\langle\varphi|\psi\rangle$ is introduced, which is [Nik95]:

1. Linear (for right multiplier):

$$\langle\varphi|\alpha_1\psi_1 + \alpha_2\psi_2\rangle = \alpha_1\langle\varphi|\psi_1\rangle + \alpha_2\langle\varphi|\psi_2\rangle;$$

2. In transposition transforms to complex conjugated:

$$\langle\varphi|\psi\rangle = \overline{\langle\psi|\varphi\rangle};$$

this implies that it is 'anti–linear' for left multiplier:

$$\langle\alpha_1\varphi_1 + \alpha_2\varphi_2\rangle = \bar{\alpha}_1\langle\varphi_1|\psi\rangle + \bar{\alpha}_2\langle\varphi_2|\psi\rangle);$$

3. Additionally it is often required, that the scalar product should be positively defined:

$$\text{for all } |\psi\rangle, \quad \langle\psi|\psi\rangle \geq 0 \qquad \text{and} \qquad \langle\psi|\psi\rangle = 0 \quad \text{iff} \quad |\psi\rangle = 0.$$

Complex conjugation of classical variables is represented as Hermitian conjugation of operators. We remind some definitions:

– two operators $\hat{f}, \hat{f}^+$ are called Hermitian conjugated (or adjoint), if

$$\langle\varphi|\hat{f}\psi\rangle = \langle\hat{f}^+\varphi|\psi\rangle \qquad (\text{for all } \varphi, \psi).$$

This scalar product is also denoted by $\langle\varphi|\hat{f}|\psi\rangle$ and called a matrix element of an operator.

– operator is Hermitian (self–adjoint) if $\hat{f}^+ = \hat{f}$ and anti–Hermitian if $\hat{f}^+ = -\hat{f}$;

– operator is unitary, if $\hat{U}^+ = \hat{U}^{-1}$; such operators preserve the scalar product:

$$\langle\hat{U}\varphi|\hat{U}\psi\rangle = \langle\varphi|\hat{U}^+\hat{U}|\psi\rangle = \langle\varphi|\psi\rangle.$$

Real classical variables should be represented by Hermitian operators; complex conjugated classical variables $(a, \bar{a})$ correspond to Hermitian conjugated operators $(\hat{a}, \hat{a}^+)$.

Multiplication of a state by complex numbers does not change the state physically.

Any Hermitian operator in Hilbert space has only real eigenvalues:

$$\hat{f}|\psi_i\rangle = f_i|\psi_i\rangle, \qquad (\text{for all } \ f_i \in \mathbb{R}).$$

Eigenvectors $|\psi_i\rangle$ form complete orthonormal basis (eigenvectors with different eigenvalues are automatically orthogonal; in the case of multiple eigenvalues one can form orthogonal combinations; then they can be normalized).

If the two operators $\hat{f}$ and $\hat{g}$ commute, i.e., $[\hat{f}, \hat{g}] = 0$ (see Heisenberg picture below), than the corresponding quantities can simultaneously have definite values. If the two operators do not commute, i.e., $[\hat{f}, \hat{g}] \neq 0$, the quantities corresponding to these operators cannot have definite values simultaneously, i.e., the general *Heisenberg's uncertainty relation* is valid:

$$(\Delta\hat{f})^2 \cdot (\Delta\hat{g})^2 \geq \frac{\hbar}{4}[\hat{f}, \hat{g}]^2,$$

where Δ denotes the deviation of an individual measurement from the mean value of the distribution. The well–known particular cases are ordinary uncertainty relations for coordinate–momentum $(q - p)$, and energy–time $(E - t)$:

$$\Delta q \cdot \Delta p_q \geq \frac{\hbar}{2}, \qquad \text{and} \qquad \Delta E \cdot \Delta t \geq \frac{\hbar}{2}.$$

For example, the rules of commutation, analogous to the classical ones written by the Poisson's brackets, are postulated for canonically–conjugate coordinate and momentum operators:

$$[\hat{q}^i, \hat{q}^j] = 0, \qquad [\hat{p}_i, \hat{p}_j] = 0, \qquad [\hat{q}^i, \hat{p}_j] = i\hbar\delta^i_j \hat{I},$$

where δ^i_j is the Cronecker's symbol. By applying the commutation rules to the system Hamiltonian $\hat{H} = \hat{H}(\hat{p}_i, \hat{q}^i)$, the *quantum Hamilton's equations* are obtained:

$$\frac{d(\hat{p}_i)}{dt} = -\frac{\partial \hat{H}}{\partial \hat{q}^i}, \qquad \text{and} \qquad \frac{d(\hat{q}^i)}{dt} = \frac{\partial \hat{H}}{\partial \hat{p}_i}.$$

A quantum state can be observed either in the *coordinate* $q-$*representation*, or in the *momentum* $p-$*representation*. In the $q-$representation, operators of coordinate and momentum have respective forms: $\hat{q} = q$, and $\hat{p}_q = -i\hbar\frac{\partial}{\partial q}$, while in the p–representation, they have respective forms: $\hat{q} = i\hbar\frac{\partial}{\partial p_q}$, and $\hat{p}_q = p_q$. The forms of the state vector $|\psi(t)\rangle$ in these two representations are mathematically related by a *Fourier–transform pair* (within the Planck constant).

Quantum Pictures

In the $q-$representation the quantum state is usually determined, i.e., the first quantization is performed, in one of the three *quantum pictures* (see e.g., [Dir82]):

1. *Schrödinger picture*,
2. *Heisenberg picture*, and
3. *Dirac interaction picture*.

These three pictures mutually differ in the time–dependence, i.e., time–evolution of the state vector $|\psi(t)\rangle$ and the Hilbert coordinate basis (q^i) together with the system operators.

1. In the *Schrödinger* (S) *picture*, under the action of the *evolution operator* $\hat{S}(t)$ the state–vector $|\psi(t)\rangle$ rotates:

$$|\psi(t)\rangle = \hat{S}(t)\,|\psi(0)\rangle,$$

and the coordinate basis (q^i) is fixed, so the operators are constant in time:

$$\hat{F}(t) = \hat{F}(0) = \hat{F},$$

and the system evolution is determined by the Schrödinger wave equation:

$$i\hbar\,\partial_t|\psi^S(t)\rangle = \hat{H}^S\,|\psi^S(t)\rangle.$$

If the Hamiltonian does not explicitly depend on time, $\hat{H}(t) = \hat{H}$, which is the case with the absence of variables of macroscopic fields, the state vector $|\psi(t)\rangle$ can be presented in the form:

$$|\psi(t)\rangle = \exp(-i\frac{E}{\hbar}t)\,|\psi\rangle,$$

satisfying the time–independent Schrödinger equation

$$\hat{H}\,|\psi\rangle = E\,|\psi\rangle,$$

which gives the eigenvalues E_m and eigenfunctions $|\psi_m\rangle$ of the Hamiltonian $\hat{H}$.

2. In the *Heisenberg* (H) *picture*, under the action of the evolution operator $\hat{S}(t)$, the coordinate basis (q^i) rotates, so the operators of physical variables evolve in time by the similarity transformation:

$$\hat{F}(t) = \hat{S}^{-1}(t)\,\hat{F}(0)\,\hat{S}(t),$$

while the state vector $|\psi(t)\rangle$ is constant in time:

$$|\psi(t)\rangle = |\psi(0)\rangle = |\psi\rangle,$$

and the system evolution is determined by the *Heisenberg equation of motion*:

$$i\hbar\,\partial_t\hat{F}^H(t) = [\hat{F}^H(t), \hat{H}^H(t)],$$

where $\hat{F}(t)$ denotes arbitrary Hermitian operator of the system, while the commutator, i.e., Poisson quantum bracket, is given by:

$$[\hat{F}(t), \hat{H}(t)] = \hat{F}(t)\,\hat{H}(t) - \hat{H}(t)\,\hat{F}(t) = \hat{\imath}K.$$

In both Schrödinger and Heisenberg picture the evolution operator $\hat{S}(t)$ itself is determined by the Schrödinger–like equation:

$$i\hbar\,\partial_t\hat{S}(t) = \hat{H}\,\hat{S}(t),$$

with the initial condition $\hat{S}(0) = \hat{I}$. It determines the Lie group of transformations of the Hilbert space $L^2(\psi)$ in itself, the Hamiltonian of the system being the generator of the group.

3. In the *Dirac interaction* (I) *picture* both the state vector $|\psi(t)\rangle$ and coordinate basis (q^i) rotate; therefore the system evolution is determined by both the Schrödinger wave equation and the Heisenberg equation of motion:

$$i\hbar\,\partial_t|\psi^I(t)\rangle = \hat{H}^I\,|\psi^I(t)\rangle, \qquad \text{and} \qquad i\hbar\,\partial_t\hat{F}^I(t) = [\hat{F}^I(t), \hat{H}^O(t)].$$

Here: $\hat{H} = \hat{H}^0 + \hat{H}^I$, where $\hat{H}^0$ corresponds to the Hamiltonian of the free fields and $\hat{H}^I$ corresponds to the Hamiltonian of the interaction.

Finally, we can show that the stationary Schrödinger equation

$$\hat{H}\,\psi = \hat{E}\,\psi$$

can be obtained from the condition for the minimum of the *quantum action*:

$$\delta S = 0.$$

The quantum action is usually defined by the integral:

$$S = \langle\psi(t)|\,\hat{H}\,|\psi(t)\rangle \; = \int \psi^* \hat{H}\psi \; dV,$$

with the additional normalization condition for the unit–probability of the psi–function:

$$\langle\psi(t)|\psi(t)\rangle \; = \int \psi^* \psi \; dV = 1.$$

When the functions ψ and ψ^* are considered to be formally independent and only one of them, say ψ^* is varied, we can write the condition for an extreme of the action:

$$\delta S = \int \delta\psi^* \hat{H}\psi \; dV - E \int \delta\psi^* \psi \; dV = \int \delta\psi^* (\hat{H}\psi - E\psi)\, dV = 0,$$

where E is a Lagrangian multiplier. Owing to the arbitrariness of $\delta\psi^*$, the Schrödinger equation $\hat{H}\psi - \hat{E}\psi = 0$ must hold.

Spectrum of a Quantum Operator

To recapitulate, each *state* of a system is represented by a *state vector* $|\psi\rangle$ with a unit–norm, $\langle\psi|\psi\rangle = 1$, in a complex Hilbert space $\mathcal{H}$, and vice versa. Each system *observable* is represented by a Hermitian operator $\hat{A}$ in a Hilbert space $\mathcal{H}$, and vice versa. A Hermitian operator $\hat{A}$ in a Hilbert space $\mathcal{H}$ has its domain $\mathcal{D}_{\hat{A}} \subset \mathcal{H}$ which must be dense in $\mathcal{H}$, and for any two state vectors $|\psi\rangle, |\varphi\rangle \in \mathcal{D}_{\hat{A}}$ holds $\langle\hat{A}\psi|\varphi\rangle = \langle\psi|\hat{A}\varphi\rangle$ (see e.g., [Mes00]).

Discrete Spectrum. A Hermitian operator $\hat{A}$ in a *finite–dimensional Hilbert space* $\mathcal{H}_d$ has a *discrete spectrum* $\{a_i,\ a \in \mathbb{R},\ i \in \mathbb{N}\}$, defined as a set of *discrete eigenvalues* a_i, for which the *characteristic equation*

$$\hat{A}|\psi\rangle = a|\psi\rangle \tag{3.5}$$

has the solution eigenvectors $|\psi_a\rangle \neq 0 \in \mathcal{D}_{\hat{A}} \subset \mathcal{H}_d$. For each particular eigenvalue a of a Hermitian operator $\hat{A}$ there is a corresponding *discrete characteristic projector* $\hat{\pi}_a = |\psi_a\rangle\,\langle\psi_a|$ (i.e., the projector to the eigensubspace of $\hat{A}$ composed of all discrete eigenvectors $|\psi_a\rangle$ corresponding to a).

Now, the *discrete spectral form* of a Hermitian operator $\hat{A}$ is defined as

$$\hat{A} = a_i \hat{\pi}_i = \sum_i a_i \left|i\right\rangle \left\langle i\right|, \qquad \text{for all } i \in \mathbb{N} \tag{3.6}$$

where a_i are different eigenvalues and $\hat{\pi}_i$ are the corresponding projectors subject to

$$\sum_i \hat{\pi}_i = \hat{I}, \qquad \hat{\pi}_i \hat{\pi}_j = \delta_{ij} \hat{\pi}_j,$$

where $\hat{I}$ is identity operator in $\mathcal{H}_d$.

A Hermitian operator $\hat{A}$ defines, with its characteristic projectors $\hat{\pi}_i$, the *spectral measure* of any interval on the real axis $\mathbb{R}$; for example, for a closed interval $[a, b] \subset \mathbb{R}$ holds

$$\hat{\pi}_{[a,b]}(\hat{A}) = \sum_{a_i \in [a,b]} \hat{\pi}_i, \tag{3.7}$$

and analogously for other intervals, $(a, b], [a, b), (a, b) \subset \mathbb{R}$; if $a_i \in [a, b] = \emptyset$ then $\hat{\pi}_{[a,b]}(\hat{A}) = 0$, by definition.

Now, let us suppose that we measure an observable $\hat{A}$ of a system in state $|\psi\rangle$. The *probability* P to get a result within the a'priori given interval $[a, b] \subset \mathbb{R}$ is given by its spectral measure

$$P([a, b], \hat{A}, \psi) = \langle \psi | \hat{\pi}_{[a,b]}(\hat{A}) | \psi \rangle. \tag{3.8}$$

As a consequence, the probability to get a discrete eigenvalue a_i as a result of measurement of an observable $\hat{A}$ equals its *expected value*

$$P(a_i, \hat{A}, \psi) = \langle \psi | \hat{\pi}_i | \psi \rangle = \langle \hat{\pi}_i \rangle,$$

where $\langle \hat{B} \rangle$ in general denotes the average value of an operator $\hat{B}$. Also, the probability to get a result a which is not a discrete eigenvalue of an observable $\hat{A}$ in a state $|\psi\rangle$ equals zero.

Continuous Spectrum. A Hermitian operator $\hat{A}$ in an *infinite–dimensional Hilbert space* $\mathcal{H}_c$ (the so–called *rigged Hilbert space*) has both a discrete spectrum $\{a_i,\ a \in \mathbb{R},\ i \in \mathbb{N}\}$ and a *continuous spectrum* $[c, d] \subset \mathbb{R}$. In other words, $\hat{A}$ has both a discrete sub–basis $\{|i\rangle : i \in \mathbb{N}\}$ and a continuous sub–basis $\{|s\rangle : s \in [c, d] \subset \mathbb{R}\}$. In this case s is called the *continuous eigenvalue* of $\hat{A}$. The corresponding characteristic equation is

$$\hat{A} |\psi\rangle = s |\psi\rangle. \tag{3.9}$$

Equation (3.9) has the solution eigenvectors $|\psi_s\rangle \neq 0 \in \mathcal{D}_{\hat{A}} \subset \mathcal{H}_c$, given by the *Lebesgue integral*

$$|\psi_s\rangle = \int_a^b \psi\left(s\right) \left|s\right\rangle ds, \qquad c \leq a < b \leq d,$$

where $\psi(s) = \langle s|\psi\rangle$ are continuous, *square integrable Fourier coefficients*,

$$\int_a^b |\psi(s)|^2 \, ds < +\infty,$$

while the continuous eigenvectors $|\psi_s\rangle$ are orthonormal,

$$\psi(t) = \langle t|\psi_s\rangle = \int_c^d \psi(s)\, \delta(s-t)\, ds, \tag{3.10}$$

i.e., normed on the Dirac $\delta-$function, with

$$\langle t|s\rangle = \delta(s-t), \qquad s,t \in [c,d].$$

The corresponding *continuous projectors* $\hat{\pi}^c_{[a,b]}(\hat{A})$ are defined as Lebesgue integrals

$$\hat{\pi}^c_{[a,b]}(\hat{A}) = \int_a^b |s\rangle\, ds\, \langle s| = |s\rangle\, \langle s|, \qquad -c \le a < b \le d. \tag{3.11}$$

In this case, projecting any vector $|\psi\rangle \in \mathcal{H}_c$ using $\hat{\pi}^c_{[a,b]}(\hat{A})$ is given by

$$\hat{\pi}^c_{[a,b]}(\hat{A})\, |\psi\rangle = \left(\int_a^b |s\rangle\, ds\, \langle s|\right) |\psi\rangle = \int_a^b \psi(s)\, |s\rangle\, ds.$$

Now, the *continuous spectral form* of a Hermitian operator $\hat{A}$ is defined as

$$\hat{A} = \int_c^d |s\rangle\, s\, ds\, \langle s|\,.$$

Total Spectrum. The *total Hilbert state–space* of the system is equal to the *orthogonal sum* of its *discrete* and *continuous subspaces*,

$$\mathcal{H} = \mathcal{H}_d \oplus \mathcal{H}_c. \tag{3.12}$$

The corresponding discrete and continuous projectors are mutually complementary,

$$\hat{\pi}_{a_i}(\hat{A}) + \hat{\pi}^c_{[c,d]}(\hat{A}) = \hat{I}.$$

Using the *closure property*

$$\sum_i |i\rangle\langle i| + \int_a^b |s\rangle\, ds\, \langle s| = \hat{I},$$

the *total spectral form* of a Hermitian operator $\hat{A} \in \mathcal{H}$ is given by

$$\hat{A} = \sum_i a_i\, |i\rangle\, \langle i| + \int_c^d |s\rangle\, s\, ds\, \langle s|\,, \tag{3.13}$$

while an arbitrary vector $|\psi\rangle \in \mathcal{H}$ is equal to

$$|\psi\rangle = \sum_i \psi_i |i\rangle + \int_c^d \psi(s)\, |s\rangle\, ds.$$

Here, $\psi_i = \langle i|\psi\rangle$ are *discrete Fourier coefficients*, while $\psi(s) = \langle s|\psi\rangle$ are continuous, square integrable, Fourier coefficients,

$$\int_a^b |\psi(s)|^{\,2}\, ds < +\infty.$$

Using both discrete and continuous Fourier coefficients, ψ_i and $\psi(s)$, the *total inner product* of $\mathcal{H}$ is defined as

$$\langle\varphi|\psi\rangle = \bar{\varphi}_i \psi_i + \int_c^d \bar{\varphi}(s)\, \psi(s)\; ds, \tag{3.14}$$

while the norm is

$$\langle\psi|\psi\rangle = \bar{\psi}_i \psi_i + \int_c^d \bar{\psi}(s)\, \psi(s)\; ds.$$

The *total spectral measure* is now given as

$$\hat{\pi}_{[a,b]}(\hat{A}) = \sum_i \hat{\pi}_i + \int_a^b |s\rangle\, ds\, \langle s|\,,$$

so the probability P to get a measurement result within the a'priori given interval $[a, b] \in \mathbb{R} \subset \mathcal{H}$ is given by

$$P([a,b], \hat{A}, \psi) = \sum_i \langle\psi|\hat{\pi}_i|\psi\rangle + \int_a^b |\psi(s)|^{\,2}\, ds, \tag{3.15}$$

where $|\psi(s)|^{\,2} = \langle\psi|s\rangle\, \langle s|\psi\rangle$ is called the *probability density*. From this the expectation value of an observable $\hat{A}$ is equal to

$$\langle\hat{A}\rangle = \sum_i a_i \langle\psi|\hat{\pi}_i|\psi\rangle + \int_a^b s\, |\psi(s)|^{\,2}\, ds = \langle\psi|\hat{A}|\psi\rangle,$$

General Representation Model

In quantum mechanics the total spectral form of the complete observable is given by relation (3.13). We can split this total spectral form into:

1. *Pure discrete spectral form,*

$$\hat{A} = \sum_i a_i\, |i\rangle\, \langle i|,$$

with its discrete eigenbasis $\{|i\rangle : i \in \mathbb{N}\}$, which is orthonormal ($\langle i|j\rangle = \delta_{ij}$) and closed ($\sum_i |i\rangle\, \langle i| = \hat{I}$); and

2. *Pure continuous spectral form*,

$$\hat{B} = \int_c^d |s\rangle\, s\, ds\, \langle s|\,,$$

with its continuous eigenbasis $\{|s\rangle : s \in [c,d] \subset \mathbb{R}\}$, which is orthonormal $(\langle s|t\rangle = \delta(s-t))$ and closed $(\int_c^d |s\rangle\, ds\, \langle s| = \hat{I})$.

The completeness property of each basis means that any vector $|\psi\rangle \in \mathcal{H}$ can be expanded/developed along the components of the corresponding basis. In case of the discrete basis we have

$$|\psi\rangle = \hat{I}\,|\psi\rangle = \sum_i |i\rangle\,\langle i|\,\psi\rangle = \sum_i \psi_i |i\rangle,$$

with discrete Fourier coefficients of the development $\psi_i = \langle i|\,\psi\rangle$.

In case of the continuous basis we have

$$|\psi\rangle = \hat{I}\,|\psi\rangle = \int_c^d |s\rangle\, ds\, \langle s|\psi\rangle = \int_c^d \psi(s)\, |s\rangle\, ds.$$

with continuous Fourier coefficients of the two development $\psi(s) = \langle s|\psi\rangle$, which are square integrable, $\int_a^b |\psi\,(s)|^{\,2}\, ds < +\infty$.

Direct Product Space

Let $\mathcal{H}_1, \mathcal{H}_2, \dots, \mathcal{H}_n$ and $\mathcal{H}$ be $n+1$ given Hilbert spaces such that dimension of $\mathcal{H}$ equals the product of dimensions of $\mathcal{H}_i$, $(i = 1, \dots, n$ in this section). We say that the *composite Hilbert space* $\mathcal{H}$ is defined as a direct product of the *factor Hilbert spaces* $\mathcal{H}_i$ and write

$$\mathcal{H} = \mathcal{H}_1 \otimes \mathcal{H}_2 \otimes ... \otimes \mathcal{H}_n$$

if there exists a one–to–one mapping of the set of all *uncorrelated vectors* $\{|\psi_1\rangle, |\psi_2\rangle, \dots, |\psi_n\rangle\}$, $|\psi_i\rangle \in \mathcal{H}_i$, with zero inner product (i.e., $\langle\psi_i|\psi_j\rangle = 0$, for $i \neq j$) – onto their direct product $|\psi_1\rangle \times |\psi_2\rangle \times \dots \times |\psi_n\rangle$, so that the following conditions are satisfied:

1. Linearity per each factor:

$$\left(\sum_{j_1=1}^{J_1} b_{j_1} |\psi_{j_1}\rangle\right) \times \left(\sum_{j_2=1}^{J_2} b_{j_2} |\psi_{j_2}\rangle\right) \times \dots \times \left(\sum_{j_n=1}^{J_n} b_{j_n} |\psi_{j_n}\rangle\right)$$
$$= \sum_{j_1=1}^{J_1} \sum_{j_2=1}^{J_2} \cdots \sum_{j_n=1}^{J_n} b_{j_1} b_{j_2} \dots b_{j_n} |\psi_{j_1}\rangle \times |\psi_{j_2}\rangle \times \dots \times |\psi_{j_n}\rangle.$$

2. Multiplicativity of scalar products of uncorrelated vectors $|\psi_i\rangle, |\varphi_i\rangle \in \mathcal{H}_i$:

$$
\begin{aligned}
&(|\psi_1\rangle \times |\psi_2\rangle \times \ldots \times |\psi_n\rangle \, , \, |\varphi_1\rangle \times |\varphi_2\rangle \times \ldots \times |\varphi_n\rangle) \\
&\quad = \langle \psi_1 | \varphi_1 \rangle \times \langle \psi_2 | \varphi_2 \rangle \times \ldots \times \langle \psi_n | \varphi_n \rangle.
\end{aligned}
$$

3. Uncorrelated vectors generate the whole composite space $\mathcal{H}$, which means that in a general case a vector in $\mathcal{H}$ equals the limit of linear combinations of uncorrelated vectors, i.e.,

$$
|\psi\rangle = \lim_{K\to\infty} \sum_{k=1}^{K} b_k |\psi_1^k\rangle \times |\psi_2^k\rangle \times \ldots \times |\psi_n^k\rangle.
$$

Let $\{|k_i\rangle\}$ represent arbitrary bases in the factor spaces $\mathcal{H}_i$. They induce the basis $\{|k_1\rangle \times |k_2\rangle \times \ldots \times |k_n\rangle\}$ in the composite space $\mathcal{H}$.

Let $\hat{A}_i$ be arbitrary operators (either all linear or all antilinear) in the factor spaces $\mathcal{H}_i$. Their direct product, $\hat{A}_1 \otimes \hat{A}_2 \otimes ... \otimes \hat{A}_n$ acts on the uncorrelated vectors

$$
\begin{aligned}
&\left(\hat{A}_1 \otimes \hat{A}_2 \otimes \ldots \otimes \hat{A}_n\right) (|\psi_1\rangle \times |\psi_2\rangle \times \ldots \times |\psi_n\rangle\,) \\
&\quad = \left(\hat{A}_1 |\psi_1\rangle\right) \times \left(\hat{A}_2 |\psi_2\rangle\right) \times \ldots \times \left(\hat{A}_n |\psi_n\rangle\right)
\end{aligned}
$$

State–Space for n Quantum Particles

Classical state–space for the system of n particles is its $6N$D phase–space $\mathcal{P}$, including all position and momentum vectors, $\mathbf{r}_i = (x, y, z)_i$ and $\mathbf{p}_i = (p_x, p_y, p_z)_i$ respectively, for $i = 1, \ldots, n$.

The *quantization* is performed as a *linear representation* of the real Lie algebra $\mathcal{L}_P$ of the phase–space $\mathcal{P}$, defined by the Poisson bracket $\{A, B\}$ of classical variables A, B – into the corresponding real Lie algebra $\mathcal{L}_H$ of the Hilbert space $\mathcal{H}$, defined by the commutator $[\hat{A}, \hat{B}]$ of skew–Hermitian operators $\hat{A}, \hat{B}$.

We start with the *Hilbert space* $\mathcal{H}_x$ for a single 1D quantum particle, which is composed of all vectors $|\psi_x\rangle$ of the form

$$
|\psi_x\rangle = \int_{-\infty}^{+\infty} \psi(x) \; |x\rangle \; dx,
$$

where $\psi(x) = \langle x|\psi\rangle$ are square integrable Fourier coefficients,

$$
\int_{-\infty}^{+\infty} |\psi(x)|^{\,2}\, dx < +\infty.
$$

The position and momentum Hermitian operators, $\hat{x}$ and $\hat{p}$, respectively, act on the vectors $|\psi_x\rangle \in \mathcal{H}_x$ in the following way:

$$
\begin{aligned}
\hat{x}|\psi_x\rangle &= \int_{-\infty}^{+\infty} \hat{x}\,\psi(x)\;|x\rangle\; dx, \qquad && \int_{-\infty}^{+\infty} |x\,\psi(x)|^{\,2}\, dx < +\infty, \\
\hat{p}|\psi_x\rangle &= \int_{-\infty}^{+\infty} -i\hbar\frac{\partial}{\partial \hat{x}}\psi(x)\;|x\rangle\; dx, \qquad && \int_{-\infty}^{+\infty} \left|-i\hbar\frac{\partial}{\partial x}\psi(x)\right|^{2} dx < +\infty.
\end{aligned}
$$

The *orbit Hilbert space* $\mathcal{H}_1^o$ for a single 3D quantum particle with the full set of compatible observable $\hat{\mathbf{r}} = (\hat{x}, \hat{y}, \hat{z}), \hat{\mathbf{p}} = (\hat{p}_x, \hat{p}_y, \hat{p}_z)$, is defined as

$$\mathcal{H}_1^o = \mathcal{H}_x \otimes \mathcal{H}_y \otimes \mathcal{H}_z,$$

where $\hat{\mathbf{r}}$ has the common generalized eigenvectors of the form

$$|\hat{\mathbf{r}}\rangle = |x\rangle \times |y\rangle \times |z\rangle\,.$$

$\mathcal{H}_1^o$ is composed of all vectors $|\psi_r\rangle$ of the form

$$|\psi_r\rangle = \int_{\mathcal{H}^o} \psi(\mathbf{r})\ |\mathbf{r}\rangle\ d\mathbf{r} = \int_{-\infty}^{+\infty}\int_{-\infty}^{+\infty}\int_{-\infty}^{+\infty} \psi(x,y,z)\ |x\rangle \times |y\rangle \times |z\rangle\ dxdydz,$$

where $\psi(\mathbf{r}) = \langle \mathbf{r}|\psi_r\rangle$ are square integrable Fourier coefficients,

$$\int_{-\infty}^{+\infty} |\psi(\mathbf{r})|^{\,2}\, d\mathbf{r} < +\infty.$$

The position and momentum operators, $\hat{\mathbf{r}}$ and $\hat{\mathbf{p}}$, respectively, act on the vectors $|\psi_r\rangle \in \mathcal{H}_1^o$ in the following way:

$$\hat{\mathbf{r}}|\psi_r\rangle = \int_{\mathcal{H}_1^o} \hat{\mathbf{r}}\,\psi(\mathbf{r})\ |\mathbf{r}\rangle\ d\mathbf{r}, \qquad \int_{\mathcal{H}_1^o} |\mathbf{r}\,\psi(\mathbf{r})|^{\,2}\, d\mathbf{r} < +\infty,$$

$$\hat{\mathbf{p}}|\psi_r\rangle = \int_{\mathcal{H}_1^o} -i\hbar\frac{\partial}{\partial\hat{\mathbf{r}}}\psi(\mathbf{r})\ |\mathbf{r}\rangle\ d\mathbf{r}, \qquad \int_{\mathcal{H}_1^o} \left|-i\hbar\frac{\partial}{\partial\mathbf{r}}\psi(\mathbf{r})\right|^2\ d\mathbf{r} < +\infty.$$

Now, if we have a system of n 3D particles, let $\mathcal{H}_i^o$ denote the orbit Hilbert space of the ith particle. Then the composite orbit state–space $\mathcal{H}_n^o$ of the whole system is defined as a direct product

$$\mathcal{H}_n^o = \mathcal{H}_1^o \otimes \mathcal{H}_2^o \otimes \ldots \otimes \mathcal{H}_n^o.$$

$\mathcal{H}_n^o$ is composed of all vectors

$$|\psi_r^n\rangle = \int_{\mathcal{H}_n^o} \psi(\mathbf{r}_1, \mathbf{r}_2, \ldots, \mathbf{r}_n)\ |\mathbf{r}_1\rangle \times |\mathbf{r}_2\rangle \times \ldots \times |\mathbf{r}_n\rangle\ d\mathbf{r}_1 d\mathbf{r}_2 \ldots d\mathbf{r}_n$$

where $\psi(\mathbf{r}_1, \mathbf{r}_2, \ldots, \mathbf{r}_n) = \langle \mathbf{r}_1, \mathbf{r}_2, \ldots, \mathbf{r}_n|\psi_r^n\rangle$ are square integrable Fourier coefficients

$$\int_{\mathcal{H}_n^o} |\psi(\mathbf{r}_1, \mathbf{r}_2, \ldots, \mathbf{r}_n)|^2\ d\mathbf{r}_1 d\mathbf{r}_2 \ldots d\mathbf{r}_n < +\infty,$$

The position and momentum operators $\hat{\mathbf{r}}_i$ and $\hat{\mathbf{p}}_i$ act on the vectors $|\psi_r^n\rangle \in \mathcal{H}_n^o$ in the following way:

$$\hat{\mathbf{r}}_i|\psi_r^n\rangle = \int_{\mathcal{H}_n^o} \{\hat{\mathbf{r}}_i\}\,\psi(\mathbf{r}_1, \mathbf{r}_2, \ldots, \mathbf{r}_n)\ |\mathbf{r}_1\rangle \times |\mathbf{r}_2\rangle \times \ldots \times |\mathbf{r}_n\rangle\ d\mathbf{r}_1 d\mathbf{r}_2 \ldots d\mathbf{r}_n,$$

$$\hat{\mathbf{p}}_i|\psi_r^n\rangle = \int_{\mathcal{H}_n^o} \left\{-i\hbar\frac{\partial}{\partial\hat{\mathbf{r}}_i}\right\}\psi(\mathbf{r}_1, \mathbf{r}_2, \ldots, \mathbf{r}_n)\,|\mathbf{r}_1\rangle \times |\mathbf{r}_2\rangle \times \ldots \times |\mathbf{r}_n\rangle\ d\mathbf{r}_1 d\mathbf{r}_2 \ldots d\mathbf{r}_n,$$

with the square integrable Fourier coefficients

$$\int_{\mathcal{H}_n^o} \left|\{\hat{\mathbf{r}}_i\}\, \psi\left(\mathbf{r}_1, \mathbf{r}_2, \ldots, \mathbf{r}_n\right)\right|^2 d\mathbf{r}_1 d\mathbf{r}_2 \ldots d\mathbf{r}_n < +\infty,$$

$$\int_{\mathcal{H}_n^o} \left|\left\{-i\hbar\frac{\partial}{\partial \mathbf{r}_i}\right\} \psi\left(\mathbf{r}_1, \mathbf{r}_2, \ldots, \mathbf{r}_n\right)\right|^2 d\mathbf{r}_1 d\mathbf{r}_2 \ldots d\mathbf{r}_n < +\infty,$$

respectively. In general, any set of vector Hermitian operators $\{\hat{\mathbf{A}}_i\}$ corresponding to all the particles, act on the vectors $|\psi_r^n\rangle \in \mathcal{H}_n^o$ in the following way:

$$\hat{\mathbf{A}}_i|\psi_r^n\rangle = \int_{\mathcal{H}_n^o} \{\hat{\mathbf{A}}_i\}\psi\left(\mathbf{r}_1, \mathbf{r}_2, \ldots, \mathbf{r}_n\right) |\mathbf{r}_1\rangle \times |\mathbf{r}_2\rangle \times \ldots \times |\mathbf{r}_n\rangle \, d\mathbf{r}_1 d\mathbf{r}_2 \ldots d\mathbf{r}_n,$$

with the square integrable Fourier coefficients

$$\int_{\mathcal{H}_n^o} \left|\left\{\hat{\mathbf{A}}_i\right\} \psi\left(\mathbf{r}_1, \mathbf{r}_2, \ldots, \mathbf{r}_n\right)\right|^2 d\mathbf{r}_1 d\mathbf{r}_2 \ldots d\mathbf{r}_n < +\infty.$$

3.1.2 Relativistic Quantum Mechanics and Electrodynamics

Difficulties of the Relativistic Quantum Mechanics

The theory outlined above is not in agreement with the Einstein's restricted *Principle of relativity*, as is at once evident from the special role played by the time t. Thus, while it works very well in the non–relativistic region of low velocities, where it appears to be in complete agreement with experiment, it can be considered only as an approximation, and one must face the task of extending it to make it conform to restricted relativity.[1] One should be prepared for possible further alterations being needed in basic physical concepts, and hence one should follow the route of first setting up the mathematical formalism and then seeking its physical interpretation.

Setting up the mathematical formalism is a fairly straightforward matter. One must first put classical Newtonian mechanics into *relativistic Hamiltonian form.* One must take into account that the various particles comprising the dynamical system interact through the medium of the electromagnetic field, and one must use Lorentz's equations of motion for them, including the damping terms which express the reaction of radiation. This is done in subsection 3.1.2 below, where, with the help of the *Dirac's electrodynamic action principle*, the equations of motion are obtained in the Hamiltonian form (3.62) with the Hamiltonians F_i, one for each particle, given by (3.61). This Hamiltonian formulation may now be made into a quantum theory by following rules

[1] General relativity (i.e., gravitation theory) need not be considered here, since gravitational effects are negligible in purely atomic theory.

which have become standardized from the non–relativistic quantum mechanics. The resulting formalism appears to be quite satisfactory mathematically, but when one proceeds to consider its physical interpretation one meets with serious difficulties [Dir26c, Dir26e, Dir32, Cha48].

Take an elementary example, that of a *free particle without spin*, moving in the absence of any field. The classical Hamiltonian for this system is the left–hand side of the equation

$$p_0^2 - p_1^2 - p_2^2 - p_3^2 - m^2 = 0, \tag{3.16}$$

where p_0 is the energy and p_1, p_2, p_3 the momentum of the particle, the velocity of light being taken as unity. Passing over to quantum theory by the standard rules, one gets from this Hamiltonian the so–called *Klein–Gordon wave equation*

$$(\hbar^2 \Box + m^2)\psi = 0, \tag{3.17}$$

where $\Box$ is the Dalambertian wave operator,

$$\Box \equiv \frac{\partial^2}{\partial x_0^2} - \frac{\partial^2}{\partial x_1^2} - \frac{\partial^2}{\partial x_2^2} - \frac{\partial^2}{\partial x_3^2}.$$

The wave function ψ here is a scalar, involving the coordinates x_1, x_2, x_3 and the time $t = x_0$ on the same footing, and so it is suitable for a relativistic theory.

If one now tries to use the old interpretation that $|\psi|^2$ is the probability per unit volume of the particle being in the neighborhood of the point $x = x_1, x_2, x_3$ at the time x_0, one immediately gets into conflict with relativity, since this probability ought to transform under *Lorentz transformations* like the time–component of a 4–vector, while $|\psi|^2$ is a scalar. Also the conservation law for total probability would no longer hold, the usual proof of it failing on account of the wave equation (3.17) not being linear in $\partial_{x_0} \equiv \partial/\partial_{x_0}$.

An important step forward was taken by [Gor26] and [Kle27], who proposed that instead of$|\psi|^2$ one should use the expression

$$\frac{1}{4\pi i}[\psi\, \partial_{x_0}\bar{\psi} \; - \bar{\psi}\, \partial_{x_0}\psi], \tag{3.18}$$

where $\bar{\psi} = \bar{\psi}(x_0, x_1, x_2, x_3)$ is the complex–conjugate wave ψ–function.

The expression (3.18) is the time component of a 4–vector. Further, it is easily verified that the divergence of this 4–vector vanishes, which gives the *conservation law* in relativistic form. Thus, (3.18) is evidently the correct mathematical form to use.

However, this form leads to trouble on the physical side, since, although it is real, it is not positive definite like $|\psi|^2$. Its employment would result in one having at times a negative probability for the particle being in a certain place.

This is not the only physical difficulty. Let us consider the energy and momentum of the particle, and take for simplicity a state for which these variables have definite values. The corresponding wave ψ–function will be of the form of plane waves,

$$\psi = \exp[-i(p_0x_0 - p_1x_1 - p_2x_2 - p_3x_3)/\hbar].$$

In order that the wave equation (3.17) may be satisfied, the energy and momentum values p_0, p_1, p_2, p_3 here must satisfy the classical equation (3.16). This equation allows of negative values for the energy p_0 as well as positive ones and is, in fact, symmetrical between positive and negative energies. The negative energies occur also in the classical theory, but do not then cause trouble, since a particle started off in a positive–energy state can never make a transition to a negative–energy one. In the quantum theory, however, such transitions are possible and do in general take place under the action of perturbing forces [Dir26c, Dir26e, Dir32].

The wave ψ–function may be transformed to the momentum and energy variables. The Klein–Gordon expression (3.18) then goes over into

$$|\psi(p_0, p_1, p_2, p_3)|^2 p_0^{-1} dp_1 dp_2 dp_3, \qquad (3.19)$$

as the probability of the momentum having a value within the small domain $dp_1dp_2dp_3$ about the value p_1, p_2, p_3, with the energy having the value p_0, which must be connected with p_1, p_2, p_3 by (3.16). The weight factor p_0^{-1} appears in (3.19) and makes it *Lorentz invariant*, since $\psi(p)$ is a scalar (it is defined in terms of $\psi(x)$ to make it so), and the differential element $p_0^{-1}dp_1dp_2dp_3$ is also Lorentz invariant. This weight factor may be positive or negative, and makes the probability positive or negative accordingly. Thus the two undesirable things, negative energy and negative probability, always occur together.

Let us pass on to another simple example, that of a *free particle with spin half a quantum.* The wave equation is of the same form (3.17) as before, but the wave ψ–function is no longer a scalar. It must have two components, or four if there is a field present, and the way they transform under Lorentz transformations is given by the general connection between the theory of angular momentum in quantum mechanics and group theory. The expression $\sum |\psi(x)|^2$, summed for the components of ψ, turns out to be the time component of a 4–vector, and further the divergence of this 4–vector vanishes. Thus it is satisfactory to use this expression as the probability per unit volume of the particle being at any place at any time. One does not now have any negative probabilities in the theory. However, the negative energies remain, as in the case of no spin.

We may go on and consider particles of higher spin. The general result is that there are always states of negative energy as well as those of positive energy. For particles whose spin is an integral number of quanta, the negative–energy states occur with a negative probability and the positive– energy ones with a positive probability, while for particles whose spin is a half–odd integral number of quanta, all states occur with a positive probability [Dir26e, Dir32].

Negative energies and probabilities should not be considered as nonsense. They are well–defined concepts mathematically, like a negative sum of money, since the equations which express the important properties of energies and probabilities can still be used when they are negative. Thus negative energies and probabilities should be considered simply as things which do not appear in experimental results. The physical interpretation of relativistic quantum mechanics that one gets by a natural development of the non–relativistic theory involves these things and is thus in contradiction with experiment. We therefore have to consider ways of modifying or supplementing this interpretation.

Particles of Half–Odd Integral Spin

Let us first consider particles with a half–odd integral spin, for which there is only the negative–energy difficulty to be removed. The chief particle of this kind for which a relativistic theory is needed is the electron, with spin half a quantum. Now electrons, and also, it is believed, all particles with a half–odd integral spin, satisfy the *Pauli's Exclusion Principle*, according to which not more than one of them can be in any quantum state.[2] With this principle there are only two alternatives for a state, either it is unoccupied or it is occupied by one particle, and a symmetry appears with respect to these two alternatives.

Dirac proposed a way of dealing with the negative–energy difficulty for electrons, based on a theory in which nearly all their negative–energy states are occupied (see [Dir36]). An unoccupied negative–energy state now appears as a 'hole' in the distribution of occupied negative–energy states and thus has a deficiency of negative energy, i.e., a positive energy. From the wave equation one finds that a hole moves in the way one would expect a positively charged electron to move. It becomes reasonable to identify the holes with the recently discovered positrons, and thus to get an interpretation of the theory involving positrons together with electrons. An electron jumping from a positive– to a negative–energy state in the theory is now interpreted as an annihilation of an electron and a positron, and one jumping from a negative– to a positive–energy state as a creation of an electron and a positron.

The theory involves an infinite density of electrons everywhere. It becomes necessary to assume that the distribution of electrons for which all positive–energy states are unoccupied and all negative–energy states occupied, what one may call the *vacuum distribution*, as it corresponds to the absence of all electrons and positrons in the interpretation, is completely unobservable. Only departures from this distribution are observable and contribute to the electric density and current which give rise to electromagnetic field in accordance with Maxwell's equations.

The above theory does provide a way out from the negative–energy difficulty, but it is not altogether satisfactory. The infinite number of electrons

[2] This principle is obtained in quantum mechanics from the requirement that wave functions shall be antisymmetric in all the particles.

that it involves requires one to deal with wave functions of very great complexity and leads to such complicated mathematics that one cannot solve even the simplest problems accurately, but must resort to crude and unreliable approximations. Such a theory is a most inconvenient one to have to work with, and on general philosophical grounds one feels that it must be wrong [Dir26c, Dir26e, Dir36].

Let us see whether one can modify the theory so as to make it possible to work out simple examples accurately, while retaining the basic idea of identifying unoccupied negative–energy states with positrons. The simple calculations that one can make involve simple wave functions, referring to only one or two electrons, and thus referring to nearly all the negative–energy states being unoccupied. The calculations therefore apply to a world almost saturated with positrons, i.e., having nearly every quantum state for a positron occupied. Such a world, of course, differs very much from the actual world. One can now calculate the probability of any kind of collision process occurring in this hypothetical world (in so far as electrons and positrons are concerned). One can deduce the probability coefficient for the process, i.e., the probability per unit number of incident particles or per unit intensity of the beam of incident particles, for each of the various kinds of incident particle taking part in the process. For this purpose one must use the laws of statistical mechanics, which tell how the probability of a collision process depends on the number of incident particles, paying due attention to the modified form of these laws arising from the Pauli's exclusion principle.

Let us now assume that probability coefficients so calculated for the hypothetical world are the same as those of the actual world. This single assumption provides a general physical interpretation for the formalism, enabling one to calculate collision probabilities in the actual world. It does not provide a complete physical theory, since it enables one to calculate only those experimental results that are reducible to collision probabilities, and some branches of physics, e.g., the structure of solids, do not seem to be so reducible. However, collision probabilities are the things for which a relativistic theory is at present most needed, and one may hope in the future to find ways of extending the scope of the theory to make it include the whole of physics.

Comparing the new theory with the old, one may say that the new assumption, identifying collision probability coefficients in the actual world with those in a certain hypothetical world, replaces the old assumption about the non–observability of the vacuum distribution of negative–energy electrons. The approximations needed for working out simple examples in the old theory are equivalent in their mathematical effect to making the new assumption; e.g., these approximations include the neglect of the Coulomb interaction between electron and positron in the calculation of the prob ability of pair creation and annihilation, and this interaction cannot appear in the new theory, since the calculation there is concerned with a one–electron system. Thus the new theory may be considered as a precise formulation of the old theory together with some general approximations needed for applying it.

The new theory for dealing with the negative–energy states of the electron may be applied to any kind of elementary particle with spin half a quantum, and probably also to particles with other half–odd integral spin values, provided, of course, they satisfy Pauli's exclusion principle. It may thus be applied to protons and neutrons. It requires for each particle the possibility of existence of an antiparticle of the opposite charge, if the original particle is charged. If the original particle is uncharged, one can arrange for the antiparticle to be identical with the original [Dir26c, Dir26e, Dir36].

Particles of Integral Spin

Most of the elementary particles of physics have half–odd integral spin, but there is the important exception of the photon (or, light–quantum), with spin one quantum, and there is the cosmic–ray particle, the meson, also probably with spin one quantum. All these kinds of particle, it is believed, satisfy the *Bose–Einstein statistics*, a statistics which allows any number of particles to be in the same quantum state with the same a priori probability.[3] For these kinds of particles the previous method of dealing with the negative–energy states is therefore no longer applicable, and there is the further difficulty of the negative probabilities.

When dealing with particles satisfying the Bose–Einstein statistics, it is useful to consider the operators corresponding to the absorption of a particle from a given state or the emission into a given state. These operators can be treated as dynamical variables, although they do not have any analogues in classical mechanics. If one works out their equations of motion and transformation equations, one finds a remarkable correspondence. The absorption operators from a set of independent states have the same equations of motion and transformation equations as the wave ψ–function representing a single particle, and similarly for the emission operators and the conjugate complex wave $\bar{\psi}$–function. Thus one can pass from a one–particle theory to a many–particle theory by making the ψ and $\bar{\psi}$ describing the one particle into *absorption and emission operators* (or anihilation and creation operators), which must satisfy the appropriate commutation relations. Such a passage is called *second quantization*.

One can get over the difficulties of negative energies and negative probabilities for Bose–Einstein particles by abandoning the attempt to get a satisfactory theory of a single particle and passing on to consider the problem of many particles, using a method given by Pauli and Weisskopf [PE34] for electrons having no spin and satisfying the Bose–Einstein statistics. [4] The method of Pauli and Wiesskopf is to work entirely with positive–energy states. The operators of absorption from and emission into negative–energy states, arising in

[3] This statistics is obtained in quantum mechanics from the requirement that wave functions shall be symmetric in all the particles.

[4] Such electrons are not known experimentally, but there is no known theoretical reason why they should not exist.

the application of second quantization to the one–electron theory, are replaced by the operators of emission into and absorption from positive–energy states of electrons with the opposite charge, respectively. This replacement does not disturb the laws of conservation of charge, energy and momentum. The resulting theory involves spinless electrons of both kinds of charge together, and leads to pair creation and annihilation, as with ordinary electrons and positrons [Dir26e, Dir26c].

The method of Pauli and Wiesskopf may be applied in a degenerate form to photons and leads to the quantum electrodynamics of Heisenberg and Pauli [HP29a, HP29b]. To take into account that photons have no charge, one must start with a one–particle theory in which the wave functions are real, so that $\bar{\psi} = \psi$. The part of the wave $\psi-$function referring to positive–energy states is then made into the absorption operators from positive–energy states, and the part referring to negative–energy states into the emission operators into positive energy states. The resulting scheme of operators, involving only positive energy photon states, may then be put into correspondence with classical electrodynamics, according to the usual laws governing the correspondence between quantum and classical theory.

It would seem that in this way the difficulties of negative energies and probabilities for Bose–Einstein particles can be overcome, but a new difficulty appears. When one tries to solve the wave equation (or the wave equations if there are several particles with their respective Hamiltonians) one gets divergent integrals in the solution, of the form, in the case of photons,

$$\int_0^\infty f(v)dv, \qquad f(v) \sim v^n \quad \text{for large } v, \tag{3.20}$$

v being the frequency of a photon. The values $1, 0$ and -1 for n are the chief ones occurring in simple examples. Thus the wave equation really has no solutions and the method fails [Dir26c, Dir26e].

Dirac had made a detailed study of the divergent integrals occurring in quantum electrodynamics and had shown [Dir36] with even values of n can be eliminated by introducing into the equations a certain limiting process, which one can justify by showing that a corresponding limiting process is needed in classical electrodynamics to get the equations of motion into Hamiltonian form (which appears according to the Dirac's electrodynamic action principle, see subsection 3.1.2 below). The divergent integrals with odd values of n remain, however, and indicate something more fundamentally wrong with the theory.

Divergent integrals are a general feature of quantum field theories, and it has usually been supposed that they should be avoided by altering the forces or the laws of interaction between the elementary particles at small distances, so as to get the integrals cut off for some high value of v. However, one can easily see that this is wrong, in the case of electrodynamics at any rate, by referring to the corresponding classical theory. The wave $\psi-$function should have its analogue in the solution of the Hamilton–Jacobi equation, in accordance with equation (3.2), but already when one tries to solve the

Hamilton–Jacobi equation of classical electrodynamics corresponding to the wave equation of Heisenberg and Pauli's quantum electrodynamics, one meets with divergent integrals. Now the classical equations of motion concerned, namely, Lorentz's equations including radiation damping, have definite solutions when treated by straightforward methods and if, on trying to get these solutions by a Hamilton–Jacobi method, one meets with divergent integrals, it means simply that the Hamilton–Jacobi method is an unsuitable one, and not that one should try to alter the physical laws of interaction to get the integrals to converge. The correspondence between the quantum and classical theories is so close that one can infer that the corresponding divergent integrals in the quantum theory must also be due to an unsuitable mathematical method.

The appearance of divergent integrals with odd $n-$values in Heisenberg and Pauli's form of quantum electrodynamics may be ascribed to the asymmetrical treatment of positive– and negative–energy photon states. If instead of using Pauli and Weisskopf's method one keeps to plain second quantization, one can build up a form of quantum electrodynamics symmetrical between positive– and negative–energy photon states [Dir26e, Dir36]. The new theory leads to similar equations as the old one, but with integrals of the type

$$\int_{-\infty}^{\infty} f(v)dv, \tag{3.21}$$

instead of (3.20), and since $f(v)$ is always a rational algebraic function, and it is reasonable on physical grounds to approach the upper and lower limits of integration in (3.21) at the same rate, the divergencies with odd $n-$values all cancel out.

Dirac had shown that the new form of quantum electrodynamics also corresponds to classical electrodynamics in accordance with the usual laws, with the exception that operators corresponding to real dynamical variables in the classical theory are no longer always self-adjoint. This exception is not important, as it rather stands apart from the general mathematical connection between quantum and classical theory. The Hamilton–Jacobi equation corresponding to the wave equation of the new quantum electrodynamics differs from that of the old one only through being expressed in terms of a different set of coordinates, but the new Hamilton–Jacobi equation can be solved without divergent integrals and is connected with a satisfactory action principle [Dir32, Dir26e, Dir36]. Thus the correspondence with classical theory of the new form of quantum electrodynamics is more far–reaching than that of the old form, which provides a strong reason for preferring the new form. It now becomes necessary to find some new physical interpretation to avoid the difficulties of negative energies and probabilities.

Let us consider in more detail the relation between the *two forms of quantum electrodynamics*. In either form the electromagnetic potentials **A** at two points **x**' and **x**" must satisfy the *commutation relations*

$$[A_\mu(\mathbf{x}'), A_\nu(\mathbf{x}")] = g_{\mu\nu}\Delta(\mathbf{x}' - \mathbf{x}"), \tag{3.22}$$

obtained from analogy with the classical theory, Δ being the four–dimensional Lorentz–invariant function introduced by Jordan and Pauli (1928), which has a singularity on the light–cone and vanishes everywhere else. In the quantum electrodynamics of Heisenberg and Pauli the **A**'s are operators referring to the absorption and emission of photons into positive energy states. Let us call such operators $\mathbf{A}^1$. One could introduce a similar set of operators referring to the absorption and emission of photons into negative–energy states. Let us call these operators $\mathbf{A}^2$. They satisfy the same commutation relations (3.22) and commute with the $\mathbf{A}^1$'s. One can now introduce a third set of operators

$$\mathbf{A}^3 = \frac{\sqrt{2}}{2}(\mathbf{A}^1 + \mathbf{A}^2),$$

which operate on wave functions referring to photons in both positive– and negative–energy states, and which satisfy the same commutation relations (3.22). The use of this third set gives the new form of quantum electrodynamics arising from plain second quantization.

The three sets of **A**'s may be expressed in terms of their Fourier components as [Dir26e, Dir32, Dir36]

$$\mathbf{A}^1(\mathbf{x}) = \int [\mathbf{R}_k \mathrm{e}^{i(k,x)} + \bar{\mathbf{R}}_k \mathrm{e}^{-i(kx)}] k_0^{-1} dk_1 dk_2 dk_3, \quad \text{with} \quad k_0 = \sqrt{k_1^2 + k_2^2 + k_3^2}, \tag{3.23}$$

where $\int$ denotes the tripple integral, $\mathbf{R}_k$ is the emission operator and $\bar{\mathbf{R}}_k$ is the absorption operator,

$$\mathbf{A}^1(\mathbf{x}) = \int [\mathbf{R}_k \mathrm{e}^{i(k,x)} + \bar{\mathbf{R}}_k \mathrm{e}^{-i(kx)}] k_0^{-1} dk_1 dk_2 dk_3, \quad \text{with} \quad k_0 = -\sqrt{k_1^2 + k_2^2 + k_3^2}, \tag{3.24}$$

$$\mathbf{A}^3(\mathbf{x}) = \frac{\sqrt{2}}{2} \sum_{k_0 = \pm\sqrt{k_1^2 + k_2^2 + k_3^2}} \int [\mathbf{R}_k \mathrm{e}^{i(k,x)} + \bar{\mathbf{R}}_k \mathrm{e}^{-i(kx)}] k_0^{-1} dk_1 dk_2 dk_3. \tag{3.25}$$

Since the three sets of **A**'s all satisfy the same commutation relations, they must correspond merely to three different representations of the same dynamical variables, and the passage from one to another must be a transformation of the linear type usual in quantum mechanics. Thus, after obtaining the divergency–free solution of the wave equation in the representation corresponding to $\mathbf{A}^3$, one could apply a transformation to get the solution in the $\mathbf{A}^1$ representation. However, the transformation would then introduce the same divergent integrals as appear with the direct solution of the wave equation in the $\mathbf{A}^1$ representation, so one would not get any further this way [Dir36].

In working with the $\mathbf{A}^3$ representation one has redundant dynamical variables. It is as though, in dealing with a system of one degree of freedom with the variables q, p, one decided to treat it as a system of two degrees–of–freedom by putting

$$q = \frac{\sqrt{2}}{2}(q_1 + q_2) \qquad \text{and} \qquad p = \frac{\sqrt{2}}{2}(p_1 + p_2).$$

This would be quite a correct procedure, but would introduce an unnecessary complication. In the case of quantum electrodynamics, the complication is a necessary one, to avoid the divergent integrals. Let us put

$$\mathcal{B}(\mathbf{x}) = \frac{\sqrt{2}}{2}[\mathbf{A}^1(\mathbf{x}) - \mathbf{A}^2(\mathbf{x})]. \tag{3.26}$$

Then the $\mathbf{B}$'s commute with the $\mathbf{A}^3$'s, and thus with all the dynamical variables appearing in the Hamiltonian, so they are the redundant variables.

To determine the significance of redundant variables in quantum mechanics one may consider a general case, and work in a representation which separates the redundant variables from the non–redundant ones. One then sees immediately that a solution of the wave equation corresponds in general, not to a single state, but to a set of states with a certain probability for each, what in the classical theory is called a *Gibbs ensemble*. The. probabilities of the various states depend on the weights attached to the various eigenvalues of the redundant variables in the wave ψ–function, these weights being arbitrary, depending on the weight factor in the representation used. If one works in a representation which does not separate the redundant and non–redundant variables, as is the case in quantum electrodynamics with the representation corresponding to the use of $\mathbf{A}^3$, the general result that wave functions represent Gibbs ensembles and not single states must still be valid. Thus one can conclude that there are no solutions of the wave equation of quantum electrodynamics representing single states, but only solutions representing Gibbs ensembles. The problem remains of interpreting the negative energies and probabilities occurring with these Gibbs ensembles.

For any $\mathbf{x}$, $\mathcal{B}(\mathbf{x})$ commutes with the Hamiltonian and is a constant of the motion. We may give it any value we like, subject to not contradicting the commutation relations. Instead of $\mathcal{B}(\mathbf{x})$ it is more convenient to work with the potential field, $\mathbf{B}(\mathbf{x})$ say, obtained from $\mathcal{B}(\mathbf{x})$ by changing the sign of all the Fourier components containing $\mathrm{e}^{ik_0x_0}$ with negative values of k_0. From (3.26), (3.23) and (3.24), we have [Dir26e, Dir36]

$$\mathbf{B}(\mathbf{x}) = \frac{\sqrt{2}}{2} \sum_{k_0=\pm\sqrt{k_1^2+k_2^2+k_3^2}} \int [\mathbf{R}_k \mathrm{e}^{i(k,x)} - \bar{\mathbf{R}}_k \mathrm{e}^{-i(kx)}] k_0^{-1} dk_1 dk_2 dk_3. \tag{3.27}$$

Let us now take $\mathbf{B}$ equal to the initial value of $\mathbf{A}^3$, a proceeding which does not contradict the commutation relations since its consequences are self–consistent. Then for the initial wave ψ–function we have

$$[\mathbf{B}(\mathbf{x}) - \mathbf{A}^3(\mathbf{x})]\psi = 0,$$

or, from (3.25) and (3.27),

$$\bar{\mathbf{R}}_k\psi = 0, \tag{3.28}$$

with k_0 either positive or negative. Thus any absorption operator applied to the initial wave ψ–function gives the result zero, which means that the corresponding state is one with no photons present.

The following natural interpretation for the wave ψ–function at some later time now appears. That part of it corresponding to no photons present may be supposed to give (through the square of its modulus) the probability of no change having taken place in the field of photons; that part corresponding to one positive–energy photon present may be supposed to give the probability of a photon having been emitted; that corresponding to one negative–energy photon present may be supposed to give the probability of a photon having been absorbed; and so on for the parts corresponding to two or more photons present. The various parts of the wave ψ–function which referred to the existence' of positive– and negative–energy photons in the old interpretation now refer to the emissions and absorptions of photons. This disposes of the negative–energy difficulty in a satisfactory way, conforming to the laws of conservation of energy and momentum. It is possible only because of the redundant variables enabling one to arrange that the initial wave ψ–function shall correspond in its entirety to no emissions or absorptions having taken place.

The interpretation is not yet complete, because the theory at present would give a negative probability for a process involving the absorption of a photon, or the absorption of any odd number of photons. To find the origin of these negative probabilities, one must study the probability distribution of the photons initially present in the Gibbs ensemble, which one can do by transforming to the representation corresponding to the $\mathbf{A}^1$ potentials. It is true that one cannot apply this transformation to a solution of the wave equation without getting divergent integrals, as has already been mentioned, but one can apply it to the initial wave ψ–function, which is of a specially simple form in the photon variables. In [Dir32, Dir26e, Dir36] it is found that the probability of there being n photons initially in any photon state is $P_n = \pm 2$, according to whether n is even or odd. Strictly, to make $\sum_{n=0}^{\infty} P_n$ converge to the limit unity, one must consider P_n as a limit,

$$P_n = 2(\epsilon - 1)^n, \tag{3.29}$$

with ϵ a small positive quantity tending to zero.

Probabilities 2 and -2 are, clearly, not physically understandable, but one can use them mathematically in accordance with the rules for working with a Gibbs ensemble. One can suppose a hypothetical mathematical world with the initial probability distribution (3.29) for the photons, and one can work out the probabilities of radiative transition processes occurring in this world. One can deduce the corresponding probability coefficients, i.e., the probabilities per unit intensity of each beam of incident radiation concerned, by using Einstein's laws of radiation. For example, for a process involving the absorption of a photon, if the probability coefficient is B, the probability of the process is

$$\sum_{n=0}^{\infty} nP_n B = -\frac{1}{2}B, \tag{3.30}$$

and for a process involving the emission of a photon, if the probability coefficient is A, the probability of the process is

$$\sum_{n=0}^{\infty}(n+1)P_n A = \frac{1}{2}A. \tag{3.31}$$

Now the probability of an absorption process, as calculated from the theory, is negative, and that for an emission process is positive, so that, equating these calculated probabilities to (3.30) and (3.31) respectively, one obtains positive values for both B and A. Generally, it is easily verified that any radiative transition probability coefficient obtained by this method is positive.

It now becomes reasonable to assume that these probability coefficients obtained for a hypothetical world are the same as those of the actual world. One gets in this way a general physical interpretation for the quantum theory of photons. When applied to elementary examples, it gives the same results as Heisenberg and Pauli's quantum electrodynamics with neglect of the divergent integrals, since the extra factor $\sqrt{2}/2$ occurring in the matrix elements of the present theory owing to the $\sqrt{2}/2$ in the right–hand side of (3.25) compensates the factor 1/2 in the right–hand side of (3.30) or (3.31). The present general method of physical interpretation is probably applicable to any kind of particle with an integral spin [Dir32, Dir26e, Dir36, Cha48].

Therefore, it appears that, whether one is dealing with particles of integral spin or of half-odd integral spin, one is led to a similar conclusion, namely, that the mathematical methods at present in use in quantum mechanics are capable of direct interpretation only in terms of a hypothetical world differing very markedly from the actual one. These mathematical methods can be made into a physical theory by the assumption that results about collision processes are the same for the hypothetical world as the actual one. One thus gets back to Heisenberg's view about physical theory, that all it does is to provide a consistent means of calculating experimental results. The limited kind of description of Nature which Schrödinger's method provides in the non–relativistic case is possible relativistically only for the hypothetical world, and even then is rather more indefinite (e.g., the principle of superposition of states no longer applies), because of the need to use a Gibbs ensemble for describing the photon distribution.

To have a description of Nature is philosophically satisfying, though not logically necessary, and it is somewhat strange that the attempt to get such a description should meet with a partial success, namely, in the non–relativistic domain, but yet should fail completely in the later development. It seems to suggest that the present mathematical methods are not final. Any improvement in them would have to be of a very drastic character, because the source of all the trouble, the symmetry between positive and negative energies arising from the association of energies with the Fourier components of functions of the time, is a fundamental feature of them [Dir32, Dir26e, Dir36, Cha48].

Dirac's Electrodynamics Action Principle

There are various forms which the action principle of classical electrodynamics may take, but most of them involve awkward conditions concerning the singularities of the field where the charged particles are situated and are not suitable for a subsequent passage to quantum mechanics.

Fokker [Fok29] set up a form of action principle which does not refer to the singularities of the field and which appears to be the best starting point for getting a quantum theory. *Fokker's action integral* may conveniently be written with the help of the δ–function as

$$S = S_1 + S_2, \qquad \text{where}$$

$$S_1 = \sum_i m_i \int ds_i \qquad \text{and} \tag{3.32}$$

$$S_2 = \sum_i \sum_{j \neq i} e_i e_j \int \int \delta(\mathbf{z}_i - \mathbf{z}_j)^2 (\mathbf{v}_i, \mathbf{v}_j) ds_i ds_j \tag{3.33}$$

Here, the scalar product notation is used as

$$(\mathbf{a}, \mathbf{b}) = a^\mu b_\mu = a_0 b_0 - a_1 b_1 - a_2 b_2 - a_3 b_3,$$

and m_i and e_i are the mass and charge of the ith particle, the 4–vector $\mathbf{z}_i$ gives the four coordinates of the point on the world–line of the ith particle whose proper–time is s_i, and $\mathbf{v}_i$ is the velocity 4–vector of the ith particle satisfying

$$\mathbf{v}_i = \frac{d\mathbf{z}_i}{ds_i}, \tag{3.34}$$

$$\mathbf{v}_i^2 = 1. \tag{3.35}$$

The integrals in (3.32–3.33) are taken along the world–lines of the particles, and the occurrence of the δ–function $\delta(\mathbf{z}_i - \mathbf{z}_j)^2$ in S_2 ensures that the only values for $\mathbf{z}_i$ and $\mathbf{z}_j$ contributing to the double integral are those for which $(\mathbf{z}_i - \mathbf{z}_j)^2 = 0$, which means that each of the points $\mathbf{z}_i, \mathbf{z}_j$ is on the past or future light–cone from the other.

The action integral as it stands is not a general one covering all possible states of motion. To make it general one must, as has been pointed out by the Dirac (1938), add to it a term of the form

$$S_3 = \sum_i e_i \int M_\mu(\mathbf{z}_i) \mathbf{v}_i^\mu ds_i. \tag{3.36}$$

The 4–vector potential $M_\mu(\mathbf{x})$ may be left for the present an arbitrary function of the field point $\mathbf{x}$.

For the purpose of deducing the equations of motion, one may take the limits of integration in the various integrals to be $-\infty$ and ∞, as was done by Fokker, but in order to introduce momenta and get the equations into Hamiltonian form one must take finite limits. Let us therefore suppose that each s_i goes from s_i^0 to s_i', and let the corresponding $\mathbf{z}_i$ and $\mathbf{v}_i$ be $\mathbf{z}_i^0$, $\mathbf{z}_i'$ and $\mathbf{v}_i^0$, $\mathbf{v}_i'$. It is desirable to restrict the initial values s_i^0 so that the points $\mathbf{z}_i^0$ all lie outside each other's light–cones, and similarly with the final values s_i'. Thus

$$(\mathbf{z}_i^0 - \mathbf{z}_j^0)^2 < 0, \qquad (\mathbf{z}_i' - \mathbf{z}_j')^2 < 0, \qquad (i \neq j). \tag{3.37}$$

Now, before making variations in S, one should replace S_1, by

$$S_1' = \sum_i m_i \int \sqrt{\mathbf{v}_i^2 ds_i}, \tag{3.38}$$

so as to make S homogeneous of degree zero in the differential elements ds_i, $\mathbf{v}_i$ counting as being of degree -1 [Dir26e]. The expression for S is then valid with s_i any parameter on the world–line of the ith particle, so that $\mathbf{v}_i$ defined by (3.34) does not necessarily satisfy (3.35).

Let us now make variations $\partial\mathbf{z}_i(s_i)$ in the world–lines of the particles, $\partial\mathbf{M}(\mathbf{x})$ in the field function $\mathbf{M}(\mathbf{x})$, and Ds_i' in the final values of the s_i, so that the end–points of the world–lines are changed by

$$D\mathbf{z}_i' = \partial\mathbf{z}_i' + \mathbf{v}_i' Ds_i', \tag{3.39}$$

$\partial\mathbf{z}_i'$ being written for $\partial\mathbf{z}_i(s_i')$. The initial values of the s_i and the initial points of the world–lines we suppose for simplicity to be fixed, since variations in them would give rise to terms of the same form as those arising from variations in the final values and would not lead to anything new.

Varying S_1' given by (3.38) and using (3.35), after the variation process, one gets with the help of (3.39),

$$S_1' = \sum_i m_i \left[-\int_{s_i^0}^{s_i'} \left(\frac{d\mathbf{v}_i}{ds_i}, \partial\mathbf{z}_i \right) ds_i + (\mathbf{v}_i', D\mathbf{z}_i') \right]. \tag{3.40}$$

To get the variation in S_2 given by (3.33) one may, owing to the symmetry between i and j in the double sum, vary only quantities involving i and multiply by 2. The result is [Dir36]

$$\begin{aligned} \partial S_2 = & \sum_i \sum_{j\neq i} e_i e_j \left\{ \int_{s_i^0}^{s_i'} \int_{s_j^0}^{s_j'} \left[\frac{\partial\delta(\mathbf{z}_i - \mathbf{z}_j)^2}{\partial\mathbf{z}_i} (\mathbf{v}_i, \mathbf{v}_j) - \frac{d}{ds_i}[\delta(\mathbf{z}_i - \mathbf{z}_j)^2 \mathbf{v}_j] \right] \partial\mathbf{z}_i ds_i ds_j \right. \\ & \left. + \int_{s_i^0}^{s_i'} \delta(\mathbf{z}_i' - \mathbf{z}_j)^2 (\mathbf{v}_j, D\mathbf{z}_i') ds_j \right\}. \end{aligned} \tag{3.41}$$

Finally, in varying S_3 given by (3.36), one has to take into account that the total variation in $\mathbf{M}$ at a point $\mathbf{z}_i(s_i)$ on the ith world–line, let us call it $D\mathbf{M}(\mathbf{z}_i)$, consists of two parts, a part $\partial\mathbf{M}(\mathbf{z}_i)$ arising from the variation in the function $\mathbf{M}(\mathbf{x})$ and equal to the value of $\partial\mathbf{M}(\mathbf{x})$ at the point $\mathbf{x}=\mathbf{z}_i$, and a part arising from the variation in $\mathbf{z}_i$, equal to $\partial\mathbf{M}/\partial\mathbf{x}$, at the point $\mathbf{x}=\mathbf{z}_i$ multiplied into $\partial\mathbf{z}_i$; thus

$$D\mathbf{M}(\mathbf{z}_i)=\partial\mathbf{M}(\mathbf{z}_i)+(\partial\mathbf{M}/\partial\mathbf{x})_{\mathbf{z}_i}\partial\mathbf{z}_i. \tag{3.42}$$

The variation in S_3 is now [Dir36]

$$\begin{aligned}\partial S_3 &=\sum_i e_i\left\{\int_{s_i^0}^{s_i'}\left[\partial M^\mu(\mathbf{z}_i)v_{\mu i}+\left(\frac{\partial M^\mu}{\partial x_\nu}\right)_{\mathbf{z}_i}v_{\mu i}\partial z_{\nu i}-\frac{dM^\mu(\mathbf{z}_i)}{ds_i}\partial z_{\mu i}\right]ds_i\right.\\ &\qquad\left.+M^\mu(\mathbf{z}_i')Dz_{\mu i}'\right\}.\end{aligned} \tag{3.43}$$

The total variation in S is given by the sum of the three expressions (3.40), (3.41) and (3.43).

By equating to zero the total coefficient of $\partial z_{\mu i}$, one gets the equation of motion of the ith particle. It is

$$\begin{aligned}&-m_i\frac{dv_i^\mu}{ds_i}+e_i\sum_{j\neq i}e_j\int_{s_j^0}^{s_j'}\left[\frac{\partial\delta(\mathbf{z}_i-\mathbf{z}_j)^2}{\partial\mathbf{z}_i}(\mathbf{v}_i,\mathbf{v}_j)-\frac{d}{ds_i}[\delta(\mathbf{z}_i-\mathbf{z}_j)^2\mathbf{v}_j]\right]ds_j\\ &\quad+e_i\left[\left(\frac{\partial M^\mu}{\partial x_\nu}\right)_{\mathbf{z}_i}v_{\mu i}-\frac{dM^\mu(\mathbf{z}_i)}{ds_i}\right]=0.\end{aligned}$$

Introducing the field function

$$A_i^\mu(\mathbf{x})=M^\mu(\mathbf{x})+\sum_{j\neq i}e_j\int_{s_j^0}^{s_j'}\partial\delta(\mathbf{x}-\mathbf{z}_j)^2v_j^\mu ds_j, \tag{3.44}$$

the above equation of motion may be written

$$m_i\frac{dv_i^\mu}{ds_i}=e_i\left[\left(\frac{\partial A_i^\mu}{\partial x_\nu}\right)_{\mathbf{z}_i}v_{\mu i}-\frac{dA^\mu(\mathbf{z}_i)}{ds_i}\right]=e_i\left[\frac{\partial A_i^\mu}{\partial x_\nu}-\frac{\partial A_i^\nu}{\partial x_\mu}\right]_{\mathbf{z}_i}v_{\mu i}. \tag{3.45}$$

It is the correct *Lorentz equation of motion* of the ith particle, provided A_i^μ is connected with the ingoing and outgoing fields and the retarded and advanced fields of the other particles by the relation, given by Dirac (1938),

$$\begin{aligned}A_i^\mu &=\frac{1}{2}\left[A_{\mathrm{in}}^\mu+A_{\mathrm{out}}^\mu\right]+\frac{1}{2}\sum_{j\neq i}\left[A_{j\mathrm{ret}}^\mu+A_{j\mathrm{adv}}^\mu\right],\qquad\text{or}\\ A_i^\mu(\mathbf{x}) &=\frac{1}{2}\left[A_{\mathrm{in}}^\mu(\mathbf{x})+A_{\mathrm{out}}^\mu(\mathbf{x})\right]+\sum_{j\neq i}e_j\int_{-\infty}^{\infty}\delta(\mathbf{x}-\mathbf{z}_j)^2v_j^\mu ds_j.\end{aligned} \tag{3.46}$$

According to (3.44) this requires (in Dirac's notation for integrals)

$$M^{\mu}(\mathbf{x}) = \frac{1}{2}\left[A^{\mu}_{\text{in}}(\mathbf{x}) + A^{\mu}_{\text{out}}(\mathbf{x})\right] + \sum_j e_j \left[\int_{-\infty}^{s_j^0} + \int_{s'_j}^{\infty}\right] \delta(\mathbf{x}-\mathbf{z}_j)^2 v_j^{\mu} ds_j, \quad (3.47)$$

Note that we are summing here over all values of j [Dir36, Dir26e], as we are dealing with a space–time region which lies inside the future light–cone from $\mathbf{z}_i^0$ and inside the past light–cone from $\mathbf{z}'_i$. By assuming that (3.47) holds throughout space–time, one gets an expression for $M^{\mu}(\mathbf{x})$ independent of i, so that the equations of motion of all the particles follow from the same Fokker's action integral.

One can now pass to the Hamiltonian formulation of the equations of motion. For each point in space–time $\mathbf{x}$, $M^{\mu}(\mathbf{x})$ may be counted as a coordinate, depending on the proper–times s_i'[5], and will have a conjugate momentum, say $K_{\mu}(\mathbf{x})$. These momenta, together with the particle momenta p_i^{μ}, are defined, as in the general theory of [Wei36], by the coefficients of $\partial M^{\mu}(\mathbf{x})$ and $Dz'_{\mu i}$ in the expression for ∂S, so that we have

$$\partial S = \sum_i p_i^{\mu} D z'_{\mu i} + \int_{-\infty}^{\infty} K_{\mu}(\mathbf{x}) \partial M^{\mu}(\mathbf{x}) dx_0 dx_1 dx_2 dx_3, \quad (3.48)$$

where the integral sign denotes the quadruple space–time integral. Comparing (3.48) with the sum of (3.40), (3.41) and (3.43), one gets [Dir26e, Dir36]

$$K_{\mu}(\mathbf{x}) = \sum_i e_i \int_{s_i^0}^{s'_i} \delta(x_0 - z_{0i})\delta(x_1 - z_{1i})\delta(x_2 - z_{2i})\delta(x_3 - z_{3i}) v_{\mu i} ds_i \quad (3.49)$$

$$\text{and} \quad p_i^{\mu} = m_i v_i'^{\mu} + e_i \left[M^{\mu}(\mathbf{z}'_i) + \frac{1}{2}\sum_j e_j \int_{s_j^0}^{s'_j} \Delta(\mathbf{z}'_i - \mathbf{z}_j + \boldsymbol{\lambda}) v_j^{\mu} ds_j \right], \quad (3.50)$$

where $\boldsymbol{\lambda}$ is a small 4–vector whose direction is within the future light–cone (so that $\boldsymbol{\lambda}^2 > 0$, $\lambda_0 > 0$), $\Delta(\mathbf{y})$ denotes the Jordan and Pauli (1928) Δ–function of any 4–vector $\mathbf{y}$, satisfying the 4D wave equation

$$\Box \Delta(\mathbf{y}) = 0 \qquad \text{which implies} \qquad \Box M^{\mu}(\mathbf{y}) = 0,$$

and related to the corresponding δ–function by

$$\Delta(\mathbf{y}) = \pm 2\delta(\mathbf{y}^2).$$

The momenta satisfy the *Poisson bracket* commutation relationships

$$[p_{\mu i}, z_{\nu j}] = g_{\mu\nu}\delta_{ij}, \quad (3.51)$$

$$[K_{\mu}(\mathbf{x}), M_{\nu}(\mathbf{x}')] = g_{\mu\nu}\delta(x_0 - x'_0)\delta(x_1 - x'_1)\delta(x_2 - x'_2)\delta(x_3 - x'_3), \quad (3.52)$$

[5] It also depends on the proper–times s_i^0, but this does not here concern us.

so that the Poisson bracket of any two momenta or of any two coordinates vanishes. Instead of $K_\mu(\mathbf{x})$ it is more convenient to work with the momentum field–function $N_\mu(\mathbf{x})$ defined by [Dir58, Dir29]

$$N_\mu(\mathbf{x}) = \frac{1}{2}\int_{-\infty}^{\infty} \Delta(\mathbf{x}-\mathbf{x}')K_\mu(\mathbf{x}')dx_0'dx_1'dx_2'dx_3', \tag{3.53}$$

$$\text{and satisfying} \qquad \Box N_\mu(\mathbf{x}) = 0. \tag{3.54}$$

Instead of (3.52) one has

$$[N_\mu(\mathbf{x}), M_\nu(\mathbf{x}')] = \frac{1}{2}g_{\mu v}\Delta(\mathbf{x}-\mathbf{x}'). \tag{3.55}$$

From (3.53) and (3.49) one gets

$$N_\mu(\mathbf{x}) = \frac{1}{2}\sum_i e_i \int_{s_i^0}^{s_i'} \Delta(\mathbf{x}-\mathbf{z}_i)v_{\mu v}ds_i, \tag{3.56}$$

so that (3.50) may be written

$$\begin{aligned} p_i^\mu &= m_i v_i'^\mu + e_i[M^\mu(\mathbf{z}_i') + N^\mu(\mathbf{z}_i' + \boldsymbol{\lambda})] \\ &= m_i v_i'^\mu + e_i A^\mu(\mathbf{z}_i'), \end{aligned} \tag{3.57}$$

$$\text{where} \qquad A^\mu(\mathbf{x}) = M^\mu(\mathbf{x}) + N^\mu(\mathbf{x}+\boldsymbol{\lambda}). \tag{3.58}$$

From (3.54) the potentials $A_\mu(\mathbf{x})$ satisfy

$$\Box A_\mu(\mathbf{x}) = 0, \tag{3.59}$$

showing that they can be resolved into waves travelling with the velocity of light, and from (3.55) it follows

$$[A_\mu(\mathbf{x}), A_\nu(\mathbf{x}')] = \frac{1}{2}g_{\mu v}[\Delta(\mathbf{x}-\mathbf{x}'+\boldsymbol{\lambda}) + \Delta(\mathbf{x}-\mathbf{x}'-\boldsymbol{\lambda})]. \tag{3.60}$$

From (3.35) and (3.57) it follows

$$F_i \equiv [p_i - e_i\mathbf{A}(\mathbf{z}_i')]^2 - m_i^2 = 0. \tag{3.61}$$

There is one of these equations for each particle. The expressions F_i may be used as Hamiltonians to determine how any dynamical variable ξ varies with the proper–times s_i', in accordance with the equations [Dir58, Dir26e, Dir82]

$$\kappa_i \frac{d\xi}{ds_i'} = [\xi, F_i], \tag{3.62}$$

were ξ is any function of the coordinates and momenta of the particles and of the fields $\mathbf{M}, \mathbf{K}, \mathbf{N}, \mathbf{A}$, and the κ's are multiplying factors not depending on ξ. Taking $\xi = z_{\mu i}'$, one finds that

$$\kappa_i = -2m_i,$$

to get agreement with (3.57). Taking $\xi = p_i^\mu$ gives one back the equation of motion (3.45) with the $\boldsymbol{\lambda}$ refinement. Taking $\xi = M_\mu(\mathbf{x})$, one gets from (3.58) and (3.55),

$$\frac{M_\mu(\mathbf{x})}{ds_i'} = e_i v_i'^\nu [M_\mu(\mathbf{x}), A_\nu(\mathbf{z}_i')] = \frac{1}{2} e_i v_{\mu i}' \Delta(\mathbf{x} - \mathbf{z}_i' - \boldsymbol{\lambda}).$$

This equation of motion for the field quantities $M_\mu(\mathbf{x})$ does not follow from the variation principle, as it involves only coordinates and velocities and not accelerations, and it has to be imposed as an extra condition in the variational method.

The above Hamiltonian formulation of the equations of classical electrodynamics may be taken over into the quantum theory in the usual way, by making the momenta into operators satisfying commutation relations corresponding to the Poisson bracket relations (3.51), (3.52). Equation (3.60) in the limit $\boldsymbol{\lambda} \to O$ goes over into the quantum equation (3.22). The Hamiltonians (3.61) provide the wave equations

$$F_i \psi = 0,$$

in which the wave ψ–function is a function of the coordinates $\mathbf{z}_i'$ of all the particles and of the field variables $M_\mu(\mathbf{x})$. One can apply the theory to spinning electrons instead of spinless particles, by modifying the Hamiltonians F_i in the appropriate way. For more details, see [Dir58, Dir26e, Dir82].

3.1.3 Feynman's Path–Integral Quantum Theory

The most complete quantum theory was developed by Richard Feynman[6] in the form of his celebrated *path–integral* and associated *Feynman diagrams*.

[6] Richard Phillips (Dick) Feynman (May 11, 1918 in Queens, New York – February 15, 1988 in Los Angeles, California) was an influential American physicist known for expanding greatly on the theory of quantum electrodynamics, particle theory, and the physics of the superfluidity of supercooled liquid helium. For his work on quantum electrodynamics, Feynman was one of the recipients of the Nobel Prize in Physics in 1965, along with Julian Schwinger and Shin-Ichiro Tomonaga; in this work, he developed a way to understand the behavior of subatomic particles using pictorial tools now called Feynman diagrams.

Feynman received a Ph.D. from Princeton University in 1942; his thesis advisor was John Archibald Wheeler. Feynman's thesis applied the principle of stationary action to problems of quantum mechanics, laying the ground work for his *path–integral* method.

He helped in the development of the atomic bomb and was later a member of the panel that investigated the Space Shuttle Challenger disaster. For all his prolific contributions, Feynman wrote only 37 research papers in his career. Apart from pure physics, Feynman is also credited with the revolutionary concept and early exploration of quantum computing, and publicly envisioning nanotechnology, the ability to create devices at the molecular scale. He held the Richard Chace Tolman professorship in theoretical physics at Caltech.

Feynman was a keen and influential popularizer of physics in both his books and lectures, notably a seminal 1959 talk on top-down nanotechnology called There's Plenty of Room at the Bottom and The Feynman Lectures on Physics, a three-volume set which has become a classic text. In his lifetime as well as in the years after his death, he became one of the most publicly known scientists of the century. Known for his insatiable curiosity, gentle wit, brilliant mind and playful temperament [1], he is also famous for his many adventures, detailed in the books Surely You're Joking, Mr. Feynman!, What Do You Care What Other People Think? and Tuva or Bust!. As well as being an inspiring lecturer, bongo player, notorious practical joker, and decipherer of Mayan hieroglyphics, Richard Feynman was, in many respects, an eccentric and a free spirit. He liked to pursue many independent paths, such as biology, art, percussion, and lockbreaking. Freeman Dyson once wrote that Feynman was "half-genius, half-buffoon", but later changed this to "all-genius, all-buffoon".

Feynman did much of his best work while at Caltech, including research in:

(i) Quantum electrodynamics. The theory for which Feynman won his Nobel Prize is known for its extremely accurate predictions. He helped develop a functional integral formulation of quantum mechanics, in which every possible path from one state to the next is considered, the final path being a sum over the possibilities.

(ii) Physics of the superfluidity of supercooled liquid helium, where helium seems to display a lack of viscosity when flowing. Applying the Schrödinger equation to the question showed that the superfluid was displaying quantum mechanical behavior observable on a macroscopic scale. This helped enormously with the problem of superconductivity.

(iii) A model of weak decay, which showed that the current coupling in the process is a combination of vector and axial. (An example of weak decay is the decay of a neutron into an electron, a proton, and an anti-neutrino.) Although E.C. George Sudharsan and Robert Marshak developed the theory nearly simultaneously, Feynman's collaboration with Murray Gell-Mann was seen as the seminal one, the theory was of massive importance, and the weak interaction was neatly described.

He also developed *Feynman diagrams*, a bookkeeping device which helps in conceptualizing and calculating interactions between particles in spacetime, notably the interactions between electrons and their antimatter counterparts, positrons. This device allowed him, and now others, to work with concepts which would have been less approachable without it, such as time reversibility and other fundamental processes. Feynman famously painted Feynman diagrams on the exterior of his van.

Feynman diagrams are now fundamental for String theory and M–theory, and have even been extended topologically. Feynman's mental picture for these diagrams started with the hard sphere approximation, and the interactions could be thought of as collisions at first. It was not until decades later that physicists thought of analyzing the nodes of the Feynman diagrams more closely. The world–lines of the diagrams have become tubes to better model the more complicated objects such as strings and M–branes.

Feynman's Sum–Over–Histories

Alternative Probability Theory

Classical Probability Concept

Recall that a *random variable* X is defined by its *distribution function* $f(x)$. Its *probabilistic description* is based on the following rules: (i) $P(X = x_i)$ is the probability that $X = x_i$; and (ii) $P(a \leq X \leq b)$ is the probability that X lies in a closed interval $[a, b]$. Its statistical description is based on: (i) μ_X or $E(X)$ is the mean of expectation of X; and (ii) σ_X is the standard deviation of X. There are two cases of random variables: discrete and continuous, each having its own probability (and statistics) theory.

Discrete random variable

Here X has only a finite (or countable) number of values $\{x_i\}$. The distribution function $f(x_i)$ has the properties:

$$P(X = x_i) = f(x_i),$$
$$f(x_i) \geq 0,$$
$$\sum_i f(x_i)\, dx = 1.$$

From his diagrams of a small number of particles interacting in spacetime, Feynman could then model all of physics in terms of those particles' spins and the range of coupling of the fundamental forces. Feynman attempted an explanation of the strong interactions governing nucleons scattering called the parton model. The parton model emerged as a rival to the quark model developed by his Caltech colleague Murray Gell-Mann. The relationship between the two models was murky; Gell–Mann referred to Feynman's partons derisively as 'put–ons'. Feynman did not dispute the quark model; for example, when the 5th quark was discovered, Feynman immediately pointed out to his students that the discovery implied the existence of a 6th quark, which was duly discovered in the decade after his death.

After the success of quantum electrodynamics, Feynman turned to quantum gravity. By analogy with the photon, which has spin 1, he investigated the consequences of a free massless spin 2 field, and was able to derive the Einstein field equation of general relativity, but little more. However, a calculational technique that Feynman developed for gravity in 1962 — 'ghosts' — later proved invaluable. In 1967 Fadeev and Popov quantized the particle behaviour of the spin 1 theories of Yang–Mills–Pauli, that are now seen to describe the weak and strong interactions, using Feynman's path integral technique. A 'ghost' is a field which is spin 0 and so should be a boson, but which is a fermion, disobeying the spin–statistics theorem. Because it does not propagate externally no effects of this are seen. Unfortunately, at this time he became exhausted by working on multiple major projects at the same time, including his Lectures in Physics. The Feynman Lectures on Physics found an appreciative audience beyond the undergraduate community.

Statistical description of X is based on the following:

$$\mu_X = E(X) = \sum_i x_i f(x_i),$$

$$\sigma_X = \sqrt{E(X^2) - \mu_X^2}.$$

Continuous random variable

Here $f(x)$ is a piecewise continuous function such that:

$$P(a \leq X \leq b) = \int_a^b f(x)\, dx,$$

$$f(x) \geq 0,$$

$$\int_{-\infty}^{\infty} f(x)\, dx = \int_{\mathbb{R}} f(x)\, dx = 1.$$

Statistical description of X is based on the following:

$$\mu_X = E(X) = \int_{-\infty}^{\infty} x f(x)\, dx,$$

$$\sigma_X = \sqrt{E(X^2) - \mu_X^2}.$$

Observe the similarity between the two descriptions. The same kind of similarity between discrete and continuous quantum spectrum stroke Dirac and Feynman when they suggested the integral approach, denoted by $\sum\!\!\!\!\!\!\int$, emphasizing both the summation over discrete spectrum and the integration over continuous one. To emphasize this similarity even further, as well as to set–up the stage for the path integral, recall the notion of the *cumulative distribution function* of a random variable X is the function $F : \mathbb{R} \to \mathbb{R}$, defined by

$$F(a) = P(X) \leq a.$$

In particular, suppose that $f(x)$ is the distribution function of X. Then

$$F(x) = \sum_{x_i \leq x} f(x_i) \qquad \text{or} \qquad F(x) = \int_{-\infty}^{\infty} f(t)\, dt,$$

according as x is a *discrete* or *continuous* random variable. In either case, $F(a) \leq F(b)$ whenever $a \leq b$. Also,

$$\lim_{x \longrightarrow -\infty} F(x) = 0 \qquad \text{or} \qquad \lim_{x \longrightarrow \infty} F(x) = 1,$$

that is, $F(x)$ is monotonic and its limit to the left is 0 and the limit to the right is 1. Furthermore,

$$P(a \leq X \leq b) = F(b) - F(a)$$

and the Fundamental Theorem of Calculus tells us that, in the continuum case,

$$f(x) = \partial_x F(x).$$

Quantum Probability Concept

An alternative concept of quantum probability is based on the following physical facts:

1. The *time–dependent Schrödinger equation* represents a *complex–valued generalization* of real–valued *Fokker–Planck equation* for describing the spatio–temporal *probability density function* for the system exhibiting *continuous–time Markov stochastic process*.
2. **Feynman's path integral** $\oint$ is a generalization of the time–dependent Schrödinger equation, including both continuous–time and discrete–time Markov stochastic processes (Markov chains, or random walks).
3. Both Schrödinger equation and path integral give 'physical description' of any system they are modelling in terms of its physical energy, instead of an abstract probabilistic description of the Fokker–Planck equation.

Therefore, the **Feynman's path integral** $\oint$, as a generalization of the time–dependent Schrödinger equation, gives a unique physical description for the general Markov stochastic process, in terms of the physically based generalized probability density functions, valid for both continuous–time and discrete–time Markov systems.

Basic consequence:

Different way for calculating probabilities. The difference is rooted in the fact that *sum of squares is different from the square of sums*, as is explained in the following text.

Namely, in Dirac–Feynman quantum formalism, each possible route from the initial system state A to the final system state B is called a *history*. This history comprises any kind of a route (see Figure 3.1), ranging from continuous and smooth deterministic (mechanical–like) paths to completely discontinues and random Markov chains (see, e.g., [Gar85]). Each history (labelled by index i) is quantitatively described by a *complex number*[7] z_i called the 'individual transition amplitude'. Its absolute square, $|z_i|^2$, is called the *individual transition probability*. Now, the *total transition amplitude* is the sum of all individual transition amplitudes, $\sum_i z_i$, called the *sum–over–histories*. The absolute square of this sum–over–histories, $|\sum_i z_i|^2$, is the *total transition probability*.

In this way, the overall probability of the system's transition from some initial state A to some final state B is given *not* by adding up the probabilities

[7] Recall that a *complex number* $z = x + \mathrm{i}y$, where $\mathrm{i} = \sqrt{-1}$ is the *imaginary unit*, x is the *real part* and y is the *imaginary part*, can be represented also in its *polar form*, $z = r(\cos\theta + \mathrm{i}\sin\theta)$, where the radius vector in the complex plane, $r = |z| = \sqrt{x^2 + y^2}$, is the *modulus* or *amplitude*, and angle θ is the *phase*; as well as in its exponential form $z = r\mathrm{e}^{\mathrm{i}\theta}$. In this way, complex numbers actually represent 2D vectors with usual vector 'head–to–tail' addition rule.

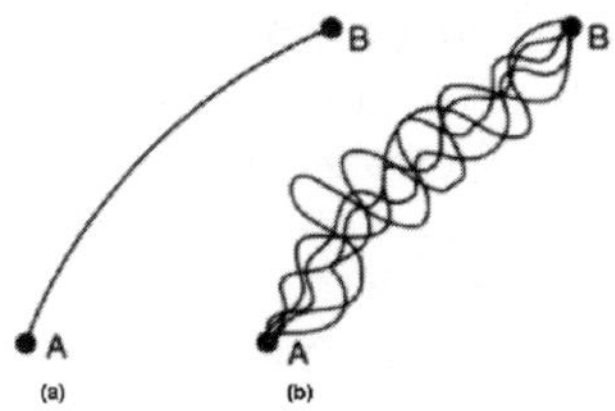

Fig. 3.1. Two ways of physical *transition* from an *initial state* A to the corresponding *final state* B. (a) Classical physics proposes a *single deterministic trajectory*, minimizing the total system's energy. (b) Quantum physics proposes a *family of Markov stochastic histories*, namely *all possible routes* from A to B, both continuous–time and discrete–time Markov chains, each giving an equal contribution to the total *transition probability*.

for each history–route, but by 'head–to–tail' adding up the sequence of amplitudes making–up each route first (i.e., performing the sum–over–histories) – to get the total amplitude as a 'resultant vector', and then squaring the total amplitude to get the overall transition probability.

Quantum Coherent States

Recall that a *quantum coherent state* is a specific kind of quantum state of the quantum harmonic oscillator whose dynamics most closely resemble the oscillating behavior of a classical harmonic oscillator. It was the first example of quantum dynamics when Erwin Schrödinger derived it in 1926 while searching for solutions of the *Schrödinger equation* that satisfy the *correspondence principle*. The quantum harmonic oscillator and hence, the coherent state, arise in the quantum theory of a wide range of physical systems. For instance, a coherent state describes the oscillating motion of the particle in a quadratic potential well. In the quantum electrodynamics and other bosonic quantum field theories they were introduced by the 2005 Nobel Prize winning work of Roy Glauber in 1963 [Gla63a, Gla63b]. Here the coherent state of a field describes an oscillating field, the closest quantum state to a classical sinusoidal wave such as a continuous laser wave.

In classical optics, light is thought of as electromagnetic waves radiating from a source. Specifically, coherent light is thought of as light that is emitted by many such sources that are in phase. For instance, a light bulb radiates light that is the result of waves being emitted at all the points along the filament. Such light is incoherent because the process is highly random in space and time. On the other hand, in a laser, light is emitted by a carefully controlled system in processes that are not random but interconnected by stimulation and the resulting light is highly ordered, or coherent. Therefore a coherent state corresponds closely to the quantum state of light emitted by an ideal laser. Semi–classically we describe such a state by an electric field oscillating as a stable wave. Contrary to the coherent state, which is the most wave–like quantum state, the *Fock state* (e.g., a single photon) is the most particle–like state. It is indivisible and contains only one quanta of energy. These two states are examples of the opposite extremes in the concept of *wave–particle duality*. A coherent state distributes its quantum–mechanical uncertainty equally, which means that the phase and amplitude uncertainty are approximately equal. Conversely, in a single–particle state the phase is completely uncertain.

Formally, the coherent state $|\alpha\rangle$ is defined to be the eigenstate of the annihilation operator a, i.e., $a|\alpha\rangle = \alpha|\alpha\rangle$. Note that since a is not Hermitian, $\alpha = |\alpha|e^{i\theta}$ is complex. $|\alpha|$ and θ are called the *amplitude* and *phase* of the state.

Physically, $a|\alpha\rangle = \alpha|\alpha\rangle$ means that a coherent state is left unchanged by the detection (or annihilation) of a particle. Consequently, in a coherent state, one has exactly the same probability to detect a second particle. Note, this condition is necessary for the coherent state's *Poisson detection statistics*. Compare this to a single–particle's Fock state: Once one particle is detected, we have zero probability of detecting another.

Now, recall that a *Bose–Einstein condensate* (BEC) is a collection of boson atoms that are all in the same quantum state. An approximate theoretical description of its properties can be derived by assuming the BEC is in a coherent state. However, unlike photons, atoms interact with each other so it now appears that it is more likely to be one of the *squeezed coherent states* (see [BSM97]). In quantum field theory and string theory, a generalization of coherent states to the case of infinitely many degrees–of–freedom is used to define a *vacuum state* with a different vacuum expectation value from the original vacuum.

Sum–Over–Histories

Recall from above that Dirac described behavior of quantum systems in terms of complex–valued *ket–vectors* $|A>$, living in the *Hilbert space* $\mathcal{H}$, and their duals, *bra–covectors* $<B|$ living in the *dual Hilbert space* $\mathcal{H}^*$. The *Hermitian inner product* of kets and bras, the *bra–ket* $<B|A>$, is a *complex number*, which is the evaluation of the ket $|A>$ by the bra $<B|$. This complex number, say $r\mathrm{e}^{\mathrm{i}\theta}$ represents the system's *transition amplitude* from its *initial state* A to its *final state* B, i.e.,

$$Transition\, Amplitude = <B|A> = r\mathrm{e}^{\mathrm{i}\theta}.$$

That is, there is a process that can mediate a transition of a system from initial state A to the final state B and the amplitude for this transition equals $<B|A> = r\mathrm{e}^{\mathrm{i}\theta}$. The absolute square of the amplitude, $|<B|A>|^2$ represents the *transition probability*. Therefore, the probability of a transition event equals the absolute square of a complex number, i.e.,

$$Transition\, Probability = |<B|A>|^2 = |r\mathrm{e}^{\mathrm{i}\theta}|^2.$$

These complex amplitudes obey the usual *laws of probability*: when a transition event can happen in alternative ways then we add the complex numbers,

$$< B_1|A_1 > + < B_2|A_2 >= r_1 e^{i\theta_1} + r_2 e^{i\theta_2},$$

and when it can happen only as a succession of intermediate steps then we multiply the complex numbers,

$$< B|A >=< B|c >< c|A >= (r_1 e^{i\theta_1})(r_2 e^{i\theta_2}) = r_1 r_2 e^{i(\theta_1+\theta_2)}.$$

In general,

1. The amplitude for n *mutually alternative processes* equals the *sum* $\sum_{k=1}^{n} r_k e^{i\theta_k}$ of the amplitudes for the alternatives; and
2. If transition from A to B occurs in a *sequence of m steps*, then the total transition amplitude equals the *product* $\prod_{j=1}^{m} r_j e^{i\theta_j}$ of the amplitudes of the steps.

Formally, we have the so–called *expansion principle*, including both products and sums,

$$< B|A >= \sum_{i=1}^{n} < B|c_i >< c_i|A > . \tag{3.63}$$

Now, *iterating* the Dirac's expansion principle (3.188) over a *complete set of all possible states* of the system, leads to the simplest form of the *Feynman's path integral* or *sum–over–histories*. Imagine that the *initial* and *final* states, A and B, are points on the vertical lines $x = 0$ and $x = n + 1$, respectively, in the $x - y$ plane, and that $(c(k)_{i(k)}, k)$ is a given point on the line $x = k$ for $0 < i(k) < m$ (see Figure 3.2). Suppose that the sum of projectors for each intermediate state is complete. Applying the completeness iteratively, we get the following expression for the transition amplitude:

$$< B|A >= \sum\sum \dots \sum < B|c(1)_{i(1)} >< c(1)_{i(1)}|c(2)_{i(2)} > \dots < c(n)_{i(n)}|A >,$$

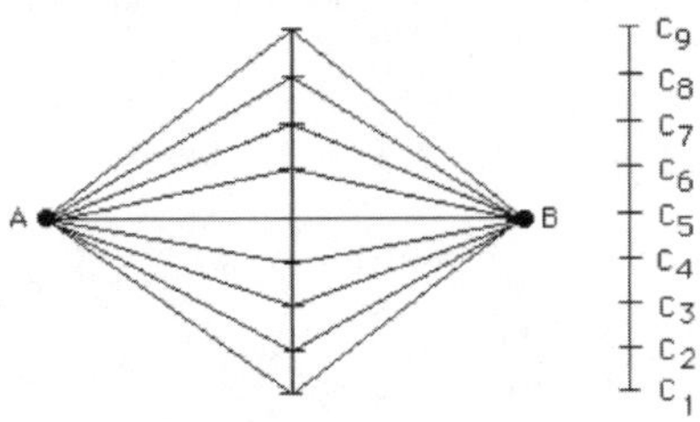

Fig. 3.2. Analysis of all possible routes from the source A to the detector B is simplified to include only double straight lines (in a plane).

where the sum is taken over all $i(k)$ ranging between 1 and m, and k ranging between 1 and n. Each term in this sum can be construed as a *combinatorial route* from A to B in the two–dimensional space of the $x-y$ plane. Thus the transition amplitude for the system going from some initial state A to some final state B is seen as a summation of contributions from *all the routes* connecting A to B.

Feynman used this description to produce his celebrated *path–integral* expression for a transition amplitude (see, e.g., [GS98, Sch81]). His path integral takes the form

$$Transition\,Amplitude = < B|A > = \oint\!\!\!\!\!\!\sum \mathcal{D}[x]\,\mathrm{e}^{\mathrm{i}\mathcal{S}[x]}, \tag{3.64}$$

where the sum–integral $\oint\!\!\!\!\!\!\sum$ is taken over all possible routes $x = x(t)$ from the initial point $A = A(t_{ini})$ to the final point $B = B(t_{fin})$, and $\mathcal{S} = \mathcal{S}[x]$ is the classical *action* for a particle to travel from A to B along a given extremal path x. In this way, Feynman took seriously Dirac's conjecture interpreting the exponential of the classical *action functional* ($\mathcal{D}\mathrm{e}^{\mathrm{i}\mathcal{S}}$), resembling a complex number ($r\mathrm{e}^{\mathrm{i}\theta}$), as an *elementary amplitude*. By integrating this elementary amplitude, $\mathcal{D}\mathrm{e}^{\mathrm{i}\mathcal{S}}$, over the infinitude of all possible histories, we get the total system's transition amplitude.[8]

Basic Form of a Path Integral

In Feynman's version of non–relativistic quantum mechanics, the time evolution $\psi(x', t') \mapsto \psi(x'', t'')$ of the wave function $\psi = \psi(x,t)$ of the elementary 1D particle may be described by the integral equation [GS98]

---

[8] For the quantum physics associated with a classical (Newtonian) particle the action S is given by the integral along the given route from a to b of the difference $T - V$ where T is the classical kinetic energy and V is the classical potential energy of the particle.
The beauty of Feynman's approach to quantum physics is that it shows the relationship between the classical and the quantum in a particularly transparent manner. Classical motion corresponds to those regions where all nearby routes contribute constructively to the summation. This classical path occurs when the *variation of the action* is null. To ask for those paths where the variation of the action is zero is a problem in the calculus of variations, and it leads directly to Newton's equations of motion (derived using the Euler–Lagrangian equations). Thus with the appropriate choice of action, classical and quantum points of view are unified.
Also, a discretization of the Schrodinger equation

$$i\hbar\frac{d\psi}{dt} = -\frac{\hbar^2}{2m}\frac{d^2\psi}{dx^2} + V\psi,$$

leads to a sum–over–histories that has a discrete path integral as its solution. Therefore, the transition amplitude is equivalent to the wave ψ. The particle travelling on the x–axis is executing a one–step *random walk*, see Figure 3.3.

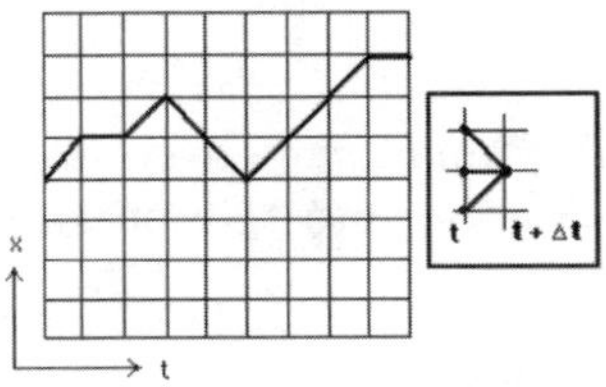

Fig. 3.3. Random walk (a particular case of Markov chain) on the x–axis.

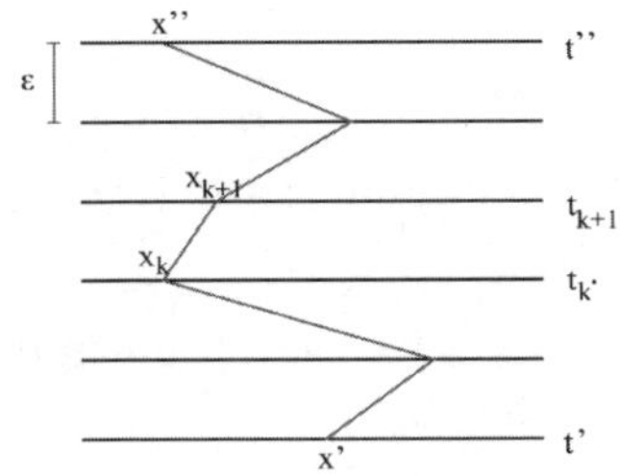

Fig. 3.4. A piecewise linear particle path contributing to the discrete Feynman propagator.

$$\psi(x'',t'') = \int_{\mathbb{R}} K(x'',x';t'',t')\,\psi(x',t'), \tag{3.65}$$

where the *propagator* or *Feynman kernel* $K = K(x'',x';t'',t')$ is defined through a limiting procedure,

$$K(x'',x';t'',t') = \lim_{\epsilon\longrightarrow 0} A^{-N} \prod_{k=1}^{N-1} \int dx_k \ \mathrm{e}^{\mathrm{i}\sum_{j=0}^{N-1} \epsilon L(x_{j+1},(x_{j+1}-x_j)/\epsilon)}. \tag{3.66}$$

The time interval $t'' - t'$ has been discretized into N steps of length $\epsilon = (t'' - t')/N$, and the r.h.s. of (3.66) represents an integral over all piecewise linear paths $x(t)$ of a 'virtual' particle propagating from x' to x'', illustrated in Figure 3.4.

The prefactor A^{-N} is a normalization and L denotes the Lagrangian function of the particle. Knowing the propagator G is tantamount to having solved the quantum dynamics. This is the simplest instance of a path integral, and is often written schematically as

$$K(x',t';x'',t'') = \sum\!\!\!\!\!\!\int \mathcal{D}[x(t)]\ \mathrm{e}^{\mathrm{i}S[x(t)]},$$

where $\mathcal{D}[x(t)]$ is a functional measure on the 'space of all paths', and the exponential weight depends on the classical action $S[x(t)]$ of a path. Recall also that this procedure can be defined in a mathematically clean way if we Wick–rotate the time variable t to imaginary values $t \mapsto \tau = \mathrm{i}t$, thereby making all integrals real [RS75].

Adaptive Path Integral

Now, we can extend the Feynman sum–over–histories (3.64), by adding the synaptic–like weights $w^i = w^i(t)$ into the measure $\mathcal{D}[x]$, to get the *adaptive path integral*:

$$Adaptive\,Transition\,Amplitude = \langle B|A \rangle_w = \not\Sigma\!\!\!\int \mathcal{D}[w,x]\, \mathrm{e}^{\mathrm{i}S[x]}, \qquad (3.67)$$

where the *adaptive measure* $\mathcal{D}[w,x]$ is defined by the weighted product (of discrete time steps)

$$\mathcal{D}[w,x] = \lim_{n\to\infty} \prod_{t=1}^{n} w^i(t)\, dx^i(t). \qquad (3.68)$$

In (3.68) the *synaptic weights* $w^i = w^i(t)$ are updated by the unsupervised *Hebbian–like learning* rule [Heb49]:

$$w^i(t+1) = w^i(t) + \frac{\sigma}{\eta}(w^i_d(t) - w^i_a(t)), \qquad (3.69)$$

where $\sigma = \sigma(t)$, $\eta = \eta(t)$ represent local *signal* and *noise* amplitudes, respectively, while superscripts d and a denote *desired* and *achieved* system states, respectively. Theoretically, equations (3.67–3.69) define an $\infty-$*dimensional complex–valued neural network*.[9] Practically, in a computer simulation we can use $10^7 \le n \le 10^8$, approaching the number of neurons in the brain. Such equations are usually solved using *Markov–chain Monte–Carlo* methods on parallel (cluster) computers (see, e.g., [WW83a, WW83b]).

Path–Integral History

Extract from Feynman's Nobel Lecture

In his Nobel Lecture, December 11, 1965, Richard (Dick) Feynman said that he and his PhD supervisor, John Wheeler, had found the *action* $A = A[x; t_i, t_j]$, directly involving the *motions of the charges only*,[10]

$$A[x;t_i,t_j] = m_i \int (\dot{x}^i_\mu \dot{x}^i_\mu)^{\frac{1}{2}}\, dt_i + \frac{1}{2} e_i e_j \int\!\!\int \delta(I^2_{ij})\, \dot{x}^i_\mu(t_i) \dot{x}^j_\mu(t_j)\, dt_i dt_j$$
$$\text{with } (i \neq j) \qquad (3.70)$$
$$I^2_{ij} = \left[x^i_\mu(t_i) - x^j_\mu(t_j)\right]\left[x^i_\mu(t_i) - x^j_\mu(t_j)\right],$$

[9] For details on complex–valued neural networks, see e.g., complex–domain extension of the standard backpropagation learning algorithm [GK92, BP92].

[10] *Wheeler–Feynman Idea* [WF49] "The energy tensor can be regarded only as a provisional means of representing matter. In reality, *matter consists of electrically charged particles.*"

where $x^i_\mu = x^i_\mu(t_i)$ is the four–vector *position* of the ith particle as a function of the proper time t_i, while $\dot{x}^i_\mu(t_i) = dx^i_\mu(t_i)/dt_i$ is the *velocity* four–vector.

The first term in the action $A[x; t_i, t_j]$ (3.70) is the integral of the proper time t_i, the *ordinary action of relativistic mechanics of free particles of mass* m_i (summation over μ). The second term in the action $A[x; t_i, t_j]$ (3.70) represents the *electrical interaction of the charges.* It is summed over each pair of charges (the factor $\frac{1}{2}$ is to count each pair once, the term $i = j$ is omitted to avoid self–action). The *interaction is a double integral over a delta function of the square of space–time interval I^2 between two points on the paths.* Thus, interaction occurs only when this interval vanishes, that is, *along light cones* (see [WF49]).

Feynman comments here: "The fact that the interaction is exactly one–half advanced and half–retarded meant that we could write such a principle of least action, whereas interaction via retarded waves alone cannot be written in such a way. So, all of classical electrodynamics was contained in this very simple form."

"...The problem is only to make a quantum theory, which has as its classical analog, this expression (3.70). Now, there is no unique way to make a quantum theory from classical mechanics, although all the textbooks make believe there is. What they would tell you to do, was find the momentum variables and replace them by $(\hbar/i)(\partial/\partial x)$, but I couldn't find a momentum variable, as there wasn't any."

"The character of quantum mechanics of the day was to write things in the famous *Hamiltonian way* (in the form of Schrödinger equation), which described how the wave function changes from instant to instant, and in terms of the Hamiltonian operator H. If the classical physics could be reduced to a Hamiltonian form, everything was all right. Now, least action does not imply a Hamiltonian form if the action is a function of anything more than positions and velocities at the same moment. If the action is of the form of the integral of the Lagrangian $L = L(\dot{x}, x)$, a function of the velocities and positions at the same time t,

$$S[x] = \int L(\dot{x}, x)\, dt, \tag{3.71}$$

then you can start with the Lagrangian L and then create a Hamiltonian H and work out the quantum mechanics, more or less uniquely. But the action $A[x; t_i, t_j]$ (3.70) involves the key variables, positions (and velocities), at two different times t_i and t_j and therefore, it was not obvious what to do to make the quantum–mechanical analogue..."

So, Feynman was looking for the action integral in quantum mechanics. He says: "...I simply turned to Professor Jehle and said, 'Listen, do you know any way of doing quantum mechanics, starting with action – where the action integral comes into the quantum mechanics?" 'No", he said, 'but Dirac has a paper in which the Lagrangian, at least, comes into quantum mechanics." What Dirac said was the following: There is in quantum mechanics a very important quantity which carries the wave function from one time to another, besides the differential equation but equivalent to it, a kind of a kernel, which we might call $K(x', x)$, which carries the wave function $\psi(x)$ known at time t, to the wave function $\psi(x')$ at time $t + \varepsilon$,

$$\psi(x', t+\varepsilon) = \int K(x', x)\, \psi(x, t)\, dx.$$

Dirac points out that this function K was analogous to the quantity in classical mechanics that you would calculate if you took the exponential of [iε multiplied by the Lagrangian $L(\dot{x}, x)$], imagining that these two positions x, x' corresponded to t and $t + \varepsilon$. In other words,

$$K(x', x) \quad \text{is analogous to} \quad \mathrm{e}^{\mathrm{i}\varepsilon L(\frac{x'-x}{\varepsilon}, x)/\hbar}.$$

So, Feynman continues: "What does he mean, they are analogous; what does that mean, *analogous*? What is the use of that?" Professor Jehle said, 'You Americans! You always want to find a use for everything!" I said that I thought that Dirac must mean that they were *equal.* 'No", he explained, 'he doesn't mean they are equal." 'Well", I said, 'Let's see what happens if we make them equal."

"So, I simply put them equal, taking the simplest example where the Lagrangian is

$$L = \frac{1}{2} M \dot{x}^2 - V(x),$$

but soon found I had to put a constant of proportionality N in, suitably adjusted. When I substituted for K to get

$$\psi(x', t+\varepsilon) = \int N \exp\left[\frac{\mathrm{i}\varepsilon}{\hbar} L(\frac{x'-x}{\varepsilon}, x)\right] \psi(x, t)\, dx \tag{3.72}$$

and just calculated things out by Taylor series expansion, *out came the Schrödinger equation.* So, I turned to Professor Jehle, not really understanding, and said, 'Well, you see, Dirac meant that they were proportional." Professor Jehle's eyes were bugging out – he had taken out a little notebook and was rapidly copying it down from the blackboard, and said, 'No, no, this is an important discovery. You Americans are always trying to find out how something can be used. That's a good way to discover things!" So, I thought I was finding out what Dirac meant, but, as a matter of fact, had made the discovery that what Dirac thought was analogous, was, in fact, equal. I had then, at least, the connection between the Lagrangian and quantum mechanics, but still with wave functions and infinitesimal times."

"It must have been a day or so later when I was lying in bed thinking about these things, that I imagined what would happen if I wanted to calculate the wave function at a finite interval later. I would put one of these factors $\mathrm{e}^{\mathrm{i}\varepsilon L}$ in here, and that would give me the wave functions the next moment, $t + \varepsilon$, and then I could substitute that back into (3.72) to get another factor of $\mathrm{e}^{\mathrm{i}\varepsilon L}$ and give me the wave function the next moment, $t + 2\varepsilon$, and so on and so on. In that way I found myself thinking of a large number of integrals, one after the other in sequence. In the integrand was the product of the exponentials, which was the exponential of the sum of terms like εL. Now, L is the Lagrangian and ε is like the time interval dt, so that if you took a sum of such terms, that's exactly like an integral. That's like Riemann's formula for the integral $\int L dt$, you just take the value at each point and add them together. We are to take the limit as $\varepsilon \to 0$. Therefore, the connection between the wave function of one instant and the wave function of another instant a finite time later could be get by an infinite number of integrals (because ε goes to zero), of exponential where S is the action expression (3.71). At last, I had succeeded in representing quantum mechanics directly in terms of the action $S[x]$."

Fully satisfied, Feynman comments: "This led later on to the idea of the *transition amplitude* for a path: that for each possible way that the particle can go from one point to another in space–time, there's an amplitude. That amplitude is **e** to the power of [i/$\hbar$ times the action $S[x]$ for the path], i.e., $\mathrm{e}^{\mathrm{i}S[x]/\hbar}$. Amplitudes from various paths superpose by addition. This then is another, a third way, of describing quantum mechanics, which looks quite different from that of Schrödinger or Heisenberg, but which is equivalent to them."

"...Now immediately after making a few checks on this thing, what we wanted to do, was to substitute the action $A[x; t_i, t_j]$ (3.70) for the other $S[x]$ (3.71). The first trouble was that I could not get the thing to work with the relativistic case of spin one–half. However, although I could deal with the matter only nonrelativistically, I could deal with the light or the photon interactions perfectly well by just putting the interaction terms of (3.70) into any action, replacing the mass terms by the non–relativistic $Ldt = \frac{1}{2}M\dot{x}^2 dt$,

$$A[x; t_i, t_j] = \frac{1}{2}\sum_i m_i \int (\dot{x}^i_\mu)^2 dt_i + \frac{1}{2}\sum_{i,j(i\neq j)} e_i e_j \int\int \delta(I^2_{ij})\, \dot{x}^i_\mu(t_i)\dot{x}^j_\mu(t_j)\, dt_i dt_j.$$

When the action has a delay, as it now had, and involved more than one time, I had to lose the idea of a wave function. That is, I could no longer describe the program as: given the amplitude for all positions at a certain time to calculate the amplitude at another time. However, that didn't cause very much trouble. It just meant developing a new idea. *Instead of wave functions we could talk about this: that if a source of a certain kind emits a particle, and a detector is there to receive it, we can give the amplitude that the source will emit and the detector receive,* $\mathrm{e}^{\mathrm{i}A[x;t_i,t_j]/\hbar}$. We do this without specifying the exact instant that the *source* emits or the exact instant that any *detector* receives, without trying to specify the state of anything at any particular time in between, but by just finding the *amplitude for the complete experiment.* And, then we could discuss how that amplitude would change if you had a scattering sample in between, as you rotated and changed angles, and so on, without really having any wave functions... It was also possible to discover what the old concepts of energy and momentum would mean with this generalized action. And, so I believed that I had a quantum theory of classical electrodynamics – or rather of this new classical electrodynamics described by the action $A[x; t_i, t_j]$ (3.70) ..."

Lagrangian Path Integral

Dirac and Feynman first developed the lagrangian approach to functional integration. To review this approach, we start with the time–dependent *Schrödinger equation*

$$\mathrm{i}\hbar\, \partial_t \psi(x,t) = -\partial_{x^2} \psi(x,t) + V(x)\, \psi(x,t)$$

appropriate to a particle of mass m moving in a potential $V(x)$, $x \in \mathbb{R}$. A solution to this equation can be written as an integral (see e.g., [Kla97, Kla00]),

$$\psi(x'', t'') = \int K(x'', t''; x', t')\, \psi(x', t')\, dx' ,$$

which represents the wave function $\psi(x'', t'')$ at time t'' as a linear superposition over the wave function $\psi(x', t')$ at the initial time t', $t' < t''$. The integral kernel $K(x'', t''; x', t')$ is known as the *propagator*, and according to Feynman [Fey48] it may be given by

$$K(x'', t''; x', t') = \mathcal{N} \int \mathcal{D}[x]\; \mathrm{e}^{(\mathrm{i}/\hbar) \int [(m/2)\, \dot{x}^2(t) - V(x(t))]\, dt},$$

which is a formal expression symbolizing an integral over a suitable set of paths. This integral is supposed to run over all continuous paths $x(t)$, $t' \leq t \leq t''$, where $x(t'') = x''$ and $x(t') = x'$ are fixed end points for all paths. Note that the integrand involves the *classical Lagrangian* for the system.

To overcome the convergence problems, Feynman adopted a *lattice regularization* as a procedure to yield well–defined integrals which was then followed by a limit as the lattice spacing goes to zero called the continuum limit. With $\varepsilon > 0$ denoting the lattice spacing, the details regarding the lattice regularization procedure are given by

$$K(x'', t''; x', t') = \lim_{\varepsilon \to 0}\, (m/2\pi \mathrm{i}\hbar\varepsilon)^{(N+1)/2} \int \cdots$$

$$\cdots \int \exp\{(\mathrm{i}/\hbar) \sum_{l=0}^{N} [(m/2\varepsilon)(x_{l+1} - x_l)^2 - \varepsilon\, V(x_l)\,]\} \prod_{l=1}^{N} dx_l ,$$

where $x_{N+1} = x''$, $x_0 = x'$, and $\varepsilon \equiv (t'' - t')/(N+1)$, $N \in \{1, 2, 3, \dots\}$. In this version, at least, we have an expression that has a reasonable chance of being well defined, provided, that one interprets the conditionally convergent integrals involved in an appropriate manner. One common and fully acceptable interpretation adds a convergence factor to the exponent of the preceding integral in the form $-(\varepsilon^2/2\hbar)\sum_{l=1}^{N} x_l^2$, which is a term that formally makes no contribution to the final result in the continuum limit save for ensuring that the integrals involved are now rendered absolutely convergent.

Hamiltonian Path Integral

It is necessary to retrace history at this point to recall the introduction of the *phase–space path integral* by Feynman [Fey51, GS98]. In Appendix B to this article, Feynman introduced a formal expression for the configuration or $q-$space propagator given by (see e.g., [Kla97, Kla00])

$$K(q'',t'';q',t') = \mathcal{M}\int \mathcal{D}[p]\,\mathcal{D}[q]\;\exp\{(\mathrm{i}/\hbar)\textstyle\int[\,p\,\dot{q} - H(p,q)\,]\,dt\}.$$

In this equation one is instructed to integrate over all paths $q(t)$, $t' \le t \le t''$, with $q(t'') \equiv q''$ and $q(t') \equiv q'$ held fixed, as well as to integrate over all paths $p(t)$, $t' \le t \le t''$, without restriction.

It is widely appreciated that the phase–space path integral is more generally applicable than the original, Lagrangian, version of the path integral. For example, the original configuration space path integral is satisfactory for Lagrangians of the general form

$$L(x) = \frac{1}{2}\,m\dot{x}^2 + A(x)\,\dot{x} - V(x)\;,$$

but it is unsuitable, for example, for the case of a relativistic particle with the Lagrangian

$$L(x) = -m\,qrt1 - \dot{x}^2$$

expressed in units where the speed of light is unity. For such a system – as well as many more general expressions – the phase–space form of the path integral is to be preferred. In particular, for the relativistic free particle, the phase–space path integral

$$\mathcal{M}\int \mathcal{D}[p]\,\mathcal{D}[q]\;\exp\{(\mathrm{i}/\hbar)\textstyle\int[\,p\,\dot{q} -\; qrtp^2 + m^2\,]\,dt\},$$

is readily evaluated and induces the correct propagator.

Feynman–Kac Formula

Through his own research, M. Kac was fully aware of *Wiener's theory of Brownian motion* and the *associated diffusion equation* that describes the corresponding *distribution function.* Therefore, it is not surprising that he was well prepared to give a path integral expression in the sense of Feynman for an equation similar to the time–dependent Schrödinger equation save for a rotation of the time variable by $-\pi/2$ in the complex plane, namely, by the change $t \to -it$ (see e.g., [Kla97, Kla00]). In particular, Kac [Kac51] considered the equation

$$\partial_t \rho(x,t) = \partial_{x^2} \rho(x,t) - V(x)\, \rho(x,t). \tag{3.73}$$

This equation is analogous to Schrödinger equation but differs from it in certain details. Besides certain constants which are different, and the change $t \to -it$, the nature of the dependent variable function $\rho(x,t)$ is quite different from the normal quantum mechanical wave function. For one thing, if the function ρ is initially real it will remain real as time proceeds. Less obvious is the fact that if $\rho(x,t) \geq 0$ for all x at some time t, then the function will continue to be nonnegative for all time t. Thus we can interpret $\rho(x,t)$ more like a probability density; in fact in the special case that $V(x) = 0$, then $\rho(x,t)$ is the probability density for a Brownian particle which underlies the *Wiener measure*. In this regard, ν is called the diffusion constant.

The fundamental solution of (3.73) with $V(x) = 0$ is readily given as

$$W(x,T;y,0) = \frac{1}{qrt2\pi\nu T} \exp\Big(- \frac{(x-y)^2}{2\nu T} \Big),$$

which describes the solution to the diffusion equation subject to the initial condition

$$\lim_{T\to 0+} W(x,T;y,0) = \delta(x-y)\,.$$

Moreover, it follows that the solution of the diffusion equation for a general initial condition is given by

$$\rho(x'',t'') = \int W(x'',t'';x',t')\, \rho(x',t')\, dx'\,.$$

Iteration of this equation N times, with $\epsilon = (t''-t')/(N+1)$, leads to the equation

$$\rho(x'',t'') = N' \int \cdots \int \mathrm{e}^{-(1/2\nu\epsilon)\sum_{l=0}^{N}(x_{l+1}-x_l)^2} \prod_{l=1}^{N} dx_l\, \rho(x',t')\, dx',$$

where $x_{N+1} \equiv x''$ and $x_0 \equiv x'$. This equation features the imaginary time propagator for a free particle of unit mass as given formally as

$$W(x'',t'';x',t') = \mathcal{N} \int \mathcal{D}[x]\, \mathrm{e}^{-(1/2\nu)\int \dot{x}^2\, dt},$$

where $\mathcal{N}$ denotes a formal normalization factor.

The similarity of this expression with the Feynman path integral [for $V(x) = 0$] is clear, but there is a profound difference between these equations.

In the former (Feynman) case the underlying measure is only *finitely additive*, while in the latter (Wiener) case the continuum limit actually defines a genuine measure, i.e., a *countably additive measure* on paths, which is a version of the famous *Wiener measure*. In particular,

$$W(x'',t'';x',t') = \int d\mu_W^\nu(x),$$

where μ_W^ν denotes a measure on continuous paths $x(t)$, $t' \leq t \leq t''$, for which $x(t'') \equiv x''$ and $x(t') \equiv x'$. Such a measure is said to be a *pinned* Wiener measure, since it specifies its path values at two time points, i.e., at $t = t'$ and at $t = t'' > t'$.

We note that Brownian motion paths have the property that with probability one they are concentrated on continuous paths. However, it is also true that the time derivative of a Brownian path is almost nowhere defined, which means that, with probability one, $\dot{x}(t) = \pm\infty$ for all t.

When the potential $V(x) \neq 0$ the propagator associated with (3.73) is formally given by

$$W(x'',t'';x',t') = \mathcal{N} \int \mathcal{D}[x] \mathrm{e}^{-(1/2\nu)\int \dot{x}^2\, dt - \int V(x)\, dt},$$

an expression which is well defined if $V(x) \geq c$, $-\infty < c < \infty$. A mathematically improved expression makes use of the Wiener measure and reads

$$W(x'',t'';x',t') = \int \mathrm{e}^{-\int V(x(t))\, dt}\, d\mu_W^\nu(x).$$

This is an elegant relation in that it represents a solution to the differential equation (3.73) in the form of an integral over Brownian motion paths suitably weighted by the potential V. Incidentally, since the propagator is evidently a strictly positive function, it follows that the solution of the differential equation (3.73) is nonnegative for all time t provided it is nonnegative for any particular time value.

Itô Formula

Itô [Ito60] proposed another version of a *continuous–time regularization* that resolved some of the troublesome issues. In essence, the proposal of Itô takes the form given by

$$\lim_{\nu\to\infty} \mathcal{N}_\nu \int \mathcal{D}[x]\; \exp\{(\mathrm{i}/\hbar) \textstyle\int [\frac{1}{2}m\dot{x}^2 - V(x)]\, dt\} \exp\{-(1/2\nu) \textstyle\int [\ddot{x}^2 + \dot{x}^2]\, dt\}.$$

Note well the alternative form of the auxiliary factor introduced as a regulator. The additional term $\ddot{x}^2$, the square of the second derivative of x, acts to smooth out the paths sufficiently well so that in the case of (21) both $x(t)$ and $\dot{x}(t)$ are continuous functions, leaving $\ddot{x}(t)$ as the term which does not exist. However, since only x and $\dot{x}$ appear in the rest of the integrand, the indicated path integral can be well defined; this is already a positive contribution all by itself (see e.g., [Kla97, Kla00]).

Standard Path–Integral Quantization

Canonical versus Path–Integral Quantization

Recall that in the usual, *canonical formulation* of quantum mechanics, the system's phase–space coordinates, q, and momenta, p, are replaced by the corresponding Hermitian operators in the Hilbert space, with real measurable eigenvalues, which obey *Heisenberg commutation relations.*

The *path–integral quantization* is instead based directly on the notion of a propagator $K(q_f, t_f; q_i, t_i)$ which is defined such that (see [Ryd96, CL84, Gun03])

$$\psi(q_f, t_f) = \int K(q_f, t_f; q_i, t_i)\, \psi(q_i, t_i)\, dq_i, \tag{3.74}$$

i.e., the wave function $\psi(q_f, t_f)$ at final time t_f is given by a Huygens principle in terms of the wave function $\psi(q_i, t_i)$ at an initial time t_i, where we have to integrate over all the points q_i since all can, in principle, send out little wavelets that would influence the value of the wave function at q_f at the later time t_f. This equation is very general and is an expression of causality. We use the normal units with $\hbar = 1$.

According to the usual interpretation of quantum mechanics, $\psi(q_f, t_f)$ is the *probability amplitude* that the particle is at the point q_f and the time t_f, which means that $K(q_f, t_f; q_i, t_i)$ is the probability amplitude for a transition from q_i and t_i to q_f and t_f. The probability that the particle is observed at q_f at time t_f if it began at q_i at time t_i is

$$P(q_f, t_f; q_i, t_i) = |K(q_f, t_f; q_i, t_i)|^2 .$$

Let us now divide the time interval between t_i and t_f into two, with t as the intermediate time, and q the intermediate point in space. Repeated application of (3.74) gives

$$\psi(q_f, t_f) = \int\int K(q_f, t_f; q, t)\, dq\, K(q, t; q_i, t_i)\, \psi(q_i, t_i)\, dq_i,$$

from which it follows that

$$K(q_f, t_f; q_i, t_i) = \int dq\, K(q_f, t_f; q, t)\, K(q, t; q_i, t_i).$$

This equation says that the transition from (q_i, t_i) to (q_f, t_f) may be regarded as the result of the transition from (q_i, t_i) to all available intermediate points

q followed by a transition from (q, t) to (q_f, t_f). This notion of *all possible paths* is crucial in the *path–integral formulation* of quantum mechanics.

Now, recall that the *state vector* $|\psi, t\rangle_S$ in the *Schrödinger picture* is related to that in the *Heisenberg picture* $|\psi\rangle_H$ by

$$|\psi, t\rangle_S = \mathrm{e}^{-\mathrm{i}Ht} |\psi\rangle_H ,$$

or, equivalently,

$$|\psi\rangle_H = \mathrm{e}^{\mathrm{i}Ht} |\psi, t\rangle_S .$$

We also define the vector

$$|q, t\rangle_H = \mathrm{e}^{\mathrm{i}Ht} |q\rangle_S ,$$

which is the Heisenberg version of the Schrödinger state $|q\rangle$. Then, we can equally well write

$$\psi(q, t) = \langle q, t\, |\psi\rangle_H . \tag{3.75}$$

By completeness of states we can now write

$$\langle q_f, t_f\, |\psi\rangle_H = \int \langle q_f, t_f\, |q_i, t_i\rangle_H \, \langle q_i, t_i\, |\psi\rangle_H \, dq_i,$$

which with the definition of (3.75) becomes

$$\psi(q_f, t_f) = \int \langle q_f, t_f\, |q_i, t_i\rangle_H \, \psi(q_i, t_i)\, dq_i.$$

Comparing with (3.74), we get

$$K(q_f, t_f; q_i, t_i) = \langle q_f, t_f\, |q_i, t_i\rangle_H .$$

Now, let us calculate the *quantum–mechanics propagator*

$$\langle q', t'\, |q, t\rangle_H = \left\langle q' |\mathrm{e}^{-\mathrm{i}H(t-t')}\, |q\right\rangle$$

using the *path–integral formalism* that will incorporate the direct quantization of the coordinates, without Hilbert space and Hermitian operators.

The first step is to divide up the time interval into $n+1$ tiny pieces: $t_l = l\varepsilon + t$ with $t' = (n+1)\varepsilon + t$. Then, by completeness, we can write (dropping the Heisenberg picture index H from now on)

$$\begin{aligned}\langle q', t'\, |q, t\rangle = \int dq_1(t_1) ... \int dq_n(t_n)\, \langle q', t'\, |q_n, t_n\rangle \times \\ \times \langle q_n, t_n\, |q_{n-1}, t_{n-1}\rangle ... \langle q_1, t_1\, |q, t\rangle .\end{aligned} \tag{3.76}$$

The integral $\int dq_1(t_1) \dots dq_n(t_n)$ is an *integral over all possible paths*, which are not trajectories in the normal sense, since there is no requirement of continuity, but rather *Markov chains*.

Now, for small ε we can write

$$\langle q', \varepsilon \, | q, 0 \rangle = \left\langle q' | \mathrm{e}^{-\mathrm{i}\varepsilon H(P,Q)} \, | q \right\rangle = \delta(q' - q) - \mathrm{i}\varepsilon \, \langle q' | H(P,Q) \, | q \rangle \, ,$$

where $H(P,Q)$ is the Hamiltonian (e.g., $H(P,Q) = \frac{1}{2}P^2 + V(Q)$, where P, Q are the momentum and coordinate operators). Then we have (see [Ryd96, CL84, Gun03])

$$\langle q' | H(P,Q) \, | q \rangle = \int \frac{dp}{2\pi} \, \mathrm{e}^{\mathrm{i}p(q'-q)} H\left(p, \frac{1}{2}(q'+q)\right).$$

Putting this into our earlier form we get

$$\langle q', \varepsilon \, | q, 0 \rangle \simeq \int \frac{dp}{2\pi} \exp\left[\mathrm{i}\left\{p(q'-q) - \varepsilon H\left(p, \frac{1}{2}(q'+q)\right)\right\}\right],$$

where the 0th order in $\varepsilon \to \delta(q'-q)$ and the 1st order in $\varepsilon \to -\mathrm{i}\varepsilon \, \langle q' | H(P,Q) \, | q \rangle$. If we now substitute many such forms into (3.76) we finally get

$$\begin{aligned} \left\langle q', t' \, | q, t \right\rangle &= \lim_{n \to \infty} \int \prod_{i=1}^{n} dq_i \prod_{k=1}^{n+1} \frac{dp_k}{2\pi} \times \qquad (3.77) \\ &\times \exp\left\{\mathrm{i} \sum_{j=1}^{n+1} [p_j(q_j - q_{j-1})] - H\left(p_j, \frac{1}{2}(q_j + q_{j+1})\right)(t_j - t_{j-1})]\right\}, \end{aligned}$$

with $q_0 = q$ and $q_{n+1} = q'$. Roughly, the above formula says to *integrate over all possible momenta and coordinate values associated with a small interval,* weighted by something that is going to turn into the *exponential of the action* $\mathrm{e}^{\mathrm{i}S}$ in the limit where $\varepsilon \to 0$. It should be stressed that the different q_i and p_k integrals are independent, which implies that p_k for one interval can be completely different from the $p_{k'}$ for some other interval (including the neighboring intervals). In principle, the integral (3.77) should be defined by *analytic continuation into the complex plane* of, for example, the p_k integrals.

Now, if we go to the differential limit where we call $t_j - t_{j-1} \equiv d\tau$ and write $\frac{(q_j - q_{j-1})}{(t_j - t_{j-1})} \equiv \dot{q}$, then the above formula takes the form

$$\langle q', t' \, | q, t \rangle = \int \mathcal{D}[p] \mathcal{D}[q] \exp\left\{\mathrm{i} \int_t^{t'} [p\dot{q} - H(p,q)] \, d\tau\right\},$$

where we have used the shorthand notation

$$\int \mathcal{D}[p]\mathcal{D}[q] \equiv \int \prod_{\tau} \frac{dq(\tau)dp(\tau)}{2\pi}.$$

Note that the above integration is an integration over the p and q values at every time τ. This is what we call a *functional integral.* We can think of a given set of choices for all the $p(\tau)$ and $q(\tau)$ as defining a *path in the* 6D *phase–space.* The most important point of the above result is that we have get an expression for a *quantum–mechanical transition amplitude* in terms of an integral involving only pure complex numbers, without operators.

We can actually perform the above integral for Hamiltonians of the type $H = H(P, Q)$. We use square completion in the exponential for this, defining the integral in the complex p plane and continuing to the physical situation. In particular, we have

$$\int_{-\infty}^{\infty} \frac{dp}{2\pi} \exp\left\{\mathrm{i}\varepsilon(p\dot{q} - \frac{1}{2}p^2]\right\} = \frac{1}{\sqrt{2\pi\mathrm{i}\varepsilon}} \exp\left[\frac{1}{2}\mathrm{i}\varepsilon\dot{q}^2\right],$$

(see [Ryd96, CL84, Gun03]) which, substituting into (3.77) gives

$$\langle q', t' \,|q, t\rangle = \lim_{n\to\infty} \int \prod_i \frac{dq_i}{\sqrt{2\pi\mathrm{i}\varepsilon}} \exp\{\mathrm{i}\varepsilon \sum_{j=1}^{n+1} [\frac{1}{2}(\frac{q_j - q_{j-1}}{\varepsilon})^2 - V(\frac{q_j + q_{j+1}}{2})]\}.$$

This can be formally written as

$$\langle q', t' \,|q, t\rangle = \int \mathcal{D}[q]\, \mathrm{e}^{\mathrm{i}S[q]},$$

where

$$\int \mathcal{D}[q] \equiv \int \prod_i \frac{dq_i}{\sqrt{2\pi\mathrm{i}\varepsilon}},$$

while

$$S[q] = \int_t^{t'} L(q, \dot{q})\, d\tau$$

is the *standard action* with the *Lagrangian*

$$L = \frac{1}{2}\dot{q}^2 - V(q).$$

Generalization to many degrees–of–freedom is straightforward:

$$\langle q_1{}'...q_N{}', t'|q_1...q_N, t\rangle = \int \mathcal{D}[p]\mathcal{D}[q] \exp\left\{\mathrm{i}\int_t^{t'} \left[\sum_{n=1}^{N} p_n\dot{q}_n - H(p_n, q_n)\right] d\tau\right\},$$

$$\text{with} \quad \int \mathcal{D}[p]\mathcal{D}[q] = \int \prod_{n=1}^{N} \frac{dq_n dp_n}{2\pi}.$$

Here, $q_n(t) = q_n$ and $q_n(t') = q_n{}'$ for all $n = 1, \ldots, N$, and we are allowing for the full Hamiltonian of the system to depend upon all the N momenta and coordinates collectively.

Basic Physical Applications of Path Integrals

(i) Consider first

$$\langle q', t'|Q(t_0)|q, t\rangle$$
$$= \int \prod dq_i(t_i) \langle q', t'|q_n, t_n\rangle \dots \langle q_{i0}, t_{i0}|Q(t_0)|q_{i-1}, t_{i-1}\rangle \dots \langle q_1, t_1|q, t\rangle ,$$

where we choose one of the time interval ends to coincide with t_0, i.e., $t_{i0} = t_0$. If we operate $Q(t_0)$ to the left, then it is replaced by its eigenvalue $q_{i0} = q(t_0)$. Aside from this one addition, everything else is evaluated just as before and we will obviously get

$$\langle q', t'|Q(t_0)|q, t\rangle = \int \mathcal{D}[p]\mathcal{D}[q]\, q(t_0) \exp\left\{ \mathrm{i} \int_t^{t'} [p\dot{q} - H(p, q)]d\tau \right\}.$$

(ii) Next, suppose we want a *path–integral expression* for $\langle q', t'|Q(t_1)Q(t_2)| q, t\rangle$ in the case where $t_1 > t_2$. For this, we have to insert as intermediate states $|q_{i1}, t_{i1}\rangle \langle q_{i1}, t_{i1}|$ with $t_{i1} = t_1$ and $|q_{i2}, t_{i2}\rangle \langle q_{i2}, t_{i2}|$ with $t_{i2} = t_2$ and since we have ordered the times at which we do the insertions we must have the first insertion to the left of the 2nd insertion when $t_1 > t_2$. Once these insertions are done, we evaluate $\langle q_{i1}, t_{i1}|\, Q(t_1) = \langle q_{i1}, t_{i1}|\, q(t_1)$ and $\langle q_{i2}, t_{i2}|\, Q(t_2) = \langle q_{i2}, t_{i2}|\, q(t_2)$ and then proceed as before and get

$$\langle q', t'|Q(t_1)Q(t_2)|q, t\rangle = \int \mathcal{D}[p]\mathcal{D}[q]\, q(t_1)\, q(t_2) \exp\left\{ \mathrm{i} \int_t^{t'} [p\dot{q} - H(p, q)]d\tau \right\}.$$

Now, let us ask what the above integral is equal to if $t_2 > t_1$? It is obvious that what we get for the above integral is $\langle q', t'|Q(t_2)Q(t_1)|q, t\rangle$. Clearly, this generalizes to an arbitrary number of Q operators.

(iii) When we enter into quantum field theory, the Q's will be replaced by fields, since it is the fields that play the role of coordinates in the 2nd quantization conditions.

Sources

The *source* is represented by modifying the Lagrangian:

$$L \to L + J(t)q(t).$$

Let us define $|0, t\rangle^J$ as the ground state (vacuum) vector (in the moving frame, i.e., with the e^{iHt} included) in the presence of the source. The required *transition amplitude* is

$$Z[J] \propto \langle 0, +\infty|0, -\infty\rangle^J ,$$

where the source $J = J(t)$ plays a role analogous to that of an electromagnetic current, which acts as a source of the electromagnetic field. In other words, we can think of the scalar product $J_\mu A^\mu$, where J_μ is the current from a scalar (or Dirac) field acting as a source of the potential A^μ. In the same way, we can always define a current J that acts as the source for some arbitrary field ϕ. $Z[J]$ (otherwise denoted by $W[J]$) is a functional of the current J, defined as (see [Ryd96, CL84, Gun03])

$$Z[J] \propto \int \mathcal{D}[p]\mathcal{D}[q] \exp\left\{ \mathrm{i} \int_t^{t'} [p(\tau)\dot{q}(\tau) - H(p,q) + J(\tau)q(\tau)] d\tau \right\},$$

with the *normalization condition* $Z[J=0]=1$. Here, the argument of the exponential depends upon the functions $q(\tau)$ and $p(\tau)$ and we then integrate over all possible forms of these two functions. So the exponential is a functional that maps a choice for these two functions into a number. For example, for a quadratically completable $H(p,q)$, the p integral can be performed as a q integral

$$Z[J] \propto \int \mathcal{D}[q] \exp\left\{ \mathrm{i} \int_{-\infty}^{+\infty} \left(L + Jq + \frac{1}{2}\mathrm{i}\varepsilon q^2 \right) d\tau \right\},$$

where the addittion to H was chosen in the form of a *convergence factor* $-\frac{1}{2}\mathrm{i}\varepsilon q^2$.

Fields

Let us now treat the *abstract scalar field* $\phi(x)$ as a coordinate in the sense that we imagine dividing space up into many little cubes and the average value of the field $\phi(x)$ in that cube is treated as a coordinate for that little cube. Then, we go through the multi–coordinate analogue of the procedure we just considered above and take the continuum limit. The final result is

$$Z[J] \propto \int \mathcal{D}[\phi] \exp\left\{ \mathrm{i} \int d^4x \left(\mathcal{L}(\phi(x)) + J(x)\phi(x) + \frac{1}{2}\mathrm{i}\varepsilon\phi^2 \right) \right\},$$

where for $\mathcal{L}$ we would employ the *Klein–Gordon Lagrangian* form. In the above, the dx_0 integral is the same as $d\tau$, while the $d^3\mathbf{x}$ integral is summing over the sub–Lagrangians of all the different little cubes of space and then taking the continuum limit. $\mathcal{L}$ is the *Lagrangian density* describing the Lagrangian for each little cube after taking the many–cube limit (see [Ryd96, CL84, Gun03]) for the full derivation).

We can now introduce *interactions*, $\mathcal{L}_I$. Assuming the simple form of the Hamiltonian, we have

$$Z[J] \propto \int \mathcal{D}[\phi] \exp\left\{ \mathrm{i} \int d^4x \left(\mathcal{L}(\phi(x)) + \mathcal{L}_I(\phi(x)) + J(x)\phi(x) \right) \right\},$$

again using the normalization factor required for $Z[J=0]=1$.

For example of Klein Gordon theory, we would use

$$\mathcal{L} = \mathcal{L}_0 + \mathcal{L}_I, \qquad \mathcal{L}_0 \frac{1}{2}[\partial_\mu \phi \partial^\mu \phi - \mu^2 \phi^2], \qquad \mathcal{L}_I = \mathcal{L}_I(\phi),$$

where $\partial_\mu \equiv \partial_{x^\mu}$ and we can freely manipulate indices, as we are working in Euclidean space $\mathbb{R}^3$. In order to define the above $Z[J]$, we have to include a convergence factor $\mathrm{i}\varepsilon\phi^2$,

$$\mathcal{L}_0 \to \frac{1}{2}\left[\partial_\mu \phi \partial^\mu \phi - \mu^2 \phi^2 + \mathrm{i}\varepsilon\phi^2\right], \qquad \text{so that}$$

$$Z[J] \propto \int \mathcal{D}[\phi] \exp\left\{\mathrm{i}\int d^4x \left(\frac{1}{2}\left[\partial_\mu \phi \partial^\mu \phi - \mu^2 \phi^2 + \mathrm{i}\varepsilon\phi^2\right] + \mathcal{L}_I(\phi(x)) + J(x)\phi(x)\right)\right\}$$

is the appropriate *generating function* in the free field theory case.

Gauges

In the path integral approach to quantization of the *gauge theory,* we implement *gauge fixing* by restricting in some manner or other the path integral over gauge fields $\int \mathcal{D}[A_\mu]$. In other words we will write instead

$$Z[J] \propto \int \mathcal{D}[A_\mu]\, \delta\,(\text{some gauge fixing condition}) \exp\{\mathrm{i}\int d^4x \mathcal{L}\,(A_\mu)\}.$$

A common approach would be to start with the *gauge condition*

$$\mathcal{L} = -\frac{1}{4} F_{\mu\nu} F^{\mu\nu} - \frac{1}{2}(\partial^\mu A_\mu)^2$$

where the electrodynamic field tensor is given by $F_{\mu\nu} = \partial_\mu A_\nu - \partial_\nu A_\mu$, and calculate

$$Z[J] \propto \int \mathcal{D}[A_\mu] \exp\left\{\mathrm{i}\int d^4x\, [\mathcal{L}(A_\mu(x)) + J_\mu(x) A^\mu(x)]\right\}$$

as the *generating function* for the *vacuum expectation* values of *time ordered products* of the A_μ fields. Note that J_μ should be conserved ($\partial^\mu J_\mu = 0$) in order for the full expression $\mathcal{L}(A_\mu) + J_\mu A^\mu$ to be *gauge–invariant* under the integral sign when $A_\mu \to A_\mu + \partial^\mu \Lambda$. For a proper approach, see [Ryd96, CL84, Gun03].

3.2 Quantum Consciousness

3.2.1 EPR Paradox and Bell's Theorem

EPR Paradox

Recall that the so–called *EPR paradox* in quantum mechanics is a *thought experiment*[11] which challenged long–held ideas about the relation between the

[11] Recall that a *thought experiment* (coined by Hans Christian Ørsted) in the broadest sense is the use of an imagined scenario to help us understand the way things

observed values of physical quantities and the values that can be accounted for by a physical theory. 'EPR' stands for A. Einstein,[12] B. Podolsky, and N. Rosen, who introduced the thought experiment in a 1935 paper [EPR35a], to argue that quantum mechanics is not a complete physical theory. It is sometimes referred to as the EPRB paradox for David Bohm, who converted the original thought experiment into something closer to being experimentally testable.

The EPR experiment yields the following dichotomy:

(i) Either the result of a measurement performed on one part A of a quantum system has a non–local effect on the physical reality of another distant part B, in the sense that quantum mechanics can predict outcomes of some measurements carried out at B, or

(ii) Quantum mechanics is incomplete in the sense that some element of physical reality corresponding to B cannot be accounted for by quantum mechanics (that is, some extra variable is needed to account for it.)

---

really are. The understanding comes through reflection on the situation. Thought experiment methodology is a priori, rather than empirical, in that it does not proceed by observation or physical experiment. Thought experiments are well–structured hypothetical questions that employ 'What if?' reasoning. Thought experiments have been used in philosophy, physics, and other fields. They have been used to pose questions in philosophy at least since Greek antiquity, some pre-dating Socrates. In physics and other sciences many famous thought experiments date from the 19th and especially the 20th Century, but examples can be found at least as early as Galileo.

Thought experiments in physics are intended to give us a priori knowledge of the natural world, rather than a priori knowledge of our concepts, as philosophy tries to do. A. Einstein and N. Tesla were famous for their thought experiment methodology.

[12] Albert Einstein (March 14, 1879 – April 18, 1955) was a German–born theoretical physicist. He is widely regarded as one of the greatest physicists who ever lived. He formulated the Special and General Relativity Theories. In addition, he made significant contributions to quantum theory and statistical mechanics. While best known for the Theory of Relativity (and specifically mass–energy equivalence, $E = mc^2$), he was awarded the 1921 Nobel Prize for Physics for his explanation of the photoelectric effect in 1905 (his 'wonderful year' or 'miraculous year') and 'for his services to Theoretical Physics'.

Following the May 1919 British solar-eclipse expeditions, whose later analysis confirmed that light rays from distant stars were deflected by the Sun's gravitation as predicted by the Field Equation of general relativity, in November 1919 Albert Einstein became world–famous, an unusual achievement for a scientist. The Times ran the headline on November 7, 1919: "Revolution in science – New theory of the Universe – Newtonian ideas overthrown." Nobel laureate Max Born viewed General Relativity as the "greatest feat of human thinking about nature;" fellow laureate Paul Dirac called it "probably the greatest scientific discovery ever made." In popular culture, the name 'Einstein' has become synonymous with great intelligence and genius.

Although originally devised as a thought experiment that would demonstrate the *incompleteness of quantum mechanics*,[13] actual experimental results ironically refutes the *principle of locality*,[14] invalidating the EPR trio's original purpose. The "spooky action at a distance" that so disturbed the

[13] Incompleteness of quantum physics is the assertion that the state of a physical system, as formulated by quantum mechanics, does not give a complete description for the system. A complete description is one which uniquely determines the values of all its measurable properties. The existence of indeterminacy for some measurements is a characteristic of quantum mechanics; moreover, bounds for indeterminacy can be expressed in a quantitative form by the *Heisenberg uncertainty principle*.

Incompleteness can be understood in two fundamentally different ways:

(i) QM is incomplete because it is not the 'right' theory; the right theory would provide descriptive categories to account for all observable behavior and not leave 'anything to chance'.

(ii) QM is incomplete, but it accurately reflects the way nature is.

Incompleteness understood as (i) is now considered highly controversial, since it contradicts the impossibility of a hidden variables theory which is shown by *Bell test experiments*. There are many variants of (ii) which is widely considered to be the more orthodox view of quantum mechanics.

[14] The *principle of locality* is that distant objects cannot have direct influence on one another: an object is influenced directly only by its immediate surroundings. This was stated as follows by Einstein in his article [Ein48].

The following idea characterizes the relative independence of objects far apart in space (A and B): external influence on A has no direct influence on B; this is known as the Principle of Local Action, which is used consistently only in field theory. If this axiom were to be completely abolished, the idea of the existence of quasi–enclosed systems, and thereby the postulation of laws which can be checked empirically in the accepted sense, would become impossible.

Local realism is the combination of the principle of locality with the 'realistic' assumption that all objects must objectively have their properties already before these properties are observed. Einstein liked to say that the Moon is 'out there' even when no one is observing it.

Local realism is a significant feature of classical mechanics, general relativity and Maxwell's theory, but quantum mechanics largely rejects this principle due the presence of distant quantum entanglements, most clearly demonstrated by the EPR paradox and quantified by Bell's inequalities. Every theory that, like quantum mechanics, is compatible with violations of Bell's inequalities must abandon either local realism or counterfactual definiteness. (The vast majority of physicists believe that experiments have demonstrated Bell's violations, but some local realists dispute the claim, in view of the recognized loopholes in the tests.) Different interpretations of quantum mechanics reject different parts of local realism and/or counterfactual definiteness.

In most of the conventional interpretations, such as the version of the *Copenhagen interpretation* and the interpretation based on *Consistent Histories*, where the wave $\psi-$function is not assumed to have a direct physical interpretation or reality it is realism that is rejected. The actual definite properties of a physical system 'do not exist' prior to the measurement and the wave $\psi-$function has a

authors of EPR consistently occurs in numerous and widely replicated experiments. Einstein never really accepted quantum mechanics as a 'real' and complete theory, struggling to the end of his career (and life) for an interpretation that could comply with his Relativity without implying "God playing dice," as he condensed his dissatisfaction with QM's intrinsic randomness and (still to be resolved) counter–intuitivity.

The EPR paradox is a paradox in the following sense: if one takes quantum mechanics and adds some seemingly reasonable conditions (referred to as locality, realism, counter factual definiteness, and completeness), then one obtains a contradiction. However, quantum mechanics by itself does not appear to be internally inconsistent, nor does it contradict relativity. As a result of further theoretical and experimental developments since the original EPR

---

restricted interpretation as nothing more than a mathematical tool used to calculate the probabilities of experimental outcomes, in agreement with positivism in philosophy as the only topic that science should discuss.

In the version of the Copenhagen interpretation where the wave ψ–function is assumed to have a physical interpretation or reality (the nature of which is unspecified), the principle of locality is violated during the measurement process via wave ψ–function collapse. This is a *nonlocal process* because Born's Rule, when applied to the system's wave function, yields a probability density for all regions of space and time. Upon measurement of the physical system, the probability density vanishes everywhere instantaneously, except where (and when) the measured entity is found to exist. This "vanishing" would be a real physical process, and clearly non-local (faster–than–light–speed), if the wave ψ–function is considered physically real and the probability density converged to zero at arbitrarily far distances during the finite time required for the measurement process.

The Bohm interpretation always wants to preserve realism, and it needs to violate the principle of locality to achieve the required correlations.

In the many–worlds interpretation realism and locality are retained but counterfactual definiteness is rejected by the extension of the notion of reality to allow the existence of parallel universes.

Because the differences between the different interpretations are mostly philosophical ones (except for the Bohm and many–worlds interpretations), the physicists usually use the language in which the important statements are independent of the interpretation we choose. In this framework, only the measurable action at a distance, a super–luminal propagation of real, physical information, would be usually considered to be a violation of locality by the physicists. Such phenomena have never been seen, and they are not predicted by the current theories (with the possible exception of the Bohm theory).

Locality is one of the axioms of relativistic quantum field theory, as required for causality. The formalization of locality in this case is as follows: if we have two observables, each localized within two distinct spacetime regions which happen to be at a space–like separation from each other, the observables must commute. This interpretation of the word 'locality' is closely related to the relativistic version in physics. In physics a solution is local if the underlying equations are either Lorentz invariant or, more generally, generally covariant or locally Lorentz invariant.

paper, most physicists today regard the EPR paradox as an illustration of how quantum mechanics violates classical intuitions, and not as an indication that quantum mechanics is fundamentally flawed.

The EPR paradox draws on a *quantum entanglement* phenomenon, to show that measurements performed on spatially separated parts of a quantum system can apparently have an instantaneous influence on one another. This effect is now known as *nonlocal behavior*[15]. In order to illustrate this, let us consider a simplified version of the EPR thought experiment due to David Bohm.

Measurements on an Entangled State

We have a source that emits pairs of electrons, with one electron sent to destination A,where there is an observer named Alice, and another is sent to destination B, where there is an observer named Bob. According to quantum mechanics, we can arrange our source so that each emitted electron pair occupies a quantum state called a spin singlet. This can be viewed as a quantum superposition of two states, which we call I and II. In state I, electron A has spin pointing upward along the $z-$axis ($+z$) and electron B has spin pointing downward along the $z-$axis ($-z$). In state II, electron A has spin $-z$ and electron B has spin $+z$. Therefore, it is impossible to associate either electron in the spin singlet with a state of definite spin. The electrons are thus said to be *entangled.*

Alice now measures the spin along the $z-$axis. She can get one of two possible outcomes: $+z$ or $-z$. Suppose she gets $+z$. According to quantum mechanics, the quantum state of the system collapses into state I (different interpretations of quantum mechanics have different ways of saying this, but

[15] A physical theory is said to 'exhibit nonlocality' if, in that theory, it is not possible to treat widely separated systems as independent. The term is most often reserved, however, for interactions that occur outside the backward light cone, i.e. super–luminal influences. Nonlocality does not imply a lack of causality only in the case when 'ethereal', not 'causal', information is transmitted between systems.

Special Relativity shows that in the case where causal information is transmitted at super–luminal rates, causality is violated. For example, if information could be exchanged at super–luminal rates, it would be possible to arrange for your grandfather to be killed before you are born, which leads to causal paradoxes. Some effects that appear nonlocal in quantum mechanics may actually obey locality, such as *quantum entanglement*. These interactions effect correlations between states of particles (expressed by a wave $\psi-$function which may be in a superposition of states), such as the infamous singlet state. Einstein criticized this interpretation of quantum mechanics on the grounds that these effects employed what he called "spooky action at a distance."

This issue is very closely related to *Bell's theorem* and the *EPR paradox*. Quantum field theory, on the other hand, which is the relativistic generalization of quantum mechanics, contains mathematical features that assure locality.

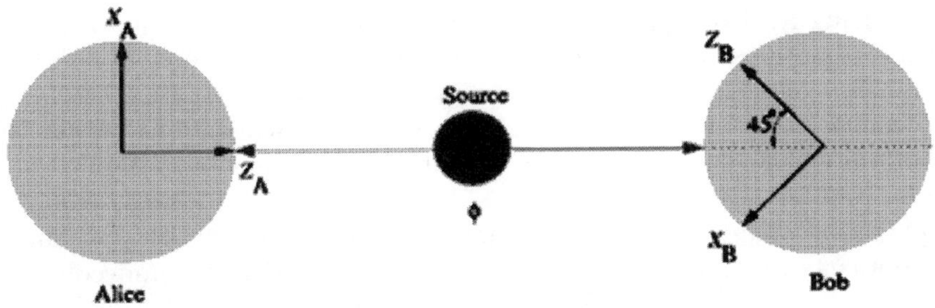

Fig. 3.5. The EPR thought experiment, performed with electrons. A source (center) sends electrons toward two observers, Alice (left) and Bob (right), who can perform spin measurements.

the basic result is the same). The quantum state determines the probable outcomes of any measurement performed on the system. In this case, if Bob subsequently measures spin along the $z-$axis, he will get $-z$ with 100% probability. Similarly, if Alice gets $-z$, Bob will get $+z$.

There is nothing special about our choice of the z axis. For instance, suppose that Alice and Bob now decide to measure spin along the $x-$axis. According to quantum mechanics, the spin singlet state may equally well be expressed as a superposition of spin states pointing in the $x-$direction. We will call these states Ia and IIa. In state Ia, Alice's electron has spin $+x$ and Bob's electron has spin $-x$. In state IIa, Alice's electron has spin $-x$ and Bob's electron has spin $+x$. Therefore, if Alice measures $+x$, the system collapses into Ia, and Bob will get $-x$. If Alice measures $-x$, the system collapses into IIa, and Bob will get $+x$.

In quantum mechanics, the $x-$spin and $z-$spin are *incompatible observables*, which means that there is a *Heisenberg uncertainty principle* operating between them: a quantum state cannot possess a definite value for both variables. Suppose Alice measures the $z-$spin and obtains $+z$, so that the quantum state collapses into state I. Now, instead of measuring the $z-$spin as well, Bob measures the $x-$spin. According to quantum mechanics, when the system is in state I, Bob's $x-$spin measurement will have a 50% probability of producing $+x$ and a 50% probability of $-x$. Furthermore, it is fundamentally impossible to predict which outcome will appear until Bob actually performs the measurement.

So how does Bob's electron know, at the same time, which way to point if Alice decides (based on information unavailable to Bob) to measure x and also how to point if Alice measures z? Using the usual Copenhagen interpretation rules that say the wave function *collapses* at the time of measurement, there must be action at a distance or the electron must know more than it is supposed to. To make the mixed part quantum and part classical descriptions of this experiment local, we have to say that the notebooks (and experimenters) are entangled and have linear combinations of + and − written in them, like *Schrödinger's Cat*.

Incidentally, although we have used spin as an example, many types of physical quantities — what quantum mechanics refers to as *quantum observables*, can be used to produce *quantum entanglement*. The original EPR paper used momentum for the observable. Experimental realizations of the EPR scenario often use the polarization of photons, because polarized photons are easy to prepare and measure.

In recent years, doubt has been cast on EPR's conclusion due to developments in understanding locality and especially quantum decoherence. The word locality has several different meanings in physics. For example, in quantum field theory *locality* means that quantum fields at different points of space do not interact with one another. However, quantum field theories that are *local* in this sense appear to violate the principle of locality as defined by EPR, but they nevertheless do not violate locality in a more general sense. A *wave ψ–function collapse*[16] can be viewed as an epiphenomenon of

[16] In certain interpretations of quantum mechanics, *wave ψ–function collapse* is one of two processes by which quantum systems apparently evolve according to the laws of quantum mechanics. It is also called collapse of the state vector or reduction of the wave packet. The reality of wavefunction collapse has always been debated, i.e., whether it is a fundamental phenomenon in its own right or just an epiphenomenon of another process (e.g., quantum decoherence).

By the time John von Neumann wrote his famous treatise 'Mathematische Grundlagen der Quantenmechanik' in 1932, the phenomenon of wave ψ–function collapse was accommodated into the mathematical formulation of quantum mechanics by postulating that there were two processes of wave ψ–function change:

(1) The probabilistic, non–unitary, non–local, discontinuous change brought about by observation and measurement, as outlined above.

(2) The deterministic, unitary, continuous time evolution of an isolated system that obeys Schrödinger's equation (or nowadays some relativistic, local equivalent).

In general, quantum systems exist in superpositions of those basis states that most closely correspond to classical descriptions, and – when not being measured or observed, evolve according to the time dependent Schrödinger equation, relativistic quantum field theory or some form of quantum gravity or string theory, which is process (2) mentioned above. However, when the wavefunction collapses – process (1) – from an observer's perspective the state seems to 'leap' or 'jump' to just one of the basis states and uniquely acquire the value of the property being measured, e.i., that is associated with that particular basis state. After the collapse, the system begins to evolve again according to the Schrödinger equation or some equivalent wave equation.

Hence, in experiments such as the double–slit experiment each individual photon arrives at a discrete point on the screen, but as more and more photons are accumulated, they form an interference pattern overall.

Consciousness causes collapse is the theory that observation by a conscious observer is responsible for the wave ψ–function collapse. It is an attempt to solve the *Wigner's friend paradox* by simply stating that collapse occurs at the first 'conscious' observer. Supporters claim this is not a revival of substance dualism, since (in a ramification of this view) consciousness and objects are entangled

quantum decoherence,[17] which in turn is nothing more than an effect of the underlying local time evolution of the wavefunction of a system and all of its environment. Since the underlying behaviour doesn't violate local causality it follows that neither does the additional effect of wavefunction collapse, whether real or apparent. Therefore, as outlined in the example above, the EPR experiment (nor any quantum experiment) does not demonstrate that FTL signalling is possible.

Resolving the Paradox

There are several ways to resolve the EPR paradox. The one suggested by EPR is that quantum mechanics, despite its success in a wide variety of experimental scenarios, is actually an incomplete theory. In other words, there is some yet undiscovered theory of nature to which quantum mechanics acts as a kind of statistical approximation (albeit an exceedingly successful one). Unlike quantum mechanics, the more complete theory contains variables corresponding to all the 'elements of reality'. There must be some unknown mechanism acting on these variables to give rise to the observed effects of 'non–commuting quantum observables', i.e., the *Heisenberg uncertainty principle*. Such a theory is called a *hidden variable theory*.[18]

Another resolution of the EPR paradox is provided by Bell's theorem.

---

and cannot be considered as separate. The consciousness causes collapse theory can be considered as a speculative appendage to almost any interpretation of quantum mechanics and many physicists reject it as unverifiable and introducing unnecessary elements into physics.

In recent decades the latter view has gained popularity.

[17] *Quantum decoherence* is the mechanism by which quantum systems interact with their environments to exhibit probabilistically additive behavior (a feature of classical physics) and give the appearance of wavefunction collapse. Decoherence occurs when a system interacts with its environment, or any complex external system, in such a thermodynamically irreversible way that ensures different elements in the quantum superposition of the system + environment's wave ψ−function can no longer interfere with each other.

Decoherence does not provide a mechanism for the actual wave function collapse; the quantum nature of the system is simply 'leaked' into the environment so that a total superposition of the wavefunction still exists, but exists beyond the realm of measurement; rather decoherence provides a mechanism for the appearance of wavefunction collapse.

Decoherence represents a major problem for the practical realization of quantum computers, since these heavily rely on the undisturbed evolution of quantum coherences.

[18] A hidden variable theory is urged by a minority of physicists who argue that the statistical nature of quantum mechanics implies that quantum mechanics is incomplete; it is really applicable only to ensembles of particles; new physical phenomena beyond quantum mechanics are needed to explain an individual event.

Bell's Theorem

Bell's theorem is the most famous legacy of the late physicist John Bell.[19] It is notable for showing that the predictions of quantum mechanics (QM) differ from those of intuition. It is simple and elegant, and touches upon fundamental philosophical issues that relate to modern physics. In its simplest form, Bell's theorem states:

No physical theory of local hidden variables can ever reproduce all of the predictions of quantum mechanics.

This theorem has even been called "the most profound in science" (Stapp, 1975). Bell's seminal 1964 paper was entitled "On the Einstein Podolsky Rosen paradox". The Einstein Podolsky Rosen paradox (EPR paradox) assumes local realism, the intuitive notion that particle attributes have definite values independent of the act of observation and that physical effects have a finite propagation speed. Bell showed that local realism leads to a requirement for certain types of phenomena that are not present in quantum mechanics. This requirement is called Bell's inequality.

Different authors subsequently derived similar inequalities, collectively termed Bell inequalities, that also assume local realism. That is, they assume that each quantum-level object has a well defined state that accounts for all its measurable properties and that distant objects do not exchange information faster than the speed of light. These well defined properties are often called hidden variables, the properties that Einstein posited when he stated his famous objection to quantum mechanics: "God does not play dice."

The inequalities concern measurements made by observers (often called Alice and Bob) on entangled pairs of particles that have interacted and then separated. Hidden variable assumptions limit the correlation of subsequent measurements of the particles. Bell discovered that under quantum mechanics this correlation limit may be violated. Quantum mechanics lacks local hidden variables associated with individual particles, and so the inequalities do not apply to it. Instead, it predicts correlation due to quantum entanglement of the particles, allowing their state to be well defined only after a measurement is made on either particle. That restriction agrees with the Heisenberg uncertainty principle, one of the most fundamental concepts in quantum mechanics.

---

[19] John S. Bell (June 28, 1928 – October 1, 1990) was a physicist who became well known as the originator of Bell's Theorem, regarded by some in the quantum physics community as one of the most important theorems of the 20th century.

In 1964, after a year's leave from CERN that he spent US, he wrote a paper [Bel64, Bel66, Bel87] entitled 'On the Einstein–Podoslky–Rosen paradox'. In this work, he showed that the carrying forward EPR's analysis [EPR35a] permits one to derive the famous inequality. What is fascinating about this inequality is that it can be derived from some quite innocent looking assumptions... ...and yet quantum mechanics itself is in conflict with it!

Per Bell's theorem, either quantum mechanics or local realism is wrong. Experiments were needed to determine which is correct, but it took many years and many improvements in technology to perform them.

Bell test experiments to date overwhelmingly show that the inequalities of Bell's theorem are violated. This provides empirical evidence against local realism and demonstrates that some of the "spooky action at a distance" suggested by the famous Einstein Podolsky Rosen (EPR) thought experiment do in fact occur. They are also taken as positive evidence in favor of QM. The principle of special relativity is saved by the no-communication theorem, which proves that the observers cannot use the inequality violations to communicate information to each other faster than the speed of light.

John Bell's papers examined both John von Neumann's 1932 proof of the incompatibility of hidden variables with QM and Albert Einstein and his colleagues' seminal 1935 paper on the subject.

Importance of the Theorem

After EPR, quantum mechanics was left in the unsatisfactory position that it was either incomplete in the sense that it failed to account for some elements of physical reality, or it violated the principle of finite propagation speed of physical effects. In the EPR thought experiment, two observers, now commonly referred to as Alice and Bob, perform independent measurements of spin on a pair of electrons, prepared at a source in a special state called a spin singlet state. It was a conclusion of EPR that once Alice measured spin in one direction (e.g. on the x axis), Bob's measurement in that direction was determined with certainty, whereas immediately before Alice's measurement, Bob's outcome was only statistically determined. Thus, either the spin in each direction is not an element of physical reality or the effects travel from Alice to Bob instantly.

In QM predictions were formulated in terms of probabilities, for example, the probability that an electron might be detected in a particular region of space, or the probability that it would have spin up or down. However, there still remained the idea that the electron had a definite position and spin, and that QM's failing was its inability to predict those values precisely. The possibility remained that some yet unknown, but more powerful theory, such as a hidden variable theory, might be able to predict these quantities exactly, while at the same time also being in complete agreement with the probabilistic answers given by QM. If a hidden variables theory were correct, the hidden variables were not described by QM and thus QM would be an incomplete theory.

The desire for a local realist theory was based on two ideas: first, that objects have a definite state that determines the values of all other measurable properties such as position and momentum and second, that (as a result of special relativity) effects of local actions such as measurements cannot travel faster than the speed of light. In the formalization of local realism used by Bell, the predictions of a theory result from the application of classical probability theory to an underlying parameter space. By a simple (but clever) argument based on classical probability he then showed that correlations between measurements are bounded in a way that is violated by QM.

Bell's theorem seemed to seal the fate of those that had local realist hopes for QM.

Bell's Thought Experiment

Bell considered a setup in which two observers, Alice and Bob, perform independent measurements on a system S prepared in some fixed state. Each observer has a detector with which to make measurements. On each trial, Alice and Bob can independently choose between various detector settings. Alice can choose a detector setting a to get a measurement A(a) and Bob can choose a detector setting b to measure B(b). After repeated trials Alice and Bob collect statistics on their measurements and correlate the results.

There are two key assumptions in Bell's analysis: (1) each measurement reveals an objective physical property of the system (2) a measurement taken by one observer has no effect on the measurement taken by the other.

In the language of probability theory, repeated measurements of system properties can be regarded as repeated sampling of random variables. One might expect measurements by Alice and Bob to be somehow correlated with each other: the random variables are assumed not to be independent, but linked in some way. Nonetheless, there is a limit to the amount of correlation one might expect to see. This is what the Bell inequality expresses.

A version of the Bell inequality appropriate for this example is given by John Clauser, Michael Horne, Abner Shimony and R. A. Holt, and is called the CHSH form,

$$\mathbf{C}[A(a),B(b)]+\mathbf{C}[A(a),B(b')]+\mathbf{C}[A(a'),B(b)]-\mathbf{C}[A(a'),B(b')]\leq 2,$$

where $\mathbf{C}$ denotes correlation.

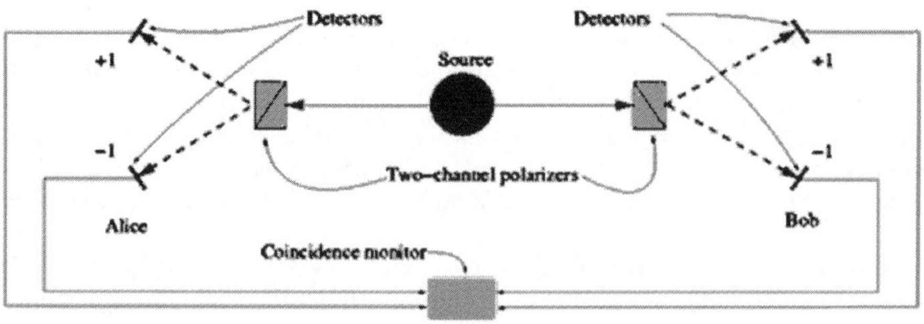

Fig. 3.6. Illustration of Bell test for spin 1/2 particles. Source produces spin singlet pair, one particle sent to Alice another to Bob. Each performs one of the two spin measurements.

Description of the Bell's Theorem

Continuing on from the situation explored in the EPR paradox, consider that again a source produces paired particles, one sent to Alice and another to Bob. When Alice and Bob measure the spin of the particles in the same axis, they will get identical results; when Bob measures at right angles to Alice's measurements they will get the same results 50% of the time, the same as a coin toss. This is expressed mathematically by saying that in the first case, their results have a correlation of 1, or perfect correlation; in the second case they have a correlation of 0; no correlation (a correlation of -1 would indicate getting opposite results the whole time).

So far, this can be explained by positing local hidden variables; each pair of particles is sent out with instructions on how to behave when measured in the $x-$axis and the $z-$axis, generated randomly. Clearly, if the source only sends out particles whose instructions are correlated for each axis, then when Alice and Bob measure on the same axis, they are bound to get identical results; but (if all four possible pairs of instructions are generated equally) when they measure on perpendicular axes they will see zero correlation.

Now consider that B rotates their apparatus (by 45 degrees, say) relative to that of Alice. Rather than calling the axes x_A, etc., henceforth we will call Alice's axes a and a', and Bob's axes b and b'. The hidden variables (supposing they exist) would have to specify a result in advance for every possible direction of measurement. It would not be enough for the particles to decide what values to take just in the direction of the apparatus at the time of leaving the source, because either Alice or Bob could rotate their apparatus by a random amount any time after the particles left the source.

Next, we define a way to 'keep score' in the experiment. Alice and Bob decide that they will record the directions they measured the particles in, and the results they got; at the end, they will tally up, scoring $+1$ for each time they got the same result and -1 for an opposite result, except that if Alice measured in a and Bob measured in b', they will score $+1$ for an opposite result and -1 for the same result. It turns out (see the mathematics below) that however the hidden variables are contrived, Alice and Bob cannot average more than 50% overall. For example, suppose that for a particular value of the hidden variables, the a and b directions are perfectly correlated, as are the a' and b' directions. Then, since a and a' are at right angles and so have zero correlation, a' and b have zero correlation, as do a and b'. The unusual 'scoring system' is designed in part to ensure this holds for all possible values of the hidden variables.

The question is now whether Alice and Bob can score higher if the particles behave as predicted by quantum mechanics. It turns out they can; if the apparatuses are rotated at 45° to each other, then the predicted score is 71%. In detail: when observations at an angle of θ are made on two entangled particles, the predicted correlation between the measurements is $\cos\theta$. In one explanation, the particles behave as if when Alice makes a measurement (in direction x, say), Bob's particle instantaneously switches to take that direction. When Bob makes a measurement, the correlation (the averaged–out value, taking +1 for the same measurement and −1 for the opposite) is equal to the length of the projection of the particle's vector onto his measurement vector; by trigonometry, $\cos\theta$. θ is 45°, and $\cos\theta$ is $\sqrt{2}/2$, for all pairs of axes except (a, b') , where they are 135° and $-\sqrt{2}/2$, but this last is taken in negative in the agreed scoring system, so the overall score is $\sqrt{2}/2$; 0.707, or 71%. If experiment shows (as it appears to) that the 71% score is attained, then hidden variable theories cannot be correct; not unless information is being transmitted between the particles faster than light, or the experimental design is flawed.

For the mathematical statement and proof of the Bell's theorem, see [Bel64, Bel66, Bel87].

3.2.2 Orchestrated Objective Reduction and Penrose Paradox

Orchestrated Objective Reduction (Orch OR) is a theory of consciousness put forth in the mid–1990s by British mathematical physicist Sir Roger Penrose and American anesthesiologist Stuart Hameroff (see [HW83, HW82]). Whereas some theories assume consciousness emerges from the brain, and among these some assume that mind emerges from complex computation at the level of synapses among brain neurons, Orch OR involves a specific form of quantum computation which underlies these neuronal synaptic activities. The proposed quantum computations occur in structures inside the brain's neurons called *microtubules*.

Now, recall from above that to make a *measurement* or *observation* of the quantum system means to concentrate on those aspects of the system that can be *simultaneously magnified* to the classical level and from which the system must then choose. Therefore, by measuring we are *disturbing the quantum system* with the *magnifying device*, which results in *decoherence*. In other words, we get classical probabilities, highly reminiscent of a standard *particle–like behavior*. The 'measurement/observation' process has caused decoherence of the wave $\psi-$function and thus led to its *collapse* to a *specific state*.

Until now, our approach to the quantum world involves two components: the one component dubbed by Penrose [Pen89, Pen94, Pen97] the $\mathbf{U}-$*part*, involves the *unitary evolution* of the system, in a *deterministic*, *continuous*, *time–symmetric* fashion, as described for example by the Schrödinger equation (3.3), i.e.,

$$\mathbf{U}: \qquad i\hbar\,\partial_t\psi(t) = \hat{H}\,\psi(t). \tag{3.78}$$

Clearly such an evolution respects *quantum coherence*, as reflected by the quantum complex superposition principle implicit in (3.78). The second component, dubbed by Penrose the $\mathbf{R}-$*part*, involves the *reduction of the state–vector* or *collapse of the wave $\psi-$function*, that enforces coexisting alternatives to resolve themselves into *actual* alternatives, one or the other,

$$\mathbf{R}: \qquad \psi = \sum_i c_{i\psi i} \longrightarrow \sum_i |c_i|^2 |\psi_i|^2, \tag{3.79}$$

where $|c_i|^2$ are classical probabilities describing *actual* alternatives. It is the **R**–part of quantum physics that introduces 'uncertainties' and 'probabilities', thus involving *discontinuous, time–asymmetric quantum jumps* and leading to gross violations of quantum coherence. It is fair to say that almost universally, when physicists talk about quantum physics, they tacitly identify it with its **U**–part only. It is the **U**–part that has absorbed all our attention for about 70 years now, and in its more advanced form, relativistic quantum field theory, has become an icon of modern physics, with spectacular success, e.g., the *Standard Model* $SU(3) \times SU(2) \times U(1)$. On the other hand, the **R**–part has been vastly and conveniently forgotten, tacitly assumed to be some mere technicality that gives us the right rules of 'measurement' or 'observation': different aspects of a quantum system are simultaneously magnified at the classical level, and between which the system must choose. This attitude has brought us finally to the *Penrose paradox*, and we need to reconsider our strategy. Actually, there is no way to deduce the **R**–part from the **U**–part, the **R**–part being a completely different procedure from the **U**–part, and effectively providing the *other 'half'* of the interpretation of quantum mechanics. It is the (**U**+**R**) parts *together* that are needed for the spectacular agreement of quantum mechanics with the observed facts. So, one is after some *new dynamics*, **N**, that provides a unified and comprehensible picture of the whole (**U** + **R**) process. In the work of [EMN92, EMN99, MN95a, MN95b, Nan95] the above approach is presented by

$$\mathbf{U} \oplus \mathbf{R} \subseteq \mathbf{N}. \tag{3.80}$$

It should be stressed that the **N**ew dynamics involved in the **N**–equation (3.80), because they have to approach at appropriate limits the **U**–equation (3.78) and the **R**–equation (3.79), i.e., almost anti–diametrical points of view, cannot be some smooth generalization of some wave dynamics. Apparently, the **N**–equation (3.80) has to contain seeds of *non–relativistic invariance* and *time asymmetry*, but in such a way that when the **R**–part or emergence of classicality is *neglected*, an approximately relativistic, time–symmetric (quantum field) theory emerges.

Orchestrated Objective Reduction

In the 1970s and 1980s Hameroff attempted to show that consciousness depends on computation within neurons in microtubules, self–assembling cylindrical polymers of the protein tubulin. Microtubules organize neuronal shape and function, e.g., forming and maintaining synapses (and help single cells like paramecium swim, find food and mates, learn and have sex without any synapses). Hameroff concluded that microtubules function as molecular–level cellular automata, and that microtubules in each neuron of the brain had the computational power of 1016 operations per second. Neuronal–level synaptic operations were regulated by these internal computations, Hameroff claimed, so attempts by artificial intelligence (AI) workers to mimic brain functions by simulating neuronal/synaptic activities would fail. Hence, as far as explaining consciousness, why we have inner experience, feelings, subjectivity, merely adding another layer of information processing within neurons in microtubules did not help.

Meanwhile Roger Penrose, famous for his work in relativity, quantum mechanics, geometry and other disciplines, had concluded for completely different reasons that AI computational approaches were inadequate to explain consciousness. In his 1989 book 'The Emperor's New Mind' Penrose used Kurt Gödel's theorem to argue that human consciousness and understanding required a factor outside algorithmic computation, and that the missing 'non–computable' factor was related to a specific type of quantum computation involving what he termed 'objective reduction' (OR), his solution to the measurement problem in quantum mechanics.

Penrose considered superposition as a separation in underlying reality at its most basic level, the Planck scale. Tying quantum superposition to general relativity, he identified superposition as spacetime curvatures in opposite directions, hence a separation in fundamental spacetime geometry. However, according to Penrose, such separations are unstable and will reduce at an objective threshold, hence avoiding multiple universes.

The threshold for Penrose OR is given by the *indeterminacy principle* $E = \hbar/t$, where E is the gravitational self–energy (i.e., the degree of spacetime separation given by the superpositioned mass), $\hbar$ is Planck's constant over 2π, and t is the time until OR occurs. Thus the larger the superposition, the faster it will undergo OR, and vice versa. Small superpositions, e.g. an electron separated from itself, if isolated from environment would require 10 million years to reach OR threshold. An isolated one kilogram object (e.g., Schrodinger's cat) would reach OR threshold in only 10–37 seconds. Penrose OR is currently being tested.

An essential feature of Penrose OR is that the choice of states when OR occurs is selected neither randomly (as are choices following measurement, or decoherence) nor completely algorithmically. Rather, states are selected by a 'non–computable' influence involving information embedded in the fundamental level of spacetime geometry at the Planck scale. Moreover, Penrose claimed that such information is Platonic, representing pure mathematical truth, aesthetic and ethical values. Plato had proposed such pure values and forms, but in an abstract realm. Penrose placed the Platonic realm in the Planck scale.

In 'The Emperor's New Mind' [Pen89], Penrose suggested (and further developed later in [Pen94, Pen97]) that consciousness required a form of *quantum computation* in the brain.

Recall that quantum computation had been suggested by Richard Feynman, Paul Benioff and David Deutsch in the 1980s. The idea is that classical information (e.g., bit states of either 1 or 0) could also be quantum superpositions of both 1 and 0 (quantum bits, or qubits). Such qubits interact and compute by nonlocal quantum entanglement, eventually being measured/observed and reducing to definite states as the solution. Quantum computations were shown to have enormous capacity if they could be constructed e.g., using qubits of ion states, electron spin, photon polarization, current in Josephson junction, quantum dots, etc. During quantum computation, qubits must be isolated from environmental interaction to avoid loss of superposition, i.e., 'decoherence'.

Penrose argued that quantum computation which terminated not by measurement, but by his version of objective reduction, constituted consciousness (allowing Platonic non–computable influences). Penrose had no definite biological qubits for such quantum computation by OR, except to suggest the possibility of superpositions of neurons both 'firing and not firing'.

Hameroff read 'The Emperor's New Mind' and suggested to Penrose that microtubules within neurons were better suited for quantum computing with OR than were superpositions of neuronal firings. The two met in the early 1990s and began to develop the theory now known as Orch OR. 'Orch' refers to orchestration, the manner in which biological conditions including synaptic–level neuronal events provide feedback to influence quantum computation with OR in microtubules [Ham87, HP96, HP93, Ham98].

The Orch OR Model

For biological qubits, Penrose and Hameroff chose conformational states of the tubulin subunit proteins in microtubules. Tubulin qubits would interact and compute by entanglement with other tubulin qubits in microtubules in the same and different neurons.

It was known that the peanut–shaped tubulin protein flexes 30 degrees, giving two different conformational shapes. Could such different states exist as superpositions, and if so, how? Penrose and Hameroff considered three possible types of tubulin superpositions: separation at the level of the entire protein, separation at the level of the atomic nuclei of the individual atoms within the proteins, and separation at the level of the protons and neutrons (nucleons) within the protein. Calculating the gravitational self–energy E of the three types, separation at the level of atomic nuclei had the highest energy, and would be the dominant factor. Penrose and Hameroff calculated E for superposition/separation of one tubulin qubit at the level of atomic nuclei in all the amino acids of the protein. They then related this to brain electrophysiology

There are claims that the best electrophysiological correlate of consciousness can be seen in the so–called *gamma–EEG waves*, synchronized oscillations in the range of 30 to 90 Hz (also known as 'coherent 40 Hz') mediated by dendritic membrane depolarizations (not axonal action potentials). This means that roughly 40 times per second (every 25 msec) neuronal dendrites depolarize synchronously throughout wide regions of brain. On the other hand, there are also claims that relaxed alpha waves (8–12 Hz) or even meditative theta waves (4–8 Hz), which have low frequency but high amplitude – are the real source of human creativity.

Using the *indeterminacy principle* $E = \hbar/t$ for OR, Penrose and Hameroff used 25 msec for t, and calculated E in terms of number of tubulins (since E was known for one tubulin). Thus they were asking: how many tubulins would be required to be in isolated superposition to reach OR threshold in 25 msec, 40 times per second, corresponding with membrane–level brain–wide effects? The answer turned out to be 2 x 1011 tubulins.

There are roughly 107 tubulins per neuron. If all tubulins in microtubules in a given neuron were involved, this would correspond with 2 x 104 (20,000) neurons. However, because dendrites are apparently more involved in consciousness than axons (which contain many microtubules), and because not all microtubules in a given dendrite are likely to be involved at any one time, an estimate of, say, 10 percent involvement gives 200,000 neurons involved in consciousness every 25 msec. These estimates (20,000 to 200,000 neurons) fit very well with others from more conventional approaches suggesting tens to hundreds of thousands of neurons are involved in consciousness at any one time.

How would microtubule quantum superpositions avoid *environmental decoherence*? Cell interiors are known to alternate between liquid phases (solution: 'sol') and quasi–solid (gelatinous: 'gel') phases due to polymerization states of the ubiquitous protein actin. In the actin-polymerized gel phase, cell water and ions are ordered on actin surfaces, so microtubules are embedded in a highly structured (i.e., non–random) medium. Tubulins are also known to have C termini 'tails', negatively charged peptide sequences extending string–like from the tubulin body into the cytoplasm, attracting positive ions and forming a plasma–like Debye layer which can also shield microtubule quantum states. Finally, tubulins in microtubules were suggested to be coherently pumped laser–like into quantum states by biochemical energy (as proposed by Herbert Frohlich).

Actin gelation cycling with 40 Hz events permits input to, and output from isolated microtubule quantum states. Thus during classical, liquid phases of actin depolymerization, inputs from membrane/synaptic inputs could 'orchestrate' microtubule states. When actin gelation occurs, quantum isolation and computation ensues until OR threshold is reached, and actin depolymerizes. The result of each OR event (in terms of patterns of tubulin states) would proceed to organize intraneuronal activities including axonal firing and synaptic modulation/learning. Each OR event (e.g., 40 per second) is proposed to be a conscious event, equivalent in philosophical terms to what philosopher Alfred North Whitehead called 'occasions of experience'.

Thus one implication of the Orch OR model is that consciousness is a sequence of discrete events, rather than a continuum. Yet conscious experience is subjectively uninterrupted, analogous to a movie appearing continuous to observers despite being a series of frames. The difference is that in

Orch OR, each conscious event is itself an intrinsic, subjective observation. Moreover the frequency of conscious events may vary, 40 Hz being an average. If someone is excited and conscious events occur more often, (e.g., at 60 Hz), then subjectively the external world seems slower, as great athletes report during peak performance. By $E = \hbar/t$, more frequent conscious events correspond with greater E, hence more tubulins/neurons per conscious events and greater intensity of experience. Thus a spectrum of conscious events may exist, similar to photons. There exists a spectrum of conscious quanta–like events ranging from longer wavelength, low intensity events (large t, low E) and shorter wavelength, higher intensity events (small t, large E).

3.2.3 Physical Science and Consciousness

Among all the human endeavors, physical science is usually considered to be the most powerful for the maximum power it endows us to manipulate the nature through an understanding of our position in it. This understanding is gained when a set of careful observations based on tangible perceptions, acquired by sensory organs and/or their extensions, is submitted to the logical analysis of human intellect as well as to the intuitive power of imagination to yield the abstract fundamental laws of nature that are not self–evident at the gross level of phenomenal existence. There exists a unity in nature at the level of laws that corresponds to the manifest diversity at the level of phenomena [Sam99].

Can consciousness be understood in this sense by an appropriate use of the methodology of science? The most difficult problem related to consciousness is perhaps, 'how to define it?'. Consciousness has remained a unitary subjective experience, its various 'components' being reflective (the recognition by the thinking subject of its own actions and mental states), perceptual (the state or faculty of being mentally aware of external environment) and a free will (volition). But how these components are integrated to provide the unique experience called 'consciousness', familiar to all of us, remains a mystery. Does it lie at the level of 'perceptions' or at the level of 'laws'? Can it be reduced to some basic 'substance' or 'phenomenon'? Can it be manipulated in a controlled way? Is there a need for a change of either the methodology or the paradigm of science to answer the above questions?

Can Consciousness be reduced to its elements?

Now, most of the successes of science over the past five hundred years or so can be attributed to the great emphasis it lays on the 'reductionist paradigm'. Following this approach, can consciousness be reduced either to 'substance' or 'phenomena' in the sense that by understanding which one can understand consciousness?

Physical Substratum

The attempts to reduce consciousness to a physical basis have been made in the following ways by trying to understand the mechanism and functioning of the human brain in various different contexts [Sam99].

- **Physics**
  The basic substratum of physical reality is the 'state' of the system and the whole job of physics can be put into a single question: *Given the initial state, how to predict its evolution at a later time?* In classical world, the state and its evolution can be reduced to events and their spatio–temporal correlations. Consciousness has no direct role to play in this process of reduction, although it is responsible to find an 'objective meaning' in such a reduction.
  But the situation is quite different in the quantum world as all relevant physical information about a system is contained in its wave $\psi-$function (or equivalently in its state vector), which is not physical in the sense of being directly measurable. Consciousness plays no role in the *deterministic and unitary Schrödinger evolution* (i.e., the $\mathbf{U}-$process of Penrose [Pen89]) that the 'un–physical' wave $\psi-$function undergoes.
  To extract any physical information from the wave $\psi-$function one has to use the *Born–Dirac rule* and thus probability enters in a new way into the quantum mechanical description despite the strictly deterministic nature of evolution of the wave $\psi-$function. The measurement process forces the system to choose an 'actuality' from all 'possibilities' and thus leads to a non–unitary collapse of the general wave $\psi-$function to an eigenstate (i.e., the $\mathbf{R}-$process of Penrose [Pen89]) of the concerned observable. The dynamics of this $\mathbf{R}-$process is not known and it is here some authors like Wigner have brought in the consciousness of the observer to cause the collapse of the wave $\psi-$function. But instead of explaining the consciousness, this approach uses consciousness for the sake of Quantum Mechanics which needs the $\mathbf{R}-$process along with the $\mathbf{U}-$process to yield all its spectacular successes.
  The $\mathbf{R}-$process is necessarily nonlocal and is governed by an irreducible element of chance, which means that the theory is not naturalistic: the dynamics is controlled in part by something that is not a part of the physical universe. Stapp [Sta95] has given a quantum–mechanical model of the brain dynamics in which this quantum selection process is a causal process governed not by pure chance but rather by a mathematically specified nonlocal physical process identifiable as the conscious process. It was reported that attempts have been made to explain consciousness by relating it to the 'quantum events', but any such attempt is bound to be futile as the concept of 'quantum event' in itself is ill–defined.
  Keeping in view the fundamental role that the quantum vacuum plays in formulating the quantum field theories of all four known basic interactions

of nature spreading over a period from the Big–Bang to the present, it has been suggested that if at all consciousness be reduced to anything 'fundamental' that should be the 'quantum vacuum' in itself. But in such an approach the following questions arise:

- 1) If consciousness has its origin in the quantum vacuum that gives rise to all fundamental particles as well as the force fields, then why is it that only living things possess consciousness?
- 2) What is the relation between the quantum vacuum that gives rise to consciousness and the space–time continuum that confines all our perceptions through which consciousness manifests itself?
- 3) Should one attribute consciousness only to systems consisting of 'real' particles or also to systems containing 'virtual' particles? Despite these questions, the idea of tracing the origin of 'consciousness' to 'substantial nothingness' appears quite promising because the properties of 'quantum vacuum' may ultimately lead us to an understanding of the dynamics of the **R**–process and thus to a physical comprehension of consciousness.
  One of the properties that distinguishes living systems from the non–living systems is their ability of self–organization and complexity. Since life is a necessary condition for possessing consciousness, can one attribute consciousness to a 'degree of complexity' in the sense that various degrees of consciousness can be caused by different levels of complexity? Can one give a suitable quantitative definition of consciousness in terms of 'entropy' that describes the 'degree of self–organization or complexity' of a system? What is the role of non-linearity and non–equilibrium thermodynamics in such a definition of consciousness? In this holistic view of consciousness what is the role played by the phenomenon of quantum nonlocality, first envisaged in EPR paper and subsequently confirmed experimentally [AGR82]? What is the role of irreversibility and dissipation in this holistic view?
- **Neurobiology**
  On the basis of the vast amount of information available on the structure and the modes of communication (neurotransmitters, neuromodulators, neurohormones) of the neuron, neuroscience has empirically found [Sam99] the neural basis of several attributes of consciousness. With the help of modern scanning techniques and by direct manipulations of the brain, neurobiologists have found out that various human activities (both physical and mental) and perceptions can be mapped into almost unique regions of the brain. Awareness, being intrinsic to neural activity, arises in higher level processing centers and requires integration of activity over time at the neuronal level. But there exists no particular region that can be attributed to have given rise to consciousness. Consciousness appears to be a collective phenomena where the 'whole' is much more than the sum of parts. Is each neuron having the 'whole of consciousness' within it, although it does work towards a particular attribute of consciousness at a time?
  Can this paradigm of finding neural correlates of the attributes of consciousness be fruitful in demystifying consciousness? Certainly not. As

it was aptly concluded [Sam99] the currently prevalent reductionist approaches are unlikely to reveal the basis of such holistic phenomenon as consciousness. There have been holistic attempts [Ham87, Pen89] to understand consciousness in terms of collective quantum effects arising in cytoskeletons and microtubules; minute substructures lying deep within the brain's neurons. The effect of general anaesthetics like chloroform ($CHCl_3$), isofluorane ($CHF_2OCHClCF_3$) etc. in switching off the consciousness, not only in higher animals such as mammals or birds but also in paramecium, amoeba, or even green slime mould has been advocated [HW83] to be providing a direct evidence that the phenomenon of consciousness is related to the action of the cytoskeleton and to microtubules. But all the implications of 'quantum coherence' regarding consciousness in such approach can only be unfolded after we achieve a better understanding of 'quantum reality', which still lies ahead of the present–day physics.

- **AI and CI**
  Can machines be intelligent? Within the restricted definition of 'artificial intelligence', the neural network approach has been the most promising one. But the possibility of realising a machine capable of artificial intelligence based on this approach is constrained at present [Sam99] by the limitations of 'silicon technology' for integrating the desired astronomical number of 'neuron–equivalents' into a reasonable compact space. Even though we might achieve such a feat in the foreseeable future by using chemical memories, it is not quite clear whether such artificially intelligent machines can be capable of 'artificial consciousness'. Because one lacks at present a suitable working definition of 'consciousness' within the framework of studies involving artificial intelligence.
  Invoking *Gödel's incompleteness theorem*, Penrose has argued [Pen89] that the technology of electronic computer–controlled robots will not provide a way to the artificial construction of an actually intelligent machine–in the sense of a machine that 'understands' what it is doing and can act upon that understanding. He maintains that human understanding (hence consciousness) lies beyond formal arguments and beyond computability i.e., in the Turing–machine–accessible sense.
  Assuming the inherent ability of quantum mechanics to incorporate consciousness, can one expect any improvement in the above situation by considering 'computation' to be a physical process that is governed by the rules of quantum mechanics rather than that of classical physics? In 'Quantum computation' [DF85] the classical notion of a Turing machine is extended to a corresponding quantum one that takes into account the quantum superposition principle. In 'standard' quantum computation, the usual rules of quantum theory are adopted, in which the system evolves according to the **U**−process for essentially the entire operation, but the **R**−process becomes relevant mainly only at the end of the

operation, when the system is 'measured' in order to ascertain either the termination or the result of the computation.
Although the superiority of the quantum computation over classical computation in the sense of complexity theory have been shown [Deu92], Penrose insists that it is still a 'computational' process since **U**−process is a computable operation and **R**−process is purely probabilistic procedure. What can be achieved in principle by a quantum computer could also be achieved, in principle, by a suitable Turing–machine–with–randomizer. Thus he concludes that even a quantum computer would not be able to perform the operations required for human conscious understanding. But we think that such a view is limited because 'computation' as a process need not be confined to a Turing–machine–accessible sense and in such situations one has to explore the power of quantum computation in understanding consciousness.

We conclude from the above discussions that the basic physical substrata to which consciousness may be reduced are 'neuron', 'event' and 'bit' at the classical level, whereas at the quantum level they are 'microtubule', 'wave ψ−function' and 'qubit'; depending on whether the studies are done in neurobiology, physics and computer science respectively [Sam99]. Can there be a common platform for these trio of substrata?

We believe the answer to be in affirmative and the first hint regarding this comes from John Wheeler's [Whe89] remarkable idea: "Every particle, every field of force, even the spacetime continuum itself, derives its function, its meaning, its very existence entirely, even if in some contexts indirectly, from the apparatus, elicited answers to yes or no questions, binary choices, bits". This view of the world refers not to an object, but to a vision of a world derived from pure logic and mathematics in the sense that an immaterial source and explanation lies at the bottom of every item of the physical world. In a recent report [Wil99] the remarkable extent of embodiment of this vision in modern physics has been discussed along with the possible difficulties faced by such a scheme. But can this scheme explain consciousness by reducing it to bits? Perhaps not unless it undergoes some modification.

Because consciousness involves an awareness of an endless mosaic of qualitatively different things, such as the color of a rose, the fragrance of a perfume, the music of a piano, the tactile sense of objects, the power of abstraction, the intuitive feeling for time and space, emotional states like love and hate, the ability to put oneself in other's position, the ability to wonder, the power to wonder at one's wondering etc. It is almost impossible to reduce them all to the 0–or–1 sharpness of the definition of 'bits'. A major part of human experience and consciousness is fuzzy and hence can not be reduced to yes or no type situations. Hence we believe that 'bit' has to be modified to incorporate this fuzzyness of the world. Perhaps the quantum superposition inherent to a 'qubit' can help. Can one then reduce the consciousness to a consistent theory of 'quantum information' based on qubits? Quite unlikely, till our knowledge of 'quantum reality' and the 'emergence of classicality from it' becomes more clear.

The major hurdles to be cleared are:

(1) Observer or Participator? In such equipment–evoked, quantum–information–theoretic approach, the inseparability of the observer from the observed will bring in the quantum measurement problem either in the form of dynamics of the **R**−process or in the emergence of classicality of the world from a quantum substratum. We first need the solutions to these long–standing problems before attempting to reduce the 'fuzzy' world of consciousness to 'qubits'.

(2) Communication? Even if we get the solutions to the above problems that enable us to reduce the 'attributes of consciousness' to 'qubits', still then the 'dynamics of the process that gives rise to consciousness' will be beyond 'quantum information' as it will require a suitable definition of 'communication' in the sense expressed by [Fol75]: "Meaning is the joint product of all evidence that is available to those who communicate." Consciousness helps us to find a 'meaning' or 'understanding' and will depend upon 'communication'. Although all 'evidence' can be reduced to qubits, 'communication' as an exchange of qubits has to be well–defined. Why do we say that a stone or a tree is unconscious? Is it because we do not know how to 'communicate' with them? Can one define 'communication' in physical terms beyond any verbal or non–verbal language? Where does one look for a suitable definition of 'communication'? Maybe one has to define 'communication' at the 'substantial nothingness' level of quantum vacuum.

(3) Time's Arrow? How important is the role of memory in 'possessing consciousness'? Would our consciousness be altered if the world we experience were reversible with respect to time? Can our consciousness ever find out why it is not possible to influence the past?

Hence we conclude that although consciousness may be beyond 'computability', it is not beyond 'quantum communicability' once a suitable definition for 'communication' is found that exploits the quantum superposition principle to incorporate the fuzzyness of our experience. Few questions arise:

(1) How to modify the qubit?

(2) Can a suitable definition of 'communication', based on immaterial entity like 'qubit' or 'modified qubit', take care of non–physical experience like dream or thoughts? We assume, being optimistic, that a suitable modification of 'qubit' is possible that will surpass the hurdles of communicability, dynamics of **R**−process and irreversibility. For the lack of a better word we will henceforth call such a modified qubit as 'Basic Entity' (BE) [Sam99].

Non–Physical Substratum

Unlike our sensory perceptions related to physical 'substance' and 'phenomena' there exists a plethora of human experiences like dreams, thoughts and lack of any experience during sleep which are believed to be non–physical in the sense that they cannot be reduced to anything basic within the confinement of space–time and causality. For example one cannot ascribe either spatiality or causality to human thoughts, dreams etc. Does one need a framework that transcends spatio–temporality to incorporate such non–physical 'events'? Or can one explain them by using BE? The following views can be taken depending on one's belief [Sam99]:

- Modified BE, or $M(BE)$
  What could be the basic substratum of these non–physical entities? Could they be understood in terms of any suitably modified physical substratum? At the classical level one might think of reducing them to 'events' which, unlike the physical events, do not have any reference to spatiality. Attempts have been made [Sam99] to understand the non–physical entities like thoughts and dreams in terms of temporal events and correlation between them. Although such an approach may yield the kinematics of these non–physical entities, it is not clear how their dynamics i.e., evolution etc., can be understood in terms of temporal component alone without any external spatial input, when in the first place they have arose from perceptions that are meaningful only in the context of spatio–temporality?. Secondly, it is not clear why the 'mental events' constructed after dropping the spatiality should require new set of laws that are different from the usual physical laws.
  At the quantum level one might try to have a suitable modification of the wave ψ–function to incorporate these non–physical entities. One may make the wave ψ–function depend on extra parameters [Sam99], either physical or non–physical, to give it the extra degrees–of–freedom to mathematically include more information. But such a wave ψ–function bound to have severe problems at the level of interpretation. For example, if one includes an extra parameter called 'meditation' as a new degree of freedom apart from the usual ones, then how will one interpret squared modulus of the wave ψ–function? It will be certainly too crude to extend the Born rule to conclude that the squared modulus in this case will give the probability of finding a particle having certain meditation value. Hence this kind of modification will not be of much help except for the apparent satisfaction of being able to write an eigenvalue equation for dreams or emotions. This approach is certainly not capable of telling how the wave ψ–function is related to consciousness, let alone a mathematical equation for the evolution of consciousness.
  If one accepts consciousness as a phenomenon that arises out of execution of processes then any suggested [Sam01] new physical basis can be shown to be redundant. As we have concluded earlier, all such possible processes and their execution can be reduced to BE and spatio–temporal correlations among BE using a suitable definition of communication.

Hence to incorporate non–physical entities as some kind of information one has to modify the BE in a subtle way. Schematically $M(BE) = BE \otimes X$, where $\otimes$ stands for a yet unknown operation and X stands for fundamental substratum of non–physical information. X has to be different from BE; otherwise it could be reducible to BE and then there will be no spatio–temporal distinction between physical and non–physical information. But, how to find out what is X? Is it evident that the laws for $M(BE)$ will be different from that for BE?

- Give up BE
  One could believe that it is the 'Qualia' that constitutes consciousness and hence consciousness has to be understood at a phenomenological level without disecting it into BE or $M(BE)$. One would note that consciousness mainly consists of three phenomenological processes that can be roughly put as retentive, reflective and creative. But keeping the tremendous progress of our physical sciences and their utility to neurosciences in view, it is not unreasonable to expect that all these three phenomenological processes, involving both human as well as animal can be understood one day in terms of $M(BE)$.
- Platonic BE
  It has been suggested [Sam99] that consciousness could be like mathematics in the sense that although it is needed to comprehend the physical reality, in itself it is not 'real'.
  The 'reality' of mathematics is a controversial issue that brings in the old debate between the realists and the constructivists whether a mathematical truth is 'a discovery' or 'an invention' of the human mind? Should one consider the physical laws based on mathematical truth as real or not?. The realist's stand of attributing a Platonic existence to the mathematical truth is a matter of pure faith unless one tries to get the guidance from the knowledge of the physical world. It is doubtful whether our knowledge of physical sciences provides support for the realist's view if one considers the challenge to 'realism' in physical sciences by the quantum world–view, which has been substantiated in recent past by experiments [AGR82] that violate Bell's inequalities.
  Even if one accepts the Platonic world of mathematical forms, this no way makes consciousness non-existent or unreal. Rather the very fact that truth of such a platonic world of mathematics yields to the human understanding as much as that of a physical world makes consciousness all the more profound in its existence.

Can Consciousness be manipulated?

Can consciousness be manipulated in a controlled manner? Experience tells us how difficult it is to control the thoughts and how improbable it is to control the dreams. We discuss below few methods prescribed by western psychoanalysis and oriental philosophies regarding the manipulation of consciousness [Sam99]. Is there a lesson for modern science to learn from these methods?

Self

The subject of 'self' is usually considered to belong to an 'internal space' in contrast to the external space where we deal with others. We will consider the following two cases here:

- Auto–suggestions
  There have been evidences that by auto–suggestions one can control one's feelings like pain and pleasure. Can one cure oneself of diseases of physical origin by auto–suggestions? This requires further investigations.
- Yoga and other oriental methods
  The eight–fold Yoga of Patanjali is perhaps the most ancient method prescribed [Iye81] to control one's thought and to direct it in a controlled manner. But it requires certain control over body and emotions before one aspires to gain control over mind. In particular it lays great stress on 'breath control' (pranayama) as a means to relax the body and to still the mind. In its later stages it provides systematic methods to acquire concentration and to prolong concentration on an object or a thought.
  After this attainment one can reach a stage where one's awareness of self and the surrounding is at its best. Then in its last stage, Yoga prescribes one's acute awareness to be decontextualized [Sam99] from all perceptions limited by spatio–temporality and thus to reach a pinnacle called (samadhi) where one attains an understanding of everything and has no doubts. In this sense the Yogic philosophy believes that pure consciousness transcends all perceptions and awareness. It is difficult to understand this on the basis of day to day experience. Why does one need to sharpen one's awareness to its extreme if one is finally going to abandon its use? How does abandoning one's sharpened awareness help in attaining a realisation that transcends spatio–temporality? Can any one realise anything that is beyond the space, time and causality? What is the purpose of such a consciousness that lies beyond the confinement of space and time?

Non–Self

The Non–Self belongs to an external world consisting of others, both living and non–living. In the following we discuss whether one can direct one's consciousness towards others such that one can affect their behavior [Sam99, Sam01].

- Hypnosis, ESP, and Paranormal
  It is a well–known fact that it is possible to hypnotize a person and then to make contact with his/her subconscious mind. Where does this subconscious lie? What is its relation to the conscious mind? The efficacy of the

method of hypnosis in curing people of deep–rooted psychological problems tells us that we are yet to understand the dynamics of the human brain fully.
The field of Para–Psychology deals with 'phenomena' like Extra Sensory Perception (ESP) and telepathy etc. where one can direct one's consciousness to gain insight into future or to influence others mind. It is not possible to explain these on the basis of the known laws of the world. It has been claimed that under hypnosis a subject could vividly recollect incidents from the previous lives including near–death and death experiences which is independent of spatio–temporality. Then, it is not clear, why most of these experiences are related to past? If these phenomena are truly independent of space and time, then studies should be made to find out if anybody under hypnosis can predict his/her own death, an event that can be easily verifiable in due course of time, unlike the recollections of past–life [Sam99].
Can mind influence matter belonging to outside of the body? The studies dubbed as Psycho–Kinesis (PK) have been conducted to investigate the 'suspect' interaction of the human mind with various material objects such as cards, dice, simple pendulum etc. An excellent historical overview of such studies leading upto the modern era is available as a review paper, titled "The Persistent Paradox of Psychic Phenomena: An Engineering Perspective," by Robert Jahn of Princeton University, published in Proc. IEEE (Feb. 1982).
The Princeton Engineering Anomalies Research (PEAR) programme of the Department of Applied Sciences and Engineering, Princeton University, has recently developed and patented a 'Field REG' (Field Random Event Generator) device which is basically a portable notebook computer with a built–in truly random number generator (based on a microelectronic device such as a shot noise resistor or a solid–state diode) and requisite software for on-line data processing and display, specifically tailored for conducting 'mind-machine interaction' studies.
After performing large number of systematic experiments over the last two decades, the PEAR group has reported [Sam99] the existence of such a consciousness related mind–machine interaction in the case of 'truly random devices'. They attribute it to a 'Consciousness Field Effect'. They have also reported that deterministic random number sequences such as those generated by mathematical algorithm or pseudo–random generators do not show any consciousness related anomalous behavior. Another curious finding is that 'intense emotional resonance' generates the effect whereas 'intense intellectual resonance' does not. It is also not clear what is the strength of the 'consciousness field' in comparison to all the four known basic force fields of nature.

One should not reject outright any phenomenon that cannot be explained by the known basic laws of nature. Because each such phenomenon holds the key to extend the boundary of our knowledge further. But before accepting these effects one should filter them through the rigors of scientific methodology. In particular, the following questions can be asked [Sam99]:

- Why are these events rare and not repeatable?
- How does one make sure that these effects are not manifestations of yet unknown facets of the known forces?
- Why is it necessary to have truly random processes? How does one make sure that these are not merely statistical artifacts?

If the above effects survive the scrutiny of the above questions (or similar ones) then they will open up the doors to a new world not yet known to science. In such a case how does one accommodate them within the existing framework of scientific methods? If these effects are confirmed beyond doubt, then one has to explore the possibility that at the fundamental level of nature, the laws are either different from the known physical laws or there is a need to complement the known physical laws with a set of non–physical laws. In such a situation, these 'suspect' phenomena might provide us with the valuable clue for modifying BE to get $M(BE)$ that is the basis of everything including both physical and mental.

Is there a need for a change of paradigm?

Although reductionist approach can provide us with valuable clues regarding the attributes of consciousness, it is the holistic approach that can only explain consciousness. But the dualism of Descartes that treats physical and mental processes in a mutually exclusive manner will not suffice for understanding consciousness unless it makes an appropriate use of complementarity for mental and physical events which is analogous to the complementarity evident in the quantum world.

Where does the brain end and the mind begin? Brain is the physical means to acquire and to retain the information for the mind to process them to find a 'meaning' or a 'structure' which we call 'understanding' that is attributed to consciousness. Whereas attributes of consciousness can be reduced to BE (or to $M(BE)$), the holistic process of consciousness can only be understood in terms of 'quantum communication', where 'communication' has an appropriate meaning. Maybe one has to look for such a suitable definition of communication at the level of 'quantum vacuum' [Sam99].

3.2.4 Quantum Brain

Biochemistry of Microtubules

Recent developments/efforts to understand aspects of the brain function at the *sub–neural* level are discussed in [Nan95]. Microtubules (MTs), protein polymers constructing the cytoskeleton of a neuron, participate in a wide variety of dynamical processes in the cell. Of special interest for this subsection is the MTs participation in bioinformation processes such as *learning* and *memory*, by possessing a well–known binary error–correcting code $[K_1(13, 2^6, 5)]$ with 64 words. In fact, MTs and DNA/RNA are *unique* cell structures that possess a code system. It seems that the MTs' code system is strongly related to a kind of *mental code* in the following sense. The MTs' periodic paracrystalline structure make them able to support a *superposition* of coherent quantum states, as it has been recently conjectured by Hameroff and Penrose [HP96], representing an *external* or *mental order*, for sufficient time needed for *efficient quantum computing.*

Living organisms are collective assemblies of cells which contain collective assemblies of organized material, including membranes, organelles, nuclei, and the *cytoplasm*, the bulk interior medium of living cells. Dynamic rearrangements of the cytoplasm within *eucaryotic cells*, the cells of all animals and almost all plants on Earth, account for their changing shape, movement, etc. This extremely important cytoplasmic structural and dynamical organization is due to the presence of networks of inteconnected protein polymers, which are referred to as the *cytoskeleton* due to their bone–like structure [HP96, Dus84]. The cytoskeleton consists of MT's, actin microfilaments, intermediate filaments and an *organizing complex*, the *centrosome* with its chief component the *centriole*, built from two bundles of microtubules in a separated **T** shape. Parallel–arrayed MTs are interconnected by cross–bridging proteins (*MT–Associated Proteins*: MAPs) to other MTs, organelle filaments and membranes to form *dynamic networks* [HP96, Dus84]. MAPs may be contractile, structural, or enzymatic. A very important role is played by contractile MAPs, like dynein and kinesin, through their participation in cell movements as well as in intra-neural, or axoplasmic transport which moves material and thus is of fundamental importance for the *maintenance* and *regulation* of *synapses* (see, e.g., [Ecc64]). The structural bridges formed by MAPs stabilize MTs and prevent their disassembly. The MT–MAP 'complexes' or *cytoskeletal networks* determine the cell architecture and dynamic functions, such a *mitosis*, or *cell division*, *growth*, *differentiation*, *movement*, and for us here the very crucial, *synapse formation and function*, all essential to the living state. It is usually said that *microtubules* are ubiquitous through the entire biology [HP96, Dus84].

MTs are hollow cylinders comprised of an exterior surface of cross–section diameter 25 nm ($1\ nm = 10^{-9}$ meters) with 13 arrays (protofilaments) of protein dimers called tubulines [Dus84]. The interior of the cylinder, of cross–section diameter 14 nm, contains *ordered water* molecules, which implies the existence of an electric dipole moment and an electric field. The arrangement of the dimers is such that, if one ignores their size, they resemble triangular lattices on the MT surface. Each dimer consists of two hydrophobic protein pockets, and has an unpaired electron. There are two possible positions of the electron, called α and β *conformations.* When the electron is in the

$\beta-$conformation there is a 29° distortion of the electric dipole moment as compared to the α conformation.

In standard models for the simulation of the MT dynamics [STZ93, SZT98], the 'physical' DOF – relevant for the description of the energy transfer – is the projection of the electric dipole moment on the longitudinal symmetry axis ($x-$axis) of the MT cylinder. The 29° distortion of the $\beta-$conformation leads to a displacement u_n along the $x-$axis, which is thus the relevant physical DOF.

There has been speculation for quite some time that MTs are involved in information processing: it has been shown that the particular geometrical arrangement (packing) of the tubulin protofilaments obeys an error–correcting mathematical code known as the $K_2(13,2^6,5)-$code [KHS93]. Error correcting codes are also used in classical computers to protect against errors while in quantum computers special error correcting algorithms are used to protect against errors by preserving quantum coherence among qubits.

Information processing occurs via interactions among the MT proto–filament chains. The system may be considered as similar to a model of *interacting Ising chains* on a triangular lattice, the latter being defined on the plane stemming from filleting open and flattening the cylindrical surface of MT. Classically, the various dimers can occur in either α or β conformations. Each dimer is influenced by the neighboring dimers resulting in the possibility of a transition. This is the basis for classical information processing, which constitutes the picture of a (classical) cellular automaton.

Kink Soliton Model of MT–Dynamics

The *quantum nature* of an MT network results from the *assumption* that each dimer finds itself in a *superposition* of α and β conformations. Viewed as a *two–state quantum mechanical system*, the MT tubulin dimers couple to conformational changes with $10^{-9}-10^{-11}$sec transitions, corresponding to an angular frequency $\omega \sim \mathcal{O}(10^{10}) - \mathcal{O}(10^{12})$ Hz [Nan95].

The *quantum computer* character of the MT network [Pen89] results from the assumption that each dimer finds itself in a superposition of α and β conformations [Ham87]. There is a macroscopic coherent state among the various chains, which lasts for $\mathcal{O}(1\,\text{sec})$ and constitutes the 'preconscious' state [Nan95]. The interaction of the chains with (non–critical stringy) quantum gravity, then, induces self–collapse of the wave function of the coherent MT network, resulting in quantum computation.

In [EMN92, EMN99, MN95a, MN95b, Nan95] the authors assumed that the collapse occurs mainly due to the interaction of each chain with quantum gravity, the interaction from neighboring chains being taken into account by including mean–field interaction terms in the dynamics of the displacement field of each chain. This amounts to a modification of the effective potential by anharmonic oscillator terms. Thus, the effective system under study is 2D, possessing one space and one time coordinate.

Let u_n be the displacement field of the nth dimer in a MT chain. The continuous approximation proves sufficient for the study of phenomena associated with energy transfer in biological cells, and this implies that one can make the replacement

$$u_n \to u(x,t), \tag{3.81}$$

with x a spatial coordinate along the longitudinal symmetry axis of the MT. There is a time variable t due to fluctuations of the displacements $u(x)$ as a result of the dipole oscillations in the dimers.

The effects of the neighboring dimers (including neighboring chains) can be phenomenologically accounted for by an effective potential $V(u)$. In the kink–soliton model[20] of ref. [STZ93, SZT98] a double–well potential was used, leading to a classical kink solution for the $u(x,t)$ field. More complicated interactions are allowed in the picture of Ellis *et al.*, where more generic polynomial potentials have been considered.

The effects of the surrounding water molecules can be summarized by a *viscous force* term that damps out the dimer oscillations,

$$F = -\gamma \partial_t u, \tag{3.82}$$

with γ determined phenomenologically at this stage. This friction should be viewed as an environmental effect, which however does not lead to energy dissipation, as a result of the non–trivial solitonic structure of the ground–state and the non–zero constant force due to the electric field. This is a well known result, directly relevant to energy transfer in biological systems.

In mathematical terms the effective equation of motion for the relevant field DOF $u(x,t)$ reads:

$$u''(\xi) + \rho u'(\xi) = P(u), \tag{3.83}$$

where $\xi = x - vt$, $u'(\xi) = du/d\xi$, v is the velocity of the soliton, $\rho \propto \gamma$ [STZ93, SZT98], and $P(u)$ is a polynomial in u, of a certain degree, stemming from the variations of the potential $V(u)$ describing interactions among the MT chains. In the mathematical literature there has been a classification of solutions of equations of this form. For certain forms of the potential the solutions include *kink solitons* that may be responsible for dissipation–free energy transfer in biological cells:

$$u(x,t) \sim c_1 \left(\tanh[c_2(x - vt)] + c_3\right), \tag{3.84}$$

where c_1, c_2, c_3 are constants depending on the parameters of the dimer lattice model. For the form of the potential assumed in the model of [STZ93, SZT98] there are solitons of the form $u(x,t) = c_1' + \frac{c_2' - c_1'}{1 + e^{c_3'(c_2' - c_1')(x - vt)}}$, where again c_i', $i = 1, \ldots 3$ are appropriate constants.

[20] Recall that kinks are solitary (non–dispersive) waves arising in various 1D (bio)physical systems.

A *semiclassical quantization* of such solitonic states has been considered by Ellis *et al.*. The result of such a quantization yields a *modified soliton equation* for the (quantum corrected) field $u_q(x,t)$ [TF91]

$$\partial_t^2 u_q(x,t) - \partial_x^2 u_q(x,t) + \mathcal{M}^{(1)}[u_q(x,t)] = 0, \tag{3.85}$$

with the notation

$$M^{(n)} = \mathrm{e}^{\frac{1}{2}(G(x,y,t)-G_0(x,y))\frac{\partial^2}{\partial z^2}} U^{(n)}(z)|_{z=u_q(x,t)}, \qquad U^{(n)} \equiv d^n U/dz^n.$$

The quantity U denotes the potential of the original soliton Hamiltonian, and $G(x,y,t)$ is a bilocal field that describes quantum corrections due to the modified boson field around the soliton. The quantities $M^{(n)}$ carry information about the quantum corrections. For the kink soliton (3.84) the quantum corrections (3.85) have been calculated explicitly in [TF91], thereby providing us with a concrete example of a large–scale quantum coherent state.

A typical propagation velocity of the kink solitons (e.g., in the model of [STZ93, SZT98]) is $v \sim 2$ m/sec, although, models with $v \sim 20$ m/sec have also been considered. This implies that, for moderately long microtubules of length $L \sim 10^{-6}$ m, such kinks transport energy without dissipation in

$$t_F \sim 5 \times 10^{-7} \text{ s}. \tag{3.86}$$

Such time scales are comparable to, or smaller in magnitude than, the decoherence time scale of the above–described coherent (solitonic) states $u_q(x,t)$. This implies the possibility that fundamental quantum mechanical phenomena may then be responsible for frictionless energy (and signal) transfer across microtubular arrangements in the cell [Nan95].

Open Liouville Neurodynamics and Self–Similarity

Recall that neurodynamics has its physical behavior both on the *macroscopic*, classical, *inter–neuronal level*, and on the *microscopic*, quantum, *intra–neuronal level*. On the *macroscopic* level, various models of neural networks (NNs, for short) have been proposed as goal–oriented models of the specific neural functions, like for instance, function–approximation, pattern–recognition, classification, or control (see, e.g., [Hay94]). In the physically–based, Hopfield–type models of NNs [Hop82), Hop84] the information is stored as a content–addressable memory in which synaptic strengths are modified after the Hebbian rule (see [Heb49]. Its retrieval is made when the network with the symmetric couplings works as the point–attractor with the fixed points. Analysis of both *activation* and *learning dynamics* of Hopfield–Hebbian NNs using the techniques of statistical mechanics [DHS91], gives us with the most important information of storage capacity, role of noise and recall performance.

On the other hand, on the general *microscopic* intra–cellular level, energy transfer across the cells, without dissipation, had been first conjectured to occur in biological matter by [FK83]. The phenomenon conjectured by them was based on their 1D superconductivity model: in 1D electron systems with holes, the formation of *solitonic structures* due to electron–hole pairing results in the transfer of electric current without dissipation. In a similar manner, Frölich and Kremer conjectured that energy in biological matter could be transferred without dissipation, if appropriate solitonic structures are formed inside the cells. This idea has lead theorists to construct various models for the energy transfer across the cell, based on the formation of *kink* classical solutions (see [STZ93, SZT98].

The interior of living cells is structurally and dynamically organized by *cytoskeletons*, i.e., networks of protein polymers. Of these structures, *microtubules* (MTs, for short) appear to be the most fundamental (see [Dus84]). Their dynamics has been studied by a number of authors in connection with the mechanism responsible for dissipation–free energy transfer. Hameroff and Penrose [Ham87] have conjectured another fundamental role for the MTs, namely being responsible for *quantum computations* in the human neurons. [Pen89, Pen94, Pen97] further argued that the latter is associated with certain aspects of quantum theory that are believed to occur in the cytoskeleton MTs, in particular quantum superposition and subsequent collapse of the wave function of coherent MT networks. These ideas have been elaborated by [MN95a, MN95b] and [Nan95], based on the quantum–gravity EMN–language of [EMN92, EMN99] where MTs have been physically modelled as non-critical (SUSY) bosonic strings. It has been suggested that the neural MTs are the microsites for the emergence of stable, macroscopic quantum coherent states, identifiable with the *preconscious states*; stringy–quantum space–time effects trigger an organized collapse of the coherent states down to a specific or *conscious state*. More recently, [TVP99] have presented the evidence for biological self–organization and pattern formation during embryogenesis.

Now, we have two space–time biophysical scales of neurodynamics. Naturally the question arises: are these two scales somehow inter-related, is there a space–time self–similarity between them?

The purpose of this subsection is to prove the formal positive answer to the self–similarity question. We try to describe neurodynamics on both physical levels by the *unique form* of a single equation, namely *open Liouville equation*: NN–dynamics using its classical form, and MT–dynamics using its quantum form in the Heisenberg picture. If this formulation is consistent, that would prove the *existence* of the *formal neurobiological space–time self–similarity.*

Hamiltonian Framework

Suppose that on the macroscopic NN–level we have a conservative Hamiltonian system acting in a $2ND$ symplectic phase–space $T^*Q = \{q^i(t), p_i(t)\}$,

$(i = 1 \dots N)$ (which is the cotangent bundle of the NN–configuration manifold $Q = \{q^i\}$), with a Hamiltonian function $H = H(q^i, p_i, t) : T^*Q \times \mathbb{R} \to \mathbb{R}$ and an inverse metric tensor g^{ij}. The conservative dynamics is defined by classical Hamiltonian canonical equations [II06b]

$$\dot{x}^i = g^{ij} p_j / m, \qquad \dot{p}_i = F_i(x). \tag{3.87}$$

Recall that within the conservative Hamiltonian framework, we can apply the formalism of classical Poisson brackets: for any two functions $A = A(q^i, p_i, t)$ and $B = B(q^i, p_i, t)$ their Poisson bracket is defined as

$$[A, B] = \left(\frac{\partial A}{\partial q^i} \frac{\partial B}{\partial p_i} - \frac{\partial A}{\partial p_i} \frac{\partial B}{\partial q^i} \right).$$

Conservative Classical System

Any function $A(q^i, p_i, t)$ is called a *constant* (or integral) of motion of the conservative system (3.87) if

$$\dot{A} \equiv \partial_t A + [A, H] = 0, \qquad \text{which implies} \qquad \partial_t A = -[A, H]. \tag{3.88}$$

For example, if $A = \rho(q^i, p_i, t)$ is a *density function* of ensemble phase–points (or, a probability density to see a state $x(t) = (q^i(t), p_i(t))$ of *ensemble* at a moment t), then equation

$$\partial_t \rho = -[\rho, H] = -\mathrm{i} L\, \rho \tag{3.89}$$

represents the *Liouville theorem*, where L denotes the (Hermitian) *Liouville operator*

$$\mathrm{i}L = [..., H] \equiv \left(\frac{\partial H}{\partial p_i} \frac{\partial}{\partial q^i} - \frac{\partial H}{\partial q^i} \frac{\partial}{\partial p_i} \right) = \mathrm{div}(\rho \dot{\mathbf{x}}),$$

which shows that the conservative Liouville equation (3.89) is actually equivalent to the mechanical *continuity equation*

$$\partial_t \rho + \mathrm{div}(\rho \dot{\mathbf{x}}) = 0. \tag{3.90}$$

Conservative Quantum System

We perform the formal quantization of the conservative equation (3.89) in the Heisenberg picture: all variables become Hermitian operators (denoted by '$\wedge$'), the symplectic phase–space $T^*Q = \{q^i, p_i\}$ becomes the Hilbert state–space $\mathcal{H} = \mathcal{H}_{\hat{q}^i} \otimes \mathcal{H}_{\hat{p}_i}$ (where $\mathcal{H}_{\hat{q}^i} = \mathcal{H}_{\hat{q}^1} \otimes ... \otimes \mathcal{H}_{\hat{q}^N}$ and $\mathcal{H}_{\hat{p}_i} = \mathcal{H}_{\hat{p}_1} \otimes ... \otimes \mathcal{H}_{\hat{p}_N}$), the classical Poisson bracket $[\,,\,]$ becomes the quantum commutator $\{\,,\,\}$ multiplied by $-\mathrm{i}/\hbar$

$$[\,,\,] \longrightarrow -\mathrm{i}\{\,,\,\} \qquad (\hbar = 1 \text{ in normal units}). \tag{3.91}$$

In this way the classical Liouville equation (3.89) becomes the *quantum Liouville equation*

$$\partial_t \hat{\rho} = \mathrm{i}\{\hat{\rho}, \hat{H}\}\,, \tag{3.92}$$

where $\hat{H} = \hat{H}(\hat{q}^i, \hat{p}_i, t)$ is the Hamiltonian evolution operator, while

$$\hat{\rho} = P(a)|\Psi_a><\Psi_a|, \qquad \text{with} \qquad \mathrm{Tr}(\hat{\rho}) = 1,$$

denotes the von Neumann *density matrix operator*, where each quantum state $|\Psi_a>$ occurs with probability $P(a)$; $\hat{\rho} = \hat{\rho}(\hat{q}^i, \hat{p}_i, t)$ is closely related to another von Neumann concept: *entropy* $S = -\,\mathrm{Tr}(\hat{\rho}[\ln\hat{\rho}])$.

Open Classical System

We now move to the *open (nonconservative) system*: on the macroscopic NN–level the *opening operation* equals to the *adding* of a *covariant* vector of external (dissipative and/or motor) forces $F_i = F_i(q^i, p_i, t)$ to (the r.h.s of) the covariant Hamiltonian *force equation*, so that Hamiltonian equations get the *open (dissipative and/or forced) form*

$$\dot{q}^i = \frac{\partial H}{\partial p_i}, \qquad \dot{p}_i = F_i - \frac{\partial H}{\partial q^i}\,. \tag{3.93}$$

In the framework of the open Hamiltonian system (3.93), dynamics of any function $A(q^i, p_i, t)$ is defined by the *open evolution equation*:

$$\partial_t A = -[A, H] + \Phi,$$

where $\Phi = \Phi(F_i)$ represents the general form of the scalar force term.

In particular, if $A = \rho(q^i, p_i, t)$ represents the density function of ensemble phase–points, then its dynamics is given by the (dissipative/forced) *open Liouville equation*:

$$\partial_t \rho = -[\rho, H] + \Phi\,. \tag{3.94}$$

In particular, the scalar force term can be cast as a linear Poisson–bracket form

$$\Phi = F_i[A, q^i]\,, \qquad \text{with} \qquad [A, q^i] = -\frac{\partial A}{\partial p_i}\,. \tag{3.95}$$

Now, in a similar way as the conservative Liouville equation (3.89) resembles the continuity equation (3.90) from continuum dynamics, also the open Liouville equation (3.94) resembles the probabilistic *Fokker–Planck equation* from statistical mechanics. If we have a ND stochastic process $x(t) = (q^i(t), p_i(t))$ defined by the vector *Itô SDE*

$$dx(t) = f(x,t)\,dt + G(x,t)\,dW,$$

where f is a ND vector function, W is a KD Wiener process, and G is a $N \times K$D matrix valued function, then the corresponding probability density function $\rho = \rho(x, t|\dot{x}, t')$ is defined by the ND Fokker–Planck equation (see, e.g., [Gar85])

$$\partial_t \rho = -\operatorname{div}[\rho\, f(x,t)] + \frac{1}{2}\frac{\partial^2}{\partial x_i \partial x_j}(Q_{ij}\, \rho), \qquad (3.96)$$

where $Q_{ij} = \big(G(x,t)\, G^T(x,t)\big)_{ij}$. It is obvious that the Fokker–Planck equation (3.96) represents the particular, stochastic form of our general open Liouville equation (3.94), in which the scalar force term is given by the (second–derivative) noise term

$$\Phi = \frac{1}{2}\frac{\partial^2}{\partial x_i \partial x_j}(Q_{ij}\, \rho) \ .$$

Equation (3.94) will represent the open classical model of our macroscopic NN–dynamics.

Continuous Neural Network Dynamics

The generalized NN–dynamics, including two special cases of *graded response neurons* (GRN) and *coupled neural oscillators* (CNO), can be presented in the form of a stochastic *Langevin rate equation*

$$\dot{\sigma}_i = f_i + \eta_i(t), \qquad (3.97)$$

where $\sigma_i = \sigma_i(t)$ are the continual neuronal variables of ith neurons (representing either membrane action potentials in case of GRN, or oscillator phases in case of CNO); J_{ij} are individual synaptic weights; $f_i = f_i(\sigma_i, J_{ij})$ are the deterministic forces (given, in GRN–case, by

$$f_i = \sum_j J_{ij} \tanh[\gamma \sigma_j] - \sigma_i + \theta_i,$$

with $\gamma > 0$ and with the θ_i representing injected currents, and in CNO–case, by

$$f_i = \sum_j J_{ij} \sin(\sigma_j - \sigma_i) + \omega_i,$$

with ω_i representing the natural frequencies of the individual oscillators); the noise variables are given by

$$\eta_i(t) = \lim_{\Delta \to 0} \zeta_i(t) \sqrt{2T/\Delta},$$

where $\zeta_i(t)$ denote uncorrelated Gaussian distributed random forces and the parameter T controls the amount of noise in the system, ranging from $T = 0$ (deterministic dynamics) to $T = \infty$ (completely random dynamics).

More convenient description of the neural random process (3.97) is provided by the Fokker–Planck equation describing the time evolution of the probability density $P(\sigma_i)$

$$\partial_t P(\sigma_i) = -\frac{\partial}{\partial \sigma_i}\left(f_i P(\sigma_i)\right) + T \frac{\partial^2}{\partial \sigma_i^2} P(\sigma_i). \qquad (3.98)$$

Now, in the case of deterministic dynamics $T = 0$, equation (3.98) can be put into the form of the conservative Liouville equation (3.89), by making the substitutions:

$$P(\sigma_i) \to \rho, \qquad f_i = \dot{\sigma}_i, \qquad [\rho, H] = \mathrm{div}(\rho\,\dot{\sigma}_i) \equiv \sum_i \frac{\partial}{\partial \sigma_i}(\rho\,\dot{\sigma}_i),$$

where $H = H(\sigma_i, J_{ij})$. Further, we can formally identify the stochastic forces, i.e., the second–order noise–term $T\sum_i \frac{\partial^2}{\partial \sigma_i^2}\rho$ with $F^i[\rho, \sigma_i]$, to get the open Liouville equation (3.94).

Therefore, on the NN–level deterministic dynamics corresponds to the conservative system (3.89). Inclusion of stochastic forces corresponds to the system opening (3.94), implying the *macroscopic arrow of time.*

Open Quantum System

By formal quantization of equation (3.94) with the scalar force term defined by (3.95), in the same way as in the case of the conservative dynamics, we get the *quantum open Liouville equation*

$$\partial_t \hat{\rho} = \mathrm{i}\{\hat{\rho}, \hat{H}\} + \hat{\Phi}, \qquad \text{with} \qquad \hat{\Phi} = -\mathrm{i}\hat{F}_i\{\hat{\rho}, \hat{q}^i\}, \tag{3.99}$$

where $\hat{F}_i = \hat{F}_i(\hat{q}^i, \hat{p}_i, t)$ represents the covariant quantum operator of external friction forces in the Hilbert state–space $\mathcal{H} = \mathcal{H}_{\hat{q}^i} \otimes \mathcal{H}_{\hat{p}_i}$.

Equation (3.99) will represent the open quantum–friction model of our microscopic MT–dynamics. Its system–independent properties are [EMN92, EMN99, MN95a, MN95b, Nan95]:

1. Conservation of probability P
$$\partial_t P = \partial_t[\mathrm{Tr}(\hat{\rho})] = 0.$$
2. Conservation of energy E, on the average
$$\partial_t \langle\langle E \rangle\rangle \equiv \partial_t[\mathrm{Tr}(\hat{\rho}\, E)] = 0.$$
3. Monotonic increase in entropy
$$\partial_t S = \partial_t[-\mathrm{Tr}(\hat{\rho} \ln \hat{\rho})] \geq 0,$$
and thus automatically and naturally implies a *microscopic arrow of time*, so essential in realistic biophysics of neural processes.

Non–Critical Stringy MT–Dynamics

In EMN–language of non–critical (SUSY) bosonic strings, our MT–dynamics equation (3.99) reads

$$\partial_t \hat{\rho} = \mathrm{i}\{\hat{\rho}, \hat{H}\} - \mathrm{i}\hat{g}_{ij}\{\hat{\rho}, \hat{q}^i\}\dot{\hat{q}}^j, \tag{3.100}$$

where the target–space density matrix $\hat{\rho}(\hat{q}^i, \hat{p}_i)$ is viewed as a function of coordinates $\hat{q}^i$ that parameterize the couplings of the generalized σ–models on the bosonic string world–sheet, and their conjugate momenta $\hat{p}_i$, while $\hat{g}_{ij} = \hat{g}_{ij}(\hat{q}^i)$ is the quantum operator of the *positive definite metric* in the space of couplings. Therefore, the covariant quantum operator of external friction forces is in EMN–formulation given as $\hat{F}_i(\hat{q}^i, \dot{\hat{q}}^i) = \hat{g}_{ij}\,\dot{\hat{q}}^j$.

Equation (3.100) establishes the conditions under which a large–scale coherent state appearing in the MT–network, which can be considered responsible for loss–free energy transfer along the tubulins.

Equivalence of Neurodynamic Forms

It is obvious that both the macroscopic NN–equation (3.94) and the microscopic MT–equation (3.99) have the same open Liouville form, which implies the arrow of time. These proves the existence of the formal neuro–biological space–time self–similarity.

In this way, we have described neurodynamics of both NN and MT ensembles, belonging to completely different biophysical space–time scales, by the unique form of open Liouville equation, which implies the arrow of time. The existence of the formal *neuro–biological self–similarity* has been proved.

Dissipative Quantum Brain Model

The *conservative brain* model was originally formulated within the framework of the quantum field theory (QFT) by [RU67] and subsequently developed in [STU78, STU79, JY95, JPY96]. The conservative brain model has been recently extended to the *dissipative quantum dynamics* in the work of G. Vitiello and collaborators [Vit95, AV00, PV99, Vit01, PV03, PV04].

The motivations at the basis of the formulation of the quantum brain model by Umezawa and Ricciardi trace back to the laboratory observations leading Lashley to remark (in 1940) that “masses of excitations ... within general fields of activity, without regard to particular nerve cells are involved in the determination of behavior” [Las42, Pri91]. In 1960’s, K. Pribram, also motivated by experimental observations, started to formulate his *holographic hypothesis*. According to W. Freeman [Fre90, Fre96, Fre00], “information appears indeed in such observations to be spatially uniform in much the way that the information density is uniform in a hologram”. While the activity of the single neuron is experimentally observed in form of discrete and stochastic pulse trains and point processes, the ‘macroscopic’ activity of large assembly of neurons appears to be spatially coherent and highly structured in phase and amplitude.

Motivated by such an experimental situation, Umezawa and Ricciardi formulated in [RU67] the quantum brain model as a many–body physics problem, using the formalism of QFT with spontaneous breakdown of symmetry (which had been successfully tested in condensed matter experiments). Such a formalism provides the only available theoretical tool capable to describe long–range correlations such as the ones observed in the brain – presenting almost simultaneous responses in several regions to some external stimuli. The understanding of these long–range correlations in terms of modern biochemical and electrochemical processes is still lacking, which suggests that these responses could not be explained in terms of single neuron activity [Pri71, Pri91].

Lagrangian dynamics in QFT is, in general, invariant under some group G of continuous transformations, as proposed by the famous Noether theorem. Now, spontaneous symmetry breakdown, one of the corner–stones of Haken's synergetics [Hak83, Hak93], occurs when the minimum energy state (the ground, or vacuum, state) of the system is not invariant under the full group G, but under one of its subgroups. Then it can be shown [IZ80, Ume93] that collective modes, the so–called Nambu–Goldstone (NG) boson modes, are dynamically generated. Propagating over the whole system, these modes are the carriers of the *long–range correlation*, in which the order manifests itself as a global property dynamically generated. The long–range correlation modes are responsible for keeping the ordered pattern: they are coherently *condensed* in the ground state (similar to e.g., in the crystal case, where they keep the atoms trapped in their lattice sites). The long–range correlation thus forms a sort of net, extending over all the system volume, which traps the system components in the ordered pattern. This explains the "holistic" macroscopic collective behavior of the system components.

More precisely, according to the *Goldstone theorem* in QFT [IZ80, Ume93], the spontaneous breakdown of the symmetry implies the existence of long–range correlation NG–modes in the ground state of the system. These modes are massless modes in the infinite volume limit, but they may acquire a finite, non-zero mass due to boundary or impurity effects [ARV02]. In the quantum brain model these modes are called dipole–wave–quanta (DWQ). The density of their condensation in the ground states acts as a *code* classifying the state and the memory there recorded. States with different code values are unitarily inequivalent states, i.e., there is no unitary transformation relating states of different codes.[21]

Now, in formulating a proper mathematical model of brain, the conservative dynamics is not realistic: we cannot avoid to take into consideration the dissipative character of brain dynamics, since the brain is an intrinsically open system, continuously interacting with the environment. As Vitiello observed in [Vit01, PV03, PV04], the very same fact of "getting an information" introduces a partition in the time coordinate, so that one may distinguish between

[21] We remark that the spontaneous breakdown of symmetry is possible since in QFT there exist infinitely many ground states or vacua which are physically distinct (technically speaking, they are "unitarily inequivalent"). In quantum mechanics (QM), on the contrary, all the vacua are physically equivalent and thus there cannot be symmetry breakdown.

before "getting the information" (the past) and *after* "getting the information" (the future): the *arrow of time* is in this way introduced.... "*Now* you know it!" is the familiar warning to mean that now, i.e. after having received a certain information, you are not the same person as before getting it. It has been shown that the psychological arrow of time (arising as an effect of memory recording) points in the same direction of the thermodynamical arrow of time (increasing entropy direction) and of the cosmological arrow of time (the expanding Universe direction) [AMV00].

The canonical quantization procedure of a dissipative system requires to include in the formalism also the system representing the environment (usually the heat bath) in which the system is embedded. One possible way to do that is to depict the environment as the time–reversal image of the system [CRV92]: the environment is thus described as the *double* of the system in the time–reversed dynamics (the system image in the mirror of time).

Within the framework of dissipative QFT, the brain system is described in terms of an *infinite collection of damped harmonic oscillators* A_κ (the simplest prototype of a dissipative system) representing the DWQ [Vit95]. Now, the collection of damped harmonic oscillators is ruled by the Hamiltonian [Vit95, CRV92]

$$\begin{aligned} &H = H_0 + H_I, \qquad \text{with} \\ &H_0 = \hbar\Omega_\kappa(A_\kappa^\dagger A_\kappa - \tilde{A}_\kappa^\dagger \tilde{A}_\kappa), \qquad H_I = \mathrm{i}\hbar\Gamma_\kappa(A_\kappa^\dagger \tilde{A}_\kappa^\dagger - A_\kappa \tilde{A}_\kappa), \end{aligned}$$

where Ω_κ is the frequency and Γ_κ is the damping constant. The $\tilde{A}_\kappa$ modes are the 'time–reversed mirror image' (i.e., the 'mirror modes') of the A_κ modes. They are the doubled modes, representing the environment modes, in such a way that κ generically labels their degrees–of–freedom. In particular, we consider the damped harmonic oscillator (DHO)

$$m\ddot{x} + \gamma\dot{x} + \kappa x = 0, \tag{3.101}$$

as a simple prototype for dissipative systems (with intention that thus get results also apply to more general systems). The damped oscillator (3.101) is a non–Hamiltonian system and therefore the customary canonical quantization procedure cannot be followed. However, one can face the problem by resorting to well known tools such as the *density matrix* ρ and the *Wigner function* $W = W(x, p, t)$.

Let us start with the special case of a *conservative particle* in the absence of friction γ, with the standard Hamiltonian,

$$H = -(\hbar\partial_x)^2/2m + V(x).$$

Recall (from the previous subsection) that the *density matrix equation of motion*, i.e., *quantum Liouville equation*, is given by

$$\mathrm{i}\hbar\dot{\rho} = [H, \rho]. \tag{3.102}$$

The density matrix function ρ is defined by

$$\langle x + \frac{1}{2}y|\rho(t)|x - \frac{1}{2}y\rangle = \psi^*(x + \frac{1}{2}y, t)\psi(x - \frac{1}{2}y, t) \equiv W(x, y, t),$$

with the associated standard expression for the *Wigner function* (see [FH65]),

$$W(p, x, t) = \frac{1}{2\pi\hbar}\int W(x, y, t)\, \mathrm{e}^{\left(-\mathrm{i}\frac{py}{\hbar}\right)} dy.$$

Now, in the coordinate x–representation, by introducing the notation

$$x_{\pm} = x \pm \frac{1}{2}y, \tag{3.103}$$

the Liouville equation (3.102) can be expanded as

$$\begin{aligned} &\mathrm{i}\hbar\, \partial_t \langle x_+|\rho(t)|x_-\rangle = \\ &\left\{-\frac{\hbar^2}{2m}\left[\partial^2_{x_+} - \partial^2_{x_-}\right] + \left[V(x_+) - V(x_-)\right]\right\}\langle x_+|\rho(t)|x_-\rangle, \end{aligned} \tag{3.104}$$

while the Wigner function $W(p, x, t)$ is now given by

$$\begin{aligned} \mathrm{i}\hbar\, \partial_t W(x, y, t) &= H_o W(x, y, t), \qquad \text{with} \\ H_o &= \frac{1}{m}p_x p_y + V(x + \frac{1}{2}y) - V(x - \frac{1}{2}y), \\ \text{and} \qquad p_x &= -\mathrm{i}\hbar\partial_x, \qquad p_y = -\mathrm{i}\hbar\partial_y. \end{aligned} \tag{3.105}$$

The new Hamiltonian H_o (3.105) may be get from the corresponding Lagrangian

$$L_o = m\dot{x}\dot{y} - V(x + \frac{1}{2}y) + V(x - \frac{1}{2}y). \tag{3.106}$$

In this way, Vitiello concluded that the density matrix and the Wigner function formalism *required*, even in the conservative case (with zero mechanical resistance γ), the introduction of a 'doubled' set of coordinates, $x_{\pm}$, or, alternatively, x and y. One may understand this as related to the introduction of the 'couple' of indices *necessary* to label the density matrix elements (3.104).

Let us now consider the case of the *particle interacting* with a *thermal bath* at temperature T. Let f denote the *random force* on the particle at the position x due to the bath. The interaction Hamiltonian between the bath and the particle is written as

$$H_{int} = -fx. \tag{3.107}$$

Now, in the *Feynman–Vernon formalism* (see [Fey72]), the *effective action* $A[x, y]$ for the particle is given by

$$A[x, y] = \int_{t_i}^{t_f} L_o(\dot{x}, \dot{y}, x, y)\, dt + I[x, y],$$

with L_o defined by (3.106) and

$$e^{\frac{i}{\hbar}I[x,y]} = \langle (e^{-\frac{i}{\hbar}\int_{t_i}^{t_f} f(t)x_-(t)dt})_- (e^{\frac{i}{\hbar}\int_{t_i}^{t_f} f(t)x_+(t)dt})_+ \rangle, \tag{3.108}$$

where the symbol $\langle . \rangle$ denotes the average with respect to the thermal bath; '$(.)_+$' and '$(.)_-$' denote time ordering and anti–time ordering, respectively; the coordinates $x_\pm$ are defined as in (3.103). If the interaction between the bath and the coordinate x (3.107) were turned off, then the operator f of the bath would develop in time according to

$$f(t) = e^{iH_\gamma t/\hbar} f e^{-iH_\gamma t/\hbar},$$

where H_γ is the Hamiltonian of the isolated bath (decoupled from the coordinate x). $f(t)$ is then the force operator of the bath to be used in (3.108).

The interaction $I[x,y]$ between the bath and the particle has been evaluated in [SVW95] for a linear passive damping due to thermal bath by following Feynman–Vernon and Schwinger [FH65]. The final result from [SVW95] is:

$$\begin{aligned} I[x,y] &= \frac{1}{2}\int_{t_i}^{t_f} dt\, [x(t)F_y^{ret}(t) + y(t)F_x^{adv}(t)] \\ &+ \frac{i}{2\hbar}\int_{t_i}^{t_f}\int_{t_i}^{t_f} dtds\, N(t-s)y(t)y(s), \end{aligned}$$

where the retarded force on y, F_y^{ret}, and the advanced force on x, F_x^{adv}, are given in terms of the retarded and advanced Green functions $G_{ret}(t-s)$ and $G_{adv}(t-s)$ by

$$F_y^{ret}(t) = \int_{t_i}^{t_f} ds\, G_{ret}(t-s)y(s), \qquad F_x^{adv}(t) = \int_{t_i}^{t_f} ds\, G_{adv}(t-s)x(s),$$

respectively. In (3.109), $N(t-s)$ is the *quantum noise* in the fluctuating random force given by

$$N(t-s) = \frac{1}{2}\langle f(t)f(s) + f(s)f(t)\rangle.$$

The real and the imaginary part of the action are given respectively by

$$\mathrm{Re}\,(A[x,y]) = \int_{t_i}^{t_f} L\, dt, \tag{3.109}$$

$$L = m\dot{x}\dot{y} - \left[V(x+\frac{1}{2}y) - V(x-\frac{1}{2}y)\right] + \frac{1}{2}\left[xF_y^{ret} + yF_x^{adv}\right], \tag{3.110}$$

$$\text{and} \qquad \mathrm{Im}\,(A[x,y]) = \frac{1}{2\hbar}\int_{t_i}^{t_f}\int_{t_i}^{t_f} N(t-s)y(t)y(s)\, dtds. \tag{3.111}$$

Equations (3.109–3.111), are *exact* results for linear passive damping due to the bath. They show that in the classical limit '$\hbar \to 0$' nonzero y yields an 'unlikely process' in view of the large imaginary part of the action implicit in (3.111). Nonzero y, indeed, may lead to a negative real exponent in the evolution operator, which in the limit $\hbar \to 0$ may produce a negligible contribution to the probability amplitude. On the contrary, at quantum level nonzero y accounts for quantum noise effects in the fluctuating random force in the system–environment coupling arising from the imaginary part of the action (see [SVW95]).

When in (3.110) we use

$$F_y^{ret} = \gamma \dot{y} \qquad \text{and} \qquad F_x^{adv} = -\gamma \dot{x} \qquad \text{we get,}$$

$$L(\dot{x}, \dot{y}, x, y) = m\dot{x}\dot{y} - V\left(x + \frac{1}{2}y\right) + V\left(x - \frac{1}{2}y\right) + \frac{\gamma}{2}(x\dot{y} - y\dot{x}). \quad (3.112)$$

By using

$$V\left(x \pm \frac{1}{2}y\right) = \frac{1}{2}\kappa\left(x \pm \frac{1}{2}y\right)^2$$

in (3.112), the DHO equation (3.101) and its complementary equation for the y coordinate

$$m\ddot{y} - \gamma\dot{y} + \kappa y = 0. \quad (3.113)$$

are derived. The y–oscillator is the time–reversed image of the x–oscillator (3.101). From the manifolds of solutions to equations (3.101) and (3.113), we could choose those for which the y coordinate is constrained to be zero, they simplify to

$$m\ddot{x} + \gamma\dot{x} + \kappa x = 0, \qquad y = 0.$$

Thus we get the classical damped oscillator equation from a Lagrangian theory at the expense of introducing an 'extra' coordinate y, later constrained to vanish. Note that the constraint $y(t) = 0$ is *not* in violation of the equations of motion since it is a true solution to (3.101) and (3.113).

Therefore, the general scheme of the dissipative quantum brain model can be summarized as follows. The starting point is that the brain is permanently coupled to the environment. Of course, the specific details of such a coupling may be very intricate and changeable so that they are difficult to be measured and known. One possible strategy is to average the effects of the coupling and represent them, at some degree of accuracy, by means of some 'effective' interaction. Another possibility is to take into account the environmental influence on the brain by a suitable *choice* of the brain vacuum state. Such a choice is triggered by the external input (breakdown of the symmetry), and it actually is the end point of the internal (spontaneous) dynamical process of the brain (self–organization). The chosen vacuum thus carries the *signature* (memory) of the reciprocal brain–environment influence at a given time under given boundary conditions. A change in the brain–environment reciprocal influence then would correspond to a change in the choice of the brain vacuum: the brain state evolution or 'story' is thus the story of the trade of the brain with the surrounding world. The theory should then provide the equations describing the brain evolution 'through the vacua', each vacuum for each instant of time of its history.

The brain evolution is thus similar to a time–ordered sequence of photograms: each photogram represents the 'picture' of the brain at a given instant of time. Putting together these photograms in 'temporal order' one gets a movie, i.e. the story (the evolution) of open brain, which includes the brain–environment interaction effects.

The evolution of a memory specified by a given code value, say $\mathcal{N}$, can be then represented as a trajectory of given initial condition running over time–dependent vacuum states, denoted by $|0(t) >_{\mathcal{N}}$, each one minimizing the free energy functional. These trajectories are known to be *classical* trajectories in the infinite volume limit: transition from one representation to another inequivalent one would be strictly forbidden in a quantum dynamics.

Since we have now two–modes (i.e., non–tilde and tilde modes), the memory state $|0(t) >_{\mathcal{N}}$ turns out to be a two–mode coherent state. This is known to be an *entangled state*, i.e., it cannot be factorized into two single–mode states, the non–tilde and the tilde one. The physical meaning of such an entanglement between non-tilde and tilde modes is in the fact that the brain dynamics is permanently a dissipative dynamics. The entanglement, which is an unavoidable mathematical result of dissipation, represents the impossibility of cutting the links between the brain and the external world.[22]

In the dissipative brain model, noise and chaos turn out to be natural ingredients of the model. In particular, in the infinite volume limit the chaotic behavior of the trajectories in memory space may account for the high perceptive resolution in the recognition of the perceptual inputs. Indeed, small differences in the codes associated to external inputs may lead to diverging differences in the corresponding memory paths. On the other side, it also happens that codes differing only in a finite number of their components (in the momentum space) may easily be recognized as being the 'same' code, which

[22] We remark that the entanglement is permanent in the large volume limit. Due to boundary effects, however, a unitary transformation could disentangle the tilde and non–tilde sectors: this may result in a pathological state for the brain. It is known that forced isolation of a subject produces pathological states of various kinds. We also observe that the tilde mode is not just a mathematical fiction. It corresponds to a real excitation mode (quasi–particle) of the brain arising as an effect of its interaction with the environment: the couples of non–tilde/tilde dwq quanta represent the correlation modes dynamically created in the brain as a response to the brain–environment reciprocal influence. It is the interaction between tilde and non–tilde modes that controls the irreversible time evolution of the brain: these collective modes are confined to live *in* the brain. They vanish as soon as the links between the brain and the environment are cut.

makes possible that 'almost similar' inputs are recognized by the brain as 'equal' inputs (as in pattern recognition).

Therefore, the brain may be viewed as a complex system with (infinitely) many macroscopic configurations (the memory states). Dissipation is recognized to be the root of such a complexity.

QED Brain

In this subsection, mainly following [Sta95], we formulate a quantum electrodynamics brain model. Recall that quantum electrodynamics (extended to cover the magnetic properties of nuclei) is the theory that controls, as far as we know, the properties of the tissues and the aqueous (ionic) solutions that constitute our brains. This theory is our paradigm basic physical theory, and the one best understood by physicists. It describes accurately, as far as we know, the huge range of actual physical phenomena involving the materials encountered in daily life. It is also related to classical electrodynamics in a particularly beautiful and useful way.

In the low–energy regime of interest here it should be sufficient to consider just the classical part of the photon interaction defined in [Sta83]. Then the explicit expression for the unitary operator that describes the evolution from time t_1 to time t_2 of the quantum electromagnetic field in the presence of a set $L = \{L_i\}$ of specified classical charged–particle trajectories, with trajectory L_i specified by the function $x_i(t)$ and carrying charge e_i, is [Sta95]

$$U[L; t_2, t_1] = \exp < a^* \cdot J(L) > \exp < -J^*(L) \cdot a > \exp[-(J^*(L) \cdot J(L)/2)],$$

where, for any X and Y,

$$< X \cdot Y > \equiv \int d^4k(2\pi)^{-4} 2\pi\delta^+(k^2) X(k) \cdot Y(k),$$

$$(X \cdot Y) \equiv \int d^4k(2\pi)^{-4} i(k^2 + i\epsilon)^{-1} X(k) \cdot Y(k),$$

and $X \cdot Y = X_\mu Y^\mu = X^\mu Y_\mu$. Also,

$$J_\mu(L; k) \equiv \sum_i -ie_i \int_{L_i} dx_\mu \exp(ikx).$$

The integral along the trajectory L_i is

$$\int_{L_i} dx_\mu \exp(ikx) \equiv \int_{t_1}^{t_2} dt(dx_{i\mu}(t)/dt) \exp(ikx).$$

The $a^*(k)$ and $a(k)$ are the photon creation and annihilation operators:

$$[a(k), a^*(k')] = (2\pi)^3 \delta^3(k - k') 2k_0.$$

The operator $U[L;t_2,t_1]$ acting on the photon vacuum state creates the coherent photon state that is the quantum–theoretic analog of the classical electromagnetic field generated by classical point particles moving on the set of trajectories $L=\{L_i\}$ between times t_1 and t_2.

The $U[L;t_2,t_1]$ can be decomposed into commuting contributions from the various values of k. The general coherent state can be written [Sta95]

$$|q,p>\equiv \exp i(<q\cdot P> - <p\cdot Q>)|0>,$$

where $|0>$ is the photon vacuum state and

$$Q(k)=(a_k+a_k^*)/\sqrt{2} \qquad \text{and} \qquad P(k)=i(a_k-a_k^*)/\sqrt{2},$$

and $q(k)$ and $p(k)$ are two functions defined (and square integrable) on the mass shell $k^2=0$, $k_0\geq 0$. The inner product of two coherent states is

$$\begin{aligned}<q,p|q',p'> = \exp -(&<q-q'\cdot q-q'> + <p-p'\cdot p-p'> \\ &+ 2i<p-p'\cdot q+q'>)/4.\end{aligned}$$

There is a decomposition of unity

$$\begin{aligned}I=&\prod d^4k(2\pi)^{-4}2\pi\delta^+(k^2)\int dq_k dp_k/\pi \\ &\times \exp(iq_kP_k-ip_kQ_k)|0_k><0_k|\exp -(iq_kP_k-ip_kQ_k).\end{aligned}$$

Here meaning can be given by quantizing in a box, so that the variable k is discretized. Equivalently,

$$I=\int d\mu(q,p)|q,p><q,p|,$$

where $\mu(q,p)$ is the appropriate measure on the functions $q(k)$ and $p(k)$. Then if the state $|\Psi><\Psi|$ were to jump to $|q,p><q,p|$ with probability density $<q,p|\Psi><\Psi|q,p>$, the resulting mixture would be [Sta95]

$$\int d\mu(q,p)|q,p><q,p|\Psi><\Psi|q,p><q,p|,$$

whose trace is

$$\int d\mu(q,p)<q,p|\Psi><\Psi|q,p>=<\Psi|\Psi>.$$

To represent the limited capacity of consciousness let us assume, in this model, that the states of consciousness associated with a brain can be expressed in terms of a relatively small subset of the modes of the electromagnetic field in the brain cavity. Let us assume that events occurring outside the brain are keeping the state of the universe outside the brain cavity in a single state, so that the state of the brain can also be represented by a single state. The brain is represented, in the path–integral method of Feynman, by a superposition of the trajectories of the particles in it, with each element of the superposition accompanied by the coherent–state electromagnetic field that this set of trajectories generates. Let the state of the electromagnetic field restricted to the modes that represent consciousness be called $|\Psi(t)>$. Using the decomposition of unity one can write

$$|\Psi(t)>= \int d\mu(q,p)|q,p><q,p|\Psi(t)> .$$

Hence the state at time t can be represented by the function $<q,p|\Psi(t)>$, which is a complex-valued function over the set of arguments $\{q_1,p_1,q_2,p_2,\dots,q_n,p_n\}$, where n is the number of modes associated with $|\Psi>$. Thus in this model the contents of the consciousness associated with a brain is represented in terms of this function defined over a $2n$D space: the ith conscious event is represented by the transition

$$|\Psi_i(t_{i+1})> \longrightarrow |\Psi_{i+1}(t_{i+1})>= P_i|\Psi_i(t_{i+1})>,$$

where P_i is a projection operator.

For each allowed value of k the pair of numbers (q_k,p_k) represents the state of motion of the kth mode of the electromagnetic field. Each of these modes is defined by a particular wave pattern that extends over the whole brain cavity. This pattern is an oscillating structure something like a sine wave or a cosine wave. Each mode is fed by the motions of all of the charged particles in the brain. Thus each mode is a representation of a certain integrated aspect of the activity of the brain, and the collection of values $q_1,p_1,\dots,p_n$ is a compact representation of certain aspects the over–all activity of the brain.

The state $|q,p>$ represents the conjunction, or collection over the set of all allowed values of k, of the various states $|q_k,p_k>$. The function

$$V(q,p,t)=<q,p|\Psi(t)><\Psi(t)|q,p>$$

satisfies $0\leq V(q,p,t)\leq 1$, and it represents, according to orthodox thinking, the 'probability' that a system that is represented by a general state $|\Psi(t)>$ just before the time t will be observed to be in the classically describable state $|q,p>$ if the observation occurs at time t. The coherent states $|q,p>$ can, for various mathematical and physical reasons, be regarded as the 'most classical' of the possible states of the electromagnetic quantum field.

To formulate a causal dynamics in which the state of consciousness itself controls the selection of the next state of consciousness one must specify a rule that determines, in terms of the evolving state $|\Psi_i(t)>$ up to time t_{i+1}, both the time t_{i+1} when the next selection event occurs, and the state $|\Psi_{i+1}(t_{i+1})>$ that is selected and actualized by that event.

In the absence of interactions, and under certain ideal conditions of confinement, the deterministic normal law of evolution entails that in each mode

k there is an independent rotation in the (q_k, p_k) plane with a characteristic angular velocity $\omega_k = k_0$. Due to the effects of the motions of the particles there will be, added to this, a flow of probability that will tend to concentrate the probability in the neighborhoods of a certain set of 'optimal' classical states $|q, p>$. The reason is that the function of brain dynamics is to produce some single template for action, and to be effective this template must be a 'classical' state, because, according to orthodox ideas, only these can be dynamically robust in the room temperature brain. According to the semi–classical description of the brain dynamics, only one of these classical–type states will be present, but according to quantum theory there must be a superposition of many such classical–type states, unless collapses occurs at lower (i.e., microscopic) levels. The assumption here is that no collapses occur at the lower brain levels: there is absolutely no empirical evidence, or theoretical reason, for the occurrence of such lower–level brain events.

So in this model the probability will begin to concentrate around various locally optimal coherent states, and hence around the various (generally) isolated points (q, p) in the $2n$D space at which the quantity [Sta95]

$$V(q, p, t) = < q, p|\Psi_i(t) >< \Psi_i(t)|q, p >$$

reaches a local maximum. Each of these points (q, p) represents a *locally–optimal solution* (at time t) to the search problem: as far as the myopic local mechanical process can see the state $|q, p>$ specifies an analog-computed 'best' template for action in the circumstances in which the organism finds itself. This action can be either intentional (it tends to create in the future a certain state of the body/brain/environment complex) or attentional (it tends to gather information), and the latter action is a special case of the former. As discussed in [Sta93], the intentional and attentional character of these actions is a consequence of the fact that the template for action actualized by the quantum brain event is represented as a projected body–world schema, i.e., as the brains projected representation of the body that it is controlling and the environment in which it is situated.

Let a certain time $t_{i+1} > t_i$ be defined by an (urgency) energy factor $E(t) = \hbar(t_{i+1} - t_i)^{-1}$. Let the value of (q, p) at the largest of the local–maxima of $V(q, p, t_{i+1})$ be called $(q(t_{i+1}), p(t_{i+1}))_{max}$. Then the simplest possible reasonable selection rule would be given by the formula

$$P_i = |(q(t_{i+1}), p(t_{i+1}))_{max} >< (q(t_{i+1}), p(t_{i+1}))_{max}|,$$

which entails that

$$\frac{|\Psi_{i+1} >< \Psi_{i+1}|}{< \Psi_{i+1}|\Psi_{i+1} >} = |(q(t_{i+1}), p(t_{i+1}))_{max} >< (q(t_{i+1}), p(t_{i+1}))_{max}|.$$

This rule could produce a tremendous speed up of the search process. Instead of waiting until all the probability gets concentrated in one state $|q, p>$, or into a set of isolated states $|q_i, p_i>$ [or choosing the state randomly, in accordance with the probability function $V(q,p,t_{i+1})$, which could often lead to a disastrous result], this simplest selection process would pick the state $|q,p>$ with the largest value of $V(q,p,t)$ at the time $t = t_{i+1}$. This process does not involve the complex notion of picking a random number, which is a physically impossible feat that is difficult even to define.

One important feature of this selection process is that it involves the state $\Psi(t)$ as a whole: the whole function $V(q,p,t_{i+1})$ must be known in order to determine where its maximum lies. This kind of selection process is not available in the semi–classical ontology, in which only one classically describable state exists at the macroscopic level. That is because this single classically describable macro–state (e.g., some one actual state $|q,p,t_{i+1}>$) contains no information about what the probabilities associated either with itself or with the other alternative possibilities would have been if the collapse had not occurred earlier, at some micro-level, and reduced the earlier state to some single classically describable state, in which, for example, the action potential along each nerve is specified by a well defined classically describable electromagnetic field. There is no rational reason in quantum mechanics for such a micro–level event to occur. Indeed, the only reason to postulate the occurrence of such premature reductions is to assuage the classical intuition that the action–potential pulse along each nerve 'ought to be classically describable even when it is not observed', instead of being controlled, when unobserved, by the local deterministic equations of quantum field theory. But the validity of this classical intuition is questionable if it severely curtails the ability of the brain to function optimally.

A second important feature of this selection process is that the actualized state Ψ_{i+1} is the state of the entire aspect of the brain that is connected to consciousness. So the feel of the conscious event will involve that aspect of the brain, taken as a whole. The 'I' part of the state $\Psi(t)$ is its slowly changing part. This part is being continually re–actualized by the sequence of events, and hence specifies the slowly changing background part of the felt experience. It is this persisting stable background part of the sequence of templates for action that is providing the over–all guidance for the entire sequence of selection events that is controlling the on–going brain process itself [Sta95].

A somewhat more sophisticated search procedure would be to find the state $|(q,p)_{max}>$, as before, but to identify it as merely a candidate that is to be examined for its concordance with the objectives imbedded in the current template. This is what a good search procedure ought to do: first pick out the top candidate by means of a mechanical process, but then evaluate this candidate by a more refined procedure that could block its acceptance if it does not meet specified criteria.

It may at first seem strange to imagine that nature could operate in such a sophisticated way. But it must be remembered that the generation of a truly random sequence is itself a very sophisticated (and indeed physically impossible) process, and that what the physical sciences have understood, so far, is only the mechanical part of nature's two–part process. Here it is the not–well–understood selection process that is under consideration. We have imposed on this attempt to understand the selection process the naturalistic requirement that the whole process be expressible in natural terms, i.e., that the universal process be a causal self–controlling evolution of the Hilbert–space state–vector in which all aspects of nature, including our conscious experiences, are efficacious.

It may be useful to describe the main features of this model in simple terms. If we imagine the brain to be, for example, a uniform rectangular box then each mode k would correspond to wave form that is periodic in all three directions: it would be formed as a combination of products of sine waves and cosine waves, and would cover the whole box–shaped brain. (More realistic conditions are needed, but this is a simple prototype.) Classically there would be an amplitude for this wave, and in the absence of interactions with the charged particles this amplitude would undergo a simple periodic motion in time. In analogy with the coordinate and momentum variables of an oscillating pendulum there are two variables, q_k and p_k, that describe the motion of the amplitude of the mode k. With a proper choice of scales for the variables q_k and p_k the motion of the amplitude of mode k if it were not coupled to the charges would be a circular motion in the (q_k, p_k)–plane. The classical theory would say that the physical system, mode k, would be represented by a point in q_k, p_k space. But quantum theory says that the physical system, mode k, must be represented by a wave (i.e., by a wave ψ–function) in (q_k, p_k) space. The reason is that interference effects between the values of this wave (function) at different points (q_k, p_k) can be exhibited, and therefore it is not possible to say the full reality is represented by any single value of (q_k, p_k): one must acknowledge the reality of the whole wave. It is possible to associate something like a 'probability density' with this wave, but the corresponding probability cannot be concentrated at a point: in units where Planck's constant is unity the bulk of the probability cannot be squeezed into a region of the (q_k, p_k) plane of area less that unity.

The mode k has certain natural states called 'coherent states', $|q_k, p_k>$. Each of these is represented in (q_k, p_k)–space by a wave function that has a 'probability density' that falls off exponentially as one moves in any direction away from the center–point (q_k, p_k) at which the probability density is maximum. These coherent states are in many ways the 'most classical' wave functions allowed by quantum theory [Gla63a, Gla63b], and a central idea of the present model is to specify that it is to one of these 'most classical' states that the mode-k component of the electromagnetic field will jump, or collapse, when an observation occurs. This specification represents a certain 'maximal' principle: the second process, which is supposed to pick out and actualize some classically describable reality, is required to pick out and actualize one of these 'most classical' of the quantum states. If this selection/actualization process really exists in nature then the classically describable states that are actualized by this process should be 'natural classical states' from some point of view. The coherent states satisfy this requirement. This strong, specific postulate should be easier to disprove, if it is incorrect, than a vague or loosely defined one.

If we consider a system consisting of a collection of modes k, then the generalization of the single coherent state $|q_k, p_k >$ is the product of these states, $|q, p >$. Classically this system would be described by specifying the values all of the classical variables q_k and p_k as functions of time. But the 'best' that can be done quantum mechanically is to specify that at certain times t_i the system is in one of the coherent states $|q, p >$. However, the equations of local quantum field theory (here quantum electrodynamics) entail that if the system starts in such a state then the system will, if no 'observation' occurs, soon evolve into a superposition (i.e., a linear combination) of many such states. But the next 'observation' will then reduce it again to some classically describable state. In the present model each a human observation is identified as a human conscious experience. Indeed, these are the same observations that the pragmatic Copenhagen interpretation of Bohr refers to, basically. The 'happening' in a human brain that corresponds to such an observation is, according to the present model, the selection and actualization of the corresponding coherent state $|q, p >$.

The quantity $V(q, p, t_{i+1})$ defined above is, according to orthodox quantum theory, the predicted probability that a system that is in the state $\Psi(t_{i+1})$ at time t_{i+1} will be observed to be in state $|q, p >$ if the observation occurs at time t_{i+1}. In the present model the function $V(q, p, t_{i+1})$ is used to specify not a fundamentally stochastic (i.e., random or chance–controlled) process but rather the causal process of the selection and actualization of some particular state $|q, p >$. And this causal process is controlled by features of the quantum brain that are specified by the Hilbert space representation of the conscious process itself. This process is a nonlocal process that rides on the local brain process, and it is the nonlocal selection process that, according to the principles of quantum theory, is required to enter whenever an observation occurs.

3.2.5 A Unified Theory of Matter and Mind

Most of the physicists today believe that we are very close to a unified theory of matter (UTM) based on $p-$brane theory, a generalization of superstring theory (see, e.g., [II06b]). This theory has been well motivated by the successes (from microchips to satellites) and the difficulties (the presence of infinities) of Quantum Gauge Field Theories (QED, Electro–Weak theory, GUT and SUSY GUT etc.) most of which are based on solid experimental evidences (measuring Lande $g-$factor of electron up to 10 significant places, prediction of Z boson etc.). This 'theory of everything ' tells us that all carbon atoms (after they are born out of nucleosynthesis in the core of stars) in this universe are the same in space and in time, and thus is unable to explain why the carbon atoms present in a lump of roughly three pound of ordinary matter the human brain – give rise to such ineffable qualities as feeling, thought, purpose, awareness and free will that are taken to be evidence for 'consciousness'.

Is Consciousness an accident caused by random evolutionary processes in the sense that it would not evolve again had the present universe to undergo a 'Big Crunch' to start with another 'Big Bang'? No. It is a fundamental property that emerges as a natural consequence of laws of nature [Sam01]. Are these laws of nature different from laws of physics? Is there a way to expand the UTM to incorporate consciousness?

Nature manifests itself not only at the gross level of phenomena (accessible to direct senses) but also at the subtle level of natural laws (accessible to 'refined' senses). Consciousness is the ability to access nature at both these levels. Hence everything in nature is conscious, but there is a hierarchy in the level of consciousness. Although animals, plants and few machines today can access to the gross level of nature, access to the subtle level seems to be purely human. In this sense a layman is less conscious compared to a scientist or an artist. (Is it possible that a level of consciousness exists compared to which a scientist or an artist of today may appear a layman?) Once everything (both matter and mind) is reduced to information it is possible to define consciousness as the capability to process information. Because any form of our access to nature is based on processing information with varying degree of complexity. The words process and complexity will be defined in the following subsection, mainly following [Sam01].

Matter

Phenomena

Some animals like dogs (in listening to ultrasound) and bats (in sensing prey through echo technique) are better equipped than humans when it comes to direct sense experiences. But unlike animals, humans have found out methods to extend their sense experiences through amplification devices like telescopes, microscopes etc. The summary [Sam89] of such sense experiences (direct and extended) acquired over the last five hundred years (equivalent to one second in a time–scale where the age of the universe is equal to a year) is that our universe extends in space from the size of an electron (10^{17} cm as probed at LEP colliders at CERN, Geneva) to size of galactic super–clusters (10^{30} cm as probed by high red–shift measurements). The theoretical possibility of Planck length (10^{33} cm) allows for further extension in space.

Our grand universe extending over more than 60 orders of magnitude in space has been constantly changing over the last 12 billion years. Living systems seem to have evolved out of non–living systems. Consciousness seems to have evolved out of living systems. Is the dramatic difference between animate and inanimate, conscious and unconscious simply a difference in their ability to process information? At the level of phenomena nature is so vast and diverse compared to the size and comprehension of human beings that one wonders how the collective inquiring human minds over the centuries could at all fathom the unity behind this diversity. This simply testifies the triumph of human mind over the physical limitations of its body. Because the human mind is capable of reaching a synthesis (an advanced form of information processing) based on careful observations of diverse phenomena [Sam01].

Inanimate

Manifold

Till Einstein, everybody thought 'absolute space' is the arena (mathematically, a manifold) on which things change with 'absolute time'. In special theory of relativity (STR) he redefined the manifold to be flat spacetime $(3+1)$ by making both space and time relative with respect to inertial observers but keeping spacetime (SpT) absolute. In general theory of relativity (GTR) he propounded this manifold to be curved spacetime that can act on matter unlike the flat spacetime. Then the Quantum Theory (QT, the standard version, not the Bohmian one) pointed out this manifold to be an abstract mathematical space called *Hilbert space* since the spacetime description of quantum processes is not available. Quantum Field Theory (QFT) that originated in a successful merge of QT and STR requires this manifold to be the Quantum Vacuum (QV). Unlike the ordinary vacuum, QV contains infinite number of 'virtual' particles that give rise to all matter and interactions. Every quantum field has its own QV and a successful amalgamation of QT and GTR (called Quantum Gravity, and yet to be achieved) may connect the curved spacetime with the QV of gravitational field. Finally, Superstring (or, $p-$brane) theory demands all fundamental entities to be strings (or $p-$branes) in 10D spacetime. This evolution in our understanding clearly shows the necessity and importance of a background manifold to formulate any scientific theory.

Basic Constituents

Energy and matter were considered to be the two basic constituents of our universe till Einstein showed their equivalence through $E=mc^2$. All forms of energy are interconvertible. Matter in all its forms (solid, liquid, gas and plasma) consists of atoms. The simplest of all atoms is the hydrogen atom that contains an electron and a proton. If one considers the (now outdated) Bohrian picture of hydrogen atom being a miniature solar system where the electron revolves around the proton in circular orbit then one realizes that most of the hydrogen atom is empty space. This means that if one blows up the hydrogen atom in imagination such that both electron and proton acquire the size of a football each then they need to be separated by 100 km or more. Hence one would think that even if 99.99 % of the hydrogen atom consists of vacuum the point–like electron and proton are material particles. But that is not so.

The elementary particle physics tells us that all the visible matter (composition of dark matter is not yet surely known) in the universe is made of six leptons and six quarks along with their antiparticles. Although they are loosely called particles they are not like the ordinary particles we experience in daily life. A 'classical' particle is localized and impenetrable whereas a 'classical' wave can be extended from minus infinity to plus infinity in space and many wave modes can simultaneously occupy the same space (like inside the telephone cable). But with the advent of quantum mechanics this seemingly contradictory differences between particle and wave lost their sharpness in the quantum world. A quantum object can simultaneously 'be' a particle and a wave until a measurement is made on it. According to QFT all fundamental entities are quantum fields (not material in the conventional sense) that are neither particle nor wave in the classical sense.

Evolution

Despite various specializations, all physical sciences share a common goal: given the complete specification of a physical system at an initial time (called the initial conditions) how to predict what will it be at a later time. To predict with exactness one needs the accurate initial conditions and the laws that govern the evolution of the system (called the dynamical laws). Either the lack of exact specifications of initial conditions or the intractability of huge number of equations (that express dynamical laws) can lead to a probabilistic description rather than a deterministic one. However, there are chaotic systems that can be both deterministic yet unpredictable because of their extreme sensitivity to initial conditions. Apart from dynamical laws there are other laws like Einstein's $E = mc^2$, etc., which do not involve time explicitly. Could there be laws at the level of initial conditions that guide us to choose a particular set over another?

A physical law is like the hidden thread (unity) of a garland with various flowers representing diverse natural phenomena. Its character seems to depend on characteristic scales (denoted by fundamental constants like Planck length, Planck's constant, and speed of light etc). Why should there be different set of laws at different scales? Most physicists believe that quantum mechanics is universal in the applicability of its laws like Schrödinger's equation and classicality of the everyday world is a limiting case. But there exists no consensus at present regarding the emergence of this limiting case.

Guiding Principles

How does one formulate these physical laws? The principle of relativistic causality helps. Do physical laws change? Is there a unity behind the diversity of laws? Is it possible to understand nature without laws? The concept of symmetry (invariance) with its rigorous mathematical formulation and generalization has guided us to know the most fundamental of physical laws. Symmetry as a concept has helped mankind not only to define 'beauty' but also to express the 'truth'. Physical laws tries to quantify the truth that appears to be 'transient' at the level of phenomena but symmetry promotes that truth to the level of 'eternity'.

Interactions

The myriad mosaic of natural phenomena is possible because, not only each fundamental entity evolves with time but also it can interact with the other basic constituents. All the physical interactions (known so far) can be put into four categories:

(i) gravitational interaction (the force that holds the universe),
(ii) electromagnetic interaction (the force that holds the atom, and hence all of us),
(iii) strong nuclear interaction (the force that holds the nucleus), and
(iv) weak nuclear interaction (the force that causes radioactive decay).

At present (ii) and (iv) are known (observationally) to be unified to a single force called Electro–Weak. Grand unified theories (GUT) and their extensions (SUSY GUT) for (ii), (iii) and (iv) do exist. Ongoing research aims to unify (i) with such theories.

According to QFT the basic matter fields interact by exchanging messenger fields (technically called gauge fields) that define the most fundamental level of communication in nature. Both matter fields and gauge fields originate in the fluctuations of QV and in this sense everything in universe including consciousness is, in principle, reducible to QV and its fluctuations. Communication in nature can happen either via the local channel mediated by gauge fields or by the nonlocal EPR [EPR35a] type channels (through entanglement) as was demonstrated by recent quantum teleportation experiments.

Composite Systems

Till the importance of quantum entanglement was realized in recent times the whole was believed to be just the sum of parts. But the whole seems to be much more than just the sum of parts in the 'quantum' world as well as classical systems having complexity. It makes quantum entanglement a very powerful resource that has been utilized in recent times for practical schemes like quantum teleportation, quantum cryptography and quantum computation. Quantum nonlocality indicates that the universe may very well be holographic in the sense that the whole is reflected in each part [Sam01].

Animate

Manifold

A $(3 + 1)-$spacetime is the manifold for all biological functions at the phenomenal level that can be explained by classical physics. If one aims to have a quantum physical explanation of certain biological functions then the manifold has to be the Hilbert space.

Basic Constituent

Cell is the basic constituent of life although the relevant information seems to be coded at the subcell (genetic) level. Neuron (or, microtubules and cytoskeletons) could be the physical substratum of brain depending on classical (or, quantum) viewpoint.

Evolution

Does biological evolution happen with respect to the physical time? If yes, then will the physical laws suffice to study biological evolution in the sense they do in chemistry? If no, then is the biological arrow of time different from the various arrows of physical time (say, cosmological or the thermodynamical arrow of time)? Is there a need for biological laws apart from physical laws to understand the functioning of biological systems?

Guiding Principles

Survivability is the guiding principles in biological systems. Organisms constantly adapt to each other through evolution, and thus organizing themselves into a delicately tuned ecosystem. Intentionality may also play a very important role in the case of more complex bio–systems.

Interactions

The interaction occurs by exchange of chemicals, electric signals, gestures, and language etc. at various levels depending upon the level of complexity involved.

Composite Systems

Composite systems are built out of the basic constituents retaining the relevant information in a holographic manner. The genetic information in the zygote is believed to contain all the details of the biology to come up later when the person grows up. The genes in a developing embryo organize themselves in one way to make a liver cell and in another way to make a muscle cell [Sam01].

Discussions

How 'material' is physical? Anything that is physical need not be 'material' in the sense we experience material things in everyday life. The concept of energy is physical but not material. Because nobody can experience energy directly, one can only experience the manifestations of energy through matter. Similarly the concept of a 'classical field' in physics is very abstract and can only be understood in terms of analogies. Still more abstract is the concept of a 'quantum field' because it cannot be understood in terms of any classical analogies. But at the same time it is a well–known fact in modern physics that all fundamental entities in the universe are quantum fields. Hence one has to abandon the prejudice that anything 'physical' has to be 'material'.

Is reductionism enough? The reductionist approach: observing a system with an increased resolution in search of its basic constituents has helped modern science to be tremendously successful. The success of modern science is the success of the experimental method that has reached an extreme accuracy and reproducibility. But the inadequacy of reductionism in physical sciences becomes apparent in two cases: emergent phenomena and quantum nonlocality. Quantum nonlocality implies a holographic universe that necessitates a holistic approach [BH93].

Though it is gratifying to discover that everything can be traced back to a small number of quantum fields and dynamical laws it does not mean that we now understand the origin of earthquakes, weather variations, the growing of trees, the fluctuations of stock market, the population growth and the evolution of life? Because each of these processes refers to a system that is complex, in the sense that a great many independent agents are interacting with each other in a great many ways. These complex systems are adaptive and undergo spontaneous self–organization (essentially nonlinear) that makes them dynamic in a qualitatively different sense from static objects such as computer chips or snowflakes, which are merely complicated. Complexity deals with emergent phenomena. The concept of complexity is closely related to that of understanding, in so far as the latter is based upon the accuracy of model descriptions of the system obtained using condensed information about it [BP97].

In this sense there are three ultimate frontiers of modern physics: the very small, the very large and the very complex. Complex systems cease to be merely complicated when they display coherent behavior involving collective organization of vast number of degrees–of–freedom. Wetness of water is a collective phenomenon because individual water molecules cannot be said to possess wetness. Lasers, *superfluidity* and *superconductivity* are few of the spectacular examples of complexity in macroscopic systems, which cannot be understood alone in terms of the microscopic constituents. In every case, groups of entities seeking mutual accommodation and self–organization somehow manage to transcend the individuality in the sense that they acquire collective properties that they might never have possessed individually. In contrast to the linear, reductionist thinking, complexity involves nonlinearity and chaos and we are at present far from understanding the complexity in inanimate processes let alone the complexity in living systems.

Emergence of Life

Is life nothing more than a particularly complicated kind of carbon chemistry? Or is it something subtler than putting together the chemical components? Do computer viruses have life in some fundamental sense or are they just pesky imitations of life? How does life emerge from the quadrillions of chemically reacting proteins, lipids, and nucleic acids that make up a living cell? Is it similar to the emergence of thought out of the billions of interconnected neurons that make up the brain? One hope to find the answer to these questions once the dynamics of complexity in inanimate systems is well understood [Sam01].

Mind

Phenomena

Mind is having three states: awake, dream, dreamless sleep. Mind is capable of free–will, self–perception (reflective) and universal perception (perceptual) in its 'awake' state. Can it be trained to have all these three attributes in the states of dream and dreamless sleep? Can there be a fourth state of mind that transcends all the above three states? Where do the brain end and the mind begin? Due to its global nature, mind cannot lie in any particular portion of the brain. Does it lie everywhere in the brain? This would require nonlocal interactions among various components of the brain. If there were no such nonlocal communication then how does the mind emerge from the brain?

Can anybody think of anything that transcends spacetime? Is mind capable of thinking something absolutely new that has not been experienced (directly or indirectly) by the body? Nobody can think of anything absolutely new. One can only think of a new way of arranging and/or connecting things that one has ever learnt. In this sense intellect is constrained by reason whereas imagination is not. But imagination is not acceptable to intellect unless it is logically consistent with what is already known. Imagination helps to see a new connection but intellect makes sure that the new connection is consistent with the old structure of knowledge. This is the way a new structure in knowledge is born and this process of acquiring larger and larger structure (hence meaning or synthesis) is the learning process. Science is considered so reliable because it has a stringent methodology to check this consistency of imagination with old knowledge.

Can one aspire to study the mind using methodology of (physical) sciences? Seeing the tremendous success of physical sciences in the external world one would think its methodology to work for understanding the inner world. It is not obvious a priori why should not QT work in this third ontology when it has worked so successfully with two different ontologies? We aim to understand nature at a level that transcends the inner and the outer worlds by synthesizing them into a more fundamental world of quantum information [Sam01].

Formalism

Manifold

A physical spacetime description of mind is not possible because thoughts that constitute the mind are acausal: it does not take 8 minutes for me to think of the sun although when I look at the sun I see how it was 8 minutes ago. We will assume that it is possible to define an abstract manifold for the space of thoughts (say, $T-$space). An element of $T-$space is a thought–state ($T-$state) and the manifold allows for a continuous change from one $T-$state to another. I presume that $T-$space is identical with mind but it need not be so if mind can exist in a thoughtless but awake state, called *turiya state*.

Basic Constituent

How does one define a $T-$state? That requires one to understand what is a 'thought'? A thought always begins as an idea (that could be based on self and universal perception) and then undergoes successive changes in that idea but roughly remaining focused on a theme. Change from one theme to another is triggered by a new idea. Hence I would suggest that the basic constituent of $T-$state is idea. An idea is like a 'snapshot' of experience complete with all sense data whereas a thought ($T-$state) is like an ensemble (where each element is not an exact replica of the other but has to be very close copy to retain the focus on the theme) of such snapshots.

Evolution

There are two types of evolution in $T-$space. First, the way an ensemble of ideas evolves retaining a common theme to produce a thought. To concentrate means to linger the focus on that theme. This evolution seems to be nonlinear and nondeterministic. Hence the linear unitary evolution of QT may not suffice to quantify this and will be perhaps best described in terms of the mathematics of self–organization and far–from–equilibrium phenomena. [23] The second type involves a change from one particular thought into another and this evolution could be linear and perhaps can be calculated through a probability amplitude description in the line of QT. Given the complete description of a thought at an initial time the refutability of any theory of mind amounts to checking how correctly it can predict the evolution of that thought at a later time.

Guiding Principles

If the guiding principle for evolution in biological world is survivability then in $T-$space it is happiness–ability. A constant pursuit of happiness (although its definition may vary from person to person) guides the change in a person's thoughts. Each and every activity (begins as mental but may or may not materialize) is directed to procure more and more happiness in terms of sensual pleasures of the body, emotional joys of the imagination and rational delights of the intellect.

Interactions

Can a thought (mind) interact with another thought (mind)? Can this interaction be similar to that between quantum fields? Perhaps yes, only if both thought and quantum fields can be reduced to the same basic entity. Then it will be possible for thought (mind) to interact with matter. What will be the messenger that has to be exchanged between interacting minds or between interacting mind and matter? This ultimate level of communication has to be at the level of QV and hence it may amount to silence in terms of conventional languages. But can any receiver (either human mind or any other mind or equipment) be made so sensitive to work with this ultimate level of communication? Interaction with the environment is believed to decohere a quantum system that causes the emergence of classicality in physical world. A completely isolated system remains quantum mechanical. Can a completely isolated mind exhibit quantum mechanical behavior in the sense of superposition and entanglement?

Composite Systems

In the $T-$Space a thought is an ensemble of ideas and a mind–state is composed of thoughts. Behavior, feeling and knowledge of self and universe are in principle reducible to composite subsets in the $T-$space [Sam01].

[23] On the practical side the time–tested techniques of Yoga teach us how to linger the focus on a theme through the practice of *concentration*, *meditation* and *samadhi*.

Discussions

Working definition of Consciousness

Consciousness (at the first level) is related to one's response R to one's environment. This response consists of two parts: habit H and learning L. Once something is learnt and becomes a habit it seems to drop out of consciousness. Once driving a bicycle is learnt one can think of something else while riding the bicycle. But if the habit changes with time then it requires conscious attention. We have defined learning earlier as a process to find commensurability of a new experience with old knowledge. One has to learn anew each time there is a change in the environment. Hence consciousness is not the response to the environment but is the time of rate of change of the response [Sam01],

$$C = \partial_t R, \qquad \text{where} \qquad R = H + L.$$

The hierarchy in consciousness depends on the magnitude of this time derivative. Everything in the universe can be fit in a scale of consciousness with unconscious and super–conscious as the limit points. It is obvious that all animals show response to their environment, so does some of the refrigerators, but there is a hierarchy in their response. Through the use of 'cresco–graph' and 'resonant cardio–graph' of J.C. Bose, one can see the response of botanical as well as inanimate world. We cannot conclude that a stone is unconscious just because we cannot communicate with it using our known means of communication. As technology progresses, we will be able to measure both the response function R and its time derivative. If this is the definition what can it tell about the future evolution of humans? My guess is that we would evolve from conscious to super–conscious in the sense that genes will evolve to store the cumulative learning of the human race.

Emergence of Consciousness

The first step in understanding consciousness consists of using reductionist method to various attributes of consciousness. A major part of the studies done by psychologists (and their equivalents doing studies on animals) and neurobiologists falls under this category. Such studies can provide knowledge about mind states (say, $M_1, M_2, M_3, \ldots$) but cannot explain the connection between these mind states with the corresponding brain states (say $B_1, B_2, \ldots$). Because this kind of dualistic model of Descartes would require to answer a) where is mind located in the brain, and b) if my mind wants me to raise my finger, how does it manage to trigger the appropriate nerves and so on in order for that to happen without exerting any known forces of nature?

To find out how the mind actually works one needs to have a theory of mind, that will relate the sequence of mental states $M_1, M_2, M_3, \ldots$ by providing laws of change (the dynamical laws for the two types of evolution discussed above) that encompass the mental realm after the fashion of the theory of matter that applies to the physical realm, with its specific laws. Such a theory of mind is possible if we synthesize the results of studies on attributes of consciousness to define the exact nature of the manifold and the basic entities of the $T-$space (or, Mind–Space). Once this is achieved then one can attempt to explain the emergence of consciousness taking clues from complexity theory in physical sciences. But such an extrapolation will make sense provided both $M-$states and $B-$states can be reduced to something fundamental that obeys laws of complexity theory. We propose in the next section that information is the right candidate for such a reduction.

Role of Indian Philosophy (IP)

(1) Unlike the Cartesian dichotomy of mind and body some schools of IP like Vaisheshika and Yoga treat both mind and body in a unified manner. Since (western) science is based on Cartesian paradigm it cannot synthesize mind and body unless it takes the clue from oriental philosophies and then blend it with its own rigorous methodology.

(2) In terms of sense awareness, awake, dream and dream–less sleep states are often called as conscious, subconscious and unconscious states. A great conceptual step taken by IP in this regard is to introduce a fourth state of mind called *turiya state* that is defined to be none of the above but a combination of all of the above states. This state is claimed to be the super–conscious state where one transcends the limitations of perceptions constrained by spacetime (3+1). Patanjali has provided very scientific and step by step instructions to reach this fourth state through samyama (concentration, meditation and samadhi are different levels of samyama). The scientific validity of this prescription can be easily checked by controlled experiments. Nobody can understand the modern physics without going through the prerequisite mathematical training. It will be foolish for any intelligent lay person to doubt the truth of modern physics without first undergoing the necessary training. Similarly one should draw conclusion about yogic methods only after disciplined practice of the eight steps of yoga.

(3) IP can provide insights regarding the role of mind in getting happiness and thus a better understanding of mind itself. Happiness lies in what the mind perceives as pleasurable and hence the true essence of happiness lies in mind and not in any external things. Once the body has experienced something mind is capable of recreating that experience in the absence of the actual conditions that gave rise to the experience in the first place. One can use this capacity of mind to create misery or ecstasy depending on one's ability to guide one's mind.

(4) There is a concept of the primordial sound in IP. Sometimes the possibility of having a universal language to communicate with everything in the universe is also mentioned. Modern physics tells us that the only universal language is at the level of gauge bosons and QV. Is there any connection between these two? Can a human mind be trained to transmit and receive at the level of QV?

(5) It is said that whole body is in the mind whereas the whole mind is not in the body. How does mind affect the body? If one believes in the answer given by IP then the results obtained in this regard by the western psychology appears to be the tip of the iceberg only. Can science verify these oriental claims through stringently controlled experiments?

Unification

Information

Information seems to be abstract and not real in the sense that, it lies inside our heads. But information can, not only exist outside the human brain (i.e., library, a CD, internet etc.) but also can be processed outside human brain (i.e., other animals, computers, etc.). Imagine a book written in a dead language, which nobody today can decipher. Does it contain information? Yes. Information exists. It does not need to be perceived or understood to exist. It requires no intelligence to interpret it. In this sense information is as real as matter and energy when it comes to the internal structure of the universe [Sto90]. But what we assume here is that information is more fundamental than matter and energy because everything in the universe can be ultimately reduced to information.

Information is neither material nor non–material. Both, quantum fields and thoughts can be reduced to information. If the human mind is not capable (by the methods known at present) of understanding this ultimate information then it is the limitation of the human mind. This may not remain so as time progresses. The whole of physical world can be reduced to information [Fri99]. Is information classical or quantum? There are enough indications from modern physics that although it can be classical at the everyday world it is quantum at the most fundamental level. The quantum information may have the advantage of describing the fuzziness of our experiences.

Formalism

Manifold

The manifold is an information field (I–field) for classical information (like that of Shannon or Fisher, etc.) Hilbert space of QT is the manifold to study quantum information. But if quantum information has to be given an ontological reality then it may be necessary for the manifold to be an extended Hilbert space.

Basic Constituent

A bit or a qubit is the basic entity of information depending on whether it is treated as classical or quantum respectively. Information can be of two types: kinetic and structural, but they are convertible to each other [Sto90].

Evolution

All organized systems contain information and addition of information to a system manifests itself by causing the system to become more organized or reorganized. The laws for evolution of information are essentially laws of organization. Are these laws different from the physical laws? Is there an equivalent in the world of information of fundamental principles like principle of least action in physical world?

Guiding Principles

Optimization seems to be the guiding principle in the world of information. What gives rise to the structure in the information such that we acquire an understanding or meaning out of it? Is there a principle of least information to be satisfied by all feasible structures?

Interactions

The interaction at the level of information has to be the ultimate universal language. What could be that language? The only fundamental language known to us is that of the gauge fields that communicate at the level of QV. Could the gauge fields serve as quanta of information? How far is this language from the conventional language? Can this help us to communicate with not only with other creatures incapable of our conventional language but also with the inanimate world? Time is not yet ripe to answer these questions.

Composite Systems

How can every composite system of information (like a gene, or a galaxy) be expressed in terms of bits or qubits? Does the holographic principle also apply to information?

Discussions

Consciousness and Information

There is no doubt that sooner or later all attributes of consciousness can be reduced to information. This is just a matter of time and progress in technology. That will complete the understanding of consciousness at the gross level of phenomena but will harbinger the understanding of consciousness at the subtle level of laws. The synthesis of the phenomenological studies of consciousness will be possible by treating information as the most basic ontological entity, which can unify mind and matter. The emergence of consciousness will be understood in terms of nonlinear, far–from–equilibrium complex processes that lead to spontaneous self–organization and adaptation of structures in the manifold of quantum information.

Consciousness will be seen as the ability to process quantum information in an effective way. Depending on the degree of complexity involved the processing would encompass activities starting from the way a planet knows which is the path of least action to the way modern supercomputers do simulations of reality to the way a scientist makes a discovery or an artist traps beauty on a canvass through the nuances of truth. The limit points of unconscious and super–conscious would correspond to the limiting cases of no information processing and infinite information processing respectively. Subconscious will be interpreted as partial information processing.

Every entity in the universe has to take a decision at every moment of time for its existence although the word existence may mean different things to different entities. The chance for continuation of existence is enhanced if the best decision on the basis of available information is taken. This is a process of optimization and the more conscious an entity is more is its ability to optimize.

Limitations of understanding

Is there any fundamental principle (or, theorem) that puts limit on the understanding of both mind and matter by reducing them to information and then applying methodology of physical sciences to understand life and consciousness as emergent phenomena? Since this approach heavily relies on mathematics the limitations of deductive logic as pointed out by Gödel in his famous incompleteness theorem may put the first limit. The second constraint may come from QT if it turns out (after having rigorous information theoretic formulation of both matter and mind) that the information related to mind is complementary to the information found in matter. I personally feel that this is quite unlikely because I believe that information at the fundamental level cannot be dualistic.

Conclusions

Unlike the Cartesian duality between mind and body, understanding consciousness requires first to understand matter and mind in a unified way. This can be achieved by giving information the most primary status in the universe. Then a generalized theory of quantum information dynamics has to be formulated (see the Table 3.1). The line of attack here involves three steps [Sam01]:

(1) understanding emergent phenomena and complexity in inanimate systems,
(2) understanding life as emergent phenomena, and
(3) understanding consciousness as emergent phenomena.

The attributes of consciousness can be understood only by a prudent application of both reductionism and holism. But the emergence of consciousness will be understood as an emergent phenomenon in the sense of structural organizations in the manifold of information to yield feasible structures through which we attribute meaning and understanding to the world.

3.2.6 Quantum Consciousness

For conscious states and brain states to mirror one another in any species, thereby establishing what von Neumann calls a psycho–physical parallelism, these intrinsically different states must evolve together and interact with one other during their time of evolution. Standard physics makes no provision for an interaction of this kind, but a quantum–mechanical opening for an objective/subjective interaction is shown to exist in [Mou95, Mou98, Mou99]. In this subsection, following this approach, we present a model of *quantum consciousness*.

Our theory of subjective evolution calls for the existence of a Central Mechanism (CM) within an evolving organism, which contains presently unknown components of the nervous system. The function of a CM is to reduce quantum–mechanical superpositions within the nervous system, and to

Table 3.1. A generalized structure of quantum information dynamics (modified and adapted from [Sam01].)

Character	Physical	Biological	Mental	Information
Manifold	SpT $(3+1)$, Hilbert Space QV, SpT (10)	SpT $(3+1)$, Hilbert Space	M–Space (Abstract Mathematical Space)	I–Field, Extended Hilbert Space
Basic Constituents	Wave ψ–function Quantum fields Strings p–branes	Cell, Neuron, Microtubule Cytoskeleton	Idea (based on self or universal perception	Bit (classical) Qubit (quantum)
Evolution	Physical Laws (mostly diff. equations)	Physical Laws Biological Laws	Laws for evolution of thought	Laws for evolution of organization
Guiding Principles	Symmetry (Group Theoretical)	Survivability Intentionality	Happiness–Ability	Optimization
Interactions	Gravity, Electro–weak, Strong Nuclear Interaction	Chemicals, Electric Signals, Language, Gesture	Primordial Sound or Vibrations	Local, and Nonlocal (EPR) channels
Composite Systems	Many–body Systems with or without interactions	Plants, Animals	Thought (Ensemble of Ideas with ordering)	Complex Systems with Hierarchy in Organization

simultaneously give rise to a conscious experience of the eigenvalues of the reduction. This accords with von Neumann's requirement that a quantum–mechanical state reduction is accompanied by an observer's conscious experience of the measured variables. At the present time, no one knows what there is about a conscious organism that gives rise to either consciousness or state reduction. We simply combined these two mysteries inside the CM, thereby placing our ignorance in a black–box so we can ask another question, namely: how do physical and mental states evolve interactively to insure the psycho–physical parallelism?

The model in [Mou95, Mou98, Mou99] requires that a conscious organism spontaneously creates a profusion of macroscopic quantum–mechanical superpositions consisting of different neurological configurations. A mechanism for this generation is proposed by H. Stapp in [Sta93]. The result is a superposition of different neurological states, each of which may be accompanied by a different subjective experience. A reduction to a single eigenstate is not assumed to be triggered microscopically along the lines of [GRW86]; but rather, it is assumed to occur in response to a macroscopic event. It occurs the moment an emerging subjective state becomes actively conscious in one of the macroscopic neurological components of a Stapp superposition. The consciousness that is associated with such a reduction is assumed to fade the moment reduction is complete, and the resulting subjective *pulse* is supposedly followed by similar pulses in rapid succession. This can make the subject aware of an apparent continuum of consciousness.

Presumably, any reduction of this kind is accompanied by a reduction of all other parts of the organism as well as all those parts of the external world that are correlated with it. This means that a second observer, coming on the heels of the first, will make an observation in agreement with the first. More formally, a measurement interaction establishes correlations between the eigenstates $|a_i\rangle$ of some apparatus (with discrete variables a_i), eigenstates of a first observer $|\Phi_i\rangle$, and eigenstates of a second observer $|\Theta_i\rangle$, such that the total state prior to reduction is given by [Mou95, Mou98, Mou99]

$$|\Psi\rangle = \Sigma_i C_i |a_i\rangle |\Phi_i\rangle |\Theta_i\rangle .$$

The coefficient C_i is the probability amplitude that the apparatus is in state $|a_i\rangle$. Let the first observer become consciously aware of the apparatus variable a_k. The resulting reduction is a projection in Hilbert space that is found by applying the projection operator of that observer $|\Phi_k\rangle\langle\Phi_k|$ to the total state.

$$|\Phi_k\rangle\langle\Phi_k||\Psi\rangle = C_k|a_k\rangle|\Phi_k\rangle|\Theta_k\rangle \qquad \text{(1st reduction)}$$

Let the second observer then become consciously aware of the apparatus variable a_m. The subsequent reduction is found by applying the projection operator of that observer $|\Phi_m\rangle\langle\Phi_m|$ to the first reduction.

$$|\Theta_m\rangle\langle\Theta_m|C_k|a_k\rangle|\Phi_k\rangle|\Theta_k\rangle = \delta_{km}C_k|a_k\rangle|\Phi_k\rangle|\Theta_m\rangle \qquad \text{(2nd reduction)}$$

Only if $m = k$ is the probability non–zero that the second observer will make a measurement. The second observer therefore confirms the results of the first observer that the apparatus has been left in the eigenstate $|a_k\rangle$.

Again, many of the particulars of a reduction (such as its nonlinearly) are ignored in this subsection so we can concentrate on the influence of subjective states on physiological states. To this end we require that 'when the emerging subjective states of a neurological superposition are different from one another, they will generally exert an influence on their relative probability amplitudes that is a function of that difference' [Mou95, Mou98, Mou99]. In particular, we imagine that when a 'painful' subjective state emerges in superposition with a 'pleasurable' subjective state, the probability amplitude of the painful

$$N \begin{cases} W\,[\text{no pain}]\ \ (0.5) \rightarrow W\,[\text{no pain}]\ (1.0) \\ C\,[\text{pain}]\ \ \ \ \ \ (0.5) \qquad\qquad\qquad\qquad (0.0) \end{cases}$$

Fig. 3.7. Nervous system of the first primitive organism (modified and adapted from [Mou99] – see text for explanation).

state will be decreased relative to the probability amplitude of the pleasurable state.

No currently known observation contradicts this conjecture, for no previously reported experiment deals specifically with the creation of different observers experiencing different degrees of pain, arising on different components of a quantum mechanical superposition.

Let N in Figure 3.7 represent the nervous system of the first primitive organism that makes a successful use of the subjective experience of 'pain'. In [Mou95, Mou98, Mou99] this creature was imagined to be a fish. It is supposed that the fish makes contact with an electric probe, at which time its nervous system splits into a superposition (*via* the Stapp mechanism) consisting of a withdrawal behavior W that is accompanied by [no pain], and a continued contact behavior C that is accompanied by [pain]. The probability of survival of each component in this highly artificial model is initially assumed to be 0.5. However, because of the hypothetical influence of subjective pain on probability amplitudes, only the withdrawal state is assumed to survive the reduction in this idealized example. State reduction in Figure 3.7 is represented by the horizontal arrow. If W is furthermore a good survival strategy from the point of view of evolution, then the association W[no pain] and C[pain] will serve the species well, whereas a wrong association W[pain] and C[no pain] will lead to its demise.

It does not matter to the above argument if the variables are 'pleasure/pain' or some other range of subjective experiences. If a subjective experience like 'A' increases the probability amplitude of an escape behavior, and if a subjective experience like 'B' diminishes the probability amplitude of that behavior, and if the escape is one that moves the creature away from something that is dangerous to its health, then a distant descendent will experience 'A' associated with life supporting escapes, and 'B' associated with life threatening failures–to–escape. It is apparent that the quality of the experience does not matter. We require only that the subjective experience in question has a predictable plus or minus effect on the probability amplitudes within a superposition, and the survival mechanisms of evolution will do the rest. They will insure that the eventual subjective life of a surviving species mirrors its experiences in a definite and predictable way thereby establishing a reliable psycho–physical parallelism.

We assume that ordinary *perception* do not have this effect. They do not give rise to the hypothetical feedback. In Figure 3.8 we imagine the existence of an externally imposed two component superposition consisting of

$$\left\{\begin{array}{ll} e^{i\varphi}e_1 & (0.5) \\ e_2 & (0.5)\end{array}\right\} N_o \begin{array}{l} \nearrow (e^{i\varphi'}eN)_1\,[x_1] \quad (0.5) \rightarrow (eN)_1\,[x_1] \quad (0.5) \\ \searrow (eN)_2\,[x_2] \quad (0.5) \rightarrow (eN)_2\,[x_2] \quad (0.5)\end{array}$$

Fig. 3.8. Pure state reduction (modified and adapted from [Mou99] – see text for explanation).

$$\left\{\begin{array}{ll} e^{i\varphi}e_1 & (0.5) \\ e_2 & (0.5)\end{array}\right\} N_o \begin{array}{l} \nearrow (e^{i\varphi'}eN)_1\,[\text{pain}] \quad (0.5) \rightarrow \\ \searrow (eN)_2\,[\text{no pain}] \quad (0.5) \rightarrow \end{array}$$

Fig. 3.9. Pure state reduction with and without a 'pain' (modified and adapted from [Mou99] – see text for explanation).

environments e_1 and e_2, which is produced by using, say, a β source. The two environments are assumed to have equal probability, and are allowed to interact with the subject's nervous system given by N_0. Before a reduction can occur, two conscious states emerge from the interaction represented by the superposition of $(eN)_1[x_1]$ and $(eN)_2[x_2]$, where the conscious part shown in brackets is the observed eigenvalue x associated with components 1 and 2. Since we require that an observer of the 'perceived' variable x cannot affect the probability of x, the pure state reduces to a mixture having the same probability as the initial superposition (horizontal arrows in Figure 3.8). State e_i represents the relevant laboratory apparatus together with the wider environment with which it is entangled. The phase angles ϕ and ϕ' are definite, but they are not localized to manageable parts of the apparatus [Mou95, Mou98, Mou99]. We call them 'arbitrary' to indicate that their values are not practically calculable, and to emphasize the lack of coherence between these 'macroscopic' components.

On the other hand, if 'pain' were the variable in Figure 3.8 rather than the externally perceived variable x, it is suggested by the hypothesis of [Mou95, Mou98, Mou99] that the resulting mixture might no longer be a 50–50 split. This possibility is represented in Figure 3.9, where the final mixture probabilities are left unspecified because they must be discovered by observation.

The Experiment

Two scalers L and R recording local background radiation are placed side–by–side in Figure 3.10. Their outputs are fed to a selector box that chooses channels L or R, depending on which is the first to record a single count after the selector has been turned on. A 20 V signal is then emitted from the output of the chosen channel. The output on the R–channel is unused, but the L–output closes a relay that puts 80 volts across two metal bars. Two seconds after the selection, an L or R–light goes on indicating which channel was selected. A finger placed across the metal bars will receive a painful 80V shock when the L–channel is selected [Mou95, Mou98, Mou99].

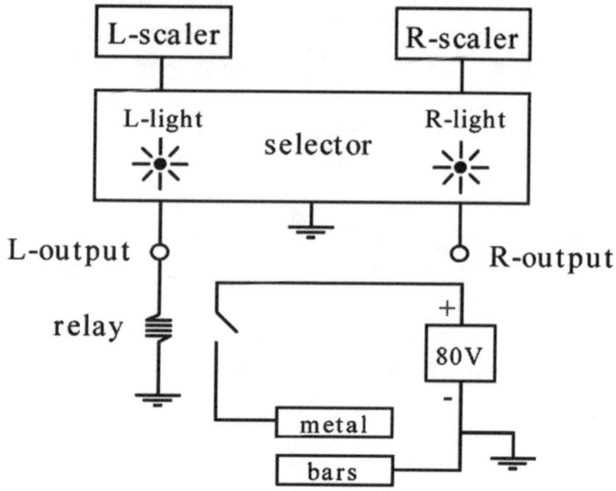

Fig. 3.10. The experimental set up (modified and adapted from [Mou99] – see text for explanation).

This apparatus allows us to carry out the experiments diagramed in Figures 3.8 and 3.9. If the selector is initiated in the absence of an observer, we say that the system will become a macroscopic superposition given by $(e^{i\phi}e_1 + e_2)$, where e_1 is the entire apparatus following an L-channel activation, and e_2 is the entire apparatus following an R–channel activation. The incoherence of the two components (represented by the arbitrary angle ϕ) is generally understood to mean that the system is indistinguishable from a classical mixture, since interference between these macroscopic components is not possible. However, for reasons given in previous papers, we claim that the final state is really an incoherent quantum–mechanical superposition rather than a classical mixture.[24] The lack of interference between the components has no bearing on our result because the hypothetical effect described here relates to, and directly affects, probability amplitudes only. The effect we are looking for should be observable with or without coherence between L and R.

If an observer is present and exposed only to the L–light or the R–light, then a reduction will occur like the one in Figure 3.8, where eigenvalues x_1 and x_2 represent a conscious experience of one or the other of those lights. If the observer is exposed only to a conscious experience of "pain or no pain" through his finger across the metal bars, then a reduction like the one in Figure 3.9

[24] The uncertainty associated with a classical mixture state represents an outsider's ignorance, whereas a pure quantum mechanical state superposition represents an uncertainty that is intrinsic to the system. Following von Neumann, we assume that the initial intrinsic uncertainty (concerning which of the scalers fires first) will remain an intrinsic uncertainty until it is reduced by 'observation'. Hence, the apparatus will remain a macroscopic pure state quantum mechanical superposition until an observation occurs.

will occur. This experiment may not appear to be quantum–mechanical, but it is quantum–mechanical by virtue of the particular hypothesis that is being tested in Figure 3.9.

The equipment in Figure 3.10 was used for a total of 2500 trials, each consisting of two parts. The experimenter's finger was first placed across the metal bars, the selector was turned on, and a 'shock' or 'no shock' was recorded before the lights were observed. In the second part of each trial the finger was replaced by an equivalent resistance, the selector was again initiated, and the appearance of the L or R channel light was recorded [Mou95, Mou98, Mou99].

Total number of trials . . . . $N = 2500$,
Number of shocks received in the first part . . . . $N_S = 1244$,
Number of times the L–light went on in the second part . . . . . $N_L = 1261$.

There are three possible outcomes of a single trial. Either the difference $N_L - N_S$ increases, or it decreases, or it remains the same. The three possibilities are represented by the variables u (increase) occurring with a probability p, and d(decrease) with a probability q, and e (remain the same) with a probability r. It was found in the experiment that $u = 632$ and $d = 615$ after 2500 trials.

If we approximate $p_0 = N_L/N$ to be the probability that the left channel fires in the second part of each trial (absent the finger), and $q_0 = 1 - p_0$ to be the probability that the right channel fires in the second part of each trial, then

$$p_0 = 1261/2500 = 0.5044 \qquad q_0 = 0.4956$$

Assuming as a null hypothesis that there is no statistical difference between the displacement of a finger across the metal bars and an equivalent resistor, we have $p = p_0 q_0$, $q = q_0 p_0$, and $r = p_0^2 + q_0^2$, giving

$$p = 0.2500 \quad q = 0.2500 \quad r = 0.5000$$

The variances of $(u + d)$ and $(u - d)$ are [Mou95, Mou98, Mou99]

$$\sigma^2(u+d) = <(u+d)^2> - <u+d>^2 = \sigma^2(u) + \sigma^2(d) + X$$
$$\sigma^2(u-d) = <(u-d)^2> - <u-d>^2 = \sigma^2(u) + \sigma^2(d) - X$$

therefore

$$\begin{aligned} \sigma^2(u-d) &= 2\sigma^2(u) + 2\sigma^2(d) - \sigma^2(u+d) \\ &= 2p(q+r)N + 2q(p+r)N - r(p+q)N \qquad \text{or} \\ \sigma(u-d) &= [[4pq + r(p+q)]N]^{1/2} = [N/2]^{1/2} = 35.4 \end{aligned}$$

Our alternative hypothesis is that $u - d$ is significantly different from 0. But from the data, $u - d = N_L - N_S = 17$ after 2500 trials, and this is well within the above the standard deviation around 0. The separate variables u and d are also within the standard deviation

$$\sigma(u) = \sigma(d) = [p(q+r)N]^{1/2} = 21.7$$

of their expected value of 625.

One can always argue that the statistics are inadequate to reveal a significant difference between u and d. However, they are sufficient to convince us that the presence of pain on one component of this externally imposed superposition has no significant effect on the outcome. We therefore conclude that the reduction in Figure 3.9 is not affected by the subjective content of the square brackets in that figure.

Bio–Active Peptides

Neurological communication depends on the diffusion of chemical neurotransmitters across the synaptic junction between neurons. There is another communication system within the body that makes use of chemicals that are produced at one site and received at another; but in this case, the distances between a production and receiver sites are macroscopic. About 95% of these chemical communicators are peptides, which are mini–proteins consisting of up to 100 amino acids having a maximum atomic mass of 10,000 u. Their classical dimensions are $\Delta \mathrm{x}$ =10 nm at most, which we assume approximates their size close to the production site [Per97]. Therefore, Heisenberg tells us that the minimum quantum–mechanical uncertainty in the velocity of one of these free peptides is $\Delta \mathrm{v}$ =0.63 mm/s. Peptides are carried through intercellular space by blood and cerebrospinal fluid. They do not move very far in a tenth of a second, but in that time the Heisenberg uncertainty in position of a peptide will be at least $\Delta \mathrm{s} = \Delta \mathrm{v} \Delta \mathrm{t} = 63$ mm. This is an enormous uncertainty of position relative to one of the peptide receptor sites which has a size similar to that of the peptide, and which is often separated from its neighbors by comparable distances. Therefore, quantum–mechanical uncertainty is an important factor in determining the probability that a given peptide is captured by a given receptor [Mou95, Mou98, Mou99].

Stapp's mechanism for introducing quantum–mechanical superpositions into the brain relies on the uncertainty in the position of calcium ions in neuron synapses. We suggest that peptides represent another possible source of super-positions that may be just as widespread. And because peptides play an important role in the chemistry of the body, they too may have a significant quantum–mechanical influence on behavior.

As with the Stapp mechanism, one might object that the uncertainty associated with the peptide's classical diffusion during its migration will overwhelm the quantum–mechanical uncertainty, or that a large number of migrating molecules will obscure all quantum–mechanical effects. However, the classical uncertainty associated with many–particle ensembles has only to do with our ignorance of initial conditions. In reality, the only uncertainties a receptor will see are those associated with an incoherent quantum–mechanical superposition of pure peptide states. This superposition will have as many components as there are peptide molecules involved. And since our hypothetical influence acts through the amplitude of these components, the presence of a large number of independent particles will only increase the hypothetical influence.

Drugs

There are many drugs that can be introduced into the body that will compete with endogenous peptides to occupy the body's receptor sites, and some of these drug molecules are small enough to have a very large quantum mechanical uncertainty of position. For this reason, peptide/drug superpositions are more promising for the purpose of experimental manipulation than calcium ion super-positions.

For example, *endorphins* are peptides that unite with special receptors to eliminate pain and/or produce euphoria. They and their receptors can be found everywhere in the body, but they are most intensely located in the limbic system of the brain. There is a drug called *naloxone* that is a strong competitor with the endorphins to occupy the same receptors, and it has the property that it reverses the analgesic/pleasurable effects of the endorphins [Per97, Sny86, Lev88] If endorphin molecules and externally administered naloxone molecules are in quantum–mechanical superposition with one another as their sizes and likely time together suggests, and if they both compete with one another for successful attachment to the same receptor site, then the ratio of endorphin attachments to naloxone attachments would (according to our hypothesis) be a function of the competing subjective states.

Evolutionary Advantage

It was pointed out in [Mou95, Mou98, Mou99] that our evolutionary mechanism of objective–subjective interaction (represented by Figure 3.7) does not insure that a creature evolving under its influence will evolve more quickly or be more successful than a creature evolving strictly as an automaton. That will be true as well of the modified model in sects. 3-5. However, it is not unreasonable to suppose that both conscious evolution and autonomic evolution might work separately and in tandem with one another. The kinds of neurological changes that are necessary for autonomic evolution might very well be independent of the kinds of neurological changes that are necessary for quantum/consciousness evolution. If that is so, and if these two processes work in tandem, then the evolution of the organism will be faster than either the autonomic route by itself, or the conscious route by itself. One would then be able to say that the introduction of consciousness as proposed here will always work to the advantage of the organism.

3.2.7 Quantum–Like Psychodynamics

In this section, which is written in the fashion of the *quantum brain*, we present the top level of natural biodynamics, using geometrical generalization of the *Feynman path integral*. To formulate the basics of *force–field psychodynamics*, we use the *action–amplitude picture* of the $BODY \rightleftarrows MIND$ adjunction:

↓ Deterministic (causal) world of *Human BODY* ↓

$$Action : S[q^n] = \int_{t_{in}}^{t_{out}} (E_k - E_p + Wrk + Src^{\pm})\, dt$$

— —

$$Amplitude : \langle out|in\rangle = \oint\!\!\!\!\!\!\sum \mathcal{D}[w_n q^n]\, \mathrm{e}^{\mathrm{i}S[q^n]}$$

↑ Probabilistic (fuzzy) world of *Human MIND* ↑

In the action integral, E_k, E_p, Wrk and $Src^{\pm}$ denote the kinetic end potential energies, work done by dissipative/driving forces and other energy sources/sinks, respectively. In the amplitude integral, the peculiar sign $\oint\!\!\!\!\!\!\sum$ denotes integration along smooth paths and summation along discrete Markov chains; i is the imaginary unit, w_n are synaptic–like weights, while $\mathcal{D}$ is the Feynman path differential (defined below) calculated along the configuration trajectories q^n. The action $S[q^n]$, through the *least action principle* $\delta S = 0$, leads to all biodynamic equations considered so far (in generalized Lagrangian and Hamiltonian form). At the same time, the action $S[q^n]$ figures in the exponent of the path integral $\oint\!\!\!\!\!\!\sum$, defining the probability transition amplitude $\langle out|in\rangle$. In this way, the whole body dynamics is incorporated in the mind dynamics. This *adaptive path integral* represents an *infinite–dimensional neural network*, suggesting an infinite capacity of human brain/mind.

For a long time the cortical systems for *language and actions* were believed to be independent modules. However, according to the recent research of [Pul05], as these systems are reciprocally connected with each other, information about language and actions might interact in distributed neuronal assemblies. A critical case is that of action words that are semantically related to different parts of the body (e.g. 'pick', 'kick', 'lick',...). The author suggests that the comprehension of these words might specifically, rapidly and automatically activate the motor system in a somatotopic manner, and that their comprehension rely on activity in the action system.

Motivational Cognition in the Life Space Foam

Applications of nonlinear dynamical systems (NDS) theory in psychology have been encouraging, if not universally productive/effective [Met97]. Its historical antecedents can be traced back to Piaget's [PHE92] and Vygotsky's [Vyg82] interpretations of the dynamic relations between action and thought, Lewinian theory of social dynamics and cognitive–affective development [Lew51, Gol67], and Bernstein's [Ber47] theory of self–adjusting, goal–driven motor action.

Now, both the original *Lewinian force–field theory* in psychology (see [Lew51, Gol67]) and modern decision–field dynamics (see [BT93, RBT01, BD02]) are based on the classical Lewinian concept of an individual's *life space.*[25] As a topological construct, Lewinian life space represents a person's psychological environment that contains *regions* separated by dynamical permeable *boundaries.* As a field construct, on the other hand, the life space is not empty: each of its regions is characterized by *valence* (ranging from positive or negative and resulting from an interaction between the person's *needs* and the dynamics of their *environment*). Need is an energy construct, according to Lewin. It creates *tension* in the person, which, in combination with other tensions, initiates and sustains behavior. Needs vary from the most primitive urges to the most idiosyncratic intentions and can be both internally generated (e.g., thirst or hunger) and stimulus–induced (e.g., an urge to buy something in response to a TV advertisement). Valences are, in essence, personal values dynamically derived from the person's needs and attached to various regions in their life space. As a field, the life space generates forces pulling the person towards positively–valenced regions and pushing them away from regions with negative valence. Lewin's term for these forces is *vectors.* Combinations of multiple vectors in the life space cause the person to move from one region towards another. This movement is termed *locomotion* and it may range from overt behavior to cognitive shifts (e.g., between alternatives in a decision–making process). Locomotion normally results in crossing the boundaries between regions. When their permeability is degraded, these boundaries become *barriers* that restrain locomotion. Life space model, thus, offers a meta–theoretical language to describe a wide range of behaviors, from goal–directed action to intrapersonal conflicts and multi–alternative decision–making.

In order to formalize the Lewinian life–space concept, a set of *action principles* need to be associated to Lewinian force–fields, (loco)motion paths (representing mental abstractions of biomechanical paths [II05]) and life space geometry. As an extension of the Lewinian concept, in this paper we introduce a new concept of *life–space foam* (LSF, see Figure 3.11). According to this new concept, Lewin's life space can be represented as a *geometrical functor* with globally smooth macro–dynamics, which is at the same time underpinned by wildly fluctuating, non–smooth, local micro–dynamics, describable by *Feynman's*: (i) *sum–over–histories* $\oint_{paths}$, (ii) *sum–over–fields* $\oint_{fields}$, and (iii) *sum–over–geometries* $\oint_{geom}$.

[25] The work presented in this subsection has been developed in collaboration with Dr. Eugene Aidman, Senior Research Scientist, Human Systems Integration, Land Operations Division, Defence Science & Technology Organisation, Australia.

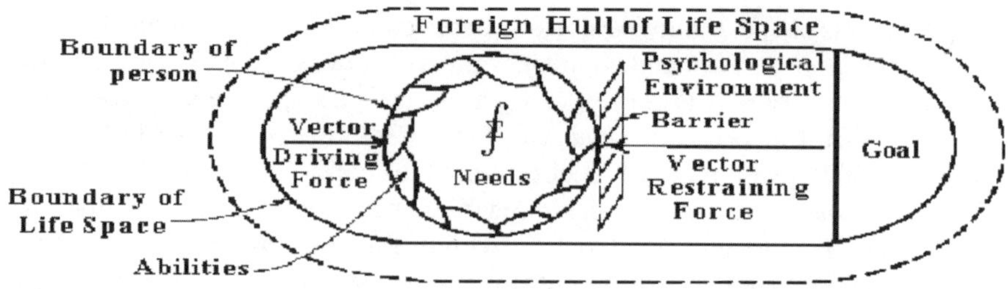

Fig. 3.11. Diagram of the *life space foam*: Lewinian life space with an adaptive path integral acting inside it and generating microscopic fluctuation dynamics.

LSF is thus a two–level *geometrodynamical functor*, representing these two distinct types of dynamics within the Lewinian life space. At its *macroscopic spatio–temporal level*, LSF appears as a 'nice & smooth' geometrical functor with globally predictable dynamics – formally, a smooth $n-$dimensional manifold M with local Riemannian metrics $g_{ij}(x)$, smooth force–fields and smooth (loco)motion paths, as conceptualized in the Lewinian theory. To model the global and smooth macro–level LSF–paths, fields and geometry, we use the general physics–like *principle of the least action.*

Now, the apparent smoothness of the macro–level LSF is achieved by the existence of another level underneath it. This *micro–level* LSF is actually a collection of wildly fluctuating force–fields, (loco)motion paths, curved regional geometries and topologies with holes. The micro–level LSF is proposed as an extension of the Lewinian concept: it is characterized by uncertainties and fluctuations, enabled by microscopic time–level, microscopic transition paths, microscopic force–fields, local geometries and varying topologies with holes. To model these fluctuating microscopic LSF–structures, we use three instances of *adaptive path integral*, defining a multi–phase and multi–path (also multi–field and multi–geometry) *transition* process from *intention* to the goal–driven *action.*

We use the new LSF concept to develop modelling framework for motivational dynamics (MD) and induced cognitive dynamics (CD).

According to Heckhausen (see [Hec77]), *motivation* can be thought of as a process of *energizing* and *directing the action.* The process of energizing can be represented by Lewin's *force–field analysis* and Vygotsky's *motive formation* (see [Vyg82, AL91]), while the process of directing can be represented by *hierarchical action control* (see [Ber47, Ber35, Kuh85]).

Motivation processes both precede and coincide with every goal–directed action. Usually these motivation processes include the sequence of the following four feedforward *phases* [Vyg82, AL91]: (*)

1. *Intention Formation* $\mathcal{F}$, including: decision making, commitment building, etc.
2. *Action Initiation* $\mathcal{I}$, including: handling conflict of motives, resistance to alternatives, etc.

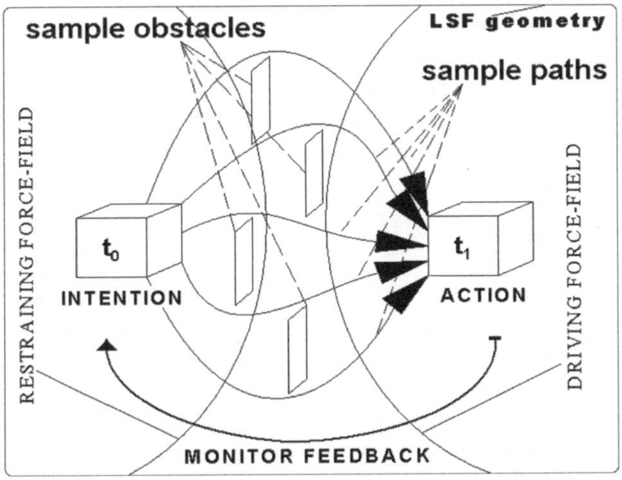

Fig. 3.12. *Transition–propagator* corresponding to each of the motivational phases $\{\mathcal{F},\mathcal{I},\mathcal{M},\mathcal{T}\}$, consisting of an ensemble of feedforward paths propagating through the 'wood of obstacles'. The paths affected by driving and restraining force–fields, as well as by the local LSF–geometry. Transition goes from *Intention*, occurring at a sample time instant t_0, to *Action*, occurring at some later time t_1. Each propagator is controlled by its own *Monitor* feedback. All together they form the transition functor $\mathcal{T}A$.

3. *Maintaining the Action* $\mathcal{M}$, including: resistance to fatigue, distractions, etc.
4. *Termination* $\mathcal{T}$, including parking and avoiding addiction, i.e., staying in control.

With each of the phases $\{\mathcal{F},\mathcal{I},\mathcal{M},\mathcal{T}\}$ in (*), we can associate a *transition propagator* – an ensemble of (possibly crossing) feedforward paths propagating through the 'wood of obstacles' (including topological holes in the LSF, see Figure 3.12), so that the complete *transition functor* $\mathcal{T}A$ is a product of propagators (as well as sum over paths). All the phases–propagators are controlled by a unique *Monitor* feedback process.

In this subsection we propose an *adaptive path integral* formulation for the motivational–transition functor $\mathcal{T}A$. In essence, we sum/integrate over different paths and make a product (composition) of different phases–propagators. Recall that this is the most general description of the general *Markov stochastic process*.

We will also attempt to demonstrate the utility of the same LSF–formalisms in representing cognitive functions, such as memory, learning and decision making. For example, in the classical *Stimulus encoding* → *Search* → *Decision* → *Response* sequence [Ste69, Ash94], the environmental input–triggered *sensory memory* and *working memory* (WM) can be interpreted as operating at the micro–level force–field under the executive control of the *Monitor* feedback, whereas *search* can be formalized as a *control* mechanism guiding retrieval from the long–term memory (LTM, itself shaped by learning) and filtering material relevant to decision making into the WM. The essential measure of these mental processes, the *processing speed* (essentially determined by Sternberg's reaction–time) can be represented by our (loco)motion speed $\dot{x}$.

Six Faces of the Life Space Foam

The LSF has three forms of appearance: $paths + field + geometries$, acting on both macro–level and micro–level, which is six modes in total. In this section, we develop three least action principles for the macro–LSF–level and three adaptive path integrals for the micro–LSF–level. While developing our psycho–physical formalism, we will address the behavioral issues of motivational fatigue, learning, memory and decision making.

General Formalism

At both macro– and micro–levels, the total LSF represents a union of transition paths, force–fields and geometries, formally written as

$$\begin{aligned} LSF_{total} &:= LSF_{paths} \bigcup LSF_{fields} \bigcup LSF_{geom} \\ &\equiv \oint_{paths} + \oint_{fields} + \oint_{geom} . \end{aligned} \tag{3.114}$$

Corresponding to each of the three LSF–subspaces in (3.114) we formulate:

1. The *least action principle*, to model deterministic and predictive, macro–level MD & CD, giving a unique, global, causal and smooth path–field–geometry on the macroscopic spatio–temporal level; and
2. Associated *adaptive path integral* to model uncertain, fluctuating and probabilistic, micro–level MD & CD, as an ensemble of local paths–fields–geometries on the microscopic spatio–temporal level, to which the global macro–level MD & CD represents both time and ensemble *average* (which are equal according to the *ergodic hypothesis*).

In the proposed formalism, transition paths $x^i(t)$ are affected by the force–fields $\varphi^k(t)$, which are themselves affected by geometry with metric g_{ij}.

Global Macro–Level of LSF_{total}. In general, at the *macroscopic* LSF–level we first formulate the *total action* $S[\Phi]$, the central quantity in our formalism that has psycho–physical dimensions of $Energy \times Time = Effort$, with immediate cognitive and motivational applications: *the greater the action – the higher the speed of cognitive processes and the lower the macroscopic fatigue* (which includes all sources of physical, cognitive and emotional fatigue that influence motivational dynamics). The action $S[\Phi]$ depends on macroscopic paths, fields and geometries, commonly denoted by an abstract field symbol Φ^i. The action $S[\Phi]$ is formally defined as a temporal integral from the *initial* time instant t_{ini} to the *final* time instant t_{fin},

$$S[\Phi] = \int_{t_{ini}}^{t_{fin}} \mathfrak{L}[\Phi]\, dt, \tag{3.115}$$

with *Lagrangian density* given by

$$\mathfrak{L}[\Phi] = \int d^n x\, \mathcal{L}(\Phi_i, \partial_{x^j}\Phi^i),$$

where the integral is taken over all n coordinates $x^j = x^j(t)$ of the LSF, and $\partial_{x^j}\Phi^i$ are time and space partial derivatives of the Φ^i–variables over coordinates.

Second, we formulate the *least action principle* as a minimal variation δ of the action $S[\Phi]$

$$\delta S[\Phi] = 0, \tag{3.116}$$

which, using techniques from the calculus of variations gives, in the form of the so–called Euler–Lagrangian equations, a shortest (loco)motion path, an extreme force–field, and a life–space geometry of minimal curvature (and without holes). In this way, we effectively derive a *unique globally smooth transition functor*

$$\mathcal{TA} : INTENTION_{t_{ini}} \Rrightarrow ACTION_{t_{fin}}, \tag{3.117}$$

performed at a macroscopic (global) time–level from some initial time t_{ini} to the final time t_{fin}.

In this way, we get macro–objects in the global LSF: a single path described Newtonian–like equation of motion, a single force–field described by Maxwellian–like field equations, and a single obstacle–free Riemannian geometry (with global topology without holes).

For example, recall that in the period 1945–1949, John Wheeler and Richard Feynman developed their *action–at–a–distance electrodynamics* [WF49], in complete experimental agreement with the classical Maxwell's electromagnetic theory, but at the same time avoiding the complications of divergent self–interaction of the Maxwell's theory as well as eliminating its infinite number of field degrees–of–freedom. In Wheeler–Feynman view, "Matter consists of electrically charged particles," so they found a form for the action directly involving the motions of the charges only, which upon variation would give the Newtonian–like equations of motion of these charges. Here is the expression for this action in the flat space–time, which is in the core of quantum electrodynamics:

$$\begin{aligned} S[x; t_i, t_j] &= \frac{1}{2} m_i \int (\dot{x}^i_\mu)^2\, dt_i + \frac{1}{2} e_i e_j \int\!\!\int \delta(I^2_{ij})\, \dot{x}^i_\mu(t_i) \dot{x}^j_\mu(t_j)\, dt_i dt_j \\ &\text{with} \\ I^2_{ij} &= \left[x^i_\mu(t_i) - x^j_\mu(t_j)\right] \left[x^i_\mu(t_i) - x^j_\mu(t_j)\right], \end{aligned} \tag{3.118}$$

where $x^i_\mu = x^i_\mu(t_i)$ is the four–vector position of the ith particle as a function of the proper time t_i, while $\dot{x}^i_\mu(t_i) = dx^i_\mu/dt_i$ is the velocity four–vector. The first term in the action (3.118) is the ordinary mechanical action in Euclidean space, while the second term defines the electrical interaction of the charges, representing the Maxwell–like field (it is summed over each pair of charges; the factor $\frac{1}{2}$ is to count each pair once, while the term $i = j$ is omitted to avoid self–action; the interaction is a double integral over a delta function of the square of space–time interval I^2 between two points on the paths; thus, interaction occurs only when this interval vanishes, that is, along light cones [WF49]).

Now, from the point of view of Lewinian geometrical force–fields and (loco)motion paths, we can give the following life–space interpretation to the Wheeler–Feynman action (3.118). The mechanical–like locomotion term occurring at the single time t, needs a covariant generalization from the flat 4D Euclidean space to the nD smooth Riemannian manifold, so it becomes (see e.g., [II06b])

$$S[x] = \frac{1}{2}\int_{t_{ini}}^{t_{fin}} g_{ij}\,\dot{x}^i\dot{x}^j\,dt,$$

where g_{ij} is the Riemannian metric tensor that generates the total 'kinetic energy' of (loco)motions in the life space.

The second term in (3.118) gives the sophisticated definition of Lewinian force–fields that drive the psychological (loco)motions, if we interpret electrical charges e_i occurring at different times t_i as motivational charges – needs.

Local Micro–Level of LSF_{total}. After having properly defined macro–level MD & CD, with a unique transition map F (including a unique motion path, driving field and smooth geometry), we move down to the *microscopic* LSF–level of rapidly fluctuating MD & CD, where we cannot define a unique and smooth path–field–geometry. The most we can do at this level of *fluctuating uncertainty*, is to formulate an adaptive path integral and calculate overall probability amplitudes for ensembles of local transitions from one LSF–point to the neighboring one. This *probabilistic transition micro–dynamics* functor is defined by a multi–path (field and geometry, respectively) and multi–phase *transition amplitude* $\langle Action|Intention\rangle$ of corresponding to the globally–smooth transition map (3.117). This absolute square of this probability amplitude gives the *transition probability* of occurring the final state of $Action$ given the initial state of $Intention$,

$$P(Action|Intention) = |\langle Action|Intention\rangle|^2.$$

The total transition amplitude from the state of $Intention$ to the state of $Action$ is defined on LSF_{total}

$$\mathcal{TA} \equiv \langle Action|Intention\rangle_{total} : INTENTION_{t_0} \Rrightarrow ACTION_{t_1}\,, \qquad (3.119)$$

given by adaptive generalization of the Feynman's path integral [FH65, Fey72, Fey98]. The transition map (3.119) calculates the *overall probability amplitude* along a multitude of wildly fluctuating paths, fields and geometries, performing the *microscopic* transition from the micro–state $INTENTION_{t_0}$ occurring at initial micro–time instant t_0 to the micro–state $ACTION_{t_1}$ at some later micro–time instant t_1, such that all micro–time instants fit inside the global transition interval $t_0, t_1, \ldots, t_s \in [t_{ini}, t_{fin}]$. It is symbolically written as

$$\langle Action|Intention\rangle_{total} := \not\int\!\!\!\!\sum \mathcal{D}[w\Phi]\, \mathrm{e}^{\mathrm{i}S[\Phi]}, \tag{3.120}$$

where the Lebesgue integration is performed over all continuous $\Phi^i_{con} = paths + field + geometries$, while summation is performed over all discrete processes and regional topologies Φ^j_{dis}). The symbolic differential $\mathcal{D}[w\Phi]$ in the general path integral (3.120), represents an *adaptive path measure*, defined as a weighted product

$$\mathcal{D}[w\Phi] = \lim_{N\to\infty} \prod_{s=1}^{N} w_s d\Phi^i_s, \qquad (i = 1, \ldots, n = con + dis), \tag{3.121}$$

which is in practice satisfied with a large N corresponding to infinitesimal temporal division of the four motivational phases (*). Technically, the path integral (3.120) calculates the *amplitude* for the transition functor $\mathcal{TA} : Intention \Rrightarrow Action.$

In the exponent of the path integral (3.120) we have the action $S[\Phi]$ and the imaginary unit $i = \sqrt{-1}$ (i can be converted into the real number -1 using the so–called *Wick rotation*, see next subsection).

In this way, we get a range of micro–objects in the local LSF at the short time–level: ensembles of rapidly fluctuating, noisy and crossing paths, force–fields, local geometries with obstacles and topologies with holes. However, by averaging process, both in time and along ensembles of paths, fields and geometries, we recover the corresponding global MD & CD variables.

Infinite–Dimensional Neural Network. The adaptive path integral (3.120) incorporates the *local learning process* according to the standard formula: $New\ Value = Old\ Value{+}Innovation$. The general *weights* $w_s = w_s(t)$ in (3.121) are updated by the $MONITOR$ feedback during the transition process, according to one of the two standard neural learning schemes, in which the micro–time level is traversed in discrete steps, i.e., if $t = t_0, t_1, \ldots, t_s$ then $t + 1 = t_1, t_2, \ldots, t_{s+1}$:

1. A *self–organized, unsupervised* (e.g., Hebbian–like [Heb49]) learning rule:

$$w_s(t+1) = w_s(t) + \frac{\sigma}{\eta}(w^d_s(t) - w^a_s(t)), \tag{3.122}$$

   where $\sigma = \sigma(t)$, $\eta = \eta(t)$ denote *signal* and *noise*, respectively, while superscripts d and a denote *desired* and *achieved* micro–states, respectively; or
2. A certain form of a *supervised gradient descent learning*:

$$w_s(t+1) = w_s(t) - \eta \nabla J(t), \tag{3.123}$$

   where η is a small constant, called the *step size*, or the *learning rate* and $\nabla J(n)$ denotes the gradient of the 'performance hyper–surface' at the $t-$th iteration.

Both Hebbian and supervised learning are used for the local decision making process (see below) occurring at the intention formation faze $\mathcal{F}$.

In this way, local micro–level of LSF_{total} represents an infinite–dimensional neural network. In the cognitive psychology framework, our adaptive path integral (3.120) can be interpreted as *semantic integration* (see [BF71, Ash94]).

Motion and Decision Making in LSF_{paths}

On the macro–level in the subspace LSF_{paths} we have the (loco)*motion action principle*

$$\delta S[x] = 0,$$

with the *Newtonian–like action* $S[x]$ given by

$$S[x] = \int_{t_{ini}}^{t_{fin}} dt\, [\frac{1}{2} g_{ij}\, \dot{x}^i \dot{x}^j + \varphi^i(x^i)], \tag{3.124}$$

where overdot denotes time derivative, so that $\dot{x}^i$ represents *processing speed*, or (loco)motion velocity vector. The first bracket term in (3.124) represents the kinetic energy T,

$$T = \frac{1}{2} g_{ij}\, \dot{x}^i \dot{x}^j,$$

generated by the *Riemannian metric tensor* g_{ij}, while the second bracket term, $\varphi^i(x^i)$, denotes the family of potential force–fields, driving the (loco)mo-tions $x^i = x^i(t)$ (the *strengths* of the fields $\varphi^i(x^i)$ depend on their positions x^i in LSF, see LSF_{fields} below). The corresponding Euler–Lagrangian equation gives the Newtonian–like equation of motion

$$\frac{d}{dt} T_{\dot{x}^i} - T_{x^i} = -\varphi^i_{x^i}, \tag{3.125}$$

(subscripts denote the partial derivatives), which can be put into the standard Lagrangian form

$$\frac{d}{dt} L_{\dot{x}^i} = L_{x^i}, \qquad \text{with} \qquad L = T - \varphi^i(x^i).$$

In the next subsection we use the micro–level implications of the action S[x] as given by (3.124), for dynamical descriptions of the local decision–making process.

On the micro–level in the subspace LSF_{paths}, instead of a single path defined by the Newtonian–like equation of motion (3.125), we have an ensemble of fluctuating and crossing paths with weighted probabilities (of the unit total sum). This ensemble of micro–paths is defined by the simplest instance of our adaptive path integral (3.120), similar to the Feynman's original *sum over histories*,

$$\langle Action|Intention\rangle_{paths} = \oint \mathcal{D}[wx]\, \mathrm{e}^{\mathrm{i}S[x]}, \tag{3.126}$$

where $\mathcal{D}[wx]$ is a functional measure on the *space of all weighted paths*, and the exponential depends on the action $S[x]$ given by (3.124). This procedure can be redefined in a mathematically cleaner way if we Wick–rotate the time variable t to imaginary values $t \mapsto \tau = it$, thereby making all integrals real:

$$\oint \mathcal{D}[wx]\, \mathrm{e}^{\mathrm{i}S[x]} \Rightarrow^{Wick} \oint \mathcal{D}[wx]\, \mathrm{e}^{-S[x]}. \tag{3.127}$$

Discretization of (3.127) gives the *thermodynamic–like partition function*

$$Z = \sum_j \mathrm{e}^{-w_j E^j/T}, \tag{3.128}$$

where E^j is the motion energy eigenvalue (reflecting each possible motivational energetic state), T is the temperature–like environmental control parameter, and the sum runs over all motion energy eigenstates (labelled by the index j). From (3.128), we can further calculate all thermodynamic–like and statistical properties of MD & CD (see e.g., [Fey72]), as for example, *transition entropy* $S = k_B \ln Z$, etc.

From cognitive perspective, our adaptive path integral (3.126) calculates all (alternative) pathways of information flow during the transition *Intention* $\rightarrow$ *Action*.

In the language of transition–propagators, the integral over histories (3.126) can be decomposed into the product of propagators (i.e., Fredholm kernels or Green functions) corresponding to the cascade of the four motivational phases (*)

$$\langle Action|Intention\rangle_{paths} = \oint dx^{\mathcal{F}} dx^{\mathcal{I}} dx^{\mathcal{M}} dx^{\mathcal{T}}\, K(\mathcal{F},\mathcal{I})K(\mathcal{I},\mathcal{M})K(\mathcal{M},\mathcal{T}), \tag{3.129}$$

satisfying the Schrödinger–like equation (see e.g., [Dir49])

$$i\,\partial_t \langle Action|Intention\rangle_{paths} = H_{Action}\, \langle Action|Intention\rangle_{paths}, \tag{3.130}$$

where H_{Action} represents the Hamiltonian (total energy) function available at the state of *Action*. Here our 'golden rule' is: the higher the H_{Action}, the lower the microscopic fatigue.

In the connectionist language, our propagator expressions (3.129–3.130) represent *activation dynamics*, to which our *Monitor* process gives a kind of *backpropagation* feedback, a version of the basic supervised learning (3.123).

Mechanisms of Decision–Making under Uncertainty. The basic question about our local decision making process, occurring under uncertainty at the intention formation faze $\mathcal{F}$, is: Which alternative to choose? (see [RBT01, Gro82, Gro99, Gro88, Ash94]). In our path–integral language this reads: Which path (alternative) should be given the highest probability weight w? Naturally, this problem is iteratively solved by the learning process (3.122–3.123), controlled by the $MONITOR$ feedback, which we term *algorithmic approach.*

In addition, here we analyze qualitative mechanics of the local decision making process under uncertainty, as a *heuristic approach.* This qualitative analysis is based on the micro–level interpretation of the Newtonian–like action $S[x]$, given by (3.124) and figuring both processing speed $\dot{x}$ and LTM (i.e., the force–field $\varphi(x)$, see next subsection). Here we consider three different cases:

1. If the potential $\varphi(x)$ is not very dependent upon position $x(t)$, then the more direct paths contribute the most, as longer paths, with higher mean square velocities $[\dot{x}(t)]^2$ make the exponent more negative (after Wick rotation (3.127)).
2. On the other hand, suppose that $\varphi(x)$ does indeed depend on position x. For simplicity, let the potential increase for the larger values of x. Then a direct path does not necessarily give the largest contribution to the overall transition probability, because the integrated value of the potential is higher than over another paths.
3. Finally, consider a path that deviates widely from the direct path. Then $\varphi(x)$ decreases over that path, but at the same time the velocity $\dot{x}$ increases. In this case, we expect that the increased velocity $\dot{x}$ would more than compensate for the decreased potential over the path.

Therefore, the most important path (i.e., the path with the highest weight w) would be one for which any smaller integrated value of the surrounding field potential $\varphi(x)$ is more than compensated for by an increase in kinetic–like energy $\frac{m}{2}\dot{x}^2$. In principle, this is neither the most direct path, nor the longest path, but rather a middle way between the two. Formally, it is the path along which the average Lagrangian is minimal,

$$< \frac{m}{2}\dot{x}^2 + \varphi(x) > \longrightarrow \min, \tag{3.131}$$

i.e., the *path that requires minimal memory* (both LTM and WM, see LSF_{fields} below) and *processing speed.* This mechanical result is consistent with the 'filter theory' of *selective attention* [Bro77], proposed in an attempt to explain a range of the existing experimental results. This theory postulates a low level filter that allows only a limited number of percepts to reach the brain at any time. In this theory, the importance of conscious, directed attention is minimized. The type of attention involving low level filtering corresponds to the concept of *early selection* [Bro77].

Although we termed this 'heuristic approach' in the sense that we can instantly feel both the processing speed $\dot{x}$ and the LTM field $\varphi(x)$ involved, there is clearly a psycho–physical rule in the background, namely the averaging minimum relation (3.131).

From the decision making point of view, all possible paths (alternatives) represent the *consequences* of decision making. They are, by default, *short–term consequences*, as they are modelled in the micro–time–level. However, the path integral formalism allows calculation of the *long–term consequences*, just by extending the integration time, $t_{fin} \to \infty$. Besides, this *averaging decision mechanics* – choosing the optimal path – actually performs the 'averaging lift' in the LSF: from micro– to the macro–level.

Force–Fields and Memory in LSF_{fields}

At the macro–level in the subspace LSF_{fields} we formulate the *force–field action principle*

$$\delta S[\varphi] = 0, \tag{3.132}$$

with the action $S[\varphi]$ dependent on Lewinian force–fields $\varphi^i = \varphi^i(x)$ ($i = 1, \dots, N$), defined as a temporal integral

$$S[\varphi] = \int_{t_{ini}}^{t_{fin}} \mathfrak{L}[\varphi]\, dt, \tag{3.133}$$

with Lagrangian density given by

$$\mathfrak{L}[\varphi] = \int d^n x\, \mathcal{L}(\varphi_i, \partial_{x^j}\varphi^i),$$

where the integral is taken over all n coordinates $x^j = x^j(t)$ of the LSF, and $\partial_{x^j}\varphi^i$ are partial derivatives of the field variables over coordinates.

On the micro–level in the subspace LSF_{fields} we have the Feynman–type *sum over fields* φ^i ($i = 1, \dots, N$) given by the adaptive path integral

$$\langle Action|Intention \rangle_{fields} = \oint\!\!\!\!\!\!\sum \mathcal{D}[w\varphi]\, \mathrm{e}^{\mathrm{i}S[\varphi]} \Rightarrow^{Wick} \oint\!\!\!\!\!\!\sum \mathcal{D}[w\varphi]\, \mathrm{e}^{-S[\varphi]}, \tag{3.134}$$

with action $S[\varphi]$ given by temporal integral (3.133). (Choosing special forms of the force–field action $S[\varphi]$ in (3.134) defines micro–level MD & CD, in the LSF_{fields} space, that is similar to standard quantum–field equations, see e.g., [II06b].) The corresponding partition function has the form similar to (3.128), but with field energy levels.

Regarding topology of the force fields, we have in place *$n-$categorical Lagrangian–field structure* on the Riemannian LSF manifold M,

$$\Phi^i : [0,1] \to M, \; \Phi^i : \Phi^i_0 \mapsto \Phi^i_1,$$

generalized from the *recursive homotopy dynamics* [II06b], using

$$\begin{aligned} \frac{d}{dt} f_{\dot{x}^i} &= f_{x^i} \longrightarrow \partial_\mu \left(\frac{\partial \mathcal{L}}{\partial_\mu \Phi^i} \right) = \frac{\partial \mathcal{L}}{\partial \Phi^i}, \\ \text{with} \quad &[x_0, x_1] \longrightarrow [\Phi^i_0, \Phi^i_1]. \end{aligned}$$

Relationship between Memory and Force–Fields. As already mentioned, the subspace LSF_{fields} is related to our *memory storage* [Ash94]. Its global macro–level represents the *long–term memory* (LTM), defined by the least action principle (3.132), related to *cognitive economy* in the model of *semantic memory* [Rat78, CQ69]. Its local micro–level represents *working memory* (WM), a limited–capacity 'bottleneck' defined by the adaptive path integral (3.134). According to our formalism, each of Miller's 7±2 units [Mil56] of the local WM are adaptively stored and averaged to give the global LTM capacity (similar to the physical notion of potential). This averaging memory lift, from WM to LTM represents *retroactive interference*, while the opposite direction, given by the path integral (3.134) itself, represents *proactive interference*. Both retroactive and proactive interferences are examples of the impact of cognitive contexts on memory. Motivational contexts can exert their influence, too. For example, a reduction in task–related recall following the completion of the task is one of the clearest examples of force–field influences on memory: the amount of details remembered of a task declines as the force–field tension to complete the task is reduced by actually completing it.

Once defined, the global LTM potential $\varphi = \varphi(x)$ is then affecting the locomotion transition paths through the path action principle (3.124), as well as general learning (3.122–3.123) and decision making process (3.131).

On the other hand, the two levels of LSF_{fields} fit nicely into the two levels of processing framework, as presented by [CL72], as an alternative to theories of separate stages for sensory, working and long–term memory. According to the *levels of processing framework*, stimulus information is processed at multiple levels simultaneously depending upon its characteristics. In this framework, our macro–level memory field, defined by the fields action principle (3.132), corresponds to the *shallow memory*, while our micro–level memory field, defined by the adaptive path integral (3.134), corresponds to the *deep memory*.

Geometries, Topologies and Noise in LSF_{geom}

On the macro–level in the subspace LSF_{geom} representing an $n-$dimensional smooth manifold M with the global Riemannian metric tensor g_{ij}, we formulate the *geometrical action principle*

$$\delta S[g_{ij}] = 0,$$

where $S = S[g_{ij}]$ is the $n-$dimensional *geodesic action* on M,

$$S[g_{ij}] = \int d^n x \sqrt{g_{ij}\, dx^i dx^j}. \tag{3.135}$$

The corresponding Euler–Lagrangian equation gives the *geodesic equation* of the *shortest path* in the manifold M,

$$\ddot{x}^i + \Gamma^i_{jk}\, \dot{x}^j\, \dot{x}^k = 0,$$

where the symbol Γ^i_{jk} denotes the so–called *affine connection* which is the source of *curvature*, which is geometrical description for *noise* (see [Ing97,

Ing98]). The higher the local curvatures of the LSF–manifold M, the greater the noise in the life space. This noise is the source of our micro–level fluctuations. It can be internal or external; in both cases it curves our micro–LSF.

Otherwise, if instead we choose an $n-$dimensional Hilbert–like action (see [MTW73]),

$$S[g_{ij}] = \int d^n x \sqrt{\det |g_{ij}|} R, \tag{3.136}$$

where R is the scalar curvature (derived from Γ^i_{jk}), we get the $n-$dimensional Einstein–like equation:

$$G_{ij} = 8\pi T_{ij},$$

where G_{ij} is the Einstein–like tensor representing geometry of the LSF manifold M (G_{ij} is the trace–reversed Ricci tensor R_{ij}, which is itself the trace of the *Riemann curvature tensor* of the manifold M), while T_{ij} is the $n-$dimensional *stress–energy–momentum* tensor. This equation explicitly states that *psycho–physics of the LSF is proportional to its geometry.* T_{ij} is important quantity, representing motivational *energy*, geometry–imposed *stress* and *momentum* of (loco)motion. As before, we have our 'golden rule': *the greater the* $T_{ij}-$*components, the higher the speed of cognitive processes and the lower the macroscopic fatigue.*

The choice between the geodesic action (3.135) and the Hilbert action (3.136) depends on our interpretation of time. If time is not included in the LSF manifold M (non–relativistic approach) then we choose the geodesic action. If time is included in the LSF manifold M (making it a relativistic–like $n-$dimensional space–time) then the Hilbert action is preferred. The first approach is more related to the information processing and the working memory. The later, space–time approach can be related to the long–term memory: we usually recall events closely associated with the times of their happening.

On the micro–level in the subspace LSF_{geom} we have the adaptive *sum over geometries*, represented by the path integral over all local (regional) Riemannian metrics $g_{ij} = g_{ij}(x)$ varying from point to point on M (modulo diffeomorphisms),

$$\langle Action|Intention\rangle_{geom} = \oint\!\!\!\!\!\!\!\!\;\Sigma\ \mathcal{D}[wg_{ij}]\, \mathrm{e}^{\mathrm{i}S[g_{ij}]} \Rightarrow^{Wick} \oint\!\!\!\!\!\!\!\!\;\Sigma\ \mathcal{D}[wg_{ij}]\, \mathrm{e}^{-S[g_{ij}]}, \tag{3.137}$$

where $\mathcal{D}[g_{ij}]$ is diffeomorphism equivalence class of $g_{ij}(x) \in M$.

To include the topological structure (e.g., a number of holes) in M, we can extend (3.137) as

$$\langle Action|Intention\rangle_{geom/top} = \sum_{\text{topol.}} \oint\!\!\!\!\!\!\!\!\;\Sigma\ \mathcal{D}[wg_{ij}]\, \mathrm{e}^{\mathrm{i}S[g_{ij}]}, \tag{3.138}$$

where the topological sum is taken over all connectedness–components of M determined by the *Euler characteristic* χ of M. This type of integral defines the *theory of fluctuating geometries*, a propagator between $(n-1)-$dimensional boundaries of the $n-$dimensional manifold M. One has to contribute a meaning to the integration over geometries. A key ingredient in doing so is to approximate (using simplicial approximation and Regge calculus [MTW73]) in a natural way the smooth structures of the manifold M by piecewise linear structures (mostly using topological simplices Δ). In this way, after the Wick–rotation (3.127), the integral (3.137–3.138) becomes a *simple statistical system*, given by partition function $Z = \sum_{\Delta} \frac{1}{C_\Delta} e^{-S_\Delta}$, where the summation is over all triangulations Δ of the manifold M, while C_T is the order of the automorphism group of the performed triangulation.

Micro–Level Geometry: the source of noise and stress in LSF. The subspace LSF_{geom} is the source of noise, fluctuations and obstacles, as well as psycho–physical stress. Its micro–level is adaptive, reflecting the human ability to efficiently act within the noisy environment and under the stress conditions. By averaging it produces smooth geometry of certain curvature, which is at the same time the smooth psycho–physics. This macro–level geometry directly affects the memory fields and indirectly affects the (loco)motion transition paths.

The Mental Force Law. As an effective summary of this section, we state that the psychodynamic transition functor $\mathcal{TA} : INTENTION_{t_{ini}} \Rightarrow ACTION_{t_{fin}}$, defined by the generic path integral (3.120), can be interpreted as a *mental force law*, analogous to our musculo–skeletal *covariant force law*, $F_i = m g_{ij} a^j$, and its associated *covariant force functor* $\mathcal{F}_* : TT^*M \to TTM$ [II05].

3.3 Quantum Computation and Chaos: Josephson Junctions

This section addresses modern electronic devices called *Josephson junctions*, which promise to be a basic building blocks of the future quantum computers. Apparently, they can exhibit chaotic behavior, both as single junctions (which have macroscopic dynamics analogous to those of the forced nonlinear oscillators), and as arrays (or ladders) of junctions, which can show high–dimensional chaos.

A *Josephson junction* is a type of electronic circuit capable of switching at very high speeds, i.e., frequency of typically $10^{10}-10^{11}$ Hz, when operated at temperatures approaching absolute zero. It is an insulating barrier separating two superconducting materials and producing the *Josephson effect*. The terms are named eponymously after British physicist Brian David Josephson, who predicted the existence of the Josephson effect in 1962 [Jos74]. Josephson junction exploits the phenomenon of *superconductivity*, the ability of certain materials to conduct electric current with practically zero resistance. Josephson junctions have important applications in quantum–mechanical circuits. They have great technological promises as amplifiers, voltage standards, detectors, mixers, and fast switching devices for digital circuits. They are used in certain specialized instruments such as highly–sensitive microwave detectors, magnetometers, and QUIDs. Finally, Josephson junctions allow the realisation of *qubits*, the key elements of *quantum computers*.

Josephson junctions have been particularly useful for experimental studies of nonlinear dynamics as the equation governing a single junction dynamics is the same as that for a pendulum [Str94]. Their dynamics can be analyzed both in a simple overdamped limit and in the more complex underdamped one, either for single junctions and for arrays of large numbers of coupled junctions.

A Josephson junction is made up of two superconductors, separated by a weak coupling non–superconducting layer, so thin that electrons can cross through the insulating barrier. It can be conceptually represented as:

$$\begin{array}{rl} Superconductor\ 1 & : \psi_1 \mathrm{e}^{i\phi_1} \\ Weak\ Coupling & \Updownarrow \\ Superconductor\ 2 & : \psi_2 \mathrm{e}^{i\phi_2} \end{array}$$

where the two superconducting regions are characterized by simple *quantum–mechanical wave functions*, $\psi_1 \mathrm{e}^{i\phi_1}$ and $\psi_2 \mathrm{e}^{i\phi_2}$, respectively. Normally, a much more complicated description would be necessary, as there are $\sim 10^{23}$ electrons to deal with, but in the superconducting ground state, these electrons form the so–called *Cooper pairs* that can be described by a single macroscopic wave function $\psi \mathrm{e}^{i\phi}$. The flow of current between the superconductors in the absence of an applied voltage is called a *Josephson current*, and the movement of electrons across the barrier is known as *Josephson tunnelling* (see Figure 3.13). Two or more junctions joined by superconducting paths form what is called a *Josephson interferometer*.

One of the characteristics of a Josephson junction is that as the temperature is lowered, superconducting current flows through it even in the absence of voltage between the electrodes, part of the *Josephson effect*. The Josephson effect in particular results from two superconductors acting to preserve their long–range order across an insulating barrier. With a thin enough barrier, the phase of the electron wave–function in one superconductor maintains a fixed relationship with the phase of the wave–function in another superconductor. This linking up of phase is called phase coherence. It occurs throughout a single superconductor, and it occurs between the superconductors in a Josephson junction. The *phase coherence*, or *long–range order*, is the essence of the Josephson effect.

While researching superconductivity, B.D. Josephson studied the properties of a junction between two superconductors. Following up on earlier work by L. Esaki and I. Giaever, he demonstrated that in a situation when there is electron flow between two superconductors through an insulating layer (in the absence of an applied voltage), and a voltage is applied, the current stops flowing and oscillates at a high frequency. The Josephson effect is influenced by magnetic fields in the vicinity, a capacity that enables the Josephson junction to be used in devices that measure extremely weak magnetic fields, such

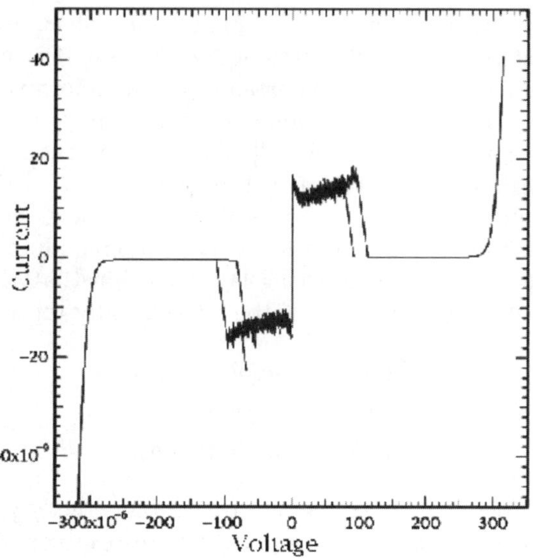

Fig. 3.13. *Josephson junction:* the current–voltage curve obtained at low temperature. The vertical portions (zero voltage) of the curve represent Cooper pair tunnelling. There is a small magnetic field applied, so that the maximum Josephson current is severely reduced. Hysteresis is clearly visible around 100 microvolts. The portion of the curve between 100 and 300 microvolts is current independent, and is the regime where the device can be used as a detector.

as superconducting quantum interference devices (SQUIDs). For their efforts, Josephson, Esaki, and Giaever shared the Nobel Prize for Physics in 1973.

The *Josephson–junction quantum computer* was demonstrated in April 1999 by Nakamura, Pashkin and Tsai of NEC Fundamental Research Laboratories in Tsukuba, Japan [NPT99]. In the same month, only about one week earlier, Ioffe, Geshkenbein, Feigel'man, Fauchère and Blatter, independently, described just such a computer in Nature [IGF99].

Nakamura, Pashkin and Tsai's computer is built around a *Cooper pair box*, which is a small superconducting island electrode weakly coupled to a bulk superconductor. Weak coupling between the superconductors creates a Josephson junction between them. Like most other junctions, the Josephson junction is also a capacitor, which is charged by the current that flows through it. A gate voltage is applied between the two superconducting electrodes. If the Cooper box is sufficiently small, e.g., as small as a quantum dot, the charging current breaks into discrete transfer of individual Cooper pairs, so that ultimately it is possible to just transfer a single Cooper pair across the junction. The effectiveness of the Cooper pair transfer depends on the energy difference between the box and the bulk and a maximum is reached when a voltage is applied, which equalizes this energy difference. This leads to *resonance* and observable *coherent quantum oscillations* [Ave99].

This contraption, like the Loss–Vincenzo quantum dot computer [LD98], has the advantage that it is controlled electrically. Unlike Loss–Vincenzo computer, this one actually exists in the laboratory. Nakamura, Pashkin and Tsai did not perform any computations with it though. At this stage it was enough of an art to observe the coherence for about 6 cycles of the Cooper pair oscillations, while the chip was cooled to about and carefully shielded from external electromagnetic radiation.

There are two general types of Josephson junctions: overdamped and underdamped. In overdamped junctions, the barrier is conducting (i.e., it is a normal metal or superconductor bridge). The effects of the junction's internal electrical resistance will be large compared to its small capacitance. An overdamped junction will quickly reach a unique equilibrium state for any given set of conditions.

The barrier of an underdamped junction is an insulator. The effects of the junction's internal resistance will be minimal. Underdamped junctions do not have unique equilibrium states, but are hysteretic.

A Josephson junction can be transformed into the so–called *Giaever tunnelling junction* by the application of a small, well defined magnetic field. In such a situation, the new device is called a superconducting tunnelling junction (STJ) and is used as a very sensitive photon detector throughout a wide range of the spectrum, from infrared to hard X–ray. Each photon breaks up a number of Cooper pairs. This number depends on the ratio of the photon energy to approximately twice the value of the gap parameter of the material of the junction. The detector can be operated as a photon-counting spectrometer, with a spectral resolution limited by the statistical fluctuations in the number of released charges. The detector has to be cooled to extremely low temperature, typically below 1 kelvin, to distinguish the signals generated by the detector from the thermal noise. Small arrays of STJs have demonstrated their potential as spectro–photometers and could further be used in astronomy [ESA05]. They are also used to perform energy dispersive X–ray spectroscopy and in principle they could be used as elements in infrared imaging devices as well [Ens05].

3.3.1 Josephson Effect and Pendulum Analog

Josephson Effect

The basic equations governing the dynamics of the Josephson effect are (see, e.g., [BP82]):

$$U(t) = \frac{\hbar}{2e}\frac{\partial \phi}{\partial t}, \qquad I(t) = I_c \sin\phi(t),$$

where $U(t)$ and $I(t)$ are the voltage and current across the Josephson junction, $\phi(t)$ is the phase difference between the wave functions in the two superconductors comprising the junction, and I_c is a constant, called the *critical current* of the junction. The critical current is an important phenomenological parameter of the device that can be affected by temperature as well as by an applied magnetic field. The physical constant $\hbar/2e$ is the magnetic flux quantum, the inverse of which is the *Josephson constant*.

The three main effects predicted by Josephson follow from these relations:

1. The DC Josephson effect. This refers to the phenomenon of a direct current crossing the insulator in the absence of any external electromagnetic field, owing to *Josephson tunnelling*. This DC Josephson current is proportional to the sine of the phase difference across the insulator, and may take values between $-I_c$ and I_c.

2. The AC Josephson effect. With a fixed voltage U_{DC} across the junctions, the phase will vary linearly with time and the current will be an AC current with amplitude I_c and frequency $2e/\hbar\, U_{DC}$. This means a Josephson junction can act as a perfect *voltage–to–frequency converter*.

3. The inverse AC Josephson effect. If the phase takes the form

$$\phi(t) = \phi_0 + n\omega t + a\sin(\omega t),$$

the voltage and current will be

$$U(t) = \frac{\hbar}{2e}\omega[n + a\cos(\omega t)], \qquad I(t) = I_c \sum_{m=-\infty}^{\infty} J_n(a)\sin[\phi_0 + (n+m)\omega t].$$

The DC components will then be

$$U_{DC} = n\frac{\hbar}{2e}\omega, \qquad I(t) = I_c J_{-n}(a)\sin\phi_0.$$

Hence, for distinct DC voltages, the junction may carry a DC current and the junction acts like a perfect *frequency–to–voltage converter*.

Pendulum Analog

To show a driven pendulum analog of a microscopic description of a single Josephson junction, we start with:

1. The *Josephson current–phase relation*

$$I = I_c \sin\phi,$$

where I_c is the *critical current*, I is the bias current, and $\phi = \phi_2 - \phi_1$ is the constant *phase difference* between the phases of the two superconductors that are weakly coupled; and

2. The *Josephson voltage–phase relation*

$$V = \frac{\hbar}{2e}\dot{\phi},$$

where $V = V(t)$ is the instantaneous voltage across the junction, $\hbar$ is the Planck constant (divided by 2π), and e is the charge on the electron.

Now, if we apply Kirhoff's voltage and current laws for the parallel RC–circuit with resistence R and capacitance C, we come to the first–order ODE

$$C\dot{V} + \frac{V}{R} + I_c \sin\phi = I,$$

which can be recast solely in terms of the phase difference ϕ as the second–order pendulum–like ODE,

$$\begin{aligned} \text{Josephson junction :} \qquad & \frac{\hbar C}{2e}\ddot{\phi} + \frac{\hbar}{2eR}\dot{\phi} + I_c \sin\phi = I, \\ \text{Pendulum :} \qquad & ml^2\ddot{\theta} + b\dot{\theta} + mgl\sin\theta = \tau. \end{aligned} \tag{3.139}$$

This mechanical analog has often proved useful in visualizing the dynamics of Josephson Junctions [Str94]. If we divide (3.139) by I_c and define a dimensionless time

$$\tau = \frac{2eI_cR}{\hbar}t,$$

we get the dimensionless oscillator equation for Josephson junction,

$$\beta\phi'' + \phi' + \sin\phi = \frac{I}{I_c}, \tag{3.140}$$

where $\phi' = d\phi/d\tau$. The dimensionless group β, defined by

$$\beta = \frac{2eI_cR^2C}{\hbar},$$

is called the McCumber parameter and represents a dimensionless capacitance.

In a simple *overdamped limit* $\beta << 1$ with *resistive loading*, the 'inertial term' $\beta\phi''$ may be neglected (as if oscillating in a highly–viscous medium), and so (3.140) reduces to a non–uniform oscillator

$$\phi' = \frac{I}{I_c} - \sin\phi, \tag{3.141}$$

with solutions approaching a stable fixed–point for $I < I_c$, and periodically varying for $I < I_c$. To find the current–voltage curve in the overdamped limit, we take the average voltage $\langle V \rangle$ as a function of the constant applied current I, assuming that all transients have decayed and the system has reached steady–state, and get

$$\langle V \rangle = I_cR \langle \phi' \rangle .$$

An overdamped *array of N Josephson Junctions* (3.141), parallel with a resistive load R, can be described by the system of first–order dimensionless ODEs [Str94]

$$\phi_k' = \Omega + a \sin \phi_k + \frac{1}{N} \sum_{j=1}^{N} \sin \phi_j, \qquad k = 1, \dots, N,$$

where

$$\Omega = I_b R_0 / I_c r, \qquad a = -(R_0 + r)/r, \qquad R_0 = R/N,$$

$$I_b = I_c \sin \phi_k + \frac{\hbar}{2eR} \dot{\phi}_k + \frac{\hbar}{2eR} \sum_{j=1}^{N} \dot{\phi}_j.$$

3.3.2 Dissipative Josephson Junction

The past decade has seen a considerable interest and remarkable activity in an area which presently is often referred to as macroscopic quantum mechanics. Specifically, one has been interested in quantum phenomena of macroscopic objects [Leg86].

In particular, macroscopic quantum tunnelling [CL81] (quantum decay of a meta–stable state), and quantum coherence [LCD87] have been studied. Soon, it became clear that dissipation has a profound influence on these quantum phenomena. Phenomenologically, dissipation is the consequence of an interaction of the object with an environment which can be thought of as consisting of infinitely many degrees of freedom. Specifically, the environmental degrees of freedom may be chosen to be harmonic oscillators such that we may consider the dissipation as a process where excitations, that are phonons, are emitted and absorbed. This, Caldeira–Leggett model has been used in [CL81] where the influence of dissipation on tunnelling has been explored.

As far as quantum coherence is concerned, the most simple system is an object with two different quantum states: it is thought to represent the limiting case of an object in a double-well potential where only the lowest energy states in each of the two wells is relevant and where the tunnelling through the separating barrier allows for transitions that probe the coherence. Since a 2–state system is equivalent to a spin–one–half problem, this standard system is often referred to by this name. In particular, with the standard coupling to a dissipative environment made of harmonic oscillators, it is called the spin–boson problem which has been studied repeatedly in the past [LCD87, SW90].

Level quantization and resonant tunnelling have been observed recently [Vaa95] in a double–well quantum–dot system. However, the influence of dissipation was not considered in this experiment. On the other hand, it seems that Josephson junctions are also suitable systems for obtaining experimental evidence pertaining to macroscopic quantum effects. In this context, evidence for level quantization and for quantum decay have been obtained [MDC85].

Recall that a Josephson junction may be characterized by a *current–phase relation*

$$I(\phi) = I_J \sin \phi, \tag{3.142}$$

where the phase ϕ is related to the voltage difference U by

$$\hbar\dot{\phi} = 2eU\,. \tag{3.143}$$

Therefore, the phase of a Josephson junction shunted by a capacitance C and biased by an external current I_x obeys a classical type of equation of motion

$$M\ddot{\phi} = -\frac{\partial V(\phi)}{\partial \phi}, \qquad \text{with the mass} \tag{3.144}$$

$$M = \left(\frac{\hbar}{2e}\right)^2 C, \qquad \text{and the potential energy} \tag{3.145}$$

$$V(\phi) = -\frac{\hbar}{2e}\left[I_J \cos\phi + I_x\phi\right]\,. \tag{3.146}$$

A widely discussed model of a dissipative object is the one where the Josephson junction is also shunted by an Ohmic resistor R. In this case, the classical equation of motion (3.144) has to be replaced by

$$M\ddot{\phi} = -\frac{\partial V(\phi)}{\partial \phi} - \eta\dot{\phi}, \qquad \eta = \left(\frac{\hbar}{2e}\right)^2 \frac{1}{R}\,. \tag{3.147}$$

The model of a dissipative environment according to the above specification has been discussed by [CL81].

The potential energy $V(\phi)$ of (3.146) displays wells at $\phi \simeq 2n\pi$ with depth shifted by an amount $\Delta \simeq (2\pi\hbar/2e)I_x$. If the wells are sufficiently deep, one needs to concentrate only on transitions between pairs of adjacent wells. Thus, one arrives at the double well problem mentioned above.

The analysis in this paper goes beyond the limiting situation where only the lowest level in each of the two wells is of importance. Roughly, this is realized when the level separation $\hbar(2E_J/M)^{1/2} \simeq (2e\hbar I_J/C)^{1/2}$ is smaller than or comparable with Δ. In particular, we will concentrate on resonance phenomena which are expected to show up whenever two levels in the adjacent wells happen to cross when the bias current I_x, that is Δ, is varied.

For such values of the bias current, there appear sharp asymmetric peaks in the current-voltage characteristic of the Josephson junction. This phenomenon has been studied by [LOS88] within the standard model in the one–phonon approximation. For bias currents that correspond to crossings of the next and next nearest levels (e.g., ground state in the left well and the first or second excited state at the right side), it is possible to neglect processes in the reverse direction provided that the temperature is sufficiently low. Thus, the restriction to a double well system receives additional support.

The transfer of the object from the left to the right potential well is accompanied by the emission of an infinite number of phonons. Therefore, in [OS94] the fact is taken into account that in the resonance region, the contribution of phonons of small energy is important as well as the contribution of resonance phonons with energy equal to the distance between levels in the wells.

Junction Hamiltonian and its Eigenstates

The model of [MS95] consists of a particle, called 'object' (coordinate R_1), which is coupled (in the sense of [CL81]) to a 'bath' of harmonic oscillators (coordinates R_j). We shall use the conventions $j \in \{2, \dots, N\}$ for the bath oscillators and $k \in \{1, \dots, N\}$ for the indices of all coordinates in the model. The double–well potential is approximated by two parabolas about the minima of the two wells.

The phase ϕ of the Josephson contact then corresponds to the object coordinate R_1 of the model, and the voltage U is related to the tunnelling rate J by $2eU = \dot{\phi} = 2\pi J$. As it has already been remarked, the current I_x is proportional to the bias Δ of the two wells. Thus, calculating the transition rate for different values of the bias Δ is equivalent to the determination of the I–V characteristics.

Specifically, following [MS95], we want to write the Hamiltonian of the model in the form

$$\hat{H} = \frac{1}{2m}\sum_k \hat{p}_k^2 + \hat{v}(\hat{R}_1) + \frac{m}{2}\sum_j \omega_j^2(\hat{R}_j - \hat{R}_1)^2,$$
$$\hat{v}(\hat{R}_1) \approx \frac{m}{2}\sum_{\pm} \Omega^2(\hat{R}_1 \pm a)^2 \pm \frac{\Delta}{2}. \tag{3.148}$$

The states for the two situations 'object in the left well' and 'object in the right well' will be denoted by $|\Lambda_L, L\rangle$ and $|\Lambda_R, R\rangle$, respectively. If one projects onto the eigenstates $|n\rangle$ of the 1D harmonic oscillator and takes into account the shift of the wells, one arrives at the following decomposition ($\phi_n(R) = \langle R|n\rangle$):

$$\langle n_L, \{R_j\}|\Lambda_L, L\rangle = \int dR_1\, \phi_{n_L}(R_1 + a)\, \phi^L_{\Lambda_L}(\{R_k\}), \tag{3.149}$$
$$\langle n_R, \{R_j\}|\Lambda_R, R\rangle = \int dR_1\, \phi_{n_R}(R_1 - a)\, \phi^R_{\Lambda_R}(\{R_k\}).$$

The situations 'object on the left' and 'object on the right' differ only by the shift and the bias of the wells. Therefore, one can find a unified representation by noting that $\phi^L_\Lambda(\{R_k\}) = \Phi_\Lambda(\{R_k + a\})$ and $\phi^R_\Lambda(\{R_k\}) = \Phi_\Lambda(\{R_k - a\})$. The eigenstates Φ_Λ are defined by the relations

$$\Phi_\Lambda(\{R_k\}) = \langle\{R_k\}|\Lambda\rangle, \qquad \hat{H}_0|\Lambda\rangle = E_\Lambda|\Lambda\rangle,$$
$$\hat{H}_0 = \frac{1}{2m}\sum_k \hat{p}_k^2 + \frac{m}{2}\sum_j \omega_j^2(\hat{R}_j - \hat{R}_1)^2 + \frac{m}{2}\,\Omega^2\hat{R}_1^2.$$

Thus, it follows from (3.149) that

$$\langle n_L, \{R_j\}|\Lambda_L, L\rangle = \langle n_L, \{R_j\}|\exp(ia\sum_j \hat{p}_j)\,|\Lambda_L>, \tag{3.150}$$
$$\langle n_R, \{R_j\}|\Lambda_R, R\rangle = \langle n_R, \{R_j\}|\exp(-ia\sum_j \hat{p}_j)\,|\Lambda_R>,$$

where we have used the shift property of the momentum operator $\hat{p}$.

The coupling of the two wells is taken into account by means of a tunnelling Hamiltonian $\hat{H}_T$ which we represent in the form

$$\langle \Lambda_L, L|\hat{H}_T|\Lambda_R, R\rangle = \int d\{R_j\} \sum_{n_L n_R} T_{n_L n_R} \langle \Lambda_L, L|n_L, \{R_j\}\rangle\langle n_R, \{R_j\}|\Lambda_R, R\rangle .$$

Using again the momentum operator, one can write

$$|x\rangle\langle x'| = \mathrm{e}^{i\hat{p}(x'-x)} |x'\rangle\langle x'| = \mathrm{e}^{i\hat{p}(x'-x)} \delta(x' - \hat{x}) .$$

From this, we conclude that

$$\begin{aligned}\langle \Lambda_L, L|\hat{H}_T|\Lambda_R, R\rangle &= \sum_{n_L n_R} T_{n_L n_R} \int dR_1 dR_1' \frac{dQ}{2\pi} \phi^*_{n_R}(R_1') \phi_{n_L}(R_1) \\ &\times \langle \Lambda_L, L|\mathrm{e}^{i\hat{p}_1(R_1'-R_1)} \mathrm{e}^{iQ(R_1'-\hat{R}_1)}|\Lambda_R, R\rangle .\end{aligned}$$

Transition Rate

The net transition rate from the left well to the right one is then in second order perturbation theory given by [MS95]

$$\begin{aligned}J &= 2\pi Z_0^{-1} \sum_{\Lambda_L,\Lambda_R} |\langle \Lambda_L, L|\hat{H}_T|\Lambda_R, R\rangle|^2 \, \delta(E_{\Lambda_L} - E_{\Lambda_R} + \Delta) \\ &\times [\mathrm{e}^{-\beta E_{\Lambda_L}} - \mathrm{e}^{\beta E_{\Lambda_R}}],\end{aligned}$$

where $Z_0 = \mathrm{Tr}\exp(-\beta H_0)$. The δ–function may be written in Fourier representation, and the fact that the E_Λ are eigen–energies of $\hat{H}_0$ serves us to incorporate the energy conservation into Heisenberg time–dependent operators $\hat{A}(t) = \exp(i\hat{H}_0 t)\hat{A}\exp(-i\hat{H}_0 t)$, i.e.,

$$\begin{aligned}&\langle \Lambda_L, L|\hat{H}_T|\Lambda_R, R\rangle \, \delta(E_{\Lambda_L} - E_{\Lambda_R} + \Delta) \\ &= \int dt \, \mathrm{e}^{i\Delta t} \langle \Lambda_L|\mathrm{e}^{-ia\sum_k \hat{p}_k(t)} \hat{H}_T(t) \, \mathrm{e}^{-ia\sum_k \hat{p}_k(t)}|\Lambda_R\rangle .\end{aligned}$$

Then, collecting our results from above we arrive at the expression

$$\begin{aligned}J &= Z_0^{-1}(1 - \mathrm{e}^{-\beta\Delta}) \int dt \, \mathrm{e}^{i\Delta t} \sum_{n_L,n_R} \sum_{\overline{n}_L,\overline{n}_R} T_{n_L,n_R} T^*_{\overline{n}_L,\overline{n}_R} \\ &\times \int \frac{dQ d\overline{Q}}{(2\pi)^2} \int dR_1 dR_1' d\overline{R}_1 d\overline{R}_1' \, \phi_{n_L}(R_1) \, \phi^*_{n_R}(R_1') \, \phi^*_{\overline{n}_L}(\overline{R}_1') \, \phi_{\overline{n}_R}(\overline{R}_1) \\ &\times \mathrm{Tr}\,\{\mathrm{e}^{-\beta\hat{H}_0} \, \mathrm{e}^{-2ia\sum_k \hat{p}_k(t)} \, \mathrm{e}^{i\hat{p}_1(t)(R_1'-R_1+2a)} \, \mathrm{e}^{-iQ(\hat{R}_1(t)-R_1')} \\ &\times \mathrm{e}^{2ia\sum_k \hat{p}_k} \, \mathrm{e}^{i\hat{p}_1(\overline{R}_1'-\overline{R}_1-2a)} \, \mathrm{e}^{-i\overline{Q}(\hat{R}_1-\overline{R}_1')}\} .\end{aligned}$$

Let us now use the relation

$$\mathrm{e}^{-i(\hat{H}_0+\hat{W})t} = \mathrm{e}^{-i\hat{H}_0 t}\,\hat{T}\,\mathrm{e}^{-i\int_0^t dt'\,\hat{W}(t')}$$

which holds for $t > 0$ when $\hat{T}$ is the *time–ordering operator* and for $t < 0$ when the anti time–ordering is used. If we define $\langle\hat{A}\rangle = \mathrm{Tr}\,\exp(-\beta\hat{H}_0)\hat{A}/Z_0$, we can write the following result for the transition rate:

$$\begin{aligned}
J &= (1-\mathrm{e}^{-\beta\Delta})\int dt\,\mathrm{e}^{i(\Delta-2m\Omega^2a^2)t}\sum_{n_L,n_R}\sum_{\overline{n}_L,\overline{n}_R} T_{n_L,n_R}T^*_{\overline{n}_L,\overline{n}_R} \\
&\times\int\frac{dQd\overline{Q}}{(2\pi)^2}\int dRdR'd\overline{R}d\overline{R}'\,\mathrm{e}^{iQ\frac{R+R'}{2}+i\overline{Q}\frac{\overline{R}+\overline{R}'}{2}} \\
&\times\phi_{n_L}(R)\,\phi^*_{n_R}(R'-2a)\,\phi^*_{\overline{n}_L}(\overline{R}')\,\phi_{\overline{n}_R}(\overline{R}-2a) \\
&\times<\hat{T}\,\exp[-iQ\hat{R}_1(t)+i\hat{p}_1(t)(R'-R)+2im\Omega^2a\int_0^t dt'\,\hat{R}_1(t') \\
&-i\overline{Q}\hat{R}_1(0)+i\hat{p}_1(0)(\overline{R}'-\overline{R})]>\,. \qquad (3.151)
\end{aligned}$$

We are now in the position to make use of the fact that the Hamiltonian is quadratic in all coordinates so that we can evaluate exactly

$$\begin{aligned}
\langle\hat{T}\mathrm{e}^{i\int dt'\,\eta(t')\hat{R}_1(t')}\rangle &= \mathrm{e}^{-\frac{i}{2}\int\int dt'dt''\eta(t')D(t',t'')\eta(t'')}, \qquad (3.152)\\
D(t',t'') &= -i\langle\hat{T}\hat{R}_1(t')\hat{R}_1(t'')\rangle\,.
\end{aligned}$$

By comparison with the last two lines in eq. (3.151), the function $\eta(t')$ is given by

$$\begin{aligned}
\eta(t') &= -Q\delta(t'-t)-\overline{Q}\delta(t')+2m\Omega^2a[\Theta(t')-\Theta(t'-t)] \\
&\quad+m(R-R')\delta'(t'-t)+m(\overline{R}-\overline{R}')\delta'(t')\,.
\end{aligned}$$

$\Theta(t)$ is meant to represent the step function. The derivatives of the δ–function arise from a partial integration of terms containing $\hat{p}(t) = md\hat{x}(t)/dt$. Note, that these act only on the coordinates but not on the step functions which arise due to the time ordering.

Moreover, the degrees of freedom of the bath can be integrated out in the usual way [CL81] leading to a dissipative influence on the object. One is then lead to the following form of the Fourier transform of $D(t,t') \equiv D(t-t')$:

$$\begin{aligned}
D(\omega) &= \frac{D^R(\omega)}{1-\exp(-\hbar\omega/k_BT)}+\frac{D^R(-\omega)}{1-\exp(\hbar\omega/k_BT)}, \qquad (3.153)\\
\left(D^R\right)^{-1}(\omega) &= m[(\omega+i0)^2-\Omega^2]+i\eta\omega,
\end{aligned}$$

where we will use a spectral density $J(\omega) = \eta\omega$, $0 \le \omega \le \omega_c$ for the bath oscillators.

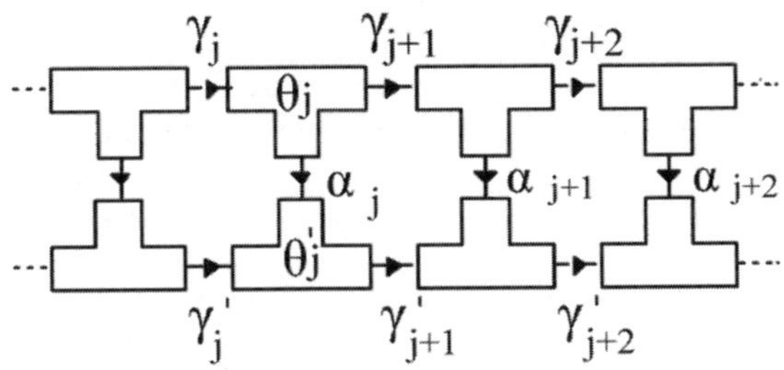

Fig. 3.14. Josephson junction Ladder (JJL).

From (3.151) and (3.152) one can conclude that the integrations with respect to $Q, \overline{Q}, R, R', \overline{R}, \overline{R}'$ can be done exactly as only Gaussian integrals are involved (note, that the eigenstates of the harmonic oscillator are Gaussian functions and derivatives of these, respectively). Therefore, for given $n_L, n_R, \overline{n}_L, \overline{n}_R$, one has to perform a 6D Gaussian integral [MS95].

3.3.3 Josephson Junction Ladder (JJL)

2D arrays of Josephson junctions have attracted much recent theoretical and experimental attention. Interesting physics arises as a result of competing vortex-vortex and vortex–lattice interactions. It is also considered to be a convenient experimental realization of the so–called *frustrated XY models*. Here, mainly following [DT95], we discuss the simplest such system, namely the *Josephson junction ladder* (JJL, see Figure 3.14) [Kar84, Gra90].

To construct the system, superconducting elements are placed at the ladder sites. Below the bulk *superconducting–normal transition* temperature, the state of each element is described by its charge and the phase of the superconducting wave ψ–function [And64]. In this section we neglect charging effects, which corresponds to the condition that $4e^2/C \ll J$, with C being the capacitance of the element and J the Josephson coupling. Let θ_j (θ'_j) denote the phase on the upper (lower) branch of the ladder at the jth rung. The Hamiltonian for the array [Tin75] can be written in terms the gauge invariant phase differences [DT95],

$$\gamma_j = \theta_j - \theta_{j-1} - (2\pi/\phi_0)\int_{j-1}^{j} A_x dx, \gamma'_j = \theta'_j - \theta'_{j-1} - (2\pi/\phi_0)\int_{j'-1}^{j'} A_x dx,$$

$$\text{and} \qquad \alpha_j = \theta'_j - \theta_j - (2\pi/\phi_0)\int_{j}^{j'} A_y dx \qquad \text{as}$$

$$H = -\sum_j (J_x \cos\gamma_j + J_x \cos\gamma'_j + J_y \cos\alpha_j), \tag{3.154}$$

where A_x and A_y are the components of the magnetic vector potential along and transverse to the ladder, respectively, and ϕ_0 the flux quantum. The sum of the phase differences around a plaquette is constrained by

$$\gamma_j - \gamma_j' + \alpha_j - \alpha_{j-1} = 2\pi(f - n_j),$$

where $n_j = 0, \pm1, \pm2, \ldots$ is the vortex occupancy number and $f = \phi/\phi_0$ with ϕ being the magnetic flux through a plaquette. With this constraint, it is convenient to write (3.154) in the form

$$H = -J\sum_j \{2\cos\eta_j \cos[(\alpha_{j-1} - \alpha_j)/2 + \pi(f - n_j)] + J_t \cos\alpha_j\},$$

$$\text{where} \qquad \eta_j = (\gamma_j + \gamma_j')/2, J = J_x \qquad \text{and} \qquad J_t = J_y/J_x \tag{3.155}$$

The Hamiltonian is symmetric under $f \to f+1$ with $n_j \to n_j+1$, and $f \to -f$ with $n_j \to -n_j$, thus it is sufficient to study only the region $0 \leq f \leq 0.5$. Since in one dimension ordered phases occur only at zero temperature, the main interest is in the ground states of the ladder and the low temperature excitations. Note that in (3.155) η_j decouples from α_j and n_j, so that all the ground states have $\eta_j = 0$ to minimize H. The ground states will be among the solutions to the current conservation equations: $\partial_{\alpha_j} H = 0$, i.e., [DT95]

$$J_t \sin\alpha_j = \sin[(\alpha_{j-1} - \alpha_j)/2 + \pi(f - n_j)] - \sin[(\alpha_j - \alpha_{j+1})/2 + \pi(f - n_{j+1})]. \tag{3.156}$$

For any given f there are a host of solutions to (3.156). The solution that minimizes the energy must be selected to get the ground state.

If one expands the inter–plaquette cosine coupling term in (3.155) about it's maximum, the discrete sine–Gordon model is obtained. A vortex ($n_j = 1$) in the JJL corresponds to a kink in the sine–Gordon model. This analogy was used by [Kar84] as an argument that this system should show similar behavior to the discrete sine–Gordon model which has been studied by several authors [AA80, CS83, PTB86]. This analogy is only valid for J_t very small so that the inter–plaquette term dominates the behavior of the system making the expansion about its maximum a reasonable assumption. However, much of the interesting behavior of the discrete sine–Gordon model occurs in regions of large J_t ($J_t \sim 1$). Furthermore, much of the work by Aubry [AA80] on the sine–Gordon model relies on the convexity of the coupling potential which we do not have in the JJL.

Following [DT95], here we formulate the problem in terms of a transfer matrix obtained from the full partition function of the ladder. The eigenvalues and eigenfunctions of the transfer matrix are found numerically to determine the phases of the ladder as functions of f and J_t. We study the properties of various ground states and the low temperature excitations. As J_t is varied, all incommensurate ground states undergo a superconducting–normal transition at certain J_t which depends on f. One such transition will be analyzed. Finally we discuss the critical current.

The partition function for the ladder, with periodic boundary conditions and $K = J/k_B T$, is

$$Z = \prod_i^N \int_{-\pi}^{\pi} \sum_{\{n_i\}} d\alpha_i d\eta_i \exp\{K(2\cos\eta_i \cos[(\alpha_{i-1} - \alpha_i)/2 + \pi(f - n_i)] + J_t \cos\alpha_i)\}.$$

The η_i can be integrated out resulting in a simple transfer matrix formalism for the partition function involving only the transverse phase differences:

$$Z = \prod_i^N \int_{-\pi}^{\pi} d\alpha_i P(\alpha_{i-1}, \alpha_i) = Tr\, \hat{P}^N.$$

The transfer matrix elements $P(\alpha, \alpha')$ are

$$P(\alpha, \alpha') = 4\pi \exp[K J_t(\cos\alpha + \cos\alpha')/2]\, I_0(2K\cos[(\alpha - \alpha')/2 + \pi f]), \quad (3.157)$$

where I_0 is the zeroth order modified Bessel function. Note that the elements of $\hat{P}$ are real and positive, so that its largest eigenvalue λ_0 is real, positive and nondegenerate. However, since $\hat{P}$ is not symmetric (except for $f = 0$ and $f = 1/2$) other eigenvalues can form complex conjugate pairs. As we will see from the correlation function, these complex eigenvalues determine the spatial periodicity of the ground states.

The two point correlation function of α_j's is [DT95]

$$\langle e^{i(\alpha_0 - \alpha_l)} \rangle = \lim_{N\to\infty} \frac{\left(\prod_i^N \int_{-\pi}^{\pi} d\alpha_i P(\alpha_{i-1}, \alpha_i)\right) e^{i(\alpha_0 - \alpha_l)}}{Z} = \sum_n c_n \left(\frac{\lambda_n}{\lambda_0}\right)^l, \quad (3.158)$$

where we have made use of the completeness of the left and right eigenfunctions. (Note that since $\hat{P}$ is not symmetric both right ψ_n^R and left ψ_n^L eigenfunctions are need for the evaluation of correlation functions.) The λ_n in (3.158) are the eigenvalues ($|\lambda_n| \geq |\lambda_{n+1}|$ and $n = 0, 1, 2, \dots$), and the constants

$$c_n = \int_{-\pi}^{\pi} d\alpha' \psi_0^L(\alpha') e^{i\alpha'} \psi_n^R(\alpha') \int_{-\pi}^{\pi} d\alpha \psi_n^L(\alpha) e^{-i\alpha} \psi_0^R(\alpha).$$

In the case where λ_1 is real and $|\lambda_1| > |\lambda_2|$, (3.158) simplifies for large l to

$$\langle e^{i(\alpha_0 - \alpha_l)} \rangle = c_0 + c_1 \left(\frac{\lambda_1}{\lambda_0}\right)^l, \quad |\lambda_1| > |\lambda_2|.$$

In the case where $\lambda_1 = \lambda_2^* = |\lambda_1| e^{i2\pi\Xi}$, (3.158) for large l is [26]

$$\langle e^{i(\alpha_0-\alpha_l)} \rangle = c_0 + \left(c_1 e^{i2\pi\Xi l} + c_2 e^{-i2\pi\Xi l}\right) \left|\frac{\lambda_1}{\lambda_0}\right|^l, \qquad \lambda_1 = \lambda_2^*.$$

There is no phase coherence between upper and lower branches of the ladder and hence no superconductivity in the transverse direction. In this case, we say that the α's are unpinned. If there exist finite intervals of α on which $\rho(\alpha) = 0$, there will be phase coherence between the upper and lower branches and we say that the α's are pinned. In term of the transfer matrix, the phase density is the product of the left and right eigenfunctions of λ_0 [GM79],

$$\rho(\alpha) = \psi_0^L(\alpha)\psi_0^R(\alpha).$$

We first discuss the case where $f < f_{c1}$. These are the *Meissner–states* in the sense that there are no vortices ($n_i = 0$) in the ladder. The ground state is simply $\alpha_i = 0$, $\gamma_j = \pi f$ and $\gamma_j' = -\pi f$, so that there is a global screening current $\pm J_x \sin \pi f$ in the upper and lower branches of the ladder [Kar84]. The phase density $\rho(\alpha) = \delta(\alpha)$. The properties of the Meissner state can be studied by expanding (3.155) around $\alpha_i = 0$,

$$H_M = (J/4)\sum_j [\cos(\pi f)(\alpha_{j-1} - \alpha_j)^2 + 2J_t\alpha_i^2].$$

The current conservation (3.156) becomes

$$\alpha_{j+1} = 2\,(1 + J_t/\cos \pi f)\,\alpha_j - \alpha_{j-1}. \tag{3.159}$$

Besides the ground state $\alpha_j = 0$, there are other two linearly independent solutions $\alpha_j = e^{\pm j/\xi_M}$ of (3.159) which describe collective fluctuations about the ground state, where

$$\frac{1}{\xi_M} = \ln\left[1 + \frac{J_t}{\cos \pi f} + \sqrt{\frac{2J_t}{\cos \pi f} + \left(\frac{J_t}{\cos \pi f}\right)^2}\,\right]. \tag{3.160}$$

ξ_M is the low temperature correlation length for the Meissner state.[27] As f increases, the Meissner state becomes unstable to the formation of vortices. A vortex is constructed by patching the two solutions of (3.159) together

[26] While the correlation length is given by $\xi = [\ln|\lambda_0/\lambda_1|]^{-1}$ the quantity $\Xi = Arg(\lambda_1)/2\pi$ determines the spatial periodicity of the state. By numerical calculation of λ_n, it is found that for f smaller than a critical value f_{c1} which depends on J_t, both λ_1 and λ_2 are real. These two eigenvalues become degenerate at f_{c1}, and then bifurcate into a complex conjugate pair [DT95].

[27] Here, $\xi_M < 1$ for $J_t \sim 1$ making a continuum approximation invalid.

using a matching condition. The energy ϵ_v of a single vortex is found to be [DT95]

$$\epsilon_v \approx [2 + (\pi^2/8)\tanh(1/2\xi_M)]\cos \pi f - (\pi + 1)\sin \pi f + 2J_t, \qquad (3.161)$$

for J_t close to one. The zero of ϵ_v determines f_{c1} which is in good agreement with the numerical result from the transfer matrix. For $f > f_{c1}$, ϵ_v is negative and vortices are spontaneously created. When vortices are far apart their interaction is caused only by the exponentially small overlap. The corresponding repulsion energy is of the order $J\exp(-l/\xi_M)$, where l is the distance between vortices. This leads to a free energy per plaquette of $F = \epsilon_v/l + J\exp(-l/\xi_M)/l$ [PTB86]. Minimizing this free energy as a function of l gives the vortex density for $f > f_{c1}$: $\langle n_j \rangle = l^{-1} = [\xi_M \ln|f_{c1} - f|]^{-1}$ where a linear approximation is used for f close to f_{c1}.

We now discuss the commensurate vortex states, taking the one with $\Xi = 1/2$ as an example. This state has many similarities to the Meissner state but some important differences. The ground state is

$$\alpha_0 = \arctan\left[\frac{2}{J_t}\sin(\pi f)\right], \qquad \alpha_1 = -\alpha_0, \; \alpha_{i\pm 2} = \alpha_i;$$
$$n_0 = 0, \; n_1 = 1, \; n_{i\pm 2} = n_i, \qquad (3.162)$$

so that there is a global screening current in the upper and lower branches of the ladder of $\pm 2\pi J(f - 1/2)/\sqrt{4 + J_t^2}$. The existence of the global screening, which is absent in an infinite 2D array, is the key reason for the existence of the steps at $\Xi = p/q$. It is easy to see that the symmetry of this vortex state is that of the (antiferromagnetic) Ising model. The ground state is two-fold degenerate. The low temperature excitations are domain boundaries between the two degenerate ground states. The energy of the domain boundary $J\epsilon_b$ can be estimated using similar methods to those used to derive (3.161) for the Meissner state. We found that $\epsilon_b = \epsilon_b^0 - (\pi^2/\sqrt{4 + J_t^2})|f - 1/2|$, where ϵ_b^0 depends only on

$$J_t = c\arctan^2(2/J_t)J_t^2\coth(1/\xi_b)/\sqrt{4 + J_t^2},$$

with c being a constant of order one and

$$\xi_b^{-1} = \ln(1 + J_t^2/2 + J_t\sqrt{1 + J_t^2/4}).$$

Thus the correlation length diverges with temperature as $\xi \sim \exp(2J\epsilon_b/k_BT)$. The transition from the $\Xi = 1/2$ state to nearby vortex states happens when f is such that $\epsilon_b = 0$; it is similar to the transition from the Meissner state to its nearby vortex states. All other steps $\Xi = p/q$ can be analyzed similarly. For comparison, we have evaluated ξ for various values of f and T from the transfer matrix and found that ξ fits $\xi \sim \exp(2J\epsilon_b/k_BT)$ (typically over several decades) at low temperature.

We now discuss the superconducting–normal transition in the transverse direction. For $J_t = 0$, the ground state has $\gamma_i = \gamma_i' = 0$ and

$$\alpha_j = 2\pi f j + \alpha_0 - 2\pi \sum_{i=0}^{i=j} n_i. \qquad (3.163)$$

The average vortex density $\langle n_j \rangle$ is f; there is no screening of the magnetic field. α_0 in (3.163) is arbitrary; the α's are unpinned for all f. The system is simply two un–coupled 1D XY chains, so that the correlation length $\xi = 1/k_BT$. The system is superconducting at zero temperature along the ladder, but not in the transverse direction. As J_t rises above zero we observe a distinct difference between the system at rational and irrational values of f. For f rational, the α's become pinned for $J_t > 0$ ($\rho(\alpha)$ is a finite sum of delta functions) and the ladder is superconducting in *both* the longitudinal and transverse directions at zero temperature. The behavior for irrational f is illustrated in the following for the state with $\Xi = a_g$, where $a_g \approx 0.381966\cdots$ is one minus the inverse of the golden mean.

Finally, we consider critical currents along the ladder. One can get an estimate for the critical current by performing a perturbation expansion around the ground state (i.e., $\{n_j\}$ remain fixed) and imposing the current constraint of $\sin\gamma_j + \sin\gamma_j' = I$. Let $\delta\gamma_j$, $\delta\gamma_j'$ and $\delta\alpha_j$ be the change of γ_j, γ_j' and α_j in the current carrying state. One finds that stability of the ground state requires that $\delta\alpha_j = 0$, and consequently $\delta\gamma_j = \delta\gamma_j' = I/2\cos\gamma_j$. The critical current can be estimated by the requirement that the γ_j do not pass through $\pi/2$, which gives $I_c = 2(\pi/2 - \gamma_{\max})\cos\gamma_{\max}$, where $\gamma_{\max} = \max_j(\gamma_j)$. In all ground states we examined, commensurate and incommensurate, we found that $\gamma_{\max} < \pi/2$, implying a finite critical current for all f. See [DT95] for more details.

Underdamped JJL

Recall that the *discrete sine–Gordon equation* has been used by several groups to model so-called hybrid Josephson ladder arrays [UMM93, WSZ95]. Such an array consists of a ladder of parallel Josephson junctions which are inductively coupled together (e.g., by superconducting wires). The sine-Gordon equation then describes the phase differences across the junctions. In an applied magnetic field, this equation predicts remarkably complex behavior, including flux flow resistance below a certain critical current, and a field–independent resistance above that current arising from so–called *whirling modes* [WSZ95]. In the flux flow regime, the fluxons in this ladder propagate as localized solitons, and the IV characteristics exhibit voltage plateaus arising from the locking of solitons to linear *spin–wave modes*. At sufficiently large values of the anisotropy parameter η_J defined later, the solitons may propagate 'ballistically' on the plateaus, i.e., may travel a considerable distance even after the driving current is turned off.

Here, mainly following [RYD96], we show that this behavior is all found in a model in which the ladder is treated as a network of coupled small junctions arranged along both the edges and the rungs of the ladder. This model is often used to treat 2D Josephson networks, and includes *no* inductive coupling between junctions, other than that produced by the other junctions. To confirm our numerical results, we derive a discrete sine–Gordon equation from our coupled–network model. Thus, these seemingly quite different models produce nearly identical behavior for ladders. By extension, they suggest that some properties of 2D arrays might conceivably be treated by a similar simplification. In simulations [Bob92, GLW93, SIT95], underdamped arrays of this type show some similarities to ladder arrays, exhibiting the analogs of both voltage steps and whirling modes.

We consider a ladder consisting of coupled superconducting grains, the ith of which has order parameter

$$\Phi_i = \Phi_0 \mathrm{e}^{i\theta_i}.$$

Grains i and j are coupled by *resistively–shunted Josephson junctions* (RSJ's) with current I_{ij}, shunt resistance R_{ij} and shunt capacitance C_{ij}, with periodic boundary conditions.

The phases θ_i evolve according to the coupled RSJ equations

$$\begin{aligned} &\hbar\dot{\theta}_i/(2e) = V_i, \\ &\quad M_{ij}\dot{V}_j = I_i^{ext}/I_c - (R/R_{ij})(V_i - V_j) - (I_{ij}/I_c)\sin(\theta_{ij} - A_{ij}). \end{aligned}$$

Here the time unit is $t_0 = \hbar/(2eRI_c)$, where R and I_c are the shunt resistance and critical current across a junction in the x–direction; I_i^{ext} is the external current fed into the ith node; the spatial distances are given in units of the lattice spacing a, and the voltage V_i in units of I_cR.

$$\begin{aligned} M_{ij} &= -4\pi eCI_cR^2/h \qquad \text{for} \quad i \neq j, \\ \text{and} \qquad M_{ii} &= -\sum_{j\neq i} M_{ij}, \end{aligned}$$

where C is the intergrain capacitance. Finally,

$$A_{ij} = (2\pi/\Phi_0)\int_i^j A \cdot dl,$$

where A is the vector potential. Following [RYD96], we assume N plaquettes in the array, and postulate a current I uniformly injected into each node on the outer edge and extracted from each node on the inner edge of the ring. We also assume a uniform transverse magnetic field $B \equiv f\phi_0/a^2$, and use the *Landau gauge* $A = -Bx\,\hat{y}$.

We now show that this model reduces approximately to a discrete sine–Gordon equation for the *phase differences*. Label each grain by (x, y) where $x/a = 0, \ldots, N-1$ and $y/a = 0, 1$. Subtracting the equation of motion for $\theta(x, 1)$ from that for $\theta(x, 2)$, and defining

$$\Psi(x) = \frac{1}{2}[\theta(x,1) + \theta(x,2)], \chi(x) = [\theta(x,2) - \theta(x,1)],$$

we get a differential equation for $\chi(x)$ which is second-order in time. This equation may be further simplified using the facts that $A_{x,y;x\pm1,y} = 0$ in the Landau gauge, and that $A_{x,1;x,2} = -A_{x,2;x,1}$, and by defining the *discrete Laplacian*

$$\chi(x+1) - 2\chi(x) + \chi(x-1) = \nabla^2\chi(x).$$

Finally, using the boundary conditions,

$$I^{ext}(x,2) = -I^{ext}(x,1) \equiv I,$$

and introducing $\phi(x) = \chi(x) - A_{x,2;x,1}$, we get

$$\begin{aligned}[1 - \eta_c^2\nabla^2]\beta\ddot{\phi} &= i - [1 - \eta_r^2\nabla^2]\dot{\phi} - \sin(\phi) + 2\eta_J^2 \\ &\times \sum_{i=\pm1} \cos\{\Psi(x) - \Psi(x+i)\} \sin\{[\phi(x) - \phi(x+i)]/2\},\end{aligned} \tag{3.164}$$

where we have defined a dimensionless current $i = I/I_{cy}$, and anisotropy factors

$$2\eta_r^2 = R_y/R_x, \qquad 2\eta_c^2 = C_x/C_y, \qquad 2\eta_J^2 = I_{cx}/I_{cy}.$$

We now neglect all combined space and time derivatives of order three or higher. Similarly, we set the cosine factor equal to unity(this is also checked numerically to be valid *a posteriori*) and linearize the sine factor in the last term, so that the final summation can be expressed simply as $\nabla^2\phi$. With these approximations, (3.164) reduces to *discrete driven sine-Gordon equation with dissipation*:

$$\beta\ddot{\phi} + \dot{\phi} + \sin(\phi) - \eta_J^2\nabla^2\phi = i, \qquad \text{where} \quad \beta = 4\pi e I_{cy} R_y^2 C_y/h. \tag{3.165}$$

Soliton Behavior

In the absence of damping and driving, the continuum version of (3.165) has, among other solutions, the sine–Gordon soliton [Raj82], given by

$$\phi_s(x,t) \sim 4\tan^{-1}\left[\exp\left\{(x - v_v t)/\sqrt{\eta_J^2 - \beta v_v^2},\right\}\right]$$

where v_v is the velocity. The phase in this soliton rises from ~0 to ~2π in a width $d_k \sim \sqrt{\eta_J^2 - \beta v_v^2}$.

The transition to the resistive state occurs at $n_{min} = 4, 2, 2, 1$ for $\eta_J^2 = 0.5, 1.25, 2.5, 5$. This can also be understood from the *kink–phason resonance* picture. To a phason mode, the passage of a kink of width d_k will appear like the switching on of a step–like driving current over a time of order d_k/v_v. The kink will couple to the phasons only if $d_k/v_v \geq \pi/\omega_1$, the half–period of the phason, or equivalently

$$\frac{1}{\sqrt{\beta}v_v} \geq \frac{\sqrt{1+\pi^2}}{\eta_J} = \frac{3.3}{\eta_J}.$$

This condition agrees very well with our numerical observations, even though it was obtained by considering soliton solutions from the continuum sine–Gordon equation.

The fact that the voltage in regime I is approximately linear in f can be qualitatively understood from the following argument. Suppose that ϕ for Nf fluxons can be approximated as a sum of well–separated solitons, each moving with the same velocity and described by

$$\phi(x,t) = \sum_{j=1}^{Nf} \phi_j, \qquad \text{where} \qquad \phi_j = \phi_s(x - x_j, t).$$

Since the solitons are well separated, we can use following properties:

$$\sin\left[\sum_j \phi_j\right] = \sum_j \sin\phi_j \qquad \text{and} \qquad \int \dot{\phi}_j \dot{\phi}_i dx \propto \delta_{ij}.$$

By demanding that the energy dissipated by the damping of the moving soliton be balanced by that the driving current provides ($\propto \int dxi\dot{\phi}(x)$), one can show that the Nf fluxons should move with the same velocity v as that for a single fluxon driven by the same current. In the *whirling regime*, the *f–independence* of the voltage can be understood from a somewhat different argument. Here, we assume a periodic solution of the form

$$\phi = \sum_j^{Nf} \phi_w(x - \tilde{v}t - j/f),$$

moving with an unknown velocity $\tilde{v}$ where $\phi_w(\xi)$ describes a whirling solution containing one fluxon. Then using the property $\phi(x + m/f) = \phi(x) + 2\pi m$, one can show that [RYD96]

$$\sin\left[\sum_j^{Nf} \phi_w(x - \tilde{v}t - j/f)\right] = \sin[Nf\phi_w(x - \tilde{v}t)].$$

Finally, using the approximate property $\phi_w(\xi) \sim \xi$ of the whirling state, one finds $\tilde{v} = v/(Nf)$, leading to an f–independent voltage.

Ballistic Soliton Motion and Soliton Mass

A common feature of massive particles is their 'ballistic motion', defined as inertial propagation after the driving force has been turned off. Such propagation has been reported experimentally but as yet has not been observed numerically in either square or triangular lattices [GLW93]. In the so–called *flux–flow regime* at $\eta_J = 0.71$, we also find no ballistic propagation, presumably because of the large pinning energies produced by the periodic lattice.

We can define the fluxon mass in our ladder by equating the *charging energy* $E_c = C/2\sum_{ij} V_{ij}^2$ to the kinetic energy of a soliton of mass M_v: $E_{kin} = \frac{1}{2}M_v v_v^2$ [GLW93]. Since E_c can be directly calculated in our simulation, while v_v can be calculated from $\langle V \rangle$, this gives an unambiguous way to determine M_v. For $\eta_J^2 = 0.5$, we find $E_c/C \sim 110(\langle V \rangle /I_c R)^2$, in the flux–flow regime. This gives $M_v^I \sim 3.4C\phi_0^2/a^2$, more than six times the usual estimate for the vortex mass in a 2D square lattice. Similarly, the vortex friction coefficient γ can be estimated by equating the rate of energy dissipation,

$$E_{dis} = 1/2\sum_{ij} V_{ij}^2/R_{ij}, \qquad \text{to} \qquad \frac{1}{2}\gamma v_v^2.$$

This estimate yields $\gamma^I \sim 3.4\phi_0^2/(Ra^2)$, once again more than six times the value predicted for 2D arrays [GLW93]. This large dissipation explains the absence of ballistic motion for this anisotropy [GLW93]. At larger values $\eta_J^2 = 5$ and 2.5, a similar calculation gives $M_v^I \sim 0.28$ and $0.34\phi_0^2/(Ra^2)$, $\gamma^I \sim 0.28$ and $0.34\phi_0^2/(Ra^2)$. These lower values of γ^I, but especially the low pinning energies, may explain why ballistic motion is possible at these values of η_J. See [RYD96] for more details.

3.3.4 Synchronization in Arrays of Josephson Junctions

The *synchronization of coupled nonlinear oscillators* has been a fertile area of research for decades [PRK01]. In particular, *Winfree–type phase models* [Win67] have been extensively studied. In 1D, a generic version of this model for N oscillators reads

$$\dot{\theta}_j = \Omega_j + \sum_{k=1}^{N} \sigma_{j,k}\Gamma\left(\theta_k - \theta_j\right), \tag{3.166}$$

where θ_j is the phase of oscillator j, which can be envisioned as a point moving around the unit circle with angular velocity $\dot{\theta}_j = d\theta_j/dt$. In the absence of coupling, this overdamped oscillator has an angular velocity Ω_j. $\Gamma(\theta_k - \theta_j)$ is the coupling function, and $\sigma_{j,k}$ describes the range and nature (e.g., attractive or repulsive) of the coupling. The special case

$$\Gamma(\theta_k - \theta_j) = \sin(\theta_k - \theta_j), \qquad \sigma_{j,k} = \alpha/N, \qquad \alpha = \text{const},$$

corresponds to the uniform, sinusoidal coupling of each oscillator to the remaining $N-1$ oscillators. This mean–field system is usually called the *globally–coupled Kuramoto model* (GKM). Kuramoto was the first to show that for this particular form of coupling and in the $N \to \infty$ limit, there is a continuous dynamical phase transition at a critical value of the coupling strength α_c and that for $\alpha > \alpha_c$ both phase and frequency synchronization appear in the system [Kur84, Str00]. If $\sigma_{j,k} = \alpha\delta_{j,k\pm1}$ while the coupling function retains the form $\Gamma(\theta_j - \theta_k) = \sin(\theta_k - \theta_j)$, then we have the so–called *locally–coupled Kuramoto model* (LKM), in which each oscillator is coupled only to its nearest neighbors. Studies of synchronization in the LKM [SSK87], including extensions to more than one spatial dimension, have shown that α_c grows without bound in the $N \to \infty$ limit [Sm88].

Watts and Strogatz introduced a simple model for tuning collections of coupled dynamical systems between the two extremes of random and regular networks [WS98]. In this model, connections between nodes in a regular array are randomly rewired with a probability p, such that $p = 0$ means the network is regularly connected, while $p = 1$ results in a random connection of nodes. For a range of intermediate values of p between these two extremes, the network retains a property of regular networks (a large clustering coefficient) and also acquires a property of random networks (a short characteristic path length between nodes). Networks in this intermediate configuration are termed *small–world networks*. Many examples of such small worlds, both natural and human–made, have been discussed [Str]. Not surprisingly, there has been much interest in the synchronization of dynamical systems connected in a small–world geometry [BP02, NML03]. Generically, such studies have shown that the presence of small–world connections make it easier for a network to synchronize, an effect generally attributed to the reduced path length between the linked systems. This has also been found to be true for the special case in which the dynamics of each oscillator is described by a Kuramoto model [HCK02a, HCK02b].

As an example of *physically–controllable systems of nonlinear oscillators*, which can be studied both theoretically and experimentally, Josephson junction (JJ) arrays are almost without peer. Through modern fabrication techniques and careful experimental methods one can attain a high degree of control over the dynamics of a JJ array, and many detailed aspects of array behavior have been studied [NLG00]. Among the many different geometries of JJ arrays, *ladder* arrays deserve special attention. For example, they have been observed to support stable time–dependent, spatially–localized states known as discrete breathers [TMO00]. In addition, the ladder geometry is more complex than that of better understood serial arrays but less so than fully two–dimensional (2D) arrays. In fact, a ladder can be considered as a special kind of 2D array, and so the study of ladders could throw some light on the behavior of such 2D arrays. Also, linearly–stable synchronization of the horizontal, or rung, junctions in a ladder is observed in the absence of a load over a wide range of dc bias currents and junction parameters (such as junction capacitance), so that synchronization in this geometry appears to be robust [TSS05].

In the mid 1990's it was shown that a serial array of zero-capacitance, i.e., overdamped, junctions coupled to a load could be mapped onto the GKM [WCS96, WCS98]. The load in this case was essential in providing an all–to–all coupling among the junctions. The result was based on an averaging process, in which (at least) two distinct time scales were identified: the 'short' time scale set by the rapid voltage oscillations of the junctions (the array was current biased above its critical current) and 'long' time scale over which the junctions synchronize their voltages. If the *resistively–shunted junction* (RSJ) equations describing the dynamics of the junctions are integrated over one cycle of the 'short' time scale, what remains is the 'slow' dynamics, describing the synchronization of the array. This mapping is useful because it allows knowledge about the GKM to be applied to understanding the dynamics of the serial JJ array. For example, the authors of [WCS96] were able, based on the GKM, to predict the level of critical current disorder the array could tolerate before frequency synchronization would be lost. Frequency synchronization, also described as entrainment, refers to the state of the array in which all junctions not in the zero–voltage state have equal (to within some numerical precision) time–averaged voltages: $(\hbar/2e)\langle\dot{\theta}_j\rangle_t$, where θ_j is the gauge–invariant phase difference across junction j. More recently, the 'slow' synchronization dynamics of finite–capacitance serial arrays of JJ's has also been studied [CS95, WS97]. Perhaps surprisingly, however, no experimental work on JJ arrays has verified the accuracy of this GKM mapping. Instead, the first detailed experimental verification of Kuramoto's theory was recently performed on systems of coupled electrochemical oscillators [KZH02].

Recently, [DDT03] showed, with an eye toward a better understanding of synchronization in 2D JJ arrays, that a ladder array of *overdamped junctions* could be mapped onto the LKM. This work was based on an averaging process, as in [WCS96], and was valid in the limits of weak critical current disorder (less than about 10%) and large dc bias currents, I_B, along the rung junctions ($I_B/\langle I_c\rangle \gtrsim 3$, where $\langle I_c\rangle$ is the arithmetic average of the critical currents of the rung junctions. The result demonstrated, for both open and periodic boundary conditions, that synchronization of the current–biased rung junctions in the ladder is well described by (3.166).

In this subsection, following [TSS05], we demonstrate that a ladder array of *underdamped junctions* can be mapped onto a second–order Winfree–type oscillator model of the form

$$a\ddot{\theta}_j + \dot{\theta}_j = \Omega_j + \sum_{k=1}^{N} \sigma_{j,k}\Gamma(\theta_k - \theta_j), \qquad (3.167)$$

where a is a constant related to the average capacitance of the rung junctions. This result is based on the *resistively & capacitively–shunted junction* (RCSJ) model and a multiple time scale analysis of the classical equations for the array. Secondly, we study the effects of *small world* (SW) connections on the synchronization of both overdamped and underdamped ladder arrays. It appears that SW connections make it easier for the ladder to synchronize, and that a Kuramoto or Winfree type model (3.166) and (3.167), suitably generalized to include the new connections, accurately describes the synchronization of this ladder.

Phase Model for Underdamped JJL

Following [TSS05] we analyze synchronization in disordered Josephson junction arrays.The ladder geometry used consists of an array with $N = 8$ plaquettes, periodic boundary conditions, and uniform dc bias currents, I_B, along the rung junctions (see Figure 3.15). The *gauge–invariant phase difference* across rung junction j is γ_j, while the phase difference across the off–rung junctions along the outer(inner) edge of plaquette j is $\psi_{1,j}(\psi_{2,j})$. The critical current, resistance, and capacitance of rung junction j are denoted I_{cj}, R_j, and C_j, respectively. For simplicity, we assume all off–rung junctions are identical, with critical current I_{co}, resistance R_o, and capacitance C_o. We also

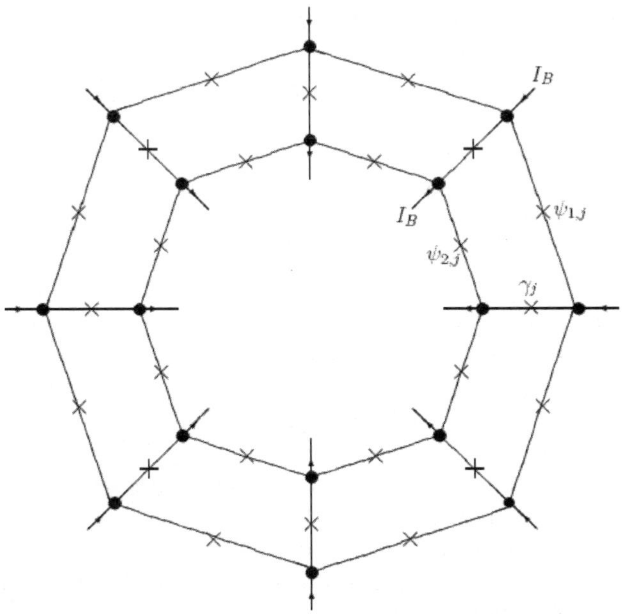

Fig. 3.15. A ladder array of Josephson junctions with periodic boundary conditions and $N = 8$ plaquettes. A uniform, dc bias current I_B is inserted into and extracted from each rung as shown. The gauge–invariant phase difference across the rung junctions is denoted by γ_j where $1 \leq j \leq N$, while the corresponding quantities for the off–rung junctions along the outer(inner) edge are $\psi_{1,j}(\psi_{2,j})$ (adapted and modified from [TSS05]).

assume that the product of the junction critical current and resistance is the same for all junctions in the array [Ben95], with a similar assumption about the ratio of each junction's critical current with its capacitance:

$$I_{cj}R_j = I_{co}R_o = \frac{\langle I_c \rangle}{\langle R^{-1} \rangle} \tag{3.168}$$

$$\frac{I_{cj}}{C_j} = \frac{I_{co}}{C_o} = \frac{\langle I_c \rangle}{\langle C \rangle}, \tag{3.169}$$

where for any generic quantity X, the angular brackets with no subscript denote an arithmetic average over the set of rung junctions,

$$\langle X \rangle \equiv (1/N) \sum_{j=1}^{N} X_j.$$

For convenience, we work with dimensionless quantities. Our dimensionless time variable is

$$\tau \equiv \frac{t}{t_c} = \frac{2e\langle I_c \rangle t}{\hbar \langle R^{-1} \rangle}, \tag{3.170}$$

where t is the ordinary time. In the following, derivatives with respect to τ will be denoted by *prime* (e.g., $\psi' = d\psi/d\tau$). The dimensionless bias current is

$$i_B \equiv \frac{I_B}{\langle I_c \rangle}, \tag{3.171}$$

while the dimensionless critical current of rung junction j is $i_{cj} \equiv I_{cj}/\langle I_c \rangle$. The McCumber parameter in this case is

$$\beta_c \equiv \frac{2e\langle I_c \rangle \langle C \rangle}{\hbar \langle R^{-1} \rangle^2}. \tag{3.172}$$

Note that β_c is proportional to the mean capacitance of the rung junctions. An important dimensionless parameter is

$$\alpha \equiv \frac{I_{co}}{\langle I_c \rangle}, \tag{3.173}$$

which will effectively tune the nearest–neighbor interaction strength in our phase model for the ladder.

Conservation of charge applied to the superconducting islands on the outer and inner edge, respectively, of rung junction j yields the following equations in dimensionless variables [TSS05]:

$$\begin{aligned} & i_B - i_{cj}\sin\gamma_j - i_{cj}\gamma_j' - i_{cj}\beta_c\gamma_j'' - \alpha\sin\psi_{1,j} - \alpha\psi_{1,j}' \\ & -\alpha\beta_c\psi_{1,j}'' + \alpha\sin\psi_{1,j-1} + \alpha\psi_{1,j-1}' + \alpha\beta_c\psi_{1,j-1}'' = 0, \end{aligned} \tag{3.174}$$

$$\begin{aligned} & -i_B + i_{cj}\sin\gamma_j + i_{cj}\gamma_j' + i_{cj}\beta_c\gamma_j'' - \alpha\sin\psi_{2,j} - \alpha\psi_{2,j}' \\ & -\alpha\beta_c\psi_{2,j}'' + \alpha\sin\psi_{2,j-1} + \alpha\psi_{2,j-1}' + \alpha\beta_c\psi_{2,j-1}'' = 0, \end{aligned} \tag{3.175}$$

where $1 \leq j \leq N$. The result is a set of $2N$ equations in $3N$ unknowns: γ_j, $\psi_{1,j}$, and $\psi_{2,j}$. We supplement (3.175) by the constraint of fluxoid quantization in the absence of external or induced magnetic flux. For plaquette j this constraint yields the relationship

$$\gamma_j + \psi_{2,j} - \gamma_{j+1} - \psi_{1,j} = 0. \tag{3.176}$$

Equations (3.175) and (3.176) can be solved numerically for the $3N$ phases γ_j, $\psi_{1,j}$ and $\psi_{2,j}$ [TSS05].

We assign the rung junction critical currents in one of two ways, randomly or nonrandomly. We generate random critical currents according to a parabolic *probability distribution function* (PDF) of the form

$$P(i_c) = \frac{3}{4\Delta^3}\left[\Delta^2 - (i_c - 1)^2\right], \tag{3.177}$$

where $i_c = I_c/\langle I_c\rangle$ represents a scaled critical current, and Δ determines the spread of the critical currents. Equation (3.177) results in critical currents in the range $1-\Delta \leq i_c \leq 1+\Delta$. Note that this choice for the PDF (also used in [WCS96]) avoids extreme critical currents (relative to a mean value of unity) that are occasionally generated by PDF's with tails. The nonrandom method of assigning rung junction critical currents was based on the expression

$$i_{cj} = 1 + \Delta - \frac{2\Delta}{(N-1)^2}\left[4j^2 - 4(N+1)j + (N+1)^2\right], \qquad 1 \leq j \leq N, \tag{3.178}$$

which results in the i_{cj} values varying quadratically as a function of position along the ladder and falling within the range $1-\Delta \leq i_{cj} \leq 1+\Delta$. We usually use $\Delta = 0.05$.

Multiple Time–Scale Analysis

Now, our goal is to derive a Kuramoto–like model for the phase differences across the rung junctions, γ_j, starting with (3.175). We begin with two reasonable assumptions. First, we assume there is a simple phase relationship between the two off–rung junctions in the same plaquette [TSS05]:

$$\psi_{2,j} = -\psi_{1,j}, \tag{3.179}$$

the validity of which has been discussed in detail in [DDT03, FW95]. As a result, (3.176) reduces to

$$\psi_{1,j} = \frac{\gamma_j - \gamma_{j+1}}{2}, \tag{3.180}$$

which implies that (3.174) can be written as

$$\begin{aligned} i_{cj}\beta_c\gamma_j'' + i_{cj}\gamma_j' + \frac{\alpha\beta_c}{2}\left[\gamma_{j+1}'' - 2\gamma_j'' + \gamma_{j-1}''\right] + \frac{\alpha}{2}\left[\gamma_{j+1}' - 2\gamma_j' + \gamma_{j-1}'\right] \\ = i_B - i_{cj}\sin\gamma_j + \alpha\sum_{\delta=\pm1}\sin\left(\frac{\gamma_{j+\delta} - \gamma_j}{2}\right). \end{aligned} \tag{3.181}$$

Our second assumption is that we can neglect the discrete Laplacian terms in (3.181), namely

$$\nabla^2\gamma'_j \equiv \gamma'_{j+1} - 2\gamma'_j + \gamma'_{j-1} \qquad \text{and} \qquad \nabla^2\gamma''_j \equiv \gamma''_{j+1} - 2\gamma''_j + \gamma''_{j-1}.$$

We find numerically, over a wide range of bias currents i_B, *McCumber parameters* β_c, and coupling strengths α that $\nabla^2\gamma'_j$ and $\nabla^2\gamma''_j$ oscillate with a time–averaged value of approximately zero. Since the multiple time scale method is similar to averaging over a fast time scale, it seems reasonable to drop these terms. In light of this assumption, (3.181) becomes

$$i_{cj}\beta_c\gamma''_j + i_{cj}\gamma'_j = i_B - i_{cj}\sin\gamma_j + \alpha\sum_{\delta=\pm1}\sin\left(\frac{\gamma_{j+\delta}-\gamma_j}{2}\right). \tag{3.182}$$

We can use (3.182) as the starting point for a multiple time scale analysis. Following [CS95] and [WS97], we divide (3.182) by i_B and define the following quantities:

$$\tilde{\tau} \equiv i_B\tau, \qquad \tilde{\beta_c} \equiv i_B\beta_c, \qquad \epsilon = 1/i_B. \tag{3.183}$$

In terms of these scaled quantities, (3.182) can be written as

$$i_{cj}\tilde{\beta_c}\frac{d^2\gamma_j}{d\tilde{\tau}^2} + i_{cj}\frac{d\gamma_j}{d\tilde{\tau}} + \epsilon i_{cj}\sin\gamma_j - \epsilon\alpha\sum_{\delta}\sin\left(\frac{\gamma_{j+\delta}-\gamma_j}{2}\right) = 1. \tag{3.184}$$

Next, we introduce a series of four (dimensionless) time scales,

$$T_n \equiv \epsilon^n\tilde{\tau}, \qquad (n = 0,1,2,3), \tag{3.185}$$

which are assumed to be independent of each other. Note that $0 < \epsilon < 1$ since $\epsilon = 1/i_B$. We can think of each successive time scale, T_n, as being 'slower' than the scale before it. For example, T_2 describes a slower time scale than T_1. The time derivatives in 3.184 can be written in terms of the new time scales, since we can think of $\tilde{\tau}$ as being a function of the four independent T_n's, $\tilde{\tau} = \tilde{\tau}(T_0, T_1, T_2, T_3)$. Letting $\partial_n \equiv \partial/\partial T_n$, the first and second time derivatives can be written as [TSS05]

$$\frac{d}{d\tilde{\tau}} = \partial_0 + \epsilon\partial_1 + \epsilon^2\partial_2 + \epsilon^3\partial_3 \tag{3.186}$$

$$\frac{d^2}{d\tilde{\tau}^2} = \partial_0^2 + 2\epsilon\partial_0\partial_1 + \epsilon^2\left(2\partial_0\partial_2 + \partial_1^2\right) + 2\epsilon^3\left(\partial_0\partial_3 + \partial_1\partial_2\right), \tag{3.187}$$

where in (3.187) we have dropped terms of order ϵ^4 and higher.

Next, we expand the phase differences in an ϵ expansion

$$\gamma_j = \sum_{n=0}^{\infty}\epsilon^n\gamma_{n,j}(T_0, T_1, T_2, T_3). \tag{3.188}$$

Substituting this expansion into (3.184) and collecting all terms of order ϵ^0 results in the expression

$$1 = i_{cj}\tilde{\beta}_c \partial_0^2 \gamma_{0,j} + i_{cj}\partial_0 \gamma_{0,j}, \tag{3.189}$$

for which we find the solution

$$\gamma_{0,j} = \frac{T_0}{i_{cj}} + \phi_j(T_1, T_2, T_3), \tag{3.190}$$

where we have ignored a transient term of the form $\mathrm{e}^{-T_0/\tilde{\beta}_c}$, and where $\phi_j(T_i)$, $(i = 1, 2, 3)$ is assumed constant over the fastest time scale T_0. Note that the expression for $\gamma_{0,j}$ consists of a rapid phase rotation described by T_0/i_{cj} and slower–scale temporal variations, described by ϕ_j, on top of that overturning. In essence, the goal of this technique is to solve for the dynamical behavior of the slow phase variable, ϕ_j. The resulting differential equation for the ϕ_j is [TSS05]:

$$\begin{aligned} \beta_c \phi_j'' + \phi_j' &= \Omega_j + K_j \sum_{\delta=\pm 1} \sin\left[\frac{\phi_{j+\delta} - \phi_j}{2}\right] + L_j \sum_{\delta=\pm 1} \sin\left[3\left(\frac{\phi_{j+\delta} - \phi_j}{2}\right)\right] \\ &+ M_j \sum_{\delta=\pm 1} \left\{\cos\left[\frac{\phi_{j+\delta} - \phi_j}{2}\right] - \cos\left[3\left(\frac{\phi_{j+\delta} - \phi_j}{2}\right)\right]\right\}, \end{aligned} \tag{3.191}$$

where Ω_j is given by the expression (letting $x_j \equiv i_{cj}/i_B$ for convenience)

$$\Omega_j = \frac{1}{x_j}\left[1 - \frac{x_j^4}{(2\beta_c^2 + x_j^2)}\right], \tag{3.192}$$

and the three coupling strengths are

$$K_j = \frac{\alpha}{i_{cj}}\left[1 + \frac{x_j^4\left(3x_j^2 + 23\beta_c^2\right)}{16\left(\beta_c^2 + x_j^2\right)^2}\right], \tag{3.193}$$

$$L_j = \frac{\alpha}{i_{cj}} \frac{x_j^4\left(3\beta_c^2 - x_j^2\right)}{16\left(\beta_c^2 + x_j^2\right)^2}, \tag{3.194}$$

$$M_j = -\frac{\alpha}{i_{cj}} \frac{x_j^5 \beta_c}{4\left(\beta_c^2 + x_j^2\right)^2}. \tag{3.195}$$

We emphasize that (3.191) is expressed in terms of the original, unscaled, time variable τ and McCumber parameter β_c.

We will generally consider bias current and junction capacitance values such that $x_j^2 \ll \beta_c^2$. In this limit, (3.193)–(3.195) can be approximated as follows [TSS05]:

$$K_j \to \frac{\alpha}{i_{cj}}\left[1 + \mathcal{O}\left(\frac{1}{i_B^4}\right)\right], \tag{3.196}$$

$$L_j \to \frac{\alpha}{i_{cj}} \left(\frac{3x_j^4}{16\beta_c^2} \right) \sim \mathcal{O}\left(\frac{1}{i_B^4} \right), \tag{3.197}$$

$$M_j \to -\frac{\alpha}{i_{cj}} \left(\frac{x_j^5}{4\beta_c^3} \right) \sim \mathcal{O}\left(\frac{1}{i_B^5} \right). \tag{3.198}$$

For large bias currents, it is reasonable to truncate (3.191) at $\mathcal{O}(1/i_B^3)$, which leaves

$$\beta_c \phi_j'' + \phi_j' = \Omega_j + \frac{\alpha}{i_{cj}} \sum_{\delta=\pm 1} \sin\left[\frac{\phi_{j+\delta} - \phi_j}{2} \right], \tag{3.199}$$

where all the cosine coupling terms and the third harmonic sine term have been dropped as a result of the truncation.

In the absence of any coupling between neighboring rung junctions ($\alpha = 0$) the solution to (3.199) is

$$\phi_j^{(\alpha=0)} = A + Be^{-\tau/\beta_c} + \Omega_j \tau,$$

where A and B are arbitrary constants. Ignoring the transient exponential term, we see that $d\phi_j^{(\alpha=0)}/d\tau = \Omega_j$, so we can think of Ω_j as the voltage across rung junction j in the un–coupled limit. Alternatively, Ω_j can be viewed as the angular velocity of the strongly–driven rotator in the un–coupled limit.

Equation (3.199) is our desired phase model for the rung junctions of the underdamped ladder [TSS05]. The result can be described as a locally–coupled Kuramoto model with a second-order time derivative (LKM2) and with junction coupling determined by α. In the context of systems of coupled rotators, the second derivative term is due to the non–negligible rotator inertia, whereas in the case of Josephson junctions the second derivative arises because of the junction capacitance. The *globally–coupled* version of the second–order Kuramoto model (GKM2) has been well studied; in this case the oscillator inertia leads to a first–order synchronization phase transition as well as to hysteresis between a weakly and a strongly coherent synchronized state [TLO97, ABS00].

Comparison of LKM2 and RCSJ Models

We now compare the synchronization behavior of the RCSJ ladder array with the LKM2. We consider frequency and phase synchronization separately. For the rung junctions of the ladder, frequency synchronization occurs when the time average voltages, $\langle v_j \rangle_\tau = \langle \phi_j' \rangle_\tau$ are equal for all N junctions, within some specified precision. In the language of coupled rotators, this corresponds to phase points moving around the unit circle with the same average angular velocity. We quantify the degree of frequency synchronization via an 'order parameter' [TSS05]

$$f = 1 - \frac{s_v(\alpha)}{s_v(0)}, \tag{3.200}$$

where $s_v(\alpha)$ is the standard deviation of the N time–average voltages, $\langle v_j \rangle_\tau$:

$$s_v(\alpha) = \sqrt{\frac{\sum_{j=1}^N \left(\langle v_j \rangle_\tau - \frac{1}{N}\sum_{k=1}^N \langle v_k \rangle_\tau\right)^2}{N-1}} \tag{3.201}$$

In general, this standard deviation will be a function of the coupling strength α, so $s_v(0)$ is a measure of the spread of the $\langle v_j \rangle_\tau$ values for N independent junctions. Frequency synchronization of all N junctions is signaled by $f = 1$, while $f = 0$ means all N average voltages have their un–coupled values.

Phase synchronization of the rung junctions is measured by the usual *Kuramoto order parameter*

$$r \equiv \frac{1}{N} \sum_{j=1}^{N} \mathrm{e}^{i\phi_j}. \tag{3.202}$$

Lastly in this subsection, we address the issue of the linear stability of the frequency synchronized states ($\alpha > \alpha_c$) by calculating their *Floquet exponents* numerically for the RCSJ model as well as analytically based on the LKM2, (3.199). The analytic technique used has been described in [TM01], giving as a result for the real part of the Floquet exponents:

$$\mathrm{Re}(\lambda_m t_c) = -\frac{1}{2\beta_c}\left[1 \pm \mathrm{Re}\sqrt{1 - 4\beta_c\left(\bar{K} + 3\bar{L}\right)\omega_m^2}\right], \tag{3.203}$$

where stable solutions correspond to exponents, λ_m, with a negative real part. One can think of the ω_m as the normal mode frequencies of the ladder. We find that for a ladder with periodic boundary conditions and N plaquettes

$$\omega_m^2 = \frac{4\sin^2\left(\frac{m\pi}{N}\right)}{1 + 2\sin^2\left(\frac{m\pi}{N}\right)}, \qquad 0 \le m \le N-1. \tag{3.204}$$

To arrive at (3.203) we have ignored the effects of disorder so that $\bar{K}$ and $\bar{L}$ are obtained from (3.193) and (3.194) with the substitution $i_{cj} \to 1$ throughout. This should be reasonable for the levels of disorder we have considered (5%). Substituting the expressions for $\bar{K}$ and $\bar{L}$ into 3.203 results in [TSS05]

$$\mathrm{Re}(\lambda_m t_c) = -\frac{1}{2\beta_c}\left[1 \pm \mathrm{Re}\sqrt{1 - 2\beta_c\alpha\left\{1 + \frac{2\beta_c^2}{\left(i_B^2\beta_c^2 + 1\right)^2}\right\}\omega_m^2}\right]. \tag{3.205}$$

We are most interested in the Floquet exponent of minimum magnitude, $\mathrm{Re}(\lambda_{\min} t_c)$, which essentially gives the lifetime of the longest–lived perturbations to the synchronized state.

'Small–World' Connections in JJL Arrays

Many properties of small world networks have been studied in the last several years, including not only the effects of network topology but also the dynamics of the node elements comprising the network [New00, Str]. Of particular interest has been the ability of oscillators to synchronize when configured in a small–world manner. Such synchronization studies can be broadly sorted into several categories [TSS05]:

(1) Work on coupled lattice maps has demonstrated that synchronization is made easier by the presence of random, *long–range connections* [GH00, BPV03].

(2) Much attention has been given to the synchronization of continuous time dynamical systems, including the first order *locally–coupled Kuramoto model* (LKM), in the presence of small–world connections [HCK02a, HCK02b, Wat99]. For example, Hong and coworkers [HCK02a, HCK02b] have shown that the LKM, which does not exhibit a true dynamical phase transition in the thermodynamic limit ($N \to \infty$) in the *pristine* case, does exhibit such a phase synchronization transition for even a small number of shortcuts. But the assertion [WC02] that any small world network can synchronize for a given coupling strength and large enough number of nodes, even when the pristine network would not synchronize under the same conditions, is not fully accepted [BP02].

(3) More general studies of synchronization in small world and scale–free networks [BP02, NML03] have shown that the small world topology does not guarantee that a network can synchronize. In [BP02] it was shown that one could calculate the average number of shortcuts per node, s_{sync}, required for a given dynamical system to synchronize. This study found no clear relation between this synchronization threshold and the onset of the small world region, i.e., the value of s such that the average path length between all pairs of nodes in the array is less than some threshold value. [NML03] studied arrays with a power–law distribution of node connectivities (scale–free networks) and found that a broader distribution of connectivities makes a network *less* synchronizable even though the average path length is smaller. It was argued that this behavior was caused by an increased number of connections on the hubs of the scale–free network. Clearly it is dangerous to assume that merely reducing the average path length between nodes of an array will make such an array easier to synchronize.

Now, regarding Josephson–junction arrays, if we have a disordered array biased such that some subset of the junctions are in the voltage state, i.e., undergoing limit cycle oscillations, the question is will the addition of random, long–range connections between junctions aid the array in attaining frequency and/or phase synchronization? Can we address this question by using the mapping discussed above between the RCSJ model for the *underdamped ladder array* and the second–order, locally–coupled Kuramoto model (LKM2). Based on the results of [DDT03], we also know that the RSJ model for an *overdamped ladder* can be mapped onto a first–order, locally–coupled Kuramoto model (LKM). Because of this mapping, the ladder array falls into category (2) of the previous paragraph. In other words, we should expect the existence of shortcuts to drastically improve the ability of ladder arrays to synchronize [TSS05].

We add connections between pairs of rung junctions that will result in interactions that are longer than nearest neighbor in range. We do so by adding two, nondisordered, off–rung junctions for each such connection. We argue that the RCSJ equations for the underdamped junctions in the ladder array can be mapped onto a straightforward variation of (3.199), in which the sinusoidal coupling term for rung junction j also includes the longer–range couplings due to the added shortcuts. Imagine a ladder with a shortcut between junctions j and l, where $l \neq j, j \pm 1$. Conservation of charge applied to the two superconducting islands that comprise rung junction j will lead to equations very similar to (3.175). For example, the analog to (3.174) will be

$$\begin{aligned} i_B - i_{cj}\sin\gamma_j - i_{cj}\gamma_j' - \beta_c i_{cj}\gamma_j'' - \alpha\sin\psi_{1,j} - \alpha\psi_{1,j}' - \beta_c\alpha\psi_{1,j}'' + \alpha\sin\psi_{1,j-1} \\ + \alpha\psi_{1,j-1}' + \beta_c\alpha\psi_{1,j-1}'' + \sum_l \left[\alpha\sin\psi_{1;jl} + \alpha\psi_{1;jl}' + \beta_c\alpha\psi_{1;jl}''\right] = 0, \end{aligned}$$

with an analogous equation corresponding to the inner superconducting island that can be generalized from (3.175). The sum over the index l accounts for all junctions connected to junction j via an added shortcut. Fluxoid quantization still holds, which means that we can augment 3.176 with

$$\gamma_j + \psi_{2;jl} - \gamma_l - \psi_{1;jl} = 0. \tag{3.206}$$

We also assume the analog of (3.179) holds:

$$\psi_{2;jl} = -\psi_{1;jl}. \tag{3.207}$$

Equations (3.206) and (3.207) allow us to write the analog to (3.180) for the case of shortcut junctions:

$$\psi_{1;jl} = \frac{\gamma_j - \gamma_l}{2} \tag{3.208}$$

Equation (3.206), in light of (3.208), can be written as

$$\begin{aligned} i_B - i_{cj}\sin\gamma_j - i_{cj}\gamma_j' - \beta_c i_{cj}\gamma_j'' + \alpha\sum_{\delta=\pm1}\sin\Big(\frac{\gamma_{j+\delta} - \gamma_j}{2}\Big) + \alpha\sum_l\sin\Big(\frac{\gamma_j - \gamma_l}{2}\Big) \\ + \frac{\alpha}{2}\nabla^2\gamma_j' + \frac{\alpha}{2}\nabla^2\gamma_j'' + \frac{\alpha}{2}\sum_l(\gamma_j' - \gamma_l') + \frac{\alpha}{2}\sum_l(\gamma_j'' - \gamma_l'') = 0, \end{aligned}$$

where the sums Σ_l are over all rung junctions connected to j via an added shortcut. As we did with the pristine ladder, we will drop the two discrete Laplacians, since they have a very small time average compared to the terms $i_{cj}\gamma_j' + i_{cj}\beta_c\gamma_j''$. The same is also true, however, of the terms $\alpha/2\sum_l(\gamma_j' - \gamma_l')$

and $\alpha/2\sum_l(\gamma_j'' - \gamma_l'')$, as direct numerical solution of the full RCSJ equations in the presence of shortcuts demonstrates. So we shall drop these terms as well. Then we have

$$i_B - i_{cj}\sin\gamma_j - i_{cj}\gamma_j' - \beta_c i_{cj}\gamma_j'' + \frac{\alpha}{2}\sum_{k\in\Lambda_j}\sin\left(\frac{\gamma_k - \gamma_j}{2}\right), \tag{3.209}$$

where the sum is over all junctions in Λ_j, which is the set of all junctions connected to junction j. From above results we can predict that a multiple time scale analysis of (3.209) results in a phase model of the form

$$\beta_c\frac{d^2\phi_j}{d\tau^2} + \frac{d\phi_j}{d\tau} = \Omega_j + \frac{\alpha}{2}\sum_{k\in\Lambda_j}\sin\left(\frac{\phi_k - \phi_j}{2}\right), \tag{3.210}$$

where Ω_j is give by (3.192). A similar analysis for the *overdamped ladder* leads to the result

$$\phi_j' = \Omega_j^{(1)} + \frac{\alpha}{2}\sum_{k\in\Lambda_j}\sin\left(\frac{\phi_k - \phi_j}{2}\right), \tag{3.211}$$

where the time-averaged voltage across each overdamped rung junction in the un–coupled limit is

$$\Omega_j^{(1)} = \sqrt{\left(\frac{i_B}{i_{cj}}\right)^2 - 1}. \tag{3.212}$$

Although the addition of shortcuts makes it easier for the array to synchronize, we should also consider the effects of such random connections on the stability of the synchronized state. The Floquet exponents for the synchronized state allow us to quantify this stability. Using a general technique discussed in [PC98], we can calculate the Floquet exponents λ_m for the LKM based on the expression

$$\lambda_m t_c = \alpha E_m^G, \tag{3.213}$$

where E_m^G are the eigenvalues of $\mathbf{G}$, the matrix of coupling coefficients for the array. A specific element, G_{ij}, of this matrix is unity if there is a connection between rung junctions i and j. The diagonal terms, G_{ii}, is merely the negative of the number of junctions connected to junction i. This gives the matrix the property $\sum_j G_{ij} = 0$. In the case of the pristine ladder, the eigenvalues of $\mathbf{G}$ can be calculated analytically, which yields Floquet exponents of the form

$$\lambda_m{}^{(p=0)}t_c = -4\alpha\sin^2\left(\frac{m\pi}{N}\right). \tag{3.214}$$

See [TSS05] for more details.

References

AA68. Arnold, V.I., Avez, A.: Ergodic Problems of Classical Mechanics. Benjamin, New York, (1968)

AA80. Aubry, S., ré, G.: Colloquium on Computational Methods in Theoretical Physics. In Group Theoretical Methods in Physics, Horowitz, Ne'eman (ed.), Ann. Israel Phys. Soc. **3**, 133–164, (1980)

Aba96. Abarbanel, H.D.I.: Analysis of Observed Chaotic Data, Springer, New York, (1996)

ABB04. Anderson, J.R., Bothell, D., Byrne, M. D., Douglass, S., Lebiere, C., Qin, Y.: An integrated theory of the mind. Psychological Review, **111**(4), 1036–1060, (2004)

Abd06. Abdi, H.: Signal detection theory. In N.J. Salkind (ed.): Encyclopedia of Measurement, Statistics. Sage, Thousand Oaks, CA, (2006)

Abd88. Abdi, H.: A generalized approach for connectionist auto-associative memories: interpretation, implications, illustration for face processing, J. Demongeot (ed.) Artificial Intelligence, Cognitive Sciences. Manchester: Manchester Univ. Press, 149–165, (1988)

ABL00. Arhem, P., Blomberg, C., Liljenstrom, H. (ed.): Disorder versus Order in Brain Function. Progress in Neural Processing 12, World Sci. Singapore, (2000)

ABR85. Anderson, J.R., Boyle, C.B., Reiser, B.J.: Intelligent tutoring systems. Science, **228**, 456–462, (1985)

ABS00. Acebrón, J.A., Bonilla, L.L., Spigler, R.: Synchronization in populations of globally coupled oscillators with inertial effects. Phys. Rev. E **62**, 3437–3454, (2000)

Ach97. Acheson, D.: From Calculus to Chaos. Oxford Univ. Press, Oxford, (1997)

AGR82. Aspect, A., Grangier, P., Roger, G.: Experimental realization of Einstein-Podolsky-Rosen-Bohm Gedankenexperiment: a new violation of Bell's inequalities. Phys. Rev. Lett., **48**, 91–94, (1982)

AHP00. Atkeson, C.G., Hale J., Pollick F., Riley M., Kotosaka S., Schaal S., Shibata T., Tevatia G., Vijayakumar S., Ude A., Kawato M.: Using humanoid robots to study human behavior. IEEE Intelligent Systems: Special Issue on Humanoid Robotics, **15**, 46–56, (2000)

AL91. Aidman, E.V., Leontiev, D.A.: From being motivated to motivating oneself: a Vygotskian perspective. Stud. Sov. Thought, **42**, 137–151, (1991)

ALL91. Amsel, E., Langer, R., Loutzenhiser, L.: Do lawyers reason differently from psychologists? A comparative design for studying expertise. In R.J. Sternberg, P.A. Frensch (eds.), Complex problem solving: Principles, mechanisms (223–250) Hillsdale, NJ: Lawr. Erl. Assoc., (1991)

Alt95. Altrock, C.V.: Fuzzy logic, NeuroFuzzy applications explained. Prentice Hall, Englewood Cliffs, N.J. (1995)

AM78. Abraham, R., Marsden, J.E.: Foundations of Mechanics (2nd ed.) Addison-Wesley, Reading, MA, (1978)

Ama77. Amari, S.: Dynamics of pattern formation in lateral-inhibition type neural fields. Biological Cybernetics, **27**, 77–87, (1977)

AMV00. Alfinito, E., Manka, R., Vitiello, G.: Vacuum structure for expanding geometry. Class. Quant. Grav. **17**, 93, (2000)

AN99. Aoyagi, T., Nomura, M.: Oscillator Neural Network Retrieving Sparsely Coded Phase Patterns. Phys. Rev. Lett. **83**, 1062–1065, (1999)

And01. Andrecut, M.: Biomorphs, program for $Mathcad^{TM}$, Mathcad Application Files, Mathsoft, (2001)

And64. Anderson, P.W.: Lectures on the Many Body Problem. E.R. Caianiello (ed.), Academic Press, New York, (1964)

And80. Anderson, J.R.: Cognitive psychology and its implications. San Francisco, Freeman, (1980)

And83. Anderson, J.R.: The Architecture of Cognition. Cambridge, Harvard Univ. Press, MA, (1983)

And84. Anderson, T.W.: An Introduction to Multivariate Statistical Analysis (2nd ed.). Wiley, New York, (1984)

And90. Anderson, J.R.: The Adaptive Character of Thought. Hillsdale and Erlbaum, NJ, (1990)

Ang92. Angluin, D.: Computational learning theory: Survey and selected bibliography. In Proc. 24 Annual ACM Symposium on Theory of Computing, 351–369, (1992)

APS98. American Psychological Association: Task force report. Gottfredson, (1998)

Arb98. Arbib, M. (ed.): Handbook of Brain Theory and Neural Networks (2nd ed.) MIT Press, Cambridge, (1998)

Arn78. Arnold, V.I.: Ordinary Differential Equations. MIT Press, Cambridge, MA, (1978)

Arn88. Arnold, V.I.: Geometrical Methods in the Theory of Ordinary differential equations. Springer, New York, (1988)

Arn89. Arnold, V.I.: Mathematical Methods of Classical Mechanics (2nd ed.). Springer, New York, (1989)

Arn92. Arnold, V.I.: Catastrophe Theory. Springer, Berlin, (1992)

Arn93. Arnold, V.I.: Dynamical systems. Encyclopaedia of Mathematical Sciences, Springer, Berlin, (1993)

Arr63. Arrow, K.J.: Social choice and individual values. Wiley, New York, (1963)

ARV02. Alfinito, E., Romei, O., Vitiello, G.: On topological defect formation in the process of symmetry breaking phase transitions. Mod. Phys. Lett. B **16**, 93, (2002)

AS02. Aguirre, J., Sanjuán, M.A.F.: Unpredictable behavior in the Duffing oscillator: Wada basins. Physica D **171**, 41, (2002)

AS72. Abramowitz, M., Stegun, I.A.: Handbook of Mathematical Functions. Dover, New York, (1972)

AS79. Anzai, K., Simon, H.A.: The theory of learning by doing. Psych. Rev., **86**, 124–140, (1979)

Ash56. Ashby, W.R.: Introduction to Cybernetics. Methuen, London, UK, (1956)

Ash94. Ashcraft, M.H.: Human Memory and Cognition (2nd ed.) HarperCollins, New York, (1994)

AV00. Alfinito, E., Vitiello, G.: Formation and life–time of memory domains in the dissipative quantum model of brain. Int. J. Mod. Phys. B, **14**, 853–868, (2000)

Ave99. Averin, D.V.: Solid-state qubits under control. Nature **398**, 748–749, (1999)

Bac96. Back, T.: Evolutionary Algorithms in Theory and Practice: Evolution Strategies, Evolutionary Programming, Genetic Algorithms, Oxford Univ. Press, (1996)

Bae91. Baesens, C.: Slow sweep through a period-doubling cascade: delayed bifurcations and renormalisation. Physica D **53**, 319, (1991)

Bai89. Bai-Iin, H.: Elentary Symbolic Dynamics and Chaos in Dissipative Systems. World Scientific, Singapore, (1989)

Bal06. Balakrishnan, J.: A geometric framework for phase synchronization in coupled noisy nonlinear systems. Phys. Rev. E **73**, 036206–036217, (2006)

Bal06. Balzac, F.: Exploring the Brain's Role in Creativity. NeuroPsychiatry Reviews **7**(1), 19–20, (2006)

BB95. Berry, D.C., Broadbent, D.E.: Implicit learning in the control of complex systems: A reconsideration of some of the earlier claims. In P.A. Frensch, J. Funke (eds.), Complex problem solving: The European Perspective (131–150) Hillsdale, NJ: Lawr. Erl. Assoc., (1995)

BBS91. Bryson, M., Bereiter, C., Scardamalia, M., Joram, E.: Going beyond the problem as given: Problem solving in expert and novice writers. In R.J. Sternberg, P.A. Frensch (eds.), Complex problem solving: Principles, mechanisms (61–84) Hillsdale, NJ: Lawr. Erl. Assoc., (1991)

BCB92. Borelli, R.L., Coleman, C., Boyce, W.E.: Differential Equations - Laboratory Workbook. Wiley, New York, (1992)

BCF02. Boffetta, G., Cencini, M., Falcioni, M., Vulpiani, A.: Predictability: a Way to Characterize Complexity. Phys. Rep., **356**, 367–474, (2002)

BD02. Busemeyer, J.R., Diederich, A.: Survey of decision field theory. Math. Soc. Sci., **43**, 345–370, (2002)

BDG93. Bielawski, S., Derozier, D., Glorieux, P.: Experimental characterization of unstable periodic orbits by controlling chaos. Phys. Rev. A, **47**, 2492, (1993)

Bea95. Bear, M.F. et al. eds.: Neuroscience: Exploring The Brain. Williams and Wilkins, Baltimore, (1995)

Bel64. Bell, J.S: On the Einstein Podolsky Rosen Paradox. Physics **1**, 195, (1964)

Bel66. Bell, J.S: On the problem of hidden variables in quantum mechanics. Rev. Mod. Phys. **38**, 447, (1966)

Bel87. Bell, J.S: Speakable and Unspeakable in Quantum Mechanics. Cambridge Univ. Press, Cambridge, (1987)

Ben84. Benettin, G.: Power law behaviour of Lyapunov exponents in some conservative dynamical systems. Physica D **13**, 211–213, (1984)

Ben95. Benz, S.P.: Superconductor-normal-superconductor junctions for programmable voltage standards. Appl. Phys. Lett. **67**, 2714–2716, (1995)

Ber35. Bernstein, N.A.: Investigations in Biodynamics of Locomotion (in Russian). WIEM, Moscow, (1935)

Ber47. Bernstein, N.A.: On the structure of motion (in Russian). Medgiz, Moscow, (1947)

Bey01. Beyer, H-G.: The Theory of Evolution Strategies. Springer, Berlin, (2001)

BF71. Bransford, J.D., Franks, J.J.: The Abstraction of Linguistic Ideas. Cogn. Psych., **2**, 331–350, (1971)

BFM97. Back, T., Fogel, D., Michalewicz, Z.: Handbook of Evolutionary Computation. Oxford Univ. Press, (1997)

BG02. Berglund, N., Gentz, B.: On the noise-induced passage through an unstable periodic orbit with additive noise. Probab. Theory Relat. Fields **122**, 341, (2002)

BG79. Barrow-Green, J.: Poincaré and the Three Body Problem. American Mathematical Society, Providence, RI, (1997)

BG96. Baker, G.L., Gollub, J.P.: Chaotic Dynamics: An Introduction (2nd ed.) Cambridge Univ. Press, Cambridge, (1996)

BGG80. Benettin, G., Giorgilli, A., Galgani, L., Strelcyn, J.M.: Lyapunov exponents for smooth dynamical systems and for Hamiltonian systems; a method for computing all of them. Part 1: theory, Part 2: numerical applications. Meccanica, **15**, 9–20, 21–30, (1980)

BGL00. Boccaletti, S., Grebogi, C., Lai, Y.-C., Mancini, H., Maza, D.: The Control of Chaos: Theory and Applications. Physics Reports **329**, 103–197, (2000)

BGS76. Benettin, G., Galgani, L., Strelcyn, J.M.: Kolmogorov Entropy and Numerical Experiments. Phys. Rev. A **14**, 2338, (1976)

BH93. Bohm, D., Hiley, B.: The Undivided Universe. Routledge, London, (1993)

BHB95. Bell, I.R., Hardin, E.E., Baldwin, C.M., Schwartz, G.E.: Increased limbic system symptomatology, sensitizability of young adults with chemical, noise sensitivities. Environmental Research. **70**, 84–97, (1995)

Bil65. Billingsley, P.: Ergodic theory and information, Wiley, New York, (1965)

Blo80. Bloom, B.S.: All Our Children Learning. McGraw-Hill, New York, (1980)

BLV01. Boffetta, G., Lacorata, G., Vulpiani, A.: Introduction to chaos and diffusion. Chaos in geophysical flows, ISSAOS, (2001)

BN01. Bastian J., Nguyenkim J.: Dendritic Modulation of Burst-Like Firing in Sensory Neurons. J. Neurophysiol. **85**, 10–22, (2001)

BNO03. Breban, R., Nusse, H.E., Ott, E.: Scaling properties of saddle–node bifurcations on fractal basin boundaries. Phys. Rev. E **68**(6), 066213.1–066213.16, (2003)

Bob92. Bobbert, P.A.: Simulation of vortex motion in underdamped two-dimensional arrays of Josephson junctions. Phys. Rev. B **45**, 7540–7543, (1992)

Bod04. Boden, M.A.: The Creative Mind: Myths and Mechanisms. Routledge, London, (2004)

BOF01. Burdet, E., Osu, R., Franklin, D., Milner, T., Kawato, M.: The central nervous system stabilizes unstable dynamics by learning optimal impedance. Nature, **414**, 446–449, (2001)

Boh92. Bohm, D.: Thought as a System. Routledge, London, (1992)

Bon73. De Bono, E.: Lateral Thinking: Creativity Step by Step. Harper, Row, (1973)

BP02. Barahona, M., Pecora, L.M.: Synchronization in Small-World Systems. Phys. Rev. Lett. **89**, 054101–054105, (2002)

BP82. Barone, A., Paterno, G.: Physics and Applications of the Josephson Effect. Wiley, New York, (1982)

BP92. Benvenuto, N., Piazza, F.: On the complex backpropagation algorithm. IEEE Trans. Sig. Proc., **40**(4), 967–969, (1992)

BP97. Badii, R. Politi, A: Complexity: Hierarchical Structures and Scaling in Physics, Cambridge Univ. Press, Cambridge, (1997)

BPS70. Baum, L.E., Petrie, T., Soules, G., Weiss, N.: A maximization technique occurring in the statistical analysis of probabilistic functions of Markov chains. Ann. Math. Statist., **41**(1), 164–171, (1970)

BPS75. Belavin, A.A., Polyakov, A.M., Swartz, A.S., Tyupkin, Yu.S.: SU(2) instantpons discovered. Phys. Lett. B **59**, 85, (1975)

BPV03. Batista, A.M., de Pinto, S.E., Viana, R.L., Lopes, S.R.: Mode locking in small-world networks of coupled circle maps. Physica A **322**, 118, (2003)

Bra01. Branke, J.: Evolutionary Optimization in Dynamic Environments. Kluwer, Dordrecht, (2001)

Bro77. Broadbent, D.E.: Levels, hierarchies and the locus of control. Quarterly J. Exper. Psychology, **29**, 181–201, (1977)

Bro86. Brooks, R.A.: A robust layered control system for a mobile robot. IEEE Trans. Rob. Aut. **2**(1), 14–23, (1986)

BS01. Baader, F., Snyder, W.: Unification Theory. In J.A. Robinson and A. Voronkov (ed.) Handbook of Automated Reasoning, Vol. 1, 447–533. Elsevier, (2001)

BS02. Beyer, H-G., Schwefel, H-P.: Evolution Strategies: A Comprehensive Introduction. J. Nat. Comp. **1**(1), 3–52, (2002)

BS77. Bhaskar, R., Simon, H.A.: Problem solving in semantically rich domains: An example from engineering thermodynamics. Cog. Sci. **1**, 193–215, (1977)

BS93. Beck, C., Schlogl, F.: Thermodynamics of chaotic systems. Cambridge Univ. Press, Cambridge, (1993)

BSM97. Breitenbach, G., Schiller, S., Mlynek, J.: Measurement of the quantum states of squeezed light. Nature, **387**, 471–475 (1997)

BT93. Busemeyer, J.R., Townsend, J.T.: Decision field theory: A dynamic–cognitive approach to decision making in an uncertain environment. Psych. Rev., **100**, 432–459, (1993)

Buc95. Buchner, A.: Theories of complex problem solving. In P. A. Frensch, J. Funke (eds.) Complex problem solving: The European Perspective (27–63) Hillsdale, NJ: Lawr. Erl. Assoc., (1995)

BY00. Bar-Yam, Y. (ed.): Unifying Themes in Complex Systems: Proc. Int. Conf. Complex Sys. Perseus Press, (2000)

BY04. Bar-Yam, Y.: Unifying Principles in Complex Systems. In Converging Technology (NBIC) for Improving Human Performance, M.C. Roco and W.S. Bainbridge (eds.), (2004)

BY97. Bar-Yam, Y.: Dynamics of Complex Systems. Perseus Books, Reading, Mas, (1997)

CA97. Chen, L., Aihara, K.: Chaos, asymptotical stability in discrete-time neural networks, Physica D **104**, 286–325, (1997)

CAM05. Cvitanovic, P., Artuso, R., Mainieri, R., Tanner, G., Vattay, G.: Chaos: Classical and Quantum. ChaosBook.org, Niels Bohr Institute, Copenhagen, (2005)

Cas92. Cassidy, D.: Uncertainty: The Life and Science of Werner Heisenberg. Freeman, New York, (1992)

CC95. Christini, D.J., Collins, J.J.: Controlling Nonchaotic Neuronal Noise Using Chaos Control Techniques. Phys. Rev. Lett. **75**, 2782–2785, (1995)

CCW03. Cohen, J., Cohen P., West, S.G., Aiken, L.S.: Applied multiple regression/correlation analysis for the behavioral sciences. (2nd ed.) Lawr. Erl. Assoc. Hillsdale, NJ, (2003)

CD98. Chen, G., Dong, X.: From Chaos to Order. Methodologies, Perspectives, Application. World Scientific, Singapore, (1998)

CE91. Cvitanovic, P., Eckhardt, B.: Periodic orbit expansions for classical smooth flows. J. Phys. A **24**, L237, (1991)

CFG81. Chi, M.T.H., Feltovich, P.J., Glaser, R.: Categorization and representation of physics problems by experts and novices. Cog. Sci. **5**, 121–152, (1981)

CFP91. Crisanti, A., Falcioni, M., Paladin, G., Vulpiani, A.: Lagrangian Chaos: Transport, Mixing, Diffusion in Fluids. Riv. Nuovo Cim. **14**, 1, (1991)

CFP94. Crisanti, A., Falcioni, M., Paladin, G., Vulpiani, A.: Stochastic Resonance in Deterministic Chaotic Systems. J. Phys. A **27**, L597, (1994)

CFR79. Clark, R.A., Ferziger, J.H., Reynolds, W.C.: Evaluation of Subgrid-Scale Turbulence Models Using an Accurately Simulated Turbulent Flow. J. Fluid. Mech. **91**, 1–16, (1979)

CG03. Carpenter, G.A., Grossberg, S.: Adaptive Resonance Theory. In M.A. Arbib (ed.) The Handbook of Brain Theory, Neural Networks, Second Edition, MIT Press, Cambridge, MA, 87–90, (2003)

CG83. Cohen, M.A., Grossberg, S.: Absolute stability of global pattern formation, parallel memory storage by competitive neural networks. IEEE Trans. Syst., Man, Cybern., **13**(5), 815–826, (1983)

CG90. Connors, B.W., Gutnick, M.J.: Intrinsic firing patterns of diverse neocortical neurons. Trends in Neuroscience, **13**, 99–104, (1990)

CGP88. Cvitanovic, P., Gunaratne, G., Procaccia, I.: Topological, metric properties of Hénon-type strange attractors. Phys. Rev. A **38**, 1503–1520, (1988)

CH94. The CHAOS Report, The Standish Group, (1994)

Cha48. Chandra, H.: Relativistic Equations for Elementary Particles. Proc. Roy. Soc. London A, **192**(1029), 195–218, (1948)

Cha97. Chalmers, D.: The Conscious Mind. Oxford Univ. Press, Oxford, (1997)

Chi79. Chirikov, B.V.: A universal instability of many-dimensional oscillator systems. Phys. Rep. **52**, 264–379, (1979)

CJP93. Crisanti, A, Jensen, M.H., Paladin, G., Vulpiani, A.: Intermittency, Predictability in Turbulence. Phys. Rev. Lett. **70**, 166, (1993)

CK97. Cutler, C.D., Kaplan, D.T.: (eds.): Nonlinear Dynamics and Time Series. Fields Inst. Comm. **11**, American Mathematical Society, (1997)

CKP02. Cho, J.-H., Ko, M.-S., Park, Y.-J., Kim, C.-M.: Experimental observation of the characteristic relations of type-I intermittency in the presence of noise. Phys. Rev. E **65**, 036222, (2002)

CL71. Cooley, W.W. Lohnes, P.R.: Multivariate Data Analysis. Wiley, New York, (1971)

CL72. Craik, F., Lockhart, R.: Levels of processing: A framework for memory research. J. Verb. Learn. Verb. Behav., **11**, 671–684, (1972)

CL81. Caldeira, A.O., Leggett, A.J.: Influence of Dissipation on Quantum Tunneling in Macroscopic Systems. Phys. Rev. Lett. **46**, 211, (1981)

CL84. Cheng, T.-P., Li, L.-F.: Gauge Theory of Elementary Particle Physics. Clarendon Press, Oxford, (1984)

Cla02a. Claussen, J.C.: Generalized Winner Relaxing Kohonen Feature Maps. arXiv cond–mat/0208414, (2002)

Cla02b. Claussen, J.C.: Floquet Stability Analysis of Ott-Grebogi-Yorke, Difference Control. arXiv:nlin.CD/0204060, (2002)

CMP98a. Claussen, J.C., Mausbach, T., Piel, A. Schuster, H.G.: Improved difference control of unknown unstable fixed–points: Drifting parameter conditions, delayed measurement. Phys. Rev. E, **58**(6), 7256–7260, (1998)

CMP98b. Claussen, J.C., Mausbach, T., Piel, A. Schuster, H.G.: Memory difference control of unknown unstable fixed–points: Drifting parameter conditions, delayed measurement. Phys. Rev. E **58**(6), 7260–7273, (1998)

Cob83. Cobb, W.A.: Recommendations for the practice of clinical neurophysiology. Elsevier, Amsterdam, (1983)

Cox92. Cox, E.: Fuzzy Fundamentals. IEEE Spectrum, 58–61, (1992)

Cox94. Cox, E.: The Fuzzy Systems Handbook. AP Professional, (1994)

CQ69. Collins, A.M., Quillian, M.R.: Retrieval Time From Semantic Memory, J. Verb. Learn., Verb. Behav., **8**, 240–248, (1969)

CRV92. Celeghini, E., Rasetti, M., Vitiello, G.: Quantum Dissipation. Annals Phys., **215**, 156, (1992)

CS00. Cristianini, N., Shawe-Taylor, J.: Support Vector Machines. Cambridge Univ. Press, Cambridge, (2000)

CS73. Chase, W.G., Simon, H.A.: Perception in chess. Cog. Psych. **4**, 55–81, (1973)

CS83. Coppersmith, S.N., Fisher, D.S.: Pinning transition of the discrete sine-Gordon equation. Phys. Rev. B **28**, 2566–2581, (1983)

CS94. Coolen, A.C.C., Sherrington, D.: Order-parameter flow in the fully connected Hopfield model near saturation. Phys. Rev. E **49** 1921–934; Erratum: Order-parameter flow in the fully connected Hopfield model near saturation, 5906, (1994)

CS95. Chernikov, A.A., Schmidt, G.: Conditions for synchronization in Josephson-junction arrays. Phys. Rev. E **52**, 3415–3419, (1995)

CS98. Claussen, J.C., Schuster, H.G.: Stability borders of delayed measurement from time–discrete systems. arXiv nlin. CD/0204031, (1998)

Cvi00. Cvitanovic, P.: Chaotic field theory: a sketch. Physica A **288**, 61–80, (2000)

Cvi91. Cvitanovic, P.: Periodic orbits as the skeleton of classical and quantum chaos. Physica, D **51**, 138, (1991)

Dor75. Dorner, D.: Wie Menschen eine Welt verbessern wollten [How people wanted to improve the world]. Bild der Wissenschaft, **12**, 48–53, (1975)

Das99. Dasgupta, D. (ed.): Artificial Immune Systems and Their Applications. Springer-Verlag, Berlin, (1999)

DBL02. Dafilis, M.P., Bourke, P.D., Liley, D.T.J., Cadusch, P.J.: Visualising Chaos in a Model of Brain Electrical Activity. Comp. Graph. **26**(6), 971–976, (2002)

DBO01. DeShazer, D.J., Breban, R., Ott, E., Roy, R.: Detecting Phase Synchronization in a Chaotic Laser Array. Phys. Rev. Lett. **87**, 044101–044105, (2001)

DCM03. Doiron, B., Chacron, M.J., Maler, L., Longtin, A., Bastian, J.: Inhibitory feedback required for network oscillatory responses to communication but not prey stimuli. Nature, **421**, 539–543, (2003)

DDT03. Daniels, B.C., Dissanayake, S.T.M., Trees, B.R.: Synchronization of coupled rotators: Josephson junction ladders and the locally coupled Kuramoto model. Phys. Rev. E **67**, 026216–026230, (2003)

Des91. Descartes, R.: Discourse on Method and Meditations on First Philosophy (tr. by D.A. Cress) Cambridge, (1991)

DF85. Deutsch, D.: Quantum theory, the Church-Turing principle and the universal quantum computer. Proc. Roy. Soc. (London), A **400**, 97–117, (1985); also, Feynman R.P., Quantum mechanical Computers, Found. of Phys., **16**(6), 507–31, (1985)

Deu92. Deutsch, D., Jozsa, R.: Rapid solution of problems by quantum computation. Proc. Roy. Soc. (London), A **439**, 553–8, (1992)

DGY97. Ding, M., Grebogi, C., Yorke, J.A.: Chaotic dynamics. In The Impact of Chaos on Science and Society. C. Grebogi, J.A. Yorke (eds.), 1–17, United Nations Univ. Press, Tokyo, (1997)

DH85. Douady, A., Hubbard, J.H.: On the dynamics of polynomial-like mappings. Ann. Sci. Ec. Norm. Sup. (Paris) **18**, 287–343, (1985)

DHS91. Domany, E., van Hemmen, J.L., Schulten, K. (eds.): Models of Neural Networks. Springer, Berlin, (1991)

Dir25. Dirac, P.A.M.: The Fundamental Equations of Quantum Mechanics. Proc. Roy. Soc. London A, **109**(752), 642–653, (1925)

Dir26a. Dirac, P.A.M.: Quantum Mechanics, a Preliminary Investigation of the Hydrogen Atom. Proc. Roy. Soc. London A, **110**(755), 561–579, (1926)

Dir26b. Dirac, P.A.M.: The Elimination of the Nodes in Quantum Mechanics. Proc. Roy. Soc. London A, **111**(757), 281–305, (1926)

Dir26c. Dirac, P.A.M.: Relativity Quantum Mechanics with an Application to Compton Scattering. Proc. Roy. Soc. London A, **111**(758), 281–305, (1926)

Dir26d. Dirac, P.A.M.: On the Theory of Quantum Mechanics. Proc. Roy. Soc. London A, **112**(762), 661–677, (1926)

Dir26e. Dirac, P.A.M.: The Physical Interpretation of the Quantum Dynamics. Proc. Roy. Soc. London A, **113**(765), 1–40, (1927)

Dir28a. Dirac, P.A.M.: The Quantum Theory of the Electron. Proc. Roy. Soc. London A, **117**(778), 610–624, (1928)

Dir28b. Dirac, P.A.M.: The Quantum Theory of the Electron. Part II. Proc. Roy. Soc. London A, **118**(779), 351–361, (1928)

Dir29. Dirac, P.A.M.: Quantum Mechanics of Many-Electron Systems. Proc. Roy. Soc. London A, **123**(792), 714–733, (1929)

Dir32. Dirac, P.A.M.: Relativistic Quantum Mechanics. Proc. Roy. Soc. London A, **136**(829), 453–464, (1932)

Dir36. Dirac, P.A.M.: Relativistic Wave Equations. Proc. Roy. Soc. London A, **155**(886), 447–459, (1936)

Dir49. Dirac, P.A.M.: The Principles of Quantum Mechanics. Oxford Univ Press, Oxford, (1949)

Dir58. Dirac, P.A.M.: Generalized Hamiltonian Dynamics. Proc. Roy. Soc. London A, **246**(1246), 326–332, (1958)

Dir82. Dirac, P.A.M.: Principles of Quantum Mechanics. (4th ed.), Oxford Univ. Press, (1982)

DLC01. Dafilis, M.P., Liley, D.T.J., Cadusch, P.J.: Robust chaos in a model of the electroencephalogram: implications for brain dynamics. Chaos **11**(3), 474–4748, (2001)

DLL02. Doiron, B., Laing, C., Longtin, A., Maler, L.: Ghostbursting: a novel neuronal burst mechanism. J. Comput. Neurosci. **12**, 5–25, (2002)

Do95. Ding, M., Ott, E.: Chaotic Scattering in Systems with More than Two Degrees of Freedom. Ann. N.Y. Acad. Sci. **751**, 182, (1995)

DSS96. Dote, Y., Strefezza, M., Suitno, A.: Neuro fuzzy robust controllers for drive systems. In Neural Networks Applications, P.K. Simpson (ed.) IEEE Tec. Upd. Ser., New York, (1996)

DT95. Denniston, C., Tang, C.: Phases of Josephson Junction Ladders. Phys. Rev. Lett. **75**, 3930, (1995)

Duf18. Duffing, G.: Erzwungene Schwingungen bei vernderlicher Eigenfrequenz. Vieweg Braunschweig, (1918)

Dum01. Dummett, M.: Origini della Filosofia Analitica. Einaudi. ISBN 88-06-15286-6, (2001)

Dun35. Duncker, K.: Zur Psychologie des produktiven Denkens [The psychology of productive thinking]. Springer, Berlin, (1935)

Dus84. Dustin, P.: Microtubules. Springer, Berlin, (1984)

DV85. Dorner, D.: Verhalten, Denken und Emotionen [Behavior, thinking, emotions]. In L.H. Eckensberger, E.D. Lantermann (eds.), Emotion und Reflexivity (157–181) Munchen, Germany: Urban, Schwarzenberg, (1985)

DW95. Dorner, D., Wearing, A.: Complex problem solving: Toward a (computer-simulated) theory. In P.A. Frensch, J. Funke (eds.) Complex problem solving: The European Perspective (65–99) Hillsdale, NJ: Lawr. Erl. Assoc., (1995)

D'Zur86. D'Zurilla, T.J.: Problem-solving therapy: a social competence approach to clinical intervention. Springer, New York, (1986)

Ebe02. Ebersole, J.S.: Current Practice of Clinical Electroencephalography. Williams, Wilkins, Lippincott, (2002)

Ecc64. Eccles, J.C.: The Physiology of Synapses. Springer, Berlin, (1964)

EHM99. Eiben, A.E., Hinterding, R., Michalewicz, Z.: Parameter Control in Evolutionary Algorithms. IEEE Trans. Evol. Comp. 3(2), 124–141, (1999)

Ein48. Einstein, A.: Quantum Mechanics and Reality (Quanten–Mechanik und Wirklichkeit). Dialectica **2**, 320–324, (1948)

Elm90. Elman, J.: Finding structure in time. Cogn. Sci. **14**, 179–211, (1990)

EMN92. Ellis, J., Mavromatos, N., Nanopoulos, D.V.: String theory modifies quantum mechanics. CERN-TH/6595, (1992)

EMN99. Ellis, J., Mavromatos, N., Nanopoulos, D.V.: A microscopic Liouville arrow of time. Chaos, Solit. Fract., **10**(2–3), 345–363, (1999)

Eng06. Engelbrecht, A.: Fundamentals of Computational Swarm Intelligence. Wiley, New York, (2006)

Ens05. Enss, C. (ed.): Cryogenic Particle Detection. Topics in Applied Physics **99**, Springer, New York, (2005)

EP98. Ershov, S.V., Potapov, A.B.: On the concept of Stationary Lyapunov basis. Physica D, **118**, 167, (1998)

EPG95. Ernst, U., Pawelzik, K., Geisel, T.: Synchronization induced by temporal delays in pulse-coupled oscillators. Phys. Rev. Lett. **74**, 1570, (1995)

EPR35. Einsten, A., Podolsky, P., Rosen, N.: Can quantum-mechanical description of physical reality be considered complete? Phys. Rev. **47**, 777–80, (1935)

EPR35a. Einstein, A., Podolsky, B., Rosen, N.: Quantum theory and Measurement. Zurek, W.H., Wheeler, J.A. (ed.), (1935)

EPR35b. Einstein, A., Podolsky, B., Rosen, N.: Can Quantum Mechanical Description of Physical Reality Be Considered Complete? Phys. Rev. **47**, 777, (1935)

EPR99. Eckmann, J.P., Pillet, C.A., Rey-Bellet, L.: Non-equilibrium statistical mechanics of anharmonic chains coupled to two heat baths at different temperatures. Commun. Math. Phys., **201**, 657–697, (1999)

ER85. Eckmann, J.P., Ruelle, D.: Ergodic theory of chaos, strange attractors, Rev. Mod. Phys., **57**, 617–630, (1985)

Erm81. Ermentrout, G.B.: The behavior of rings of coupled oscillators. J. Math. Biol. **12**, 327, (1981)

ES03. Eiben, A.E., Smith, J.E.: Introduction to Evolutionary Computing. Springer, New York, (2003)

ESA05. European Space Agency. Payload and Advanced Concepts: Superconducting Tunnel Junction (STJ) February 17, (2005)

ESH98. Elson, R.C., Selverston, A.I., Huerta, R. *et al.*: Synchronous Behavior of Two Coupled Biological Neurons. Phys. Rev. Lett. **81**, 5692–5695, (1998)

Eys69. Eysenck, H.J., Eysenck, S.B.G.: Personality Structure and Measurement. Routledge, London, (1969)

Eys76. Eysenck, H.J., Eysenck, S.B.G.: Psychoticism as a Dimension of Personality. Hodder and Stoughton, London, (1976)

Eys92a. Eysenck, H.J.: A reply to Costa, McCrae. P or A, C - the role of theory. Personality and Individual Differences, **13**, 867–868, (1992)

Eys92b. Eysenck, H.J.: Four ways five factors are not basic. Personality, Individual Differences, **13**, 667–673, (1992)

Fol75. Follesdal, D.: Meaning, Experience. In ‘Mind and Language’, ed. S. Guttenplan. Oxford, Clarendon, 25–44, (1975)

Fan85. Fancher, R.: The Intelligence Men: Makers of the IQ Controversy. W.W. Norton and Company, New York, (1985)

Fat19. Fatou, P.: Sur les équations fonctionnelles. Bull. Soc. math. France **47**, 161–271, (1919)

Fat22. Fatou, P.: Sur les fonctions méromorphes de deux variables, Sur certaines fonctions uniformes de deux variables. C.R. Acad. Sc. Paris **175**, 862–65, 1030–33, (1922)

Fei78. Feigenbaum, M.J.: Quantitative universality for a class of nonlinear transformations. J. Stat. Phys. **19**, 25–52, (1978)

Fei79. Feigenbaum, M.J.: The universal metric properties of nonlinear transformations. J. Stat. Phys. **21**, 669–706, (1979)

Fer99. Ferber, J.: Multi-Agent Systems. An Introduction to Distributed Artificial Intelligence. Addison-Wesley, Reading, MA, (1999)

Fey48. Feynman, R.P.: Space–time Approach to Nonrelativistic Quantum Mechanics. Rev. Mod. Phys. **20**, 367–387, (1948)

Fey51. Feynman, R.P.: An Operator Calculus having Applications in Quantum Electrodynamics. Phys. Rev., **84**, 108–128, (1951)

Fey51. Feynman, R.P.: An Operator Calculus Having Applications in Quantum Electrodynamics. Phys. Rev. **84**, 108–128, (1951)

Fey72. Feynman, R.P.: Statistical Mechanics, a Set of Lectures. WA Benjamin, Inc., Reading, Massachusetts, (1972)

Fey98. Feynman, R.P.: Quantum Electrodynamics. Advanced Book Classics, Perseus Publishing, (1998)

FH65. Feynman, R.P., Hibbs, A.R.: Quantum Mechanics, Path Integrals, McGraw-Hill, New York, (1965)

Fit61. FitzHugh, R.A.: Impulses, physiological states in theoretical models of nerve membrane. Biophys J., **1**, 445–466, (1961)

FK83. Frolich, H., Kremer, F.: Coherent Excitations in Biological Systems. Springer, New York, (1983)

FKH04. Fee, M.S., Kozhevnikov, A.A., Hahnloser, R.H.: Neural mechanisms of vocal sequence generation in the songbird. Ann. N.Y. Acad. Sci. **1016**, 153–170, (2004)

Fly04. Flynn, T.: J.P. Sartre. In The Stanford Encyclopedia of Philosophy, Stanford, (2004)

Fog98. Fogel, D. (ed.): Evolutionary Computation: The Fossil Record, IEEE Press, New York, (1998)

Fok29. Fokker, A.D.: Z. Phys., **58**, 386–393, (1929)

FOW66. Fogel, L.J., Owens, A.J., Walsh, M.J.: Artificial Intelligence through Simulated Evolution. Wiley, New York, (1966)

FPP86. Farmer, J.D., Packard, N., Perelson, A.: The immune system, adaptation, machine learning, Physica D, **22**, 187–204, (1986)

Fre00. Freeman, W.J.: Neurodynamics: An exploration of mesoscopic brain dynamics. Springer, Berlin, (2000)

Fre90. Freeman, W.J.: On the problem of anomalous dispersion in chaotic phase transitions of neural masses, its significance for the management of perceptual information in brains. In H.Haken, M.Stadler (ed.) Synergetics of cognition 45, 126–143. Springer Verlag, Berlin, (1990)

Fre91. Freeman, W.J.: The physiology of perception. Sci. Am., **264**(2), 78–85, (1991)

Fre92. Freeman, W.J.: Tutorial on neurobiology: from single neurons to brain chaos. Int. J. Bif. Chaos. **2**(3), 451–82, (1992)

Fre96. Freeman, W.J.: Random activity at the microscopic neural level in cortex sustains is regulated by low dimensional dynamics of macroscopic cortical activity. Int. J. of Neural Systems 7, 473, (1996)

Fri94. Fritzke, B.: Fast learning with incremental RBF networks. Neural Processing Letters, **1**, 1–5, (1994)

Fri99. Frieden R.B.: Physics from Fisher Information: A Unification, Cambridge Univ. Press, (1999)

FS91. Frensch, P.A., Sternberg, R.J.: Skill-related differences in game playing. In R.J. Sternberg, P.A. Frensch (eds.) Complex problem solving: Principles, mechanisms (pp. 343–381) Hillsdale, NJ: Lawr. Erl. Assoc., (1991)

FS92. Freeman, J.A., Skapura, D.M.: Neural Networks: Algorithms, Applications and Programming Techniques. Addison-Wesley, Reading, MA, (1992)

Fuk75. Fukushima, K.: Cognitron: a self–organizing multilayered neural network. Biol. Cyb., **20**, 121–136, (1975)

Fun95. Funke, J.: Solving complex problems: Human identification, control of complex systems. In R.J. Sternberg, P.A. Frensch (eds.) Complex problem solving: Principles, mechanisms (185–222) Hillsdale, NJ: Lawr. Erl. Assoc., (1995)

FW95. Filatrella, G., Wiesenfeld, K.: Magnetic-field effect in a two-dimensional array of short Josephson junctions. J. Appl. Phys. **78**, 1878–1883, (1995)

FY83. Fujisaka, H., Yamada, T.: Amplitude Equation of Higher-Dimensional Nikolaevskii Turbulence. Prog. Theor. Phys. **69**, 32, (1983)

GA99. Gallagher, R., Appenzeller, T.: Beyond Reductionism. Science **284**, 79, (1999)

Gar85. Gardiner, C.W.: Handbook of Stochastic Methods for Physics, Chemistry, Natural Sciences, (2nd ed.) Springer-Verlag, New York, (1985)

GGS81. Guevara, M.R., Glass, L., Shrier, A.: Phase locking, period-doubling bifurcations and irregular dynamics in periodically stimulated cardiac cells. Science **214**, 1350–53, (1981)

GH00. Gade, P.M., Hu, C.K.: Synchronous chaos in coupled map lattices with small-world interactions. Phys. Rev. E **62**, 6409–6413, (2000)

GH83. Guckenheimer, J., Holmes, P.: Nonlinear Oscillations, Dynamical Systems and Bifurcations of Vector Fields. Springer-Verlag, Berlin, (1983)

GK92. Georgiou, G.M., Koutsougeras, C.: Complex domain backpropagation. IEEE Trans. Circ. Sys., **39**(5), 330–334, (1992)

GK96. Gomi, H., Kawato, M.: Equilibrium-point control hypothesis examined by measured arm-stiffness during multi-joint movement. Science, **272**, 117–120, (1996)

Gla63a. Glauber, R.J.: The Quantum Theory of Optical Coherence. Phys. Rev. **130**, 2529–2539, (1963)

Gla63b. Glauber, R.J.: Coherent and Incoherent States of the Radiation Field. Phys. Rev. **131**, 2766–2788, (1963)

Gle87. Gleick, J.: Chaos: Making a New Science. Penguin–Viking, New York, (1987)

GLW93. Geigenmuller, U., Lobb, C.J., Whan, C.B.: Friction and inertia of a vortex in an underdamped Josephson array. Phys. Rev. B **47**, 348–358, (1993)

GM96. Gray, C.M., McCormick, D.A.: Chattering cells: superficial pyramidal neurons contributing to the generation of synchronous oscillations in the visual cortex. Science. **274**(5284), 109–113, (1996)

GM79. Guyer, R.A., Miller, M.D.: Commensurability in One Dimension at $T \neq 0$. Phys. Rev. Lett. **42**, 718–722, (1979)

GN90. Gaspard, P., Nicolis, G.: Transport properties and Lyapunov exponents and entropy per unit time. Phys. Rev. Lett. **65**, 1693–1696, (1990)

Gol01. Goldreich, O.: Foundations of Cryptography, Vol. 1: Basic Tools. Cambridge Univ. Press, Cambridge, (2001)

Gol67. Gold, E.: Language Identification in the Limit. Information and Control, **10**, 447–474, (1967)

Gol89. Goldberg, D.E.: Genetic Algorithms in Search, Optimization and Machine Learning. Kluwer Acad. Pub., Boston, MA, (1989)

Gol99. Goldberger, A.L.: Nonlinear Dynamics, Fractals, and Chaos Theory: Implications for Neuroautonomic Heart Rate Control in Health, Disease. In: Bolis CL, Licinio J, eds. The Autonomic Nervous System. World Health Organization, Geneva, (1999)

GOP84. Grebogi, C., Ott, E., Pelikan, S., Yorke, J.A.: Strange attractors that are not chaotic. Physica D **3**, 261–268, (1984)

Gor26. Gordon, W.: Z. Phys., **40**, 117–133, (1926)

Got96. Gottlieb, H.P.W.: Question #38. What is the simplest jerk function that gives chaos? Am. J. Phys., **64**(5), 525, (1996)

GOY87. Grebogi, C., Ott, E., Yorke, J.A.: Chaos, strange attractors, and fractal basin boundaries in nonlinear dynamics. Science, **238**, 632–637, (1987)

Goz83. Gozzi, E.: Functional-integral approach to Parisi-Wu stochastic quantization: Scalar theory. Phys. Rev. D **28**, 1922, (1983)

GP83. Grassberger, P., Procaccia, I.: Measuring the strangeness of strange attractors, Physica D **9**, 189–208, (1983)

GP83a. Grassberger, P., Procaccia, I.: Measuring the Strangeness of Strange Attractors. Phys. D **9**, 189–208, (1983)

GP83b. Grassberger, P., Procaccia, I.: Characterization o f Strange Attractors. Phys. Rev. Lett. **50**, 346–349, (1983)

Gra90. Granato, E.: Phase transitions in Josephson-junction ladders in a magnetic field. Phys. Rev. B **42**, 4797–4799, (1990)

Gro69. Grossberg, S.: Embedding fields: A theory of learning with physiological implications. J. Math. Psych. **6**, 209–239, (1969)

Gro82. Grossberg, S.: Studies of Mind and Brain. Kluwer, Dordrecht, Holland, (1982)

Gro87. Grossberg, S.: Competitive learning: from interactive activation to adaptive resonance. Cog. Sci. **11**, 23–63, (1987)

Gro88. Grossberg, S.: Neural Networks and Natural Intelligence. MIT Press, Cambridge, MA, (1988)

Gro99. Grossberg, S.: How does the cerebral cortex work? Learning, attention and grouping by the laminar circuits of visual cortex. Spatial Vision **12**, 163–186, (1999)

GRW86. Ghirardi, G.C., Rimini, A., Weber, T.: Unified dynamics for microscopic and macroscopic systems. Phys. Rev. D, **34**(2), 470–491, (1986)

GS98. Grosche, C., Steiner, F.: Handbook of Feynman path integrals. Springer tracts in modern physics 145, Springer, Berlin, (1998)

GSD92. Garfinkel, A., Spano, M.L., Ditto, W.L., Weiss, J.N.: Controlling cardiac chaos. Science **257**, 1230–1235, (1992)

Gui67. Guilford, J.P.: The Nature of Human Intelligence. McGraw-Hill, New York, (1967)

Gun03. Gunion, J.F.: Class Notes on Path-Integral Methods. U.C. Davis, 230B, (2003)

Gut90. Gutzwiller, M.C.: Chaos in Classical and Quantum Mechanics. Springer, New York, (1990)

Gut98. Gutkin, B.S., Ermentrout, B.: Dynamics of membrane excitability determine interspike interval variability: A link between spike generation mechanisms, cortical spike train statistics. Neural Comput., **10**(5), 1047–1065, (1998)

GZ94. Glass, L., Zeng, W.: Bifurcations in flat-topped maps and the control of cardiac chaos. Int. J. Bif. Chaos **4**, 1061–1067, (1994)

Hak83. Haken, H.: Synergetics: An Introduction (3rd ed.). Springer, Berlin, (1983)

Hak91. Haken, H.: Synergetic Computers and Cognition. Springer-Verlag, Berlin, (1991)

Hak93. Haken, H.: Advanced Synergetics: Instability Hierarchies of Self-Organizing Systems and Devices (3nd ed.) Springer, Berlin. (1993)

Hak96. Haken, H.: Principles of Brain Functioning: A Synergetic Approach to Brain Activity, Behavior and Cognition. Springer, Berlin, (1996)

Hak00. Haken, H.: Information, Self-Organization: A Macroscopic Approach to Complex Systems. Springer, Berlin, (2000)

Hak02. Haken, H.: Brain Dynamics, Synchronization and Activity Patterns in Pulse-Codupled Neural Nets with Delays and Noise. Springer, New York, (2002)

Ham87. Hameroff, S.R.: Ultimate Computing: Biomolecular Consciousness and Nanotechnology. North-Holland, Amsterdam, (1987)

Ham98. Hameroff, S.: Quantum computation in brain microtubules? The Penrose-Hameroff Orch OR model of consciousness. Philos. Trans. R. Soc. London Ser. A **356**, 1869–1896, (1998)

Har75. Harris, R.J.: A Primer of Multivariate Statistics, Acad. Press, New York, (1975)

Har92. Harris-Warrick, R.M. (ed.): The Stomatogastric Nervous System. MIT Press, Cambridge, MA, (1992)

Har96. Hart, W.D.: Dualism. In A Companion to the Philosophy of Mind, Blackwell, Oxford, (1996)

Hay91. Haykin, S.: Adaptive Filter Theory. Prentice–Hall, Englewood Cliffs, (1991)

Hay94. Haykin, S.: Neural Networks: A Comprehensive Foundation. Macmillan, (1994)

HBB96. Houk, J.C., Buckingham, J.T., Barto, A.G.: Models of the cerebellum, motor learning. Behavioral, Brain Sciences, **19**(3), 368–383, (1996)

HCK02a. Hong, H., Choi, M.Y., Kim, B.J.: Synchronization on small-world networks. Phys. Rev. E **65**, 26139, (2002)

HCK02b. Hong, H., Choi, M.Y., Kim, B.J.: Phase ordering on small-world networks with nearest-neighbor edges. Phys. Rev. E **65**, 047104, (2002)

HCL74. Hornby, A.S., Cowie, A.P., Lewis, J.W.: Oxford Advanced Learner's Dicitionary of Current English. Oxford Univ. Press, (1974)

HCT97. Hall, K., Christini, D.J., Tremblay, M., Collins, J.J., Glass, L., Billette, J.: Dynamic Control of Cardiac Alternans. Phys. Rev. Lett. **78**, 4518–4521, (1997)

Heb49. Hebb, D.O.: The Organization of Behavior. Wiley, New York, (1949)

Hec77. Heckhausen, H.: Achievement motivation, its constructs: a cognitive model. Motiv. Emot, **1**, 283–329, (1977)

Hec87. Hecht-Nielsen, R.: Counterpropagation networks. Applied Optics, **26**(23), 4979–4984, (1987)

Hec90. Hecht-Nielsen, R.: NeuroComputing. Addison–Wesley, Reading, (1990)

Heg77. Hegel, G.W.F.: Phenomenology of the Spirit. (Translated by A.V. Miller with analysis of the text, foreword by J. N. Findlay). Clarendon Press, Oxford, (1977)

Heg91. Hegarty, M.: Knowledge, processes in mechanical problem solving. In R.J. Sternberg, P.A. Frensch (eds.), Complex problem solving: Principles, mechanisms (253–285) Hillsdale, NJ: Lawr. Erl. Assoc., (1991)

Hen66. Hénon, M.: Sur la topologie des lignes de courant dans un cas particulier. C.R. Acad. Sci. Paris A, **262**, 312–314, (1966)

Hen69. Hénon, M.: Numerical study of quadratic area preserving mappings. Q. Appl. Math. **27**, (1969)

Hen76. Hénon, M.: A two-dimensional mapping with a strange attractor. Com. Math. Phys. **50**, 69–77, (1976)

Hew69. Hewitt, C.: PLANNER: A Language for Proving Theorems in Robots. IJCAI, (1969)

HH52. Hodgkin, A.L., Huxley, A.F.: A quantitative description of membrane current and application to conduction and excitation in nerve. J. Physiol., **117**, 500–544, (1952)

HH97. Harmon, M.E., Harmon, S.S.: Reinforcement learning: A tutorial. Tec. Rep. Wright Laboratory, Wright State Univ., (1997)

HHH99. Hirai, K., Hirose, M., Haikawa, Takenaka, T.: The Development of Honda Humanoid Robot. Proc. of the IEEE Int. Conf. on Robotics and Automation, Leuven, Belgium, 1321–1326, (1999)

HI01. Hoppensteadt, F.C., Izhikevich, E.M.: Canonical Neural Models. In Arbib MA (ed.) Brain Theory, Neural Networks (2nd ed.) MIT press, Cambridge, MA, (2001)

HI97. Hoppensteadt, F.C., Izhikevich, E.M.: Weakly Connected Neural Networks. Springer, New York, (1997)

HI99. Hoppensteadt, F.C., Izhikevich, E.M.: Oscillatory Neurocomputers With Dynamic Connectivity, Phys. Rev. Let., **82**(14), 2983–86, (1999)

Hil00. Hilfer, R. (ed.): Applications of Fractional Calculus in Physics. World Scientific, Singapore, (2000)

Hil94. Hilborn, R.C.: Chaos and Nonlinear Dynamics: An Introduction for Scientists, Engineers. Oxford Univ. Press, Oxford, (1994)

HK87. Heppner, P.P., Krauskopf, C.J.: An information-processing approach to personal problem solving. The Counseling Psychologist, **15**, 371–447, (1987)

HL84. Horsthemke, W., Lefever, R.: Noise–Induced Transitions. Springer, Berlin, (1984)

HL86a. Hale, J.K., Lin, X.B.: Symbolic dynamics and nonlinear semiflows. Ann. Mat. Pur. Appl. **144**(4), 229–259, (1986)

HL86b. Hale, J.K., Lin, X.B.: Examples of transverse homoclinic orbits in delay equations. Nonlinear Analysis **10**, 693–709, (1986)

HL93. Hale, J.K., Lunel, S.M.V.: Introduction to Functional Differential Equations. Springer, New York, (1993)

HO96. Hunt, B.R., Ott, E.: Optimal periodic orbits of chaotic systems occur at low period. Phys. Rev. E **54**, 328–337, (1996)

Hod64. Hodgkin, A.L.: The Conduction of the Nervous Impulse. Liverpool Univ. Press, Liverpool, (1964)

HOG88. Hsu, G.H., Ott, E., Grebogi, C.: Strange Saddles and Dimensions of their Invariant Manifolds. Phys. Lett. A **127**, 199–204, (1988)

Hol92. Holland, J.H.: Adaptation in Natural, Artificial Systems (2nd ed.) MIT Press, Cambridge, MA, (1992)

Hol95. Holland, J.H.: Hidden order: how adaptation builds complexity. Addison-Wesley, New York, (1995)

Hop82). Hopfield, J.J.: Neural networks, physical systems with emergent collective computational activity. Proc. Natl. Acad. Sci. USA., **79**, 2554–2558, (1982)

Hop82. Hopfield, J.J.: Neural networks, physical systems with emergent collective computational abilities. Proc. Natl. Acad. Sci. USA **79**, 2554, (1982)

Hop84. Hopfield, J.J.: Neurons with graded response have collective computational properties like those of two–state neurons. Proc. Natl. Acad. Sci. USA, **81**, 3088–3092, (1984)

Hou79. Houk, J.C.: Regulation of stiffness by skeletomotor reflexes. Ann. Rev. Physiol., **41**, 99–123, (1979)

HOY96. Hunt, B.R., Ott, E., Yorke, J.A.: Fractal dimensions of chaotic saddles of dynamical systems. Phys. Rev. E **54**, 4819–4823, (1996)

HP29a. Heisenberg, W., Pauli, W.: Z. Phys., **56**, 1–61, (1929)

HP29b. Heisenberg, W., Pauli, W.: Z. Phys., **59**, 168–190, (1929)

HP93. Hameroff, S.R., Penrose, R.: Conscious events as orchestrated spacetime selections. Journal of Consciousness Studies, **3**(1), 36–53, (1996)

HP96. Hameroff, S.R., Penrose, R.: Orchestrated reduction of quantum coherence in brain microtubules: A model for consciousness. In: Hameroff, S. R., Kaszniak, A.W., Scott, A.C. Eds.: Toward a Science of Consciousness: the First Tucson Discussion, Debates, 507–539. MIT Press, Cambridge, MA, (1996)

HP96. Hawking, S.W., Penrose, R.: The Nature of Space and Time. Princetone Univ. Press, Princetone, NJ, (1996)

HS74. Hirsch, M.W., Smale, S.: Differential Equations, Dynamical Systems and Linear Algebra. Academic Press, New York, (1974)

HT85. Hopfield, J.J., Tank, D.W.: Neural computation of decisions in optimisation problems. Biol. Cybern., **52**, 114–152, (1985)

HU79. Hopcroft, J.E., Ullman, J.D.: Introduction to Automata Theory, Languages and Computation. Addison-Wesley, New York, (1979)

Hun01. Hunt, E.L.: Multiple views of multiple intelligence. (Review of Intelligence Reframed: Multiple Intelligences for the 21st Century.) Contemp. Psych. **46**, 5–7, (2001)

Hux898. Huxley, T.H.: On the Hypothesis that Animals are Automata, its History. Reprinted in Method, Results: Essays by Thomas H. Huxley. D. Appleton, Company, New York (1898)

HW82. Hameroff, S.R., Watt, R.C.: Information processing in microtubules. J. Theo. Bio. **98**, 549–561, (1982)

HW83. Hameroff, S.R., Watt, R.C.: Do anesthetics act by altering electron mobility? Anesth. Analg., **62**, 936–40, (1983)

IB05. Ivancevic, V., Beagley, N.: Brain-like functor control-machine for general humanoid biodynamics. Int. J. Math., Math. Sci. **11**, 1759–1779, (2005)

IDW03. Izhikevich, E.M., Desai, N.S., Walcott, E.C., Hoppensteadt, F.C.: Bursts as a unit of neural information: selective communication via resonance. Trends in Neurosci., **26**, 161–167, (2003)

IGF99. Ioffe, L.B., Geshkenbein, V.B., Feigel'man, M.V., Fauchère, A.L., Blatter, G.: Environmentally decoupled sds-wave Josephson junctions for quantum computing. Nature **398**, 679–681, (1999)

II05. Ivancevic, V., Ivancevic, T.: Human–Like Biomechanics. Springer, Dordrecht, (2005)

II06a. Ivancevic, V., Ivancevic, T.: Natural Biodynamics. World Scientific, Singapore, (2006)

II06b. Ivancevic, V., Ivancevic, T.: Geometrical Dynamics of Complex Systems. Springer, Dordrecht, (2006)

IJB99a. Ivancevic, T., Jain, L.C., Bottema, M.: A New Two-feature GBAM-Neurodynamical Classifier for Breast Cancer Diagnosis. In Proc. from KES'99, IEEE Press, USA, (1999)

IJB99b. Ivancevic, T., Jain, L.C., Bottema, M.: A New Two–Feature FAM–Matrix Classifier for Breast Cancer Diagnosis, Proc. from KES'99, IEEE Press, USA, (1999)

IKM03. Imamizu, H., Kuroda, T., Miyauchi, S., Yoshioka, T., Kawato, M.: Modular organization of internal models of tools in the human cerebellum. Proc Natl Acad Sci USA., **100**, 5461–5466, (2003)

IMT00. Imamizu, H., Miyauchi, S., Tamada, T., Sasaki, Y., Takino, R., Puetz, B., Yoshioka, T., Kawato, M.: Human cerebellar activity reflecting an acquired internal model of a novel tool. Nature, **403**, 192–195, (2000)

Ing97. Ingber, L.: Statistical mechanics of neocortical interactions: Applications of canonical momenta indicators to electroencephalography. Phys. Rev. E, **55**(4), 4578–4593, (1997)

Ing98. Ingber, L.: Statistical mechanics of neocortical interactions: Training, testing canonical momenta indicators of EEG. Mathl. Computer Modelling **27**(3), 33–64, (1998)

IP01a. Ivancevic, V., Pearce, C.E.M.: Topological duality in humanoid robot dynamics. ANZIAM J. **43**, 183–194, (2001)

IP01b. Ivancevic, V., Pearce, C.E.M.: Poisson manifolds in generalized Hamiltonian biomechanics. Bull. Austral. Math. Soc. **64**, 515–526, (2001)

IS01. Ivancevic, V., Snoswell, M.: Fuzzy-stochastic functor machine for general humanoid robot dynamics. IEEE Trans. Syst., Man, Cybern. B **31**(3), 319–330, (2001)

Isi89. Isidori, A.: Nonlinear Control Systems, An Introduction (2nd ed.). Springer, Berlin, (1989)

Ito60. Ito, K.: Wiener Integral, Feynman Integral. Proc. Fourth Berkeley Symp. Math., Stat., Prob., **2**, 227–238, (1960)

Iva02. Ivancevic, V.: Generalized Hamiltonian biodynamics, topology invariants of humanoid robots. Int. J. Math., Math. Sci. **31**(9), 555–565, (2002)

Iva04. Ivancevic, V.: Symplectic rotational geometry in human biomechanics, SIAM Rev., **46**(3), 455–474, (2004)

Iva06a. Ivancevic, V.: Lie-Lagrangian model for realistic human bio-dynamics. Int. J. Hum. Rob., **3**(2), 205–218, (2006)

Iva06b. Ivancevic, V.: Dynamics of Humanoid Robots: Geometrical, Topological Duality. In Biomathematics: Modelling, Simulation, ed. J.C. Misra, World Scientific, Singapore, (2006)

Iye81. Iyengar, B.K.S.: Light on Yoga. Unwin Publishers, London, (1981)

IZ80. Itzykson, C., Zuber, J.: Quantum field theory. McGraw-Hill, New York, (1980)

Izh00. Izhikevich, E.M.: Neural Excitability, Spiking and Bursting. International Journal of Bifurcation and Chaos, **10**, 1171–1266, (2000)

Izh01. Izhikevich, E.M.: Synchronization of Elliptic Bursters, SIAM Rev., **43**(2), 315–344, (2001)

Izh01. Izhikevich, E.M.: Resonate-and-fire neurons. Neu. Net. **14**, 883–894, (2001)

Izh04. Izhikevich, E.M.: Which model to use for cortical spiking neurons? IEEE Trans. Neu. Net. **15**, 1063–1070, (2004)

Izh07. Izhikevich, E.M.: Dynamical Systems in Neuroscience: The Geometry of Excitability and Bursting. The MIT Press, Cambridge, MA, (2007)

Izh99a. Izhikevich, E.M.: Class 1 neural excitability, conventional synapses, weakly connected networks and mathematical foundations of pulse-coupled models. IEEE Trans. Neu. Net., **10**, 499–507, (1999)

Izh99b. Izhikevich, E.M.: Weakly Connected Quasiperiodic Oscillators, FM Interactions and Multiplexing in the Brain. SIAM J. Appl. Math., **59**(6), 2193–2223, (1999)

Jac82. Jackson, F.: Epiphenomenal Qualia. Reprinted in Chalmers and David (ed., 2002). Philosophy of Mind: Classical, Contemporary Readings. Oxford Univ. Press, Oxford, (1982)

JBO97. Just, W., Bernard, T., Ostheimer, M., Reibold, E., Benner, H.: Mechanism of time–delayed feedback control. Phys. Rev. Lett., **78**, 203–206, (1997)

Joh72. Johnson, D.M.: Systematic introduction to the psychology of thinking. Harper, Row, (1972)

Jos74. Josephson, B.D.: The discovery of tunnelling supercurrents. Rev. Mod. Phys. **46**(2), 251–254, (1974)

JPY96. Jibu, M., Pribram, K.H., Yasue, K.: From conscious experience to memory storage and retrieval: the role of quantum brain dynamics, boson condensation of evanescent photons, Int. J. Mod. Phys. B, **10**, 1735, (1996)

Jul18. Julia, G.: Mémoires sur l'itération des fonctions rationelles. J. Math. **8**, 47–245, (1918)

Jun80. Jung, C.G.: Psychology and Alchemy. Princeton Univ. Press., Princeton, New Jersey, (1980)

Jun93. Jung, P.: Periodically driven stochastic systems. Phys. Reports **234**, 175, (1993)

JV72. Juricic, D, Vukobratovic, M.: Mathematical Modeling of Biped Walking Systems. ASME Publ. 72-WA/BHF-13, (1972)

JY95. Jibu, M., Yasue, K.: Quantum brain dynamics and consciousness. John Benjamins, Amsterdam, (1995)

Kac51. Kac, M.: On Some Connection between Probability Theory, Differential and Integral Equations. Proc. 2nd Berkeley Sympos. Math. Stat. Prob., 189–215, (1951)

Kar84. Kardar, M.: Free energies for the discrete chain in a periodic potential and the dual Coulomb gas. Phys. Rev. B **30**, 6368–6378, (1984)

Kas02. Kasabov, N.: Evolving connectionist systems: Methods and applications in bioinformatics, brain study and intelligent machines. Springer, London, (2002)

Kaw99. Kawato, M.: Internal models for motor control and trajectory planning. Current Opinion in Neurobiology, **9**, 718–727, (1999)

Kay91. Kay, D.S.: Computer interaction: Debugging the problems. In R.J. Sternberg, P.A. Frensch (eds.) Complex problem solving: Principles, mechanisms (317–340) Hillsdale, NJ: Lawr. Erl. Assoc., (1991)

KBA99. Koza, J.R., Bennett, F.H., Andre, D., Keane, M.A.: Genetic Programming III: Darwinian Invention, Problem Solving. Morgan Kaufmann, (1999)

KG85. Kantz, H., Grassberger, P.: Repellers, semi-attractors and long-lived chaotic transients. Physica D **17**, 75–86, (1985)

Khi57. Khinchin, A.I.: Mathematical foundations of Information theory. Dover, (1957)

KHS93. Koruga, D.L., Hameroff, S.I., Sundareshan, M.K., Withers, J., Loutfy, R.: Fullerence C60: History, Physics, Nanobiology and Nanotechnology. Elsevier Science Pub, (1993)

Kim95a. Kim, J.: Problems in the Philosophy of Mind. Oxford Companion to Philosophy. Ted Honderich (ed.) Oxford Univ. Press, Oxford, (1995)

Kim95b. Kim, J.: Mind–Body Problem. Oxford Companion to Philosophy. Ted Honderich (ed.) Oxford Univ. Press, Oxford, (1995)

KK00. Kye, W.-H., Kim, C.-M.: Characteristic relations of type-I intermittency in the presence of noise. Phys. Rev. E **62**, 6304–6307, (2000)

KKS03. Koza, J.R., Keane, M.A., Streeter, M.J., Mydlowec, W., Yu, J., Lanza, G.: Genetic Programming IV: Routine Human-Competitive Machine Intelligence. Kluwer, Dordrecht, (2003)

Kla00. Klauder, J.R.: Beyond Conventional Quantization. Cambridge Univ. Press, Cambridge, (2000)

Kla97. Klauder, J.R.: Understanding Quantization. Found. Phys. **27**, 1467–1483, (1997)

Kle27. Klein, O.: Z. Phys., **41**, 407–442, (1927)

Kli00. Kline, P.: A Psychometrics Primer. Free Assoc. Books, London, (2000)

KLR03. Kye, W.-H., Lee, D.-S., Rim, S., Kim, C.-M., Park, Y.-J.: Periodic Phase Synchronization in coupled chaotic oscillators. Phys. Rev. E **68**, 025201–025205(R), (2003)

KM78a. Kim, J., Mueller, C.W.: Introduction to factor analysis: What it is and how to do it. Thousand Oaks, CA: Sage Publications, Quantitative Applications in the Social Sciences Series, 13, (1978)

KM78b. Kim, J., Mueller, C.W.: Factor Analysis: Statistical methods and practical issues. Thousand Oaks, CA: Sage Publications, Quantitative Applications in the Social Sciences Series, 14, (1978)

KMM94. Konen, W., Maurer, T., von der Malsburg, C.: A fast dynamic link matching algorithm for invariant pattern recognition. Neural Networks, **7**, 1019–1030, (1994)

KMY84. Kaplan, J.L., Mallet-Paret, J., Yorke, J.A.: The Lyapunov dimension of a nowhere dierentiable attracting torus. Ergod. Th. Dynam. Sys. **4**, 261 (1984)

KN00. Kotz, S., Nadarajah, S.: Extreme Value Distributions. Imperial College Press, London, (2000)

Koe64. Koestler, A. The Act of Creation. Penguin, London, (1964)

Koh82. Kohonen, T.: Self–Organized Formation of Toplogically Correct Feature Maps. Biological Cybernetics **43**, 59–69, (1982)

Koh88. Kohonen, T.: Self Organization, Associative Memory. Springer, (1988)

Koh91. Kohonen, T.: Self–Organizing Maps: Optimization Approaches. In: Artificial Neural Networks, ed. T. Kohonen etal. North-Holland, Amsterdam, (1991)

Kos86. Kosko, B.: Fuzzy Cognitive Maps. Int. J. Man-Mach. Stud. **24**, 65–75, (1986)

Kos88. Kosko, B.: Bidirectional Associative Memory. IEEE Trans. Sys. Man Cyb. **18**, 49–60, (1988)

Kos92. Kosko, B.: Neural Networks, Fuzzy Systems, A Dynamical Systems Approach to Machine Intelligence. Prentice–Hall, New York, (1992)

Kos93. Kosko, B.: Fuzzy Thinking. Disney Books, Hyperion, (1993)

Kos96. Kosko, B.: Fuzzy Engineering. Prentice Hall, New York, (1996)

Kos99. Kosko, B.: The Fuzzy Future: From Society, Science to Heaven in a Chip. Random House, Harmony, (1999)

Koz92. Koza, J.R.: Genetic Programming: On the Programming of Computers by Means of Natural Selection. MIT Press, Cambridge, MA, (1992)

Koz95. Koza, J.R.: Genetic Programming II: Automatic Discovery of Reusable Programs. MIT Press, Cambridge, MA, (1995)

KP95. Kocarev, L., Parlitz, U.: General Approach for Chaotic Synchronization with Applications to Communication. Phys. Rev. Lett. **74**, 5028–5031, (1995)

KP96. Kocarev, L., Parlitz, U.: Generalized Synchronization, Predictability and Equivalence of Unidirectionally Coupled Dynamical Systems. Phys. Rev. Lett. **76**, 1816–1819, (1996)

KS02. Kasabov, N., Song, Q.: Denfis: Dynamic evolving neural fuzzy inference systems and its application for time series prediction. IEEE Trans. Fuz. Sys. **10**(2), 144–154, (2002)

KT01. Kye, W.-H., Topaj, D.: Attractor bifurcation and on-off intermittency. Phys. Rev. E **63**, 045202–045206(R), (2001)

KT73. Krall, N.A., Trivelpiece, A.W.: Principles of Plasma Physics. McGraw-Hill, New York, (1973)

Kuh85. Kuhl, J.: Volitional Mediator of cognition-Behaviour consistency: Self-regulatory Processes, action versus state orientation (101–122) In: J. Kuhl, S. Beckman (eds.) Action control: From Cognition to Behaviour. Springer, Berlin, (1985)

Kur84. Kuramoto, Y.: Chemical Oscillations. Waves, Turbulence. Springer, New York, (1984)

Kuz95. Kuznetsov, Y.A.: Elements of Applied Bifurcation Theory. Applied Mathematical Sciences **112**, Springer-Verlag, Berlin, (1995)

KY75. Kaplan, J.L., Yorke, J.A.: On the stability of a periodic solution of a differential delay equation. SIAM J. Math. Ana. **6**, 268–282, (1975)

KY79. Kaplan, J.L., Yorke, J.A.: Numerical Solution of a Generalized Eigenvalue Problem for Even Mapping. Peitgen, H.O., Walther, H.O. (Eds.): Functional Differential Equations, Approximations of Fixed Points, Lecture Notes in Mathematics, **730**, 228–256, Springer, Berlin, (1979)

KY79. Kaplan, J.L., Yorke, J.A.: Preturbulence: a regime observed in a fluid flow of Lorenz. Commun. Math. Phys. **67**, 93–108, (1979)

KY91. Kennedy, J., Yorke, J.A.: Basins of Wada. Physica D **51**, 213–225, (1991)

KYR98. Kim, C.M., Yim, G.S., Ryu, J.W., Park, Y.J.: Characteristic Relations of Type-III Intermittency in an Electronic Circuit. Phys. Rev. Lett. **80**, 5317–5320, (1998)

KZH02. Kiss, I.Z., Zhai, Y., Hudson, J.L.: Emerging coherence in a population of chemical oscillators. Science **296**, 1676–1678, (2002)

Lai94. Lai, Y.-C.: Controlling chaos. Comput. Phys., **8**, 62–67, (1994)

Lak03. Lakshmanan, M., Rajasekar, S: Nonlinear Dynamics: Integrability, Chaos and Patterns, Springer-Verlag, New York, (2003)

Lak97. Lakshmanan, M.: Bifurcations, Chaos, Controlling and Synchronization of Certain Nonlinear Oscillators. In Lecture Notes in Physics, **495**, 206, Y. Kosmann-Schwarzbach, B. Grammaticos, K.M. Tamizhmani (ed.), Springer-Verlag, Berlin, (1997)

Lam02. Lampinen, J.: A Constraint Handling Method for the Differential Evolution Algorithm. In: Sincak P., Vascak J., Kvasnicka V., Pospichal J. (eds.) Intelligent Technologies – Theory, Applications, 152–158. IOS Press, (2002)

Las42. Lashley, K.S.: The problem of cerebral organization in vision. In Biological Symposia, VII, Visual mechanisms, 301–322. Jaques Cattell Press, Lancaster, (1942)

LCD02. Liley, D.T.J., Cadusch, P.J., Dafilis MP.: A spatially continuous mean field theory of electrocortical activity. Comp. Neu. Sys. **13**(1), 67–113, (2002)

LCD87. Leggett, A.J., Chakravarty, S., Dorsey, A.T., Fisher, M.P.A., Chang, A., Zwerger, W.: Dynamics of the dissipative two-state system. Rev. Mod. Phys. **59**, 1, (1987)

LCW99. Liley, D.T.J., Cadusch, P.J., Wright, J.J.: A continuum theory of electrocortical activity. Neurocom. **26**, 795–800, (1999)

LD98. Loss, D., DiVincenzo, D.P.: Quantum computation with quantum dots. Phys. Rev. A **57**(1), 120–126, (1998)

LEA00. Lehnertz, K., Elger, C.E., Arnhold, J., Grassberger, P. (ed.): Chaos in Brain. World Scientific, Singapore, (2000)

Lee90. Lee, C.C.: Fuzzy Logic in Control Systems. IEEE Trans. Sys., Man, Cybern., **20**(2), 404–435, (1990)

Leg86. Leggett, A.J.: In The Lesson of Quantum Theory. Niels Bohr Centenary Symposium 1985; J. de Boer, E. Dal, O. Ulfbeck (ed.) North Holland, Amsterdam, (1986)

Lei714. Leibniz, G.W.: Monadology, (1714)

Lev88. Levinthal, C.F.: Messengers of Paradice, Opiates and the Brain. Anchor Press, Freeman, New York, (1988)

Lev92. Levy, S.: Artificial Life: A Report from the Frontier Where Computers Meet Biology. Vintage Books: Random House, New York, (1992)

Lew51. Lewin, K.: Field Theory in Social Science. Univ. Chicago Press, Chicago, (1951)

Lew97. Lewin, K.: Resolving Social Conflicts: Field Theory in Social Science, American Psych. Assoc., New York, (1997)

Lis97. Lisman, J.: Bursts as a unit of neural information: making unreliable synapses reliable. Trends in Neurosci. **20**, 38–43, (1997)

LL91. Lesgold, A., Lajoie, S.: Complex problem solving in electronics. In R. J. Sternberg, P.A. Frensch (eds.), Complex problem solving: Principles, mechanisms (287–316) Hillsdale, NJ: Lawr. Erl. Assoc., (1991)

Lor63. Lorenz, E.N.: Deterministic Nonperiodic Flow. J. Atmos. Sci., **20**, 130–141, (1963)

LOS88. Larkin, A.I., Ovchinnikov, Yu.N., Schmid, A.: Physica B **152**, 266, (1988)

LP95. Lebovitz, N.R., Pesci, A.I.: Dynamics bifurcation in Hamiltonian systems with one degree of freedom. SIAM J. Appl. Math. **55**, 1117–1133, (1995)

LS04. Lesica, N.A, Stanley, G.B.: Encoding of Natural Scene Movies by Tonic and Burst Spikes in the Lateral Geniculate Nucleus, J. Neurosci. 24, 10731–10740, (2004)

Lug02. Luger, G.F.: Artificial Intelligence: Structures and Strategies for Complex Problem Solving. Pearson Educ (4th ed.) Ltd, Harlow, UK, (2002)

LVB93. Lades, M., Vorbruggen, J.C., Buhmann, J., Lange, J.C. von der Malsburg, C., Wurtz, R.P., Konen, W.: Distortion invariant object recognition in the dynamic link architecture. IEEE Transactions on Computers, 42(3), 300–311, (1993)

LY85a. Ledrappier, F., Young, L.-S.: The metric entropy of diffeomorphisms I. Characterization of measures satisfying Pesin's entropy formula. Ann. of Math, 122, 509–539, (1985)

LY85b. Ledrappier, F., Young, L.-S.: The metric entropy of diffeomorphisms II. Relations between entropy, exponents and dimension. Ann. of Math. (2), 122(3), 540–574, (1985)

Mac59. Mach, E.: The Analysis of Sensations and the Relation of Physical to the Psychical. (5th ed.) Dover, New York, (1959)

Mal798. Malthus, T.R.: An essay on the Principle of Population. Originally published in 1798. Penguin, (1970)

Mal85. Von der Malsburg, C.: Nervous structures with dynamical links. Ber. Bunsenges. Phys. Chem., 89, 703–710, (1985)

Mal88. Von der Malsburg, C.: Pattern recognition by labelled graph matching. Neural Networks, **7**, 1019–1030, (1988)

Man80a. Mandelbrot, B.: Fractal aspects of the iteration of $z \mapsto \lambda z(1-z)$ for complex λ, z, Annals NY Acad. Sci. **357**, 249–259, (1980)

Man80b. Mandelbrot, B.: The Fractal Geometry of Nature. WH Freeman, Co., New York, (1980)

Mar99. Marsden, J.E.: Elementary Theory of Dynamical Systems. Lecture notes. CDS, Caltech, (1999)

May73. May, R.M. (ed.): Stability and Complexity in Model Ecosystems. Princeton Univ. Press, Princeton, NJ, (1973)

May76. May, R.: Simple Mathematical Models with Very Complicated Dynamics. Nature, **261**(5560), 459–467, (1976)

May76. May, R.M. (ed.): Theoretical Ecology: Principles and Applications. Blackwell Sci. Publ. (1976)

May92. Mayer, R.E.: Thinking, problem solving and cognition. Second edition. New York: W. H. Freeman, Company, (1992)

MB98. Meyer, D.A., Brown, T.A.: Statistical mechanics of voting. Phys. Rev. Lett. **81**, 1718–1721, (1998)

MDC85. Martinis, J.M., Devoret, M.H., Clarke, J.: Energy-Level Quantization in the Zero-Voltage State of a Current-Biased Josephson Junction. Phys. Rev. Lett. **55**, 1543–1546, (1985)

Mes00. Messiah, A.: Quantum Mechanics (two volumes bound as one). Dover Pubs, (2000)

Met97. Metzger, M.A.: Applications of nonlinear dynamical systems theory in developmental psychology: Motor and cognitive development. Nonlinear Dynamics, Psychology, Life Sciences, **1**, 55–68, (1997)

MF04. Michalewicz, Z., Fogel, D.: How to Solve It: Modern Heuristics, 2nd ed., Springer-Verlag, (2004)

MGO85. McDonald, S.W., Grebogi, C., Ott, E., Yorke, J.A.: Fractal basin boundaries. Physica D **17**, 125–153, (1985)

Mic06. Michalewicz, Z.: Adaptive Business Intelligence. Talk presented at DSTO-Adelaide, (2006)

Mic99. Michalewicz, Z.: Genetic Algorithms + Data Structures = Evolution Programs. Springer-Verlag, Berlin, (1999)

Mil56. Miller, G.A.: The magical number seven, plus or minus two: Some limits on our capacity for processing information, Psych. Rev., **63**, 81–97, (1956)

Mil99. Milnor, J.: Periodic Orbits, Externals Rays and the Mandelbrot Set. Stony Brook IMS Preprint # 1999/3, (1999)

Mit96. Mitchell, M.: An Introduction to Genetic Algorithms. MIT Press, Cambridge, MA, (1996)

ML81. Morris, C., Lecar, H.: Voltage oscillations in the barnacle giant muscle fiber. Biophys. J., **35**, 193–213, (1981)

MN95a. Mavromatos, N.E., Nanopoulos, D.V.: A Non-critical String (Liouville) Approach to Brain Microtubules: State Vector reduction and Memory coding, Capacity. ACT-19/95, CTP-TAMU-55/95, OUTP-95-52P, (1995)

MN95b. Mavromatos, N.E., Nanopoulos, D.V.: Non-Critical String Theory Formulation of Microtubule Dynamics and Quantum Aspects of Brain Function. ENSLAPP-A-524/95, (1995)

Moo89. Moore, W.: Schr'odinger: Life and Thought. Cambridge Univ. Press, Cambridge, (1989)

Mos73. Moser, J.: Stable and Random Motions in Dynamical Systems. Princeton Univ. Press, Princeton, (1973)

Mos96. Mosekilde, E.: Topics in Nonlinear Dynamics: Application to Physics, Biology and Economics. World Scientific, Singapore, (1996)

Mou95. Mould, R.: The inside observer in quantum mechanics. Found. Phys. **25**(11), 1621–1629, (1995)

Mou98. Mould, R.: Consciousness and Quantum Mechanics. Found. Phys. **28**(11), 1703–1718, (1998)

Mou99. Mould, R.: Quantum Consciousness. Found. Phys. **29**(12), 1951–1961, (1999)

MP43. McCulloch W., Pitts W.: A logical calculus of the ideas imminent in the nervous activity. Bull. Math. Biophys. **5**, 115–133, 1943

MP69. Minsky, M., Papert, S.: Perceptrons. MIT Press, Cambridge, MA, (1969)

MRJ06. Masgrau, L., Roujeinikova, A., Johannissen, L.O., *et al.*: Atomic Description of an Enzyme Reaction Dominated by Proton Tunneling. Science, **312**, 237–241, (2006)

MS95. M'ullers, J., Schmid, A.: Resonances in the current-voltage characteristics of a dissipative Josephson junction. cond-mat/9508035, (1995)

MSM05. Michalewicz, Z., Schmidt, M., Michalewicz, M., Chiriac, C.: A Decision-Support System based on Computational Intelligence: A Case Study. IEEE Intelligent Systems, 20(4), 44–49, (2005)

MTW04. Markram, H, Toledo-Rodriguez, M, Wang, Y, Gupta, A, Silberberg, G, Wu, C.: Interneurons of the neocortical inhibitory system. Nature Review Neuroscience, **5**,793–807, (2004)

MTW73. Misner, C.W., Thorne, K.S., Wheeler, J.A.: Gravitation. Freeman, San Francisco, (1973)

Mur02. Murray, J.D.: Mathematical Biology, Vol. I: An Introduction (3rd ed.), Springer, New York, (2002)

MW00. Mansfield A.J., Wayman J.L.: Best practices in testing, reporting performance of biometric devices, Issue 1, Report for CESG, BiometricsWorking Group, February (2000)

MW02. Mansfield A.J., Wayman J.L., Best practices in testing, reporting performance of biometric devices, NPL Report CMSC 14/02, Center for Mathematics, Scientific Computing, National Physical Laboratory, August (2002)

MWH01. Michel, A.N., Wang, K., Hu, B.: Qualitative Theory of Dynamical Systems (2nd ed.) Dekker, New York, (2001)

Nag74. Nagel, T.: What is it like to be a bat? Philos. Rev. **83**, 435–456, (1974)

Nan95. Nanopoulos, D.V.: Theory of Brain Function, Quantum Mechanics and Superstrings. CERN-TH/95128, (1995)

NAY60. Nagumo, J., Arimoto, S., Yoshizawa, S.: An active pulse transmission line simulating 1214-nerve axons, Proc. IRL **50**, 2061–2070, (1960)

Nay73. Nayfeh, A.H.: Perturbation Methods. Wiley, New York, (1973)

New00. Newman, M.E.J.: Models of the small world. J. Stat. Phys. **101**, 819, (2000)

NF91. Nitta, T., Furuya, T.: A complex back-propagation learning. Trans. Inf. Proc. Soc. Jpn., **32**(10), 1319–1329, (1991)

Nik95. Nikitin, I.N.: Quantum string theory in the space of states in an indefinite metric. Theor. Math. Phys. **107**(2), 589–601, (1995)

Nit00. Nitta, T.: An analysis on fundamental structure of complex-valued neuron. Neu. Proc. Let., **12**(3), 239–246, (2000)

Nit04. Nitta, T.: Reducibility of the Complex-valued Neural Network. Neu. Inf. Proc., **2**(3), 53–56, (2004)

Nit97. Nitta, T.: An extension of the back-propagation algorithm to complex numbers. Neu. Net., **10**(8), 1392–1415, (1997)

NLG00. Newrock, R.S., Lobb, C.J., Geigenm´uller, U., Octavio, M.: Solid State Physics. Academic Press, San Diego, Vol. 54, (2000)

NML03. Nishikawa, T., Motter, A.E., Lai, Y.C., Hoppensteadt, F.C.: Heterogeneity in Oscillator Networks: Are Smaller Worlds Easier to Synchronize? Phys. Rev. Lett. **91**, 014101, (2003)

NNM00. Nagao, N., Nishimura, H., Matsui, N.: A Neural Chaos Model of Multistable Perception. Neural Processing Letters **12**(3): 267–276, (2000)

Nor99. Normile, D.: Complex Systems: Building Working Cells 'in Silico'. Science, **284**, 80, (1999)

NOY95. Nusse, H.E., Ott, E., Yorke, J.A.: Saddle-Node Bifurcations on Fractal Basin Boundaries. Phys. Rev. Lett. **75**(13), 2482, (1995)

NPT99. Nakamura, Y., Pashkin, Yu.A., Tsai, J.S.: Coherent control of macroscopic quantum states in a single-Cooper-pair box, Nature, **398**, 786–788, (1999)

NR02. Neiman, A.B., Russell, D.F.: Synchronization of Noise-Induced Bursts in Noncoupled Sensory Neurons. Phys. Rev. Lett. **88**, 138–103, (2002)

NS72. Newell, A., Simon, H.A.: Human problem solving. Englewood Cliffs, NJ: Prentice-Hall, (1972)

NS90. Nijmeijer, H., van der Schaft, A.J.: Nonlinear Dynamical Control Systems, Springer, New York, (1990)

Nun00. Nunez, P.L.: Toward a quantitative description of largescale neocortical dynamic function, EEG. Beh. Brain Sci. **23**, 371–437, (2000)

Nun81. Nunez, P.L.: Electric fields of the brain: the neurophysics of EEG. Oxford Univ. Press, New York, (1981)

Nus87. Nusse, H.E.: Asymptotically Periodic Behaviour in the Dynamics of Chaotic Mappings. SIAM J. Appl. Math. **47**, 498, (1987)

NY89. Nusse, H.E., Yorke, J.A.: A procedure for finding numerical trajectories on chaotic saddles. Physica D **36**, 137, (1989)

NY92. Nusse, H.E., Yorke, J.A.: The equality of fractal dimension and uncertainty dimension for certain dynamical systems. Commun. Math. Physics **150**, 1, (1992)

OCD04. Oswald, A.M., Chacron, M.J., Doiron, B., Bastian, J., Maler, L.: Parallel processing of sensory input by bursts, isolated spikes. J Neurosci. **24**(18), 4351–62, (2004)

OGY90. Ott, E., Grebogi, C., Yorke, J.A.: Controlling chaos. Phys. Rev. Lett., **64**, 1196–1199, (1990)

Oja82. Oja, E.: A simplified neuron modeled as a principal component analyzer. J. Math. Biol. **15**, 267–273, (1982)

OS94. Ovchinnikov, Yu.N., Schmid, A.: Resonance phenomena in the current-voltage characteristic of a Josephson junction. Phys. Rev. B **50**, 6332–6339, (1994)

OSB02. Ott, E., So, P., Barreto, E., Antonsen, T.: The onset of synchronization in systems of globally coupled chaotic and periodic oscillators. Physica D, 173(1–2), 29–51, (2002)

Ose68. Oseledets, V.I.: A Multiplicative Ergodic Theorem: Characteristic Lyapunov Exponents of Dynamical Systems. Trans. Moscow Math. Soc., **19**, 197–231, (1968)

Ott89. Ottino, J.M.: The kinematics of mixing: stretching, chaos and transport. Cambridge Univ. Press, Cambridge, (1989)

Ott93. Ott, E.: Chaos in dynamical systems. Cambridge Univ. Press, Cambridge, (1993)

PC05. Park, J., Chung, W-K.: Geometric Integration on IEEE Trans. Robotics, **21**(5), 850–863, (2005)

PC90. Pecora, L.M., Carroll, T.L.: Synchronization in chaotic systems. Phys. Rev. Lett. **64**, 821–824, (1990)

PC91. Pecora, L.M., Carroll, T.L.: Driving systems with chaotic signals. Phys. Rev. A **44**, 2374–2383, (1991)

PC98. Pecora, L.M., Carroll, T.L.: Master stability functions for synchronized coupled systems. Phys. Rev. Lett. **80**, 2109–2112, (1998)

PD95. Pritchard, W.S., Duke, D.W.: Measuring ‘Chaos’ in the brain: a tutorial review of EEG dimension estimation. Brain Cog., **27**, 353–97, (1995)

PE02. Popper, Karl, Eccles, John: The Self and Its Brain. Springer, (2002)

PE34. Pauli, W., Weisskopf, V.: ʹUber die Quantisierung der skalaren relativistischen. Helv. Phys. Acta, **7**, 708–731, (1934)

Pe89. Penrose, R.: The Emperor's New Mind. Oxford Univ. Press, Oxford, (1989)

PEL00. Principe, J., Euliano, N., Lefebvre, C.: Neural and Adaptive Systems: Fundamentals Through Simulations. Wiley, New York, (2000)

Pel85. Pelikan, S.: A dynamical meaning of fractal dimension. Trans. Am. Math. Soc. **292**, 695–703, (1985)

Pen89. Penrose, R.: The Emperor's New Mind, Oxford Univ. Press, Oxford, (1989)

Pen94. Penrose, R.: Shadows of the Mind. Oxford Univ. Press, Oxford, (1994)

Pen97. Penrose, R.: The Large, the Small and the Human Mind. Cambridge Univ. Press, (1997)

Per97. Pert, C.B.: Molecules of Emotion. Scribner, New York, (1997)

Pes76. Pesin, Ya.B.: Invariant manifold families which correspond to non-vanishing characteristic exponents. Izv. Akad. Nauk SSSR Ser. Mat. **40**(6), 1332–1379, (1976)

Pes77. Pesin, Ya.B.: Lyapunov Characteristic Exponents, Smooth Ergodic Theory. Russ. Math. Surveys, **32**(4), 55–114, (1977)

Pet02. Petras, I.: Control of Fractional-Order Chua's System. J. El. Eng. **53**(7-8), 219–222, (2002)

Pet93. Peterson, I.: Newton's Clock: Chaos in the Solar System. W.H. Freeman, San Francisco, (1993)

Pet96. Petrov, V., Showalter, K.: Nonlinear Control from Time-Series. Phys. Rev. Lett. **76**, 3312, (1996)

Pet99. Petras, I.: The Fractional-order controllers: Methods for their synthesis, application. J. El. Eng. **9-10**, 284–288, (1999)

PGO89. Park, B.-S., Grebogi, C., Ott, E., Yorke, J.A.: Scaling of fractal basin boundaries near intermittency transitions to chaos. Phys. Rev. A **40**(3), 1576–1581, (1989)

PGY06. Politi, A., Ginelli, F., Yanchuk, S., Maistrenko, Y.: From synchronization to Lyapunov exponents, back. arXiv:nlin.CD/0605012, (2006)

PHE92. Piaget, J., Henriques, G., Ascher, E.: Morphisms and categories. Erlbaum Associates, Hillsdale, NJ, (1992)

PHV86. Posch, H.A., Hoover, W.G., Vesely, F.J.: Canonical Dynamics of the Nosé Oscillator: Stability, Order and Chaos. Phys. Rev. A, **33**(6), 4253–4265, (1986)

Pic86. Pickover, C.A.: Computer Displays of Biological Forms Generated From Mathematical Feedback Loops. Computer Graphics Forum, **5**, 313, (1986)

Pic87. Pickover, C.A.: Mathematics, Beauty: Time–Discrete Phase Planes Associated with the Cyclic System. Computer Graphics Forum, **11**, 217, (1987)

Pine97. Pinel, J.P.: Psychobiology. Prentice Hall, New York, (1997)

Pink97. Pinker, S. How the Mind Works. Norton, New York, (1997)

PJ01. Pauli, W., Jung, C.G.: Atom and Archetype, The Pauli/Jung Letters, 1932–1958. (Meier, C.A. ed.) Princeton Univ. Press., Princeton, New Jersey, (2001)

PJ55. Pauli, W., Jung, C.G.: The Interpretation of Nature and the Psyche. Random House. (1955)

PK97. Pikovsky, A., Kurth, J.: Coherence Resonance in a Noise-Driven Excitable Systems. Phys. Rev. Lett. **78**, 775–778, (1997)

PM80. Pomeau, Y., Manneville, P.: Intermittent transition to turbulence in dissipative dynamical systems. Commun. Math. Phys. **74**(2), 189–197, (1980)

PMW00. Phillips, P.J., Martin A., Wilson C.L., Przybocki M.: An introduction to evaluating biometric systems. IEEE Computer, 56–63, February (2000)

Pol45. Polya, G. How to Solve It. Princeton Univ. Press, Princeton, NJ, (1945)

Pop00. Pope, S.B.: Turbulent Flows. Cambridge Univ. Press, Cambridge, (2000)

POR97. Pikovsky, A., Osipov, G., Rosenblum, M., Zaks, M., Kurths, J.: Attractor–repeller collision and eyelet intermittency at the transition to phase synchronization. Phys. Rev. Lett. **79**, 47–50, (1997)

Pri71. Pribram, K.H.: Languages of the brain. Prentice-Hall, Englewood Cliffs, N.J., (1971)

Pri91. Pribram, K.H.: Brain and perception. Lawrence Erlbaum, Hillsdale, N.J., (1991)

PRK01. Pikovsky, A., Rosenblum, M., Kurths, J.: Synchronization: A Universal Concept in Nonlinear Sciences. Cambridge Univ. Press, Cambridge, UK, (2001)

PS90. Preparata, F.P., Shamos, M.I.: Computational geometry. Springer, New York, 1990.

PT93. Palis, J., Takens, F.: Hyperbolicity, sensitive-chaotic dynamics at homoclinic bifurcations. Cambridge Univ. Press, (1993)

PTB86. Pokrovsky, V.L., Talapov, A.L., Bak, P.: In Solitons, 71–127, edited by Trullinger, Zakharov, Pokrovsky. Elsevier Science, (1986)

Pul05. Pulverm'uller, F.: Brain mechanicsms kinking language and action. Nature Rev. Neurosci. **6**, 576–582, (2005)

Put93. Puta, M.: Hamiltonian Mechanical Systems and Geometric Quantization, Kluwer, Dordrecht, (1993)

PV03. Pessa, E., Vitiello, G.: Quantum noise, entanglement and chaos in the quantum field theory of mind/brain states, Mind, Matter, **1**, 59–79, (2003)

PV04. Pessa, E., Vitiello, G.: Quantum noise induced entanglement and chaos in the dissipative quantum model of brain, Int. J. Mod. Phys. B, **18** 841–858, (2004)

PV88. Palis, J., Viana, M.: Continuity of Hausdorff dimension and limit capacity. Lecture Notes in Math. 1331, 150–161, Springer Verlag, Berling, (1988)

PV99. Pessa, E., Vitiello, G.: Quantum dissipation and neural net dynamics. Biolectrochem. Bioener., **48**, 339–342, (1999)

Pyr92. Pyragas, K.: Continuous control of chaos, by self-controlling feedback. Phys. Lett. A, **170**, 421–428, (1992)

Pyr95. Pyragas, K.: Control of chaos via extended delay feedback. Phys. Lett. A, **206**, 323–330, (1995)

PZR97. Pikovsky, A., Zaks, M., Rosenblum, M., Osipov, G., Kurths, J.: Phase synchronization of chaotic oscillations in terms of periodic orbits. Chaos **7**, 680, (1997)

Rab89. Rabiner, L.R.: A Tutorial on Hidden Markov Models and Selected Applications in Speech Recognition. Proceedings of the IEEE, 77(2), 257–286, February (1989)

Raj82. Rajaraman, R.: Solitons, Instantons. North-Holland, Amsterdam, (1982)

RAO00. Robert, C., Alligood, K.T., Ott, E., Yorke, J.A.: Explosions of chaotic sets. Physica D **144**, 44 (2000)

Rat78. Ratcliff, R.: A theory of memory retrieval. Psych. Rev., **85**, 59–108, (1978)

Rav02. Raven, J.: Intelligence, Engineered Invisibility and the Destruction of Life on Earth. In: R.K. McKinzey (ed.) WebPsychEmpiricist, WPE, (2002)

RBT01. Roe, R.M., Busemeyer, J.R., Townsend, J.T.: Multi-alternative decision field theory: A dynamic connectionist model of decision making. Psych. Rev., **108**, 370–392, (2001)

RG98. Rao, A.S., Georgeff, M.P.: Decision Procedures for BDI Logics. Journal of Logic and Computation, **8**(3), 292–343, (1998)

RGL99. Rodriguez, E., George, N., Lachaux, J., Martinerie, J., Renault, B., Varela, F.: Long-distance synchronization of human brain activity. Nature, **397**, 430, (1999)

RGS99. Reinagel, P., Godwin, D., Sherman, S.M., Koch, C.: Encoding of visual information by LGN bursts. J Neurophysiol. **81**, 2558–69, (1999)

RH89. Rose, R.M., Hindmarsh, J.L.: The assembly of ionic currents in a thalamic neuron. I The three-dimensional model. Proc. R. Soc. Lond. B, **237**, 267–288, (1989)

Rho61. Rhodes, M.: An analysis of creativity. Phi Delta Kappan **42**, 305–311, (1961)

Rin85. Rinzel, J.: Bursting oscillations in an excitable membrane model. In: Sleeman B.D., Jarvis R.J., eds. Ordinary, partial Differential Equations. Proceedings of the 8th Dundee Conference. Lecture Notes in Mathematics, 1151. Springer, Berlin, (1985)

Rit02. Ritchey, T.: General Morphological Analysis: A general method for non-quantified modelling. http://www.swemorph.com/ma.html, (2002)

RN03. Russel, S., Norvig, P.: Artificial Intelligence: A Modern Approach. Prentice Hall, New Jersey, (2003)

Rob79. Robbins, K.: Periodic solutions and bifurcation structure at high r in the Lorenz system. SIAM J. Appl. Math. **36**, 457–472, (1979)

Rob83. Robinson, H.: Aristotelian dualism. Oxford Studies in Ancient Philosophy 1, 123–44, (1983)

Rob95. Robinson, C.: Dynamical Systems. CRC Press. Boca Raton, FL, (1995)

ROH98. Rosa, E., Ott, E., Hess, M.H.: Transition to Phase Synchronization of Chaos. Phys. Rev. Lett. **80**, 1642–1645, (1998)

Ros58b. Rosenblatt F.: The perceptron: a probabilistic model for information storage and organization in the brain. Physiol. Rev., **65**, 386–408, (1958)

Rot01. Roth, G.: The brain and its reality. Cognitive Neurobiology and its philosophical consequences. Frankfurt a.M.: Aufl. Suhrkamp, (2001)

RPG94. Rasmussen, J., Pejtersen, A.M., Goodstein, L.P.: Cognitive Systems Engineering. Wiley, New York, (1994)

RPK96. Rosenblum, M., Pikovsky, A., Kurths, J.: Phase synchronization of chaotic oscillators. Phys. Rev. Lett. **76**, 1804, (1996)

RPK97. Rosenblum, M., Pikovsky, A., Kurths, J.: From Phase to Lag Synchronization in Coupled Chaotic Oscillators. Phys. Rev. Lett. **78**, 4193–4196, (1997)

RPW97. Robinson, P.A., Rennie, C.J., Wright, J.J.: Propagation and stability of waves of electrical activity in the cerebral cortex. Phys. Rev. E 56(1), 826–40, (1997)

RS75. Reed, M., Simon, B.: Methods of modern mathematical physics, Vol. 2: Fourier analysis, self-adjointness. Academic Press, San Diego, (1975)

RSE82. van Rotterdam, A., Lopes da Silva, F.H., van den Ende, J., Viergever, M.A., Hermans, A.J.: A model of the spatialtemporal characteristics of the alpha rhythm. Bul. Mat. Bio **44**(2), 283–305, (1982)

RU67. Ricciardi, L.M., Umezawa, H.: Brain physics and many-body problems, Kibernetik, **4**, 44, (1967)

Rul01. Rulkov, N.F.: Regularization of Synchronized Chaotic Bursts. Phys. Rev. Lett. **86**, 183–186, (2001)

Rus18. Russell, B.: Mysticism, Logic and Other Essays, London: Longmans, Green, (1918)

Ryd96. Ryder, L.: Quantum Field Theory. Cambridge Univ. Press, (1996)

RYD96. Ryu, S., Yu, W., Stroud, D.: Dynamics of an underdamped Josephson-junction ladder. Phys. Rev. E **53**, 2190–2195, (1996)

Ryl49. Ryle, G.: The Concept of Mind. Chicago: Chicago Univ. Press, (1949)

Sam01. Samal, M.K.: Speculations on a Unified Theory of Matter and Mind. Proc. Int. Conf. Science, Metaphysics: A Discussion on Consciousness, Genetics. NIAS, Bangalore, India, (2001)

Sam89. Davies, P.: The New Physics. Cambridge Univ. Press, Cambridge (1989)

Sam95. Samuels, A.: Jung and the Post-Jungians. Routledge, London, (1985)

Sam99. Samal, M.K.: Can Science 'explain' Consciousness? Proc. Nat. Conf. Scientific, Philosophical Studies on Consciousness, ed. Sreekantan, B.V. *et al.*, NIAS, Bangalore, India, (1999)

SB98. Sutton, R.S., Barto, A.G.: Reinforcement Learning: An Introduction. MIT Press, Cambridge, MA, (1998)

SC91. Stanovich, K.E., Cunningham, A.E.: Reading as constrained reasoning. In R.J. Sternberg, P.A. Frensch (eds.), Complex problem solving: Principles, mechanisms (3–60) Hillsdale, NJ: Lawr. Erl. Assoc., (1991)

Sch02. Schmaltz, T.: Nicolas Malebranche, The Stanford Encyclopedia of Philosophy, Stanford, (2002)

Sch06. Scholarpedia, http://www.scholarpedia.org, (2006)

Sch81. Schulman, L.S.: Techniques and Applications of Path Integration. New York: Wiley, (1981)

Sch85. Schoenfeld, A.H.: Mathematical problem solving. Orlando, FL: Academic Press, (1985)

Sch88. Schuster, H.G. (ed.): Handbook of Chaos Control. Wiley-VCH, (1999)

Sch94. Schiff S.J. *et al.*: Controlling chaos in the brain. Nature **370**, 615–620, (1994)

Sch98. Schlacher, K.: Mathematical Strategies Common to Mechanics, Control, Zeits. Math. Mech., **78**(11), 723–730 (1998)

Sea80. John Searle: Minds, Brains and Programs. Behav. Brain Sci. **3**(3), 417–457, (1980)

Ser99. Service, R.F.: Complex Systems: Exploring the Systems of Life. Science **284**, 80, (1999)

SF91. Sternberg, R.J., Frensch, P.A. (eds.): Complex problem solving: Principles, mechanisms. Hillsdale, NJ: Lawr. Erl. Assoc., (1991)

Sha06. Sharma, S.: An Exploratory Study of Chaos in Human-Machine System Dynamics. IEEE Trans. SMC B. **36**(2), 319–326, (2006)

She01. Sherman, S.M.: Tonic, burst firing: dual modes of thalamocortical relay. Trends in Neuroscience, 24, 122–126, (2001)

Si89. Sastri, S.S., Isidori, A.: Adaptive control of linearizable systems. IEEE Trans. Aut. Con., **34**(1), 1123–1131, (1989)

Sim51. Simpson, E.H.: The Interpretation of Interaction in Contingency Tables. J. Roy. Stat. Soc. B **13**, 238–241, (1951)

Sio05. Sioutis, C.: Reasoning, learning for intelligent agents. PhD thesis, Univ. SA, Adelaide, SA, (2005)

SIT95. Shea, H.R., Itzler, M.A., Tinkham, M.: Inductance effects and dimensionality crossover in hybrid superconducting arrays Phys. Rev. B **51**, 12690–12697, (1995)

Siv77. Sivashinsky, G.I.: Nonlinear analysis of hydrodynamical instability in laminar flames – I. Derivation of basic equations. Acta Astr. **4**, 1177, (1977)

SJD94. Schiff, S.J., Jerger, K., Duong, D.H., Chang, T., Spano, M.L., Ditto, W.L.: Controlling chaos in the brain. Nature, **370**, 615–620, (1994)

SKG93. Shidara, M., Kawano, K., Gomi, H., Kawato, M.: Inverse-dynamics model eye movement control by Purkinje cells in the cerebellum. Nature, **365**, 50–52, (1993)

Ski72. Skinner, B.F.: Beyond Freedom, Dignity. New York: Bantam/Vintage Books, (1972)

SL00. Sprott, J.C., Linz, S.J.: Algebraically Simple Chaotic Flows. Int. J. Chaos Theory, Appl., **5**(2), 3–22, (2000)

SL99. Sternberg, R.J., Lubart, T.I.: The Concept of Creativity: Prospects and Paradigms, ed. Sternberg, R.J. Handbook of Creativity. Cambridge Univ. Press, Cambridge, (1999)

SM00. Smierzchalski, R., Michalewicz, Z.: Modeling of Ship Trajectory in Collision Situations by an Evolutionary Algorithm. IEEE Trans. Evol. Comp. 4(3), 227–241, (2000)

Sm88. Strogatz, S.H., Mirollo, R.E.: Phase-locking and critical phenomena in lattices of coupled nonlinear oscillators with random intrinsic frequencies. Physica D **31**, 143, (1988)

SM91. Sokol, S.M., McCloskey, M.: Cognitive mechanisms in calculation. In R.J. Sternberg, P.A. Frensch (eds.), Complex problem solving: Principles, mechanisms (85–116) Hillsdale, NJ: Lawr. Erl. Assoc., (1991)

Sno75. Snodgrass, J.G.: Psychophysics. In: Experimental Sensory Psychology. B Scharf. (ed.), 17–67, (1975)

Sny86. Snyder, S.H.: Drugs and the Brain. Scientific American Library, W.H. Freeman, Co., New York, (1986)

SO00. Sweet D., Ott E.: Fractal dimension of higher-dimensional chaotic repellors. Physica D **139**(1), 1–27, (2000)

Spa82. Sparrow, C.: The Lorenz Equations: Bifurcations, Chaos and Strange Attractors. Springer, New York, (1982)

Spi670. Spinoza, B.: Tractatus Theologico–Politicus (A Theologico-Political Treatise), (1670)

Spr93a. Sprott, J.C.: Automatic Generation of Strange Attractors. Comput. Graphics, **17**(3), 325–332, (1993)

Spr93b. Sprott, J.C.: Strange Attractors: Creating Patterns in Chaos. M&T Books, New York, (1993)

Spr94. Sprott, J.C.: Some Simple Chaotic Flows. Phys. Rev. E, **50**(2), R647–R650, (1994)

Spr97. Sprott, J.C.: Some Simple Chaotic Jerk Functions. Am. J. Phys., **65**(6), 537–543, (1997)

SRK98. C. Schafer, M.G. Rosenblum, J. Kurths, H.-H Abel: Heartbeat Synchronized with Ventilation. Nature **392**, 239–240 (1998)

SRV04. Stoico, C., Renzi, D., Vericat, F.: From Darwin to Sommerfeld: Genetic algorithms and the electron gas. arXiv:cond-mat/0412239, (2004)

SS01. Scholkopf, B., Smola, A.: Leaning with Kernels. MIT Press, Cambridge, MA, (2001)

SSK87. Sakaguchi, H., Shinomoto, S., Kuramoto, Y.: Local and global self-entrainments in oscillator-lattices. Prog. Theor. Phys. **77**, 1005–1010, (1987)

Sta83. Stapp, H.P.: Exact solution of the infrared problem. Phys. Phys. Rev. D **28**, 1386–1418, (1983)

Sta93. Stapp, H.P.: Mind, Matter and Quantum Mechanics. Spinger-Verlag, Heidelberg, (1993)

Sta95. Stapp, H.P.: Chance, Choice and Consciousness: The Role of Mind in the Quantum Brain. arXiv:quant-ph/9511029, (1995)

Ste69. Sternberg, S.: Memory-scanning: Mental processes revealed by reaction-time experiments. Am. Sci., **57**(4), 421–457, (1969)

Ste95. Sternberg, R.J.: Conceptions of expertise in complex problem solving: A comparison of alternative conceptions. In P.A. Frensch, J. Funke (eds.), Complex problem solving: The European Perspective (295–321) Hillsdale, NJ: Lawr. Erl. Assoc., (1995)

Sto05. Stoljar, D.: Physicalism. The Stanford Encyclopedia of Philosophy, Stanford, (2005)

Sto90. Stonier, T.: Information and the internal structure of the Universe. Springer, New York, (1990)

Str. Strogatz, S.H.: Exploring complex networks. Nature, **410**, 268, (2001)

Str00. Strogatz, S.H.: From Kuramoto to Crawford: exploring the onset of synchronization in populations of coupled oscillators. Physica D, **143**, 1–20, (2000)

Str94. Strogatz, S.: Nonlinear Dynamics and Chaos. Addison-Wesley, Reading, MA, (1994)

STU78. Stuart, C.I.J., Takahashi, Y., Umezawa, H.: On the stability, non-local properties of memory, J. Theor. Biol. **71**, 605–618, (1978)

STU79. Stuart, C.I.J., Takahashi, Y., Umezawa, H.: Mixed system brain dynamics: neural memory as a macroscopic ordered state, Found. Phys. **9**, 301, (1979)

STZ93. Satarić, M.V., Tuszynski, J.A., Zakula, R.B.: Kinklike excitations as an energy-transfer mechanism in microtubules. Phys. Rev. E, **48**, 589–597, (1993)

SUK01. Su, H, Alroy G, Kirson ED, Yaari Y.: Extracellular calcium modulates persistent sodium current-dependent burst-firing in hippocampal pyramidal neurons. J. Neurosci. **21**, 4173–4182, (2001)

SVW95. Srivastava, Y.N., Vitiello, G., Widom, A.: Quantum dissipation and quantum noise, Annals Phys., **238**, 200, (1995)

SW90. Sassetti, M., Weiss, U.: Universality in the dissipative two-state system. Phys. Rev. Lett. **65**, 2262–2265, (1990)

SWV99. Srivastava, Y.N., Widom, A., Vitiello, G.: Quantum measurements, information and entropy production. Int. J. Mod. Phys. B13, 3369–3382, (1999)

SZT98. Satarić, M.V., Zeković, S., Tuszynski, J.A., Pokorny, J.: Ḿossbauer effect as a possible tool in detecting nonlinear excitations in microtubules. Phys. Rev. E **58**, 6333–6339, (1998)

Tan93. Tanaka, K.: Neuronal mechanisms of object recogntition. Science, **262**, 685–688, (1993)

Tay88. Taylor, C.W.: Various approaches to, definitions of creativity, ed. Sternberg, R.J. The nature of creativity: Contemporary psychological perspectives. Cambridge Univ. Press, Cambridge, (1988)

TF91. Tsue, Y., Fujiwara, Y.: Time-Dependent Variational Approach to (1+1)-Dimensional Scalar-Field Solitons, Progr. Theor. Phys. **86**(2), 469–489, (1991)

Tho75. Thom, R.: Structural Stability and Morphogenesis. Addison–Wesley, Reading, (1975)

Tin75. Tinkham, M.: Introduction to Superconductivity. McGraw-Hill, New York, (1975)

TLO97. Tanaka, H.A., Lichtenberg, A.J., Oishi, S.: First Order Phase Transition Resulting from Finite Inertia in Coupled Oscillator Systems. Phys. Rev. Lett. **78**, 2104–2107, (1997)

TLO97. Tanaka, H.A., Lichtenberg, A.J., Oishi, S.: Self-synchronization of coupled oscillators. with hysteretic response. Physica D **100**, 279, (1997)

TM01. Trees, B.R., Murgescu, R.A.: Phase locking in Josephson ladders and the discrete sine-Gordon equation: The effects of boundary conditions, current-induced magnetic fields. Phys. Rev. E **64**, 046205–046223, (2001)

TMO00. Tras, E., Mazo, J.J., Orlando, T.P.: Discrete Breathers in Nonlinear Lattices: Experimental Detection in a Josephson Array. Phys. Rev. Lett. **84**, 741, (2000)

TRW98. Tass, P., Rosenblum, M.G., Weule, J., Kurths, J. *et al.*: Detection of $n : m$ Phase Locking from Noisy Data: Application to Magnetoencephalography. Phys. Rev. Lett. **81**, 3291–3294, (1998)

TS01. Thompson, J.M.T., Stewart, H.B.: Nonlinear Dynamics, Chaos: Geometrical Methods for Engineers, Scientists. Wiley, New York, (2001)

TS91. Thompson, J.M.T., Soliman, M.S.: Indeterminate jumps to resonance from a tangled saddle-node bifurcation. Proc. R. Soc. Lond. A **432**, 101–111, (1991)

TSS05. Trees, B.R., Saranathan, V., Stroud, D.: Synchronization in disordered Josephson junction arrays: Small-world connections and the Kuramoto model. Phys. Rev. E **71**, 016215–016235, (2005)

TVP99. Tabony, J., Vuillard, L., Papaseit, C.: Biological self-organisation, pattern formation by way of microtubule reaction-diffusion processes. Adv. Complex Syst. **2**(3), 221–276, (1999)

Ume93. Umezawa, H.: Advanced field theory: micro, macro and thermal concepts. American Institute of Physics, New York, (1993)

UMM93. Ustinov, A.V., Cirillo, M., Malomed, B.A.: Fluxon dynamics in one-dimensional Josephson-junction arrays. Phys. Rev. B **47**, 8357–8360, (1993)

Vaa95. Van der Vaart, N.C. *et al.*: Resonant Tunneling Through Two Discrete Energy States. Phys. Rev. Lett. **74**, 4702–4705, (1995)

Vap95. Vapnik, V.: The Nature of Statistical Learning Theory. Springer, New York, (1995)

Vap98. Vapnik, V.: Statistical Learning Theory. Wiley, New York, (1998)

VB04. Vukobratović, M., Borovac, B.: Zero–Moment Point: Thirty–five Years of its Life. Int. J. Hum. Rob., **1**(1), 157–173, (2004)

VBB05. Vukobratovic M., Borovac B., Babkovic K.: Contribution to the study of anthropomorphism of humanoid robots. Int. J. Hum. Rob. **2**(3), (2005)

VBP06. Vukobratovic, M., Borovac, B., Potkonjak, V.: Towards a Unified Understanding of Basic Notions and Terms in Humanoid Robotics. Int. J. Hum. Rob. (to appear), (2006)

VBS90. Vukobratović, M., Borovac, B., Surla, D., Stokić, D.: Biped Locomotion – Dynamics, Stability, Control and Application. Springer-Verlag, Berlin, (1990)

Ver838. Verhulst, P.F.: Notice sur la loi que la population pursuit dans son accroissement. Corresp. Math. Phys. **10**, 113–121, (1838)

Ver845. Verhulst, P.F.: Recherches Mathematiques sur La Loi D'Accroissement de la Population (Mathematical Researches into the Law of Population Growth Increase) Nouveaux Memoires de l'Academie Royale des Sciences et Belles-Lettres de Bruxelles, **18**(1), 1–45, (1845)

VHC73. Vukobratović, M., Hristić, D., Ćirić, V., Zečević, M.: Analysis of Energy Demand Distribution Within Anthropomorphic System, Trans. of ASME, Series G, J. Dyn. Sys. Meas. Con., **17**, 191–242, (1973)

Vir00. Virgin, L.N.: Introduction to Experimental Nonlinear Dynamics. Cambridge Univ. Press, Cambridge, (2000)

Vit01. Vitiello, G.: My Double Unveiled. John Benjamins, Amsterdam, (2001)

Vit95. Vitiello, G.: Dissipation and memory capacity in the quantum brain model, Int. J. Mod. Phys. B, **9**, 973–989, (1995)

VJ69. Vukobratović, M., Juričić, D: Contribution to the Synthesis of Biped Gait. IEEE Trans. Biomed. Eng., **16**, 1, (1969)

Vos99. Vose, M.D.: The Simple Genetic Algorithm: Foundations and Theory. MIT Press, Cambridge, MA, (1999)

Vuk73. Vukobratovic, M.: How to Control the Artificial Anthropomorphic Systems. IEEE Trans. Sys. Man. Cyber. SMC-3, 497–507, (1973)

Vuk75. Vukobratovic, M.: Legged Locomotion Systems and Anthropomorphic Mechanisms. Mihajlo Pupin Institute, Belgrade, (1975); also published in Japanese, Nikkan Shimbun Ltd. Tokyo, in Russian, MIR, Moscow, 1976; in Chinese, Beijing 1983.

VWL91. Voss, J.F., Wolfe, C.R., Lawrence, J.A., Engle, R.A.: From representation to decision: An analysis of problem solving in international relations. In R. J. Sternberg, P.A. Frensch (eds.), Complex problem solving: Principles, mechanisms (119–158) Hillsdale, NJ: Lawr. Erl. Assoc., (1991)

Vyg82. Vygotsky, L.S. Historical meaning of the Psychological crisis. Collected works. Vol. 1. Pedag. Publ., Moscow, (1982)

Wag91. Wagner, R.K.: Managerial problem solving. In R. J. Sternberg, P. A. Frensch (eds.), Complex problem solving: Principles, mechanisms (159–183) Hillsdale, NJ: Lawr. Erl. Assoc., (1991)

Wat90. Watson, L.T.: Globally convergent homotopy algorithms for nonlinear systems of equations. Nonlinear Dynamics, **1**, 143–191, (1990)

Wat99. Watts, D.J.: Small Worlds. Princeton Univ. Press, Princeton, (1999)

WBI. Weng, G., Bhalla, U.S., Iyengar, R.: Complexity in Biological Signaling Systems. Science, **284**, 92, (1999)

WC02. Wang, X.F., Chen, G.: Synchronization in small-world dynamical networks. Int. J. Bifur. Chaos **12**, 187, (2002)

WCS96. Wiesenfeld, K., Colet, P., Strogatz, S.H.: Synchronization Transitions in a Disordered Josephson Series Array. Phys. Rev. Lett. **76**, 404–407, (1996)

WCS98. Wiesenfeld, K., Colet, P., Strogatz, S.H.: Frequency locking in Josephson arrays: Connection with the Kuramoto model. Phys. Rev. E **57**, 1563–1569, (1998)

Wei36. Weiss, P.: Proc. Roy. Soc., A **156**, 192–220, (1936)

Wer89. Werbos, P.J.: Backpropagation, neurocontrol: A review and prospectus. In IEEE/INNS Int. Joint Conf. Neu. Net., Washington, D.C., **1**, 209–216, (1989)

Wer90. Werbos, P.: Backpropagation through time: what it does, how to do it. Proc. IEEE, **78** (10), (1990)

WF49. Wheeler, J.A., Feynman, R.P.: Classical Electrodynamics in Terms of Direct Interparticle Action. Rev. Mod. Phys. **21**, 425–433, (1949)

Whe89. Wheeler, J.A.: Information, Physics and Quantum: the Search for the Links. Proc. 3rd Int. symp. Foundations of Quantum Mechanics, Tokyo, 354–368, (1989)

Wig90. Wiggins, S.: Introduction to Applied Dynamical Systems, Chaos. Springer, New York, (1990)

Wik05. Wikipedia, the free encyclopedia. http://wikipedia.org, (2005)

Wil99. Wilczek, F.: Getting its from bits, Nature **397**, 303–306, (1999)

Win67. Winfree, A.T.: Biological rhythms and the behavior of populations of coupled oscillators. J. Theor. Biol. **16**, 15, (1967)

Win80. Winfree, A.T.: The Geometry of Biological Time. Springer, New York, (1980)

Wis95. Wiskott, L.: Labeled Graphs and Dynamic Link Matching for Face Recognition, Scene Analysis. PhD thesis, Fakultat f́ur Physik und Astronomie, Ruhr-Universitat Bochum, D-44780 Bochum, (1995)

WK02. Womack, M.D., Khodakhah, K.: Active contribution of dendrites to the tonic and trimodal patterns of activity in cerebellar Purkinje neurons. J Neurosci. **15**(24), 10603–10612, (2002)

WK98. Wolpert, D., Kawato, M.: Multiple paired forward, inverse models for motor control. Neural Networks, **11**, 1317–1329, (1998)

WM06. Wojtusiak, J., Michalski, R.S.: The LEM3 Implementation of Learnable Evolution Model and Its Testing on Complex Function Optimization Problems, Proceedings of Genetic, Evolutionary Computation Conference, GECCO 2006, Seattle, WA, (2006)

Wol02. Wolfram, S.: A New Kind of Science. Wolfram Media, (2002)

Wol84. Wolfram, S.: Cellular Automata as Models of Complexity. Nature, **311**, 419–424, (1984)

Woo00. Wooldridge, M: Reasoning about rational agents. MIT Press, Boston, MA, (2000)

WS97. Watanabe, S., Swift, J.W.: Stability of periodic solutions in series arrays of Josephson junctions with internal capacitance. J. Nonlinear Sci. **7**, 503, (1997)

WS98. Watts, D.J., Strogatz, S.H.: Collective dynamics of small-world networks. Nature **393**, 440, (1998)

WSS85. Wolf, A., Swift, J.B., Swinney, H.L., Vastano, J.A.: Determining Lyapunov Exponents from a Time Series. Physica D, **16**(3), 285–317, (1985)

WSZ95. Watanabe, S., Strogatz, S.H., van der Zant, H.S.J., Orlando, T.P.: Whirling Modes, Parametric Instabilities in the Discrete Sine-Gordon Equation: Experimental Tests in Josephson Rings. Phys. Rev. Lett. **74**, 379–382, (1995)

WW83a. Wehner, M.F., Wolfer, W.G.: Numerical evaluation of path–integral solutions to Fokker–Planck equations. I., Phys. Rev. A **27**, 2663–2670, (1983)

WW83b. Wehner, M.F., Wolfer, W.G.: Numerical evaluation of path–integral solutions to Fokker–Planck equations. II. Restricted stochastic processes, Phys. Rev. A, **28**, 3003–3011, (1983)

WZ98. Waelbroeck, H., Zertuche, F.: Discrete Chaos. J. Phys. A **32**, 175, (1998)

WZZ03. Williams, R.H., Zimmerman, D.W., Zumbo, B.D., Ross, D.: Charles Spearman: British Behavioral Scientist. Human Nature Review. **3**, 114–118, (2003)

Yag87. Yager, R.R.: Fuzzy Sets and Applications: Selected Papers by L.A. Zadeh, Wiley, New York, (1987)

YAS96. Yorke, J.A., Alligood, K., Sauer, T.: Chaos: An Introduction to Dynamical Systems. Springer, New York, (1996)

Yeo92. Yeomans, J.M.: Statistical Mechanics of Phase Transitions. Oxford Univ. Press, Oxford, (1992)

YL97. Yalcinkaya, T., Lai, Y.-C.: Phase Characterization of Chaos. Phys. Rev. Lett. **79**, 3885–3888, (1997)

YML00. Yanchuk, S., Maistrenko, Yu., Lading, B., Mosekilde, E.: Effects of a parameter mismatch on the synchronizatoin of two coupled chaotic oscillators. Int. J. Bifurcation, Chaos **10**, 2629–2648, (2000)

Zad65. Zadeh, L.A.: Fuzzy sets. Inform. Contr. **8**, 338–353, (1965)

Zad78. Zadeh, L.A.: Fuzzy sets as a basis for a theory of possibility. Fuzzy Sets, Systems, **1**(1), 3–28, (1978)

Zha84. Zhadin, M.N.: Rhythmic processes in the cerebral cortex. J. The. Bio. 108, 565–95, (1984)

Zwi69. Zwicky, F.: Discovery, Invention and Research – Through the Morphological Approach. Macmillian, Toronto, (1969)

Index

Printing: Krips bv, Meppel
Binding: Stürtz, Würzburg